For All Practical Purposes

PROJECT DIRECTOR
Solomon Garfunkel,
Consortium for Mathematics and its Applications

CONTRIBUTING AUTHORS

PART I MANAGEMENT SCIENCE
Joseph Malkevitch, York College, CUNY

PART II STATISTICS: THE SCIENCE OF DATA
David S. Moore, Purdue University

PART III VOTING AND SOCIAL CHOICE
Alan D. Taylor, Union College
Bruce P. Conrad, Temple University
Steven J. Brams, New York University

PART IV FAIRNESS AND GAME THEORY
Alan D. Taylor, Union College
Bruce P. Conrad, Temple University
Steven J. Brams, New York University

PART V THE DIGITAL REVOLUTION
Joseph Gallian, University of Minnesota–Duluth

PART VI ON SIZE AND GROWTH
Paul J. Campbell, Beloit College

PART VII YOUR MONEY AND RESOURCES
Paul J. Campbell, Beloit College

FOR ALL PRACTICAL PURPOSES

Mathematical Literacy In Today's World

SEVENTH EDITION

W. H. FREEMAN AND COMPANY
New York

Publisher: Craig Bleyer
Acquisitions Editor: Terri Ward
Senior Marketing Manager: Sarah Martin
Senior Developmental Editor: Mary J. Maechling
Assistant Editor: Brendan Cady
Media Editor: Victoria Anderson
Associate Editor: Amy Shaffer
Photo Editor and Researcher: Bianca Moscatelli
Cover and Text Designer: Blake Logan, Cambraia Fernandes
Project Editor: Vivien Weiss
Illustrations: Fine Line Illustrations
Illustration Cordinator: Shawn Churchman
Production Coordinator: Paul W. Rohloff
Composition: Matrix Publishing Services
Printing and Binding: Quebecor

Library of Congress Control Number: 2005933990

Chapter Openers 1: Thinkstock/Picturequest; 2: Massachusetts Bay Transportation Authority; 3: Najlah Feanny/Corbis; 4: Fogstock LLC/Index Stock Imagery; 5: Jean Louis Batt/Getty Images; 6: Michael Newman/PhotoEdit; 7: Spencer Grant/PhotoEdit; 8: Cut and Deal/Punchstock; 9: Brooks Kraft/Corbis; 10: Stock Connection/Picturequest; 11: Robert Galbraith/Reuters/Corbis; 12: Joe Raedle/Getty Images; 13: Digital Vision/ Picturequest; 14: Luke Frazza/Corbis; 15: Corbis/Punchstock; 16: mediacolor's/Alamy; 17: NASA; 18: Robert F. Walters; 19: Phil Schermeister/CORBIS; 20: Peter Aprahamian/CORBIS; 21: Chuck Savage/CORBIS; 22: Mason Morfit/ Getty Images; 23: Paul Harris/Getty Images

Printed in the United States of America

Second printing

W. H. Freeman and Company

41 Madison Avenue

New York, NY 10010

Houndmills, Basingstoke RG21 6XS, England

www.whfreeman.com

Brief Contents

Contents

NASA/JPL

Daemmrich AS/A. Wade/
Corbis Sygma

Gleb Garanich/Reuters/CORBIS

United Artists/The Kobal
Collection

75468 61617

Peter Mason/Getty Images

Joseph Sohm; ChromoSohm Inc./CORBIS

Creatas Images/Punchstock

Preface

To the Student

For All Practical Purposes represents our effort to bring the excitement of contemporary mathematical thinking to the nonspecialist. Environmental and economic decisions dominate modern life, and behind these decisions are fundamental principles of science, technology, and mathematics. We want you to have an awareness of these fundamental yet accessible principles. In science and industry, mathematical models are the main tools for analyzing and solving problems that arise. In this book, our goal is to convey the power of mathematics by showing you the great variety of problems that can be modeled and solved by quantitative means.

There are many ways to talk about why mathematics and its applications matter. You will hear expressions such as "mathematical literacy" or "quantitative literacy." They mean, essentially, that math is important. It is important in explaining the way the world works. It is important because knowing it can make your life easier. In other words, it can help to explain how your world works. We created this course and this book because we know that not everyone looks at mathematics in this way.

In school, you spent a great deal of time learning the tools of mathematics—how to manipulate symbols, how to solve equations. In this course, you will spend time learning the uses of mathematics and the power of mathematics to help us to understand so many different parts of our everyday lives and the world itself. We hope that this exploration will give you a broader sense of what our subject is about and why we wanted you to take a math course every year you were in school. It's "for all practical purposes," because, in a sense, you've learned to hammer nails and saw wood. Now we're going to build houses.

Enjoy!

To the Instructor

Because For All Practical Purposes stresses the connections between contemporary mathematics and modern society, our text must be flexible enough to accommodate new ideas in mathematics and their new applications to our daily lives. We maintain this flexibility in the seventh edition.

Our primary goal for this edition was to improve the ease of use for instructors and students alike.

Our first efforts were to improve the flow of the text through enhanced design.

- New Part Openers offer an introductory narrative that ties together chapters within each part by briefly discussing the overriding theme.

More subtle changes that further improve the flow of the text include:

- New titles that call out definitions, procedures, and formulas
- Examples that are numbered and titled.
- Exercises and end-of-chapter material, reorganized to be more open, effective, and accessible.

Second, we worked to create a consistent style between chapters. Together the authors and editors smoothed individual writing styles into one cohesive FAPP voice to avoid digressions and jumps in conceptual level and in complexity of notation within and between chapters.

We have significantly **reordered and restructured some parts** of the book to improve the relationships between chapters. These changes will streamline your lesson planning and improve the ease-of-use for you and your students. The changes are listed in the order they appear in the text.

PART I   MANAGEMENT SCIENCE Chapter 4, Linear Programming, was revised to focus on linear programming and offers new applications on delivering commodities.

PART II   STATISTICS: THE SCIENCE OF DATA Content within the chapters was reorganized so that important statistical ideas that students need to know are moved forward and followed by more advanced concepts.

PART III   VOTING AND SOCIAL CHOICE and **PART IV   FAIRNESS AND GAME THEORY** Part III has been significantly revised and reordered to increase the manageability of this material. With the new organization, the less complicated material is separated and brought forward into Part III: Voting and Social Choice and followed by the more complex topics in Part IV: Fairness and Game Theory. This approach keeps the conceptual foundation of the material.

NEW! **Chapter 10: The Manipulability of Voting Systems** Chapter 10 dovetails with and then extends the coverage about manipulating elections (how it works and what does not work).

PART V   THE DIGITAL REVOLUTION Chapters have been significantly revised to reflect our fast-moving society and business practices. Material on the Internet, the Web, and Logic (formerly Chapter 11 in the sixth edition) has been integrated into Chapter 17, now titled Information Science.

PART VI   ON SIZE AND GROWTH Chapters have been blended to create more even sections to ease reading and assigning chapters. The seventh edition offers a new section on Fractals (section 19.6) and an expanded section on quasicrystals (section 20.6).

PART VII YOUR MONEY AND RESOURCES These chapters, which focus on growth and growth models, have been revised to reflect current financial applications including new interest rates and lending practices. For example, there is now more information on adjustable-rate mortgages and low-APR car loans.

Sequencing of the end-of-chapter material has been changed to improve the overall flow of the material.

- **Skills Check Exercises have doubled.** You will now find 20 conceptual and computational exercises that are designed to help students master basic skills before proceeding to the rest of the exercises.

- **Exercise sets are renewed and refreshed** by updating old problems and adding completely new exercises.

- **Writing Projects are updated and simplified** to allow for shorter writing assignments. This flexibility will better accommodate your "Writing Across the Curriculum" goals as well as address any limitations in time.

- **Suggested Readings are streamlined** to focus on the most important works.

Focus on Accuracy

For this edition we implemented a detailed accuracy checking plan to sustain the quality of the exercises and improve the solutions. To this end, we are very fortunate to welcome **John Samons of Florida Community College at Jacksonville** to the FAPP team. He tirelessly worked with the authors to ensure accuracy in this edition of the text. John also collaborated with supplements author, **Heidi Howard of Florida Community College at Jacksonville**, to ensure both accuracy and consistency between the text and supplements package.

For All Practical Purposes has also enjoyed a particular devotion to its mission from several dedicated professors over the course of its lifetime, which has now passed the 20-year mark. We have worked hard on this edition to address their concerns as well as those expressed by our newest friends. In particular, we would like to thank Valerie Morgan-Krick and Karen Clark, both of Tacoma Community College; Scott Inch of Bloomsburg University; and Dr. Joe Marasco.

Media and Supplements

The media and supplements package for the seventh edition has been updated to reflect changes in the book. Both instructors and students will benefit from the innovative materials available to them.

Student Resources

Student Study Guide, ISBN: 0-7167-6946-8
Heidi Howard, Florida Community College at Jacksonville
Offers study tips and tools to help students gain a better understanding of course material.

NEW! Student Solutions Manual, ISBN: 0-7167-0010-7
Heidi Howard, Florida Community College at Jacksonville
Contains full, worked solutions to the odd-numbered problems in the text.

Web Site: www.whfreeman.com/fapp7e
For students, the site serves as an online study guide, featuring:

- **Interactive Applets.** Selected exercises in the text, indicated by an icon, ask students to manipulate data or graphics to better understand the mathematical concepts by using an applet.
- **Online Quizzes.** This self-grading quiz engine generates up to 20 questions for each chapter.
- **Video Clips** excerpted from the acclaimed Annenberg/CPB Project, *For All Practical Purposes*, provide students with visual depictions of key concepts.
- **Flashcards.** This self-grading feature helps students master key terms.
- **Suggested Web Sites.** Continually updated annotated Web links support the coverage of specific topics in every chapter. Instructors are invited to submit links to be included.
- **Writing Projects.** These projects, which require mathematically reasoned responses, complement projects in the text.
- **Spreadsheet Projects.** Excel spreadsheets can be downloaded for select chapters.

Instructor Resources

Instructor's Manual with Full Solutions, ISBN: 0-7167-6947-6
Heidi Howard, Florida Community College at Jacksonville
Includes teaching support for each chapter *and* full solutions for all problems in the text.

NEW! Teaching Guide for First-Time Instructors, ISBN: 0-7167-0128-6
Heidi Howard, Florida Community College at Jacksonville
This new guide for new instructors, adjuncts, and teaching assistants will help make planning your course and teaching with FAPP easier and more effective. Ideas set forth in this guide will also offer fresh perspective and ideas to experienced instructors.

Enhanced Instructor's Resource CD-ROM (IRCD), ISBN: 0-7167-6940-9
Created to help instructors develop lecture presentations, course Web sites, and other resources, this CD-ROM allows instructors to **search** and **export** all the resources contained below by key term or chapter:

- All text images
- Applets, movies, flahcards, spreadsheet projects, self-quizzes available on the Web site
- Instructor's Manual with full solutions

Assessment

Test Bank

CD-ROM (Windows and Mac): 0-7167-6941-7
Printed: 0-7167-6945-X
John Emert, Ball State University
The *Test Bank* offers 75 multiple-choice and fill-in-the-blank questions and 35 short-answer questions per chapter. The easy-to-use CD includes Windows and Mac versions on a single disc, in a format that lets you add, edit, and re-sequence questions to suit your needs.

Course Management

WebCT and Blackboard
All the book's Web and testing materials are compatible with WebCT and Blackboard. We offer the electronic content as a service to adopters; please contact your local sales representative.

Acknowledgements

For All Practical Purposes continues to evolve in great part due to our many friends and colleagues who have offered suggestions, comments, and corrections. We are grateful to them all.

Scott Balcomb, *St. Joseph's University*
Nancy Balle, *Ball State University*
Richard Bedient, *Hamilton College*
Rebecca Bergs, *Ball State University*
Terence R. Blows, *Northern Arizona University*
Terry Boyd, *University of Indianapolis*
Linda Braddy, *East Central University*
Barry Brunson, *Western Kentucky University*
Paul Buckelew, *Oklahoma City Community College*
Annette M. Burden, *Youngstown State University*
Shana Calaway, *Shoreline Community College*
Tim Carroll, *Eastern Michigan University*
G. Andy Chang, *Youngstown State University*
Leo Chouinard, *University of Nebraska–Lincoln*
Karen Clark, *Tacoma Community College*

Valerie Morgan-Crick, *Tacoma Community College*
Greg Crow, *Point Loma Nazarene University*
Sloan Despeaux, *Western Carolina University*
Rob Donnelly, *Murray State University*
Daniel Dreibelbis, *University of North Florida*
Gina Poore Dunn, *Lander University*
Nancy Eaton, *University of Rhode Island*
John W. Emert, *Ball State University*
Sandra Fillebrown, *Saint Joseph's University*
W. Bart Frye, *Ball State University*
Martha Gady, *Whitworth College*
Monica Pierri-Galvao, *Gannon University*
Steve Gendler, *Clarion University*
Marty Getz, *University of Alaska, Fairbanks*
T. R. Hamlett, *East Central University*

Mohammad Halim, *Ball State University*

Peter Johnson, *Auburn University*

W. T. Kiley, *George Mason University*

Julie Killingbeck, *Ball State University*

Nancy Kitt, *Ball State University*

Samuel Kohn, *Thomas Edison State College*

Kathy Lewis, *State University of New York, Oswego*

Monica Liddle, *State University of New York, Delhi*

Jay Malmstrom, *Oklahoma City Community College*

Barbara Margoulius, *Cleveland State University*

Mary T. McMahon, *North Central College*

Christopher McCord, *University of Cincinnati*

Steve Morics, *University of Redlands*

Dean Morrow, *Washington and Jefferson College*

Anne Marie Mosher, *St. Louis Community College*

Ellen Mulqueeny, *Cleveland State University*

Mika Munakata, *Montclair State University*

Chris Oehrlein, *Oklahoma City Community College*

Steven Ohs, *Western Michigan University*

Patricia Parkison, *Ball State University*

Deb Pearson, *Ball State University*

Marilyn Reba, *Clemson University*

Leo Robinson, *Ball State University*

Robin Ruffato, *Ball State University*

Steven Schecter, *North Carolina State*

Brian Siebenaler, *Ball State University*

Samuel Bruce Smith, *St. Joseph's University*

Patricia Stanley, *Ball State University*

James D. Stoops, *Ball State University*

Tamas Szabo, *Weber State University*

Robert Terrell, *Cornell University*

Helen Thorwarth, *Northern Kentucky University*

David Urion, *Winona State University*

Bonnie Wachhaus, *Messiah College*

W. D. Wallis, *Southern Illinois University*

Kim Ward, *Eastern Connecticut State University*

John Weglarz, *Kirkwood Community College*

Gideon Weinstein, *Montclair State University*

Liz Whittern, *Ball State University*

Meredith Wort, *East Central University*

Christian Yankov, *Eastern Connecticut State University*

Janet Yi, *Ball State University*

John Zerger, *Catawba College*

In addition, Paul Campbell thanks Friedrich Pukelsheim of the Mathematics Institute of the University of Augsburg, Germany, for his gracious hosting in 2005 during the revision of chapters in this edition.

We owe our appreciation to the people at W. H. Freeman and Company who participated in the development and production of this edition. We wish especially to thank the editorial staff for their tireless efforts and support. Among them are Terri Ward, Acquisitions Editor; Mary J. Maechling, Senior Development Editor; Brendan Cady, Assistant Editor; Vivien Weiss, Project Editor; Paul W. Rohloff, Production Manager; Blake Logan, Designer, with Cambraia Fernandes; Bianca Moscatelli, Photo Editor; Victoria Anderson, Media Editor; Amy Shaffer, Associate Editor; Scott Amerman and Trumbull Rogers, copy editors; and Carol Dean, proofreader. We would also like to extend our appreciation to outside Developmental Editor Donald Gecewicz.

The efforts of the COMAP staff must also be recognized. We thank our production and administrative staff, George Ward and Roland Cheney, and Laurie Aragon, the manager.

To everyone who helped make our purposes practical, we offer our appreciation for an exciting and exhilarating time.

Solomon Garfunkel, COMAP

Management Science

Recently, the National Aeronautics and Space Administration (NASA) landed two mobile robotic units (*Spirit* and *Opportunity*) on Mars and placed one of the most sophisticated spacecraft ever launched, *Cassini*, into an orbit around the planet Saturn. The *Huygens* spacecraft deployed by *Cassini* reached the surface of Saturn's moon Titan and sent back images of what appear to be channels formed by liquid methane. The Mars rovers have sent back to earth remarkable pictures of the complex and varied landscapes and rocks on the surface of Mars. Amazingly, *Cassini*, launched in 1997, arrived right on schedule in 2004 for its close encounter with Saturn! Exploring the world around us has occupied humankind from the days when the known world did not include the Americas, to Neil Armstrong's first steps on the moon, to robot spacecraft such as *Cassini, Spirit,* and *Opportunity*.

To accomplish tasks of these dimensions, with as little cost as possible, NASA, a government agency, and its many private subcontractors, have had to engage in complex planning, coordination of the allocation of resources, detailed scheduling, and cost-minimization procedures. The many workers who were involved in these dramatic success stories put mathematics to work for them. They used mathematical tools that have been developed over more than 100 years and now constitute the part of mathematics known as *operations research,* the subject of the first four chapters of this book.

Collectively, the chapters that follow show the range of mathematical ideas that are being used by government, business, and individuals that help make our lives more satisfying and enjoyable, while at the same time demonstrating ideas that are intrinsically nifty. ■

1

Urban Services

The underlying theme of management science, also called **operations research,** is finding the best method for solving some problem—what mathematicians call the **optimal solution.** In some cases, the goal may be to finish a job as quickly as possible. In other situations, the objective might be to maximize profit or minimize cost. In this chapter, our goal is to save time in traversing a street network while checking parking meters, delivering mail, removing snow, or collecting bottles for recycling.

Let's begin by assisting the parking department of a city government. Most cities and many small towns have parking meters that must be regularly checked for parking violations or emptied of coins. We will use an imaginary town to show how management science techniques can help to make parking control more efficient.

1.1 Euler Circuits

The street map in Figure 1.1 is typical of many towns across the United States, with streets, residential blocks, and a village green. Our job, or that of the commissioner of parking, is to find the most efficient route for the parking-control officer, who travels on foot, to check the meters in an area. Efficient routes save money. Our map shows only a small area, allowing us to start with an easy problem. But the problem occurs on a larger scale in all cities and towns and, for larger areas, there are almost unlimited possibilities for parking-control routes.

The commissioner has two goals in mind: (1) The parking-control officer must cover all the sidewalks that have parking meters without retracing any more steps than are necessary; and (2) the route should end at the same point at which

3

FIGURE 1.1 A street
map for part of a town.

it began, perhaps where the officer's patrol car is parked. To be specific, suppose
there are only two blocks that have parking meters, the two lightly shaded blocks
that are side by side toward the top of Figure 1.1. Suppose further that the parking-control officer must start and end at the upper left corner of the left-hand
block. You might enjoy working out some routes by trial and error and evaluating their good and bad points. We are going to leave this problem for the moment and establish some concepts that will give us a better method than trial and
error to deal with this problem.

Describing a Graph

A **graph** is a finite set of dots and connecting links. The dots are called **vertices** (a single dot is called a **vertex**), and the links are called **edges.** Each edge
must connect two different vertices. A **path** is a connected sequence of edges
showing a route on the graph that starts at a vertex and ends at a vertex; a
path is usually described by naming in turn the vertices visited in traversing
it. A path that starts and ends at the same vertex is called a **circuit.** A graph
can represent our city map, a communications network, or even a system of
air routes.

EXAMPLE 1   Parts of a Graph

We can use the graph in Figure 1.2 to help explain these technical terms. The
graph shown has 5 vertices and 7 edges. The vertices represent cities, and the
edges represent nonstop airline routes between them. We see that there is a nonstop flight between Berlin and Rome, but no such flight between New York and
Berlin. There are several paths that describe how a person might travel with this

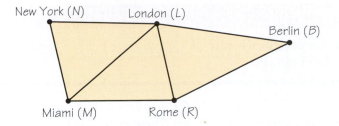

FIGURE 1.2 The edges of the graph show nonstop routes that an airline might offer.

airline from New York to Berlin. The path that seems most direct is New York, London, Berlin, but New York, Miami, Rome, Berlin is also a path. We can describe these two paths as *NLB* and *NMRB*. Another path would be New York, Miami, Rome, London, Berlin, which can be written as *NMRLB*. An example of a circuit is Miami, Rome, London, Miami. It is a circuit because the path starts and ends at the same vertex. This circuit can best be described in symbols by *MRLM.* Another example of a circuit in this graph would be *LRBL,* which is the circuit involving the cities London, Rome, Berlin, and back to London. In this chapter, we are especially interested in circuits, just as we are in real life. Most of us end our day in the same location where we start it—at home! ◼

Returning to the case of parking control in Figure 1.1, we can use a graph to represent the whole territory to be patrolled: Think of each street intersection as a vertex and each sidewalk that contains a meter as an edge, as in Figure 1.3. Notice in Figure 1.3b that the width of the street separating the blocks is not explicitly represented; it has been shrunk to nothing. In effect, we are simplifying our problem by ignoring any distance traveled in crossing streets.

The numbered sequence of edges in Figure 1.4a shows one circuit that covers all the meters (note that it is a circuit because its path returns to its starting point). However, one edge is traversed three times. Figure 1.4b shows another solution that is better because its circuit covers every edge (sidewalk) exactly once. In Figure 1.4b, no edge is covered more than once, or *deadheaded* (a term borrowed from shipping, which means making a return trip without a load).

(a)

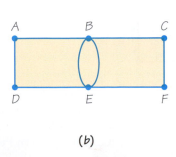

(b)

FIGURE 1.3 (a) A graph superimposed upon a street map. The edges show which sidewalks have parking meters. (b) The same graph enlarged.

:::::: Euler Circuit

A circuit that covers each edge of a graph once but not more than once is called an **Euler circuit**.

Figure 1.4b shows an Euler circuit. These circuits get their name from the great eighteenth-century mathematician Leonhard Euler (pronounced *oy′ lur*), who first studied them (see Spotlight 1.1). Euler was the founder of the theory of graphs, or graph theory. One of his first discoveries was that some graphs have no Euler circuits at all. For example, in the graph in Figure 1.5b, it would be impossible to start at one point and cover all the edges without retracing some steps: If we try to start a circuit at the leftmost vertex, we discover that once we have left the vertex, we have "used up" the only edge meeting it. We have no way to return to our starting point except to reuse that edge. But this is not allowed in an Euler circuit. If we try to start a circuit at one of the other two vertices, we likewise can't complete it to form an Euler circuit.

As mentioned in Spotlight 1.2, realistic problems of this type will involve larger neighborhoods that might require the use of a computer. In addition, there may be other complications that might take us beyond the simple mathematics we want to stick to.

Because we are interested in finding circuits, and Euler circuits are the most efficient ones, we will want to know how to find them. If a graph has no Euler circuit, we will want to develop the next best circuits, those having minimum deadheading. These topics make up the rest of this chapter.

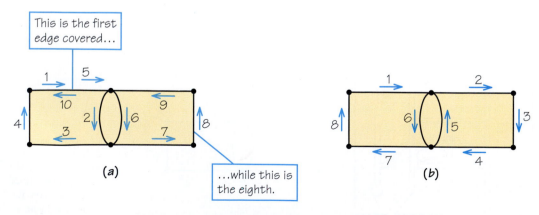

FIGURE 1.4 (a) A circuit and (b) an Euler circuit.

SPOTLIGHT 1.1   Leonhard Euler

Leonhard Euler (1707–1783) was one of those rare individuals who was remarkable in many ways. He was extremely prolific, publishing over 500 works in his lifetime. But he wasn't devoted just to mathematics; he was a people person, too. He was extremely fond of children and had thirteen of his own, of whom only five survived childhood. It is said that he often wrote difficult mathematical works with a child or two in his lap.

Human interest stories about Euler have been handed down through two centuries. He was a prodigy at doing complex mathematical calculations under less than ideal conditions, and he continued to do them even after he became totally blind later in life. His blindness diminished neither the quantity nor the quality of his output. Throughout his life, he was able to mentally calculate in a short time what would have taken ordinary mathematicians hours of pencil-and-paper work. A contemporary claimed that Euler could calculate effortlessly, "just as men breathe, as eagles sustain themselves in the air." His collected works are not yet fully published.

Euler invented the idea of a graph in 1736 when he solved a problem in "recreational mathematics." He showed that it was impossible

Leonhard Euler

(Portrait by Emanuel Handmann, Bildnis des Mathematikers, 1753, Oeffentliche Kunstsammlung Basel, Kunstmuseum.)

to stroll a route visiting the seven bridges of the German town of Königsberg exactly once. Ironically, in 1752 he discovered that three-dimensional polyhedra obey the remarkable formula $V - E + F = 2$ (that is, number of vertices − number of edges + number of faces = 2) but failed to give a proof because he did not analyze the situation using graph theory methods.

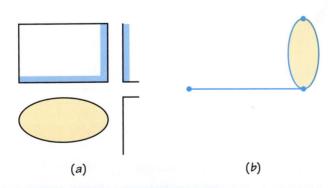

FIGURE 1.5 (a) The three shaded sidewalks cannot be covered by an Euler circuit. (b) The graph of the shaded sidewalks in part (a).

(a)   (b)

Thomas Magnanti, professor of operations research and management, heads the Department of Management Science at MIT's Sloan School of Management. Here are some of his observations:

Thomas Magnanti
(Courtesy of Thomas Magnanti.)

Typically, a management science approach has several different ingredients. One is just structuring the problem—understanding that the problem is an Euler circuit problem or a related management science problem. After that, one has to develop the solution methods.

But one should also recognize that you don't just push a button and get the answer. In using these underlying mathematical tools, we never want to lose sight of our common sense, of understanding, intuition, and judgment. The computer provides certain kinds of insights. It deals with some of the combinatorial complexities of these problems very nicely. But a model such as an Euler circuit can never capture the full essence of a decision-making problem.

Typically, when we solve the mathematical problem, we see that it doesn't quite correspond to the real problem we want to solve. So we make modifications in the underlying model. It is an interactive approach, using the best of what computers and mathematics have to offer and the best of what we, as human beings, with our own decision-making capabilities, have to offer.

1.2 Finding Euler Circuits

Now that we know what an Euler circuit is, we are faced with two obvious questions:

1. Is there a way to tell by calculation, not by trial and error, if a graph has an Euler circuit?
2. Is there a method, other than trial and error, for finding an Euler circuit when one exists?

Euler investigated these questions in 1735 by using the concepts of valence and connectedness.

Valence

The **valence** of a vertex in a graph is the number of edges meeting at the vertex.

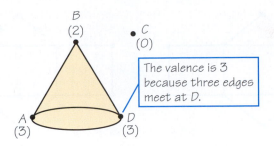

FIGURE 1.6 Valences of vertices.

The valence is 3 because three edges meet at D.

Figure 1.6 illustrates the concept of valence, with vertices A and D having valence 3, vertex B having valence 2, and vertex C having valence 0. Isolated vertices such as vertex C are an annoyance in Euler circuit theory. Because they don't occur in typical applications, we henceforth assume that our graphs have no vertices of valence 0.

Figure 1.3b has four vertices of valence 2, namely, A, C, F, and D. This graph also has two vertices, B and E, of valence 4. Notice that each vertex has a valence that is an even number. We'll soon see that this is very significant.

Connected Graph

A graph is said to be **connected** if for every pair of its vertices there is at least one path connecting the two vertices.

Given a graph, if we can find even one pair of vertices not connected by a path, then we say that the graph is not connected. For example, the graph in Figure 1.7 is not connected because we are unable to join A to D with a path of edges. However, the graph does consist of two "pieces" or connected components, one containing the vertices A, B, F, and G, the other containing C, D, and E. A connected graph will contain a single connected component. Notice that the parking-control graph of Figure 1.3b is connected.

We can now state Euler's theorem, his simple answer to the problem of detecting when a graph G has an Euler circuit:

1. If G is connected and has all valences even, then G has an Euler circuit.

2. Conversely, if G has an Euler circuit, then G must be connected and all its valences must be even numbers.

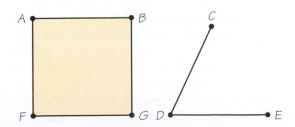

FIGURE 1.7 A nonconnected graph.

FIGURE 1.8 (a) A graph having (b) an Euler circuit.

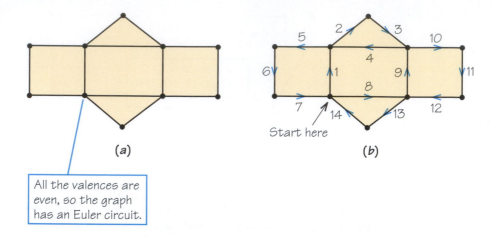

(a)

(b)

Start here

All the valences are even, so the graph has an Euler circuit.

Because the parking-control graph of Figure 1.3b conforms to the connectedness and even-valence conditions, Euler's theorem tells us that it has an Euler circuit. We already have found an Euler circuit for Figure 1.4b by trial and error. For a very large graph, however, trial and error may take a long time. It is usually quicker to check whether the graph is connected and even-valent than to find out if it has an Euler circuit.

Once we know there is an Euler circuit in a certain graph, how do we find it? Many people find that, after a little practice, they can find Euler circuits by trial and error, and they don't need detailed instructions on how to proceed. At this point you should see if you can develop this skill by trying to find Euler circuits in Figure 1.8a, Figure 1.9a, and Figure 1.10. In doing your experiments, draw your graph in ink and the circuit in pencil so you can erase if necessary. Make your graph big and clear so you won't get confused.

If you would like more guidance on how to find an Euler circuit without trial and error, here is a method that works: Never use an edge that is the only link between two parts of the graph that still need to be covered. Figure 1.9b illustrates this. Here we have started the circuit at *A* and gotten to *D* via *B* and *C*, and we want to know what to do next. Going to *B* would be a bad idea because the uncovered part of the graph would then be disconnected into left and right portions. You will never be able to get from the left part back to the right part because you have just used the last remaining link between these parts. Therefore, you should stay on the right side and finish that before using the edge from *D* to *E*. This kind of thinking needs to be applied every time you need to choose a new edge.

Let's see how this works, starting at the beginning at *A*. From vertex *A* there are two possible edges, and neither of them disconnects the unused portion of the graph. Thus, we could have gone either to the left or down. Having gone down to *B*, we now have three choices, none of which disconnects the unused part of the graph. After choosing to go from *B* to *C*, we find that any of the three choices at *C* is acceptable. Can you complete the Euler circuit? Figure 1.9c shows one of many ways to do this.

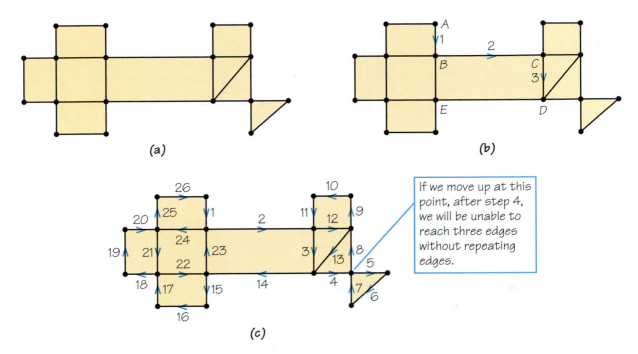

FIGURE 1.9 (a) A graph that has an Euler circuit. (b) A critical junction in finding an Euler circuit in this graph, starting from vertex *A*. (c) A description of a full Euler circuit for this graph.

The method just described leaves many edge choices up to you. When there are many acceptable edges for your next step, you can pick one at random. You might even flip a coin. When computers carry out algorithms of this sort, they use random-number generators, which mimic the flipping of a coin.

Finding an Euler Circuit

EXAMPLE 2

Check the valences of the vertices and the connectivity of the graph in Figure 1.8a to verify that the graph does have an Euler circuit. Now try to find an Euler circuit for that graph. You can start at any vertex. When you are done, compare your solution with the Euler circuit given in Figure 1.8b. If your path covers each edge exactly once and returns to its original vertex (is a circuit), then it is an Euler circuit, even if it is not the same as the one we give. ■

Proving Euler's Theorem

We'll start by proving that if a graph has an Euler circuit, then it must have only even-valent vertices and it must be connected. Let *X* be any vertex of the graph. We will show that the edges at *X* can be paired up, and this will prove that the valence is even. Every edge at *X* is used by the Euler circuit as an outgoing edge (leaving from *X*) or an incoming edge (arriving at *X*). If the Euler circuit starts at *X*, then pair up the first edge used by the circuit with the last one (when the circuit returns to *X* for the last time). In addition, each other edge at *X* that is

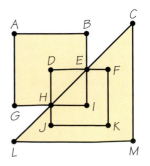

FIGURE 1.10 A graph with an Euler circuit.

FIGURE 1.11 An
Euler circuit starting and
ending at A.

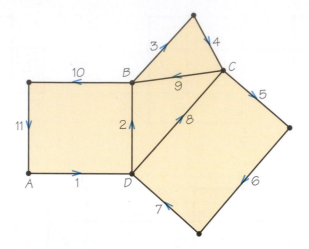

used by the circuit as an incoming edge will be paired with the outgoing edge
that is used next. Because all edges at X are used by the Euler circuit, none more
than once, this pairs up the edges.

But what if X is not the start of the Euler circuit? Then do the pairing like
this: The first incoming edge at X is paired with the outgoing one used next, the
second incoming edge at X is paired with the outgoing one used next, and so on.
For example, in Figure 1.11 at vertex B, we would pair up edges 2 and 3 and
edges 9 and 10. At vertex C, we would pair up edges 4 and 5 and edges 8 and 9.
Can you see how the pairings would work at D? How about vertex A?

In studying this particular example, you might think it would be simpler to
count the edges at a vertex to see that the valence is even. True, but our pairing
method works for a graph about which we know nothing except that it has an
Euler circuit.

To see that a graph with an Euler circuit is connected, note that by following the Euler circuit around we can get from edge to any other edge (it covers
them all) using a portion of the Euler circuit. Because every vertex is on an edge
(there are no vertices of 0 valence), we can get from any vertex to any other using a portion of the Euler circuit.

So far, this is not a complete proof of Euler's theorem. We also need to prove
that if a graph has all vertices even-valent and is connected, then an Euler circuit
can be found for it.

1.3 Beyond Euler Circuits

Now let's see what Euler's theorem tells us about the three-block neighborhood
with parking meters, represented by dots in Figure 1.12a. Figure 1.12b shows the
corresponding graph. (Because we use edges to represent only sidewalks along
which the officer must walk, the sidewalk with no meters is not represented by
any edge in the graph.) This graph has odd valences (at vertices C and G), so
Euler's theorem tells us that there is no Euler circuit for this graph.

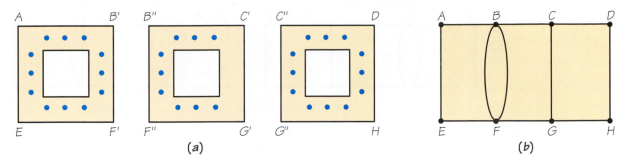

FIGURE 1.12 (a) A street network and (b) its graphic representation. Locations such as *B'* and *B"*, *C'* and *C"*, *F'* and *F"*, and *G'* and *G"* are merged to form the vertices *B*, *C*, *F*, and *G*.

Because we must reuse some edges in this graph to cover all edges in a circuit, for efficiency we need to keep the total length of reused edges to a minimum. This type of problem, in which we want to minimize the length of a circuit by carefully choosing which edges to retrace, is often called the **Chinese postman problem** (like parking-control routes, mail routes need to be efficient). The problem was first studied by the Chinese mathematician Meigu Guan in 1962—hence the name. The remainder of this chapter is dedicated to solving the Chinese postman problem and discussing applications besides parking control.

Solving the Chinese Postman Problem

In a realistic Chinese postman problem, we need to consider the lengths of the sidewalks, streets, or whatever the edges represent, because we want to minimize the total length of the reused edges. However, to simplify things at the start, we can suppose that all edges represent the same length. (This is often called the *simplified* Chinese postman problem.) In this case, we need only count reused edges and need not add up their lengths. To solve the problem, we want to find a circuit that covers each edge and that has the minimal number of reuses of edges already covered.

To follow the procedure we are going to develop, look at the graph in Figure 1.13a, which is the same graph as in Figure 1.12b. This graph has no Euler circuit, but there is a circuit that has only one reuse of an edge (*CG*), namely, *ABCDHGCGFBFEA*. Let's draw this circuit so that when edge *CG* is about to be reused, we install a new, extra, blue edge in the graph for the circuit to use. By duplicating edge *CG*, we can avoid reusing the edge. To duplicate an edge, we must add an edge that joins the two vertices that are already joined by the edge we want to duplicate. (It makes no sense to join vertices that are not already connected by an edge, because such edges would not represent sidewalk sections with meters; see Figure 1.15.)

We have now created the graph of Figure 1.13b. In the graphs we draw, the edges that are added will be shown in color to distinguish them from the original edges, which are shown in black. (You may want to use a similar scheme to

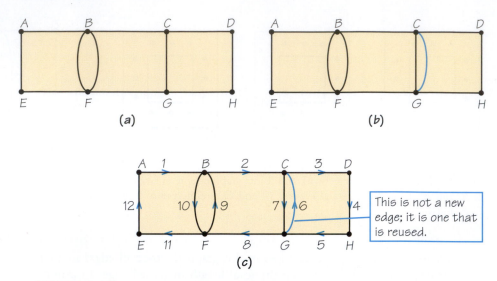

FIGURE 1.13 Making a circuit by reusing an edge.

help you remember which edges are the originals and which are duplicated in the graphs you draw.) In the graph of Figure 1.13b, the original circuit can be traced as an Euler circuit, using the new edge when needed. The circuit is shown in Figure 1.13c. Our theory will be based on using this idea in reverse, as follows:

1. Take the given graph and add edges by duplicating existing ones, until you arrive at a graph that is connected and even-valent. Note that after a graph is *eulerized*, the new graph produced will have an Euler circuit.
2. Find an Euler circuit on the eulerized graph.
3. "Squeeze" this Euler circuit from the eulerized graph onto the original graph by reusing an edge of the original graph each time the circuit on the eulerized graph uses an added edge.

Eulerizing a Graph

Adding edges to a graph to make all valences even is called **eulerizing** the graph.

EXAMPLE 3   Eulerizing a Graph

Suppose we want to eulerize the graph of Figure 1.14a. When we eulerize a graph, we first locate the vertices with odd valence. The graph in Figure 1.14a has two, *B* and *C*. Next, we add one end of an edge at each such vertex, matching up the

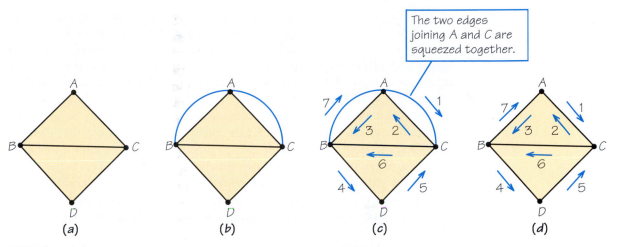

The two edges joining A and C are squeezed together.

FIGURE 1.14 Eulerizing a graph.

new edge with an existing edge in the original graph. Figure 1.14b shows one way to eulerize the graph. Note that B and C have even valence in the second graph. After eulerization, each vertex has even valence. To see an Euler circuit on the eulerized graph in Figure 1.14c, simply follow the edges in numerical order and in the direction of the arrows, beginning and ending at vertex A. The final step, shown in Figure 1.14d, is to "squeeze" our Euler circuit onto the original graph. There are two reuses of previously covered edges. Notice that each reuse of an edge corresponds to an added edge. ■

In the previous example, we noticed that we could count how many reuses we needed by counting added edges. This is generally true in this type of problem: *If you add the new edges correctly, the number of reuses of edges equals the number of edges added during eulerization.*

Adding new edges correctly means adding only edges that are duplicates of existing edges. Doing this makes the rule, just stated in italics, always true, and so it is easy to count the needed reuses.

To see why we add only duplicate edges, examine Figure 1.15a. We need to alter the valences of vertices X and Y by adding edges so that they become evenvalent. Adding one long edge from X to Y (Figure 1.15b) might seem like

(a)

(b)

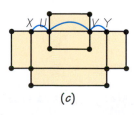

(c)

FIGURE 1.15
Eulerizing when the vertices are more than one edge apart.

an attractive idea, but adding this edge is equivalent to asking a snowplow, say, to get from X to Y without moving along existing streets. At times it is necessary to traverse sections of the graph that have been previously traversed. This is the significance of the duplicated edges. Here the structure of the graph forces us to repeat some edges. We can get away with no fewer than three repeats—the three edges XU, UV, and VY (Figure 1.15c). The duplicated edges are shown in color.

Now that we have learned to eulerize, the next step is to try to get a best eulerization we can—one with the fewest added edges. It turns out that there are many ways to eulerize a graph. It is even possible that the smallest number of added edges can be achieved with two different eulerizations. This is the reason we use the phrase "a best eulerization" rather than "the best eulerization." Remember, we want a best eulerization because this enables us to find the circuit for the original graph that has the minimum number of reuses of edges.

EXAMPLE 4   A Better Eulerization

In Figure 1.16a, we begin with the same graph as in Figure 1.14, but we eulerize it in a different way—by adding only one edge (see Figure 1.16b). Figure 1.16c shows an Euler circuit on the eulerized graph, and in Figure 1.14d we see how it is squeezed onto the original graph. There is only one reuse of an edge, because we added one edge during eulerization. ▪

The solution in Figure 1.16 is better than the solution in Figure 1.14 because one reuse is better than two. These examples suggest the following addition to our solution procedure: Try to find the eulerization with the smallest number of added edges. This extra requirement makes the problem both more interesting and more difficult. For large graphs, a best eulerization may not be obvious. We can try out a few and pick the best among the ones we find, but there may be an even better one that our haphazard search does not turn up.

A systematic procedure for finding a best eulerization does exist, but the process is complicated. There is an especially easy technique for eulerizing the following special category of networks often found in our neighborhoods.

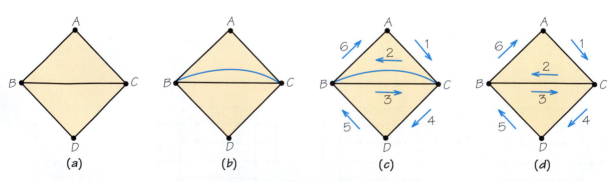

FIGURE 1.16 A better eulerization of Figure 1.14.

Rectangular Network

If a street network is composed of a series of rectangular blocks that form a large rectangle a certain number of blocks high and a certain number of blocks wide, the network is called *rectangular*.

Examples of rectangular street networks (a 3-by-3, a 3-by-4, and a 4-by-4) are shown in Figure 1.17. The graph on the right in each pair shows a best eulerization for the rectangular street network on the left. There appear to be three different eulerization patterns, depending upon whether the rectangle height and width in the original graph are odd or even numbers. In Figure 1.17a, both lengths are 3, both odd; in Figure 1.17b, one length is odd (3) and one is even (4); in Figure 1.17c, both lengths are 4, an even number.

Although the patterns appear different, one technique can be used to create all of them. This technique can be thought of as involving an "edge walker" who walks around the outer boundary of the large rectangle in some direction, say, clockwise. He starts at any corner, say, the upper left corner. As he goes around, he adds edges by the following rules. When he comes to an odd-valent vertex, he links it to the next vertex with an added edge. This next vertex now becomes either even or odd. If it became even, he skips it and continues around, looking for an odd vertex. If it became odd (this could happen only at a corner of the big rectangle), the edge walker links it to the next vertex and then checks this

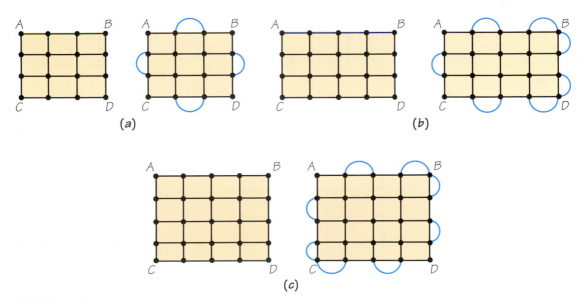

FIGURE 1.17 Eulerizations of three rectangular networks.

vertex to see whether it is even or odd. Each of the three parts of Figure 1.17 has been eulerized by this method.

In a street network that is not rectangular, the eulerization process is started by locating all the vertices with odd valence and then pairing these vertices with each other and finding the length of the shortest path between each pair. We look for the shortest paths because each edge on the connecting paths will be duplicated. The idea is to choose the pairings cleverly so that the sum of the lengths of those paths is the smallest it can be. With a little practice, most people can find a best or nearly best eulerization using only this idea together with trial and error and some ingenuity.

Finding Good Eulerizations

Suppose we want a perfect procedure for eulerizing a graph. What theoretical ideas and methods could we use to build such a tool?

One building block we could use is a method for finding the shortest path between two given vertices of a graph. For example, let us focus on vertices X and Y in Figure 1.18a. Both have odd valence. We can connect them with a pat-

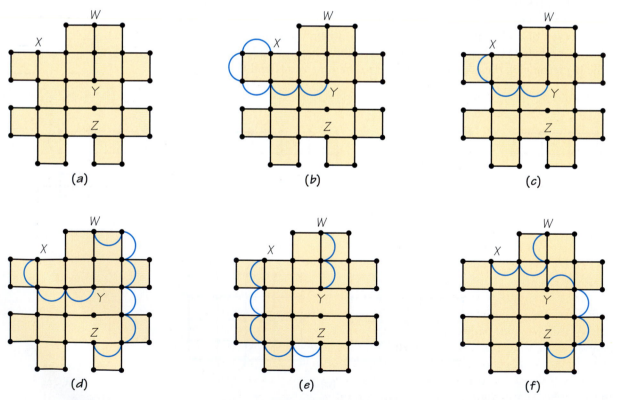

FIGURE 1.18 Choosing among eulerizations.

tern of duplicate edges, as in Figure 1.18b. The cost of this is the length of the path we duplicated from X to Y. A shorter path from X to Y, such as the one shown in Figure 1.18c, would be better. Fortunately, the *shortest-path problem* has been well studied, and we have many good procedures for solving it exactly, even in large, complex graphs.

But there is more to eulerizing the graph in Figure 1.18a than dealing with X and Y. Notice that we have odd valences at Z and W. Should we connect X and Y with a path, and then connect Z and W, as in Figure 1.18d? Or should we connect X to Z and Y to W, as in Figure 1.18e? Another alternative is to use connections X to W and Y to Z, as in Figure 1.18f. It turns out that the alternatives in both Figures 1.18e and 1.18f are preferable to the one in Figure 1.18d, because they involve seven added edges, whereas Figure 1.18d uses nine.

Remember that when an unweighted graph is eulerized in an optimal way, then the total cost of traversing each edge at least once can be found by adding the total number of edges in the graph to the number of edges that are reused (duplicated). Small problems involving eulerization can be carried out by trial-and-error methods. Unfortunately, although there is an algorithm that can be applied to find the best eulerization for large problems, the details of this algorithm are quite complex. However, the procedure works not only for graphs without weights but also for graphs with weights on the edges.

1.4 Urban Graph Traversal Problems

Euler circuits and eulerizing have many more practical applications than just checking parking meters. Almost anytime services must be delivered along streets or roads, our theory can make the job more efficient. Examples include collecting garbage, salting icy roads, plowing snow, inspecting railroad tracks, and reading electric meters (see Spotlight 1.3).

Each of these problems has its own special requirements that may call for modifications in the theory. For example, in the case of garbage collection, the edges of our graph will represent streets, not sidewalks. If some of the streets are one-way, we need to put arrows on the corresponding edges, resulting in a directed graph, or **digraph.** The circuits we seek will have to obey these arrows. In the case of salt spreaders and snowplows, each lane of a street needs to be modeled as a direct edge, as shown in Figure 1.19. Note that the arrows on the map and digraph are not in color because these arrows denote restrictions in traversal possibilities, not parts of circuits.

Like salt spreaders, street-sweeping trucks can travel in only one lane at a time and must obey the direction of traffic. Street sweepers, however, have an additional complication: parked cars. It is very difficult to clean the street if cars are parked along the curb. Yet for overall efficiency, those who are responsible for routing street sweepers want to interfere with parking as little as possible. The common solution is to post signs specifying times when parking is prohibited. Because the parking-time factor is a constraint on street sweeping, it is important to find not only an Euler circuit, or a circuit with very few duplications, but also

(a)

(b)

FIGURE 1.19 (a) Salt-spreading route, where each west–west street has two traffic lanes in the same direction, and (b) an appropriate digraph model.

SPOTLIGHT 1.3   Israel Electric Company Reduces Meter-Reading Task

The Beersheba branch of Israel's major electric company wanted to make the job of meter reading more efficient. When the branch managers decided to minimize the number of people required to read the electric meters in the houses of one particular neighborhood, they set a precedent by applying management science. Formerly, each person's route had been worked out by trial and error and intuition, with no help from mathematics. The whole job required 24 people, each doing a part of the neighborhood in a five-hour shift.

At first, it looks as though one would find a more efficient way of doing the work the same way as in the Chinese postman problem, but there are two important differences. First, the neighborhood was big enough to negate any possibility of having only one route assigned to one person. Instead, it was necessary to find a number of routes that, taken together, covered all the edges (sidewalks). Second, a meter reader who was done with a route was allowed to return home directly. Thus, there was no reason for the individual routes to return to their starting points; therefore, routes could be paths instead of circuits.

The Beersheba researchers found solutions to these problems by modifying the basic ideas we have described in this chapter. They managed to cover the neighborhood with 15 five-hour routes, a 40% reduction of the original 24 five-hour routes. Altogether, these routes involve a total of 4338 minutes of walking time, of which 41 minutes (less than 1%) is deadheading.

a circuit that visits streets when they are free of cars. Once again, mathematicians have developed techniques to handle this constraint.

Finally, because towns and cities of any size will have more than one street sweeper, parking officer, or garbage truck, a single best route may not suffice. Instead, it becomes necessary to divide the territory into multiple routes. The general goal is to find optimal solutions while taking into account traffic direction, number of lanes, parking-time restrictions, and divided routes (see Figure 1.20).

Management science makes all this possible. For example, a pilot study done in the 1970s in New York City showed that applying these techniques to street sweepers in just one district could save about $30,000 per year. With 57 sanitation districts in New York, this would amount to a savings of more than $1.5 million in a single year. To translate these totals into 2005 terms, simply multiply by 5. In addition, the same principles could be extended to garbage collection, parking control, and other services carried out on street networks.

This plan was not adopted when first proposed. Because city services take place in a political context, several other factors come into play. For example, union leaders try to protect the jobs of city workers, bureaucrats might try to keep their departmental budgets high, and elected politicians rarely want to be accused of cutting the jobs of their constituents. Thus, political obstacles can overrule management science. As mentioned in Spotlight 1.2, such human factors often

(a)

(b)

FIGURE 1.20
(a) Residential neighborhoods, whether they be in cities or the suburbs, require many services such as mail delivery, garbage collection, street sweeping, meter reading, or sewage systems. The mathematical techniques of operations research make it possible to provide these services as cheaply as possible. When optimal solutions to providing such services can be found, everyone is a winner. *(Brand X Pictures/ Picturequest.)*
(b) Computers can be used to extract the essential information needed to solve routing problems from photographs.

arise when applying management science. Perhaps a more acceptable street-sweeping plan would have been devised for New York City if more attention had been paid to the human factors earlier.

Despite the complications of real-world problems, management science principles provide ways to understand these problems by using graphs as models. We can reason about the graphs and then return to the real-world problem with a workable solution. The results we get can have a lasting effect on the efficiency and economic well-being of any organization or community.

▓▓▓▓ REVIEW VOCABULARY

Chinese postman problem The problem of finding a circuit on a graph that covers every edge of the graph at least once and that has the shortest possible length.

Circuit A path that starts and ends at the same vertex.

Connected graph A graph is connected if it is possible to reach any vertex from any specified starting vertex by traversing edges.

Digraph A graph in which each edge has an arrow indicating the direction of the edge. Such directed edges are appropriate when the relationship is "one-sided" rather than symmetric (for instance, one-way streets as opposed to regular streets).

Edge A link joining two vertices in a graph.

Euler circuit A circuit that traverses each edge of a graph exactly once.

Eulerizing Adding new edges to a graph so as to make a graph that possesses an Euler circuit.

Graph A mathematical structure in which points (called vertices) are used to represent things of

interest and in which links (called edges) are used to connect vertices, denoting that the connected vertices have a certain relationship.

Management science A discipline in which mathematical methods are applied to management problems in pursuit of optimal solutions that cannot readily be obtained by common sense.

Operations research Another name for management science.

Optimal solution When a problem has various solutions that can be ranked in preference order (perhaps according to some numerical measure of "goodness"), the optimal solution is the best-ranking solution.

Path A connected sequence of edges in a graph.

Valence (of a vertex) The number of edges touching that vertex.

Vertex A point in a graph where one or more edges end.

✔ SKILLS CHECK

1. What is the valence of vertex A in the graph below?

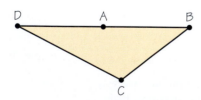

(a) 3
(b) 2
(c) 4

2. The valences of the vertices in the accompanying graph listed in decreasing order are

(a) 1, 1, 1, 2, 3, 3, 4, 5.
(b) 5, 4, 4, 2, 1, 1, 1.
(c) 5, 4, 3, 3, 2, 1, 1, 1.

3. The graph shown below is not connected because

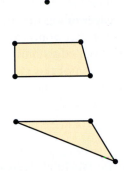

(a) it has all even-valent vertices.
(b) it consists of two parts.
(c) it consists of three parts.

4. A vertex in the graph below that is even-valent is

(a) C.
(b) A.
(c) E.

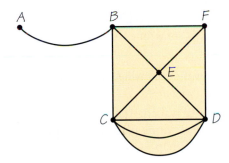

5. The graph below has

(a) four vertices and six edges.
(b) four vertices and four edges.
(c) five vertices and six edges.

6. Which of the following statements is true about a *path*?

(a) A path always forms a circuit.
(b) A path is always connected.
(c) A path can visit any vertex only once.

7. If a graph consists of four vertices and every pair of vertices is connected by a single edge, how many edges are in the graph?

(a) Four
(b) Five
(c) Six

8. It is not possible for a graph to have five vertices of valence 3 and six vertices of valence 4 because

(a) there are no graphs with exactly 11 vertices.
(b) a graph cannot have an even number of 4-valent vertices.
(c) a graph cannot have an odd number of odd-valent vertices.

9. If a graph is connected and has six vertices, what can be said about the number of edges in the graph?

(a) There are at least five edges in the graph.
(b) There are exactly five edges in the graph.
(c) There are at least six edges in the graph.

10. For which of the situations below is it most desirable to find an Euler circuit or an efficient eulerization of the graph?

(a) Sweeping the sidewalks of a small town
(b) Planning a new highway
(c) Planning a parade route in Muncie, Indiana

11. Consider the path represented by the sequence of numbered edges on the graph below. Which statement is correct?

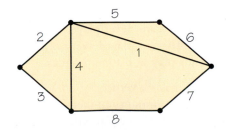

(a) The sequence of numbered edges forms an Euler circuit.

(b) The sequence of numbered edges traverses each edge exactly once but is not an Euler circuit.

(c) The sequence of numbered edges forms a circuit but not an Euler circuit.

12. What is the minimum number of duplicated edges needed to create a good eulerization for the graph below?

(a) Four edges
(b) Three edges
(c) Two edges

13. For the graph below, which statement is correct?

(a) The graph has an Euler circuit.
(b) One new edge is required to eulerize this graph.
(c) Three new edges are required to eulerize this graph.

14. Suppose each vertex of a graph represents a baseball team and each edge represents a game played by two baseball teams. If the resulting graph is not connected, which of the following statements must be true?

(a) At least one team never played a game.
(b) At least one team played every other team.
(c) The teams play in distinct leagues.

15. If a graph has six vertices of odd valence, what is the absolute minimum number of edges that must be added (duplicated) to eulerize the graph?

(a) Zero
(b) Two
(c) Three

16. Suppose the edges of a graph represent streets that must be plowed after a snowstorm. To eulerize the graph, four edges must be added. The real-world interpretation of this is that

(a) four streets will not be plowed.
(b) four streets will be traversed twice.
(c) four new streets would be built.

17. If the vertices of a graph represent cities and the edges of a graph represent flight routes for a particular airline, then which of the objects below best models a pilot's daily schedule?

(a) A path
(b) An Euler circuit
(c) A graph

18. If the valences of the vertices of a graph are added together, what can be said about their sum?

(a) The sum is always even.
(b) The sum is always odd.
(c) The sum is sometimes even and sometimes odd.

19. Which of the following situations is best modeled by a *digraph*?

(a) A system of walking trails
(b) An electrical wiring plan for a home
(c) A bus route map

20. Suppose a civic club offers several craft courses, and each club member can choose to participate in up to two different courses. Let each vertex of a graph represent one of these courses and each edge represent a club member who wants to take the two courses represented by the vertices at its endpoints. What can be said about the vertices in the resulting graph whose valence is zero?

(a) There are no vertices whose valence is zero.
(b) These vertices represent courses that can occur at the same time without displeasing any club member.
(c) These vertices represent the least popular courses.

EXERCISES   ■ Challenge   ◆ Discussion

Basic Concepts

1. In the graph below, the vertices represent cities and the edges represent roads connecting them. What are the valences of the vertices in this graph? (Keep in mind that *E* is part of the graph.) What might the valence of city *E* be showing about the geography?

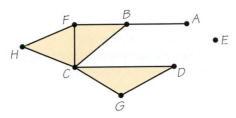

2. In the two graphs below, the vertices represent cities and the edges represent roads connecting them. In which graphs could a person located in city *A* choose any other city and then find a sequence of roads to get from *A* to that other city?

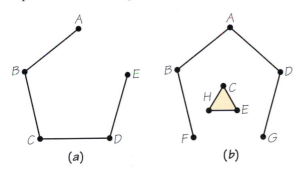

(a)   (b)

3. (a) Is the figure below a graph? explain your answer.

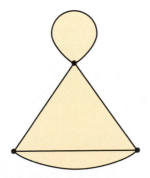

(b) The graph below has edges that "cross" at points that are not vertices of the graph. Which edges are these?

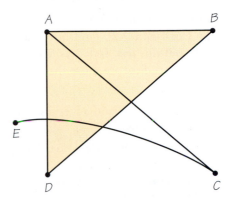

(c) How many vertices and edges are there in the graph above?

4. The graph below shows the stores and roads connecting them in a small shopping mall.

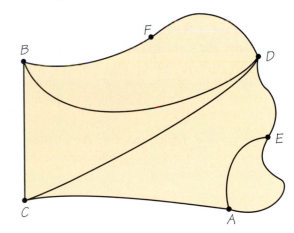

(a) How many stores does the mall have?
(b) How many roads connect up the stores in the mall?
(c) Write down a path from *C* to *F*.
(d) Write down a path from *E* to *B*.

5. Refer to the figure in Exercise 4.

(a) Write down a circuit the includes the vertices *C* and *D* but does not start or end at either of these vertices.

(b) If two paths are considered different if they use different edges, write down:

  (i) two different paths from *B* to *D*.
  (ii) three different paths from *C* to *F*.
  (iii) a circuit that has four edges.

6. Jack and Jill are located in Miami and want to fly to Berlin (see Figure 1.2). Find three paths for them to carry out this trip. Jack says that a path can repeat edges so there are more than three. Is Jack right?

7. (a) How many vertices and edges does the graph in Figure 1.6 have?

(b) How many vertices and edges does the graph in Figure 1.7 have?

(c) How many vertices and edges does the graph in Figure 1.8a have?

8. (a) Add up the numbers you get for the valences of the vertices in Figure 1.6.

(b) Add up the numbers you get for the valences of the vertices in Figure 1.7.

(c) Add up the numbers you get for the valences of the vertices in Figure 1.8a.

(d) Describe the pattern you see in the answers you got for parts (a) through (c).

(e) Show that the pattern describes a fact that is true for any graph. (*Hint:* How many endpoints does an edge have?)

9. In the graph in Figure 1.8a, find the smallest possible number of edges you could remove that would disconnect the graph.

10. In the graphs in Figure 1.17, find the smallest possible number of edges you could remove that would disconnect the graph.

11. Draw a graph with six vertices that is connected where

(a) each vertex has valence 3.
(b) each vertex has valence 4.

12. (a) Draw a connected graph with six vertices, all of whose vertices have valence 2.

(b) Draw a disconnected graph with 6 vertices, all of whose vertices have valence 2.

(c) Do all graphs with 6 vertices having valence 2 have the same number of edges?

13. Is it possible that a street network gives rise to a disconnected graph? If so, draw such a network of blocks and streets and parking meters (in the style of Figure 1.12a). Then draw the disconnected graph it gives rise to.

14. (a) Draw a graph where every vertex has valence of at least 3 but where removing a single edge disconnects the graph.

(b) In what urban settings might a road network be represented by a graph that has an edge whose removal would disconnect the graph?

15. (a) Find a graph where the valences of the seven vertices of the graph are 1, 2, 2, 3, 3, 3, 4.

(b) Find another graph with the same valences as above that is "different" from the one you found for part (a).

16. For some services provided along streets, it may matter whether the roads are one-way or two-way. Give some examples where the street directions do and do not matter for our graph model analysis.

◆ **17.** A postal worker is supposed to deliver mail on all streets represented by edges in the graph below by traversing each edge exactly once. The first day the worker traverses the numbered edges in the order shown in (a), but the supervisor is not satisfied—why? The second day the worker follows the path indicated in (b), and the worker is unhappy—why? Is the original job description realistic? Why?

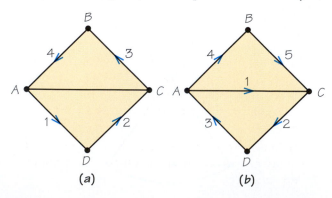

(a)   (b)

18. For the street network in Exercise 17, draw the graph that would be useful for routing a snowplow. Assume that all streets are two-way, one lane in each direction, and that you need to pass down each lane separately.

19. Find an efficient route for the snowplow to follow in the graph you drew in Exercise 18.

20. (a) Give examples of services that could be performed by a vehicle that moved in the direction of traffic down either lane of a two-way street.

(b) Give examples of services that would probably require a vehicle to travel down each of the lanes of a two-way street (in the direction of traffic for that lane) to perform the service.

21. For the street network shown below, draw the graph that would be useful for finding an efficient route for checking parking meters. (*Hint:* Notice that not every sidewalk has a meter; see Figure 1.12.)

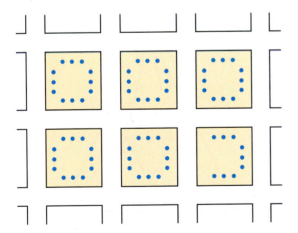

22. (a) For the street network in Exercise 21, draw a graph that would be useful for routing a garbage truck. Assume that all streets are two-way and that passing once down a street suffices to collect from both sides.

(b) Do the same problem on the assumption that one pass down the street suffices to collect from only one side.

23. (a) In the graph below, find the largest number of paths from *A* to *F* that do not have any edges in common.

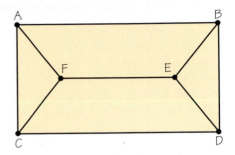

(b) Verify that the largest number of paths with no edges in common between any pair of vertices in this graph is the same.
(c) Why might one want to be able to design graphs such that one can move between two vertices of the graph using paths that have no edges in common?

Euler Circuits

24. Examine the paths represented by the numbered sequences of edges in both parts of the figure below. Determine whether each path is a circuit. If it is a circuit, determine if it is an Euler circuit.

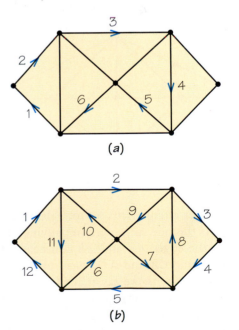

25. In Figure 1.13c, suppose we started an Euler circuit using this sequence of edges: 6, 7, 8, 9 (ignore existing arrows on the edges). What does our guideline for finding Euler circuits tell you *not* to do next?

26. In Figure 1.8b, suppose we started an Euler circuit using this sequence of edges: 14, 13, 8, 1, 4 (ignore existing arrows on the edges). What does our guideline for finding Euler circuits tell you *not* to do next?

27. Find an Euler circuit on the graph of Figure 1.15c (including the blue edges).

28. Find Euler circuits in the right-hand graphs in Figures 1.17a and 1.17b.

29. In the graph below, we see a territory for a parking-control officer that has no Euler circuit. How many sidewalks (edges) need to be dropped in order to enable us to find an Euler circuit? What effect would this have in the associated real-world situation?

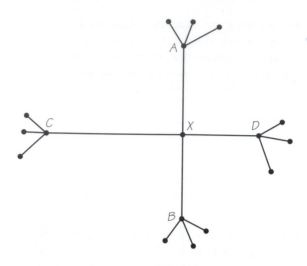

Suppose *G* is a four-valent graph such as that in the diagram below. Is it possible to find an Euler circuit of this graph that never cuts through any vertex? Explain why it might be desirable to find an Euler circuit of this special kind in an applied situation.

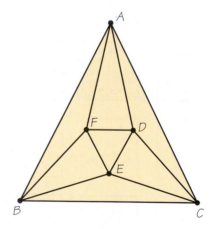

30. An Euler circuit visits a four-valent vertex *X*, such as the one in the accompanying graph, by using the edges *AX* and *XB* consecutively, and then using *CX* and *XD* consecutively. When this happens, we say that the Euler circuit cuts through at *X*.

Eulerization and Squeezing

31. Find an Euler circuit on the eulerized graph (b) of the following figure. Use it to find a circuit on the original graph (a) that covers all edges and reuses edges only five times.

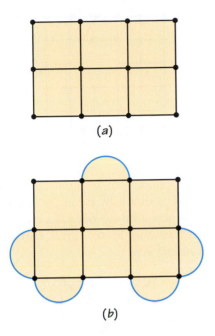

(a)

(b)

32. Squeeze the circuit shown in graph (a) below onto graph (b). Show your answers by writing numbered arrows on the edges and by listing a sequence of vertices (for example, *ABEB . . . A*).

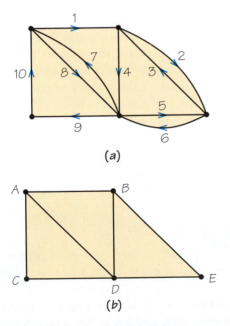

(a)

(b)

Then squeeze the circuit shown in graph (c) onto graph (d). Show your answers by writing numbered arrows on the edges and by listing a sequence of vertices.

(c)

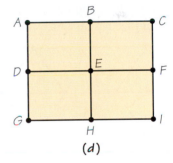

(d)

33. For the graph below:

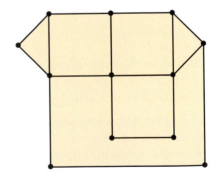

(a) Determine the minimum number of edges that have to be removed for the resulting graph to have all even-valent vertices.

(b) Does the graph you obtain in part (a) have an Euler circuit?

For the graph below:

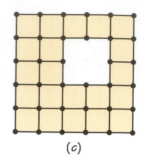

(c)

(c) Determine the minimum number of edges that have to be removed for the resulting graph to have all even-valent vertices.

(d) Does the graph you obtain in part (c) have an Euler circuit?

34. In the graph below, add one or more edges to produce a graph that has an Euler circuit.

35. Eulerize these rectangular street networks using the same patterns that would be used by the edge walker described in the text.

(a) A 5 × 5 rectangle
(b) A 4 × 5 rectangle
(c) A 6 × 6 rectangle
(d) Can you find an eulerization with nine added edges for a 2-by-7-block rectangular street network? Can you do better than nine added edges?

36. Find good eulerizations for these graphs, using as few duplicated edges as you can. See "Finding Good Eulerizations" for hints.

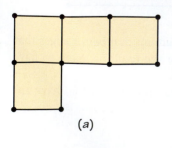

(a)   (b)

37. A college campus has a central square with sides arranged as shown by the edges in the graph below. Show how all these sidewalks can be traversed at least once in a tour that starts and ends at the same vertex.

38. The figure below shows a river, some islands, and bridges connecting the islands and riverbanks. A charity is sponsoring a race in which entrants have to start at *A*, go over each bridge at least once, and end at *A*. Draw a graph that would be useful for finding a route that requires the least recrossing of bridges. Show what that route would be. (*Historical note:* This situation resembles the one that inspired Leonhard Euler's 1736 "recreational mathematics" problem that resulted in the first work in graph theory.)

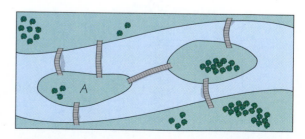

39. Find a circuit in the accompanying graph that covers every edge and has as few reuses as possible.

40. In the figure below, all blocks are 1000-by-1000 feet, except for the middle column of blocks, which are 1000-by-4000 feet. Find a circuit of minimum total length that covers all edges.

41. In the figure below, all blocks are 1000-by-1000 feet, except for the middle column of blocks, which are 1000-by-4000 feet. Find a circuit of minimum total length that covers all edges.

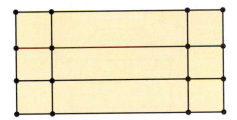

42. (a) Find the cheapest route in the graph below, where one starts at vertex *A*, finishes at vertex *A*, and traverses each edge at least once. The cost of a route is computed by summing the numbers along the edges that one uses.

(b) How many edges are repeated in the minimal-cost route?

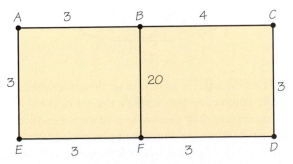

(c) Discuss the implications of this example for the relation between finding good eulerizations of graphs and the problem of finding cheap routes that start and end at the same vertex and traverse each edge at least once.

(d) The physical edge with cost 20 in the diagram is not physically longer than other edges with lower costs attached to them. Explain why in an urban setting it might make sense to assign two stretches of street of similar length very different "costs" for traversing them.

(e) What are some different meanings that "weights" (for example, traffic volume) potentially assigned to edges in a graph might have in an urban setting?

43. Which graphs (see figures below) have Euler circuits? In the ones that do, find the Euler circuits by numbering the edges in the order the Euler circuit uses them. For the ones that don't, explain why no Euler circuit is possible.

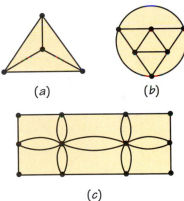

44. Eulerize the graph below by using four new edges. Find an Euler circuit in the eulerized graph and use that circuit to find a circuit of the original graph that covers all edges but reuses edges only four times. How many different ways can the four edges be chosen?

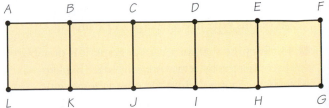

■ **45.** A graph *G* represents a street network to be traveled by a postal worker who must traverse every street twice, once for each side of the street. In graph *G*, the edges represent sidewalks. Does such a graph always have an Euler circuit? Explain your answer.

46. In the graph below, find a circuit that covers every edge and has as few reuses as possible.

47. (a) Find the best eulerizations you can for the two graphs below.

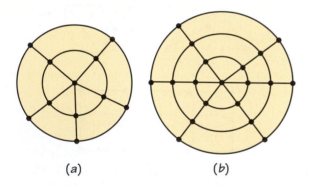

(a)   (b)

(b) Graph (a) can be thought of as having five rays and two circles, and graph (b) as having six rays and three circles. Draw a graph with four rays and four circles and find the best eulerization you can for this graph.
(c) Find a "formula" involving *r* and *s* for the smallest number of edges needed to eulerize a graph of this type having *r* rays and *s* circles.

■ **48.** Suppose that for a certain graph it is possible to disconnect it by removing one edge. Explain why such a graph (before the edge is removed)

must have at least one vertex of odd valence. (*Hint:* Show that it cannot have an Euler circuit.)

49. Can you draw a graph with six vertices where the valence of each vertex is 5?

50. Each graph below represents the sidewalks to be cleaned in a fancy garden (one pass over a sidewalk will clean it). Can the cleaning be done using an Euler circuit? If so, show the circuit in each graph by numbering the edges in the order the Euler circuit uses them. If not, explain why no Euler circuit is possible.

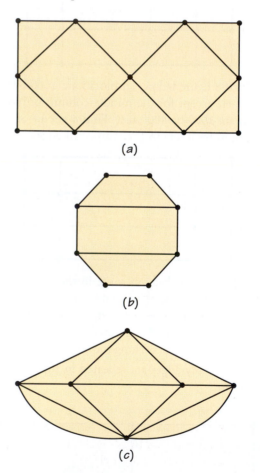

(a)

(b)

(c)

■ **51.** If an edge is added to an already existing graph, connecting two vertices already in the graph, explain why the number of vertices with odd valence has the same parity before and after.

(This means if it was even before, it is even after, while if it was odd before, it remains odd.)

■ **52.** Any graph can be built in the following fashion: Put down dots for the vertices, then add edges connecting the dots as needed. When you have put down the dots, and before any edges have been added, is the number of vertices with odd valence an even number or an odd number? What is the number of vertices with odd valence when all the edges have been added (see Exercise 51)?

53. Draw the graph for the parking-control territory shown in the figure below. Label each vertex with its valence and determine if the graph is connected.

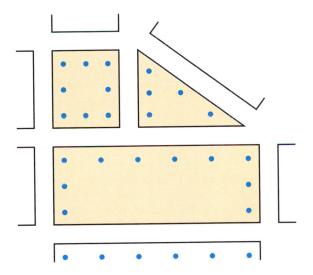

54. If a rectangular street network is *r* blocks by *s* blocks, find a formula for the minimum number of edges that must be added to eulerize a graph representing the network in terms of *r* and *s*. (*Hint:* Treat the case $r = 1$ separately. Test your formula with the cases 6 blocks by 5 blocks, 6 blocks by 6 blocks, and 5 blocks by 3 blocks.)

◆ **55.** The word *valence* is also used in chemistry. Find out what it means in chemistry and explain how this usage is similar to the use we make of it here.

■ **56.** For the street network below, draw a graph that represents the sidewalks with meters. Then find the minimum-length circuit that covers all sidewalks with meters. If you drew the graph as we recommended, you would find that the shortest circuit has length 18 (it reuses every edge).

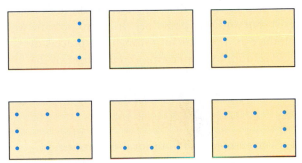

But the meter checker comes to you and says: "I don't know anything about your theories, but I have found a way to cover the sidewalks with meters using a circuit of length 10. My trick is that I don't rule out walking on sidewalks with no meters." Explain what he means and discuss whether his strategy can be used in other problems.

57. Each edge of the accompanying graph represents a two-lane highway. A grass-mowing machine is located at *A*, and its operator has the job of cutting the grass along each of the edges of road shown. Find a tour for the mowing machine that begins and ends at *A*. Find such a tour that begins and ends at *A* and, as the mowing is done, moves along the edge of the road in the same direction as the traffic is going.

APPLET EXERCISES

To do these exercises, go to
www.whfreeman.com/fapp7e.

Eulerizing a Graph

We learned that if a graph has exactly two vertices with odd valences, then an Euler circuit does not exist—but an Euler path does. It is also possible to produce an Euler circuit through the process of eulerization, by duplicating certain edges of the graph. But how many duplications are necessary to obtain an Euler circuit? Investigate this problem and more general related topics using the *Eulerizing a Graph* applet.

Euler Circuits

We know that if all the vertices have even valence, then an Euler circuit exists. Try your hand at finding such circuits in the *Euler Circuit* applet.

 ## WRITING PROJECTS

1. Write a memo to your local department of parking control (or police department) in which you suggest that management science techniques like the ones in this chapter be used to plan routes. Assume that the person to whom you are writing is not extensively trained in mathematics but is willing to read through some technical material, provided you make it seem worth the trouble.

2. Do the same as in Writing Project 1, but to the department in charge of spreading salt on roads after snowstorms.

3. If you were making a recommendation to the mayor of New York City concerning proposed new street-sweeping routes, designed using the theory of this chapter, would you recommend that the changes be adopted or not? Write a memo that outlines the pros and cons as fairly as you can, and then conclude with your recommendation.

SUGGESTED READINGS

BELTRAMI, EDWARD J., *Models for Public Systems Analysis*, Academic Press, New York, 1977. This book gives a good overview of the way that operations research has provided and continues to provide new tools for solving societal problems. Among the ideas discussed are police patrol tactics, organization of emergency services, and scheduling. Some of the mathematics used is advanced.

MALKEVITCH, JOSEPH, and WALTER MEYER, *Graphs, Models and Finite Mathematics*, Prentice-Hall, Englewood Cliffs, NJ, 1974. This introductory book includes much of the same material as presented here but provides more details of the proofs and uses somewhat different algorithms for solving the problems involved.

The following books treat many of the topics discussed here as well as shortest-path problems and matching problems, and they formulate some problems in more realistic terms:

ROBERTS, FRED S., AND BARRY TESMAN, *Applied Combinatorics*, Second Edition, Pearson Prentice Hall, Upper Saddle River, NJ, 2004.

TUCKER, ALAN. *Applied Combinatorics*, Third Edition, Wiley, New York, 1995.

SUGGESTED WEB SITES

www.hsor.org/what_is_or.cfm This site discusses the history of operations research (OR) and some of the areas where OR is being applied. Be sure to follow the "Networks Routing" link to see applications of the Chinese postman problem.

www.maths.may.ie/images/konig.html This site treats problems involving the traversing of edges in graphs, including the case where one does not start and end at the same vertex.

www-gap.dcs.st-and.ac.uk/~history/ Mathematicians/Euler.html This essay discusses the numerous contributions that Euler made to mathematics, and provides biographical information about him.

Business Efficiency

In the previous chapter, we saw that there was an easy way of telling whether a connected graph has a circuit that traverses each of the edges of a graph exactly once—for example, a route for a snowplow that covers the streets of a section of a town only once. However, the situation changes radically if we make a seemingly small change in the problem: When is it possible to find a route along distinct edges of a graph that visits each *vertex* once and only once in a simple circuit? Perhaps there has been a hurricane and it is important to check whether or not the storm sewers at every corner in town are clogged.

This problem is called the *Hamiltonian circuit problem,* and, like the Euler circuit problem, it is a graph theory problem. The Hamiltonian circuit problem has many applications. Suppose inspections or deliveries need to be made at each vertex (rather than along each edge) of a graph. An "efficient" tour of the graph would be a route that started and ended at the same vertex and passed through all the vertices without reuse, or repetition; that is, the route would be a Hamiltonian circuit. Such routes would be useful for inspecting traffic signals or for delivering mail to drop-off boxes, which hold heavy loads of mail so that urban postal carriers do not have to carry them long distances.

For example, the wiggly line in Figure 2.1a shows a circuit we can take to tour that graph, visiting each vertex once and only once. This tour can be written *ABDGIHFECA.* Note that another way of writing the same circuit would be *EFHIGDBACE.* A different circuit visiting each vertex once and only once would be *CDBIGFEHAC* (Figure 2.1b). Do not be confused because *C* is written twice when we write down this list of vertices. We can think of the circuit as starting at any of its vertices, but we do start and end at the same vertex.

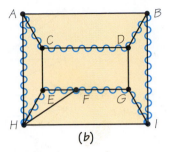

FIGURE 2.1 Wiggly edges illustrate Hamiltonian circuits.

Hamiltonian Circuit

A tour, like the ones marked by wiggly edges in Figure 2.1, that starts at a vertex of a graph and visits each vertex once and only once, returning to where it started, is called a **Hamiltonian circuit.**

2.1 Hamiltonian Circuits

The concept is named for the Irish mathematician William Rowan Hamilton (1805–1865), who was one of the first to study it. We now know that the concept was discovered somewhat earlier by Thomas Kirkman (1806–1895), a British minister with a penchant for mathematics.

The concepts of Euler and Hamiltonian circuits are similar in that both forbid reuse: the Euler circuit of edges, the Hamiltonian circuit of vertices. However, it is far more difficult to determine which connected graphs admit a Hamiltonian circuit than to determine which connected graphs have Euler circuits. As we saw in Chapter 1, looking at the valences of vertices tells us whether a connected graph has an Euler circuit, but we have no such simple method for telling whether or not a graph has a Hamiltonian circuit.

Some special classes of graphs are known to have Hamiltonian circuits, and some special classes of graphs are known to lack them. For example, here is a method for constructing an infinite family of graphs where each graph in the family cannot have a Hamiltonian circuit. Construct a vertical column of m vertices and a parallel column of n vertices, where m is bigger than n, as shown in Figure 2.2. The figure illustrates the typical case where $m = 4$ and $n = 2$. Now join each vertex on the left in the figure to every vertex on the right. As m and n vary, we get a family of different graphs.

No graph obtained in this manner can have a Hamiltonian circuit. If a Hamiltonian circuit existed, it would have to alternately include vertices on the left and right of the figure. This is not possible because the number of vertices on the left and right are not the same. It is unlikely that a method will ever be found to easily determine whether or not a graph has a Hamiltonian circuit. If Hamiltonian circuits were easy to find in any graph at all, many applied problems could be solved in a less costly way.

It may seem that delivering mail in a cheap and timely manner should not be that hard. However, finding the optimal way to deliver mail over a variety of environments, rural, suburban, and urban, is very complex. How should a large geographic area be divided into smaller sections? Should each mail carrier use a truck as a "depot" to resupply small amounts of mail for delivery or should there be deposit boxes on street corners? Mathematics can be used to get answers to such questions. *(Lawrence Migdale/Photo Researchers, Inc.)*

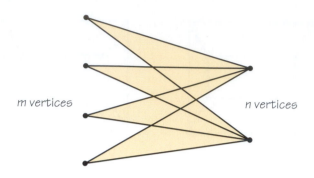

FIGURE 2.2 An example of one graph from a family of graphs that has no Hamiltonian circuit. The number of vertices *m* on the left is chosen to be greater than the number of vertices *n* on the right. The case $m = 4$ and $n = 2$ is shown.

The Hamiltonian circuit problem itself has many applications. This is not unusual in mathematics. Often mathematics used to solve a particular real-world problem leads to new mathematics that suggests applications to other real-world situations. One class of problems to which we can apply Hamiltonian circuits is vacation planning.

EXAMPLE 1   Vacation Planning

Let's imagine that you are a college student studying in Chicago. During spring break you and a group of friends have decided to take a car trip to visit other friends in Minneapolis, Cleveland, and St. Louis. There are many choices as to the order of visiting the cities and returning to Chicago, but you want to design a route that minimizes the distance you have to travel. Presumably, you also want a route that cuts costs, and you know that minimizing distance will minimize the cost of gasoline for the trip. Similar problems with different complications would arise for bus, railroad, or airplane trips.

Express mail and parcel post delivery companies need to make complicated patterns of deliveries and pick-ups. To do this they need to know driving distances between the various geographical locations involved. Using this information, together with driving times, they can use mathematics to cut costs and to make the pick-ups and deliveries on time. *(© Rhoda Sidney/PhotoEdit.)*

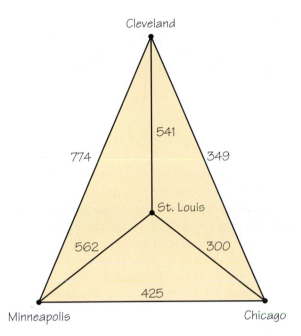

FIGURE 2.3 Road mileages between four cities.

Imagine now that the local automobile club has provided you with the intercity driving distances between Chicago, Minneapolis, Cleveland, and St. Louis. We can construct a graph model with this information, representing each city by a vertex and the legs of the journey between the cities by edges joining the vertices. To complete the model, we add a number called a **weight** to each graph edge, as in Figure 2.3. In this example, the weights represent the distances between the cities, each of which corresponds to one of the endpoints of the edges in the graph. (In other examples the weight might represent a cost, time, or profit.) We want to find a minimum-cost tour that starts and ends in Chicago and visits each other city only once. Using our earlier terminology, what we wish to find is a **minimum-cost Hamiltonian circuit**—a Hamiltonian circuit with the lowest possible sum of the weights of its edges.

Finding a Minimum-Cost Hamiltonian Circuit

How can we determine which Hamiltonian circuit has minimum cost? There is a conceptually easy **algorithm**, or mechanical step-by-step process, for solving this problem:

1. Generate all possible Hamiltonian tours (starting from Chicago).
2. Add up the distances on the edges of each tour.
3. Choose the tour of minimum distance.

FIGURE 2.4 First stage in finding vacation-planning routes.

Steps 2 and 3 of the algorithm are straightforward. Thus, we need worry only about step 1, generating all the possible Hamiltonian circuits in a systematic way. To find the Hamiltonian tours, we will use the **method of trees**, as follows. Starting from Chicago, we can choose any of the three cities to visit after leaving Chicago. The first stage of the enumeration tree is shown in Figure 2.4. If Minneapolis is chosen as the first city to visit, then there are two possible cities to visit next, Cleveland and St. Louis. The possible branchings of the **tree** at this stage are shown in Figure 2.5. In this second stage, however, for each choice of first city to visit, there are two choices from this city to the second city to visit. This would lead to the diagram in Figure 2.6.

Having chosen the order of the first two cities to visit, and knowing that no revisits (reuses) can occur in a Hamiltonian circuit, there is only one choice left for the next city. From this city we return to Chicago. The complete tree diagram showing the third and fourth stages for these routes is given in Figure 2.7. Notice, however, that because we can traverse a circular tour in either of two directions, the paths shown in the tree diagram of Figure 2.7 do not correspond to different Hamiltonian circuits. For example, the leftmost path (C–M–S–CL–C) and the rightmost path (C–CL–S–M–C) represent the same Hamiltonian circuit. Thus, among what appear to be six different paths in the tree diagram, only three in fact correspond to different Hamiltonian circuits. These three distinct Hamiltonian circuits are shown in Figure 2.8.

Note that in generating the Hamiltonian circuits we disregard the distances involved. We are concerned only with the different patterns of carrying out the visits. To find the optimal route, however, we must add up the distances on the edges to get each tour's length. Figure 2.8 shows that the optimal tour is Chicago, Minneapolis, St. Louis, Cleveland, Chicago. The length of this tour is 1877 miles, which saves 163 miles over the longest choice of tour. ■

FIGURE 2.5 Part of the second stage in finding vacation-planning routes.

FIGURE 2.6 Complete second stage in finding vacation-planning routes.

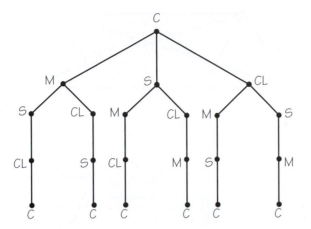

FIGURE 2.7
Completed enumeration
of routes using the
method of trees for the
vacation-planning
problem.

The method of trees is not always as easy to use as our example suggests. Instead of doing our analysis for four cities, consider the general case of *n* cities. The graph model similar to that in Figure 2.3 would consist of a weighted graph with *n* vertices, with every pair of vertices joined by an edge. Such a graph is called **complete** because the edge between any pair of vertices is present in the graph. A complete graph with five vertices is illustrated in Figure 2.9.

Fundamental Principle of Counting

How many Hamiltonian circuits are in a complete graph of *n* vertices? We can solve this problem by using the same type of analysis that we used in the method

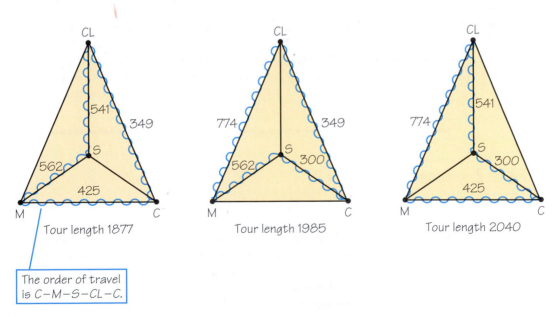

FIGURE 2.8 The three Hamiltonian circuits for the vacation-planning problem of Figure 2.3.

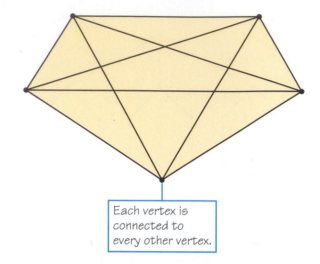

FIGURE 2.9 A complete graph with five vertices. Every pair of vertices is joined by an edge.

Each vertex is connected to every other vertex.

of trees. The method of trees is a visual application of the fundamental principle of counting, a procedure for counting outcomes in multistage processes. Using this procedure, we can count how many patterns occur in a situation by looking at the number of ways in which the component parts can occur. For example, if Jack has 9 shirts and 4 pairs of trousers, he can wear $9 \times 4 = 36$ shirt–pants outfits. Each shirt can be worn with any of the pants. (This can be verified by drawing a tree diagram, but such a diagram is cumbersome for big numbers.)

The Fundamental Principle of Counting

In general, the **fundamental principle of counting** can be stated this way. If there are a ways of choosing one thing, b ways of choosing a second after the first is chosen, . . . , and z ways of choosing the last item after the earlier choices, then the total number of choice patterns is $a \times b \times c \times \cdots \times z$.

EXAMPLE 2   Counting

Here are some examples of how to use the fundamental principle of counting:

1. In a restaurant there are 4 kinds of soup, 12 entrees, 6 desserts, and 3 drinks. How many different four-course meals can a patron choose from? The four choices can be made in 4, 12, 6, and 3 ways, respectively. Hence, applying the fundamental principle of counting, there are $4 \times 12 \times 6 \times 3 = 864$ possible meals.

2. In a state lottery a contestant gets to pick a four-digit number that contains no zero followed by an uppercase or lowercase letter. How many such

sequences of digits and a letter are there? Each of the four digits can be chosen in 9 ways (that is, 1, 2, . . . , 9), and the letter can be chosen in 52 ways (that is, A, B, . . . , Z plus a, b, . . . , z). Hence, there are $9 \times 9 \times 9 \times 9 \times 52 = 341{,}172$ possible patterns.

3. A corporation is planning a musical logo consisting of four different ordered notes from the scale C, D, E, F, G, A, and B. How many logos are there to choose from? The first note can be chosen in 7 ways, but because reuse is not allowed, the next note can be chosen in only 6 ways. The remaining two notes can be chosen in 5 and 4 ways, respectively. Using the fundamental principle of counting, $7 \times 6 \times 5 \times 4 = 840$ musical logos are possible. If reuse of notes is allowed, $7 \times 7 \times 7 \times 7 = 2401$ logos are possible. ▪

Let's now return to the problem of enumerating Hamiltonian circuits for the complete graph with n vertices. The city visited first after the home city can be chosen in $n - 1$ ways, the next city in $n - 2$ ways, and so on, until only one choice remains. Using the fundamental principle of counting, there are $(n - 1)! = (n - 1)(n - 2) \times \cdots \times 3 \times 2 \times 1$ routes. The exclamation mark in $(n - 1)!$ is read "factorial" and is shorthand notation for the product $(n - 1)(n - 2) \times \cdots \times 3 \times 2 \times 1$. For example, $5! = 5 \times 4 \times 3 \times 2 \times 1 = 120$.

As we saw in Figure 2.7, pairs of routes correspond to the same Hamiltonian circuit because one route can be obtained from the other by traversing the cities in reverse order. Thus, although there are $(n - 1)!$ possible routes, there are only half as many, or $(n - 1)!/2$, different Hamiltonian circuits. Now, if we have only a few cities to visit, $(n - 1)!/2$ Hamiltonian circuits can be listed and examined in a reasonable amount of time. Analysis of a six-city problem would require generation of $(6 - 1)!/2 = 5!/2 = 120/2 = 60$ tours. But for, say, 25 cities, $24!/2$ is approximately 3×10^{23}. Even if these tours could be generated at the rate of 1 million a second, it would take 10 billion years to generate them all. Because it would take so long to solve large vacation-planning problems using this method, it is sometimes referred to as the **brute force method** (that is, trying all the possibilities).

2.2 Traveling Salesman Problem

If the only benefit were saving money and time in vacation planning, the difficulty of finding a minimum-cost Hamiltonian circuit in a complete graph with n vertices for large values of n would not be of great concern. However, the problem we are discussing is one of the most common in *operations research*, the branch of mathematics concerned with getting governments and businesses to operate more efficiently. This problem is usually called the **traveling salesman problem (TSP)** because of its early formulation: Determine the trip of minimum cost that a salesperson can make to visit the cities in a sales territory, starting and ending the trip in the same city.

Many situations require solving a TSP:

1. A lobster fisherman has set out traps at various locations and wishes to pick up his catch.

2. The telephone company wishes to pick up the coins from its pay telephone booths. (To avoid the high cost of picking up these coins, phone companies in many countries have adopted a system that uses prepurchased phone cards to operate phones. This means that there are no coins to collect.)

3. The electric (or gas) company needs to design a route for its meter readers.

4. A minibus must pick up six day campers, deliver them to camp, and later in the day return them home.

5. In drilling holes in a series of plates, the drill press operator (perhaps a robot) must drill the holes in a predetermined order.

6. Physical records generated at automated teller machine (ATM) locations—as backup in case of failure of the electronic systems—must be picked up periodically.

7. A limousine service with a van located at an airport must pick up five customers and deliver them to the airport in time to catch their flights.

Perhaps surprisingly, TSP problems are also solved regularly in the design of computer chips. The components must be located so that the machines involved in the assembly can insert them on the chips as efficiently as possible. Because many chips are manufactured, even a small improvement in the time needed to make a chip can save a lot of money.

The meaning of *cost* can vary from one formulation of a TSP to another. We can measure cost as distance, airplane ticket prices, time, or any other factor that is to be optimized. In many situations, the TSP arises as a subproblem of a more complicated problem. For example, a supermarket chain may have a very large number of stores to be served from a single large warehouse. If there are fewer trucks than stores, the stores must be grouped into clusters so that one truck can serve each cluster. If we then solve the TSP for every truck, we can minimize total costs for the supermarket chain. Similar vehicle-routing problems—for dial-a-ride services that transport senior citizens to activity centers, for example, or that deliver children to their schools or camps—often involve solving the TSP as a subproblem.

2.3 Helping Traveling Salesmen

Because the traveling salesman problem arises so often in situations where the associated complete graphs would be very large, we must find a faster method than the brute force method we have described. We need to look at our original problem in Figure 2.3 and try to find an alternative algorithm for solving it. Recall that our goal is to find the minimum-cost Hamiltonian circuit.

Nearest-Neighbor Algorithm

Starting from the home city, first visit the nearest city, then visit the nearest city that has not already been visited. We return to the start city when no other choice is available. This approach is called the **nearest-neighbor algorithm.**

Applying this algorithm to the TSP in Figure 2.3 quickly leads to the tour of Chicago, St. Louis, Cleveland, Minneapolis, and Chicago, with a length of 2040 miles. Here is how this tour is determined. Because we are starting in Chicago, there is a choice of going to a city that is 425, 300, or 349 miles away. Because the smallest of these numbers is 300, we next visit St. Louis, which is the nearest neighbor of Chicago not already visited. At St. Louis, we have a choice of visiting next cities that are 541 or 562 miles away. Hence, Cleveland, which is nearer (541 miles), is visited. To complete the tour, we visit Minneapolis and return to Chicago, thereby adding 774 and 425 miles to the length of the tour.

The nearest-neighbor algorithm is an example of a **greedy algorithm,** because at each stage a best (greedy) choice, based on an appropriate criterion, is made. Unfortunately, this is not the optimal tour, which we saw was C–M–S–CL–C, for a total length of 1877 miles. Making the best choice at each stage may not yield the best "global" solution. However, even for a large TSP, one can always find a nearest-neighbor route quickly.

Figure 2.10 again illustrates the ease of applying the nearest-neighbor algorithm, this time to a weighted complete graph with five vertices. Starting at vertex

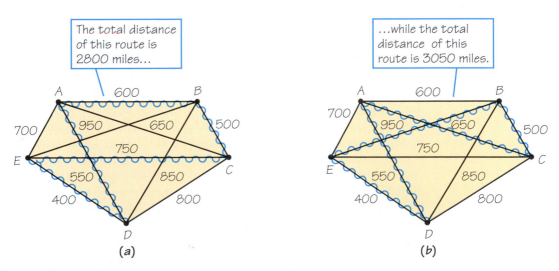

The total distance of this route is 2800 miles...

...while the total distance of this route is 3050 miles.

FIGURE 2.10 (a) A weighted complete graph with five vertices that illustrates the use of the nearest-neighbor algorithm (starting at *A*). (b) TSP tour generated by the nearest-neighbor algorithm (starting at *B*).

A, we get the tour *ADECBA* (cost 2800) (Figure 2.10a). Note that the nearest-neighbor algorithm starting at vertex *B* yields the tour *BCADEB* (cost 3050) (Figure 2.10b).

This example illustrates that a nearest-neighbor tour can be computed for each vertex of the complete graph being considered and that different nearest-neighbor tours can be obtained starting at different vertices. Thus, even though we may seek a tour starting at a particular vertex—say, *A* in Figure 2.10—because a Hamiltonian circuit can be thought of as starting at any of its vertices, we can just as easily apply the nearest-neighbor procedure starting at vertex *B* (rather than at *A*). The Hamiltonian circuit we get can still be thought of as beginning at vertex *A* rather than *B*. Even for complete graphs with a large number of vertices, it would still be faster to apply the nearest-neighbor algorithm for each vertex and pick the cheapest of the tours generated (though such a tour might not be optimal) than to apply the brute force method.

Sorted-Edges Algorithm

We might start by sorting or arranging the edges of the complete graph in order of increasing cost (or, equivalently, arranging the intercity distances in order of increasing distance). Then at each stage we can select that edge of least cost that (1) never requires that three used edges meet at a vertex (because a Hamiltonian circuit uses up exactly two edges at each vertex) and that (2) never closes up a circular tour that doesn't include all the vertices. This algorithm is called the **sorted-edges algorithm.**

Applying the sorted-edges algorithm to the TSP in Figure 2.3 works as follows: First, the six weights on the edges listed in increasing order would be 300, 349, 425, 541, 562, and 774. Because the cheapest edge in this sorted list is 300, this is the first edge we select for the tour we are building. Next we add the edge with weight 349 to the tour. The next-cheapest edge would be 425, but using this edge together with those already selected would result in having three edges at a vertex (Figure 2.11a), which is not consistent with having a Hamiltonian circuit. Hence, we do not use this edge. The next-cheapest edge, 541, used together with the edges already selected, would create a circuit (see Figure 2.11b) that does not include all the vertices. Thus, this edge, too, would be skipped over. However, we are able to add the edges 562 and 774 without either creating a circuit shorter than one including all the vertices or having three edges at a vertex. Hence, the tour we arrive at is Chicago, St. Louis, Minneapolis, Cleveland, and Chicago. Again, this solution is not optimal because its length is 1985. Note that this algorithm, like the nearest-neighbor, is greedy.

Although the edges selected by applying the sorted-edges method to the example in Figure 2.3 are connected to each other at every stage, this does not always happen. For example, if we apply the sorted-edges algorithm to the graph

FIGURE 2.11 (a) When three shortest edges are added in order of increasing distance, three edges at a vertex are selected, which is not allowed as part of a Hamiltonian circuit. (b) When the edges of distances 300, 349, and 541 are selected, a circuit that does not include all vertices results.

You can't have 3 edges meeting at one vertex...

...and you can't have a circuit that doesn't include all the vertices.

in Figure 2.10a, we build up the tour first with edge *ED* (400) and then edge *BC* (500), which do not touch. The edges that are then selected are *AD, AB,* and *EC,* giving the circuit *EDABCE,* which is the same as the nearest-neighbor circuit starting at vertex *A.*

Many "quick-and-dirty" methods for solving the TSP have been suggested; while some methods give an optimal solution in some cases, none of these methods guarantees an optimal solution. Surprisingly, most experts believe that no efficient method that guarantees an optimal solution for the TSP will ever be found (see Spotlight 2.1).

Recently, mathematical researchers have adopted a somewhat different strategy for dealing with TSP problems. If finding a fast algorithm to generate optimal solutions for large problems is unlikely, perhaps we can show that the quick-and-dirty methods, usually called **heuristic algorithms,** come close enough to giving optimal solutions to be important for practical use. For example, suppose we could prove that the nearest-neighbor heuristic was never off by more than 25% in the worst case or by more than 15% in the average case. For a medium-sized TSP, we would then have to choose whether to spend a lot of time (or money) to find an optimal solution or instead to use a heuristic algorithm to obtain a fairly good solution. Investigators at AT&T Research have developed many remarkably good heuristic algorithms. The best-known guarantee for a heuristic algorithm for a TSP is that it yields a cost no worse than one and a half times the optimal cost. Interestingly, this heuristic algorithm involves solving a Chinese postman problem (see Chapter 1), for which a "fast" algorithm is known to exist.

Throughout our discussion of the TSP, we have concentrated on the goal of minimizing the cost (or time) of a tour that visited each of a variety of sites once and only once. One of the things that makes mathematical modeling exciting, however, is the subtle issues that arise in specific real-world situations (or that provide a contrast between seemingly similar situations). For example, suppose

SPOTLIGHT 2.1   NP-Complete Problems

Steven Cook, a computer scientist at the University of Toronto, showed in 1971 that certain computational problems are equivalently difficult. This class of problems, now referred to as **NP-complete problems,** has the following characteristic: If a "fast" algorithm for solving one of these problems could be found, then a fast method would exist for all these problems.

In this context, "fast" means that as the size n of the problem grows (the number of cities gives the problem size in the traveling salesman problem), the amount of time needed to solve the problem grows no more rapidly than a polynomial function in n. (A polynomial function has the form $a_k n^k + a_{k-1} n^{k-1} + \cdots + a_1 n + a_0$.) On the other hand, if it could be shown that any problem in the class of NP-complete problems required an amount

of time that grows faster than any polynomial (an exponential function, such as 3^n, is an example of a function that grows faster than any polynomial) as the problem size increased, then all problems in the NP-complete class would share this characteristic. The TSP, along with a wide variety of other practical problems, is known to be NP-complete. It is widely believed that large versions of none of these problems can be solved quickly. Furthermore, the security of some recent cryptographical systems relies on the hope that large NP-complete problems are actually as time consuming to solve as they appear to be. The Clay Foundation is offering a $1 million prize for determining whether NP-complete problems are truly computationally hard. The prize is still unclaimed!

the TSP situation is that of picking up day campers and taking them to and from the camp. The camp wants to minimize the total length of time that the bus needs to pick up the campers. The parents of the campers, however, may want to minimize the time their children spend on the bus. For some problems, the tour that minimizes the mean (average) time that a child spends on the bus may not be the same tour that minimizes the total time of the tour. (Specifically, if the bus first picks up the child who lives the farthest from the camp, and then picks up the other children, this may yield a relatively short time on the bus for the kids but a relatively long time for the tour itself.) It is these subtleties between problems that mathematicians return to examine at a later time, after the basic structure of the main problem itself is well understood. It is in this way that mathematics continues to grow, explore new ideas, and find new applications.

2.4 Minimum-Cost Spanning Trees

The traveling salesman problem is but one of many graph theory optimization problems that have grown out of real-world problems in both government and industry. Here is another.

EXAMPLE 3   Pictaphone Service

Imagine that Pictaphone service (telephone service that provides a video image of the callers) is to be set up on an experimental basis among five cities. The

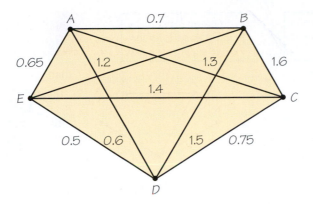

FIGURE 2.12 Costs (in millions of dollars) of installing Pictaphone service among five cities.

graph in Figure 2.12 shows the possible links that might be included in the Pictaphone network, with each edge showing the cost in millions of dollars to create that particular link. To send a Pictaphone message between two cities, a direct communication link is not necessary because it is possible to send a message indirectly via another city. Thus, in Figure 2.12, sending a message from A to C could be achieved by sending the message from A to B, from B to E, and from E to C, provided the links AB, BE, and EC are part of the network. We assume that the cost of relaying a message, compared with the cost of the direct communication link, is so small that we can neglect this amount. The problem that concerns us, therefore, is to provide service between any pair of cities in a way that minimizes the total cost of the links.

Our first guess at a solution is to put in the cheapest possible links between cities first, until all cities could send messages to any other city. Such an approach

Some videophone (pictaphone) and video conferencing requires the creation of specialized networks for which the techniques of finding a minimum-cost spanning tree are required. *(Digital Vision/Punchstock.)*

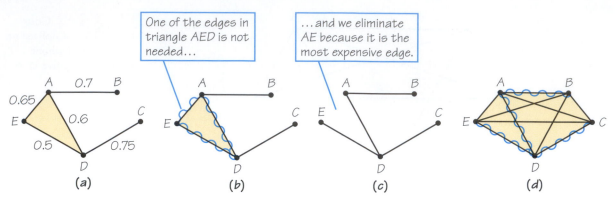

One of the edges in triangle AED is not needed...

...and we eliminate AE because it is the most expensive edge.

FIGURE 2.13 (a) Cities are linked in order of increasing cost until all cities are connected. (b) Circuit in part (a) highlighted. (c) Most expensive link in circuit in part (a) deleted. (d) Highlighted edges show, as a subgraph of the original graph, those links connecting the cities with minimum cost, obtained using Kruskal's algorithm.

would be analogous to the sorted-edges method that was used to study the traveling salesman problem. In our example, if the cheapest links are added until all cities are joined, we obtain the connections shown in Figure 2.13a.

The links were added in the order *ED, AD, AE, AB, DC*. However, because this graph contains the circuit *ADEA* (wiggly edges in Figure 2.13b), it has redundant edges: We can still send messages between any pair of cities using relays after omitting the most expensive edge in the circuit—*AE*. After deleting an edge of a circuit, a message can still be relayed among the cities of the circuit by sending signals the long way around. After *AE* is deleted, messages from *A* to *E* can be sent via *D* (Figure 2.13c). These ideas constitute a procedure developed by Joseph Kruskal (AT&T Research) in 1956.

Kruskal's Algorithm

Kruskal's algorithm: Add links in order of cheapest cost so that no circuits form and so that every vertex belongs to some link added (Figure 2.13d).

In Kruskal's procedure, as in the sorted-edges method for the TSP, the edges that are added need not be connected to each other until the end. A subgraph formed in this way will be a **tree**; that is, it will consist of one piece and contain no circuits. It will also include all the vertices of the original graph. A subgraph that is a tree and that contains all the vertices of the original graph is called a **spanning tree** of the original graph.

To understand these concepts better, consider the graph *G* in Figure 2.14a. The wiggly edges in Figure 2.14b would constitute a subgraph of *G* that is a tree (because it is connected and has no circuit), but this tree would not be a span-

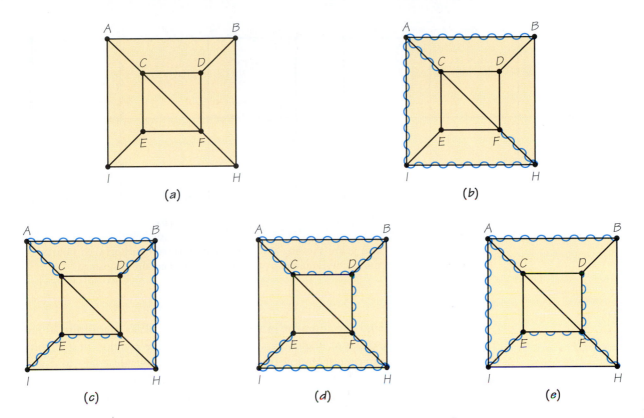

FIGURE 2.14 (a) A graph to help illustrate the concept of a spanning tree. (b) The wiggly edges are a tree, but not a spanning tree, because vertices D and E are not part of the tree. (c) The vertices of the graph are, however, endpoints of wiggly edges. (d) The wiggly edges are not a tree, because they contain the edges of the circuit $BDCAB$. All the vertices of the graph are, however, endpoints of wiggly edges. (e) The wiggly edges form a tree and include all of the vertices of the graph as endpoints of wiggly edges.

ning tree of G because the vertices D and E would not be included. On the other hand, the wiggly edges in Figure 2.14c and 2.14d show subgraphs of G that include all the vertices of G but are not trees because the first is not connected and the second contains a circuit. Figure 2.14e shows a spanning tree of G; the wiggly edges are connected and contain no circuit, and every vertex of the original graph is an endpoint of some wiggly edge. ■

Finding a **minimum-cost spanning tree**—that is, a spanning tree whose edge weights sum to a minimum value—solves the Pictaphone problem. Note that having a different goal in the Pictaphone problem led to a different mathematical question from that of finding a Chinese postman tour or TSP tour. This application required that we find a minimum-cost spanning tree. In Figure 2.15a we have a graph model showing the costs of putting in roads to connect new houses

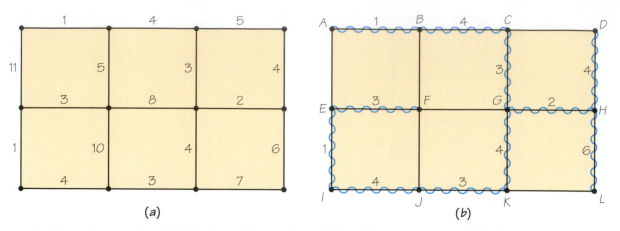

FIGURE 2.15 (a) A graph showing costs for construction of roads between houses. (b) Wiggly edges show a minimum-cost spanning tree for the graph in part (a).

in a suburban land-development project. Applying Kruskal's algorithm–adding the edges in the order of increasing cost, but avoiding the creation of a circuit– yields a minimum-cost spanning tree, indicated by the wiggly edges in Figure 2.15b. This tree is the cheapest one that makes it possible to drive between any pair of homes, though the driving distance between some of the homes will be relatively large, because only roads corresponding to wiggly edges will be built.

Remember that the weights on the edges of the graph in Figure 2.15a represent the costs of building roads, not the driving distances between the houses. Note that Figure 2.15a is not a complete graph, one in which all possible edges are included. Edges that correspond to roads that would be economically prohibitive to build have not been shown in the graph model. Also, in Figure 2.15b, the two edges of weight 5 (shown in Figure 2.15a) do not become part of the minimum-cost spanning tree, because they would create circuits with edges already chosen.

Although Kruskal's algorithm worked in our example, how do we know that the spanning tree found by this algorithm will always achieve the minimum possible cost? While this sounds very plausible, our experience with the TSP should suggest caution. Remember that for the TSP, the sorted-edges algorithm did not necessarily give an optimal solution even though it is a greedy algorithm like Kruskal's. On what basis should we have more faith in Kruskal's algorithm?

Kruskal proposed his algorithm as a way to solve a pure mathematics problem put forward by Czechoslovakian mathematician Otakar Borůvka. In mathematics it is surprising but not uncommon to find that ideas used to solve problems with no apparent application often turn out to have many real-world uses. Kruskal's solution to the problem of finding a minimum-cost spanning tree in a graph with weights is a good example of this phenomenon.

Kruskal showed that the greedy algorithm described does yield the minimum answer, and his work led to applications of these and related ideas in designing

SPOTLIGHT 2.2 AT&T Manager Explains How Long-Distance Calls Run Smoothly

Although long-distance calls are now routine, it takes great expertise and careful planning for a company like AT&T to handle its vast amounts of telephone traffic. Rich Wetmore was district manager of AT&T's Communications Network Operations Center in Bedminster, New Jersey. Here are his responses to questions about how AT&T handles its huge volumes of long-distance traffic and how it tracks its operations to keep things running smoothly.

How do you make sure that a customer doesn't run into a delayed signal when attempting a long-distance call?

We monitor the performance of our AT&T network by displaying data collected from all over the country on a special wallboard. The wallboard is configured to tell us if a customer's call is not going through because the network doesn't have enough capacity to handle it.

That's when we step in and take control to correct the problem. The typical control we use is to reroute the call. Instead of sending the customer's call directly to its destination, we'll route it via a third city—to someplace else in the country that has the capacity to complete the call.

It would seem that routing via another city would take longer. Is the customer aware of this process?

Routing a call via a third city is entirely transparent [imperceptible] to the customer. I'm an expert about the network, and even when I make a phone call, I have no idea how that individual call was routed. It's transparent both in terms of how far away the other person sounds and in how quickly the telephone call gets set up. With the signaling network we use, it takes milliseconds for switching systems to "talk" to each other to set up a call. So the fact that you are involved in a third switch in some distant city is something you would never know.

You want to be sure to keep costs down while supplying enough service to customers. So how do you balance company benefits with customer benefit?

In terms of making the network efficient, we want to do two things. First, we want our customers to be happy with our service and for all their calls to go through, which means we must build enough capacity in the network to allow that to happen. Second, we want to be efficient for stockholders and not spend more money than we need to for the network to be at the optimum size.

There are basically two costs in terms of building the network. There is the cost of switching systems and the cost of the circuits that connect the switching systems. Basically, you can use operations-research techniques and mathematics to determine cost trade-offs. It may make sense to build direct routing between two switching systems and use a lot of circuits, or maybe to involve three switching systems, with fewer circuits between the main two, and so on.

minimum-cost computer networks, phone connections, and road and railway systems. For additional discussion of operations research in the communications industry, see Spotlight 2.2. To explore how one can reconstruct full information from partial information using the tree concept, see Spotlight 2.3.

SPOTLIGHT 2.3 Common Ancestors?

In the study of ancient manuscripts, different manuscripts of the same book are available, even though the original manuscript upon which they are based has been lost. Examples of this include Euclid's *Elements* and Chaucer's *Canterbury Tales*. What interests scholars is reconstructing the relationships between the manuscripts and the common ancestors of the manuscripts, even when some of the ancestors are now missing.

Similarly, perceptual psychologists may be interested in which colors people perceive as being closely related and comparing these perceptions with those of people who are color-blind. Finally, in studying different species, biologists are interested in determining which species are more closely related to each other, including species known only in fossil form, and constructing a "tree" of life that shows which species were ancestors of others.

Reconstructions of this kind are made possible by using graph theory, specifically using the graph theory concept of a tree. The value of the graph theory in these and many other situations lies in using the distance between pairs of vertices in the tree as a way of reflecting the closeness of the relationships that pairs of manuscripts, pairs of colors, or pairs of species have. The distance between two vertices in a tree is the sum of the weights along the one path that joins the two vertices. If there are no weights on the edges, the distance is the number of edges in the path. In some reconstruction problems, a special vertex of the tree called the *root* is singled out. This root

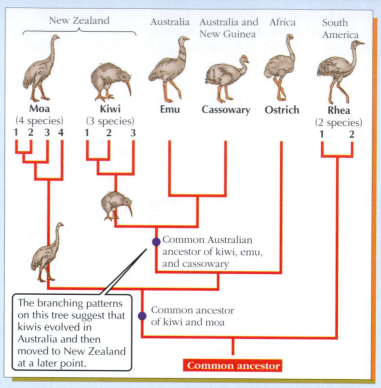

The branching patterns on this tree suggest that kiwis evolved in Australia and then moved to New Zealand at a later point.

Common Australian ancestor of kiwi, emu, and cassowary

Common ancestor of kiwi and moa

Common ancestor

plays the role of the original common ancestors, and distances to the root are of critical interest.

In the case of species, trees of family relatedness were traditionally constructed based on similarities of bones and physical appearance. With the discovery of molecular biology, many new avenues have been opened up. We can now draw trees of relatedness based on an organism's genetic material, DNA, or the proteins that the DNA codes for. The traditional trees based on physical traits often show different species as being more closely related than trees based on newer molecular biological approaches. These differences focus scholars on how to resolve the discrepancies and thereby reach a deeper understanding of the unity of life.

In our discussion of routing problems in graphs, we have not touched on one of the most obvious ones: finding the path between two specified, distinct vertices while keeping the sum of the weights of the edges in the path as small as possible. (Here there is no need to cover all vertices or to cover all edges.) We have seen that the weights on the edges have many possible interpretations, including time, distance, and cost. The following are some of the many possible applications:

1. Design routes to be used by an ambulance, police car, or fire engine to get to an emergency as quickly as possible.
2. Design delivery routes that minimize gasoline use.
3. Design routes to bring soldiers to the front as quickly as possible.
4. Design a route for a truck carrying nuclear waste.

Many people have found it valuable to consult the Web to get directions and time estimates for driving to a location they wish to visit. The software that provides this information relies on algorithms that compute the shortest-path route in an appropriate weighted graph, which involves distances or times.

The need to find shortest paths seems natural. Next we investigate a situation in which finding a *longest* path is the right tool.

2.5 Critical-Path Analysis

Mathematics can confirm the obvious in certain situations while showing that our intuition is wrong in other circumstances. Our next group of applications will illustrate this point.

A characteristic of American life is its fast pace. People are interested in getting things done quickly and efficiently. This means that when you take your car in to be repaired before going to work, you want to know for sure that the repairs will be done when you pick the car up. You want the trains and the bus that take you to your doctor's appointment to run on time. When you arrive at the doctor's office, you want a nurse to be free to take a blood sample and a throat culture. You want your outpatient appointment for an X-ray at the local hospital to occur on schedule. You want the X-ray to be interpreted quickly and the results reported back to your internist.

Scheduling machines and people is a big part of modern life. It is involved in running a school, a hospital, an airline, or in landing a person on Mars, and modern mathematics plays a big part in solving scheduling problems.

Part of what makes scheduling complicated is that the tasks that make up a job usually cannot be done in a random order. For example, to make Thanksgiving dinner you must buy and prepare the turkey before putting it in the oven, and you must set the table before serving the food.

If the tasks cannot be performed in a random order, we can specify the order in an **order-requirement digraph**. The term *digraph* is short for "directed

FIGURE 2.16 A typical order-requirement digraph.

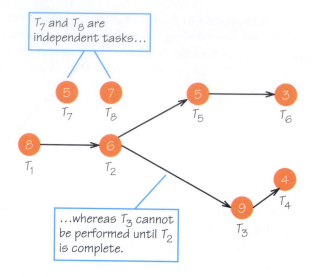

T_7 and T_8 are independent tasks...

...whereas T_3 cannot be performed until T_2 is complete.

graph." A digraph is a geometrical tool similar to a graph except that each edge has an arrow on it to indicate a direction for that edge. Digraphs can be used to illustrate that traffic on a street must go in one direction or that certain tasks in a job must be completed before other tasks. A typical example of an order-requirement digraph is shown in Figure 2.16. There is a vertex in this digraph for each task. If one task must be done immediately before another, we draw a directed edge, or arrow, from the prerequisite task to the subsequent task. The numbers within the circles representing vertices are the times it takes to complete the tasks. In Figure 2.16 there is no arrow from T_1 to T_5 because task T_2 intervenes. Also, T_1, T_7, and T_8 have no tasks that must precede them. Hence, if there are at least three processors (such as people or machines) available, tasks T_1, T_7, and T_8 can be worked on simultaneously at the start of the job.

Let's investigate a typical scheduling problem faced by a business.

EXAMPLE 4   Turning a Plane Around

Consider an airplane that carries both freight and passengers. The plane must have its passengers and freight unloaded and new passengers and cargo loaded before it can take off again. Also, the cabin must be cleaned before departure can occur. Thus, the job of "turning the plane around" requires completion of five tasks:

Task A	Unload passengers	13 minutes
Task B	Unload cargo	25 minutes
Task C	Clean cabin	15 minutes
Task D	Load new cargo	22 minutes
Task E	Load new passengers	27 minutes

Turning a plane around, which involves such tasks as refueling, unloading, and then reloading cargo and passengers, entails very careful scheduling to avoid time slippage. *(David Butow/CORBIS SABA.)*

The order-requirement digraph for the problem of turning an airplane around is shown in Figure 2.17. The presence or absence of an edge in the order-requirement digraph depends on the analysis made as part of the modeling process for the problem. It seems natural that we need an arrow between task *A* and task *C,* because the passengers have to be unloaded before the cabin is cleaned. Other arrows may not seem natural—say, perhaps the arrow from task *B* (unload the cargo) to task *E* (load new passengers). This arrow may be due to government rules or requirements.

What matters is that the mathematics of solving the problem does not depend on the reason that the order-requirement digraph looks the way that it does. The person solving the problem constructs the order-requirement digraph and then the mathematical techniques we will develop can be applied, regardless of whether or not some business faced with a similar problem might model the problem in a different way. Because we want to find the earliest completion time, it might seem that finding the shortest path in the digraph (path *BD* with time length 25 + 22 = 47) would solve the problem. But this approach shows the danger of ignoring the relationship between the mathematical model (the digraph) and the original problem.

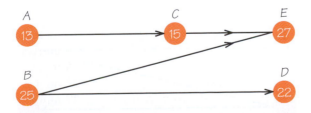

FIGURE 2.17 An order-requirement digraph for turning an airplane around after landing.

The time required to complete all the tasks, *A* through *E,* must be at least as long as the time necessary to do the tasks on any particular path. Consider the path *BD,* which has length $25 + 22 = 47$. Recall that here *length* of a path refers to the sum of the times of the tasks that lie along the path. Because task *B* must be done before task *D* can begin, the two tasks *B* and *D* cannot be completed before time 47. Hence, even if work on other tasks (such as *A, C,* and *E*) proceeds during this period, all the tasks cannot be finished before the tasks on path *BD* are finished. The same statement is true for every other path in the order-requirement digraph. Thus, the earliest completion time actually corresponds to the length of the longest path. In the airplane example, this earliest completion time is 55 ($= 13 + 15 + 27$) minutes, corresponding to the path *ACE.* We call *ACE* the *critical path* because the times of the tasks on this path determine the earliest completion time.

:::::: Critical Path

A **critical path** in an order-requirement digraph is a longest path. The length is measured in terms of summing the task times of the tasks making up the path.

Note that if none of these tasks could go on simultaneously, the time to complete all the tasks would be $13 + 25 + 15 + 22 + 27 = 102$ minutes. However, even though some tasks may be performed simultaneously, the fact that the length of the critical path is 55 means that completion of the tasks in less than 55 minutes is not possible. Only by speeding up the times to complete the critical-path tasks themselves can a completion time less than 55 time units be achieved.

Suppose it were desirable to speed the turnaround of the plane to less than 55 minutes. One way to do this might be to build a second jetway to help unload passengers more quickly. For example, we could unload passengers (task *A*) in 7 minutes instead of 13. However, reducing task *A* to 7 minutes does not reduce the completion time by 6 minutes, because in the new digraph (Figure 2.18) *ACE* is no longer the critical (longest) path. The longest path is now *BE,* which has a length of 52 minutes. Thus, shortening task *A* by 6 minutes results in only a 3-minute saving in completion time. This may mean that building a new jetway is uneconomical.

FIGURE 2.18 An order-requirement digraph for turning an airplane around in reduced time due to construction of a new jetway.

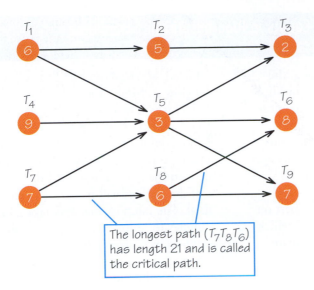

FIGURE 2.19 An order-requirement digraph with 12 paths, to examine how to find the length of the longest path.

The longest path ($T_7T_8T_6$) has length 21 and is called the critical path.

Note also that shortening the time to complete tasks that are not on the original critical path *ACE* will not shorten the completion time at all. Speeding tasks on the critical path will shorten completion time of the job only up to the point where a new critical path is created. Also note that a digraph may have more than one longest path. ■

Not all order-requirement digraphs are as simple as the one shown in Figure 2.17. The order-requirement digraph in Figure 2.19 has 12 paths, which can be found by exhaustive search. Examples of such paths are $T_1T_2T_3$, $T_1T_5T_9$, $T_4T_5T_9$, and $T_7T_5T_3$. (Although we have not discussed them here, fast algorithms for finding longest and shortest paths in graphs are known.) The critical path is $T_7T_8T_6$ (length 21), and the earliest completion time for all nine tasks is time 21.

These examples are typical of many scheduling problems that occur in practice (see Spotlight 2.4). Perhaps the most dramatic use of critical-path analysis is in the construction trades. No major new building project is now carried out without a critical-path analysis first being performed to ensure that the proper personnel and materials are available at the right times in order to have the project finished as quickly as possible. Many such problems are too large and complicated to be solved without the aid of computers.

The critical-path method was popularized and came into wider use as a consequence of the *Apollo* project. This project, which aimed at landing a man on the moon within 10 years of 1960, was one of the most sophisticated projects in planning and scheduling ever attempted. The dramatic success of the project can be attributed partly to the use of critical-path ideas and the related program evaluation and review technique (PERT), which helped keep the project on schedule.

In Chapter 3, we will see how mathematical ideas drawn from outside of graph theory can be used to gain insight into scheduling problems.

SPOTLIGHT 2.4   Every Moment Counts in Rigorous Airline Scheduling

When people think of airline scheduling, the first thing that comes to mind is how quickly a particular plane can safely reach its destination. But using ground time efficiently is just as important to an airline's timetable as the time spent in flight. Bill Rodenhizer was the manager of control operations for an airline that provided shuttle service between Boston and New York. He is considered to be an expert on airplane turnaround time, the process by which an airplane is prepared for almost immediate takeoff once it has landed. He tells us how this well-orchestrated effort works:

Scheduling, to the airline, is just about the whole ball game. Everything is scheduled right to the minute. The whole fleet operates on a strict schedule. Each of the departments responsible for turning around an aircraft has an allotted period of time in which to perform its function. Manpower is geared to the amount of ground time scheduled for that aircraft. This would be adjusted during off-weather or bad-weather days or during heavy air-traffic delays.

Most of our aircraft in Boston are scheduled for a 42- to 65-minute ground time. Boston is the end of the line, so it is a "terminating and originating station." In plain talk, that means almost every aircraft that comes in must be fully unloaded, refueled, serviced, and dispatched within roughly an hour's time.

This is how the process works: In the larger aircraft, it takes passengers roughly 20 minutes to load and 20 minutes to unload. During this period, we will have completely cleaned the aircraft and unloaded the cargo, and the caterers will have taken care of the food. The ramp service may take 20 to 30 minutes to unload the baggage, mail, and cargo from underneath the plane, and it will take the same amount of time to load it up again. We double-crew those aircraft with heavier weights so that the workload will fit the time it takes passengers to load and unload upstairs.

While this has been going on, the fueler has fueled the aircraft. As to repairs, most major maintenance is done during the midnight shift, when [most of our] several hundred aircraft are inactive. We all work under a very strict time frame.

New security requirements in the wake of the World Trade Center attack (9/11/2001) have increased the difficulty of adhering to timetables in operating shuttle services between East Coast cities such as New York and Boston. This makes it even more important to use analytical tools in keeping operations on schedule.

REVIEW VOCABULARY

Algorithm A step-by-step descriptions of how to solve a problem.

Brute force method The method that solves the traveling salesman problem (TSP) by enumerating all the Hamiltonian circuits and then selecting the one with minimum cost.

Complete graph A graph in which every pair of vertices is joined by an edge.

Critical path The longest path in an order-requirement digraph. The length of this path gives the earliest completion time for all the tasks making up the job consisting of the tasks in the digraph.

Fundamental principle of counting A method for counting outcomes of multistage processes.

Greedy algorithm An approach for solving an optimization problem, where at each stage of the algorithm the best (or cheapest) action is taken. Unfortunately, greedy algorithms do not always lead to optimal solutions.

Hamiltonian circuit A circuit using distinct edges of a graph that starts and ends at a particular vertex of the graph and visits each vertex once and only once. A Hamiltonian circuit can start at any one of its vertices.

Heuristic algorithm A method of solving an optimization problem that is "fast" but does not guarantee an optimal answer to the problem.

Kruskal's algorithm An algorithm developed by Joseph Kruskal (AT&T Research) that solves the minimum-cost spanning-tree problem by selecting edges in order of increasing cost, but in such a way that no edge forms a circuit with edges chosen earlier. It can be proved that this algorithm always produces an optimal solution.

Method of trees A visual method of carrying out the fundamental principle of counting.

Minimum-cost Hamiltonian circuit A Hamiltonian circuit in a graph with weights on the edges, for which the sum of the weights of the edges of the Hamiltonian circuit is as small as possible.

Minimum-cost spanning tree A spanning tree of a weighted connected graph having minimum cost. The cost of a tree is the sum of the weights on the edges of the tree.

Nearest-neighbor algorithm An algorithm for attempting to solve the TSP that begins at a "home" vertex and visits next that vertex not already visited that can be reached most cheaply. When all other vertices have been visited, the tour returns to home. This method may not give an optimal answer.

NP-complete problems A collection of problems, which includes the TSP, that appear to be very hard to solve quickly for an optimal solution.

Order-requirement digraph A directed graph that shows which tasks precede other tasks among the collection of tasks making up a job.

Sorted-edges algorithm An algorithm for attempting to solve the TSP where the edges added to the circuit being built up are selected in order of increasing cost, but no edge is added that would prevent a Hamiltonian circuit's being formed. These edges must all be connected at the end, but not necessarily at earlier stages. The tour obtained may not have the lowest possible cost.

Spanning tree A subgraph of a connected graph that is a tree and includes all the vertices of the original graph.

Traveling salesman problem (TSP) The problem of finding a minimum-cost Hamiltonian circuit in a complete graph where each edge has been assigned a cost (or weight).

Tree A connected graph with no circuits.

Weight A number assigned to an edge of a graph that can be thought of as a cost, distance, or time associated with that edge.

✔ SKILLS CHECK

1. Which of the following describes a Hamiltonian circuit for the graph below?

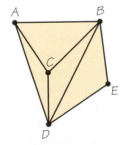

(a) *ABCDEA*
(b) *ABEDCBDAC*

(c) *ACDEBA*

2. Using the nearest-neighbor algorithm and starting at vertex *A*, find the cost of the Hamiltonian circuit for the graph below.

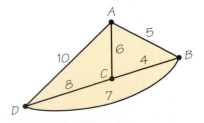

(a) 25

(b) 27

(c) 26

3. The nearest-neighbor traveling salesman tour for the following graph starting at *B* is

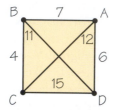

(a) *BCDAB*.

(b) *BDCAB*.

(c) *BCADB*.

4. Using the sorted-edges algorithm, find the cost of the Hamiltonian circuit for the graph below.

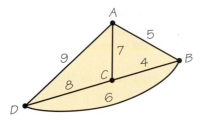

(a) 25

(b) 26

(c) Another answer

5. Suppose that after a hurricane, a van is dispatched to pick up five nurses at their homes and bring them to work at the local hospital. Which of these techniques is most likely to be useful in solving this problem?

(a) Finding an Euler circuit in a graph

(b) Solving a TSP (traveling salesman problem)

(c) Finding a minimum-cost spanning tree in a graph

6. The graph shown below has

(a) no Hamiltonian circuit and no Euler circuit.

(b) an Euler circuit and a Hamiltonian circuit.

(c) no Hamiltonian circuit, but it has an Euler circuit.

7. When the sorted-edges method and nearest-neighbor method are applied to a complete graph on seven vertices with nonnegative weights,

(a) both methods always give the same optimal answer.

(b) both methods always give the same answer but that answer may not be optimal.

(c) neither method may give an optimal answer.

8. Assuming a graph with *E* edges and *V* vertices has a Hamiltonian circuit, which of the following statements *must* be true?

(a) The graph has only one Hamiltonian circuit.

(b) The Hamiltonian circuit has exactly *V* edges.

(c) The Hamiltonian circuit has at least *E*/2 edges.

9. Paul has packed four ties, three shirts, and two pairs of pants for a trip. How many different outfits can he create if he never wears a tie?

(a) Fewer than 10

(b) Between 10 and 25

(c) More than 25

10. Janet can design a lunch by selecting one of three meats, one of three salads, and one of six vegetables. How many different lunches can be designed?

(a) Fewer than 16

(b) Between 17 and 50

(c) More than 50

11. An ice-cream shop offers 3 types of cones, 20 flavors, and 4 different toppings (crushed peanuts, crushed almonds, chocolate bits, or corn flakes). If a customer is allergic to nuts, how many different choices can she choose from?

(a) 240

(b) 120

(c) 25

12. If a three-character password system must begin with a lowercase letter of the English alphabet followed by two decimal digits that may be repeated, the number of different possible passwords is

(a) 2600.

(b) 46.

(c) 2340.

13. Using Kruskal's algorithm, find the minimum-cost spanning tree for the graph below. Which statement is true?

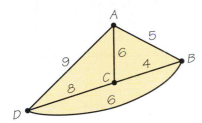

(a) Edges AC and BD are included in the minimum-cost spanning tree.

(b) Edges AB and BD are included in the minimum-cost spanning tree.

(c) Edges CD and BD are included in the minimum-cost spanning tree.

14. Assume that every edge of a graph G has a different cost. If Kruskal's algorithm is used to find the minimum-cost spanning tree T for graph G, which of the following statements *must* be true?

(a) Any other spanning tree for graph G will have more edges than T.

(b) Any other spanning tree for graph G will have a greater cost than T.

(c) The edge of graph G having greatest weight is included in T.

15. Assuming a graph with E edges and V vertices has a minimum-cost spanning tree T, which of the following statements *must* be true?

(a) The tree T has exactly V edges.

(b) The tree T includes every minimum-cost edge.

(c) The graph is connected.

16. If a graph contains a circuit, which of the following statements is true?

(a) The graph cannot be a tree.

(b) The graph must have the same number of vertices as edges.

(c) The graph is not connected.

17. Which of the following statements about algorithms is true?

(a) An algorithm *always* finds the *best* solution.

(b) An algorithm *sometimes* finds the *best* solution.

(c) An algorithm *always* finds a *best* or *nearly best* solution.

18. Assume a job has an order-requirement digraph with five tasks whose critical path is 25 minutes in length. Based on this information, what can be said about the tasks?

(a) Each task takes exactly 5 minutes.

(b) Some task takes 25 minutes.

(c) The five tasks in total take at least 25 minutes.

19. Assume a job has an order-requirement digraph consisting of six tasks ranging in time from 5 to 12 minutes. The earliest possible completion time is at least

(a) 12 minutes.

(b) 30 minutes.

(c) 51 minutes.

20. What is the earliest completion time (in minutes) for a job with the following order-requirement digraph?

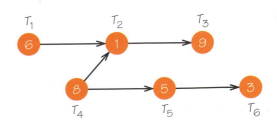

(a) 16 minutes

(b) 17 minutes

(c) 18 minutes

EXERCISES   ■ Challenge   ◆ Discussion

Hamiltonian Circuits

1. For the accompanying graphs (a) through (e), write a Hamiltonian circuit starting at X_5.

(a)

(b)

(c)

(d)

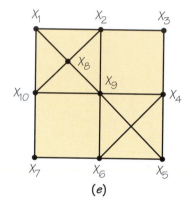

(e)

2. If the edge X_2X_3 is erased from each of the graphs in Exercise 1, does the resulting graph still have a Hamiltonian circuit?

3. (a) If the vertex X_6 and the edges attached to X_6 are removed from the graphs in Exercise 1, do the new graphs that result still have Hamiltonian circuits?

(b) If you think of the graphs in Exercise 1 as communications networks, what interpretation might be given to the "removal" of a vertex and the edges attached as described in part (a)?

◆ **4.** Give examples of real-world situations that can be modeled using a graph and for which finding a Hamiltonian circuit in the graph would be of interest.

5. Suppose two Hamiltonian circuits are considered different if the collections of edges that they use are different. How many other Hamiltonian circuits can you find in the graph in Figure 2.1 that are different from the two discussed?

6. For each of the following graphs, add wiggly edges to indicate a Hamiltonian circuit.

(a)

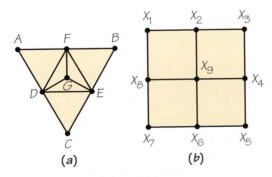

(b)

7. (a) Neither of the following graphs has a Hamiltonian circuit. Is it possible to add a single new edge to these graphs to obtain a new graph that has a Hamiltonian circuit?

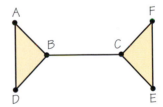

(a)   (b)

(b) Find an example of a graph that has no Hamiltonian circuit and will still have no

Hamiltonian circuit no matter what single edge is added to it.

8. Explain why the graph below has no Hamiltonian circuit.

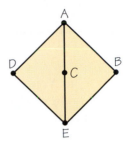

9. Use the graph shown in Exercise 8 to help you construct a connected graph for which every vertex has valence exactly 3 and that does not have a Hamiltonian circuit.

■ **10.** Explain why the tour *ABCFECBDA* is not a Hamiltonian circuit for the graph below. Does this graph have a Hamiltonian circuit?

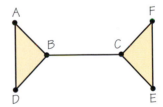

■ **11.** Do the following graphs have Hamiltonian circuits? If not, can you demonstrate why not?

(a)

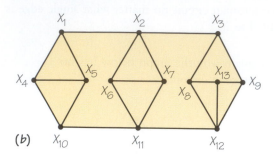

(b)

12. If an edge from X_2 to X_5 is added to each graph in Exercise 11, do the new graphs that result have a Hamiltonian circuit?

13. For each of the following graphs, determine whether there is a Hamiltonian circuit.

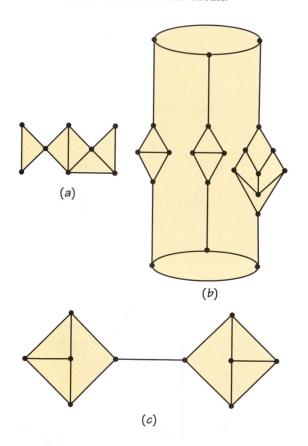

(a)

(b)

(c)

■ **14. (a)** The graph below is known as a four spokes and three concentric circles graph. What

conditions on m and n guarantee that an m spokes and n concentric circles graph has a Hamiltonian circuit? (Assume $m \geq 2$, $n \geq 1$.)

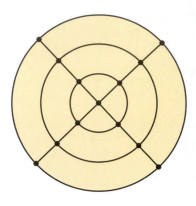

(b) The graph below is known as a 3×4 grid graph. What conditions on m and n guarantee that an $m \times n$ grid graph has a Hamiltonian circuit?

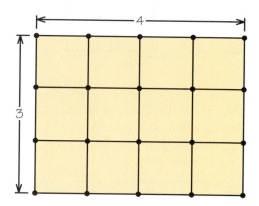

Can you think of a real-world situation in which finding a Hamiltonian circuit in an $m \times n$ grid graph would represent a solution to the problem? If an $m \times n$ grid graph has no Hamiltonian circuit, can you find a tour that repeats a minimum number of vertices and starts and ends at the same vertex?

15. A Hamiltonian path in a graph is a tour of the vertices of the graph that visits each vertex once and only once and starts and ends at different vertices.

(a) For each of the graphs shown in Exercise 11, does the graph have a Hamiltonian path?

(b) Does each of these graphs have a Hamiltonian path that starts at X_1 and ends at X_2?

(c) Describe three real-world situations where finding a Hamiltonian path in a graph would be required.

16. Using the terminology of Exercise 15, draw a graph that has

(a) a Hamiltonian path but no Hamiltonian circuit.

(b) an Euler circuit but no Hamiltonian path.

(c) a Hamiltonian path but no Euler circuit.

17. To practice your understanding of the concepts of Euler circuits and Hamiltonian circuits, determine for the following graphs (a) through (d) whether there is an Euler circuit and/or a Hamiltonian circuit. If so, write it down.

(c)

(d)

(a)

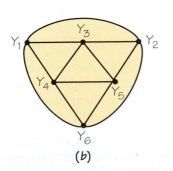

(b)

■ **18.** The n-dimensional cube is obtained from two copies of an $(n - 1)$-dimensional cube by joining corresponding vertices. (The process is illustrated for the 3-cube and the 4-cube in the following figure.) Find formulas for the number of vertices and the number of edges of an n-cube. Can you show that every n-cube has a Hamiltonian circuit? [*Hint:* Show that if you know how to find a Hamiltonian circuit on an $(n - 1)$-cube, then you can use two copies of this to build a Hamiltonian circuit on an n-cube.]

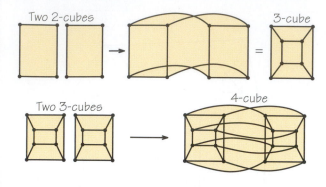

Two 2-cubes   3-cube

Two 3-cubes   4-cube

19. If an edge is added from the vertex with subscript 1 to the vertex with subscript 5 in each graph in Exercise 17, which of the resulting graphs will have Hamiltonian circuits and which will have Euler circuits?

20. Find a family of graphs of which none have Hamiltonian circuits but for which adding a single edge to the first graph in the family creates a Hamiltonian circuit, adding two edges to the second graph in the family creates a Hamiltonian circuit, and so forth.

21. A Hamiltonian path in a graph is a tour of the vertices that visits each vertex once and only once and that starts and ends at different vertices.
(a) Draw an example of a graph that has no Hamiltonian path and where all the vertices are 3-valent.
(b) Draw a graph that has no Hamiltonian path but that does have an Euler circuit.
(c) By analogy with the Hamiltonian path, develop a definition of "Euler path."

Counting Problems

22. To encourage her son to try new things, a mother offers to take him for a dish of ice cream with a topping once a week, for as many weeks as he does not get the same choice as on a previous occasion. If the store offers 12 flavors and six toppings, for how many weeks will she have to do this if her son never picks either of the two types of chocolate ice cream or the three types of nut topping that the store carries?

23. A large corporation has found that it has "outgrown" its current code system for routing interoffice mail. The current system places a code of three ordered, distinct nonzero digits on the mail. The new proposal calls for the use of two ordered capital letters. Is the new system an improvement? If so, how many more locations will the new system enable the company to encode over the current system?

24. A lottery game requires that a person select an upper- or lowercase letter followed by five different two-digit numbers (where the digits cannot both be zero). How many different ways are there to fill out a lottery ticket?

25. (a) In designing a security system for its accounts, a bank asks each customer to choose a five-digit number, all the digits to be distinct and nonzero. How many choices can a customer make?
(b) A suitcase with a liquid-crystal display allows one to unlock it with a specific combination of three capital letters that are not necessarily different. How many choices would a thief have to go through to be sure that all the possibilities had been tried? How does this compare to a "standard" combination lock?

26. (a) When going outside on a cold winter day, Jill can choose from three winter coats, four wool scarves, four pairs of boots, and three ski hats. How many outfits might her friends see her in?
(b) If Jill always insists on wearing her green wool scarf, how many outfits might her friends see her in?

27. The notes C, D, E, F, G, A, and B are to be used to form an ordered five-note musical logo. In how many ways can this be done if **(a)** no note can be repeated; **(b)** notes can be repeated; **(c)** notes can be repeated but all the notes cannot be the same?

28. Repeat Exercise 27a, except that exactly one of the notes in the musical logo must be a sharp and the note chosen to be sharped cannot appear elsewhere (for example, BCD#AG, where D# denotes D sharp).

◆ **29. (a)** In New York State, one type of license plate has three letters followed by three numbers.

Suppose the digits from 0, 1, . . . , 9 can be used, except that all three digits cannot be zero, and that any letter from A to Z (repeats allowed) can be used. How many plates are possible?

(b) Investigate what schemes for license plates are used in your state and determine how many different plates are possible.

30. A restaurant offers 4 soups, 10 entrees, and 8 desserts. How many different choices for a meal can a customer make if one selection is made from each category? If 3 of the desserts are pies and the customer will never order pie, how many meals can the customer choose?

31. In the last several years, heavily populated regions that had had only one area code have had to be divided into service areas with more than one area code. What is the largest number of different phone numbers that can be served using one area code? If an area code cannot begin with a zero, how many different area codes are possible?

32. (a) A credit-card company makes it easier for customers to memorize their PIN (personal identification number) by using a four-digit PIN that consists of three different digits selected from 0, 1, 2, . . . , 9 where one of the digits must be a zero, another is a nonzero digit that is repeated, and another is a digit different from these two. How many different PINs of this kind are there?

(b) How many PINs are possible if there are no restrictions on repeats of the 10 possible digits that can be used?

Traveling Salesman Problems

33. Draw complete graphs with four, five, and six vertices. How many edges do these graphs have? Can you generalize to n vertices? How many TSP tours would these graphs have? (Tours yielding the same Hamiltonian circuit are considered the same.)

34. Calculate the values of 5!, 6!, 7!, 8!, 9!, and 10!. Then find the number of TSP tours in the complete graph with nine vertices.

35. The following table shows the mileage between four cities: Springfield, Ill. (S); Urbana, Ill. (U); Effingham, Ill. (E); and Indianapolis, Ind. (I).

	E	I	S	U
E	–	147	92	79
I	147	–	190	119
S	92	190	–	88
U	79	119	88	–

(a) Represent this information by drawing a weighted complete graph on four vertices.
(b) Use the weighted graph in part (a) to find the cost of the three distinct Hamiltonian circuits in the graph. (List them starting at U.)
(c) Which circuit gives the minimum cost?
(d) Would there be any different in parts (b) and (c) if the start vertex were at I?
(e) If one applies the nearest-neighbor method starting at U, what circuit would be obtained? Does the answer change if one applies the nearest-neighbor algorithm starting at S? At E? At I?
(f) If one applies the sorted-edges method, what circuit would be obtained? Does one get the optimal answer?

36. After a party at her house, Francine (F) has agreed to drive Mary (M), Rachel (R), and Constance (C) home. If the times (in minutes) to drive between her friends' homes are shown below, what route gets Francine back home the quickest?

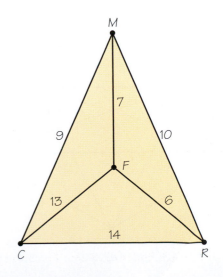

37. In Exercise 36, what route would Francine have to follow to get home as quickly as possible, assuming she promised to drive Mary home first?

38. In Exercise 36, Francine is planning to deliver her friends home and then spend the night at Rachel's house. What would her fastest route be?

39. Starting from the location where she moors her boat (*M*), a fisherwoman wishes to visit three areas—*A*, *B*, and *C*—where she has set fishing nets. If the times (in minutes) between the locales are given in the figure below, what route to visit the three sites and return to the mooring place would be optimal?

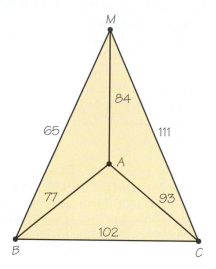

40. (a) For the two complete graphs that follow, find the costs of the nearest-neighbor tour starting at *B* and of the tour generated by the sorted-edges algorithm.

(a)

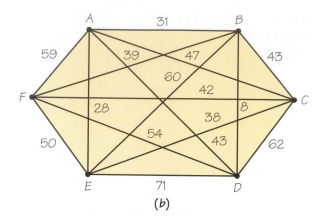

(b)

(b) How many Hamiltonian circuits would have to be examined to find a shortest route for part (a) by the brute force method?
(c) Invent an algorithm different from the sorted-edges and nearest-neighbor algorithms that is easy to apply for finding TSP solutions.

41. An airport limo must take its six passengers from the airport to different downtown hotels. Is this a traveling salesman problem, a Chinese postman problem, or an Euler circuit problem?

42. For each of the following graphs with weights, apply the nearest-neighbor method (starting at vertex *A*) and the sorted-edges method to find (it is hoped) a cheap tour.

(a)

(b)

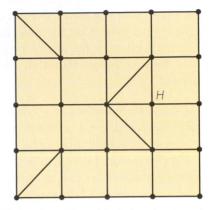

Assume that at equally spaced intervals along the blocks in this graph there are storm sewers that must be inspected after each thunderstorm to see whether they are clogged. Is this a Hamiltonian circuit problem, an Euler circuit problem, or a Chinese postman problem? Find an optimal tour to do this inspection.

44. (a) Solve the six-city TSP shown in the diagram using the nearest-neighbor algorithm starting at vertex A and starting at vertex B.
(b) Apply the sorted-edges method.

(c)

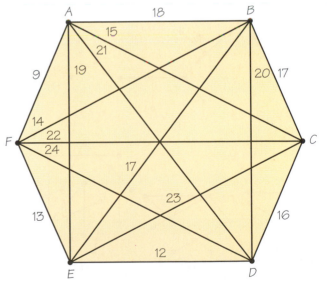

43. The following figure represents a town where there is a sewer located at each corner (where two or more streets meet). After every thunderstorm, the department of public works wishes to have a truck start at its headquarters (at vertex H) and make an inspection of sewer drains to be sure that leaves are not clogging them. Can a route start and end at H that visits each corner exactly once? (Assume that all the streets are two-way streets.) Does this problem involve finding an Euler circuit or a Hamiltonian circuit?

■ **45.** Construct an example of a complete graph of five vertices, with distinct weights on the edges for which the nearest-neighbor algorithm starting at a particular vertex and the sorted-edges algorithm yield different solutions for the traveling salesman problem. Can you find a five-vertex complete graph with weights on the edges in which the optimal solution, the nearest-neighbor solution, and the sorted-edges algorithm solution are all different?

■ **46.** If the brute force method of solving a 20-city TSP is employed, use a calculator to determine how many Hamiltonian circuits must be examined. How long would it take to determine the minimum-cost tour if the cost of tours could be computed at the rate of 1 billion per second? (Convert your answer to years by seeing how many years are equivalent to a billion seconds!)

47. Suppose one has found an optimal tour for a given 10-city TSP problem to have weight 4200. Now suppose the weights on the edges of the complete graph are increased by 30. What can you say about the optimal tour and its weight?

Trees and Spanning Trees

48. For each graph below, explain why it is or is not a tree.

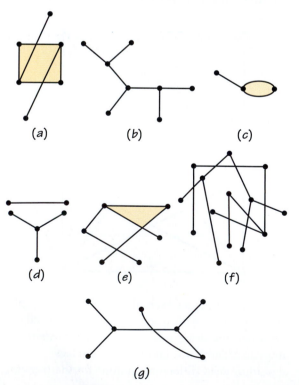

49. For each of the diagrams below, explain why the wiggly edges are not

(a) a spanning tree.

(b) a Hamiltonian circuit.

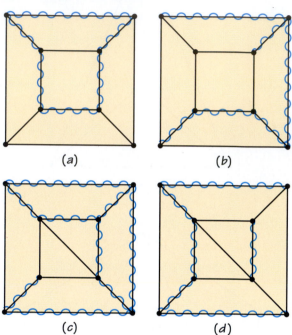

50. Find all the spanning trees in the graphs below.

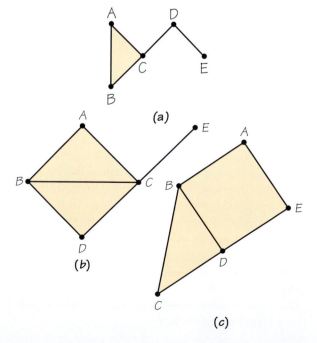

51. Use Kruskal's algorithm to find a minimum-cost spanning tree for the following graphs (a), (b), (c), and (d). In each case, what is the cost associated with the tree?

(a)

(b)

(c)

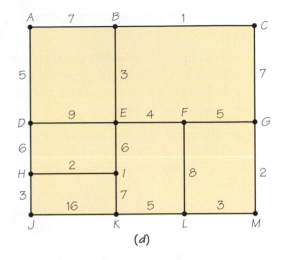

(d)

52. A connected graph G has 20 vertices. How many edges does a spanning tree of G have? How many vertices does a spanning tree of G have? What can one say about the number of edges G has?

53. A connected graph H has a spanning tree with 25 edges. How many vertices does the spanning tree have? How many vertices does H have? What can one say about the number of edges H has?

54. A large company wishes to install a pneumatic tube system that would enable small items to be sent between any of 10 locales, possible by relay. If the nonprohibitive costs (in \$100) are shown in the graph model below, between which sites should the tube be installed to minimize the total cost?

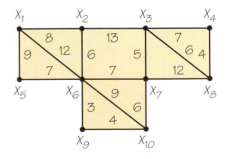

55. If the weight of each edge in Exercise 54 is increased by 3, will the tree that achieves minimum cost for the new collection of weights be the same as the one that achieves minimum cost for the original set of weights?

◆ **56.** Give examples of real-world situations that can be modeled using a weighted graph and for which finding a minimum-cost spanning tree for the graph would be of interest.

■ **57.** Can Kruskal's algorithm be modified to find a maximum-weight spanning tree? Can you think of an application for finding a maximum-weight spanning tree?

◆ **58.** Find the cost of providing a relay network between the six cities with the largest populations in your home state, using the road distances between the cities as costs. Does it follow that the same solution would be obtained if air distances were used instead?

■ **59.** Would there ever be a reason to find a minimum-cost spanning tree for a weighted graph in which the weights on some of the edges were negative? Would Kruskal's algorithm still apply?

■ **60.** Suppose G is a graph such that all the weights on its edges are different numbers. Show that there is a unique minimum-cost spanning tree.

61. Two spanning trees of a (weighted) graph are considered different if they use different edges. Show that the graph below has different minimum-cost spanning trees, though all these different trees have the same cost.

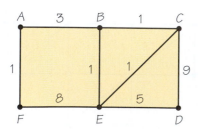

62. Let G be a graph with weights assigned to each edge. Consider the following algorithm:

(a) Pick any vertex V of G.

(b) Select an edge E with a vertex at V that has a minimum weight. Let the other endpoints of E be W.

(c) Contract the edge VW so that edge VW disappears and vertices V and W coincide (see the following figures).

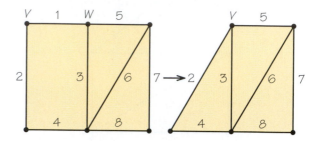

If in the new graph two or more edges join a pair of vertices, delete all but the cheapest. Continue to call the new vertex V.

(d) Repeat steps (b) and (c) until a single point is obtained. The edges selected in the course of this algorithm (called Prim's algorithm) form a minimum-cost spanning tree. Apply this algorithm to the following graphs.

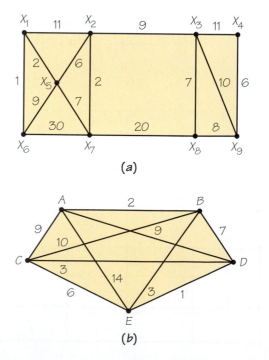

(a)

(b)

63. Determine whether each of the following statements is true or false for a minimum-cost spanning tree T for a weighted connected graph G:

(a) T contains a cheapest edge in the graph.
(b) T cannot contain a most expensive edge in the graph.
(c) T contains one fewer edge than there are vertices in G.
(d) There is some vertex in T to which all others are joined by edges.
(e) There is some vertex in T that has valence 3.

■ **64.** In the following graphs, the number in the circle for each vertex is the cost of installing equipment at the vertex if relaying must be done at the vertex, while the number on an edge indicates the cost of providing service between the endpoints of the edge.

In each case, find the minimum cost (allowing relays) for sending messages between any pair of vertices, taking vertex relay costs into account.

(a)

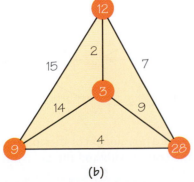

(b)

Would your answer be different if vertex relay costs were neglected? (*Warning:* Kruskal's algorithm cannot be used to answer the first question. This problem illustrates the value of having an algorithm overrelying on "brute force.")

65. (a) Show that for each edge of graph J below there is a spanning tree of J that avoids that edge.
(b) For each spanning tree that you found in graph J, count the number of vertices and edges. Do you notice any pattern?
(c) For graph H below and each edge e in the graph, is there a spanning tree that does not include edge e of H?

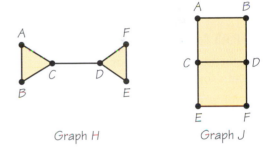

Graph H   Graph J

66. (a) The table shown gives the "closeness" or distance values between four objects. Construct a four-vertex tree with weights on its edges such that the distances between pairs of vertices of the tree (as measured by the sum of the weights on the path in the tree between these vertices) give rise to this table.

	A	B	C	D
A	0	3	10	14
B	3	0	7	11
C	10	7	0	4
D	14	11	4	0

(b) Produce several real-world contexts that might give rise to the situation described here.

67. The figure below represents four objects using a tree with weights on the edges. Construct a table with four rows and four columns recording how "close" pairs of vertices in the tree are to each other. To find how close a pair of objects is, add together the weights along the path that joins these two objects.

Critical-Path Analysis

68. Find the earliest completion time and critical paths for the order-requirement digraphs below.

(a)

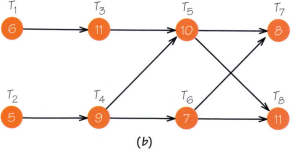

(b)

69. Find the earliest completion time and critical paths for the following order-requirement digraphs.

(a)

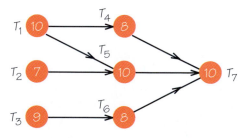

(b)

70. Construct an example of an order-requirement digraph with two different critical paths.

71. In the order-requirement digraph below, determine which tasks, if shortened, would reduce the earliest completion time and which would not. Then find the earliest completion time if task T_5 is reduced to time length 7. What is the new critical path?

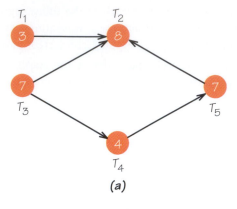

72. For the order-requirement digraph in Exercise 71, find the critical path and the task(s) in the critical path whose time, when reduced the least, creates a new critical path.

73. To build a new addition on a house, the following tasks must be completed:

(a) Lay foundation.
(b) Erect sidewalls.
(c) Erect roof.
(d) Install plumbing.
(e) Install electric wiring.
(f) Lay tile flooring.
(g) Obtain building permits.
(h) Put in door that connects new room to existing house.
(i) Install track lighting on ceiling.
(j) Install wall air-conditioner.

Construct reasonable time estimates for these tasks and a reasonable order-requirement digraph. What is the fastest time in which these tasks can be completed?

74. At a large toy store, scooters arrive unassembled in boxes. To assemble a scooter, the following tasks must be performed:

TASK 1. Remove parts from the box.
TASK 2. Attach wheels to the footboard.
TASK 3. Attach vertical housing.
TASK 4. Attach handlebars to vertical housing.
TASK 5. Put on reflector tape.
TASK 6. Attach bell to handlebars.
TASK 7. Attach decals.
TASK 8. Attach kickstand.
TASK 9. Attach safety instructions to handlebars.

Give reasonable time estimates for these tasks and construct a reasonable order-requirement digraph. What is the earliest time by which these tasks can be completed?

75. Construct an order-requirement digraph with six tasks that has three critical paths of length 24.

APPLET EXERCISES

To do these exercises, go to www.whfreeman.com/fapp7e.

1. TSP: Nearest-Neighbor Algorithm. There is an extended version of the nearest-neighbor algorithm, in which you compare the total distances of the Hamiltonian circuits produced by applying the ordinary nearest-neighbor algorithm starting at each of the vertices of the graph (rather than just a specific one). Explore the effectiveness of this algorithm using the *TSP: Nearest-Neighbor* applet.

2. TSP: Sorted-Edges Algorithm. Go to the *TSP: Sorted Edges* applet, where you can apply the sorted-edges algorithm to see if it solves the traveling salesman problem for the following graphs (and others):

(a)

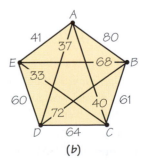

(b)

3. Kruskal's Algorithm. Go to the *Kruskal's Algorithm* applet, where you can apply Kruskal's algorithm to find the minimum-cost spanning trees in the following graphs (and others):

(a)

(b)

 # WRITING PROJECTS

1. Write an essay about a variety of situations in which you are personally involved for which a solution of the TSP is (perhaps implicitly) required. Explain under what circumstances it might be valuable to carry out a formal mathematical solution to such TSPs rather than use an ad hoc solution.

2. Construct an example, of the kind suggested on page 48, that shows that in a situation where three day campers must be picked up and brought to camp, it may make a difference if the optimization criterion is minimizing distance traveled by the camp bus versus minimizing average time that the children spend on the bus.

3. Determine the six largest cities in the state in which you live. By consulting a road atlas (or by some other means) construct the graph that represents the road distances between your hometown and these six other cities. Now apply (a) the nearest-neighbor method, (b) the sorted-edges method, and (c) the nearest neighbor from each city, and pick the minimum tour method to solve the associated TSP. Do you have reason to believe that the answers you get might include an optimum solution among them?

SUGGESTED READINGS

BODIN, LAWRENCE. Twenty years of routing and scheduling, *Operations Research*, 38 (1990): 571–579. A survey of real-world situations where routing and scheduling were used, written by a pioneer in this area.

DOLAN, ALAN, and JOAN ALDUS. *Networks and Algorithms: An Introductory Approach*, Wiley, Chichester, England, 1993. An excellent introduction to graph theory algorithms.

GUSFIELD, DAN. *Algorithms on Strings, Trees, and Sequences*, Cambridge University Press, New York, 1997. Details applications of graph theory in pattern recognition and reconstruction problems.

JONES, NEIL C., and PAVEL A. PEVZNER, *An Introduction to Bioinformatics Algorithms,* MIT Press, Cambridge, Mass., 2004. This book has material on how graph theory ideas, particularly those related to Hamiltonian circuits, are being used in molecular genetics and computational biology.

LAWLER, EUGENE, J. LENSTRA, RINNOY KAN, and D. SHMOYS, eds. *The Traveling Salesman Problem,* Prentice-Hall, Englewood Cliffs, N.J., 1985. Includes survey and technical articles on all aspects of the TSP.

LUCAS, WILLIAM, FRED ROBERTS, and ROBERT THRALL, eds. *Discrete and Systems Models,* vol. 3: *Modules in Applied Mathematics,* Springer-Verlag, New York, 1983. Chapter 6, "A Model for Municipal Street Sweeping Operations," by A. Tucker and L. Bodin, describes street-sweeping and related models in detail. Other chapters detail many recent applications of mathematics.

ROBERTS, FRED S., and BARRY TESMAN, *Applied Combinatorics,* 2nd ed., Pearson Prentice Hall, Upper Saddle River, N.J., 2004. The material on network-optimization problems is excellent.

ROBERTS, FRED. *Graph Theory and Its Applications to Problems of Society,* Society for Industrial and Applied Mathematics, Philadelphia, 1978. A very readable account of how graph theory is finding a wide variety of applications.

SUGGESTED WEB SITES

www.tsp.gatech.edu This site provides a detailed history and many applications of the TSP.

www.geocities.com/satishgoda/html/mst.html This Web page provides a collection of useful resources related to minimum-cost spanning trees and graph theory.

www-gap.dcs.st-and.ac.uk/~history/ Mathematicians/Hamilton.html This site provides biographical information about William Rowan Hamilton, for whom Hamiltonian circuits are named.

Planning and Scheduling

In a society as complex as ours, everyday problems such as providing services efficiently and on time require accurate planning of both people and machines. Take the example of a hospital in a major city. Around-the-clock scheduling of nurses and doctors must be provided to guarantee that people with particular expertise are available during each shift. The operating rooms must be scheduled in a manner flexible enough to deal with emergencies. Equipment used for X-ray, CT, or MRI scans must be scheduled for maximum efficiency.

Although many scheduling problems are often solved on an ad hoc basis, we can also use mathematical ideas to gain insight into the complications that arise in scheduling. The ideas we develop in this chapter have practical value in a relatively narrow range of applications, but they shed light on many characteristics of more realistic, and hence more complex, scheduling problems.

3.1 Scheduling Tasks

Assume that a certain number of identical **processors** (machines, humans, or robots) work on a series of tasks that make up a job. Associated with each task is a specified amount of time required to complete the task. For simplicity, we assume that any of the processors can work on any of the tasks. Our problem, known as the **machine-scheduling problem,** is to decide how the tasks should be scheduled so that the completion time for the tasks collectively is as early as possible.

Even with these simplifying assumptions, complications in scheduling will arise. Some tasks may be more important than others and perhaps should be scheduled first. When "ties" occur, they must be resolved by special rules. As an example, suppose we are scheduling patients to be seen in a hospital emergency

Hospitals are increasingly making use of mathematical techniques applied to scheduling problems. To make efficient use of one or more operating rooms requires the complicated assembly of a team of doctors, nurses, equipment, and support staff. Mathematical techniques for scheduling have made it possible to carry out more operations in less time. *(Eyewire/ Punchstock.)*

room staffed by one doctor. If two patients arrive simultaneously, one with a bleeding foot, the other with a bleeding arm, which patient should be processed first? Suppose the doctor treats the arm patient first, and while treatment is going on, a person in cardiac arrest arrives. Scheduling rules must establish appropriate priorities for cases such as these.

Another common complication arises with jobs consisting of several tasks that cannot be done in an arbitrary order. For example, if the job of putting up a new house is treated as a scheduling problem, the task of laying the foundation must precede the task of putting up the walls, which in turn must be completed before work on the roof can begin.

Assumptions and Goals

To simplify our analysis, we need to make clear and explicit assumptions:

1. If a processor starts work on a task, the work on that task will continue without interruption until the task is completed.

2. No processor stays voluntarily idle. In other words, if there is a processor free and a task available to be worked on, then that processor will immediately begin work on that task.

3. The requirements for ordering the tasks are given by an order-requirement digraph. (A typical example is shown in Figure 3.1, with task times highlighted within each vertex. The ordering of the tasks imposed by the order-requirement digraph represents constraints of physical reality. For example, you cannot fly a plane until it has taken fuel on board.)

4. The tasks are arranged in a **priority list** that is independent of the order requirements. (The priority list is a ranking of the tasks according to some criterion of "importance.")

FIGURE 3.1 A typical order-requirement digraph.

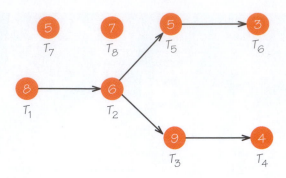

EXAMPLE 1   Home Construction

Let's see how these assumptions might work for an example involving a home construction project. In this case, the processors are human workers with identical skills. Assumption 1 means that once a worker begins a task, the work on this task is finished without interruption. Assumption 2 means that no worker will stay idle if there is some task for which the predecessors are finished. Assumption 3 requires that the ordering of the tasks be summarized in an order-requirement digraph. This digraph would code facts such as that the site must be cleared before the foundation is begun. Assumption 4 requires that the tasks be ranked in a list from some perspective, perhaps a subjective view.

The task with highest priority rank is listed first in the list, followed left to right by the other tasks in priority rank. The priority list might be based on the size of the payments made to the construction company when a task is completed, even though these payments have no relation to the way the tasks must be done, as indicated in the order-requirement digraph. Alternatively, the priority list might reflect an attempt to find an algorithm to schedule the tasks needed to complete the whole job more quickly. ■

When considering a scheduling problem, there are various goals we might want to achieve. Among these are:

Goal 1. Minimizing the completion time of the job.
Goal 2. Minimizing the total time that processors are idle.
Goal 3. Finding the minimum number of processors necessary to finish the job by a specified time.

In the context of the construction example, goal 1 would complete the home as quickly as possible. Goal 2 would ensure that workers, who are perhaps paid by the hour, were not paid for doing nothing. One way of accomplishing this would be to hire one fewer worker even if it means the house takes longer to finish. Goal 3 might be reasonable if the family wants the house done before the first day of school, even if they have to pay a lot more workers to get the house done by this time.

For now we will concentrate on goal 1, finishing all the tasks at the earliest possible time. Note, however, that optimizing with respect to one goal may not optimize with respect to another. Our discussion here goes beyond what was discussed in Chapter 2 (see section 2.4) by dealing with how to assign tasks in a job to the processors that do the work.

List-Processing Algorithm

The scheduling problem we have described sounds more complicated than the traveling salesman problem (TSP). Indeed, like the TSP, it is known to be NP-complete. This means that it is unlikely that anyone will ever find a computationally fast algorithm that can find an optimal solution. Thus, we will be content to seek a solution method that is computationally fast and gives only approximately optimal answers.

List-Processing Algorithm and Ready Task

The algorithm we use to schedule tasks is the **list-processing algorithm.** In describing it, we will call a task **ready** at a particular time if all its predecessors as indicated in the order-requirement digraph have been completed at that time. In Figure 3.1 at time 0, the ready tasks are T_1, T_7, and T_8, while T_2 cannot be ready until 8 time units after T_1 is started. The algorithm works as follows: At a given time, assign to the lowest-numbered free processor the first task on the priority list that is *ready* at that time and that hasn't already been assigned to a processor.

In applying this algorithm, we will need to develop skill at coordinating the use of the information in the order-requirement digraph and the priority list. It will be helpful to cross out the tasks in the priority list as they are assigned to a processor to keep track of which tasks remain to be scheduled.

Let's apply this algorithm to one possible priority list—T_8, T_7, T_6, . . . , T_1—using two processors and the order-requirement digraph in Figure 3.1. The result is the schedule shown in Figure 3.2, where idle processor time is indicated by white. How does the list-processing algorithm generate this schedule?

Machine 1 is idle because it cannot begin T_2 until T_1 is complete.

FIGURE 3.2 The schedule produced by applying the list-processing algorithm to the order-requirement digraph in Figure 3.1 using the list T_8, T_7, . . . , T_1.

Because T_8 (task 8) is first on the priority list and ready at time 0, it is assigned to the lowest-numbered free processor, processor 1. Task 7, next on the priority list, is also ready at time 0 and thus is assigned to processor 2. The first processor to become free is processor 2 at time 5. Recall that by assumption 1, once a processor starts work on a task, its work cannot be interrupted until the task is complete. Task 6, the next unassigned task on the list, is not ready at time 5, as can be seen by consulting Figure 3.1. The reason task 6 is not ready at time 5 is that task 5 has not been completed by time 5. In fact, at time 5, the only ready task on the list is T_1, so that task is assigned to processor 2. At time 7, processor 1 becomes free, but no task becomes ready until time 13.

Thus, processor 1 stays idle from time 7 to time 13. At this time, because T_2 is the first ready task on the list not already scheduled, it is assigned to processor 1. Processor 2, however, stays idle because no other ready task is available at this time. The remainder of the scheduling shown in Figure 3.2 is completed in this manner.

We can summarize this procedure as follows:

List-Processing Algorithm

As the priority list is scanned from left to right to assign a task to a processor at a particular time, we pass over tasks that are not ready to find ones that are ready. If no task can be assigned in this manner, we keep one or more processors idle until such time that, reading the priority list from the left, there is a ready task not already assigned. After a task is assigned to a processor, we resume scanning the priority list, starting over at the far left, for unassigned tasks.

When Is a Schedule Optimal?

The schedule in Figure 3.2 has a lot of idle time, so it may not be optimal. Indeed, if we apply the list-processing algorithm for two processors to another possible priority list $T_1, \ldots, T_8$, using the digraph in Figure 3.1, the resulting schedule is that shown in Figure 3.3.

Here are the details of how this schedule was arrived at. Remember that we must coordinate the list $T_1, T_2, \ldots, T_8$ with the information in the order-requirement digraph shown in Figure 3.3a. At time 0, task T_1 is ready, so this task is assigned to processor 1. However, at time 0, tasks $T_2, T_3, \ldots, T_6$ are not ready because their predecessors are not done. For example, T_2 is not ready at time 0 because T_1, which precedes it, is not done at time 0. The first ready task on the list, reading from left to right, that is not already assigned is T_7, so task T_7 gets assigned to processor 2. Both processors are now busy until time 5, at which point processor 2 becomes idle (Figure 3.3b).

Tasks T_1 and T_7 have been assigned. Reading from left to right along the list, the first task not already assigned whose predecessors are done by time 5 is T_8,

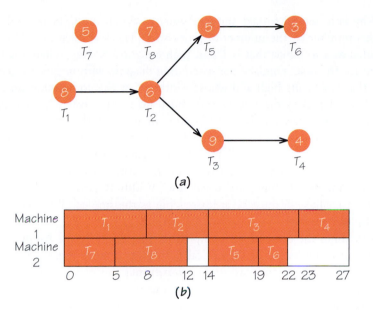

(a)

FIGURE 3.3 (a) A typical order-requirement digraph (repeat of Figure 3.1). (b) The schedule produced by applying the list-processing algorithm to the order-requirement digraph in Figure 3.3a using the list $T_1, T_2, \ldots, T_8$.

(b)

so this task is started at time 5 on processor 2; processor 2 will continue to work on this task until time 12, because the task time for this task is 7 time units. At time 8, processor 1 becomes free, and reading the list from left to right we find that T_2 is ready (because T_1 has just been completed). Thus, T_2 is assigned to processor 1, which will stay busy on this task until time 14. At time 12, processor 2 becomes free, but the tasks that have not already been assigned from the list, T_3, T_4, T_5, T_6 are not ready, because they depend on T_2 being completed before these tasks can start. Thus, processor 2 must stay idle until time 14. At this time, T_3 and T_5 become ready. Since both processors 1 and 2 are idle at time 14, the lower numbered of the two, processor 1, gets to start on T_3 because it is the first ready task left to be assigned on the list scanned from left to right. Task T_5 gets assigned to processor 2 at time 14. The remaining tasks are assigned in a similar manner.

The schedule shown in Figure 3.3b is optimal because the path T_1, T_2, T_3, T_4, with length 27, is the critical path in the order-requirement digraph. As we saw in Chapter 2, the earliest completion time for the job made up of all the tasks is the length of the longest path in the order-requirement digraph.

There is another way of relating optimal completion time to the completion time that is yielded by the list-processing algorithm. Suppose that we add all the task times given in the order-requirement digraph and divide by the number of processors. The completion time using the list-processing algorithm must be at least as large as this number. For example, the task times for the order-requirement digraph in Figure 3.3a sum to 47. Thus, if these tasks are scheduled on two processors, the completion time is at least $\frac{47}{2} = 23.5$ (in fact, 24, because the list-processing algorithm applied to integer task times must yield an integer solution), while for three processors the completion time is at least $\frac{47}{3}$ (in fact, 16).

Why is it helpful to take the total time to do all the tasks in a job and divide this number by the number of processors? Think of each task that must be scheduled as a rectangle that is 1 unit high and t units wide, where t is the time allotted for the task. Think of the scheduling diagram with m processors as a rectangle that is m units high and whose width, W, is the completion time for the tasks. The scheduling diagram is to be filled up by the rectangles that represent the tasks. How small can W be? The area of the rectangle that represents the scheduling diagram must be at least as large as the sum of all the rectangles representing tasks that are "packed" into it. The area of the scheduling diagram rectangle is mW. The combined areas of all the tasks, plus the area of rectangles corresponding to idle time, will equal mW. Width W is smallest when the idle time is zero. Thus, W must be at least as big as the sum of all the task times divided by m.

Sometimes the estimate for completion time given by the list-processing algorithm from the length of the critical path gives a more useful value than the approach based on adding task times. Sometimes the opposite is true. For the order-requirement digraph in Figure 3.1, except for a schedule involving one processor, the critical-path estimate is superior. For some scheduling problems, both these estimates may be poor.

The number of priority lists that can be constructed if there are n tasks is $n!$ and can be computed using the fundamental principle of counting. For example, for eight tasks, $T_1, \ldots, T_8$, there are $8 \times 7 \times 6 \cdots \times 1 = 40{,}320$ possible priority lists. For different choices of the priority list, the list-processing algorithm may schedule the tasks, subject to the constraints of the order-requirement digraph, in different ways. More specifically, two different lists may yield different completion times or the same completion time, but the order in which the tasks are carried out will be different. It is also possible that two different lists produce identical ordering of the assignments of the tasks to processors and completion times. A little later we will see a method that can be used to select a list that, if we are lucky, will give a schedule with a relatively good completion time. In fact, no method is known, except for very specialized cases, of how to choose a list that can be guaranteed to produce an optimal schedule when the list algorithm is applied to it.

Strange Happenings

The list-processing algorithm involves four factors that affect the final schedule. The answer we get depends on the following:

1. Times of the tasks
2. Number of processors
3. Order-requirement digraph
4. Ordering of the tasks on the list

To see the interplay of these four factors, consider another scheduling problem, this time asociated with the order-requirement digraph shown in Figure 3.4

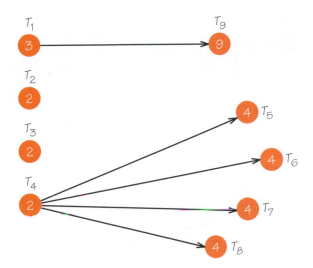

FIGURE 3.4 An order-requirement digraph designed to help illustrate some paradoxical behavior produced by the list-processing algorithm.

(the highlighted numbers are task time lengths). The schedule generated by the list-processing algorithm applied to the list $T_1, T_2, \ldots, T_9$, using three processors, is given in Figure 3.5.

Treating the list $T_1, \ldots, T_9$ as fixed, how might we make the completion time earlier? Our alternatives are to pursue one or more of these strategies:

1. Reduce task times.
2. Use more processors.
3. "Loosen" the constraints of the order-requirement digraph.

Let's consider each alternative in turn, changing one feature of the original problem at a time, and see what happens to the resulting schedule. If we use strategy 1 and reduce the time of each task by one unit, we would expect the completion time to go down. Figure 3.6 shows the new order-requirement digraph, and Figure 3.7 shows the schedule produced for this problem, using the list-processing algorithm with three processors applied to the list $T_1, \ldots, T_9$.

The completion time is now 13—longer than the completion time of 12 for the case (Figure 3.5) with longer task times. Here is something unexpected! Let's explore further and see what happens.

FIGURE 3.5 The schedule produced by applying the list-processing algorithm to the order-requirement digraph in Figure 3.4 using the list $T_1, T_2, \ldots, T_9$ with three processors.

FIGURE 3.6 The order-requirement digraph obtained from the one in Figure 3.4 by reducing by one unit each of the task times shown there.

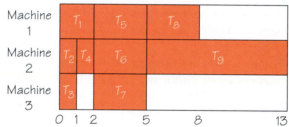

FIGURE 3.7 The schedule produced by applying the list-processing algorithm to the order-requirement digraph in Figure 3.6 using the list $T_1, T_2, \ldots, T_9$ with three processors.

Next we consider strategy 2, increasing the number of machines. Surely this should speed matters up. When we apply the list-processing algorithm to the original graph in Figure 3.4, using the list $T_1, \ldots, T_9$ and four machines, we get the schedule shown in Figure 3.8. The completion time is now 15—an even later completion time than for the previous alteration.

Finally, we consider strategy 3, trying to shorten completion time by erasing all constraints (edges with arrows) in the order-requirement digraph shown in Figure 3.4. By increasing flexibility of the ordering of the tasks, we might guess we could finish our tasks more quickly. Figure 3.9 shows the schedule using the list

FIGURE 3.8 The schedule produced by applying the list-processing algorithm to the order-requirement digraph in Figure 3.4 using the list $T_1, T_2, \ldots, T_9$ with four processors.

FIGURE 3.9 The schedule produced by applying the list-processing algorithm to the order-requirement digraph in Figure 3.4, modified by erasing all its directed edges, using the list $T_1, T_2, \ldots, T_9$ with three processors.

$T_1, \ldots T_9$—now it takes 16 units! This is the worst of our three strategies to reduce completion time.

The failures we have seen here are surprising at first glance, but they are typical of what can happen when a situation is too complex to analyze with naïve intuition. The value of using mathematics rather than intuition or trial and error to study scheduling and other problems is that it points out flaws that can occur in unguarded intuitive reasoning.

It is tempting to believe that we can make an adjustment in the rules for scheduling that we adopted to avoid the paradoxical behavior that has just been illustrated. Unfortunately, operations research experts have shown that there are no "simple fixes." This means that, in practice, for large scheduling problems such as those that face our hospitals and transportation system, finding the best solution to a particular scheduling problem cannot be guaranteed (see Spotlight 3.1).

3.2 Critical-Path Schedules

In our discussion so far, we have acted as though the priority list used in applying the list-processing algorithm was given to us in advance based on external considerations. Let's now consider the question of whether there is a systematic method of *choosing* a priority list that yields optimal or nearly optimal schedules. We will show how to construct a specific priority list based on this principle, to which the list-processing algorithm can then be applied.

Recall from our discussion of critical-path analysis in Chapter 2 that no matter how a schedule is constructed, the finish time cannot be earlier than the length of the longest path in the order-requirement digraph. This suggests that we should try to schedule first those tasks that occur early in long paths, because they might be a bottleneck for the other tasks.

Scheduling Two Processors

EXAMPLE 2

To illustrate this method, consider the order-requirement digraph in Figure 3.10a. Suppose we wish to schedule these tasks on two processors. Initially, there are two critical paths of length 64: T_1, T_2, T_3 and T_1, T_4, T_3. Thus, we place T_1 first on the priority list. With T_1 "gone," there is a new critical path of length 60

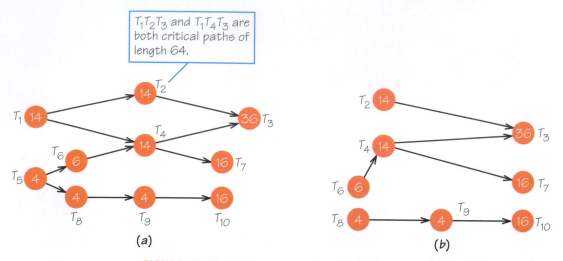

FIGURE 3.10 (a) An order-requirement digraph used to illustrate the critical-path scheduling method. (b) Residual order-requirement digraph after tasks T_1 and T_5 have been removed.

(T_5, T_6, T_4, T_3) that starts with T_5, so T_5 is placed second on the priority list. At this stage, with T_1 and T_5 removed, we have the residual order-requirement digraph shown in Figure 3.10b. In this diagram there are paths of length 50 (T_2, T_3), 56 (T_6, T_4, T_3), 36 (T_6, T_4, T_7), and 24 (T_8, T_4, T_{10}). Because T_6 heads the path that is currently longest in length, it is placed third in the priority list. Once T_6 is removed from Figure 3.10b, there is a tie for which is the longest path remaining, because both T_2, T_3 and T_4, T_3 are paths of length 50.

When there is a tie between two longest paths, we place next on the priority list in the lowest-numbered task heading a longest path. In the example shown here, this means that T_2 is placed next on the priority list, to be followed by T_4. Continuing in this fashion, we obtain the priority list $T_1, T_5, T_6, T_2, T_4, T_3, T_8, T_9, T_7, T_{10}$. Note that the order of T_7 and T_{10} was decided using the rule for breaking ties. The list-processing algorithm is now applied using this priority list and the order-requirement digraph in Figure 3.10a. We obtain the schedule in Figure 3.11. ∎

FIGURE 3.11 The optimal schedule produced by applying the critical-path scheduling method to the order-requirement digraph in Figure 3.10. The list used was T_1, T_5, T_6, $T_2, T_4, T_3, T_8, T_9, T_7, T_{10}$.

The city of New York depends on a public transportation system of subways and roads to bring hundreds of thousands of people who live in the four outer boroughs (Queens, Brooklyn, the Bronx, and Staten Island) into Manhattan to work and "play." New York City also has a communication system of telephones, radio and television stations, and computer networks. These systems speed information between New York's citizens and people outside the city and around the world. The area in southern Manhattan, in the vicinity of the World Trade Center (WTC), was a center for banking, insurance, financial markets, and domestic and international commerce. The attack on the World Trade Center on September 11, 2001, disrupted these networks and markets but did not destroy them, partly because the principles of operations research and management science were used in the design and development of these systems over a long period of time.

The diagram below shows a very simple subway (train) system between an eastern and a western terminus.

(AP/World Wide Photo.)

Western terminus (W)

Eastern terminus (E)

There are two tracks, each dedicated for use by westbound or eastbound trains to run between the two termini. The only place where trains can be turned around is at these termini. Simple graph theory tells us that in such a system, if a vertex is "destroyed" or out of service, or an edge is "destroyed" or out of service, the system totally breaks down. However, the simple provision that trains can be turned around at U, even though this is usually only one stop on the way from W to E, gives much greater flexibility to the system if

there is a water main break, or a gas leak, etc. Thanks to simple principles of this kind and creating routes that use independent lines with many transfer points, New Yorkers were able to use the subway system in a flexible way after the World Trade Center disaster. In the days right after the WTC collapsed, trains were not allowed past the geographic area near the WTC for fear that the tunnels' structural foundation had been weakened and that subway vibrations could cause the collapse of damaged buildings. After it was ascertained that running the subways was safe both for partially damaged buildings and for the subways themselves, routes were altered several times to give rescue workers and people returning to their daily routines maximum support. One line's tunnels did collapse, and several stations had to be closed for extended periods, but due to the redundancy and flexibility of the design of the system, a remarkable amount of service was quickly restored.

Unfortunately, with one major exception, the antennae that supported New York's broadcast television system were located atop one of the towers. The exception is WCBS, Channel 2, which has its antenna atop the Empire State Building. This meant that weeks after the WTC was destroyed, New Yorkers who did not have cable television were able to watch only one major channel and a few minor ones. The cable television system was largely unaffected.

Good planning and wise application of the principles of management science make it possible to minimize the effects of natural and manmade disasters.

:::::: Critical-Path Scheduling

The **critical-path scheduling** algorithm applies the list-processing algorithm using the priority list L obtained as follows:

1. Find a task that heads a critical (longest) path in the order-requirement digraph. If there is a tie, choose the task with the lower number.
2. Place the task found in step 1 next on the list L. (The first time through the process this task will head the list.)
3. Remove the task found in step 1 and the edges attached to it from the current order-requirement digraph, obtaining a new (modified) order-requirement digraph.
4. If there are no vertices left in the new order-requirement digraph, the procedure is complete; if there are vertices left, go to step 1.

This procedure will terminate when all the tasks in the original order-requirement digraph have been placed on the list L.

The preceding example shows that critical-path scheduling can sometimes yield optimal solutions. Unfortunately, this algorithm does not always perform well. For example, the critical-path method employing four processors applied to the order-requirement digraph shown in Figure 3.12 yields the list T_1, T_8, T_9, T_{10}, T_{11}, T_5, T_6, T_7, T_{12}, T_2, T_3, T_4 and then the schedule in Figure 13.3. (Note that T_5, T_6, T_7 are thought of as heading paths of length 10.) In fact, there can be no worse schedule than this one. An optimal schedule is shown in Figure 3.14.

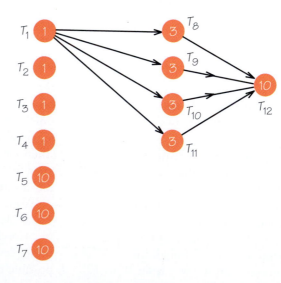

FIGURE 3.12 An order-requirement digraph used to illustrate how poorly the critical-path scheduling method can sometimes behave.

FIGURE 3.13 The schedule produced by applying the critical-path scheduling method to the order-requirement digraph in Figure 3.12 using four processors. The list used was $T_1, T_8, T_9, T_{10}, T_{11}, T_5, T_6, T_7, T_{12}, T_2, T_3, T_4$.

Many of the results we have examined so far are negative because we are dealing with a general class of problems that defy our using computationally efficient algorithms to find an optimal schedule. But we can close on a more positive note. Consider an arbitrary order-requirement digraph, but assume all the tasks take equal time. It turns out that we can always construct an optimal schedule using two processors in this situation. Ironically, we can choose among many algorithms to produce these optimal schedules. The algorithms are easy to understand (though not easy to prove optimal) and have all been discovered since 1969! Many people think that mathematics is a subject that is no longer alive, and that all its ideas and methods were discovered hundreds of years ago—but as we have just seen, this is not true. In fact, more new mathematics has been discovered and published in the last 30 years than during any previous 30-year period.

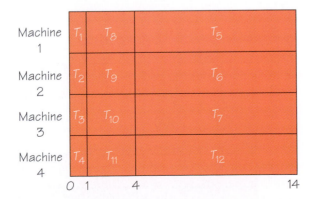

FIGURE 3.14 An optimal schedule for the order-requirement digraph in Figure 3.12 using four processors.

3.3 Independent Tasks

Mathematicians suspect that no computationally efficient algorithm for solving general scheduling problems optimally will ever be found. Owing to our limited success in designing algorithms for finding optimal schedules for general order-requirement digraphs, we will consider a special class of scheduling problems for which the order-requirement digraph has no edges. In this case, we say that the tasks are *independent* of one another, because they can be performed in any order. (No edges in the order-requirement digraph indicates that no tasks need to precede others; that is, the tasks can be done in any order.) In this section we consider the problem of scheduling **independent tasks.**

Geometrically, we can think of the independent tasks as rectangles of height 1 whose lengths are equal to the time length of the task. Finding an optimal schedule amounts to packing the task rectangles into a longer rectangle whose height equals the number of machines. For example, Figure 3.15 shows two different ways to schedule tasks of length 10, 4, 5, 9, 7, 7 on two machines. (For convenience, the rectangles in the case of independent tasks are labeled with their task times rather than their task numbers.) Scheduling basically means efficiently packing the task rectangles into the machine rectangle. Finding the optimal answer among all possible ways to pack these rectangles is like looking for a needle in a haystack. The list-processing algorithm produces a packing, but it may not be a good one.

There are two approaches we can consider. To study **average-case analysis,** we might ask if the average (mean) of the completion times arrived at by using the list-processing algorithm with all the possible different lists is close to the optimal possible completion time. To study **worst-case analysis,** we might ask *How far from optimal is a schedule obtained using the list-processing algorithm with one particular priority list?* What is being contrasted with these two points of view is that an algorithm may work well most of the time (give an answer close to optimal) even

FIGURE 3.15 (a) A nonoptimal way to schedule independent tasks of time lengths 10, 4, 5, 9, 7, 7 using two processors. (b) An optimal way to schedule independent tasks of time lengths 10, 4, 5, 9, 7, 7 using two processors.

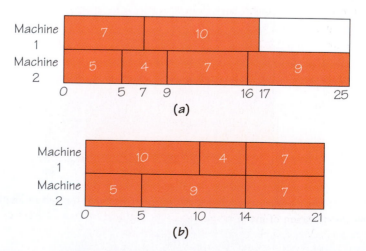

though there may be a few cases in which it performs very badly. Average-case analysis is amenable to mathematical solution but requires methods of great sophistication.

Decreasing-Time Lists

Is there some way of choosing a priority list that consistently yields relatively good schedules? The surprising answer is yes! The idea is that when long tasks appear toward the end of the list, they often seem to "stick out" on the right end, as in Figure 3.15a. This suggests that before one tries to schedule a collection of tasks, the tasks should be placed in a list where the longest tasks are listed first.

:::::: Decreasing-Time-List Algorithm

The list-processing algorithm applied to a list of task times arranged in order of nonincreasing size is called the **decreasing-time-list algorithm**.

If we apply it to the set of tasks listed previously (10, 4, 5, 9, 7, 7), we obtain the times 10, 9, 7, 7, 5, 4 and the schedule (packing) shown in Figure 3.16. This packing is again optimal, but it is different from the optimal scheduling in Figure 3.15b. It is worth noting that the decreasing-time list and the list obtained by the critical-path method discussed earlier will coincide in the case of independent tasks. The decreasing-time list can also be constructed for the case in which the tasks are not independent. For general order-requirement digraphs, the decreasing-time list does not produce particularly good schedules.

It is important to remember that the decreasing-time-list algorithm does not *guarantee* optimal solutions. This can be seen by scheduling the tasks with times 11, 10, 9, 6, 4 (Figure 3.17). The schedule has a completion time of 21. However, the rearranged list 9, 4, 6, 11, 10 yields the schedule in Figure 3.18, which finishes at time 20. This solution is obviously optimal because the machines finish at the same time and there is no idle time. Note that when tasks are independent, if there are *m* machines available, the completion time cannot be less than the sum of the task times divided by *m*.

FIGURE 3.16 The optimal schedule resulting from applying the decreasing-time-list algorithm to a collection of independent tasks. The list used, written in terms of task times only, is 10, 9, 7, 7, 5, 4.

FIGURE 3.17 The nonoptimal schedule resulting from applying the decreasing-time-list algorithm to a collection of independent tasks. The list used, written in terms of task times only, is 11, 10, 9, 6, 4.

FIGURE 3.18 The optimal schedule resulting from applying the list-processing algorithm to a collection of independent tasks. The list used, written in terms of task times only, is 9, 4, 6, 11, 10.

A modern copy shop provides a wide array of services ranging from copying a few sheets for a "drop in" customer, to printing elaborate reports for small businesses, to publishing monographs and advertising flyers. Using mathematical scheduling techniques can save time and cost by ensuring the many tasks are completed most efficiently. (Stockbyte Gold/Alamy.)

Photocopy Shop and Data Entry Problems

EXAMPLE 3

Imagine a photocopy shop with three photocopiers. Photocopying tasks that must be completed overnight are accepted until 5 P.M. The tasks are to be done in any manner that minimizes the finish time for all the work. Because this problem involves scheduling machines for independent tasks, the decreasing-time-list algorithm would be a good heuristic to apply.

For another example, consider a data entry pool at a large corporation or college, where individual entry tasks can be assigned to any data entry specialist. In this setting, however, the assumption that the data entry workers are identical in skill is less likely to be true. Hence, the tasks might have different times with different processors. This phenomenon, which occurs in real-world scheduling problems, violates one of the assumptions of our mathematical model. ∎

3.4 Bin Packing

Suppose you plan to build a wall system for your books, CDs, DVDs, and stereo set. It requires 24 wooden shelves of various lengths: 6, 6, 5, 5, 5, 4, 4, 4, 4, 2, 2, 2, 2, 3, 3, 7, 7, 5, 5, 8, 8, 4, 4, and 5 feet. The lumberyard, however, sells wood only in boards of length 9 feet. If each board costs $8, what is the minimum cost to buy sufficient wood for this wall system?

Because all shelves required for the wall system are shorter than the boards sold at the lumberyard, the largest number of boards needed is 24, the precise number of shelves needed for the wall system. Buying 24 boards would, of course, be a waste of wood and money because several of the shelves you need could be cut from one board. For example, pieces of length 2, 2, 2, and 3 feet can be cut from one 9-foot board.

To be more efficient, we think of the boards as bins of capacity W (9 feet in this case) into which we will pack (without overlap) n weights (in this case, lengths) whose values are $w_1, \ldots, w_n$, where each $w_i \leq W$. We wish to find the minimum number of bins into which the weights can be packed. In this formulation, the problem is known as the **bin-packing problem.**

Bin-Packing Problem

The bin-packing problem involves finding the minimum number of bins of weight capacity W into which weights $w_1, w_2, \ldots, w_n$ (each less than or equal to W) can be packed without exceeding the capacity of the bins.

At first glance, bin-packing problems may appear unrelated to the machine-scheduling problems we have been studying. However, there is a connection.

Let's suppose we want to schedule independent tasks so that each machine working on the tasks finishes its work by time W. Instead of fixing the number of machines and trying to find the earliest completion time, we must find the minimum number of machines that will guarantee completion by the fixed completion time (W). Despite this similarity between the machine-scheduling problem and the bin-packing problem, the discussion that follows will use the traditional terminology of bin packing.

By now, it should come as no surprise to learn that no one knows a fast algorithm that always picks the optimal (smallest) number of bins (boards). In fact, the bin-packing problem belongs to the class of NP-complete problems (see Spotlight 2.1), which means that most experts think it unlikely that any fast optimal algorithm will ever be found.

Bin-Packing Heuristics

We will think of the items to be packed, in any particular order, as constituting a list. In what follows we will use the list of 24 shelf lengths given for the wall system. We will consider various **heuristic** algorithms, namely, methods that can be carried out quickly but cannot be guaranteed to produce optimal results. Probably the easiest approach is simply to put the weights into the first bin until the next weight won't fit, and then start a new bin. (Once you open a new bin, don't use leftover space in an earlier, partially filled bin.) Continue in the same way until as many bins as necessary are used.

The resulting solution is shown in Figure 3.19. This algorithm, called **next fit (NF),** has the advantage of not requiring knowledge of all the weights in advance. Only the remaining space in the bin currently being packed must be remembered. The disadvantage of this heuristic is that a bin packed early on may have had room for small items that come later in the list.

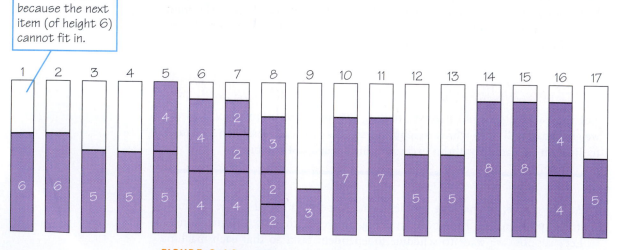

Bin 1 is closed because the next item (of height 6) cannot fit in.

FIGURE 3.19 The list 6, 6, 5, 5, 5, 4, 4, 4, 4, 2, 2, 2, 2, 3, 3, 7, 7, 5, 5, 8, 8, 4, 4, 5 packed in bins using next fit.

Our wish to avoid permanently closing a bin too early suggests a different heuristic—**first fit (FF):** Put the next weight into the first bin already opened that has room for this weight. If no such bin exists, start a new bin. Note that a computer program to carry out first fit would have to keep track of how much room was left in all the previously opened bins. For the 24 wall-system shelves, the first-fit algorithm would generate a solution that uses only 14 bins (see Figure 3.20) instead of the 17 bins generated by the next-fit algorithm.

If we are keeping track of how much room remains in each unfilled bin, we can put the next item to be packed into the bin that currently has the most room available. This heuristic will be called **worst fit (WF).** The name *worst fit* refers to the fact that an item is packed into a bin with the most room available, that is, into which it fits "worst," rather than into a bin that will leave little room left over after it is placed in that bin ("best fit"). The solution generated by this approach looks the same as that shown in Figure 3.20. Although this heuristic also leads to 14 bins, the items are packed in a different order. For example, the first item of size 2, the tenth item in the list, is put into bin 6 in worst fit, but into bin 1 in first fit.

Decreasing-Time Heuristics

One difficulty with all three of these heuristics is that large weights that appear late in the list can't be packed efficiently. Therefore, we should first sort the items to be packed in order of decreasing size, assuming that all items are known in advance. We can then pack large items first and the smaller items into leftover spaces. This approach yields three new heuristics: **next-fit decreasing (NFD), first-fit decreasing (FFD),** and **worst-fit decreasing WFD).** Here is the original list sorted by decreasing size: 8, 8, 7, 7, 6, 6, 5, 5, 5, 5, 5, 5, 4, 4, 4, 4, 4, 4, 3, 3, 2, 2, 2, 2. Packing using first-fit-decreasing order yields the solution in Figure 3.21. This solution uses only 13 bins.

Is there any packing that uses only 12 bins? No. In Figure 3.21, there are only 2 free units (1 unit each in bins 1 and 2) of space in the first 12 bins, but

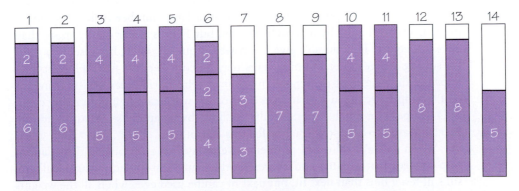

FIGURE 3.20 The list 6, 6, 5, 5, 5, 4, 4, 4, 4, 2, 2, 2, 2, 3, 3, 7, 7, 5, 5, 8, 8, 4, 4, 5 packed in bins using first fit. Worst fit would yield a packing that would look identical.

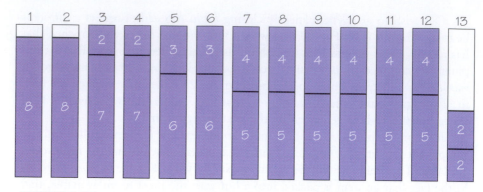

FIGURE 3.21 The bin packing resulting from applying first-fit decreasing to the wall-system numbers. The list involved, which uses the original list sorted in decreasing order, is 8, 8, 7, 7, 6, 6, 5, 5, 5, 5, 5, 5, 4, 4, 4, 4, 4, 4, 3, 3, 2, 2, 2, 2.

4 occupied units (two 2's) in bin 13. We could have predicted this by dividing the total length of the shelves (110) by the capacity of each bin (board): $\frac{110}{9} = 12\frac{2}{9}$. Thus, no packing could squeeze these shelves into 12 bins—there would always be at least 2 units left over for the 13th bin. (In Figure 3.21, there are 4 units in bin 13 because of the 2 wasted empty spaces in bins 1 and 2.) Even if this division created a zero remainder, there would still be no guarantee that the items could be packed to fill each bin without wasted space. For example, if the bin capacity is 10 and there are weights of 6, 6, 6, 6, and 6, the total weight is 30; dividing by 10, we get 3 bins as the minimum requirement. Clearly, however, 5 bins are needed to pack the five 6's.

None of the six heuristic methods shown will necessarily find the optimal number of bins for an arbitrary problem. How can we decide which heuristic to use? One approach is to see how far from the optimal solution each method might stray.

Various formulas have been discovered to calculate the maximum discrepancy between what a bin-packing algorithm actually produces and the best possible result. For example, in situations where a large number of bins are to be packed, FF can be off by as much as 70%, but FFD is never off by more than 22%. Of course, FFD doesn't give an answer as quickly as FF, because extra time for sorting a large collection of weights may be considerable. Also, FFD requires knowing the whole list of weights in advance, whereas FF does not. It is important to emphasize that a 22% margin of error is a worst-case figure. In many cases, FFD will perform much better. Results obtained by computer simulation indicate excellent average-case performance for this algorithm.

When solving real-world problems, we always have to look at the relationship between mathematics and the real world. Thus, first-fit decreasing usually results in fewer bins than next fit, but next fit can be used even when all the weights are not known in advance. Next fit also requires much less computer storage than first fit, because once a bin is packed, it need never be looked at again.

SPOTLIGHT 3.2 Using Mathematical Tools

No Euler circuit
Nonconnected
Even-valent

No Euler circuit
Connected
Not even-valent

The tools of a carpenter include the saw, T square, level, and hammer. A mathematician also requires tools of the trade. Some of these tools are the proof techniques that enable verification of mathematical truths. Another set of tools consists of strategies to sharpen or extend the mathematical truths already known. For example, suppose that if A and B hold, then C is true. What happens if only A holds? Will C still be true? Similarly, if only B holds, will C still be true?

This type of thinking is of value because such questions will result either in more general cases where C holds or in examples showing that B alone and/or A alone can't imply C. For example, we saw that if a graph G is connected (hypothesis A) and even-valent (hypothesis B), then G has a tour of its edges using each edge only once (conclusion C). If either hypothesis is omitted, the conclusion fails to hold. The figures illustrate this point. On the left is an even-valent but nonconnected graph; on the right, a connected graph with two odd-valent vertices. Neither graph has an Euler circuit.

Here is another way that a mathematician might approach extending mathematical knowledge. If A and B imply C, will A and B imply both C and D, where D extends the conclusion of C? For example, not only can we prove that a connected, even-valent (hypotheses A and B) graph has an Euler circuit, but we can also show that the first edge of the Euler circuit can be chosen arbitrarily (conclusions C and D). It turns out that being able to specify the first two edges of the Euler circuit may not always be possible. Mathematicians are trained to vary the hypotheses and conclusions of results they prove, in an attempt to clarify and sharpen the range of applicability of the results.

We have seen that machine scheduling and bin packing are probably computationally difficult to solve because they are NP-complete. A mathematician could then try to find the simplest version of a bin-packing problem that would still be NP-complete: What if the items to be packed can have only eight weights? What if the weights are only 1 and 2? Asking questions like these is part of the mathematician's craft. Such questions help to extend the domain of mathematics and hence the applications of mathematics.

Fine-tuning of the conditions of the actual problem often results in better practical solutions and in interesting new mathematics as well. See Spotlight 3.2 for a discussion of some of the tools mathematicians use to verify and even extend mathematical truths.

3.5 Resolving Conflict via Coloring

In attempting to understand situations that involve scheduling, one might desire to achieve a wide variety of goals. For example, in certain types of scheduling problems, as we have seen here, one is interested in optimization issues. What is the earliest completion time for getting a collection of tasks done on two identical processors? However, in other situations a different goal may arise. For example, in sports, consider a league of baseball teams. Each team has to play some games during the day, some at night, some at home, and some away from home. In the interests of *equity,* it may be desirable for each team to play the same number of day games and night games both at home and away against each of the other teams in the league. If, for example, team A plays 8 games away against team B and 2 games at home against B, then if A wins both home games but loses 7 out of 8 away games, it may appear that B had an advantage due to the way its games against A were scheduled.

Another goal of scheduling, other than optimization and equity, may be to prevent conflicts from occurring. We can use our knowledge of graph theory to solve some interesting scheduling problems where the goal is "conflict resolution." For example, at most colleges, every semester and summer session final examinations must be scheduled. From the point of view of students and faculty both, it would be desirable to schedule these examinations so that (1) no two ex-

Graph theory can be used to resolve scheduling conflicts that occur in trying to provide students access to limited database or computer resources. *(Bananastock/Picturequest.)*

aminations are scheduled at the same time when a student is enrolled in both of the courses and (2) the examinations are scheduled in as "compact" a way as possible, that is, in as few time slots or days as possible. The administration of the college may share the desire for these two features and want still another property for the scheduling: (3) no more than five examinations are scheduled for any time slot. The reason for a condition such as the last might be that during the summer only five rooms with reliable enough air conditioning are available (or there might be only five rooms large enough to hold all the students taking the common final for multiple-section courses).

Scheduling Examinations

EXAMPLE 4

Small State is offering eight courses during its summer session. The table shows with an X which pairs of courses have two or more students in the same course. Only two air-conditioned lecture halls are available for use at any one time. To design an efficient way to schedule the final examinations, we can represent the information in this table by using a graph, as shown in Figure 3.22a. In the graph, courses are represented by vertices and two courses are joined by an edge if there is any student enrolled in both courses.

	F	M	H	P	E	I	S	C
French (F)		X		X	X	X		X
Mathematics (M)	X				X	X		
History (H)						X	X	X
Philosophy (P)	X							X
English (E)	X	X					X	
Italian (I)	X	X	X		X			X
Spanish (S)			X			X		
Chemistry (C)	X		X	X				

The graph theory problem we are faced with is the following: Can we assign labels to the vertices of the graph in such a way that vertices that are joined by an edge get different labels? We think of the labels as the time slots the courses are assigned for final examinations. Traditionally, in graph theory such labels are referred to as *colors*. In this language we seek to color the vertices of the graph so that vertices that are joined by an edge get different colors. Such a coloring is called a **vertex coloring**.

Vertex Coloring

The vertex coloring problem for a graph requires assigning each vertex of the graph a color (label) such that two vertices joined by an edge are assigned different colors.

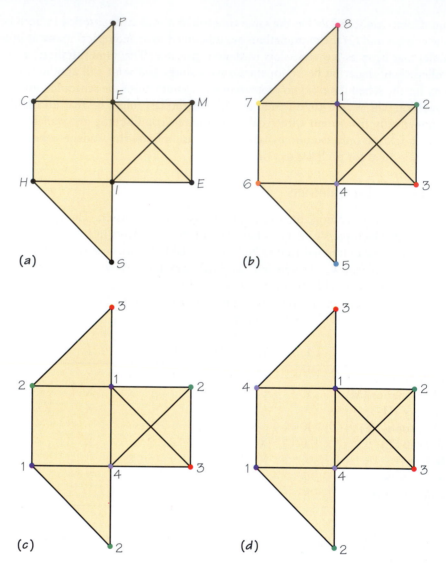

FIGURE 3.22 (a) A graph used to represent conflict information about courses. When two courses have a common student, an edge is drawn between the vertices that represent these courses. (b) A coloring of the scheduling graph with 8 colors, representing 8 time slots. Using this coloring would lead to a schedule where 8 time slots are used to schedule the examinations. This number is far from optimal. (c) A coloring of the scheduling graph with 4 colors. This translates into a way of scheduling the examinations during 4 time slots, and it is not possible to design a schedule with fewer time slots. However, this schedule calls for the use of three different rooms, because three examinations are scheduled during time slot 2. (d) A coloring of the scheduling graph with 4 colors. This means that the examinations can be scheduled in 4 time slots. However, because each color appears only twice, all the examinations can be scheduled in two air-conditioned rooms.

Figure 3.22b shows one way to color the vertices of the graph so that each vertex gets a different color. Note that numbers are being used to represent the different colors. This solution is not very valuable, however, because it means that each course must be given its own time slot.

To minimize the number of time slots used, we assign colors so that no two vertices that are joined by an edge get the same color. Thus, vertices *F, M, I,* and *E* must get four different colors. These four colors can then be used to color the remaining vertices, ensuring that no two connected vertices have the same color.

The coloring in Figure 3.22c is a major improvement over the one in Figure 3.22b. It uses only four colors. In fact, this is the smallest number of colors that can be used. To see this, notice that the vertices *F, M, I, E* in Figure 3.22a are all joined by edges to each other. Thus, in any coloring of this graph they would require four different colors. The improved coloring in Figure 3.22c was found by trial and error.

Chromatic Number

The **chromatic number** is the minimum number of colors needed to label the vertices of a graph so that no two vertices of the graph joined by an edge get the same color.

The examination graph we have been studying has chromatic number 4; hence, we can schedule the eight examinations in four time slots without a conflict. Notice, however, that the coloring in Figure 3.22c schedules three different courses for the time slot corresponding to color 2. This means that not enough rooms with air conditioning will be available. Is there a way to recolor the graph with four colors so that each of the four colors is used only twice? Figure 3.22d shows that the answer is yes.

Thus, we are able to schedule the eight final examinations in four time slots, using only two air-conditioned rooms, and no student will have a conflict under this schedule! ■

Realistic problems in scheduling government committees, high school and university final examinations, and job interviews (see Spotlight 3.3) are usually so large that graph coloring algorithms have to be incorporated into elaborate software packages to solve them.

Mathematicians have examined many kinds of coloring problems. One can study problems that involve the coloring of the edges of a graph rather than its vertices. Using techniques that have emerged from the study of coloring problems, problems involving such diverse contexts as scheduling government committees, using runways at airports efficiently, assigning frequencies for use by mobile pagers and cell phones, and designing timetables for public transportation have been solved—all these benefits from a problem that at first glance looks as if it belongs to recreational mathematics!

A group of companies is coming to campus for job interviews. The companies have been assigned the number of hours they need to cover the number of students they hope to interview during a block of consecutive hours wherein a representative from each company can hold interviews. Due to the fact that classes are going on at the same time, five departmental conference rooms have been made available to the companies to conduct their interviews.

The interviews will follow the school's regular hourly periods, which start at 9 A.M. and end at 4 P.M. (Companies will be scheduled for continuous interviews during lunch-hour times. Interviews cannot be scheduled beyond the end of the period that starts at 4 P.M. and ends at 5 P.M.)

	Company	Time Slot Requested
A	(Apricot Computers)	7
B	(Big Green)	1
C	(Challenge Insurance)	4, 5
D	(Daisy Printers)	7, 8
E	(Earnest Engine)	4, 5, 6
F	(Flexible Systems)	2, 3
G	(Gutter Leaders)	1, 2
H	(Halley's Combs)	6, 7
I	(Indelible Ink Corporation)	7, 8
J	(Jay's Produce)	4, 5
K	(Kelly's Detective Agency)	2, 3
L	(Large Clothes)	4, 5, 6
M	(Metropolitan TV)	1, 2
N	(Nationwide Bank)	4, 5, 6, 7

Look at the list of time blocks that the companies requested (where 1 = 9–10 A.M., . . . , 8 =

4–5 P.M.). Is it possible to accommodate all the companies that wish to do interviewing in the five rooms available while meeting their desired schedule times?

Problems of this kind seem simple enough, and you should try your hand at solving this particular one, for which a schedule does exist! However, this situation is not simple at all. The following facts are known about problems of this kind.

Fact 1. Suppose there are i interviews, p time periods, and r rooms where interviews can be scheduled. Each interviewer has specified periods during which he or she wishes to conduct interviews. Is it possible to design a schedule that meets the desired specifications? It turns out that this problem is NP-complete (see Spotlight 2.1), that is, it belongs to a large group of problems for which, among other things, the fastest known algorithms run very slowly on large-problem versions.

Fact 2. The problem just described remains NP-complete even for the case where only three rooms have to be scheduled ($p = 3$).

The moral is *surprisingly simple:* Scheduling problems are very hard to solve.

However, the situation is not always as hopeless as it might seem. If you look at the list of time requests for the corporations, you will note that, not surprisingly, each company has requested a contiguous block of times. It turns out that when this condition holds, it is possible to determine whether there is a feasible schedule using an algorithm that works relatively quickly.

::::::::::: REVIEW VOCABULARY

Average-case analysis The study of the list-processing algorithm (more generally, any algorithm) from the point of view of how well it performs in all the types of problems it may be used for and seeing on average how well it does. *See also* worst-case analysis.

Bin-packing problem The problem of determining the minimum number of containers of capacity W into which objects of size $w_1, \ldots, w_n$ ($w_i \leq W$) can be packed.

Chromatic number The chromatic number of a graph G is the minimum number of colors (labels) needed in any vertex coloring of G.

Critical-path scheduling A heuristic algorithm for solving scheduling problems where the list-processing algorithm is applied to the priority list obtained by listing next in the priority list a task that heads a longest path in the order-requirement digraph. This task is then deleted from the order-requirement digraph, and the next task placed in the priority list is obtained by repeating the process.

Decreasing-time-list algorithm The heuristic algorithm that applies the list-processing algorithm to the priority list obtained by listing the tasks in decreasing order of their time length.

First fit (FF) A heuristic algorithm for bin packing in which the next weight to be packed is placed in the lowest-numbered bin already opened into which it will fit. If it fits in no open bin, a new bin is opened.

First-fit decreasing (FFD) A heuristic algorithm for bin packing where the first-fit algorithm is applied to the list of weights sorted so that they appear in decreasing order.

Heuristic algorithm An algorithm that is fast to carry out but that doesn't necessarily give an optimal solution to an optimization problem.

Independent tasks Tasks are independent when there are no edges in the order-requirement digraph.

List-processing algorithm A heuristic algorithm for assigning tasks to processors: Assign the first ready task on the priority list that has not already been assigned to the lowest-numbered processor that is not working on a task.

Machine scheduling problem The problem of assigning tasks to processors so as to complete the tasks by the earliest time possible.

Next fit (NF) A heuristic algorithm for bin packing in which a new bin is opened if the weight to be packed next will not fit in the bin that is currently being filled; the current bin is then closed.

Next-fit decreasing (NFD) A heuristic algorithm for bin packing where the next-fit algorithm is applied to the list of weights sorted so that they appear in decreasing order.

Priority list An ordering of the collection of tasks to be scheduled for the purpose of attaining a particular scheduling goal. One such goal is minimizing completion time when the list algorithm is applied.

Processor A person, machine, robot, operating room, or runway with time that must be scheduled.

Ready task A task is called ready at a particular time if its predecessors, as given by the order-requirement digraph, have been completed by that time.

Vertex coloring A vertex coloring of a graph G is an assignment of labels, which can be thought of as "colors," to the vertices of G so that vertices joined by an edge get different labels (colors).

Worst-case analysis The study of the list-processing algorithm (more generally, any algorithm) from the point of view of how well it performs on the hardest problems it may be used on. *See also* average-case analysis.

Worst fit (WF) A heuristic algorithm for bin packing in which the next weight to be packed is placed into the open bin with the largest amount of room remaining. If the weight fits in no open bin, a new bin is opened.

Worst-fit decreasing (WFD) A heuristic algorithm for bin packing where the worst-fit algorithm is applied to the list of weights sorted so that they appear in decreasing order.

SKILLS CHECK

1. Given the order-requirement digraph below (time in minutes) and the priority list T_1, T_2, T_3, T_4, T_5, T_6, apply the list-processing algorithm to construct a schedule using two processors. How much time does the resulting schedule require?

(a) 11 minutes
(b) 13 minutes
(c) 14 minutes

2. What is the minimum time required to complete 8 independent tasks with a total task time of 64 minutes on 4 machines?

(a) Less than 8 minutes
(b) Between 8 and 10 minutes
(c) More than 12 minutes

3. Suppose that a crew can complete in a minimum amount of time the job whose order-requirement digraph is shown below. If task T_2 is shortened from 5 minutes to 2 minutes, then what is the maximum amount by which the completion time for the entire job can be shortened?

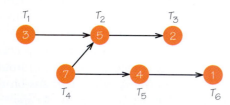

(a) 3 minutes
(b) 2 minutes
(c) 1 minute

4. The following digraph cannot be an order-requirement digraph because

(a) no vertex has three edges that enter that particular vertex.
(b) it has a directed circuit.
(c) all the tasks require the same time to complete.

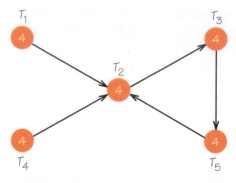

5. If all the paths in an order-requirement digraph go through a single vertex X, then

(a) the order-requirement digraph has no critical path.
(b) all the paths must be critical.
(c) any increase in the time necessary to finish the task associated with vertex X will mean the whole job will take that much longer to finish.

6. Assume an order-requirement digraph requires 15 minutes when scheduled on two machines. Based on this information, when the digraph is instead scheduled on three machines, how much time will be required?

(a) Exactly 10 minutes
(b) At least 10 minutes
(c) Exactly 15 minutes

7. Which statement is true for the following digraph?

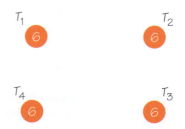

(a) This digraph cannot be the order-requirement digraph for a scheduling problem because the digraph has no (directed) edges.
(b) This digraph cannot be the order-requirement digraph for a scheduling problem because it is not

allowed for all the tasks to have the same time length.

(c) This digraph can be the order-requirement digraph for a scheduling problem.

8. A critical path list associated with the order-requirement digraph below is

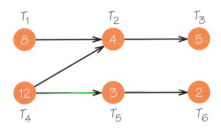

(a) $T_4, T_1, T_3, T_2, T_5, T_6$.
(b) $T_4, T_1, T_2, T_3, T_5, T_6$.
(c) $T_1, T_2, T_3, T_4, T_5, T_6$.

9. Use the decreasing-time-list algorithm to schedule these independent tasks on two machines: 6 minutes, 7 minutes, 4 minutes, 3 minutes, 6 minutes. How much time does the resulting schedule require?

(a) 13 minutes
(b) 14 minutes
(c) More than 14 minutes

10. Assume an order-requirement digraph has a critical path with length 20 minutes. Based on this information, when the tasks are scheduled on two machines, how much time will be required?

(a) Exactly 10 minutes
(b) Exactly 20 minutes
(c) At least 20 minutes

11. Assume a job consists of six independent tasks ranging in time from 2 to 10 minutes and totaling 27 minutes. Efficiently scheduled on three machines, how much time will the job require?

(a) Exactly 9 minutes
(b) Exactly 10 minutes
(c) More than 10 minutes

12. Which of the following statements about the decreasing-time-list algorithm is true?

(a) The decreasing-time-length algorithm *always* yields an *optimal* schedule.
(b) The decreasing-time-length algorithm *often* yields an *optimal* schedule.
(c) The decreasing-time-length algorithm *always* yields an *optimal* or *near-optimal* schedule.

13. A radio announcer has 10 songs of various lengths to schedule into several segments. The announcer must identify the station at least once every 15 minutes, so the segments cannot be longer than 15 minutes. This job can be solved using the

(a) list-processing algorithm for independent tasks.
(b) critical-path scheduling algorithm.
(c) first-fit algorithm for bin packing.

14. Use the first-fit (FF) bin-packing algorithm to pack the following weights into bins that can hold no more than 10 lb: 6 lb, 7 lb, 4 lb, 3 lb, 6 lb. How many bins are required?

(a) 3 bins
(b) 4 bins
(c) 5 bins

15. Use the worst-fit-decreasing (WFD) bin-packing algorithm to pack the following weights into bins that can hold no more than 10 lb: 6 lb, 7 lb, 4 lb, 3 lb, 6 lb. How many bins are holding a full 10 lb?

(a) 0 bins
(b) 1 bin
(c) 2 bins

16. The first-fit decreasing (FFD) bin-packing algorithm is applied to the weight list 1, 2, 3, 4, 5, 5, 6, 8 for packing into bins of capacity 10. The item of weight 2 is packed into which bin, when the packed bins are numbered from left to right?

(a) The first bin
(b) The second bin
(c) The fourth bin

17. A vertex coloring seeks to color the vertices of a graph to ensure which of the following traits?

(a) Every color is used.

(b) Every edge connects vertices of the same color.

(c) Vertices of the same color are never connected by an edge.

18. Assume the 8 corners of a cube represent vertices of a graph and the 12 edges of a cube represent the cube's edges. What is the chromatic number of this graph?

(a) 2

(b) 3

(c) More than 3

19. The minimum number of colors needed to color the vertices of the accompanying graph is

(a) 4.

(b) 2.

(c) 3.

20. A graph that has a circuit length 3 can always be colored with

(a) no more than four colors.

(b) exactly three colors.

(c) no fewer than three colors.

EXERCISES   ■ Challenge   ◆ Discussion

Scheduling

1. List as many scheduling situations as you can for these environments:

(a) Your school

(b) Bus station

(c) Fast food restaurant

(d) Your home

(e) A 911 dispatch center

(f) Hospital

(g) Firehouse

(h) Bookstore

(i) Police station

2. For the situation where a family with three children is preparing a Thanksgiving meal for 8 guests, list tasks that must be completed and the types of processors that are involved. Can any of these tasks be done simultaneously?

◆ **3.** To get to a ski resort for a weekend vacation, Jocelyn must accomplish a variety of things. She will leave work early at 1 P.M. and must get to the airport to be on a 5 P.M. shuttle to Boston. She then hopes to take a bus to get to the resort. Discuss some of the tasks that must be accomplished to get Jocelyn to the resort by 10 P.M. What are the different types of processors that are involved in getting these tasks done? Can any of these tasks be done simultaneously?

4. Consider the following order-requirement digraph:

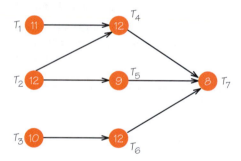

(a) Find the length of the critical path.

(b) Schedule these seven tasks on two processors using the list algorithm and the lists:

  (i) $T_1, T_2, T_3, T_4, T_5, T_6, T_7$

  (ii) $T_2, T_1, T_3, T_6, T_5, T_4, T_7$

(c) Does either list lead to a completion time that equals the length of the critical path?

(d) Show that no list can ever lead to a completion time equal to the length of the critical path (providing the schedule uses two processors).

5. (a) Use the following order-requirement digraph to schedule the 6 tasks $T_1, T_2, T_3, T_4, T_5, T_6$ on two processors with the priority lists:

  (i) $T_1, T_2, T_3, T_4, T_5, T_6$

  (ii) $T_1, T_6, T_3, T_5, T_4, T_2$

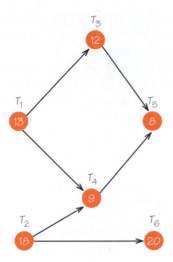

(b) Are either of the schedules produced from these lists optimal? If not, can you find a priority list that will result in an optimal schedule?

(c) Find the critical path and its length. Explain why no schedule has earliest completion time equal to the length of the critical path.

6. (a) Repeat Exercise 5, but interchange the task times of tasks T_2 and T_6.

(b) How does the completion time for an optimum schedule for this situation compare with the optimum schedule for Exercise 5?

7. Use the list-processing algorithm to schedule the tasks in the following order-requirement digraph on

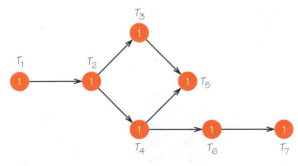

(a) two processors using the list $T_1, \ldots, T_7$.
(b) two processors using the list $T_1, T_2, T_3, T_4, T_6, T_5, T_7$.
(c) Is either of the schedules that you obtain optimal?

(d) Will adding a third processor enable the tasks to be finished earlier?
(e) Which tasks in this order-requirement digraph can be shortened and not affect the completion time of all the tasks?

◆ 8. Discuss scheduling problems for which it is not reasonable to assume that once a processor starts a task, it will always complete that task, before it works on any other task. Give examples for which this approach would be reasonable.

9. For the accompanying order-requirement digraph, apply the list-processing algorithm, using three processors for lists (a) through (c). How do the completion times obtained compare with the length of the critical path?

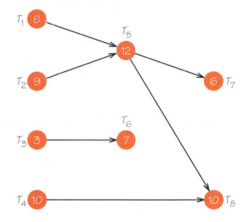

(a) $T_1, T_2, T_3, T_4, T_5, T_6, T_7, T_8$
(b) $T_1, T_3, T_5, T_7, T_2, T_4, T_6, T_8$
(c) $T_8, T_6, T_4, T_2, T_1, T_3, T_5, T_7$

Using the List-Processing Algorithm

10. (a) Can you find an order-requirement digraph with four tasks for which every priority list used to schedule the tasks on two machines assigns task T_4 to machine 1 at time 0?

(b) Can you choose the order-requirement digraph in part (a) so that machine 2 stays idle for all lists from time 0 to time 3?

11. Can you give examples of scheduling problems for which it seems reasonable to assume that all the task times are the same?

12. Use the list-processing algorithm to schedule the tasks in the following order-requirement digraph on

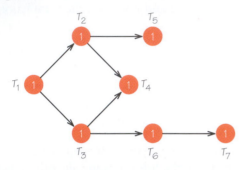

(a) two processors using the list $T_1, \ldots, T_7$.
(b) two processors using the list $T_1, T_2, T_3, T_4, T_6, T_5, T_7$.
(c) Is either of the schedules that you obtain optimal?

13. Can you find a list that gives rise to the optimal schedule shown in Figure 3.14 for the order-requirement digraph in Figure 3.12?

14. Consider the following order-requirement digraph:

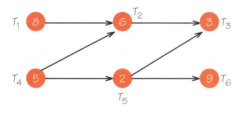

(a) Find the critical path(s).
(b) Schedule these tasks on one processor using the critical-path scheduling method.
(c) Schedule these tasks on one processor using the priority list obtained by listing the tasks in order of decreasing time.
(d) Does either of these schedules have idle time? How do their completion times compare?
(e) If two different schedules have the same completion time, what criteria can be used to say one schedule is superior to the other?

(f) Schedule these tasks on two processors using the order-requirement digraph shown and the priority list from part (b).
(g) Does the schedule produced in part (f) finish in half the time that the schedule in part (b) did, which might be expected, since the number of processors has doubled?
(h) Schedule the tasks on (i) one processor and (ii) two processors (using the decreasing-time list), assuming that each task time has been reduced by one. Do the changes in completion time agree with your expectations?

15. (a) Can all the processors being used to schedule tasks be simultaneously idle at a time before the completion time of a collection of tasks scheduled using the list-processing algorithm?

(b) Explain why the list-processing algorithm cannot give rise to the schedule below, regardless of what priority list was used to schedule the tasks on the two processors.

(c) Construct an order-requirement digraph and a priority list that will give rise to the following schedule on two processors.

16. To prepare a meal quickly involves carrying out the tasks shown (time lengths in minutes) in the following order-requirement digraph:

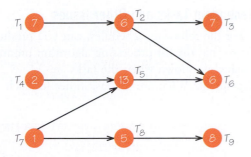

(a) If Mike prepares the meal alone, how long will it take?

(b) If Mike can talk Mary into helping him prepare the meal, how long will it take them if the tasks are scheduled using the list $T_5, T_9, T_1, T_3, T_2, T_6, T_8, T_4, T_7$ and the list-processing algorithm?

(c) If Mike can talk Mary and Jack into helping him prepare the meal, how long will it take if the tasks are scheduled using the same list as in part (b)?

(d) What would be a reasonable set of criteria for choosing a priority list in this situation?

17. (a) Making use of the order-requirement digraph below, determine at time 0 which tasks are ready.

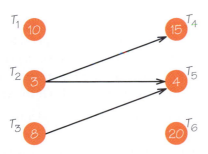

(b) What is special about tasks T_1 and T_6?

(c) What is the critical path, and what is its length?

(d) Schedule the tasks on three processors with the priority list $T_1, \ldots, T_6$.

(e) Is the schedule found in part (d) optimal?

(f) Schedule the tasks on three processors using the priority list $T_6, \ldots, T_1$.

(g) Is the schedule found in part (f) optimal?

(h) Can you find a priority list that yields an optimal schedule?

18. (a) In Exercise 17, what priority list would be used if you applied the critical-path scheduling method?

(b) Use this priority list to schedule the tasks on three processors. Is this schedule optimal?

(c) How does this schedule compare with the schedules that you found using the lists in Exercise 17?

19. Consider the order-requirement digraph below. Suppose one plans to schedule these tasks on two identical processors.

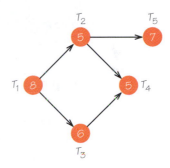

(a) How many different priority lists can be used to schedule the tasks?

(b) Can all these priority lists lead to different schedules? If not, why not?

(c) Can an optimal schedule have no idle time? Can you give two different reasons why an optimal schedule must have some idle time?

(d) Is there any list that produces a schedule where the second processor has no idle time?

20. (a) In Exercise 19, how many different lists are there that do not list T_1 first?

(b) Would it make any sense not to list T_1 first in a list?

(c) Construct a list and schedule the tasks on two processors.

(d) Can you find another list that leads to a different completion time than the schedule you found for part (c)?

(e) Find a list that leads to an optimal schedule.

21. Can you find an order-requirement digraph with five tasks for which every possible list yields exactly the same schedule?

22. Can you find an order-requirement digraph such that the schedule corresponding to every list is different?

23. At a large toy store, scooters arrive unassembled in boxes. To assemble a scooter, the following tasks must be performed:

Task 1. Remove parts from the box.

Task 2. Attach wheels to the footboard.

Task 3. Attach vertical housing.

Task 4. Attach handlebars to vertical housing.

Task 5. Put on reflector tape.

Task 6. Attach bell to handlebars.

Task 7. Attach decals.

Task 8. Attach kickstand.

Task 9. Attach safety instructions to handlebars.

(a) Give reasonable time estimates for these tasks and construct a reasonable order-requirement digraph. What is the earliest time by which these tasks can be completed?

(b) Schedule this job on two processors (humans) using the decreasing-time-list algorithm.

24. If two schedules for the same number of processors have the same completion time, can one schedule have more idle time than the other?

25. Could the schedule below be obtained by applying the list-scheduling algorithm to some order-requirement digraph?

26. Could the schedule below be obtained by applying the list-scheduling algorithm to some order-requirement digraph?

Independent Tasks and Other Issues

27. For the following schedules, can you produce a list so that the list-processing algorithm produces the schedule shown when the tasks are independent? What are the task times for each task?

(a)

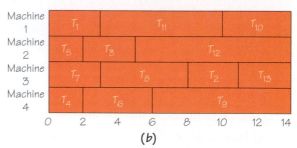

(b)

◆ **28.** Once an optimal schedule has been found for independent tasks (see diagrams in Exercise 27), usually the scheduling of the tasks can be rearranged and the same optimal time achieved. One can, among other things, reorder the tasks done by a particular processor. Discuss criteria that might be used to implement the rearrangement process.

29. The task times of eight independent tasks T_1 to T_8 are 1, 2, 3, 4, 5, 6, 7, 8.

(a) Schedule the tasks on two processors using the lists (i) $T_1, T_2, \ldots, T_8$ and (ii) $T_8, T_7, \ldots, T_1$.

(b) Is either of the schedules you get in part (a) optimal? If not, find a list that gives an optimal schedule.

30. Repeat Exercise 29, but schedule the tasks (with the same lists) on three processors. If the schedules you get are not optimal, find a list that gives an optimal schedule.

◆ **31.** Discuss different criteria that might be used to construct a priority list for a scheduling problem.

◆ **32.** Some scheduling projects have due dates for tasks (times by which a given task should be completed) and release dates (times before which a task cannot have work begun on it). Give examples of circumstances where these situations might arise.

33. Using the lists you found in Exercise 27 and the task times you computed for those independent tasks, schedule the tasks for (a) on four processors and the tasks for (b) on five processors. Can you see why for any schedule you may produce for (a) on four processors and (b) on five processors there must be some idle time for one or more processors?

34. Given the following order-requirement digraph:

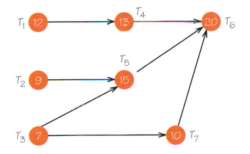

(a) Use the list-processing algorithm to schedule these seven tasks on two processors using these lists:

  (i) $T_1, T_3, T_7, T_2, T_4, T_5, T_6$
  (ii) $T_1, T_3, T_2, T_4, T_5, T_6, T_7$
  (iii) The list obtained by listing the tasks in order of decreasing time
(b) Try to determine if any of the resulting schedules are optimal.
(c) Schedule the tasks using the critical-path scheduling method. Try to determine if this schedule is optimal.

35. Repeat the questions in Exercise 34 using the order-requirement digraph obtained by erasing all the (directed) edges shown there. How do the schedules you get compare with the ones you originally got?

36. (a) Find the completion time for independent tasks of length 8, 11, 17, 14, 16, 9, 2, 1, 18, 5, 3, 7, 6, 2, 1 on two processors, using the list-processing algorithm.

(b) Find the completion time for the tasks in part (a) on two processors, using the decreasing-time-list algorithm.

(c) Does either algorithm give rise to an optimal schedule?

(d) Repeat for tasks of lengths 19, 19, 20, 20, 1, 1, 2, 2, 3, 3, 5, 5, 11, 11, 17, 18, 18, 17, 2, 16, 16, 2.

37. A photocopy shop must schedule independent batches of documents to be copied. The times for the different sets of documents are (in minutes): 12, 23, 32, 13, 24, 45, 23, 23, 14, 21, 34, 53, 18, 63, 47, 25, 74, 23, 43, 43, 16, 16, 76.

(a) Construct a schedule using the list-processing algorithm on three machines.
(b) Construct a schedule using the list-processing algorithm on four machines.
(c) Repeat parts (a) and (b), but use the decreasing-time-list algorithm.
(d) Suppose union regulations require that an 8-minute rest period be allowed for any photocopy task over 45 minutes. Use the decreasing-time-list algorithm, with the preceding times modified to take into account the union requirement, to schedule the tasks on three human-operated machines.

38. Find a list that produces the following optimal schedule when the list-processing algorithm is applied to this list. (Assume the tasks are independent.)

What completion time and schedule are obtained when the decreasing-time-list algorithm is applied to this list?

39. Can you think of situations other than those mentioned in the text where scheduling independent tasks on processors occurs?

40. Can you think of real-world scheduling situations in which all the tasks have the same time and are independent? Find an algorithm for solving this problem optimally. (If there are n independent tasks of time length k, when will all the tasks be finished?)

41. Show that when tasks to be scheduled are independent, the critical-path method and the decreasing-time-list method are identical.

Bin Packing

42. Two wooden wall systems are to be made of pieces of wood with lengths shown in the accompanying diagram. If wood is sold in 10-foot planks and can be cut with no waste, what number of boards would be purchased if one uses the first-fit-decreasing, next-fit-decreasing, and worst-fit-decreasing heuristics, respectively?

In solving this problem, does it make a difference if the 10-foot horizontal shelves and 6-foot vertical boards employ single-length pieces as compared with using pieces of boards that add up to 10- and 6-foot lengths?

43. It takes 4 seconds to photocopy one page. Manuscripts of 10, 8, 15, 24, 22, 24, 20, 14, 19, 12, 16, 30, 15, and 16 pages are to be photocopied. How many photocopy machines would be required, using the first-fit-decreasing algorithm, to guarantee that all manuscripts are photocopied in 2 minutes or less? Would the solution differ if worst-fit decreasing were used?

44. A radio station's policy allows advertising breaks of no longer than 2 minutes, 15 seconds. Using first-fit and first-fit-decreasing algorithms, determine the minimum number of breaks into which the following ads will fit (lengths given in seconds): 80, 90, 130, 50, 60, 20, 90, 30, 30, 40.

Can you find the optimal solution? Do the same for these ad lengths: 60, 50, 40, 40, 60, 90, 90, 50, 20, 30, 30, 50.

45. Fiberglass insulation comes in 36-inch precut sections. A plumber must install insulation in a basement on piping that is interrupted often by joints. The distances between the joints on the stretches of pipe that must be insulated are 12, 15, 16, 12, 9, 11, 15, 17, 12, 14, 17, 18, 19, 21, 31, 7, 21, 9, 23, 24, 15, 16, 12, 9, 8, 27, 22, 18 inches. How many precut sections would he have to use to provide the insulation if he bases his decision on

(a) next fit?
(b) next-fit decreasing?
(c) worst fit?
(d) worst-fit decreasing?

46. The files that a company has for its employees dealing with utilities occupy 100, 120, 60, 90, 110, 45, 30, 70, 60, 50, 40, 25, 65, 25, 55, 35, 45, 60, 75, 30, 120, 100, 60, 90, 85 sectors. If, after operating systems are installed, a disk can store up to 480 sectors, determine the number of disks needed to store the utilities if each of these heuristics is used to pack the disk with files:

(a) NF
(b) NFD
(c) FF
(d) FFD

47. Advertisements for the TV show Q are permitted to last up to a total of 8 minutes, and each group of ads can last up to 2 minutes. If the ads slated for Q last 63, 32, 11, 19, 24, 87, 64, 36, 27, 42, 63 seconds, determine if FF and FFD yield acceptable configurations for the ads.

48. Consider the heuristic for packing bins known as *best fit*. Keep track of how much room remains in each unfilled bin and put the next item to be packed into that bin that would leave the least room left over after the item is put into the bin. (For example, suppose that bin 4 had 6 units left, bin 7 had 5 units left, and bin 9 had 8 units left. If the next item in the list had size 5, then first fit would place this item in bin 4, worst fit would place the item in bin 9, while best fit would place the item in bin 7.) If there is a tie, place the item into the bin with the lowest number. Apply this heuristic to the list 8, 7, 1, 9, 2, 5, 7, 3, 6, 4, where the bins have capacity 10.

◆ **49.** We have described two algorithms for bin packing called worst fit and best fit (see page 99 and Exercise 48). The words *best* and *worst* have connotations in English. However, the performance of algorithms depends on their merits as algorithms, not on the names we give them.

(a) On the basis of experiments you perform with the best-fit and worst-fit algorithms, which one do you think is the "better" of the two?
(b) Can you construct an example where worst fit uses fewer bins than best fit?

50. The best-fit heuristic (see Exercise 48) also has a "decreasing" version, where the list is first sorted in decreasing order. Using bins of capacity 10, apply the best-fit heuristic and its decreasing version to the following list: 6, 9, 5, 8, 3, 2, 1, 9, 2, 7, 2, 5, 4, 3, 7, 6, 2, 8, 3, 7, 1, 6, 4, 2, 5, 3, 7, 2, 5, 2, 3, 6, 2, 7, 1, 3, 5, 4, 2, 6.

■ **51.** One pianist's recording of the complete Mozart piano sonatas takes the following times (given in minutes and seconds): 13:46, 6:15, 3:29, 5:37, 7:52, 2:55, 5:00, 4:28, 4:21, 7:39, 7:55, 6:42, 4:23, 3:52, 4:21, 4:20, 5:46, 6:29, 5:34, 6:23, 6:39, 7:19, 5:54, 6:54, 2:58, 5:22, 1:42, 5:00, 1:29, 5:47, 7:30, 8:19, 4:44, 4:57, 4:09, 14:31, 3:55, 4:04, 4:01, 6:06, 6:50, 5:27, 4:28, 5:40, 2:52, 5:16, 5:34, 3:10, 7:22, 4:40, 3:08, 6:32, 4:47, 6:59, 5:38, 7:57, 3:38. If the maximum time that can be recorded on a compact disc is 70:30, can all the music be performed on four compact discs? Can all the music be performed on five compact discs?

■ **52.** In the wall-system example in the text, first fit and worst fit required equal numbers of bins (see Figure 3.20). Can you find an example where first fit and worst fit yield different numbers of bins? Can you find an example where first fit,

worst fit, and next fit yield answers with different numbers of bins?

◆ **53.** A common suggestion for heuristics for the bin-packing problem with bins of capacity W involves finding weights that sum to exactly W. Discuss the pros and cons of a heuristic of this type.

■ **54.** A recording company wishes to record all the Beethoven string quartets (16 quartets, each consisting of several consecutive parts called movements) on LPs. It wishes to complete the project on as few records as possible. Recording can be done on two sides as long as the movements are consecutive. Is this an example of a bin-packing problem? (Defend your answer.) If the project were to record the quartets on (standard) tape cassettes or compact discs, would your answer be different?

55. Give examples where it would be realistic to keep bins open as more items "arrive" to be packed, rather than to close a bin permanently based on some criterion.

56. Give examples where it would be unrealistic to keep bins open as more items "arrive" to be packed, rather than to close a bin permanently based on some criterion.

57. A data entry group must handle 30 (independent) tasks that will take the following amounts of time (in minutes) to type: 25, 18, 13, 19, 30, 32, 12, 36, 25, 17, 18, 26, 12, 15, 31, 18, 15, 18, 16, 19, 30, 12, 16, 15, 24, 16, 27, 18, 9, 14. Using these times as a priority list:

(a) Use the list-processing algorithm to find the completion time for scheduling tasks with four secretaries. Also, solve with five secretaries.
(b) Repeat the scheduling using the decreasing-time-list algorithm.
(c) Can you show that any of the schedules that you get are optimal?

If one needs to finish the typing in one hour:

(d) Use the FFD heuristic to find how many typists would be needed.

(e) Repeat for the NFD and WFD heuristics.
(f) Can you show that any of the solutions you get are optimal?

58. Find the minimum number of bins necessary to pack items of size 8, 5, 3, 4, 3, 7, 8, 8, 6, 5, 3, 2, 1, 2, 1, 2, 1, 3, 5, 2, 4, 2, 6, 5, 3, 4, 2, 6, 7, 7, 8, 6, 5, 4, 6, 1, 4, 7, 5, 1, 2, 4 in bins of capacity (a) through (d) using the first-fit and first-fit-decreasing algorithms. Can you determine if any of the packings you get are optimal?

(a) 9
(b) 10
(c) 11
(d) 12

■ **59.** Two-dimensional bin packing refers to the problem of packing rectangles of various sizes into a minimum number of $m \times n$ rectangles, with the sides of the packed rectangles parallel to those of the containing rectangle.

(a) Suggest some possible real-world applications of this problem.
(b) Devise a heuristic algorithm for this problem.
(c) Give an argument to show that the problem is at least as hard to solve as the usual bin-packing problem.
(d) If you have $1 \times m$ rectangles with total area W to be packed into a single rectangle of area $p \times q = W$, can the packing always be accomplished?

◆ **60.** In what situations would packing bins of different capacities be the appropriate model for real-world situations? Suggest some possible algorithms for this type of problem.

■ **61.** Find an example of weights that, when packed into bins using first fit, use fewer bins than the number of bins used when the first-fit algorithm is applied with the first weight on the list removed.

◆ **62.** Formulate "paradoxical" situations for bin packing that are analogous to those we found for scheduling processors.

Coloring Problems and Conflict Resolution

63. For each of the graphs below:

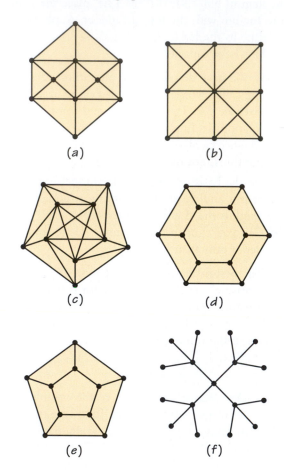

(a)

(b)

(c)

(d)

(e)

(f)

(a) Color the vertices (if possible) with three different colors.
(b) Color the vertices (if possible) with four different colors.
(c) Find the chromatic number of the graph.

64. For each of the following graphs:

(a)

(b)

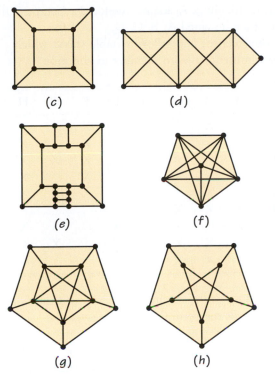

(c)   (d)

(e)   (f)

(g)   (h)

(a) Color the vertices (if possible) with two different colors.
(b) Color the vertices (if possible) with three different colors.
(c) Find the chromatic number of the graph.

65. The owner of a new pet store wishes to display tropical fish in display tanks. The following table shows the incompatibilities between the species, in the sense that an X indicates that it is unwise to allow those species in the row and column that meet at the X to be in the same tank.

	A	B	C	D	E	F	G	H	I
A						X	X		X
B			X					X	
C		X			X			X	
D					X	X		X	
E			X	X			X		
F	X			X			X		X
G	X				X	X		X	X
H		X	X	X			X		
I	X					X	X		

(a) What is the minimum number of tanks needed to display all the fish she wishes to sell?
(b) Display the species so that the number of species in each tank is as nearly equal as possible.

66. The managers of a zoo are planning to open a small satellite branch. The animals are to be in enclosures in which compatible animals are displayed together. The accompanying table indicates those pairs of animals that are compatible. (Thus, an X in a particular row and column means that the animals that label this row and column *can* share an enclosure.)

	A	B	C	D	E	F	G	H	I	J
A	X	X			X	X	X	X		
B	X	X			X	X	X		X	X
C			X		X	X	X			
D	X			X	X	X	X		X	X
E	X	X	X	X	X			X	X	
F	X	X	X	X			X	X	X	X
G	X	X	X	X		X	X	X		
H					X	X	X	X		
I		X		X	X	X			X	
J		X		X		X				X

(a) What is the minimum number of enclosures needed to avoid housing incompatible animals in the same enclosure?
(b) Is it possible to enclose the animals in such a way that each enclosure contains the same number of animals?
(c) Why might that be desirable? Why might this approach to grouping the animals not be ideal?

67. The nine standing committees of a state legislature are designing a schedule for when the committees can meet. The matrix shown in the following table has an X in a position where the committees corresponding to the row and column have a common member and, hence, should not be scheduled to meet at the same hour. The committees involved are Agriculture (A), Commerce (C), Consumer Affairs (CA), Education (E), Forests (F), Health (H), Justice (J), Labor (L), and Rules (R).

	A	C	CA	E	F	H	J	L	R
A		X	X			X			
C	X		X	X	X				
CA	X	X					X		X
E		X			X	X			
F		X		X		X	X		
H	X			X	X			X	
J			X		X			X	X
L						X	X		X
R			X				X	X	

(a) Draw a graph that will be of value in determining the minimum number of time slots the committees can meet in without any legislator having to be in two places at one time.

(b) What is the minimum number of time slots in which the committees can be scheduled without a conflict?

(c) How many different rooms are needed at any time that a committee is scheduled to meet? (Why might this issue matter?)

68. Determine the minimum number of colors, and how often each color is used, in a vertex coloring of the graphs below.

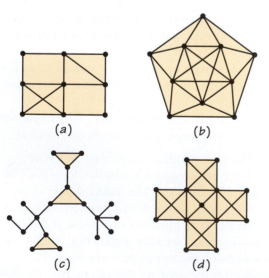

(a)   (b)

(c)   (d)

69. The faculty–student governing council at All State College has nine standing committees (such as Curriculum, Academic Standards, Campus Life) that are designed A, B, C, D, . . . , I for convenience. The following table shows which committees have no member in common.

	A	B	C	D	E	F	G	H	I
A		X		X		X	X		X
B	X				X	X		X	X
C				X		X	X	X	X
D	X		X			X		X	
E		X					X	X	X
F	X	X	X	X					
G	X		X		X			X	
H		X	X	X	X		X		X
I	X	X	X		X			X	

(a) What is the minimum number of time slots in which all the committee meetings can be scheduled?

(b) How many rooms are needed during each time slot to accommodate the committees that are scheduled to meet in that time slot?

70. When two towns are within 145 miles of each other, the frequency used by a certain type of emergency response system for the towns requires that they be on different frequencies to avoid possible interference with each other. The table below shows the mileage distances between six towns.

	E	F	G	I	S	T
Evansville (E)		290	277	168	303	133
Ft. Wayne (F)	290		132	83	79	201
Gary (G)	277	132		153	58	164
Indianapolis (I)	168	83	153		140	71
South Bend (S)	303	79	50	140		196
Terre Haute (T)	113	201	164	71	196	

(a) What would be the minimum number of frequencies that are needed for each town to have its emergency broadcasts not conflict with those of any other town using this system?

(b) How many different towns would be assigned to each frequency used?

71. Show that the vertices of any tree can be colored with two colors.

72. Can you find a family of graphs H_n that require n colors to color their vertices?

73. The edge-coloring number of a graph G is the minimum number of colors needed to color the edges of G so that edges that share a common vertex get different colors. Determine the edge-coloring number for each of the graphs in Exercise 63. Can you make a conjecture about the value of the minimum number of colors needed to color the edges of any graph?

74. Can you think of any applications that require determining the minimum number of colors needed to color the edges of a graph?

75. When a graph has been drawn on a piece of paper so that edges meet only at vertices, the graph divides the paper up into regions called *faces*. The faces include one called the "infinite" face, which surrounds the whole graph. The face-coloring number of a graph G (which can be drawn in this special way) is the minimum number of colors needed to color the faces of G so that two faces that share an edge receive different colors. (Note that if two faces meet only at a vertex, they can be colored the same color.)

(a) Determine the minimum number of colors needed to color the faces of the following graphs. In each case, remember to color the infinite face, which is labeled I (for "infinite").

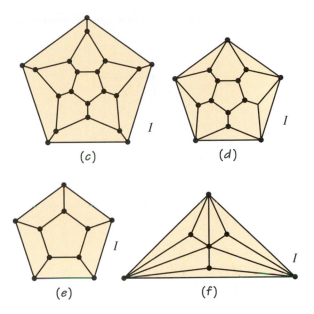

(c)   (d)

(e)   (f)

(b) Can you think of an application of the problem of coloring the faces of a graph with a minimum number of colors?

76. Each vertex in the graph below represents a child who attends a day care center. An edge between two children indicates these children tend to cause problems when they are in the same play group. What is the minimum number of play groups that will ensure that no conflicts arise? Can conflict-free play groups with the same number of children in each group be formed?

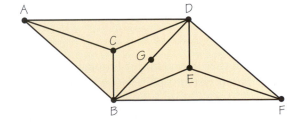

77. A company sells herbs, each of which requires a certain level of proper watering. The following graph is constructed by having one vertex for each type of herb. The vertices representing two herbs are joined by an edge if they must have different levels of watering. What is the minimum number of terrariums that the herbs can be displayed in so

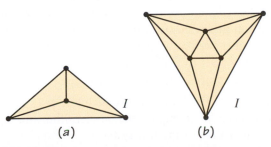

(a)   (b)

that herbs in the same terrarium can be watered at the same level?

78. The company in Exercise 77 is disappointed by the minimum number of terrariums needed to display the herbs with the proper watering requirements. One company employee suggests that if the information about watering requirements is altered for a single pair of herbs (e.g., a single edge is erased from the diagram), then the number of terrariums needed will be reduced by 1. Is this true?

 ## APPLET EXERCISES

To do these exercises, go to
www.whfreeman.com/fapp7e.

Graph Coloring

Solving a scheduling problem such as the one below can be accomplished by constructing a related graph and then coloring it in a way that adjacent vertices have different colors. Explore the problem of graph coloring in the *Graph Coloring* applet.

Scheduling

A mathematics department has seven faculty committees–A, B, C, D, E, F and G. Because there is overlap in the composition of the committees, the chairman of the department is attempting to work out a schedule that will avoid conflicts

among the committees. The following chart indicates the overlapping committee structure:

	A	B	C	D	E	F	G
A		X		X		X	
B	X		X			X	
C		X			X		X
D	X						X
E			X			X	X
F	X	X			X		
G			X	X	X		

Help the chairman arrange a schedule without conflicts in the *Scheduling* applet.

 ## WRITING PROJECTS

1. Scheduling is important for hospitals, schools, transportation systems, police services, and fire services. Pick one of these areas and write about the different scheduling situations that come up, types of processors, and extent to which the assumptions of the list-processing model hold for the area you pick.

2. Compare and contrast the basic scheduling problem we investigated with the scheduling version of the bin-packing problem.

3. One of the oversimplifications made in our discussion of scheduling was that there were no "due dates" involved for the tasks making up a job. Develop an algorithm for solving a scheduling problem under the assumption that each task has a due date as well as a time length. You will probably want to decide on a penalty amount that will occur when a due date is exceeded.

4. Consider the problem of scheduling tasks on a single machine. Design different algorithms for achieving different goals. You will probably wish to assume that each task has a due date such that if the task is not finished by this date, some penalty payment must be made.

5. Discuss the role of graph colorings for scheduling committee meetings so as to avoid conflicts. Research whether or not these ideas are used in the legislature of your home state.

SUGGESTED READINGS

BRUCKER, P. *Scheduling Algorithms*, 4th ed., Springer-Verlag, Heidelberg, Germany, 2004. A detailed mathematical look at scheduling.

GRAHAM, RONALD. Combinatorial scheduling theory, in Lynn Steen (ed.), *Mathematics Today*, Springer-Verlag, New York, 1978, pp. 183–211. This essay on scheduling is one of many excellent accounts of recent developments in mathematics in this book.

GRAHAM, RONALD. The combinatorial mathematics of scheduling. *Scientific American*, March 1978, pp. 124–132. A very readable introduciton to scheduling and bin packing.

JENSEN, T. R., and BJARNE TOFT. *Graph Coloring Problems*, Wiley, New York, 1995. A detailed summary of what is known about coloring problems and many questions that await answering.

LAWLER, E., et al. Sequencing and scheduling algorithms and complexity, in S. C. Graves et al. (eds.), *Handbooks in OR and MS*, vol. 4, Elsevier, New York, 1993, pp. 445–522. A recent survey of results about scheduling.

LEUNG, JOSEPH Y-T., *Handbook of Scheduling*, Chapman & Hall/CRC, Boca Raton, Florida, 2004. This book has an encyclopedic treatment of scheduling algorithms and the great variety of situations where mathematical analysis has assisted schedulers, ranging from sports to hospitals.

PARKER, R. GARY, *Deterministic Scheduling Theory*, Chapman & Hall, London, 1995. A wide-ranging look at scheduling methods and their applications.

SUGGESTED WEB SITES

www.ctl.ua.edu/math103/scheduling/ schedmnu.htm This site provides an overview of scheduling as discussed in this chapter.

www.ams.org/featurecolumn/archive/ machines1.html, www.ams.org/featurecolumn/ archive/packings1.html, www.ams.org/ featurecolumn/archive/bins1.html These Web pages describe mathematical aspects of machine scheduling and bin packing and give a discussion of the relationship between these two mathematical problems.

www.ie.bilkent.edu.tr/~ie672/docs/ resources.html This Web page contains links to many aspects of scheduling theory, including research on the frontier.

Linear Programming

A manager's job often calls for making very complicated decisions. One set of decisions involves planning what products the business is to make and determining what resources are needed. In the modern business world, diversification of products provides a company with stability in a climate of changing tastes and needs. So it is not surprising that companies would produce many products, some of which share resource needs. For example, any bakery uses many resources—like butter, sugar, eggs, and flour—to make its products such as cookies, cakes, pies, and breads.

Resources can include more than just raw materials. A labor force with appropriate skills, farmland, time, and machinery are also resources. Typically, resources are limited: A farmer owns only so much land; there are only so many hours in a day; in a year of drought the wheat crop is very small. Resource availability is also limited by location and competition.

Because resources are limited, management faces important questions: How should the available resources be shared among the possible products? One goal of management is to maximize profit. How can that determine how much of each product should be produced? There are usually so many alternative product mixes that it is impossible to evaluate them all individually. Despite this complexity, millions of dollars may ride on management's decision.

Many business and government agencies must deal with supply-and-demand problems. The general idea is that goods or services can be provided by different providers to individuals or businesses who need these goods or services. There are varying costs to the suppliers to provide different recipients with these goods or services. The goal is to find how to meet the demands for the supplies as cheaply as possible. For example, what is the cheapest way for a company with several oil refineries to provide oil distributors, in many different geographical locations, with the oil they need?

In this chapter, we learn about **linear programming,** a management science technique that helps a business allocate the resources it has on hand to make a particular mix of products that will maximize profit. Linear programming is a tool for maximizing or minimizing a quantity, typically a profit or a cost, subject to constraints. The technique is so powerful that linear programming is said to account for over 50% and perhaps as much as 90% of all computing time used for management decisions in business.

Linear programming is an example of "new" mathematics. It came into being, along with many other management science techniques, during and shortly after World War II, in the 1940s. It is quite young as intellectual ideas go. Yet, during its short history, linear programming has changed the way businesses and governments make decisions, from "seat-of-the-pants" methods based on guesswork and intuition to using an algorithm based on available data and guaranteed to produce an optimal decision.

Linear programming is but one operations research tool belonging to a family of tools known as mathematical programming. Another such tool is integer programming. The difference between linear programming and integer programming is that for linear programming, the quantities being studied can take on values such as $\pi = 3.14159\ldots$ or $7\frac{1}{8}$; in integer programming, the values are confined

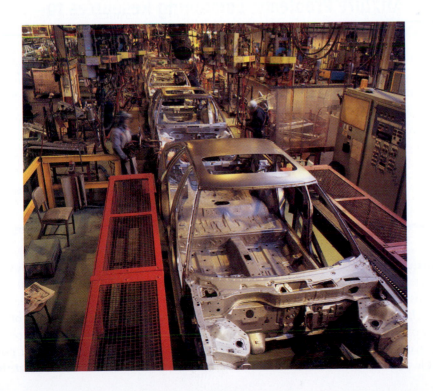

The assembly of an automobile requires many complicated steps and processes. Using linear-programming techniques enables the robots and humans to carry out their tasks faster and more accurately than would be possible without the use of mathematics. This makes American cars more competitive and of a higher quality than otherwise would be the case. (*Digital Vision/Getty Images.*)

to whole numbers such as 8, 50, or 1,102,362. Whole numbers are conceptually easier than the broader group consisting of all numbers that can be represented by decimals (1.32, 1.455555 . . .).

Yet integer-programming problems have proved much harder to solve. In the discussions that follow, we often describe "relaxed" versions of integer-programming problems as linear-programming problems. For example, it would make no sense to produce 3.24 dolls to sell. So, strictly speaking, we must find an optimum whole number of dolls to produce. If we are "lucky," the linear-programming problem associated with an integer-programming problem has an integer solution. In this case, we have also found the correct answer to the integer-programming problem. Some other examples that fall into this category are discussed below.

Linear programming has saved businesses and governments billions of dollars. Of all the management science techniques presented in this book, linear programming is far and away the most frequently used. It can be applied in a variety of situations, in addition to the one we study in this chapter. Some of the problems studied in Chapters 1, 2, and 3 can be viewed as linear-programming problems, and examples of other uses are in Spotlight 4.1. Linear programming is an excellent example of a mathematical technique useful for solving many different kinds of problems that at first do not seem to be similar problems at all. It has been suggested that without linear programming, management science would not exist.

4.1 Mixture Problems: Combining Resources to Maximize Profit

In this chapter, we study how to use linear programming to solve a special kind of problem—a mixture problem. Realistic versions of such problems would be much more involved. Our discussion is designed to give you the flavor of what is actually done.

Mixture Problem

In a **mixture problem,** limited resources are combined into products so that the profit from selling those products is a maximum.

Mixture problems are widespread because nearly every product in our economy is created by combining resources. A typical example would be how different kinds of aviation fuel are manufactured using different kinds of crude oil.

Let's analyze small versions of the kinds of problems that might confront a toy or a beverage manufacturer. Both manufacturers can sell many different products on which each company can make a profit. There could be dozens of possible products and many resources. A manufacturer must periodically look at the

SPOTLIGHT 4.1   Case Studies in Linear Programming

Linear programming is not limited to mixture problems. Here are two case studies that do not involve mixture problems, yet where applying linear-programming techniques produced impressive savings:

▶ The Exxon Corporation spends several million dollars per day running refineries in the United States. Because running a refinery takes a lot of energy, energy-saving measures can have a large effect. Managers at Exxon's Baton Rouge plant had over 600 energy-saving projects under consideration. They couldn't implement them all because some conflicted with others, and there were so many ways of making a selection from the 600 that it was impossible to evaluate all selections individually.

   Exxon used linear programming to select an optimal configuration of about 200 projects, resulting in millions of dollars in savings.

▶ American Edwards Laboratories uses heart valves from pigs to produce artificial heart valves for human beings. Pig heart valves come in different sizes. Shipments of pig heart valves often contain too many of some sizes and too few of others. However, each supplier tends to ship roughly the same imbalance of valve sizes on every order, so the company can expect consistently different imbalances from the different suppliers. Thus, if they order shipments from all the suppliers, the imbalances could cancel each other out in a fairly predictable way. The amount of cancellation will depend on the sizes of the individual shipments. Unfortunately, there are too many combinations of shipment sizes to consider all combinations individually.

   American Edwards used linear programming to figure out which combination of shipment sizes would give the best cancellation effect. This reduced the company's annual cost by $1.5 million.

quantities and prices of resources and then determine which products should be produced in which quantities in order to gain the greatest, or optimum, profit. This is an enormous task that usually requires a computer to solve.

What does it mean to find a solution to a linear-programming mixture problem? A solution to a mixture problem is a production policy that tells us how many units of each product to make.

Optimal Production Policy

An **optimal production policy** has two properties. First it is possible; that is, it does not violate any of the limitations under which the manufacturer operates, such as availability of resources. Second, the optimal production policy gives the maximum profit.

Common Features of Mixture Problems

Although our first mixture problem (Example 1) has only two products and one resource, it does contain the essential features that are common to *all* mixture problems:

▶ *Resources*. Definite resources are available in limited, known quantities for the time period in question. The resource in Example 1 is containers of plastic.

▶ *Products*. Definite products can be made by combining, or mixing, the resources. In Example 1, the products are skateboards and dolls.

▶ *Recipes*. A recipe for each product specifies how many units of each resource are needed to make one unit of that product. Each skateboard in Example 1 uses five units of plastic, and each doll uses two units.

▶ *Profits*. Each product earns a known profit per unit. (We assume that every unit produced can be sold. More complicated mathematical models, which we will not discuss here, are needed if we want to consider the possibility of items being produced but not sold.)

▶ *Objective*. The objective in a mixture problem is to find how much of each product to make so as to maximize the profit without exceeding any of the resource limitations.

The examples we show are not designed to be realistic. Rather, our goal is to demonstrate how ideas whose roots are in basic algebra and geometry can solve, when scaled up to realistic versions, problems that save Americans much time and money and make our government and American businesses more efficient.

EXAMPLE 1   Making Skateboards and Dolls

A toy manufacturer can manufacture only skateboards, only dolls, or some mixture of skateboards and dolls. Skateboards require five units of plastic and can be sold for a profit of $1, while dolls require two units of plastic and can be sold for a $0.55 profit. If 60 units of plastic are available, what numbers of skateboards and/or dolls should be manufactured for the company to maximize its profit?

Attacking this and other mixture problems requires the carrying out of a series of steps that get at the essence of the problem.

As a first step, we need to take the "verbal" information that we have been given and display it in a form that makes it easier to convert into the mathematics necessary to solve the problem. This is done by making a **mixture chart** for the information we are given (see Figure 4.1).

In the rows of this chart we display the products we want to make, and in the column of the chart we display the resources and the profit margin information that is available. In this case we have two products, so we have two rows. We have one resource, which accounts for one column.

The other column is reserved for profit information. Since this information is used in a somewhat different way from the information about the re-

FIGURE 4.1 Mixture chart for Example 1.

PRODUCTS	RESOURCE(S) Containers of Plastic 60	PROFIT
Skateboards (x units)	5	$1.00
Dolls (y units)	2	$0.55

FIGURE 4.1 Mixture chart for Example 1.

sources, we will separate the resource column(s) from the profit column by a double bar.

Since we do not know the number of skateboards the company should make, we will use a letter x to represent the unknown number of skateboard units that the company might manufacture. Similarly, y will represent the number of dolls that the company might manufacture. We enter these letters as part of the labels of the rows of our table.

We can now enter the numbers about resources in the columns based on the information we have been given. In this case, there is one resource, plastic. Thus, for the five units of plastic needed for a skateboard we record a 5 in the Skateboards row and the Containers of Plastic column. Similarly, we enter a 2 in the second row and first column because dolls need two units of plastic for each doll. Since we have 60 units of plastic available, we display this fact by placing the number 60 at the top of this column. We complete the table with the information about profit. We enter $1 in the Skateboards row and Profit column and $0.55 in the Dolls row and Profit column. ■

Making a Mixture Chart

EXAMPLE 2

Make a mixture chart to display this situation. A clothing manufacturer has 60 yards of cloth available to make shirts and decorated vests. Each shirt requires 3 yards of material and provides a profit of $5. Each vest requires 2 yards of material and provides a profit of $3.

SOLUTION: See the mixture chart in Figure 4.2. ■

PRODUCTS	RESOURCE(S) Yards of Cloth 60	PROFIT
Shirts (x units)	3	$5
Vests (y units)	2	$3

FIGURE 4.2 Mixture chart for the clothing manufacturer.

Translating Mixture Charts into Mathematical Form

Consider again the mixture chart in Figure 4.1. What can we say about the numbers of skateboards and dolls that might be manufactured? Clearly we cannot make negative numbers of skateboards or dolls. Since we are using the letter x to represent the number of skateboards we plan to make, we can write down the algebraic expression that $x \geq 0$. Here we are using the standard symbol $\geq$ for "greater than or equal to."

Algebraic expressions that involve the symbol $\geq$ or its companion symbols $\leq$ (less than or equal to), $>$ (greater than), and $<$ (less than) are known as *inequalities*. We can also write down an inequality for y dolls we plan to make, based on the fact that we cannot make a negative number of dolls. Thus, we must have $y \geq 0$. We will use the phrase **minimum constraints** for these two inequalities, $x \geq 0$ and $y \geq 0$, which say simply that one cannot manufacture negative numbers of objects.

However, we also have only a limited number of units of plastic available. How can we represent this information? Consulting the mixture table, we see that we need five units of plastic for every skateboard we make. Thus, we will need $5x$ (5 times x) units of plastic for the skateboards we make. Similarly, we will need $2y$ (2 times y) units of plastic for the dolls we make. Hence we will need $5x + 2y$ units of plastic for the mixture of skateboards and dolls we make. We added the $5x$ and $2y$, because we need to find the total plastic used when we make a mixture of skateboards and dolls.

Reading from the table, we see that we are limited in having only 60 units of plastic. So we can express the constraint imposed by the limited number of units of plastic by writing that $5x + 2y \leq 60$. Here we use the symbol for less than or equal to, $\leq$, to express the fact that we cannot use more than the amount of plastic we have available.

Notice that all of the numbers in this equation can be obtained from a column of the mixture chart. One of the reasons we construct a mixture chart is that it helps us speed up the conversion into equalities of the information about the problem we wish to solve.

In addition to the resource inequalities (of which realistic problems will often have hundreds), there is one additional algebraic expression, this time an equality, that the mixture table allows us to create. Using the mixture table, we can compute the profit that will be produced when we manufacture different mixtures. For each skateboard, we make a profit of $1, so if x skateboards are made, the profit is $1x$ (1 times x). For each doll made, the profit is $0.55. So if y dolls are made, the profit is $0.55y$ (0.55 times y). Denoting by P the total profit from making x skateboards and y dolls, we get the equation

$$P = 1x + 0.55y$$

Note that unlike the situation for the resources where we got an inequality, here we get an expression for what the profit will be as we vary the numbers of skateboards and dolls manufactured. Our goal is to find which values of x and y (skateboards and dolls) make this profit as large as possible.

Revisiting Our Clothing Manufacturer

EXAMPLE 3

We can also translate the information in the mixture chart shown in Figure 4.2 into inequalities and an equation for expressing the profit in terms of how many shirts and vests are produced. Using the first column of the mixture chart and the fact that for cloth there are only 60 units available, we can write

$$3x + 2y \leq 60$$

And using the last column, we get the following expression for the profit P:

$$P = 5x + 3y$$

Now that we have the information from the original problems represented in mathematical terms, we will return our attention to finding a solution to the problems.

Finding the best (largest profit) mixture of skateboards and/or dolls to make can be carried out in two phases.

1. Determine those mixtures of skateboards and/or dolls that can be manufactured subject to the limited resources that are available. This step involves finding the **feasible set** for the mixture problem.

Feasible Set or Feasible Region

The **feasible set,** also called the **feasible region,** for a linear-programming problem is the collection of all physically possible solution choices that can be made.

We can use a geometric diagram such as the one in Figure 4.3 to help us understand the feasible set of options that the manufacturer of skateboards and dolls has available.

2. Determine how to pick out, from the feasible set, the mixture (or mixtures) that gives rise to the largest profit. ■

Representing the Feasible Region with a Picture

After we have constructed inequalities using a mixture chart or have the inequalities that must be obeyed for a more general linear-programming problem, we can draw a helpful picture to visualize the choices to be made in solving the problem. This picture will show in a convenient way the different choices that are available in solving the linear-programming problem at hand. To get the picture that will help us, we need to draw graphs of the inequalities associated with the linear-programming problem.

To draw the graph of an inequality, let's first review how to draw the graph of the equation of a straight line. Remember that two points can be used to uniquely determine a straight line. Let's use the equation associated with the less-than-or-equal-to inequality

$$5x + 2y \leq 60$$

namely,

$$5x + 2y = 60$$

There are two points that are easy to find on this line. When $x = 0$, this gives rise to one point on the line, and when $y = 0$, we can find another point. (See Figure 4.3.) When $x = 0$, if we substitute this value in the equation $5x + 2y = 60$, we get $5(0) + 2y = 60$. Solving this equation, we discover that $y = 30$. Similarly, if we substitute $y = 0$ in the equation $5x + 2y = 60$, we get $5x + 2(0) = 60$, from which we conclude that $x = 12$. We now have two points $(0, 30)$ and $(12, 0)$ that lie on the line $5x + 2y = 60$.

We are using the usual convention that when we write a pair such as $(3, 10)$, we are describing a point that has $x = 3$ and $y = 10$. We always list the x-value first in such a pair and the y-value second. Furthermore, when a point (x, y) is represented in a diagram, larger values of x are shown farther to the right (east) and larger value of y are shown farther up (north).

Using the two points we find on the line $5x + 2y = 60$, we can draw the graph shown in Figure 4.3a, where we have also displayed the point $(3, 10)$. How do we know that $(3, 10)$ is not on the line? We can see this by replacing x by 3 and y by 10 in the equation $5x + 2y = 60$, getting $5(3) + 2(10) = 15 + 20 = 35$. To have been on the line, we would have to have had the value 60. Furthermore, we know that the point $(3, 10)$ is below the line $5x + 2y = 60$ because when

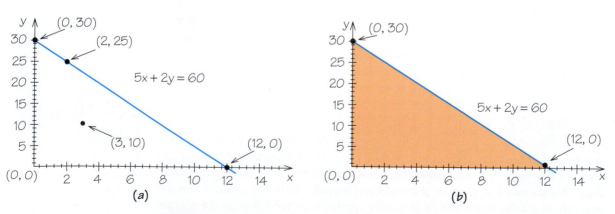

FIGURE 4.3 The feasible region for Example 3. (a) Graph of $5x + 2y = 60$. (b) Shading of the half-plane $5x + 2y < 60$, and where $x \geq 0$, $y \geq 0$.

$x = 3$, and we replace x by this value in the equation $5x + 2y = 60$, we get $5(3) + 2y = 60$, which means that $y = 45/2$. Since $45/2$ is greater than 10, the y-value for the point $(3, 10)$, we conclude that $(3, 10)$ is below the line.

Now that we know what the graph of the equation $5x + 2y = 60$ looks like, we can think through where points (x, y) that satisfy $5x + 2y < 60$ are located. The points that are either on the line $5x + 2y = 60$ or satisfy $5x + 2y < 60$ will satisfy $5x + 2y \leq 60$.

Any line, for example, $5x + 2y = 60$, divides the xy-plane into three parts: those points on the line, and the points in one of two half-planes. In one of these half-planes we have the points for which $5x + 2y < 60$, and in the other we have the points for which $5x + 2y > 60$. How can we tell which of the two half-planes is above the line $5x + 2y = 60$ and which is below?

The key is the use of a test point (x, y) that is not on the line and whose half-planes we wish to distinguish. We saw above that $(3, 10)$ is not on the line $5x + 2y = 60$ and is below the line. This enables us to see that the half-plane for which $5x + 2y < 60$ consists of the points below the line $5x + 2y = 60$.

To complete the drawing of the points that are feasible for the skateboard and dolls manufacturing problem, we also have to know which points satisfy the constraints that state that the number of skateboards produced x cannot be negative ($x \geq 0$) and the number of dolls produced y cannot be negative ($y \geq 0$). Each of these inequalities corresponds to a half-plane, and we can again test which of the half-planes associated with the line $x = 0$ is determined by $x \geq 0$.

This can be done using the point $(3, 10)$ as a test point again. Since $x = 3$ is greater than 0, $x \geq 0$ determines the half-plane to the right of the line $x = 0$ (the x-axis). Similarly, using the point $(3, 10)$ to test which half-plane is determined by the line $y = 0$ (associated with the inequality $y \geq 0$), we see that since $y = 10$ is greater than 0, we are dealing with the half-plane above the line $y = 0$ (the y-axis).

Putting this information together leads us to the conclusion that the collection of points (x, y) that meets the three inequalities involved ($x \geq 0, y \geq 0, 5x + 2y \leq 60$) corresponds to the shaded region in Figure 4.3b.

Note that since the minimality conditions are always present in the kind of linear-programming problems we are dealing with, the points that are feasible for these problems are always in the upper right region (quadrant) that is created by the x-axis and y-axis. Next, we draw the feasible region for the clothing manufacturing problem.

Drawing a Feasible Region

EXAMPLE 4

In the earlier clothing manufacturer example, we developed a resource constraint of $3x + 2y \leq 60$. Draw the feasible region corresponding to that resource constraint, using the reality minimums of $x \geq 0$ and $y \geq 0$.

SOLUTION: First we find the two points where the line, $3x + 2y = 60$, crosses the axes. When $x = 0$, we get $3(0) + 2y = 60$, giving $y = \frac{60}{2} = 30$, yielding the point $(0, 30)$. For $y = 0$, we get $3x + 2(0) = 60$, or $x = \frac{60}{3} = 20$, so we have the point $(20, 0)$. We draw the line connecting those points. Testing the point $(0, 0)$, we

FIGURE 4.4 Feasible region for the clothing manufacturer.

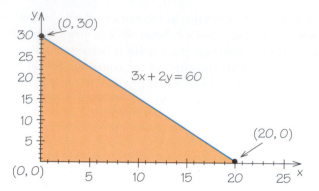

find that the down side of the line we have drawn corresponds to $3x + 2y < 60$. The feasible region is shown in Figure 4.4. ■

4.2 Finding the Optimal Production Policy

Our next step is that we still must find the *optimal production policy*, a point within that region that gives a maximum profit. There are a lot of points in that region. If you consider points with only whole numbers as values for x or y, there are many points, but in fact either x or y or both of them could be some fractional number. There are so many points in this feasible region that to consider the profit at each one of them would require us to calculate profits from now until we grow very old, and still the calculations would not be done. Here is where the genius of the linear-programming technique comes in, with the *corner point principle*, which we define in terms of our mixture problems.

Corner Point Principle

The **corner point principle** states that in a linear-programming problem, the maximum value for the profit formula always corresponds to a **corner point** of the feasible region.

The corner point principle is probably the most important insight into the theory of linear programming. The geometric nature of this principle explains the value of creating a geometric model from the data in a mixture chart.

The corner point principle gives us the following method to solve a linear-programming problem:

1. Determine the corner points of the feasible region.
2. Evaluate the profit at each corner point of the feasible region.
3. Choose the corner point with the highest profit as the production policy.

TABLE 4.1	Calculation of the Profit Formula for Skateboards and Dolls

Corner Point	Value of the Profit Formula: $\$1.00x + \$0.55y$
(0, 0)	$\$1.00(0) + \$0.55(0) = \$0.00 + \$0.00 = \$0.00$
(0, 30)	$\$1.00(0) + \$0.55(30) = \$0.00 + \$16.50 = \$16.50$
(12, 0)	$\$1.00(12) + \$0.55(0) = \$12.00 + \$0.00 = \$12.00$

Let's look at the feasible region we drew in Figure 4.3. It is a triangle having three corners, namely, (0, 0), (0, 30), and (12, 0). Now all we need to do is find out which of these three points gives us the highest value for the profit formula, which in this problem is $\$1.00x + \$0.55y$. We display our calculations in Table 4.1. The maximum profit for the toy manufacturer is $\$16.50$, and that happens if the manufacturer makes 0 skateboards and 30 dolls. The point (0, 30) is called the *optimal production policy*.

Optimal Production Policy

An **optimal production policy** corresponds to a corner point of the feasible region where the profit formula has a maximum value.

Finding the Optimal Production Policy

EXAMPLE 5

Our analysis of the clothing manufacturer problem resulted in a feasible region with three corner points (0, 0), (0, 30), and (20, 0). Which of these maximizes the profit formula, $\$5x + \$3y$, and what does that corner represent in terms of how many shirts and vests to manufacture?

SOLUTION: The evaluation of the profit formula at the corner points is shown in Table 4.2. The maximum profit of $\$100$ occurs at the corner point (20, 0), which represents making 20 shirts and no vests. ∎

TABLE 4.2	Evaluating the Profit Formula in the Clothing Example

Corner Point	Value of the Profit Formula: $\$5x + \$3y$
(0, 0)	$\$5(0) + \$3(0) = \$0 + \$0 = \$0$
(0, 30)	$\$5(0) + \$3(30) = \$0 + \$90 = \$90$
(20, 0)	$\$5(20) + \$3(0) = \$100 + \$0 = \$100$

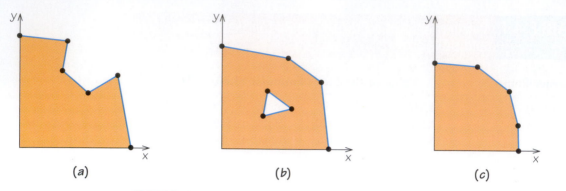

(a)   (b)   (c)

FIGURE 4.5 A feasible region may not have (a) dents or (b) holes. Graph (c) shows a typical feasible region.

General Shape of Feasible Regions

The shape of a feasible region for a linear-programming mixture problem has some important characteristics, without which the corner point principle would not work:

1. The feasible region is a polygon in the first quadrant, where both $x \geq 0$ and $y \geq 0$. This is because the minimum constraints require that both x and y be nonnegative.
2. The region is a polygon that has neither dents (as in Figure 4.5a) nor holes (as in Figure 4.5b). Figure 4.5c is a typical example. Such polygons are called *convex*.

The Role of the Profit Formula: Skateboards and Dolls

In practice, there are often different amounts of resources available in different time periods. The selling price for the products can also change. For example, if competition forces us to cut our selling price, the profit per unit can decrease. To maximize profit, it is usually necessary for a manufacturer to redo the mixture problem calculations whenever any of the numbers change.

Suppose that business conditions change and now the profits per skateboard and doll are, respectively, $1.05 and $0.40. Let us keep everything else about the skateboards and dolls problem the same. The change in profits would give us a new profit formula of $1.05x + $0.40y. When we evaluate the new profit formula at the corner points, we get the results shown in Table 4.3. This time the optimal production policy, the point that gives the maximum value for the profit formula, is the point (12, 0). To get the maximum profit of $12.60, the toy manufacturer should now make 12 skateboards and 0 dolls.

We see from this example that the shape of the feasible region, and thus the corner points we test, are determined by the constraint inequalities. The profit formula is used to choose an optimal point from among the corner points, so it is not surprising that different profit formulas might give us different optimal production policies.

TABLE 4.3	A Different Profit Formula: Skateboards and Dolls

Corner Point	Value of the Profit Formula: $\$1.05x + \$0.40y$
(0, 0)	$\$1.05(0) + \$0.40(0) = \$0.00 + \$0.00 = \$0.00$
(0, 30)	$\$1.05(0) + \$0.40(30) = \$0.00 + \$12.00 = \$12.00$
(12, 0)	$\$1.05(12) + \$0.40(0) = \$12.60 + \$0.00 = \$12.60$

We started the exploration of skateboard and doll production with the idea that a toy manufacturer has a product line with either one to two products. But both linear-programming solutions we have found tell the manufacturer that to maximize profit, make just one product. This is probably not an acceptable result for the manufacturer, who might want to produce both products for business reasons other than profit, such as establishing brand loyalty. And it certainly would be very difficult for the manufacturer to be ready to switch back and forth between producing either skateboards or dolls every time the profit formula changed. Linear programming is a flexible enough technique that it can accommodate the desire for there to be both products in the optimal production policy. This is done by specifying that there be nonzero minimum quantities for each period.

Summary of the Pictorial Method Using a Feasible Region

Let's stop and summarize the steps we are following to find the optimal production policy in a mixture problem:

1. Read the problem carefully. Identify the resources and the products.
2. Make a mixture chart showing the resources (associated with limited quantities), the products (associated with profits), the recipes for creating the products from the resources, the profit from each product, and the amount of each resource on hand. If the problem has nonzero minimums, include a column for those as well.
3. Assign an unknown quantity, x or y, to each product. Use the mixture chart to write down the resource constraints, the minimum constraints, and the profit formula.
4. Graph the line corresponding to each resource constraint and determine which side of the line is in the feasible region. If there are nonzero minimum constraints, graph lines for them also, and determine which side of each is in the feasible region. Sketch the feasible region by finding the common points in the half-planes from all the resource constraints plus the minimum constraints. (This process is called finding the "intersection" of the half-planes.)
5. Find the coordinates of all the corner points of the feasible region. Some of these may have been calculated so that you can graph the individual

lines. Proceed in order around the boundary of the feasible region. Be sure that every point you consider is part of the feasible region.

6. Evaluate the profit formula for each of the corner points. The production policy that maximizes profit is the one that gives the biggest value to the profit formula.

Two Products and Two Resources: Skateboards and Dolls

We return to the toy manufacturer, now to consider two limited resources instead of one. The second limited resource will be time, the number of person-minutes available to prepare the products. Suppose that there are 360 person-minutes of labor available and that making one skateboard requires 15 person-minutes and making one doll requires 18 person-minutes. We will continue to use the original figures regarding containers of plastic, the first of our two profit formulas, and to keep the problem relatively simple, we use the zero minimum constraints: $x \geq 0$ and $y \geq 0$. We need a new mixture chart. In general, we will include a column for minimums in a mixture chart only if there are any nonzero minimum constraints. In Figure 4.6 we have the mixture chart for this problem. Using the mixture chart, we can write the two resource constraints:

$$5x + 2y \leq 60 \qquad \text{for containers of plastic}$$

and

$$15x + 18y \leq 360 \qquad \text{for person-minutes}$$

We can also write the profit formula: $\$1.00x + \$0.55y$.

The half-plane corresponding to the plastic resource is shown in Figure 4.7a. We now need to graph the half-plane corresponding to the time constraint. We find where the line $15x + 18y = 360$ intersects the two axes by substituting first $x = 0$ and then $y = 0$ into that equation.

The line corresponding to the time constraint contains the two points (0, 20) and (24, 0). When we substitute the point (0, 0) into the inequality $15x + 18y <$

RESOURCE(S)

PRODUCTS		Containers of Plastic 60	Person-minutes 360	PROFIT
	Skateboards (x units)	5	15	$1.00
	Dolls (y units)	2	18	$0.55

FIGURE 4.6 Mixture chart for Skateboards and Dolls (two resources).

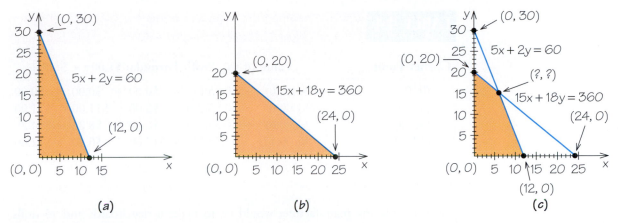

FIGURE 4.7 Feasible region for Skateboards and Dolls (two resources). (a) Half-plane for the plastic resource constraint. (b) Half-plane for the time resource constraint. (c) Intersection for the two half-planes.

360, we get $15(0) + 18(0) < 360$, or $0 < 360$, which is true, so $(0, 0)$ is on the side of the line that we shade. Putting all this together, we get the half-plane in Figure 4.7b as the correct half-plane for the time resource constraint.

We are not permitted to exceed the supply of even a single resource. Therefore, the feasible region must be made up of points that are shaded twice—both in the half-plane for the plastic resource constraint, shown in Figure 4.7a, and in the half-plane for the time resource constraint in Figure 4.7b. The procedure with several half-plane constraints is that we build our feasible region by finding the intersection, or overlap, of the individual half-planes in the problem. In Figure 4.7c we show the result of intersecting the half-planes from the two resource constraints. Because this problem has minimums that are zeroes, the shaded region in Figure 4.7c is in fact the feasible region for the problem.

The next step that we need to carry out to use the pictorial method for solving this problem is to find the corner points of the feasible region. This is done by using the algebra necessary to solve two equations in two unknowns. This leads to the points $(0, 0)$, $(0, 20)$, $(6, 15)$, and $(12, 0)$. Three of these points have a zero value for one or more of the unknowns, so the calculations are easy.

To find the coordinates of the point $(6, 15)$, it is necessary to solve for the point that satisfies both of the equations $5x + 2y = 60$ and $15x + 18y = 360$. One way to do this is to multiply the first of these equations by -3, obtaining the equation $-15x - 6y = -180$. When this is added to the second equation, the x term "drops out" and we can solve $12y = 180$, to get $y = 15$. Now it is an easy matter to substitute this value into either of the original equations to get the x value of 6.

We are ready to finish the problem. In Table 4.4 we have evaluated the profit formula at the four corner points of the feasible regions. The optimal production

TABLE 4.4	The Profit at the Four Corner Points

Corner Point	Value of the Profit Formula: $\$1.00x + \$0.55y$
(0, 0)	$\$1.00(0) + \$0.55(0) = \$0.00 + \$0.00 = \$0.00$
(0, 20)	$\$1.00(0) + \$0.55(20) = \$0.00 + \$11.00 = \$11.00$
(6, 15)	$\$1.00(6) + \$0.55(15) = \$6.00 + \$8.25 = \$14.25$
(12, 0)	$\$1.00(12) + \$0.55(0) = \$12.00 + \$0.00 = \$12.00$

policy for the toy manufacturer would be to make 6 skateboards and 15 dolls, for a maximum profit of $14.25.

Here is another mixture problem example of how the pictorial method using a feasible region works from start to finish.

EXAMPLE 6   Mixtures of Two Fruit Juices: Beverages

A juice manufacturer produces and sells two fruit beverages: 1 gallon of cranapple is made from 3 quarts of cranberry juice and 1 quart of apple juice; and 1 gallon of appleberry is made from 2 quarts of apple juice and 2 quarts of cranberry juice. The manufacturer makes a profit of 3 cents on a gallon of cranapple and 4 cents on a gallon of appleberry. Today, there are 200 quarts of cranberry juice and 100 quarts of apple juice available. How many gallons of cranapple and how many gallons of appleberry should be produced to obtain the highest profit without exceeding available supplies? We use zeroes as "reality minimums." The mixture chart for this problem is shown in Figure 4.8.

For each resource, we develop a resource constraint reflecting the fact that the manufacturer cannot use more of that resource than is available. The number of quarts of cranberry juice needed for x gallons of cranapple is $3x$. Similarly, $2y$ quarts of cranberry are needed for making y gallons of appleberry. So if the manufacturer makes x gallons of cranapple and y gallons of appleberry, then $3x +$

FIGURE 4.8 A mixture chart for Example 6.

	RESOURCE(S)		
PRODUCTS	Cranberry 200 quarts	Apple 100 quarts	PROFIT
Cranapple (x gallons)	3 quarts	1 quart	3 cents/gallon
Appleberry (y gallons)	2 quarts	2 quarts	4 cents/gallon

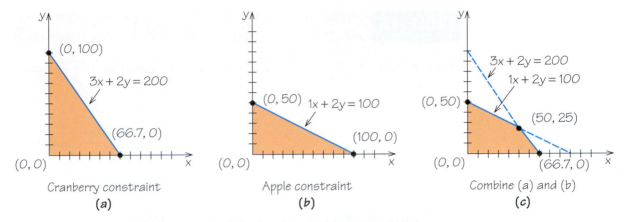

FIGURE 4.9 Feasible region for Example 6.

$2y$ quarts of cranberry juice will be used. Because there are only 200 quarts of cranberry available, we get the cranberry resource constraint $3x + 2y \le 200$. Note that the numbers 3, 2, and 200 are all in the Cranberry column. We get another resource constraint from the column for the apple juice resource: $1x + 2y \le 100$. We also have these minimum constraints: $x \ge 0$ and $y \ge 0$.

Finally, we have the profit formula. Because $3x$ is the profit from making x units of cranapple and $4y$ is the profit from making y units of appleberry, we get the profit formula $3x + 4y$.

We summarize our analysis of the juice mixture problem. Maximize the profit formula, $3x + 4y$, given these constraints:

cranberry: $3x + 2y \le 200$
apple: $1x + 2y \le 100$
minimums: $x \ge 0$ and $y \ge 0$

Remember, in a mixture problem, our job is to find a production policy (x, y), that makes all the constraints true and maximizes the profit.

Figure 4.9a shows the result of graphing the constraint associated with the cranberry resource, while Figure 4.9b shows the result of graphing the constraint associated with the apple resource, taking into account that the amounts of these resources used cannot be negative. When these two diagrams are superimposed, we get the diagram in Figure 4.9c. Now, to carry out the pictorial method, we need to find the profits associated with the four corner points shown. This is done in Table 4.5.

When we evaluate the profit formula at the four corner points, we see that the optimal production policy is to make 50 gallons of cranapple and 25 gallons of appleberry for a profit of 250 cents. ∎

TABLE 4.5	Finding the Optimal Production Policy for Beverages
Corner Point	**Value of the Profit Formula: $3x + 4y$ cents**
(0, 0)	$3(0)$ $+ 4(0)$ $= 0$ cents
(0, 50)	$3(0)$ $+ 4(50) = 200$ cents
(50, 25)	$3(50)$ $+ 4(25) = 250$ cents
(66.7, 0)	$3(66.7) + 4(0)$ $= 200$ cents (rounded)

4.3 Why the Corner Point Principle Works

In finding solutions to our mixture problems, we have been using the corner point principle, which says that the highest profit value on a polygonal feasible region is always at a corner point. A feasible region has infinitely many points, making it impossible to compute the profit for each point. The corner point principle gives us a finite set of points, making the calculation possible.

You can visualize a mathematical proof of the corner point principle by imagining that each point of the plane is a tiny light bulb that is capable of lighting up. For the juice mixture example, whose feasible region is shown in Figure 4.9c, imagine what would happen if we ask this question: Will all points with profit = 360 please light up? What geometric figure do these lit-up points form?

In algebraic terms, we can restate the profit question in this way: Will all points (x, y) with $3x + 4y = 360$ please light up? As it happens, this version of the profit question is one mathematicians learned to answer hundreds of years before linear programming was born.

The points that light up make a straight line because $3x + 4y = 360$ is the equation of a straight line. Furthermore, it is a routine matter to determine the exact position of the line. We call this line the **profit line** for 360; it is shown in Figure 4.10. For numbers other than 360, we would get different profit lines. Unfortunately, there are no points on the profit line for 360 that are feasible, that is, which lie in the feasible region. Therefore, the profit of 360 is impossible. *If the profit line corresponding to a certain profit doesn't touch the feasible region, then that profit isn't possible.*

Because 360 is too big, perhaps we should ask the profit line for a more modest amount, say, 160, to light up. You can see that the new profit line of 160 in Figure 4.10 is parallel to the first profit line and closer to the origin. This is no accident: All profit lines for the profit formula $3x + 4y$ have the same coefficients for x and y—namely, 3 for x and 4 for y. Because the slope of the line is determined by those coefficients, they all have the same slope. Changing the profit value from 360 to 160 has the effect of changing where the line intersects the y-axis, but it does not affect the slope.

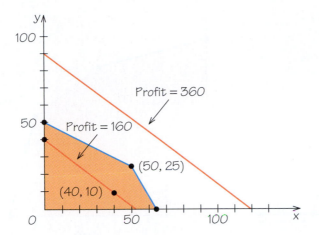

FIGURE 4.10 The profit line for 360 lies outside the feasible region, whereas the profit line for 160 passes through the region.

The most important feature of the profit line for 160 is that it has points in common with the interior of the feasible region. For example, (40, 10) is on that profit line because $3(40) + 4(10) = 160$; in addition, (40, 10) is a feasible point. This means that it is possible to make 40 gallons of cranapple and 10 gallons of appleberry and that if we do so, we will have a profit of 160.

Can we do better than a 160 profit? As we slowly increase our desired profit from 160 toward 360, the location of the profit line that lights up shifts smoothly upward away from the origin. As long as the line continues to cross the feasible region, we are happy to see it move away from the origin, because the more it moves, the higher the profit represented by the line. We would like to stop the movement of the line at the last possible instant, while the line still has one or more points in common with the feasible region. It should be obvious that this will occur when the line is just touching the feasible region either at a corner point (Figure 4.11a) or along a line segment joining two corners (Figure 4.11b). That point or line segment corresponds to the production policy or policies with the maximum achievable profit. This is just what the corner point principle says: The maximum profit always occurs at a corner or along an edge of the feasible region.

Adding Nonzero Minimums: Beverages   EXAMPLE 7

Suppose that in Example 6 the profit for cranapple changes from 3 cents per gallon to 2 cents and the profit of appleberry changes from 4 cents per gallon to 5 cents. You can verify that this change moves the optimal production policy to the point (0, 50)—no cranapple is produced. This result is not surprising: Appleberry is giving a higher profit and the policy is to produce as much of it as possible. But suppose the manufacturer wants to incorporate nonzero minimums into the linear-programming specifications so that there will always be both cranapple, x, and appleberry, y, produced. Specifically, they decide that $x \geq 20$

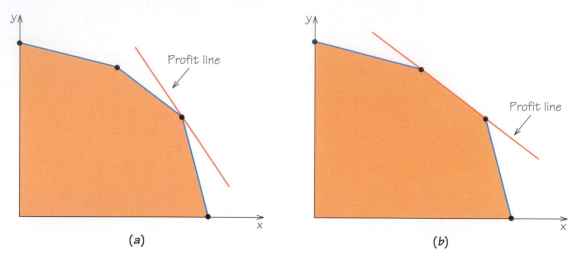

FIGURE 4.11 The highest profit will occur when the profit is just touching the feasible region, either (a) at the corner point or (b) along a line segment.

and $y \geq 10$ are desirable minimums. Figure 4.12 is the mixture chart showing the new profit formula and the nonzero minimums along with the unchanged rest of the beverage problem.

The feasible region for Beverages, Example 6, is shown in Figure 4.13a. The feasible region for Beverages, Example 7, is shown in Figure 4.13b. You can verify that, starting at the lower left corner of the new feasible region and moving clockwise around its boundary, we have corner points (20, 10), (20, 40), (50, 25), and (60, 10). (One of those points was also a corner point of the old feasible region. Can you explain why?) Table 4.6 shows the evaluation of the profit formula at these corner points. For this modified problem the optimal production policy is to produce 20 gallons of cranapple and 40 of appleberry for a maximum profit of 240 cents.

		RESOURCE(S)		
	Cranberry Juice 200 quarts	Apple Juice 100 quarts	MINIMUMS	PROFIT
Cranapple (x gallons)	3	1	20	2 cents
Appleberry (y gallons)	2	2	10	5 cents

PRODUCTS

FIGURE 4.12 Mixture chart for Example 7.

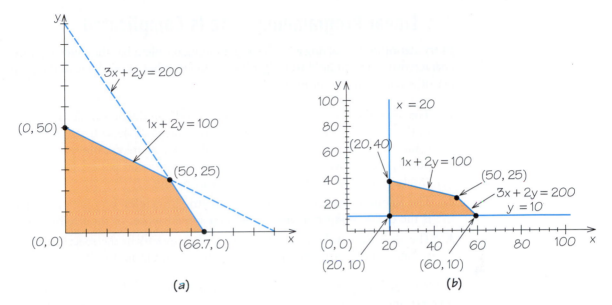

FIGURE 4.13 Feasible region for Examples 6 and 7. (a) Zero minimums. (b) Nonzero minimums.

One final note about this solution concerns the resources. The point (20, 40) is on the resource constraint line for the apple juice resource, so it represents using up all the available apple juice. We can see this by substituting into the apple juice resource constraint: $1(20) + 2(40) = 100$ is true. However, (20, 40) is *below* the line for the cranberry juice resource, indicating that there will be *slack*, or leftover, amounts of cranberry juice. Specifically, substituting (20, 40) into the cranberry juice constraint gives $3x + 2y = 3(20) + 2(40) = 60 + 80 = 140$, which is 60 quarts less than the 200 quarts available. The slack is 60 quarts of cranberry juice. Dealing with slack can be an important consideration for manufacturers. Can you see why? ■

TABLE 4.6	Profit Evaluation for Beverages
Corner Point	**Value of the Profit Formula: $2x + 5y$**
(20, 10)	$2(20) + 5(10) = \quad 40 + \quad 50 = \quad 90$ cents
(20, 40)	$2(20) + 5(40) = \quad 40 + 200 = 240$ cents
(50, 25)	$2(50) + 5(25) = 100 + 125 = 225$ cents
(60, 10)	$2(60) + 5(10) = 120 + \quad 50 = 170$ cents

4.4 Linear Programming: Life Is Complicated

Every algorithm for solving a linear-programming problem has the following three characteristics, which hold true regardless of the number of products or the number of resources in the problem:

1. The algorithm can distinguish between "good" production policies—those in the feasible set that satisfy all the constraints—and those that violate some constraint(s) and are thus not feasible. There are usually many good points, each of which corresponds to some production policy; for example, "Make x units of product 1 and y units of product 2."
2. The algorithm makes use of some geometric principles—one such principle is the corner point principle—to select a special subset of the feasible set.
3. The algorithm evaluates the profit formula at points in the special subset to find which corner point actually gives the maximum profit.

The various algorithms for linear programming differ in how they find the feasible set and in how quickly the algorithm finds the production policy—corner point—that gives the optimal profit. The better algorithms are faster at finding the optimal production policy.

In practical linear-programming problems, the feasible region will not be as simple as the ones we have examined here. There are two ways the feasible region can be more complex:

1. Sometimes, as in Figure 4.14, we have a great many corners. The more corners there are, the more calculations we need to determine the coordinates of all of them and the profit at each one. The number of corners literally can exceed the number of grains of sand on the earth. Even with the fastest computer, computing the profit of every corner is impossible.
2. It is not possible to visualize the feasible region as a part of two-dimensional space when there are more than two products. Each product is represented by an unknown, and each unknown is represented by a dimension of space. If we have 50 products, we would need 50 dimensions and couldn't visualize the feasible region.

Another type of complication can occur even in simple two-dimensional regions: Corner points can have fractional coordinates, not the integer ones we see in the specially constructed problems in this text. Making 3.75 skateboards and 5.45 dolls is not possible. Integer programming, a special type of linear programming, is used when it is not possible to use fractional answers.

FIGURE 4.14 A feasible region with many corners.

The Simplex Method

Several methods are used for the typically large linear-programming problems solved in practice. The oldest method is the **simplex method,** which is still the most commonly used. Devised by the American mathematician George Dantzig

SPOTLIGHT 4.2 Father of Linear Programming Recalls Its Origins

George Dantzig, who died in 2005, spent most of his career as a professor of operations research and computer science at Stanford University. He is credited with inventing the linear-programming technique called the simplex method. Since its invention in the 1940s, the simplex method has provided solutions to linear-programming problems that have saved both industry and the military time and money. Here Dantzig talks about the background of his famous technique:

Initially, all the work we did had to do with military planning. During World War II, we were planning on a very extensive scale. The civilian population and the military were all performing scheduling and planning tasks, perhaps on a larger scale than at any time in history. And this was the case up until about 1950. From 1950 on, the whole emphasis shifted from military planning to practical planning for the civilian population, and industry picked it up.

The first areas of industry to use linear programming were the petroleum refineries. They used it for blending gasoline. Nowadays, all of the refineries in the world (except for one) use linear programming methods. They are one of the biggest users of it, and it's been picked up by every other industry you can think of—the forestry industry, the steel industry—you could fill up a book with all the different places it's used.

The question of why linear programming wasn't invented before World War II is an interesting one. In the postwar period, various technologies just evolved that had never been there before. Computers were one example. These technologies were talked about before. You can go back in history and you'll find papers on them, but these were isolated cases that never went anywhere. . . .

The problems we solve nowadays have thousands of equations, sometimes a million variables. One of the things that still amazes me is to see a program run on the computer—and to see the answer come out. If we think of the number of combinations of different solutions that we're trying to choose the best of, it's akin to the stars in the heavens. Yet we solve them in a matter of moments. This, to me, is staggering. Not that we can solve them—but that we can solve them so rapidly and efficiently.

The simplex method has been used now for roughly 50 years. There has been steady work going on trying to use different versions of the simplex method, nonlinear methods, and interior methods. It has been recognized that certain classes of problems can be solved much more rapidly by special algorithms than by using the simplex method. If I were to say what my field of specialty is, it is in looking at these different methods and seeing which are more promising than others. There's a lot of promise in this—there's always something new to be looked at.

(see Spotlight 4.2), this ingenious mathematical invention makes it possible to find the best corner point by evaluating only a tiny fraction of all the corners. With the use of the simplex method, a problem that might be impossible to solve if each corner point had to be checked can be solved in a few minutes or even a few seconds on a typical business computer.

The operation of the simplex method may be likened to the behavior of an ant crawling on the edges of a polyhedron (a solid with flat sides) looking for an

FIGURE 4.15 The simplex method can be compared to an ant crawling along the edges of a polyhedron, looking for the "target"—the optimal corner point.

optimal corner point—one that gives the highest profit (Figure 4.15). The ant cannot see where the optimal corner is. As a result, if it were to wander along the edges randomly, it might take a long time to reach that corner. The ant will do much better if it has a temperature clue to let it know it is getting warmer (closer to the optimal corner) or colder (farther from the optimal corner).

Think of the simplex method as a way of calculating these temperature hints. We begin at any corner. All neighboring corners are evaluated to see which ones are warmer and which are colder. A new corner is chosen from among the warmer ones, and the evaluation of neighbors is repeated—this time checking neighbors of the new corner. The process ends when we arrive at a corner all of whose neighbors are colder than it is.

Part of what the simplex method has going for it is that it works faster in practice than its worst-case behavior would lead us to believe. Although mathematicians have devised artificial cases for which the simplex method bogs down in unacceptable amounts of arithmetic, the examples arising from real applications are never like that. This may be the world's most impressive counterexample to Murphy's law, which says that if something can go wrong, it will.

Although the simplex method usually avoids visiting every corner, it may require visiting many intermediate ones as it moves from the starting corner to the optimal one. The simplex method has to search along edges on the boundary of the polyhedron. If it happens that there are a great many small edges lying between the starting corner and the optimal one, the simplex method must operate like a slow-moving bus that stops on every block.

Many computer programs are available that will use the simplex method to produce an optimal production policy if we just supply the computer with the constraint inequalities and profit formula. Simplex method programs can be found in a variety of places, among which are spreadsheets, packages of mathematics programs designed for business applications or finite mathematics courses, and large "all-purpose" mathematics packages. A graphical solution is possible only for problems limited to two products; these special exercises involve more than two products.

An Alternative to the Simplex Method

In 1984, Narendra Karmarkar (see Figure 4.16), a mathematician working at Bell Laboratories, devised an alternative method for linear programming that finds the optimal corner point in fewer steps than the simplex algorithm by making use of search routes through the interior of the feasible region. The applications of Karmarkar's algorithm are important to a lot of industries, including telephone communications and the airlines (see Spotlight 4.3). Routing millions of long-distance calls, for example, means deciding how to use the resources of long-distance landlines, repeater amplifiers, and satellite terminals to best advantage. The problem is similar to the juice company's need to find the best use of its stocks of juice to create the most profitable mix of products.

Many airlines use software based on Karmarkar's algorithm to reduce fuel costs and deal with delays caused by storms.

In the 1980s, scientists at Bell Labs applied Karmarkar's algorithm to a problem of unprecedented complexity: deciding how to economically build telephone

FIGURE 4.16
Narendra Karmarkar, a researcher at AT&T Bell Laboratories, invented a powerful new linear-programming algorithm that solves many complex linear-programming problems faster and more efficiently than any previous method. *(Courtesy of AT&T Labs.)*

links between cities so that calls can get from any city to any other, possibly being relayed through intermediate cities. Figure 4.17 shows one such linking. The number of possible linkings is unimaginably large, so picking the most economical one is difficult. For any given linking, there is also the problem of deciding how to economically route calls through the network to reach their destinations.

Now, similar approaches are being used to route email packets and phone calls over the Internet.

FIGURE 4.17 A map of the United States showing one conceivable network of major communication lines connecting major cities. Routing millions of calls over this immense network requires sophisticated linear-programming techniques and high-speed computers. *(Courtesy of AT&T Labs.)*

SPOTLIGHT 4.3

SPOTLIGHT 4.3 Finding Fast Algorithms Means Better Airline Service

Linear-programming techniques have a direct impact on the efficiency and profitability of major airlines. Thomas Cook, director of operations research at American Airlines, was interviewed in 1985 concerning his ideas on why optimal solutions are essential to his business:

Finding an optimal solution means finding the best solution. Let's say you are trying to minimize a cost function of some kind. For example, we may want to minimize the excess costs related to scheduling crews, hotels, and other costs that are not associated with flight time. So we try to minimize that excess cost, subject to a lot of constraints, such as the amount of time a pilot can fly, how much rest time is needed, and so forth.

An optimal solution, then, is either a minimum-cost solution or a maximizing solution. For example, we might want to maximize the profit associated with assigning aircrafts to the schedule; so we assign large aircraft to high-need segments and small aircraft to low-load segments.

The simplex method, which was developed some 50 years ago by George Dantzig, has been very useful at American

Airlines and, indeed, at a lot of large businesses. The difference between his method and Narendra Karmarkar's is speed. Finding fast solutions to linear-programming problems is also essential. With an algorithm like Karmarkar's, which is 50 to 100 times faster than the simplex method, we could do a lot of things that we couldn't do otherwise. For example, some applications could be real-time applications, as opposed to batch applications. So instead of running a job overnight and getting an answer the next morning, we could actually key in the data or access the database, generate the matrix, and come up with a solution that could be implemented a few minutes after keying in the data.

A good example of this kind of application is what we call a major weather disruption. If we get a major weather disruption at one of the hubs, such as Dallas or Chicago, then a lot of flights may get canceled, which means we have a lot of crews and airplanes in the wrong places. What we need is a way to put that whole operation back together again so that the crews and airplanes are in the right places. That way, we minimize the cost of the disruption as well as passenger inconvenience.

4.5 A Transportation Problem: Delivering Perishables

A supermarket chain gets bread deliveries from a bakery chain that does its baking in different places. Each supermarket store needs a certain number of loaves each day, and the supplier bakes in total enough breads to exactly meet the demands. Figure 4.18 shows the cost to ship a loaf from a particular baking location to the store involved. How many breads should be shipped from each locale to each of the stores to stay within the demands and to minimize the cost?

Similarly, after a long holiday weekend, a car rental company will have extra cars in some cities and too few cars in other cities. It is faced with the problem of reshuffling the cars at minimal cost so that each city has the right number of

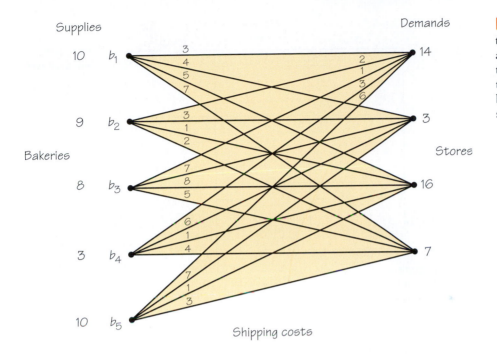

FIGURE 4.18 A graph theory representation of a supply-and-demand transportation problem that involves shipping breads from bakeries to stores.

cars. Problems such as these go under the general name of **transportation problems** and they form a special class of linear-programming problems that can be solved by a specialized method.

Delivering Bread

EXAMPLE 8

Imagine we have three bakeries and three stores, though the ideas we develop will also solve problems where the number of stores and bakeries need not be the same. The three stores require 3 dozen, 7 dozen, and 1 dozen loaves of bread, respectively, while the three bakeries can supply 8 dozen, 1 dozen, and 2 dozen loaves, respectively. The information given so far can be displayed in Figure 4.19, where the "suppliers" are represented by the rows of the table (labeled with Roman numerals) and the "demanders" are represented by the columns (labeled with Hindu-Arabic numerals).

The numbers of breads available and the numbers being required are shown on the right side and bottom of the table and will be referred to as **rim conditions**. Each entry of the table shown in Figure 4.19 is known as a cell. It is convenient to have a name for each of these cells. For example, the cell in the third row and second column will be denoted (III, 2). The first number always corresponds to a row, the second to a column. Thus cell (I, 2) refers to bakery I and store 2.

In deciding which bakeries should ship to which stores, it seems natural to take into account the costs of shipping a dozen breads from a particular bakery to a particular store. If bakery I is farther from store 2 than is bakery II, it seems

152   PART I   Management Science

FIGURE 4.19 A
representation of a
specific problem
involving meeting the
demands of three stores
for breads from the
supplies available at
three bakeries. Shipping
costs between bakeries
and stores are also
shown.

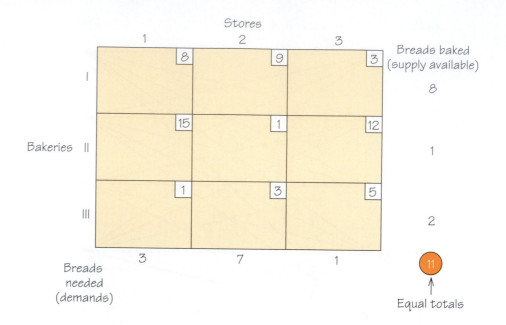

reasonable that the shipping cost for I will be higher than for II when shipping
to that particular store.

However, the costs of shipping may also involve time considerations. (The
distance to a store may be shorter, but it may be that this route is a very slow
one.) Also, it may take extra time for a truck coming from I to park when making the next delivery.

The numbers we use in our diagrams are "aggregate" costs. The nice thing
about what we are doing is that the solution method works independently of the
way the costs are computed or arrived at. These costs (see Figure 4.19) are shown
in the upper-right-hand corner of a cell. Thus the 9 shown in the cell (I, 2) means
that it costs nine units to ship a bread from bakery I to store 2. Our goal will be
to supply the stores with the breads they require from the supplies available at
the bakeries so that the total cost of providing the breads to the stores is as small
as possible (a minimum). ■

The tools for solving transportation problems like these were developed during World War II in conjunction with trying to get supplies from different ports
in the United States to different ports in Europe (mostly the United Kingdom)
in as efficient a manner as possible. (The U.S. ports were like the bakeries, and
the British ports were like the stores that needed the breads.)

We can think of finding a solution to a problem like this as a special kind
of linear-programming problem, because we can express the objective of minimizing the cost using a linear relationship. The constraints that express that the
rim conditions are met can also be expressed using linear equations. However, it
turns out there are algorithms that make it possible to solve problems of this kind

that are rather larger than general linear-programming problems that can be solved by hand. These algorithms are intuitively appealing.

We can divide the problem of finding a solution to a transportation problem into two phases, as we did for general linear-programming problems. First, find a solution that is feasible (that is, a solution that does not violate any of the constraints of the problem). Second, if the current solution is not optimal, we move to a better one. Thus, we will first find a solution that meets the constraints and then try to find an improved solution. If there is no better solution than the one we have, under suitable circumstances we show that there is never a better one. Thus, the solution that we have found is an optimal solution. We will work our way through a simple example as being typical of what is required in general transportation problems.

Let's turn to the table shown in Figure 4.20, where certain numbers have been inserted with circles around them.

Tableau

A table showing costs and rim conditions for a transportation problem is known as a **tableau**.

When we see a circled number such as the 6 in row I, column 2, this means that we plan to ship six breads to store 2 from bakery I. Similarly, the circled number 1 in row III and column 3 means we plan to ship one bread from bakery III to store 3. The cells that have no circled numbers are thought of as

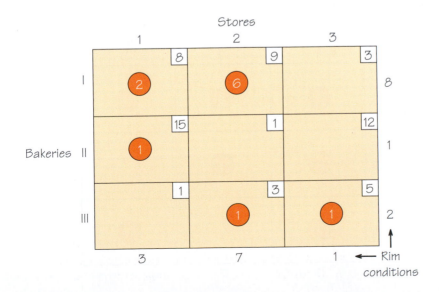

FIGURE 4.20 A possible solution to meeting the needs of three stores for bread from supplies available at three bakeries. The circled numbers show the amounts shipped from the bakeries to the stores.

having zero entries; no breads are being shipped between these stores and these bakeries. Note, for example, that the row sum of the circled numbers in the first row is 8. This means that all the breads available at bakery I are being shipped to some store.

Similarly, the fact that circled entries in column 2 add up to 7 means that all the breads needed by store 2 are being supplied to it. You can verify that all the row sums and column sums add to exactly the numbers that we want to ship from each bakery to each store. Note that 11 breads have been shipped by the bakeries and received by the stores. When this happens, the circled numbers are said to be a *feasible* solution to the problem.

How much will it cost to ship these amounts of breads (see Figure 4.20) to the stores? The number can be computed by multiplying the circled numbers by the cost shown in the associated cell. For example, to ship two breads from bakery I to store 1 costs 2(8) = 16 because the cost associated with the cell in which the 2 appears is 8. The cost of shipping six breads from bakery I to store 2 is 6(9) = 54. To get the total cost of this "shipment plan," we sum all the shipped amounts by the associated costs to get

$$2(8) + 6(9) + 1(15) + 1(3) + 1(5) = 16 + 54 + 15 + 3 + 5 = 93$$

However, at this point we do not know if there is a cheaper way to ship the breads to the stores. Notice that the number of cells with circled numbers is exactly equal to the number of rows m plus the number of columns n minus 1. This is the general pattern with transportation problems. Cells that are used for shipping are circled. On occasion, we ship a zero amount because the procedure works only when $m + n - 1$ cells are circled.

If we look at the pattern of circled numbers in the tableau in Figure 4.21, we see that there is a difficulty even though 11 breads are involved (the sum of all the circled numbers).

FIGURE 4.21 The circled numbers are not a possible solution to this transportation problem because the rim conditions are not satisfied.

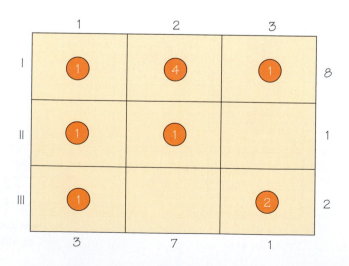

The numbers in the first row add to 6, which means that there will be breads left over at bakery I that have not been shipped. In row 2 the sum of the circled numbers is 2, but this means that something is wrong. How can bakery II, which has a supply of only one bread, ship two breads? Furthermore, column 3 sums to 3, which means that three breads have been shipped to store 1 despite the fact that it only requested one bread! These facts add up to the realization that this assignment of numbers to the cells violates the rules we are requiring. This proposed shipment plan also violates our rule that we are not allowed to circle more than five cells.

How can we find a solution that meets the constraints of the problem (the rim conditions)? We will show two ways to do this. The first is "fast and dirty" but typically does not find a very good solution with which to start. The second (developed in the exercises) usually gives a better "initial" solution but is a little harder to carry out.

This pair of approaches displays a common tension in problem solving: the ease of getting started but requiring more work later on, or more work at the start, which often proves to be a good investment of extra effort because less work is needed to find an optimal solution. If we know in advance the method being chosen to solve a problem, we can often find an example where this particular method does poorly. Mathematicians work hard to find methods that work well on the kinds of problems that come up in genuine applications.

Northwest Corner Rule

The easier approach involves what is called the **Northwest Corner Rule (NCR).** This rule is simple because it is based on the geometry of the table that is involved and does not even look at the costs associated with the cells in the table, which in the long run cannot be a good idea, because these costs come into play when trying to get an optimal solution.

How does the Northwest Corner Rule work? The algorithm carries out the following procedure until exactly one cell remains in the "altered tableau."

1. Locate that cell of the current tableau that is as far to the top and to the left as possible (that is, in the northwest corner). Ship via this cell the smaller of the two rim values (call the value s) associated with the row and column of this cell. (Indicate that this cell is being used by putting a circle around the entry in the tableau.)

2. Cross out the row or column that had rim value s and reduce the other rim value for this cell by s.

3. When a single cell remains, there will be a tie for the rim conditions of both the row and column involved, and this amount is entered into the cell and circled.

Note that it is possible (when there is more than one cell at the start) for there to be a tie when step 1 above is applied. In this case we simultaneously fulfill the rim conditions for a row and column. If this happens we can always choose to cross out, say, the column (not both the row and column) and reduce to 0 the rim condition for the row involved. Now when the algorithm is applied, one has a rim value of 0 for the northwest corner cell. This now requires that 0 be shipped via that cell. Even though this will not change the cost, it is necessary to put a zero in this cell and circle it. Here we have usually designed the examples to avoid ties so as to make it easier to get the essential ideas across.

Applying the Northwest Corner Rule to our original tableau (see Figure 4.19), we get the sequence of tableaux in Figure 4.22 as we cross out the rows or columns, where for clarity the costs associated with the cells are suppressed. The last diagram in the sequence shows the results on the original tableau, with the cost restored. Note that, at the steps in between, the costs played no role. It is a good idea to check that the circled numbers in each row and column really add up to the rim value for that row and column and that exactly $m + n - 1$ cells are filled.

We can now compute the cost of the associated solution that we have found (feasible solution), which obeys the rim conditions. As we did previously, we add up the cost multiplied by the amount shipped for each cell with a circled entry. We get the following calculation:

$$3(8) + 5(9) + 1(1) + 1(3) + 1(5) = 78$$

This shows a cost that is smaller than the solution we found earlier. That solution involved a cost of 93. But is this solution the cheapest one? Since finding this feasible solution did not make use of the costs on the cells, it suggests that it is not very likely.

Improving the Feasible Solution

The next phase of the transportation problem algorithm attempts to answer the question of how to tell if the feasible solution found by using the Northwest Corner Rule is the best. If this solution is not the best, we should be able to find a way to improve it.

Suppose we decided to ship an additional bread from bakery II to store 3. Now, this would violate the fact that we had shipped exactly the right numbers of breads before this new additional shipment, so we have shipped one bread too many from bakery II. We can compensate for this by reducing from 1 to 0 the bread shipped from bakery II to store 2. But this now means that store 2 has not gotten all the breads it needs. We can take care of this by shipping one more bread from bakery III to store 2, but again we now have one extra bread shipped from bakery III. We can compensate for this by reducing the number of breads shipped via cell (III, 3) (that is, from bakery III to store 3). This step will ensure that the rim conditions will hold for the circled numbers. This is because we have located a circuit–(II, 3), (II, 2), (III, 2), (III, 3), (II, 3)–where, if we increase and decrease the breads alternately going around that circuit, we maintain the rim conditions (see Figure 4.23).

Check for yourself that the tableau on the right in Figure 4.23 with the circled entries meets the rim conditions. To the left we show the circuit of cells with

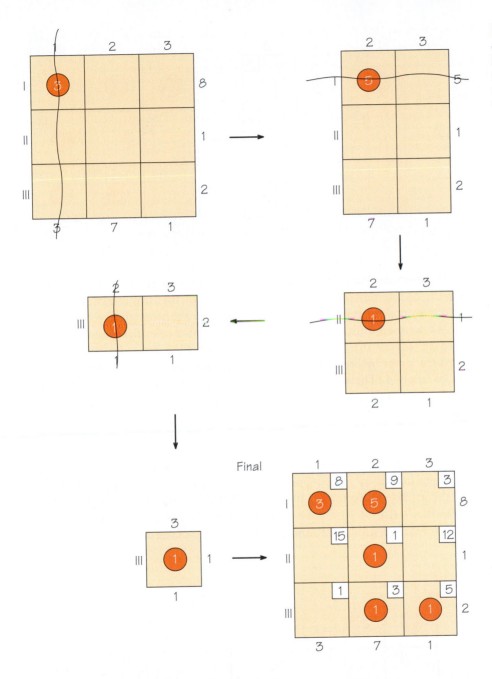

FIGURE 4.22 The construction of an initial solution to a transportation problem using the Northwest Corner Rule.

plus and minus signs (+ and −) where we have increased the amounts in the cells with + and decreased the amounts in the cells with − by 1 unit. (Note that to keep a total of five cells circled, we have set one of the circled cells to 0, because when we reduce the amounts of bread shipped in cells (II, 2) and (III, 3), we get a tie of value 0.)

FIGURE 4.23 An illustration of how to take a current solution to a transportation problem and get an improved cheaper solution that still meets the rim conditions.

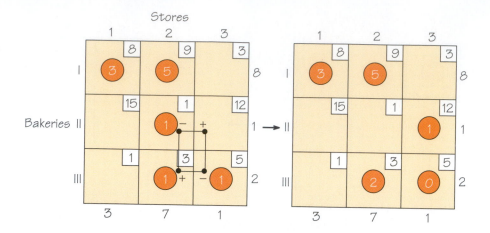

We now have to ask whether this new solution is cheaper or more expensive than the one we started with. We can figure out whether this is a better or worse solution by tracking the costs of moving from the previous solution to the new one.

We went around a circuit where we increased a cost, decreased a cost (because we reduced the number of breads in that cell), increased a cost, and then decreased a cost before coming back to where we started, having traversed a circuit of length 4. The net effect of this collection of cost changes is $+12 - 1 + 3 - 5 = +9$. Thus, these changes, while producing a new feasible solution, give a more costly solution!

Perhaps increasing the amount shipped via a different circuit of cells would be better. Suppose we try the same process for cell (I, 3) (Figure 4.24)—that is, increase the shipping of breads from bakery I to store 3. To see if this is worthwhile, check the circuit formed by shipping more via cell (I, 3). It takes a bit of practice to find the circuit this cell forms. In the case of cell (I, 3), the circuit we get is (I, 3), (I, 2), (III, 2), (III, 3), (I, 3). The cost of moving around this circuit is $+3 - 9 + 3 - 5 = -8$. We will refer to this number as the *indicator value* for this cell.

FIGURE 4.24 Using a cell with a negative indicator value, we can find a cheaper way of meeting the demands from the supplies.

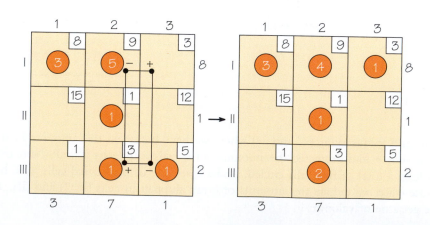

::::::: Indicator Value

The **indicator value of a cell** C (not currently a circled cell) is the cost change associated with increasing or decreasing the amounts shipped in a circuit of cells starting at C. It is computed by summing with alternating signs the costs of the cells in the circuit.

This means that we can lower the cost of shipping breads by using a different pattern of meeting the demands from the supplies. We have computed the saving for shipping one bread more via cell (I, 3), but perhaps we might be able to save even more by shipping even more breads via this cell. To determine whether we could, we look at the circuit that begins at cell (I, 3). To maintain a feasible solution, we have to increase the amounts shipped via some cells of this circuit and decrease the amounts shipped via others. Since we cannot decrease the amount shipped via any cell below zero, the minimum value of any cell that must be reduced is the maximum amount that can be shipped via cell (I, 3). In this case, it means that only one bread can be shipped via cell (I, 3), thereby lowering the cost from the previous solution by 8.

When we looked to improve the solution shown in Figure 4.22, we have now seen that by shipping via cell (I, 3) we can get a better solution. However, there might be several cells in the solution shown in Figure 4.22 that would lead to improvement. Which one should we choose? The answer is that we should adopt a greedy point of view. If there are several cells with a negative indicator value, pick the one that is "most negative" to improve the solution.

Given a current feasible solution (one that satisfies the rim condition), we check each cell that does not have a circled number for improvement if we ship via that cell. If a cell leads to a positive indicator value with the circuit associated with it, no improvement is possible. If a cell has a negative indicator value associated with the circuit for that cell, we can get an improvement. We select as the cell to increase that cell with the largest negative indicator value. We now have a new feasible solution that is cheaper than the one we started with and can repeat our procedure just described starting from this new feasible solution.

It turns out that there was no better cell than (I, 3) (using this greedy approach) to get an improved solution. We will take the current best solution and see if we can improve it further. It turns out that for the current tableau (Figure 4.24), all the cells have a positive indicator except for cell (III, 1):

Indicator for cell (III, 1): $+1 - 3 + 9 - 8 = -1$

Since the minimum of the circled numbers in the cell with a negative label is 2 in cell (III, 2), we can increase by 2 the amount shipped in cell (III, 1) and get a new solution as shown in Figure 4.25.

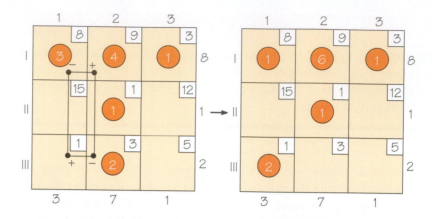

Now, for this tableau, all the empty cells have positive indicator values.

Indicator (II, 1): $+15 - 1 + 9 - 8 = 15$

Indicator (II, 3): $+12 - 3 + 9 - 1 = 17$

Indicator (III, 2): $+3 - 9 + 8 - 1 = 1$

Indicator (III, 3): $+5 - 3 + 8 - 1 = 9$

This means that the current solution is optimal. The cost of this solution is

$$1(8) + 6(9) + 1(3) + 1(1) + 2(1) = 8 + 54 + 3 + 1 + 2 = 68$$

It turns out that if all the cells associated with a feasible solution have positive indicator values, then the solution one has reached is always optimal. (Cells with zero indicator value show that there are other solutions that achieve the same optimal value.)

::::::: Theorem: How to Recognize an Optimal Solution

We are given a transportation problem with m suppliers and n demanders where the amount of the supplies equals the amount of demands. A collection of $m + n - 1$ of circled cells is optimal (that is, the circled cells determine a minimum cost solution) if the indicator value associated with the empty cells is positive. If some indicator cells are positive and some are zero, there are multiple solutions for an optimal value.

This theorem is the analog of the result for linear programming that states that if a corner point is feasible, and if no neighbor of the corner point has a better value of the objective function, then the corner point we are at is already an optimal one. Note that there may be other optimal solutions that use a different number of cells than $m + n - 1$, but we can never do any better in terms of the cheapness of a solution than what we have described above.

For those interested in the exciting fact that one piece of mathematics is often useful for other mathematics, we see an example of that here. The reason an empty cell gives rise to a unique circuit with which we can try to improve the current solution of a transportation problem results from the fact that when an edge not in a tree is added to a tree, it creates a unique circuit (see Chapter 3). Since we have m rows and n columns, a tree associated with a graph on $m + n$ vertices has $m + n - 1$ edges, exactly the number of cells we need to fill in a transportation problem!

4.6 Improving on the Current Solution

We have now described a method guaranteed to find an optimal solution to a transportation problem.

The Stepping Stone Method

The **stepping stone method** consists of taking some feasible solution of a transportation problem and improving this solution, if it is not optimal, by shipping an additional amount using a cell with a negative indicator value.

We will work out another small example to illustrate the technique of applying the Northwest Corner Rule to get an initial solution, and then improving this solution if it is not optimal. Again, we do so by computing the indicator values of the cells and improving the current solution by shipping using a cell with a negative indicator value.

We start with an initial tableau where there are two mines that can supply ore to three companies that extract ore. There are 10 units of ore being mined and the extractors need 10 units to run at full capacity. The initial tableau for the problem is displayed in Figure 4.26.

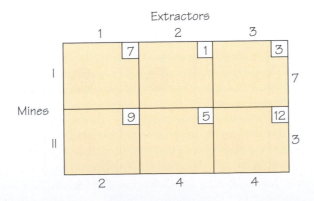

FIGURE 4.26 A transportation problem where two mines supply ore to three companies that extract metal from the ore. The shipping costs are indicated.

FIGURE 4.27 The
Northwest Corner Rule
has been used to find a
possible way to meet the
demands from the
supplies for the tableau
in Figure 4.26.

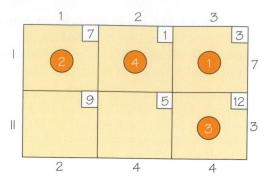

Using the Northwest Corner Rule we find an initial feasible solution as shown in Figure 4.27. When applying the Northwest Corner Rule we eliminate a row or column as follows: first column 1, then column 2, then row I, and we are now left with a single cell. The cost of the feasible solution shown is

$$2(7) + 4(1) + 1(3) + 3(12) = 14 + 4 + 3 + 36 = 57$$

The two empty cells we have are (II, 1) and (II, 2). We compute the indicator value for each of these cells:

Indicator for cell (II, 1): $+9 - 12 + 3 - 7 = -7$
Indicator for cell (II, 2): $+5 - 12 + 3 - 1 = -5$

Since cell (II, 1) has a more negative indicator value, we can reduce the cost more by using that cell. Increasing by 2 (since this is the minimum of circled numbers with negative signs in the computation of the indicator) the amount of metal shipped via cell (II, 1) and cell (I, 3) and reducing by 2 the amount in cells (I, 1) and (II, 3), we obtain the new tableau in Figure 4.28. This has cost

$$4(1) + 3(3) + 2(9) + 1(12) = 4 + 9 + 18 + 12 = 43$$

Note that as a partial check on our work, if we multiply the indicator (−7) by 2, this is −14 and 57 − 43 = 14, so we reduced the cost of our first solution by 14, as expected.

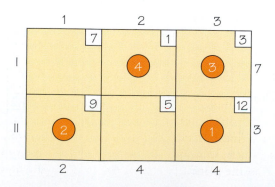

FIGURE 4.28 An
improved solution based
on the negative indicator
value for cell (II, 1) in
Figure 4.27.

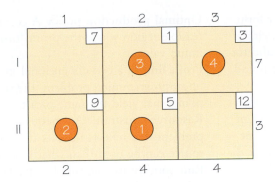

FIGURE 4.29 An improved solution, which turns out to be optimal, based on the negative indicator value for cell (II, 2) in Figure 4.28.

We now repeat this procedure for this new tableau. We must compute the indicator value of cells (I, 1) and (II, 2).

Indicator for cell (I, 1): $+7 - 9 + 12 - 3 = +7$
Indicator for cell (II, 2): $+5 - 12 + 3 - 1 = -5$

Thus, it turns out that we can increase by 1 the amount shipped by (II, 2), and get the tableau in Figure 4.29.

From this tableau, we need to compute the indicator values for the cells (I, 1) and (II, 3). We obtain

Indicator for cell (I, 1): $+7 - 9 + 5 - 1 = +2$
Indicator for cell (II, 3): $+12 - 3 + 1 - 5 = +5$

Not surprisingly, the cell (II, 2) has a positive indicator value because in the previous tableau, that cell was the one that, when we shipped less via it, enabled us to reduce the cost. The fact that both of these indicator values are positive means that the current shipping schedule is an optimal one; that is, using a shipping schedule that ships via only four cells, we cannot find any other solution with the same value.

Transportation problems arise in a very large range of situations including shipping milk from dairies to supermarkets, vegetables to health food stores, and vitamins to your local drug store. The next time you sit down to breakfast, think about how many mathematics problems were solved for you to have a healthy breakfast!

⬤⬤⬤⬤ REVIEW VOCABULARY

Corner point principle The principle states that there is a corner point of the feasible region that yields the optimal solution.

Feasible point A possible solution (but not necessarily the best) to a linear-programming problem. With just two products, we can think of a feasible point as a point on the plane.

Feasible region The set of all feasible points, that is, possible solutions to a linear-programming problem. For problems with just two products, the feasible region is a part of the plane. Also called **feasible set.**

Indicator value of a cell The change in cost due to shipping an increased or decreased amount, using the cells in a transportation tableau that form a circuit consisting of circled cells together with a selected cell that is not circled. When an indicator value is negative, a cheaper solution can be found by shipping using this cell.

Linear programming A set of organized methods of management science used to solve problems of finding optimal solutions, while at the same time respecting certain important constraints. The mathematical formulations of the constraints in linear-programming problems are linear equations and inequalities. Mixture problems are usually solved by some type of linear programming.

Minimum constraint An inequality in a mixture problem that gives a minimum quantity of a product. Negative quantities can never be produced.

Mixture chart A table displaying the relevant data in a linear-programming mixture problem. The table has a row for each product and a column for each resource, for any nonzero minimums, and for the profit.

Mixture problem A problem in which a variety of resources available in limited quantities can be combined in different ways to make different products. It is usually desired to find the way of combining the resources that produces the most profit.

Northwest Corner Rule (NCR) A method for finding an initial but rarely optimal solution to a transportation problem starting from a tableau with rim conditions. The amounts to be shipped between the suppliers and demanders are indicated by circling numbers in the cells in the tableau. The number of cells circled after applying the method will equal the number of rows plus the number of columns minus 1. The method depends on locating at each stage the "northwest corner" of the original tableau or a part of it.

Optimal production policy A corner point of the feasible region where the profit formula has a maximum value.

Profit line In a two-dimensional, two-product, linear-programming problem, the set of all feasible points that yield the same profit.

Resource constraint An inequality in a mixture problem that reflects the fact that no more of a resource can be used than what is available.

Rim conditions The supplies available (listed in a column at the right of a transportation tableau) and demands required (listed in a row at the bottom of a transportation tableau) in a transportation problem. The supplies available are usually taken to exactly meet the demands required.

Simplex method One of a number of algorithms for solving linear-programming problems.

Stepping stone method A method for solving a transportation problem that improves the current solution, when it is not optimal, by increasing the amount shipped using a cell with a negative indicator value.

Tableau A table for a transportation problem indicating the supplies available and demands required, as well as the cost of shipping from a supplier to a demander. The amounts to be shipped from different suppliers to different users are indicated by circled cells in the tableau. The number of such circled cells is always the number of rows plus the number of columns diminished by 1 for the tableau.

Transportation problem A special type of linear-programming problem where we have sources of supplies and users of, or demand for, these supplies. There is a cost to ship an item from a supplier to a demander. The goal is to minimize the total shipping cost to meet the demands from the supplies.

✔ SKILLS CHECK

1. Where do the lines $6x + 2y = 26$ and $2x + 3y = 18$ intersect?

(a) At the point (3, 4)

(b) At the point (6, 2)

(c) At the point (3, 2)

2. Where do the lines $x + 3y = 12$ and $y = 3$ intersect?

(a) At the point (9, 5)
(b) At the point (9, 3)
(c) At the point (3, 3)

3. Which of these points lie in the region
$4x + 3y \leq 24$, $x \geq 0$, $y \geq 0$?

(a) Points (5, 2) and (3, 4)
(b) Points (2, 5) and (3, 4)
(c) Points (5, 2) and (2, 5)

4. A tart requires 3 oz of fruit and 2 oz of dough;
a pie requires 13 oz of fruit and 7 oz of dough.
There are 140 oz of fruit and 90 oz of dough
available. Each tart earns 6 cents profit; each pie
earns 25 cents profit. What are the resource
inequalities of this situation?

(a) $3x + 2y \leq 140$
$13x + 7y \leq 90$
$x \geq 0$, $y \geq 0$
(b) $3x + 13y \leq 140$
$2x + 7y \leq 90$
$x \geq 0$, $y \geq 0$
(c) $3x + 2y \leq 6$
$13x + 7y \leq 25$
$x \geq 0$, $y \geq 0$

5. Producing a bench (x) requires 2 boards, and
producing a table (y) requires 5 boards. There are
25 boards available. What is the resource
constraint for this situation?

(a) $x + y \leq 25$
(b) $2x + 5y \leq 25$
(c) $5x + 2y \leq 25$

6. A tart requires 3 oz of fruit and 2 oz of dough;
a pie requires 13 oz of fruit and 7 oz of dough.
There are 140 oz of fruit and 90 oz of dough
available. Each tart earns 6 cents profit; each pie
earns 25 cents profit. What is the profit formula for
this situation?

(a) $P = 140x + 90y$
(b) $P = 70/3x + 18/5y$
(c) $P = 6x + 25y$

7. Graph the feasible region identified by the
following inequalities:

$2x + 4y \leq 20$

$4x + 2y \leq 16$
$x \geq 0$, $y \geq 0$

Which of these points is *not* in the feasible region
of the graph drawn?

(a) (2, 4)
(b) (1, 1)
(c) (10, 0)

8. Suppose the feasible region has four corners, at
points (0, 0), (4, 0), (0, 3), and (3, 2). If the profit
formula is $3x - 2y$, what is the maximum profit
possible?

(a) $0
(b) $5
(c) $12

9. Suppose the feasible region has four corners, at
points (0, 0), (4, 0), (0, 3), and (3, 2). For which of
these profit formulas is the profit maximized by
producing a mix of products?

(a) $x - $y
(b) $x + $2y
(c) $2x - $y

10. The shaded region in the accompanying
diagram is an example of a region

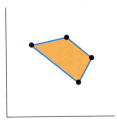

(a) whose area is not bounded.
(b) that is not convex.
(c) that is bounded by straight-line segments.

11. Suppose the feasible region has five corners, at
points (1, 1), (2, 1), (3, 2), (2, 4), (1, 5). If the
profit formula is $5x - $3y, which point
maximizes the profit?

(a) (2, 1)
(b) (3, 2)
(c) (1, 5)

12. Suppose the feasible region has five corners, at points (1, 1), (2, 1), (3, 2), (2, 4), (1, 5). If the profit formula is $5x − $3y, what is the maximum profit possible?

(a) $13

(b) $10

(c) $9

13. Suppose the feasible region has five corners, at points (1, 1), (2, 1), (3, 2), (2, 4), (1, 5). Which of these points is *not* in the feasible region?

(a) (1, 3)

(b) (2, 2)

(c) (0, 0)

14. The corner point method cannot be applied to find the optimal answer for the value of the profit $P = 3x + 7y$, where the feasible region is shown in the diagram, because the feasible region

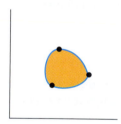

(a) has "infinite" area.

(b) is nonconvex.

(c) is bounded by curves rather than by straight-line segments.

15. Consider the feasible region identified by the inequalities $x \geq 0$, $y \geq 0$, $3x + y \leq 10$, $x + 2y \leq 5$.

Which point is *not* a corner of the region?

(a) (0, 2.5)

(b) (3, 1)

(c) (5, 0)

16. How does the line representing the maximum feasible profit intersect the feasible region?

(a) No points of intersection

(b) Only one point of intersection

(c) At least one point of intersection, and sometimes more than one point of intersection

17. When the Northwest Corner Rule is applied to the accompanying transportation problem tableau, the cells that remain empty are

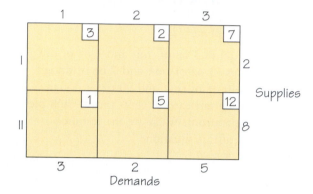

(a) cell (II, 2) and cell (I, 2).

(b) cell (I, 2) and cell (II, 3).

(c) cell (I, 2) and cell (I, 3).

18. The circled cells in the accompanying tableau give a solution that satisfies the rim conditions. The cost associated with this solution is

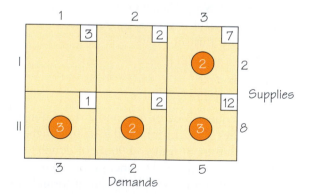

(a) 57.

(b) 63.

(c) 62.

19. The indicator value associated with cell (I, 2) of the accompanying tableau is

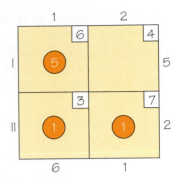

(a) −6.

(b) +6.

(c) 0.

20. The circled cells in the accompanying tableau satisfy the rim conditions. When the indicator value for cell (I, 2) is computed,

(a) the result being positive, the current solution is optimal.

(b) the result being positive, this tableau has no minimal cost solution.

(c) the result being negative means this tableau does not give a minimal cost solution.

EXERCISES   ■ Challenge   ◆ Discussion

Graphing

1. Using intercepts, the points where the lines cross the axes, graph each line.

(a) $2x + 3y = 12$ (d) $7x + 4y = 42$

(b) $3x + 5y = 30$ (e) $x = -3$

(c) $4x + 3y = 24$ (f) $y = 6$

2. Graph both lines on the same axes. Put a dot where the lines intersect. Use algebra to find the x- and y-coordinates of the point of intersection.

(a) $4x + 3y = 18$ and $x = 0$

(b) $5x + 3y = 45$ and $y = -5$

3. Graph both lines on the same axes. Put a dot where the lines intersect. Use algebra to find the x- and y-coordinates of the point of intersection.

(a) $x + y = 10$ and $x + 2y = 14$

(b) $y - 2x = 0$ and $x = 4$

Resource-Constraint Inequalities

4. Graph the line and half-plane corresponding to the inequality, a typical constraint from a mixture problem.

(a) $x \geq 8$ (c) $5x + 3y \leq 15$

(b) $y \geq 5$ (d) $4x + 5y \leq 30$

5. Graph the line and half-plane corresponding to the inequality, a typical constraint from a mixture problem.

(a) $x \geq 4$ (c) $3x + 2y \leq 18$

(b) $y \geq 9$ (d) $7x + 2y \leq 42$

In Exercises 6–8, for each description, write one or more appropriate resource-constraint inequalities. The unknown to use for each product is given in parentheses.

6. (a) One bridesmaid's bouquet (x) requires 2 roses, and one corsage (y) requires 4 roses. There are 28 roses available.

(b) Maintaining a large tree (x) takes 2 hours of pruning time and 30 minutes of shredder time; maintaining a small tree (y) takes 30 minutes of pruning time and 15 minutes of shredder time. There are 40 hours of pruning time and 2 hours of shredder time available.

7. (a) Manufacturing one package of hot dogs (x) requires 6 oz of beef, and manufacturing one

package of bologna (y) requires 4 oz of beef. There are 300 oz of beef available.
(b) It takes 30 ft of 12-in. board to make one bookcase (x); it takes 72 ft of 12-in. board to make one table (y). There are 420 ft of 12-in. board available.

8. Manufacturing one salami (x) requires 12 oz of beef and 4 oz of pork. Manufacturing one bologna (y) requires 10 oz of beef and 3 oz of pork. There are 40 lb of beef and 480 oz of pork available.

Graphing the Feasible Region

Graph the feasible region, label each line segment bounding it with the appropriate equation, and give the coordinates of every corner point.

9. $x \geq 0; y \geq 0; x + 2y \leq 12$

10. $x \geq 0; y \geq 0; 2x + y \leq 10$

11. $x \geq 0; y \geq 0; 2x + 5y \leq 60$

12. $x \geq 10; y \geq 0; 3x + 5y \leq 120$

13. $x \geq 0; y \geq 4; x + y \leq 20$

14. $x \geq 2; y \geq 6; 3x + 2y \leq 30$

Finding Points in the Feasible Region

Determine whether the points (2, 4) and/or (10, 6) are points of the given feasible regions of:

15. Exercises 9, 11, and 13.

16. Exercises 10, 12, and 14.

17. In the toy problem, x represents the number of skateboards and y the number of dolls. Using the version of that problem whose feasible region is presented in Figure 4.3b, with the profit formula $\$2.30x + \$3.70y$, write a sentence giving the maximum profit and describing the production policy that gives that profit.

18. In the toy problem, x represents the number of skateboards and y the number of dolls. Using the version of that problem whose feasible region is presented in Figure 4.3b, with the profit formula $\$5.50x + \$1.80y$, write a sentence giving the maximum profit and describing the production policy that gives that profit.

19. Graph both lines on the same axes. Put a dot where the lines intersect. Use algebra to find the x- and y-coordinates of the point of intersection.
(a) $5x + 4y = 22$ and $5x + 10y = 40$
(b) $x + y = 7$ and $3x + 4y = 24$

Graphing the Feasible Region

Graph the feasible region, label each line segment bounding it with the appropriate equation, and give the coordinates of every corner point.

20. $x \geq 0; y \geq 0; 3x + y \leq 9; x + 2y \leq 8$

21. $x \geq 0; y \geq 0; 2x + y \leq 4; 4x + 4y \leq 12$

22. $x \geq 0; y \geq 2; 5x + y \leq 14; x + 2y \leq 10$

23. $x \geq 4; y \geq 0; 5x + 4y \leq 60; x + y \leq 13$

24. Determine whether the points (4, 2) and/or (1, 3) are points of the given feasible regions of Exercises 21 and 23.

Linear Programming and Its Geometry

25. Find the maximum value of P where $P = 3x + 2y$ subject to the constraints $x \geq 3, y \geq 2, x + y \leq 10, 2x + 3y \leq 24$.

26. Find the maximum value of P where $P = 3x - 2y$ subject to the constraints $x \geq 2, y \geq 3, 3x + y \leq 18, 6x + 4y \leq 48$.

27. Find the maximum value of P where $P = 5x + 2y$ subject to the constraints $x \geq 2, y \geq 4, x + y \leq 10$.

28. Given profit $P = 21x + 11y$ subject to the constraints $x \geq 0, y \geq 0, 7x + 4y \leq 13$:
(a) Graph the feasible region.
(b) Determine a corner point where there is an optimal solution.

(*Warning:* The corner point where the optimal solution occurs will not have integer values for both x and y.)

29. (a) Referring to Exercise 28, use the usual rounding rule to round the x-coordinate and the y-coordinate of the point where the optimal linear-programming solution occurs. Call the point with these coordinates Q.

(b) Determine if Q's coordinates define a feasible point by checking them against the constraints.
(c) Evaluate the profit value P at point Q. How does the profit value compare with the point where the optimal value occurred in Exercise 28?
(d) Let R be the point with coordinates $(0, 3)$. Is R in the feasible region? Evaluate P at point R and compare the result with the answer at Q and where the optimum linear-programming value occurred.
(e) Explain the significance of the situation here for solving maximization problems where $P = ax + by$ (a and b are known in advance) is subject to linear constraints but where the variables must be nonnegative integers rather than arbitrary nonnegative decimal numbers.

Mixture Problems and Linear Programming

Exercises 30–41 each have several steps leading to a complete solution to a mixture problem. Practice a specific step of the solution algorithm by working out just that step for several problems. The steps are:

(a) Make a mixture chart for the problem.
(b) Using the mixture chart, write the profit formula and the resource- and minimum-constraint inequalities.
(c) Draw the feasible region for those constraints and find the coordinates of the corner points.
(d) Evaluate the profit information at the corner points to determine the production policy that best answers the question.
(e) (Requires technology) Compare your answer with the one you get from running the same problem on a simplex algorithm computer program.

30. A clothing manufacturer has 600 yd of cloth available to make shirts and decorated vests. Each shirt requires 3 yd of material and provides a profit of $5. Each vest requires 2 yd of material and provides a profit of $2. The manufacturer wants to guarantee that under all circumstances there are minimums of 100 shirts and 30 vests produced. How many of each garment should be made to maximize profit? If there are no minimum quantities, how, if at all, does the optimal production policy change?

31. A car maintenance shop must decide how many oil changes and how many tune-ups can be scheduled in a typical week. The oil change takes 20 min, and the tune-up requires 100 min. The maintenance shop makes a profit of $15 on an oil change and $65 on a tune-up. What mix of services should the shop schedule if the typical week has available 8000 min for these two types of services? How, if at all, do the maximum profit and optimal production policy change if the shop is required to schedule at least 50 oil changes and 20 tune-ups?

32. A clerk in a bookstore has 90 min at the end of each workday to process orders received by mail or on voice mail. The store has found that a typical mail order brings in a profit of $30 and a typical voice-mail order brings in a profit of $40. Each mail order takes 10 min to process and each voice-mail order takes 15 min. How many of each type of order should the clerk process? How, if at all, do the maximum profit and optimal processing policy change if the clerk must process at least three mail orders and two voice-mail orders?

33. In a certain medical office, a routine office visit requires 5 min of doctors' time and a comprehensive office visit requires 25 min of doctors' time. In a typical week, there are 1800 min of doctors' time available. If the medical office clears $30 from a routine visit and $50 from a comprehensive visit, how many of each should be scheduled per week? How, if at all, do the maximum profit and optimal production policy change if the office is required to schedule at least 20 routine visits and 30 comprehensive ones?

34. A bakery makes 600 specialty breads—multigrain or herb—each week. Standing orders from restaurants are for 100 multigrain breads and 200 herb breads. The profit on each multigrain bread is $8 and on herb bread, $10. How many breads of each type should the bakery make in order to maximize profit? How, if at all, do the maximum profit and optimal production policy change if the bakery has no standing orders?

35. A student has decided that passing a mathematics course will, in the long run, be twice as valuable as passing any other kind of course. The student estimates that to pass a typical math course will require 12 hr a week to study and do homework. The student estimates that any other course will require only 8 hr a week. The student has available 48 hr for study per week. How many of each kind of course should the student take? (*Hint:* The profit could be viewed as 2 "value points" for passing a math course and 1 "value point" for passing any other course.) How, if at all, do the maximum value and optimal course mix change if the student decides to take at least two math courses and two other courses?

Exercises 36–41 require finding the point of intersection of two lines, each corresponding to a resource constraint.

36. The firm websareus creates and maintains Web sites for client companies. There are two types of Web sites: "Hot" sites change their layout frequently but keep their content for long times; "cool" sites keep their layout for a while but frequently change their content. To maintain a hot site requires 1.5 hr of layout time and 1 hr for content changes. To maintain a cool site requires 1 hr of layout time and 2 hr for content changes. Every day, websareus has available 12 hr for layout changes and 16 hr for content changes. Net profit is $50 for a set of changes on a hot site and $250 for a set of changes on a cool site. In order to maximize profit, how many of each type of site should websareus maintain daily? How, if at all, do the maximum profit and optimal policy change if the company must maintain at least two hot and three cool sites daily?

37. A paper recycling company uses scrap cloth and scrap paper to make two different grades of recycled paper. A single batch of grade A recycled paper is made from 25 lb of scrap cloth and 10 lb of scrap paper, whereas one batch of grade B

recycle paper is made from 10 lb of scrap cloth and 20 lb of scrap paper. The company has 100 lb of scrap cloth and 120 lb of scrap paper on hand. A batch of grade A paper brings a profit of $500, whereas a batch of grade B paper brings a profit of $250. What amounts of each grade should be made? How, if at all, do the maximum profit and optimal production policy change if the company is required to produce at least one batch of each type?

38. Jerry Wolfe has a 100-acre farm that he is dividing into one-acre plots, on each of which he builds a house. He then sells the house and land. It costs him $20,000 to build a modest house and $40,000 to build a deluxe house. He has $2,600,000 to cover these costs. The profits are $25,000 for a modest house and $60,000 for a deluxe house. How many of each type of house should he build to maximize profit? How, if at all, do the maximum profit and optimal production policy change if Wolfe is required to build at least 20 of each type of house?

39. The maximum production of a soft-drink bottling company is 5000 cartons per day. The company produces regular and diet drinks, and must make at least 600 cartons of regular and 1000 cartons of diet per day. Production costs are $1.00 per carton of regular and $1.20 per carton of diet. The daily operating budget is $5400. How many cartons of each type of drink should be produced if the profit is $0.10 per regular and $0.11 per diet? How, if at all, do the maximum profit and optimal bottling policy change if the company has no minimum required production?

40. Wild Things raises pheasants and partridges to restock the woodlands and has room to raise 100 birds during the season. The cost of raising one bird is $20 per pheasant and $30 per partridge. The Wildlife Foundation pays Wild Things for the birds; the latter clears a profit of $14 per pheasant and $16 per partridge. Wild Things has $2400 available to cover costs. How many of each type of bird should they raise? How, if at all, do the maximum profit and optimal restocking policy change if Wild Things is required to raise at least 20 pheasants and 10 partridges?

41. Lights Afire makes desk lamps and floor lamps, on which the profits are $2.65 and $4.67, respectively. The company has 1200 hr of labor and $4200 for materials each week. A desk lamp takes 0.8 hr of labor and $4 for materials; a floor lamp takes 1.0 hr of labor and $3 for materials. What production policy maximizes profit? How, if at all, do the maximum profit and optimal production policy change if Lights Afire wants to produce at least 150 desk lamps and 200 floor lamps per week?

In Exercises 42–45, there are more than two products in the problem. Although you cannot solve these problems using the two-dimensional graphical method, you can follow these steps:

(a) Make a mixture chart for each problem.
(b) Using the mixture chart, write the resource- and minimum-constraint inequalities. Also write the profit formula.
(c) (Requires technology) If you have a simplex method program available, run the program to obtain the optimal production policy.

42. A toy company makes three types of toys, each of which must be processed by three machines: a shaper, a smoother, and a painter. Each Toy A requires 1 hr in the shaper, 2 hr in the smoother, and 1 hr in the painter, and brings in a $4 profit. Each Toy B requires 2 hr in the shaper, 1 hr in the smoother, and 3 hr in the painter, and brings in a $5 profit. Each Toy C requires 3 hr in the shaper, 2 hr in the smoother, and 1 hr in the painter, and brings in a $9 profit. The shaper can work at most 50 hr per week, the smoother 40 hr, and the painter 60 hr. What production policy would maximize the toy company's profit?

43. A rustic furniture company handcrafts chairs, tables, and beds. It has three workers, Chris, Sue, and Juan. Chris can work only 80 hr per month, but Sue and Juan can each put in 200 hr. Each of

these artisans has special skills. To make a chair takes 1 hr of Chris's time, 3 from Sue, and 2 from Juan. A table needs 3 hr from Chris, 5 from Sue, and 4 from Juan. A bed requires 5 hr from Chris, 4 from Sue, and 8 from Juan. Even artisans are concerned about maximizing their profit, so what product mix should they stick with if they get $100 profit per chair, $250 per table, and $350 per bed?

44. A candy manufacturer has 1000 lb of chocolate, 200 lb of nuts, and 100 lb of fruit in stock. The Special Mix requires 3 lb of chocolate, 1 lb each of nuts and fruit, and it brings in $10. The Regular Mix requires 4 lb of chocolate, 0.5 lb of nuts, and no fruit, and brings in $6. The Purist Mix requires 5 lb of chocolate, no nuts or fruit, and brings in $4. How many boxes of each type should be produced to maximize profit?

45. A gourmet coffee distributor has on hand 17,600 lb of African coffee, 21,120 oz of Brazilian coffee, and 12,320 oz of Colombian coffee. It sells four blends–Excellent, Southern, World, and Special–on which it makes these per-pound profits, respectively: $1.80, $1.40, $1.20, and $1.00. One pound of Excellent is 16 oz of Colombian; it is not a blend at all. One pound of Southern consists of 12 oz of Brazilian and 4 oz of Colombian. One pound of World requires 6 oz of African, 8 of Brazilian, and 2 of Colombian. One pound of Special is made up of 10 oz of African and 6 of Brazilian. What product mix should the gourmet coffee distributor prepare in order to maximize profit?

In Exercises 46 and 47, use the fact that the corner point approach can also solve minimization problems to minimize the given expression for cost C.

46. Minimize C given by $C = 4x + 7y$ over the feasible region for Exercise 25.

47. Minimize C given by $C = 3x + 11y$ over the feasible region for Exercise 26.

48. Show by example that a feasible region that has the nonnegativity constraints $x \geq 0$, $y \geq 0$, and $x + y \leq 0.5$ can have no feasible points with integer coordinates other than $(0, 0)$.

49. Courtesy Calls makes telephone calls for businesses and charities. A profit of $0.50 is made for each business call and $0.40 for each charity call. It takes 4 min (on average) to make a business call and 6 min (on average) to make a charity call. If there are 240 min of calling time to be distributed each day, how should that time be spent so that Courtesy Calls makes a maximum profit? What changes, if any, occur in the maximum profit and optimal production policy if every day they must make at least 12 business and 10 charity calls?

50. A refinery mixes high-octane and low-octane fuels to produce regular and premium gasolines. The profits per gallon on the two gasolines are $0.30 and $0.40, respectively. One gallon of premium gasoline is produced by mixing 0.5 gal of each of the fuels. One gallon of regular gasoline is produced by mixing 0.25 gal of high octane with 0.75 gal of low octane. If there are 500 gal of high octane and 600 gal of low octane available, how many gallons of each gasoline should the refinery make? How, if at all, do the maximum profit and optimal production policy change if the refinery is required to produce at least 100 gal of each gasoline?

51. A toy manufacturer makes bikes, for a profit of $12, and wagons, for a profit of $10. To produce a bike requires 2 hr of machine time and 4 hr painting time. To produce a wagon requires 3 hr machine time and 2 hr painting time. There are 12 hr of machine time and 16 hr of painting time available per day. How many of each toy should be produced to maximize profit? How, if at all, do the maximum profit and optimal production policy change if the manufacturer must daily produce at least two bikes and two wagons?

52. Apply the Northwest Corner Rule to the following transportation problem tableaux.

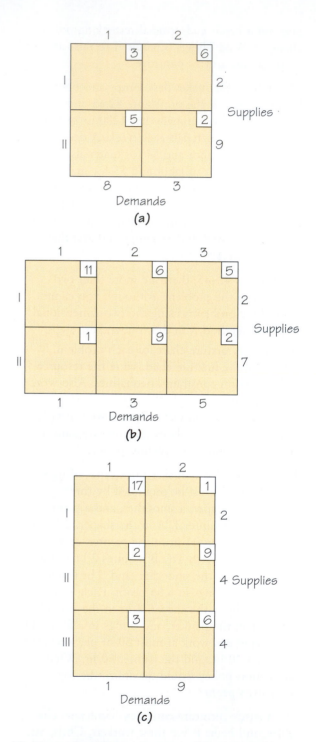

(a)

(b)

(c)

(d) For each tableau, give a possible real-world setting for the problem.

(e) For each tableau, find the cost of shipping using the cells that were circled when you used the Northwest Corner Rule.

53. (a) Apply the Northwest Corner Rule to the accompanying tableau, which arose from meeting the demands of fruit stands for peaches from supplies available from local orchards.

(b) Determine the cost associated with the solution you found.
(c) Compute the indicator value for each noncircled cell.

54. The accompanying tableau represents the shipping costs and supply-and-demand constraints for supplies of purified water to be shipped to companies that resell the water to office buildings.

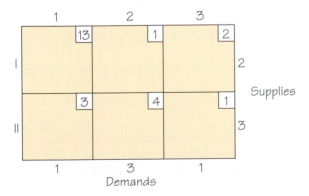

(a) Find the Northwest Corner Rule initial solution.
(b) Determine the indicator value for each noncircled cell.

(c) Is the current solution optimal? If not, find a cheaper solution.

55. The accompanying tableau arose by applying the Northwest Corner Rule.

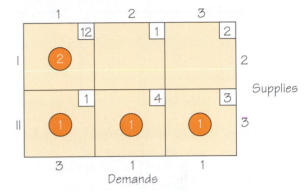

The accompanying graph was constructed so that there is one edge for each circled vertex in the tableau above.

(a) Verify that the graph is a tree.
(b) Show that for each empty cell in the tableau, adding to the graph the unique edge that corresponds to the empty cell creates one circuit.
(c) Show that this circuit corresponds to the one used to find the indicator value of the empty cell.

56. (a) For each row of the following tableau, compute the minimum cost for that row. Now select the row R that among all the rows has the smallest row minimum. In a way similar to the Northwest Corner Rule, use the cheapest cell in row R and ship as much as possible via that cell, crossing out a row or a column, and adjust the rim conditions and repeat the process. This is known as the *minimum row entry method*. Use the minimum row entry method to find an initial solution to the following transportation problem, which shows the costs of returning rental cars from cities that have more cars than necessary to cities that have too few cars.

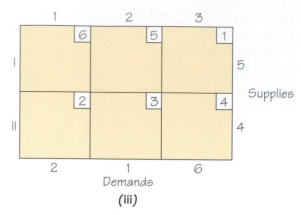

(iii)

(b) Compute the cost of the solution you find using the minimum row entry method.

(c) Compare the cost found in part (b) with the cost of the initial solution obtained using the Northwest Corner Rule.

57. (a) For each of the following tableaux, find an initial solution using the Northwest Corner Rule.

(b) If the solution you find using the Northwest Corner Rule is not optimal, then apply the stepping stone algorithm to find an optimal solution.

58. (a) Apply the Northwest Corner Rule to the following tableau.

(b) Determine the cost associated with the solution you found.

(c) Compute the indicator value for each noncircled cell.

✎ WRITING PROJECTS

1. Interview a local businessperson who is in charge of deciding the product mix for a business. Must this business take into consideration situations other than minimum and resource constraints? If so, what are these considerations? Find out what methods the person uses to make production policy decisions. Is linear programming used? Are other methods used? If so, what are they? Write a report of your findings, and add some of your own conclusions about the usefulness of linear programming for this business.

2. In economics, it is often useful to distinguish between a firm that has a monopoly (for example, is the only supplier of a product) and firms that supply only a small share of the market. How would the presence of a monopoly affect the relation between production and price? Would the presence of a monopoly tend to ensure the fixed-profit assumption of linear programming, or would it make it more likely that the interplay of supply and demand would have to be considered in order to have a truly realistic model? Write an essay addressing these issues.

SUGGESTED READINGS

ANDERSON, DAVID R., DENNIS J. SWEENEY, and THOMAS A. WILLIAMS. *An Introduction to Management Science: Quantitative Approaches to Decision Making*, West, St. Paul, Minn., 1985. A business management text with seven chapters on linear programming.

DOLAN, ALAN, and JOAN ALDUS, *Networks and Algorithms: An Introductory Approach*, Wiley, New York, 1993. A graph theoretical approach to network optimization problems, including the transportation problem.

GASS, SAUL I. *Decision Making, Models, and Algorithms*, Krieger, Melbourne, Fla., 1991. This book demonstrates how to use linear programming and related ideas to solve a variety of industrial and governmental problems.

HARDWICK, I., *Decision and Discrete Mathematics*, Albion Publishing, Chichester, England, 1996. A survey of situations that can be modeled using graphs in the area of operations research. It treats both the simplex method for solving linear-programming problems and the transportation problem.

Note: Simplex software can be found in *Maple* (keyword is *simplex*), *Mathematica* (keyword is *Linear Programming*), in both *Lotus 1-2-3* and *MSExcel* via *Solver,* and in other software packages, especially those intended for quantitative mathematics courses focusing on business applications.

SUGGESTED WEB SITES

www.informs.org This Web site is maintained by the Institute for Operations Research and the Management Sciences, the main professional organization in these fields in the United States. It contains information on (and/or links to) news items about operations research and management science and employment opportunities and summer internships; it also has a student newsletter. Much of the material is written in a nontechnical style.

www.hsor.org/what_is_or.cfm?name= linear_programming This Web page discusses how linear programming fits into the broader subject of operations research.

www-gap.dcs.st-and.ac.uk/~history/ Mathematicians/Dantzig_George.html This site contains biographical information about George Dantzig, who, by developing the simplex method, greatly expanded the use and applicability of linear programming.

www-unix.mcs.anl.gov/otc/Guide/faq/ linear-programming-faq.html This site is the "frequently asked questions" section of an online newsgroup for people interested in linear programming.

en.wikipedia.org/wiki/Linear_programming This Web page outlines the theory of linear programming.

fraud

ncerned are you that you will encounter problems
ht prevent you from voting?

- Very 10%
- Somewhat 19%
- Not very 25%

dwork: Oct 28-30
ple: 920
gin of error +/- 4%

SOURCE: NY Times/CBS poll

Statistics: The Science of Data

What CDs are big sellers this week? When you buy a CD, the checkout scanner probably reports your choice to a company that tallies sales and reports the winners. Are there genetic differences between two related types of cancer? To find out, biologists use "microarrays" to report the activity of thousands of genes at once. Checkout scanners and microarrays produce immense amounts of *data,* numerical facts. So do opinion polls, medical studies, and even the sports pages. *Statistics* is the science of collecting, organizing, and interpreting data.

Chapters 5 and 6 concern *data analysis,* the art of seeing what data say. We learn from data by making graphs and doing calculations, guided by principles that help us decide what graphs to make, what to look for in our graphs, and what calculations are helpful based on what we see. Sometimes we want to know more: An opinion poll or a medical study looks at only some people, but we want conclusions that apply to all voters or all patients. This is called *statistical inference,* because we infer conclusions about a large group from data on a small part of the group. Chapter 7 discusses inference from beginning to end: from how to produce data when we have inference in mind to how to say just how much confidence we can have in our conclusions. Confidence, uncertainty, risk, chance—the mathematics that describes all these ideas is *probability theory,* the topic of Chapter 8. Probability is the mathematics behind statistical inference, but that's just a small part of its usefulness. ■

Exploring Data: Distributions

A flood of data is a prominent feature of modern society. Data are essential for making decisions in almost every area of life and work. Like other great floods, the flood of numbers threatens to overwhelm us. We must control the flood by careful organization and interpretation. A corporate database, for example, contains an immense volume of data—on employees, sales, inventories, customer accounts, equipment, taxes, and other topics. These data are useful only if we can organize them and present them so that their meaning is clear. The penalties for ignoring data can be severe—several banks have suffered billion-dollar losses from unauthorized trades in financial markets by their employees, trades that were hidden in a mass of data that the banks' management did not examine carefully.

This chapter and the next show you how to use graphs and numerical summaries to work with data. Always remember that although knowing which tools to use and how to use them is important, interpreting your work is even more important. Data come from the real world, and in the end, your goal is to use data to learn something about the real world.

Any set of data contains information about some group of *individuals*. The information is organized in *variables*.

::::::: Individuals and Variables

Individuals are the objects described by a set of data. Individuals may be people, but they may also be animals or things.

A **variable** is any characteristic of an individual. A variable can take different values for different individuals.

179

FIGURE 5.1 Part of a data set displayed by the Excel spreadsheet program.

	A	B	C	D	E	F
1	SEX	HAND	HEIGHT	STUDY	COINS	
2	F	L	65	200	50	
3	M	L	72	30	35	
4	M	R	62	95	35	
5	F	L	64	120	0	
6	M	R	63	220	0	
7	F	R	58	60	76	
8	F	R	67	150	215	
9						

Sheet1 / Sheet2 / Sheet3

EXAMPLE 1 Data from a Student Questionnaire

Figure 5.1 is a small part of a data set that describes the students in a large statistics class. The data come from anonymous responses to a class questionnaire. Each row records data on one *individual,* that is, one student. Each column contains the values of one *variable* for all the individuals. There are five variables. Sex (female or male) and handedness (left-handed or right-handed) are variables that classify the individuals but don't take numerical values. The remaining three variables do take numerical values. They are height in inches, minutes spent studying on a typical weeknight, and "How many cents in coins (not bills) are you carrying?"

Most data tables follow this format—each row is an individual, and each column is a variable. This data set appears in a *spreadsheet* program that has rows and columns ready for your use. Spreadsheets are commonly used to enter and transmit data, and spreadsheet programs also have functions for basic statistics.

Knowing the context of the data—that these are student responses to a class questionnaire—helps us make sense of them. For example, one student claimed to study 30,000 minutes on a typical night. We know that this is a joke. ■

Statistical tools and ideas help us examine data in order to describe their main features. This examination is called **exploratory data analysis.** Like an explorer crossing unknown lands, we want first to simply describe what we see. In this chapter and the next, we use both numbers and graphs to explore data. Here are two principles that provide the tactics for exploratory analysis of data.

Exploring Data

1. Begin by examining each variable by itself. Then move on to study the relationships among the variables.
2. Begin with a graph or graphs. Then add numerical summaries of specific aspects of the data.

These principles also organize the material in Chapters 5 and 6. In this chapter, we look at data on a single variable. Chapter 6 moves on to relations among several variables. In each chapter, we first display data in graphs, then add numerical summaries.

5.1 Displaying Distributions: Histograms

Data analysis begins with graphical displays of the values of a single variable. For example, you may want to compare the study times claimed by female and male students. Because individual study times vary so much, we are interested in the *distribution* of study time for female and male students.

Distribution of a Variable

The **distribution** of a variable tells us what values the variable takes and how often it takes these values.

Numerical variables often take many values. A graph of the distribution is clearer if nearby values are grouped together. The most common graph of the distribution of one numerical variable is a **histogram.**

Making a Histogram

EXAMPLE 2

Every 10 years, the Census Bureau tries to contact every household in the United States. One of the most striking findings of the 2000 census was the growth of the Hispanic population of the United States. Table 5.1 presents the percent of adult (age 18 and over) residents in each of the 50 states who identified themselves in the 2000 census as "Spanish/Hispanic/Latino." The *individuals* in this data set are the 50 states. The *variable* is the percent of Hispanics in a state's population. To make a histogram of the distribution of this variable, proceed as follows:

Step 1. Choose the classes. Divide the range of the data into classes of equal width. The data in Table 5.1 range from 0.7 to 42.1, so we decide to choose these classes:

$$0.0 \leq \text{percent Hispanic} < 5.0$$
$$5.0 \leq \text{percent Hispanic} < 10.0$$
$$\vdots$$
$$40.0 \leq \text{percent Hispanic} < 45.0$$

Be sure to specify the classes precisely so that each individual falls into exactly one class. A state with 4.9% Hispanic residents would fall into the first class, but a state with 5.0% falls into the second.

| TABLE 5.1 | Percent of Adult Population of Hispanic Origin, by State (2000) | | | | |

State	Percent	State	Percent	State	Percent
Alabama	1.5	Louisiana	2.4	Ohio	1.9
Alaska	4.1	Maine	0.7	Oklahoma	5.2
Arizona	25.3	Maryland	4.3	Oregon	8.0
Arkansas	2.8	Massachusetts	6.8	Pennsylvania	3.2
California	32.4	Michigan	3.3	Rhode Island	8.7
Colorado	17.1	Minnesota	2.9	South Carolina	2.4
Connecticut	9.4	Mississippi	1.3	South Dakota	1.4
Delaware	4.8	Missouri	2.1	Tennessee	2.0
Florida	16.8	Montana	2.0	Texas	32.0
Georgia	5.3	Nebraska	5.5	Utah	9.0
Hawaii	7.2	Nevada	19.7	Vermont	0.9
Idaho	7.9	New Hampshire	1.7	Virginia	4.7
Illinois	10.7	New Jersey	13.3	Washington	7.2
Indiana	3.5	New Mexico	42.1	West Virginia	0.7
Iowa	2.8	New York	15.1	Wisconsin	3.6
Kansas	7.0	North Carolina	4.7	Wyoming	6.4
Kentucky	1.5	North Dakota	1.2		

Step 2. Count the individuals in each class. Here are the counts:

Class	Count	Class	Count	Class	Count
0.0 to 4.9	27	15.0 to 19.9	4	30.0 to 34.9	2
5.0 to 9.9	13	20.0 to 24.9	0	35.0 to 39.9	0
10.0 to 14.9	2	25.0 to 29.9	1	40.0 to 44.9	1

Step 3. Draw the histogram. First mark the scale for the variable whose distribution you are displaying on the horizontal axis. That's the percent of a state's population who are Hispanic. The scale runs from 0 to 45 because that is the span of the classes we chose. The vertical axis contains the scale of counts. Each bar represents a class. The base of the bar covers the class, and the bar height is the class count. There is no horizontal space between the bars unless a class is empty, so that its bar has height zero. Figure 5.2 is our histogram. ■

The bars of a histogram should cover the entire range of values of a variable. When the possible values of a variable have gaps between them, extend the bases of the bars to meet halfway between two adjacent possible values. For example,

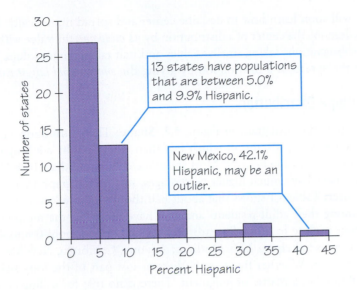

FIGURE 5.2 Histogram of the percent of Hispanics among the adult residents of the states.

13 states have populations that are between 5.0% and 9.9% Hispanic.

New Mexico, 42.1% Hispanic, may be an outlier.

in a histogram of the ages in years of university faculty, the bars representing 25 to 29 years and 30 to 34 years would meet at 29.5.

Our eyes respond to the *area* of the bars in a histogram. Because the classes are all the same width, area is determined by height and all classes are fairly represented. There is no one right choice of the classes in a histogram. Too few classes will give a "skyscraper" graph, with all values in a few classes with tall bars. Too many will produce a "pancake" graph, with most classes having one or no observations. Neither choice will give a good picture of the shape of the distribution. You must use your judgment in choosing classes to display the shape. Statistics software will choose the classes for you. The computer's choice is usually a good one, but you can change it if you want.

5.2 Interpreting Histograms

Making a statistical graph is not an end in itself. The purpose of the graph is to help us understand the data. After you make a graph, always ask, "What do I see?" Once you have displayed a distribution, you can see its important features as follows.

Examining a Distribution

In any graph of data, look for the **overall pattern** and for striking **deviations** from that pattern.

You can describe the overall pattern of a distribution by its **shape**, **center**, and **spread**.

An important kind of deviation is an **outlier**, an individual value that falls outside the overall pattern.

We will soon learn how to describe center and spread numerically. For now, you can describe the center of a distribution by its *midpoint*, the value with roughly half the observations taking smaller values and half taking larger values. You can describe the spread of a distribution by giving the *smallest and largest values.*

▦ EXAMPLE 3   Describing a Distribution

Look again at the histogram in Figure 5.2. **Shape:** The distribution has a *single peak*, which represents states in which less than 5% of adults are Hispanic. The distribution is *skewed to the right*. Most states have no more than 10% Hispanics, but some states have much higher percentages, so that the graph trails off to the right. **Center:** Table 5.1 shows that about half the states have less than 4.7% Hispanics among their adult residents and half have more. So the midpoint of the distribution is close to 4.7%. **Spread:** The spread is from about 0% to 42%, but only four states fall above 20%. **Outliers:** Arizona, California, New Mexico, and Texas stand out. Whether these are outliers or just part of the long right tail of the distribution is a matter of judgment. There is no rule for calling an observation an outlier. Once you have spotted possible outliers, look for an explanation. Some outliers are due to mistakes, such as typing 4.2 as 42. Other outliers point to the special nature of some observations. These four states are heavily Hispanic by history and location. ■

When you describe a distribution, concentrate on the main features. Look for major peaks, not for minor ups and downs in the bars of the histogram. Look for clear outliers, not just for the smallest and largest observations. Look for rough *symmetry* or clear *skewness.*

▦ Symmetric and Skewed Distributions

A distribution is **symmetric** if the right and left sides of the histogram are approximately mirror images of each other.

A distribution is **skewed to the right** if the right side of the histogram (containing the half of the observations with larger values) extends much farther out than the left side. It is **skewed to the left** if the left side of the histogram extends much farther out than the right side.

Some variables have distributions with predictable shapes. Many biological measurements on specimens from the same species and sex—lengths of bird bills, heights of young women—have symmetric distributions. Money amounts, on the other hand, usually have right-skewed distributions. There are many moderately priced houses, for example, but the few very expensive mansions give the distribution of house prices a strong right skew. Here are more examples of describing the overall pattern of a histogram.

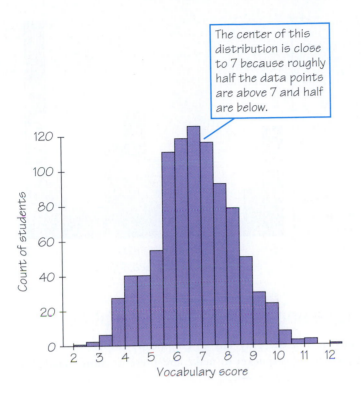

> The center of this distribution is close to 7 because roughly half the data points are above 7 and half are below.

Iowa Test Scores

EXAMPLE 4

Figure 5.3 displays the scores of all 947 seventh-grade students in the public schools of Gary, Indiana, on the vocabulary part of the Iowa Test of Basic Skills. The distribution is *single-peaked* and *symmetric*. In mathematics, the two sides of symmetric patterns are exact mirror images. Real data are almost never exactly symmetric. We are content to describe Figure 5.3 as symmetric. The center (half above, half below) is close to 7. This is a seventh-grade reading level. The scores range from 2.0 (second-grade level) to 12.1 (twelfth-grade level). ■

College Tuition

EXAMPLE 5

Jeanna plans to attend college in her home state of Massachusetts. She looks up the tuition and fees for the 2003–2004 academic year for all 56 four-year colleges in Massachusetts (omitting art schools and other special colleges). Figure 5.4 is a histogram of the data. As is often the case, we can't call this irregular distribution either symmetric or skewed. It does show two separate *clusters* of colleges, 11 with tuition less than $9000 and the remaining 45 costing more than $13,000. Clusters suggest that two types of individuals are mixed in the data set. In fact, the histogram distinguishes the 11 state colleges in Massachusetts from the 45 private colleges, which charge much more. ■

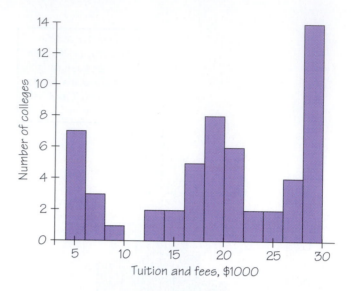

FIGURE 5.4 Histogram of the tuition and fees charged by four-year colleges in Massachusetts.

5.3 Displaying Distributions: Stemplots

Histograms are not the only way to graphically display distributions. For small data sets, a *stemplot* is quicker to make and presents more detailed information.

Stemplot

To make a **stemplot**:

1. Separate each observation into a **stem** consisting of all but the final (rightmost) digit and a **leaf**, the final digit. Stems may have as many digits as needed, but each leaf contains only a single digit.
2. Write the stems in a vertical column with the smallest at the top, and draw a vertical line at the right of this column.
3. Write each leaf in the row to the right of its stem, in increasing order out from the stem.

EXAMPLE 6   Making a Stemplot

For the "percent Hispanic" percents in Table 5.1, take the whole-number part of the percent as the stem and the final digit (tenths) as the leaf. The Massachusetts entry, 6.8%, has stem 6 and leaf 8. Wyoming, at 6.4%, places leaf 4 on the same stem. These are the only observations on this stem. We then arrange the leaves in order, as 48, so that 6|48 is one row in the stemplot. Figure 5.5 is the complete stemplot for the data in Table 5.1. To save space, we left out California,

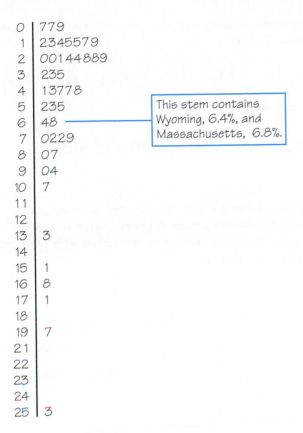

```
 0 | 779
 1 | 2345579
 2 | 00144889
 3 | 235
 4 | 13778
 5 | 235
 6 | 48 ————————
 7 | 0229
 8 | 07
 9 | 04
10 | 7
11 |
12 |
13 | 3
14 |
15 | 1
16 | 8
17 | 1
18 |
19 | 7
21 |
22 |
23 |
24 |
25 | 3
```

This stem contains Wyoming, 6.4%, and Massachusetts, 6.8%.

High 32.0 32.4 42.1

FIGURE 5.5 Stemplot of the percent of Hispanics among the adult residents of the states.

Texas, and New Mexico, which have stems 32 and 42. These observations are listed as "High" below the stemplot. ▪

A stemplot looks like a histogram turned on end. Comparing the stemplot in Figure 5.5 with the histogram in Figure 5.2 reveals the strengths and weaknesses of stemplots. The stemplot, unlike the histogram, preserves the actual value of each observation. But you can choose the classes in a histogram, whereas the classes (the stems) of a stemplot are forced on you. Whether the large number of classes in Figure 5.5 is an improvement over Figure 5.2 is a matter of taste. Stemplots do not work well for large data sets like the 947 Iowa Test scores in Figure 5.3, because each stem must hold a large number of leaves.

When the observed values have many digits, it is often best to *round* the numbers to just a few digits before making a stemplot. For example, a stemplot of data like

3.468   2.567   2.981   1.095 . . .

would have very many stems and no leaves or just one leaf on most stems. You can round these data to

$$3.5 \qquad 2.6 \qquad 3.0 \qquad 1.1 \ldots$$

before making a stemplot.

5.4 Describing Center: Mean and Median

Interested in a sporty car? Worried that it may use too much gas? The Environmental Protection Agency lists most such vehicles in its "minicompact" or "two-seater" categories. Table 5.2 gives the city and highway gas mileage for cars in these groups. (The mileages are for the basic engine and transmission combination for each car.) We want to compare two-seaters with minicompacts and city mileage with highway mileage.

TABLE 5.2 Fuel Economy (Miles per Gallon) for Model Year 2004 Vehicles

Two-Seater Cars			Minicompact Cars		
Model	City	Highway	Model	City	Highway
Acura NSX	17	24	Aston Martin Vanquish	12	19
Audi TT Roadster	20	28	Audi TT Coupe	21	29
BMW Z4 Roadster	20	28	BMW 325CI	19	27
Cadillac XLR	17	25	BMW 330CI	19	28
Chevrolet Corvette	18	25	BMW M3	16	23
Dodge Viper	12	20	Jaguar XK8	18	26
Ferrari 360 Modena	11	16	Jaguar XKR	16	23
Ferrari Maranello	10	16	Lexus SC 430	18	23
Ford Thunderbird	17	23	Mini Cooper	25	32
Honda Insight	60	66	Mitsubishi Eclipse	23	31
Lamborghini Gallardo	9	15	Mitsubishi Spyder	20	29
Lamborghini Murcielago	9	13	Porsche Cabriolet	18	26
Lotus Esprit	15	22	Porsche Turbo 911	14	22
Maserati Spyder	12	17			
Mazda Miata	22	28			
Mercedes-Benz SL500	16	23			
Mercedes-Benz SL600	13	19			
Nissan 350Z	20	26			
Porsche Boxster	20	29			
Porsche Carrera 911	15	23			
Toyota MR2	26	32			

We start with graphs. Figure 5.6 is a stemplot of the highway mileages of the 21 two-seater cars. As is often the case when there are only a few observations, the shape of the distribution is irregular. The most striking feature is an extreme high outlier, listed separately at the bottom of the stemplot. The Honda Insight (66 miles per gallon) is a hybrid gas-electric car that doesn't belong in the same category as the 20 gasoline-powered two-seater cars.

Numerical summaries make the comparisons we want more specific. Numerical description of a distribution begins with a measure of its center or average. The two common measures of center are the *mean* and the *median*. The mean is the "average value" and the median is the "middle value." These are two different ideas for "center," and the two measures behave differently. We need precise recipes for the mean and the median.

13	0
14	
15	0
16	00
17	0
18	
19	0
20	0
21	
22	0
23	000
24	0
25	00
26	0
27	
28	000
29	0
30	
31	
32	0

High 66

FIGURE 5.6 Stemplot of the highway gas mileages of two-seater cars. The Honda Insight (66 miles per gallon) is an extreme outlier.

The Mean $\bar{x}$

To find the **mean** of a set of observations, add their values and divide by the number of observations. If the n observations are $x_1, x_2, \ldots, x_n$, their mean is

$$\bar{x} = \frac{x_1 + x_2 + \cdots x_n}{n}$$

The bar over the x indicates the mean of all the x-values. Pronounce the mean $\bar{x}$ as "x-bar." This notation is very common. When writers who are discussing data use $\bar{x}$ or $\bar{y}$, they are talking about a mean.

Calculating the Mean

EXAMPLE 7

The mean highway mileage for the 21 two-seaters in Table 5.2 is

$$\bar{x} = \frac{x_1 + x_2 + \cdots + x_n}{n}$$

$$= \frac{24 + 28 + 28 + \cdots + 32}{21}$$

$$= \frac{518}{21} = 24.7 \text{ miles per gallon (mpg)}$$

In practice, you can key the data into your calculator and hit the $\bar{x}$ key. You don't have to actually add and divide. But you should know that this is what the calculator is doing. ■

We said that the Honda Insight doesn't belong with the other cars. If we exclude the Insight, the mean highway mileage drops to 22.6 mpg. The single outlier adds more than 2 mpg to the mean highway mileage. This illustrates an important weakness of the mean as a measure of center: *The mean is sensitive to the influence of a few extreme observations.* These may be outliers, but a skewed distribution that has no outliers will also pull the mean toward its long tail.

We have used the midpoint of a distribution as an informal measure of center. The *median* is the formal version of the midpoint, with a specific rule for calculation.

The Median *M*

The **median** *M* is the midpoint of a distribution, the number such that half the observations are smaller and the other half are larger. To find the median of a distribution:

1. Arrange all observations in order of size, from smallest to largest.
2. If the number *n* of observations is odd, the median *M* is the center observation in the ordered list. The location of the median is found by counting $(n + 1)/2$ observations up from the bottom of the list.
3. If the number *n* of observations is even, the median *M* is the average of the two center observations in the ordered list. The location of the median is again $(n + 1)/2$ from the bottom of the list.

Be sure to write down each individual observation in the data set, even if several observations repeat the same value. And be sure to arrange the observations in order of size before locating the median. Note that the recipe $(n + 1)/2$ gives the *position* of the median in the ordered list of observations, *not* the median itself.

EXAMPLE 8   Calculating the Median

To find the median highway mileage for 2004 model two-seater cars, arrange the data in increasing order:

13 15 16 16 17 19 20 22 23 23 **23** 24 25 25 26 28 28 28 29 32 66

The median is the bold **23**, the 11th observation in the ordered list. You can find the median by eye—there are 10 observations to the left and 10 to the right. Or you can use the recipe $(n + 1)/2 = 22/2 = 11$ to locate the median in the list.

What happens if we drop the Honda Insight? The remaining 20 cars have highway mileages:

13 15 16 16 17 19 20 22 23 **23** **23** 24 25 25 26 28 28 28 29 32

Because the number of observations $n = 20$ is even, there is no center observation. There is a center pair–the bold pair of 23s have 9 observations to their left and 9 to their right. The median M is the mean of the center pair, which is 23. The recipe $(n + 1)/2 = 21/2 = 10.5$ for the position of the median in the list says that the median is at location "ten and one-half," that is, halfway between the 10th and 11th observations. ■

You see that the median resists the influence of extreme observations better than the mean. Removing the Honda Insight did not change the median at all. Even if we mistakenly enter the Insight's mileage as 660 rather than 66, the median remains 23. The very high value is simply one observation to the right of center. The *Mean and Median* applet (at www.whfreeman.com/fapp) is an excellent way to compare the resistance of M and $\bar{x}$. See Applet Exercises 1 and 2.

SPOTLIGHT 5.1 Think!

Computer software will do all of your statistical graphs and calculations for you. But you can think and the computer can't. Here are some examples of the importance of thinking as well as calculating.

What's not on the plot matters Abraham Wald (1902–1950), like many statisticians, worked on war problems during World War II. Wald invented some statistical methods that were military secrets until the war ended. Here is one of his simpler contributions. Asked where extra armor should be added to airplanes, Wald studied the location of enemy bullet holes in planes returning from combat. He plotted the locations on an outline of the plane. As data accumulated, most of the outline filled up. Put the armor in the few spots with no bullet holes, said Wald. That's where bullets hit the planes that didn't make it back.

What are your units? Not paying attention to units of measurement can get you into trouble. In 1999, the *Mars Climate Orbiter* spacecraft burned up in the Martian atmosphere. It was supposed to be 93 miles (150 kilometers) above the planet, but was in fact only 35 miles (57 kilometers) up. It seems that Lockheed Martin, which built the *Orbiter*, specified important measurements in English units (pounds, miles). The National Aeronautics and Space Administration team who flew the spacecraft thought the numbers were in metric system units (kilograms, kilometers). There went $125 million.

What are we counting? A news report says that 22% of American high school students smoke. That sounds serious. But what counts as "smoking"? It turns out that this is the percent who smoked at least once in the past month. If we say that a smoker is someone who smoked on at least 20 of the past 30 days, only 9.7% of high school students smoke. The number you get depends on how you define your variable—and you can make a problem sound more or less serious by changing the definition. Interest groups do this all the time, so pay attention.

The median and mean are the most common measures of the center of a distribution. The mean and median of a symmetric distribution are close together. If the distribution is exactly symmetric, the mean and median are exactly the same. In a skewed distribution, the mean is farther out in the long tail than is the median. For example, the distribution of house prices is skewed to the right. There are many moderately priced houses and a few very expensive mansions. The mansions pull the mean up but do not affect the median. The mean price of existing single-family houses sold in March 2005 was $247,000, but the median price for these same houses was only $193,000.

5.5 Describing Spread: The Quartiles

The mean and median provide two different measures of the center of a distribution. But a measure of center alone can be misleading. Two neighborhoods with median house price $193,000 are very different if one has both mansions and modest homes and the other has little variation among houses. We are interested in the spread or variability of house prices as well as their centers. *The simplest useful numerical description of a distribution consists of both a measure of center and a measure of spread.*

One way to measure spread is to give the smallest and largest observations. For example, the percents of Hispanics in the states range from 0.7% in Maine and West Virginia to 42.1% in New Mexico. These observations show the full spread of the data, but they may be outliers. We can improve our description of spread by also looking at the spread of the middle half of the data. The *quartiles* mark out the middle half. Count up the ordered list of observations, starting from the smallest. The *first quartile* lies one-quarter of the way up the list. The *third quartile* lies three-quarters of the way up the list. In other words, the first quartile is larger than 25% of the observations, and the third quartile is larger than 75% of the observations. The second quartile is the median, which is larger than 50% of the observations. That is the idea of quartiles. We need a rule to make the idea exact. The rule for calculating the quartiles uses the rule for the median.

The Quartiles Q_1 and Q_3

To calculate the **quartiles**:

1. Arrange the observations in increasing order and locate the median M in the ordered list of observations.
2. The **first quartile** Q_1 is the median of the observations whose position in the ordered list is to the left of the location of the overall median.
3. The **third quartile** Q_3 is the median of the observations whose position in the ordered list is to the right of the location of the overall median.

Calculating Quartiles

EXAMPLE 9

The highway mileages of the 20 gasoline-powered two-seater cars, arranged in increasing order, are:

13 15 16 16 17 19 20 22 23 23 | 23 24 25 25 26 28 28 28 29 32

The median is midway between the center pair of observations. We have marked its position in the list by |. The first quartile is the median of the 10 observations to the left of the position of the median. Check that its value is $Q_1 = 18$. Similarly, the third quartile is the median of the 10 observations to the right of the |. Check that $Q_3 = 27$.

When there is an odd number of observations, the median is the unique center observation, and the rule for finding the quartiles excludes this center value. The highway mileages of the 13 minicompact cars in Table 5.2 are (in order):

19   22   23   23   23   26   **26**   27   28   29   29   31   32

The median is the bold **26**. The first quartile is the median of the 6 observations falling to the left of this point in the list, $Q_1 = 23$. Similarly, $Q_3 = 29$. ■

Some software packages use a slightly different rule to find the quartiles, so computer results may be a bit different from your own work. Don't worry about this. The differences will always be too small to be important.

5.6 The Five-Number Summary and Boxplots

We started by using the smallest and largest observations to indicate the spread of a distribution. These single observations tell us little about the distribution as a whole, but they give information about the tails of the distribution that is missing if we know only Q_1, M, and Q_3. To get a quick summary of both center and spread, combine all five numbers.

The Five-Number Summary

The **five-number summary** of a distribution consists of the smallest observation, the first quartile, the median, the third quartile, and the largest observation, written in order from smallest to largest. In symbols, the five-number summary is

Minimum Q_1 M Q_3 Maximum

FIGURE 5.7 Boxplots of the highway and city gas mileages for cars classified as two-seaters and as minicompacts by the Environmental Protection Agency. These boxplots are drawn vertically, but it is equally correct to draw them horizontally.

These five numbers offer a reasonably complete description of center and spread. The five-number summaries for highway gas mileages are

$$13 \quad 18 \quad 23 \quad 27 \quad 32$$

for two-seaters and

$$19 \quad 23 \quad 26 \quad 29 \quad 32$$

for minicompacts. The median describes the center of the distribution. The quartiles show the spread of the center half of the data. The minimum and maximum show the full spread of the data. The five-number summary leads to another visual representation of a distribution, the *boxplot*. Figure 5.7 shows boxplots for both city and highway gas mileages for our two groups of cars.

Boxplot

A **boxplot** is a graph of the five-number summary.

▶ A central box spans the quartiles Q_1 and Q_3.
▶ A line in the box marks the median M.
▶ Lines extend from the box out to the smallest and largest observations.

Because boxplots show less detail than histograms or stemplots, they are best used for side-by-side comparison of more than one distribution, as in Figure 5.7. When you look at a boxplot, first locate the median, which marks the center of the distribution. Then look at the spread. The quartiles show the spread of the middle half of the data, and the extremes (the smallest and largest observations) show the spread of the entire data set. We see at once that city mileages are lower than highway mileages. The minicompact cars have slightly higher median gas mileages than the two-seaters and their mileages are markedly less variable. In particular, the low gas mileages of the Ferrari and Lamborghinis in the two-seater group pull the group minimum down.

5.7 Describing Spread: The Standard Deviation

Although the five-number summary is the most generally useful numerical description of a distribution, it is not the most common. That distinction belongs to the combination of the mean with the *standard deviation*. The mean, like the median, is a measure of center. The standard deviation, like the quartiles and extremes in the five-number summary, measures spread. The standard deviation and its close relative, the *variance*, measure spread by looking at how far the observations are from their mean.

Understanding the Standard Deviation   EXAMPLE 10

A person's metabolic rate is the rate at which the body consumes energy. Metabolic rate is important in studies of weight gain, dieting, and exercise. Here are the metabolic rates of 7 men who took part in a study of dieting. (The units are calories per 24 hours. These are the same calories used to describe the energy content of foods.)

1792   1666   1362   1614   1460   1867   1439

Figure 5.8 displays the data as points above the number line, with their mean marked by an asterisk (*). The arrows mark two of the deviations from the mean. These deviations show how spread out the data are about their mean. Some of the deviations are positive and some are negative. Squaring the deviations makes them all positive. Observations far from the mean in either direction will have

FIGURE 5.8 The variance and standard deviation measure spread by looking at the deviations of observations from their mean.

large positive squared deviations. So a reasonable measure of spread is the average of the squared deviations. This average is called the *variance*. The variance is large if the observations are widely spread about their mean. It is small if the observations are all close to the mean.

But the variance has the wrong units: If we measure metabolic rate in calories, the variance of the metabolic rates is in squared calories. Taking the square root of the variance gets us back to calories. The square root of the variance is the *standard deviation*. ■

The Standard Deviation s

The **variance** s^2 of a set of observations is an average of the squares of the deviations of the observations from their mean. In symbols, the variance of n observations $x_1, x_2, \ldots, x_n$ is

$$s^2 = \frac{(x_1 - \bar{x})^2 + (x_2 - \bar{x})^2 + \cdots + (x_n - \bar{x})^2}{n - 1}$$

The **standard deviation** s is the square root of the variance s^2.

Notice that it is usual to divide by $n - 1$ rather than by n when we average the squared deviations to get the variance. The reason is a bit subtle: The deviations $x_i - \bar{x}$ always add to exactly 0, so if you know the value of all but one, you know them all. Only $n - 1$ of the squared deviations can vary freely, and we average by dividing the total by $n - 1$. Many calculators offer a choice between dividing by n and dividing by $n - 1$, so be sure to use $n - 1$.

In practice, use your calculator or software to obtain the standard deviation from keyed-in data. Doing an example step by step, however, will help you understand how the variance and standard deviation work.

EXAMPLE 11   Calculating the Standard Deviation

To find the standard deviation of the 7 metabolic rates, first find the mean:

$$\bar{x} = \frac{1792 + 1666 + 1362 + 1614 + 1460 + 1867 + 1439}{7}$$

$$= \frac{11,200}{7} = 1600 \text{ calories}$$

The deviations shown in Figure 5.8 are the starting point for calculating the variance and the standard deviation.

Observations x_i	Deviations $x_i - x$			Squared Deviations $(x_i - x)^2$		
1792	$1792 - 1600$	$=$	192	192^2	$=$	36,864
1666	$1666 - 1600$	$=$	66	66^2	$=$	4,356
1362	$1362 - 1600$	$=$	-238	$(-238)^2$	$=$	56,644
1614	$1614 - 1600$	$=$	14	14^2	$=$	196
1460	$1460 - 1600$	$=$	-140	$(-140)^2$	$=$	19,600
1867	$1867 - 1600$	$=$	267	267^2	$=$	71,289
1439	$1439 - 1600$	$=$	-161	$(-161)^2$	$=$	25,921
	sum	$=$	0	sum	$=$	214,870

The variance is the sum of the squared deviations divided by 1 less than the number of observations:

$$s^2 = \frac{214{,}870}{6} = 35{,}811.67$$

The standard deviation is the square root of the variance:

$$s = \sqrt{35{,}811.67} = 189.24 \text{ calories}$$

If the 7 observations had the same mean 1600 calories, but were more spread out, their deviations from 1600 would be larger. The variance and standard deviation would then be larger. Remember that s^2 and s get larger as the spread of the observations about the mean gets larger. ■

More important than the details of hand calculation are the properties that determine the usefulness of the standard deviation:

▶ s measures spread about the mean $\bar{x}$. Use s to describe the spread of a distribution only when you use $\bar{x}$ to describe the center.

▶ $s = 0$ only when there is *no spread*. This happens only when all observations have the same value. Otherwise $s > 0$. As the observations become more spread out about their mean, s gets larger.

▶ s has the same units of measurement as the original observations. For example, if you measure metabolic rates in calories, both the mean $\bar{x}$ and the standard deviation s are also in calories.

▶ The use of squared deviations makes s even more sensitive than $\bar{x}$ to a few extreme observations. For example, dropping the Honda Insight from our list of two-seater cars cuts the standard deviation more than half, from 10.8 mpg with the Insight to 5.3 mpg without it. Distributions with outliers and strongly skewed distributions have large standard deviations. The number s does not give much helpful information about such distributions.

We now have a choice between two descriptions of the center and spread of a distribution: the five-number summary, or $\bar{x}$ and s. Because $\bar{x}$ and s are sensitive to extreme observations, they can be misleading when a distribution is strongly skewed or has outliers. In fact, because the two sides of a skewed distribution have different spreads, no single number such as s describes the spread well. The five-number summary, with its two quartiles and two extremes, does a better job.

Choosing a Summary

The five-number summary is usually better than the mean and standard deviation for describing a skewed distribution or a distribution with outliers. Use $\bar{x}$ and s only for reasonably symmetric distributions that are free of outliers.

Although the standard deviation is widely used, it is not a natural or convenient measure of the spread of a distribution. The real reason for the popularity of the standard deviation is that it is the natural measure of spread for *normal distributions*, an important class of distributions that we will meet next.

Do remember that a graph gives the best overall picture of a distribution. Numerical measures of center and spread report specific facts about a distribution, but they do not describe its entire shape. Numerical summaries do not disclose the presence of clusters, for example. *Always start with a graph of your data.*

5.8 Normal Distributions

We now have a kit of graphical and numerical tools for describing distributions. What is more, we have a clear strategy for exploring data on a single numerical variable:

1. Always plot your data: make a graph, usually a histogram or a stemplot.
2. Look for the overall pattern (shape, center, spread) and for striking deviations such as outliers.
3. Calculate a numerical summary to briefly describe center and spread.

Here is one more step to add to this strategy:

4. Sometimes the overall pattern of a large number of observations is so regular that we can describe it by a smooth curve.

Figure 5.3 is a histogram of the Iowa Test vocabulary scores of 947 seventh-grade students. The histogram is symmetric, is single-peaked, and has a quite regular shape. In Figure 5.9, we draw a smooth curve through the tops of the histogram bars to describe the shape. The curve is an idealized description of the

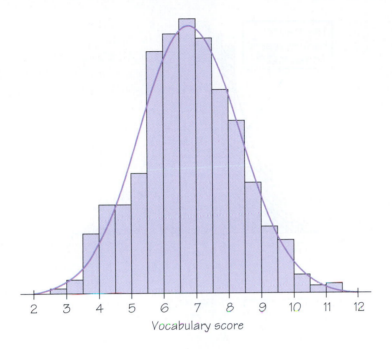

FIGURE 5.9 Histogram of the vocabulary scores of all seventh-grade students in Gary, Indiana. The smooth curve shows the overall shape of the distribution.

distribution. It gives a compact picture of the overall pattern of the data but ignores minor irregularities as well as any outliers. The curve in Figure 5.9 is a *normal curve*. A distribution whose shape is described by a normal curve is a *normal distribution.*

From Histogram to Normal Curve

EXAMPLE 12

You can think of a normal curve as a smoothed-out histogram. Our eyes respond to the *areas* of the bars in a histogram. The bar areas represent proportions of the observations. Figure 5.10a is a copy of Figure 5.9 with the leftmost bars shaded. The area of the shaded bars in Figure 5.10a represents the students with vocabulary scores 6.0 or lower. There are 287 such students, who make up the proportion 287/947 = 0.303 of all Gary seventh graders.

Now look at the curve drawn through the bars. In Figure 5.10b, the area under the curve to the left of 6.0 is shaded. We know that the areas of histogram bars represent proportions of all the observations, but we don't worry about the actual area. We can draw a histogram taller or shorter as we wish. In moving from histogram bars to a smooth curve, we no longer allow this. Adjust the scale of the graph so that *the total area under the curve is exactly 1.* The total area represents the proportion 1, that is, all the observations. Now areas under the curve actually *are* proportions of the observations. This curve is a normal curve. The shaded area under the normal curve in Figure 5.10b is the proportion of students with score 6.0 or lower. This area is 0.293, only 0.010 away from the histogram result. You see that areas under the normal curve give quite good approximations of areas given by the histogram. ■

FIGURE 5.10a The proportion of scores less than or equal to 6.0 from the histogram is 0.303.

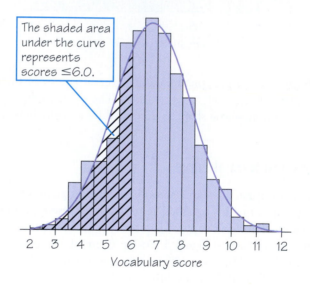

The shaded bars represent scores ≤6.0.

Vocabulary score

FIGURE 5.10b The proportion of scores less than or equal to 6.0 from the normal curve is 0.293.

The shaded area under the curve represents scores ≤6.0.

Vocabulary score

EXAMPLE 13   Heights of Young Women

The distribution of the heights of young American women is approximately described by the normal curve in Figure 5.11. The proportion of young women who are between 60 inches (5 feet) and 65 inches tall is given by the area under the curve between 60 and 65. This area is about 0.54, so 54% of young women are between 60 and 65 inches tall. This is exactly correct for the normal distribution, and approximately correct for the actual heights of young women. ■

⬚ Normal Distributions

The distribution of a variable tells us what values the variable takes and how often it takes these values. A normal distribution is described by a normal curve. The area under the curve above any interval of values tells us what proportion of all values of the variable lie in that interval. The total area under any normal curve is exactly 1.

Normal distributions play a large role in statistics, but they are rather special and not at all "normal" in the sense of being average or natural. Normal curves can be specified exactly by an equation, but we will be content with pictures. Figure 5.12 shows two normal curves. All normal curves have the same overall **shape.** They are symmetric and bell-shaped, with tails that fall off rapidly from a central peak. The **center** of the symmetric normal curve is the center of the distribution in several senses. It is the mean of the distribution. It is also the median in the sense that half the observations (half the area under the curve) lie on each side of the center.

What about the **spread** of a normal curve? Normal curves have the special property that their spread is completely measured by a single number, the standard deviation. We have learned how to calculate the standard deviation from a set of observations. For normal distributions, the standard deviation, like the

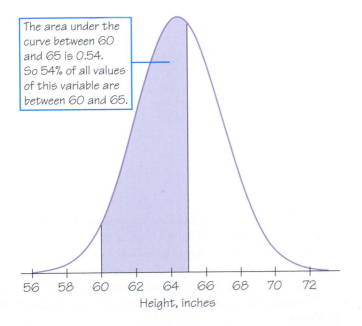

The area under the curve between 60 and 65 is 0.54. So 54% of all values of this variable are between 60 and 65.

Height, inches

FIGURE 5.11 Areas under a normal curve describe a normal distribution. This normal curve describes the distribution of heights of young women.

FIGURE 5.12 Two normal curves with the same mean and different standard deviations. The standard deviation for each curve is the distance from the center (the mean) to the change-of-curvature point on one side of the center.

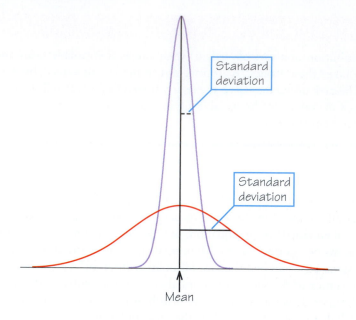

FIGURE 5.12 Two normal curves with the same mean and different standard deviations. The standard deviation for each curve is the distance from the center (the mean) to the change-of-curvature point on one side of the center.

mean, can be found directly from the curve. Here's how. Imagine that you are skiing down a mountain that has the shape of a normal curve. At first, you descend at an ever-steeper angle as you go out from the peak:

Fortunately, before you find yourself going straight down, the slope begins to grow flatter rather than steeper as you go out and down:

The points at which this change of curvature takes place are located one standard deviation on either side of the mean. You can feel the change as you run your finger along a normal curve, and so find the standard deviation. Try it with the two normal curves in Figure 5.12. Normal curves with the same standard deviation have exactly the same shape. Changing the mean just moves the center of the curve to a new location. Changing the standard deviation changes the spread of the curve, as Figure 5.12 shows.

Mean and Standard Deviation of Normal Distributions

The shape of a normal distribution is completely described by two numbers, the mean and the standard deviation.

The mean of a normal distribution is at the center of symmetry of the normal curve. The standard deviation is the distance from the center to the change-of-curvature points on either side.

The mean and median of a skewed distribution are not equal. Figure 5.13, for example, shows a *right-skewed distribution*. The right tail of the curve is much longer than the left. We know that the mean, more than the median, is pulled out by the long right tail.

We have often used the quartiles to indicate the spread of a distribution. Because the standard deviation completely describes the spread of any normal distribution, it tells us where the quartiles are. Here are the facts.

Quartiles of Normal Distributions

The first quartile of any normal distribution is located 0.67 standard deviation below the mean; the third quartile is 0.67 standard deviation above the mean.

Heights of Young Women

EXAMPLE 14

The distribution of heights of young women is approximately normal with mean 64.5 inches and standard deviation 2.5 inches. Figure 5.14 shows this normal curve. The quartiles lie 0.67 standard deviation, or

$$(0.67)(2.5) = 1.7 \text{ inches}$$

on either side of the mean. The first quartile is $64.5 - 1.7$, or 62.8 inches. The third quartile is $64.5 + 1.7$, or 66.2 inches. The middle 50% of women's heights lie between 62.8 inches and 66.2 inches. This is exactly true for the normal distribution with mean 64.5 inches and standard deviation 2.5 inches. It is approximately

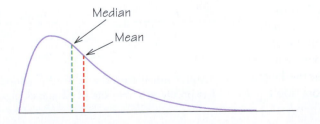

FIGURE 5.13 This smooth curve describes a right-skewed distribution. The mean is pulled toward the long tail more than the median.

FIGURE 5.14 The quartiles of a normal distribution are located 0.67 standard deviation on either side of the mean. For this normal curve, the mean is 64.5 inches and the standard deviation is 2.5 inches.

Third quartile

25% of young women are taller than 66.2 inches

Height, inches

true for the actual heights of young women because the distribution of heights is approximately normal. ■

SPOTLIGHT 5.2 Density Estimation

Smooth curves that describe the overall pattern of distributions of data are called *density curves.* Normal curves are one type of density curve. There are many other types used for different purposes. However, you don't have to call for a specific type such as the normal curves. Clever software for "density estimation" will calculate a density curve to describe any set of observations you give it.

The figure shows a strongly skewed distribution, the survival times of 72 guinea pigs in a medical experiment. Two graphs of the distribution are overlaid: a histogram and a density curve produced by software from the data. The histogram and density curve agree on the overall shape and on the "bumps" in the long right tail. The density curve shows a higher single peak as a main feature of the distribution. The histogram divides the observations near the peak between two bars, thus reducing the height of the peak. Because density estimators don't

depend on dividing the data into classes, as histograms do, many statisticians prefer them when they need a picture of a distribution.

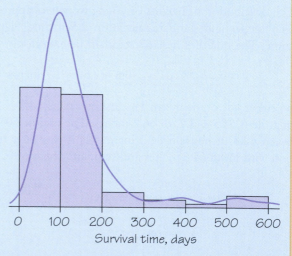

Survival time, days

Density estimation software fits this smooth curve to data on the survival time of 72 guinea pigs.

Why are the normal distributions important in statistics? Here are two reasons. First, normal distributions are good descriptions for some distributions of *real data*. Distributions that are often close to normal include scores on tests taken by many people (such as SAT exams and many psychological tests), repeated careful measurements of the same quantity, and characteristics of biological populations (such as heights of young women and yields of corn). Second, normal distributions are good approximations to the results of many kinds of *chance outcomes*, such as tossing a coin many times. We will return to normal curves in Chapter 8, when we study probability, the mathematics of chance. However, many sets of data do not follow a normal distribution. Most income distributions, for example, are skewed to the right and so are not normal. Nonnormal data, like nonnormal people, not only are common but are sometimes more interesting than their normal counterparts.

5.9 The 68–95–99.7 Rule

One consequence of the fact that the mean and standard deviation completely specify a normal distribution is that all normal distributions are the same when we record observations in terms of how many standard deviations they lie from the mean. Here is an important rule based on this fact.

The 68–95–99.7 Rule for Normal Distributions

According to the **68–95–99.7 rule**, in any normal distribution:

▶   68% of the observations fall within 1 standard deviation of the mean.

▶   95% of the observations fall within 2 standard deviations of the mean.

▶   99.7% of the observations fall within 3 standard deviations of the mean.

Figure 5.15 illustrates the 68–95–99.7 rule. By remembering these three numbers, you can think about normal distributions without making detailed calculations. For more detailed information, you can use tables or software that give areas under normal curves, but the 68–95–99.7 rule is adequate for our purposes.

Heights of Young Women

EXAMPLE 15

The heights of women between the ages of 18 and 24 are roughly normally distributed, with mean 64.5 inches and standard deviation 2.5 inches. Two standard deviations is 5 inches for this distribution. The 95 part of the 68–95–99.7 rule

FIGURE 5.15 The
68–95–99.7 rule for
normal distributions.

says that the middle 95% of young women are between 64.5 − 5 and 64.5 + 5 inches tall, that is, between 59.5 inches and 69.5 inches. This fact is exactly true for an exactly normal distribution. It is approximately true for the heights of young women because the distribution of heights is approximately normal.

The other 5% of young women have heights outside the range from 59.5 to 69.5 inches. Because the normal distributions are symmetric, half of these women are on the tall side. So the tallest 2.5% of young women are taller than 69.5 inches. ■

EXAMPLE 16   SAT Scores

The distribution of scores on tests such as the SAT college entrance examination is close to normal. SAT scores are adjusted so that the mean score is about $\mu = 500$ and the standard deviation is about $\sigma = 100$. This information allows us to answer many questions about SAT scores.

▶   *How high must a student score to fall in the top 25%?*

The third quartile is $(0.67)(100) = 67$ points above the mean. So scores above 567 are in the top 25%.

▶   *What percent of scores fall between 200 and 800?*

Scores of 200 and 800 are 3 standard deviations on either side of the mean. The 99.7 part of the 68–95–99.7 rule says that 99.7% of all scores lie in this range. (In fact, 200 and 800 are the lowest and highest scores that are reported on the SAT. The few scores higher than 800 are reported as 800.)

▶   *What percent of scores are above 700?*

A score of 700 is 2 standard deviations above the mean. By the 95 part of the 68–95–99.7 rule, 95% of all scores fall between 300 and 700 and 5% fall below 300 or above 700. Because normal curves are symmetric, half of this 5% are above 700. So a score above 700 places a student in the top 2.5% of test-takers.

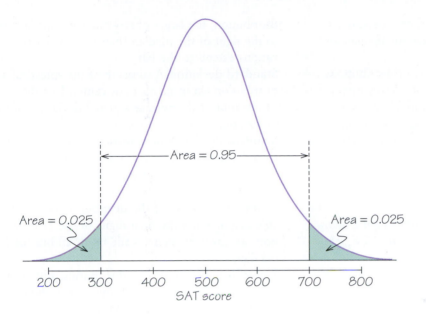

Sketching a normal curve with the points 1, 2, and 3 standard deviations from the mean marked can help you use the 68–95–99.7 rule. Figure 5.16 shows the distribution of SAT scores with the areas needed to find the percent of scores above 700. ■

The 68–95–99.7 rule allows you to find selected areas under a normal curve—areas for outcomes bounded by 1, 2, or 3 standard deviations away from the mean. You can use software, a graphing calculator, or the *Normal Curve* applet to find any area under a normal curve. See Applet Exercises 3 to 5 for use of the applet.

REVIEW VOCABULARY

Boxplot A graph of the five-number summary. A box spans the quartiles, with an interior line marking the median. Lines extend out from this box to the extreme high and low observations.

Distribution The pattern of outcomes of a variable. The distribution describes what values the variable takes and how often each value occurs.

Exploratory data analysis The practice of examining data for overall patterns and special features, without necessarily seeking answers to specific questions.

Five-number summary A summary of a distribution that gives the median, the first and third quar-

tiles, and the largest and smallest observations.

Histogram A graph of the distribution of outcomes (often divided into classes) for a single variable. The height of each bar is the number of observations in the class of outcomes covered by the base of the bar. All classes should have the same width.

Individuals The people, animals, or things described by a data set.

Mean The ordinary arithmetic average of a set of observations. To find the mean, add all the observations and divide the sum by the number of observations summed.

Median The midpoint of a set of observations. Half the observations fall below the median and half fall above.

Normal distributions A family of distributions that describe how often a variable takes its values by areas under a curve. The normal curves are symmetric and bell-shaped. A specific normal curve is completely described by giving its mean and its standard deviation.

Outlier A data point that falls clearly outside the overall pattern of a set of data.

Quartiles The first quartile of a distribution is the point with 25% of the observations falling below it; the third quartile is the point with 75% below it.

68–95–99.7 rule In any normal distribution, 68% of the observations lie within 1 standard deviation on either side of the mean; 95% lie within 2 standard deviations of the mean; and 99.7% lie within 3 standard deviations of the mean.

Skewed distribution A distribution in which observations on one side of the median extend notably farther from the median than do observations on the other side. In a right-skewed distribution, the larger observations extend farther to the right of the median than the smaller observations extend to the left.

Standard deviation A measure of the spread of a distribution about its mean as center. It is the square root of the average squared deviation of the observations from their mean.

Standard deviation of a normal curve The standard deviation of a normal curve is the distance from the mean to the change-of-curvature points on either side.

Stemplot A display of the distribution of a variable that attaches the final digits of the observations as leaves on stems made up of all but the final digit.

Symmetric distribution A distribution with a histogram or stemplot in which the part to the left of the median is roughly a mirror image of the part to the right of the median.

Variable Any characteristic of an individual.

Variance A measure of the spread of a distribution about its mean. It is the average squared deviation of the observations from their mean. The square root of the variance is the standard deviation.

✔ SKILLS CHECK

1. Here are the first lines of a professor's data set at the end of a mathematics course:

Name	Major	Points	Grade
ADVANI, SURA	COMM	397	B
BARTON, DAVID	HIST	323	C
BROWN, ANNETTE	BIOL	446	A
CHIU, SUN	PSYC	405	B
CORTEZ, MARIA	PSYC	461	A

The individuals in these data are

(a) the students.
(b) the total points.
(c) the course grades.

Figure 5.4 is a histogram of the tuition and fee charges for the 2003–2004 academic year for 56 four-year colleges in Massachusetts. Exercises 2 and 3 are based on this histogram.

2. The number of colleges with tuition and fee charges covered by the leftmost bar in the histogram is

(a) 7.
(b) 8.
(c) 14.

3. The leftmost bar in the histogram covers tuition and fee charges ranging from about

(a) $3000 to $7000.
(b) $4000 to $6000.
(c) $4500 to $5500.

4. Here are the IQ test scores of 10 randomly chosen fifth-grade students:

145 139 126 122 125 130 96 110 118 118

To make a stemplot of these scores, you would use as stems

(a) 0 and 1.
(b) 09, 10, 11, 12, 13, and 14.
(c) 96, 110, 118, 122, 125, 126, 130, 139, and 145.

5. You look at real estate ads for houses in Sarasota, Florida. There are many houses ranging from $200,000 to $400,000 in price. The few houses on the water, however, have prices up to $15 million. The distribution of house prices will be

(a) skewed to the left.
(b) roughly symmetric.
(c) skewed to the right.

6. Exercise 4 gives the IQ test scores of 10 students. The mean of these scores is

(a) 122.9.
(b) 123.4.
(c) 136.6.

7. The median of the 10 IQ test scores in Exercise 4 is

(a) 125.
(b) 123.5.
(c) 122.9.

8. If a distribution is skewed to the right,

(a) the mean is less than the median.
(b) the mean and median are equal.
(c) the mean is greater than the median.

9. The first quartile of the 10 IQ test scores in Exercise 4 is

(a) 114.
(b) 118.
(c) 130.

10. What percent of the observations in a distribution lie between the first quartile and the third quartile?

(a) 25%
(b) 50%
(c) 75%

11. The five-number summary of the 10 IQ scores in Exercise 4 is

(a) 96, 114, 125, 134.5, 145.

(b) 96, 118, 122.9, 130, 145.
(c) 96, 118, 123.5, 130, 145.

12. To make a boxplot of a distribution, you must know

(a) all of the individual observations.
(b) the mean and the standard deviation.
(c) the five-number summary.

13. The standard deviation of the 10 IQ scores in Exercise 4 (use your calculator) is

(a) 13.23.
(b) 13.95.
(c) 194.6.

14. What are all the values that a standard deviation s can possibly take?

(a) $0 \le s$
(b) $0 \le s \le 1$
(c) $-1 \le s \le 1$

15. You have data on the weights in grams of 5 baby pythons. The mean weight is 31.8 and the standard deviation of the weights is 2.39. The correct units for the standard deviation are

(a) no units—it's just a number.
(b) grams.
(c) grams squared.

16. Which of the following is least affected if an extreme high outlier is added to your data?

(a) The median
(b) The mean
(c) The standard deviation

17. To completely specify the shape of a normal distribution, you must give

(a) the mean and the standard deviation.
(b) the five-number summary.
(c) the mean and the median.

18. Which of these variables is least likely to have a normal distribution?

(a) Income per person for 150 different countries
(b) Lengths of 50 newly hatched pythons
(c) Heights of 100 white pine trees in a forest

19. The length of human pregnancies from conception to birth varies according to a distribution that is approximately normal with

mean 266 days and standard deviation 16 days. 95% of all pregnancies last between

(a) 250 and 282 days.
(b) 234 and 298 days.
(c) 218 and 314 days.

20. The scale of scores on an IQ test is approximately normal with mean 100 and standard deviation 15. The organization MENSA, which calls itself "the high IQ society," requires an IQ score of 130 or higher for membership. What percent of adults would qualify for membership?

(a) 95%
(b) 5%
(c) 2.5%

EXERCISES   ■ Challenge   ◆ Discussion

Some exercises require use of a calculator (or software) that will find mean and standard deviation from keyed-in data.

1. Here is a small part of a data set that describes the fuel economy (in miles per gallon) of 2004 model motor vehicles:

◆ **3.** Sketch a histogram for a distribution that is skewed to the left. Suppose that you and your friends emptied your pockets of coins and recorded the year marked on each coin. The distribution of dates would be skewed to the left. Explain why.

Make and Model	Vehicle Type	Transmission Type	Number of Cylinders	City mpg	Highway mpg
⋮					
Acura NSX	Two-seater	Automatic	6	17	24
BMW 330I	Compact	Manual	6	20	30
Cadillac Seville	Midsize	Automatic	8	18	26
Ford F150 2WD	Standard pickup truck	Automatic	6	16	19
⋮					

(a) What are the individuals in this data set?
(b) For each individual, what variables are given? For which of these variables would a histogram be helpful?

Displaying Distributions: Histograms

◆ **2.** Figure 5.17 is a histogram of the lengths of words used in Shakespeare's plays. Because there are so many words in the plays, the vertical axis of the graph is the percent that are of each length, rather than the count. What is the overall shape of this distribution? What does this shape say about word lengths in Shakespeare? Do you expect other authors to have word-length distributions of the same general shape? Why?

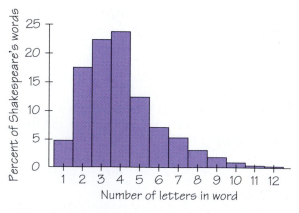

FIGURE 5.17 Histogram of the lengths of words used in Shakespeare's plays, for Exercise 2.

TABLE 5.3	Carbon Dioxide Emissions, Metric Tons per Person						
Country	CO_2	Country	CO_2	Country	CO_2	Country	CO_2
Algeria	2.3	Germany	10.0	Mexico	3.7	South Africa	8.1
Argentina	3.9	Ghana	0.2	Morocco	1.0	Spain	6.8
Australia	17.0	India	0.9	Myanmar	0.2	Sudan	0.2
Bangladesh	0.2	Indonesia	1.2	Nepal	0.1	Tanzania	0.1
Brazil	1.8	Iran	3.8	Nigeria	0.3	Thailand	2.5
Canada	16.0	Iraq	3.6	Pakistan	0.7	Turkey	2.8
China	2.5	Italy	7.3	Peru	0.8	Ukraine	7.6
Colombia	1.4	Japan	9.1	Philippines	0.9	United Kingdom	9.0
Congo	0.0	Kenya	0.3	Poland	8.0	United States	19.9
Egypt	1.7	Korea, North	9.7	Romania	3.9	Uzbekistan	4.8
Ethiopia	0.0	Korea, South	8.8	Russia	10.2	Venezuela	5.1
France	6.1	Malaysia	4.6	Saudi Arabia	11.0	Vietnam	0.5

4. Make a histogram of the highway gas mileages of the two-seater cars in Table 5.2. Use classes with width 5 mpg. Do you prefer the histogram or the stemplot of the same data in Figure 5.6? Why?

5. Burning fuels in power plants or motor vehicles emits carbon dioxide (CO_2), which contributes to global warming. Table 5.3 displays CO_2 emissions per person from countries with population at least 20 million.

(a) Why do you think we choose to measure emissions per person rather than total CO_2 emissions for each country?

(b) Display the data of Table 5.3 in a histogram. Describe the shape, center, and spread of the distribution. Which countries are outliers?

■ **6.** A survey of a large college class asked the following questions:

  1. Are you female or male? (In the data, male = 0, female = 1.)

  2. Are you right-handed or left-handed? (In the data, right = 0, left = 1.)

  3. What is your height, in inches?

4. How many minutes do you study on a typical weeknight?

Figure 5.18 shows histograms of the student responses, in scrambled order and without scale markings. Which histogram goes with each variable? Explain your reasoning.

Displaying Distributions: Stemplots

7. The population of the United States is aging, though less rapidly than in other developed countries. Figure 5.19 is a stemplot of the percents of residents aged 65 and over in the 50 states, according to the 2000 census. The stems are whole percents and the leaves are tenths of a percent.

(a) There are two outliers: Alaska has the lowest percent of older residents, and Florida has the highest. What are the percents for these two states?

(b) Ignoring Alaska and Florida, describe the shape, center, and spread of this distribution.

8. People with diabetes must monitor and control their blood glucose level. The goal is to maintain "fasting plasma glucose" between about 90 and

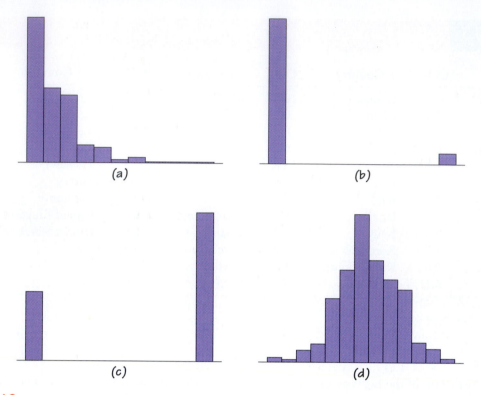

FIGURE 5.18 Match each histogram with its variable, for Exercise 6.

130 milligrams per deciliter (mg/dl). Here are the fasting plasma glucose levels for 18 diabetics enrolled in a diabetes control class, five months after the end of the class:

141	158	112	153	134	95
96	78	148	172	200	271
103	172	359	145	147	255

Round these values to the nearest 10 and then drop the zero. For example, 141 rounds to 14 and 158 rounds to 16. Make a stemplot of the rounded data. Describe the main features of the distribution. Are there outliers? How well is the group as a whole achieving the goal for controlling glucose levels?

9. The Survey of Study Habits and Attitudes (SSHA) is a psychological test that evaluates college students' motivation, study habits, and attitudes toward

```
 5 | 7
 6 |
 7 |
 8 | 5
 9 | 679
10 | 6
11 | 0223367 7
12 | 001111 3445789
13 | 000 1 2 2 33345568
14 | 034579
15 | 36
16 |
17 | 6
```

FIGURE 5.19 Stemplot of the percentages of residents aged 65 and over in the 50 states, for Exercise 7.

school. A private college gives the SSHA to 18 of its incoming first-year women students. Their scores are:

154	109	137	115	152	140
154	178	101	103	126	126
137	165	165	129	200	148

Make a stemplot of these data. The overall shape of the distribution is irregular, as often happens when only a few observations are available. Are there any outliers? About where is the center of the distribution (the score with half the scores above it and half below)? What is the spread of the scores (ignoring any outliers)?

10. In 1798 the English scientist Henry Cavendish measured the density of the earth in a careful experiment with a torsion balance. Here are his 29 repeated measurements of the same quantity (the density of the earth relative to that of water) made with the same instrument. [S. M. Stigler, Do robust estimators work with real data? *Annals of Statistics*, 5 (1977): 1055–1078.]

5.50	5.61	4.88	5.07	5.26	5.55
5.36	5.29	5.58	5.65	5.57	5.53
5.62	5.29	5.44	5.34	5.79	5.10
5.27	5.39	5.42	5.47	5.63	5.34
5.46	5.30	5.75	5.68	5.85	

Make a stemplot of the data. Describe the distribution: Is it approximately symmetric or distinctly skewed? Are there gaps or outliers?

Describing Center: Mean and Median

11. Here are the scores of 18 first-year college women on the Survey of Study Habits and Attitudes (SSHA):

154	109	137	115	152	140
154	178	101	103	126	126
137	165	165	129	200	148

(a) Find the mean score from the formula for the mean. That is, add the 18 scores, record the sum,

and divide by 18. Then enter the data into your calculator and use the calculator's $\bar{x}$ button to obtain the mean. Verify that you get the same result.

(b) Your stemplot of the scores (Exercise 9) suggests that the score 200 is an outlier. Use your calculator to find the mean for the 17 observations that remain when you drop the outlier. How does the outlier change the mean?

12. The Major League Baseball single-season home run record is held by Barry Bonds of the San Francisco Giants, who hit 73 in 2001. Here are Bonds's home run totals from 1986 (his first year) to 2003:

16	25	24	19	33	25
34	46	37	33	42	40
37	34	49	73	46	45

(a) Make a stemplot of the data. Bonds's record year is a high outlier.

(b) Find his career mean and median number of home runs. How do these change when you drop the record 73? What general fact about the mean and median does your result illustrate?

◆ **13.** The distribution of income in the United States is skewed to the right. In 2002, the mean and median incomes of American households were $42,409 and $57,852. Which of these numbers is the mean and which is the median? Explain your reasoning.

◆ **14.** A report on the assets of American households says that the median net worth of households headed by someone aged less than 35 years is $11,600. The mean net worth of these same young households is $90,700. What explains the difference between these two measures of center?

15. Give an example of a small set of data for which the mean lies in the top 25% of the observations.

■ **16.** Create a set of five positive numbers (repeats allowed) that have median 10 and mean 7.

What thought process did you use to create your numbers?

Quartiles, the Five-Number Summary, and Boxplots

17. The stemplot in Figure 5.19 displays the distribution of the percents of residents aged 65 and over in the 50 states. Stemplots help you find the five-number summary because they arrange the observations in increasing order. Give the five-number summary of this distribution.

18. Here are the percents of the popular vote won by the successful candidate in each of the presidential elections from 1948 to 2004:

Year	1948	1952	1956	1960	1964
Percent	49.6	55.1	57.4	49.7	61.1

Year	1968	1972	1976	1980	1984
Percent	43.4	60.7	50.1	50.7	58.8

Year	1988	1992	1996	2000	2004
Percent	53.9	43.2	49.2	47.9	51.2

(a) Make a stemplot of the winners' percents.
(b) What is the median percent of the vote won by the successful candidate in presidential elections?
(c) Call an election a landslide if the winner's percent falls at or above the third quartile. Find the third quartile. Which elections were landslides?

19. Find the five-number summary of the city gas mileages for the 20 gas-powered two-seater cars in Table 5.2. (That is, omit the Honda Insight.) Your result should match the boxplot for these mileages in Figure 5.7.

20. Find the five-number summary of Cavendish's measurements of the density of the earth in Exercise 10. How is the symmetry of the distribution reflected in the five-number summary?

21. Table 5.3 gives carbon dioxide (CO_2) emissions per person for countries with population at least 20 million. The distribution is strongly skewed to the right. The United States and several other countries appear to be high outliers. Give the five-number summary. Explain why this summary suggests that the distribution is right-skewed.

22. Figure 5.4 is a histogram of the tuition and fees charged by the 56 four-year colleges in the state of Massachusetts. Here are those charges (in dollars), arranged in increasing order. [Data for 2003, from the College Board Web site, www.collegeboard.com.]

4,123	4,186	4,324	4,342	4,557
4,884	5,397	6,129	6,963	6,972
8,232	13,584	13,612	15,500	15,934
16,230	16,696	16,700	17,044	17,500
18,550	18,750	19,145	19,300	19,410
19,700	19,700	19,910	20,234	20,400
20,640	20,875	21,165	21,302	22,663
23,550	24,324	25,840	26,965	27,522
27,544	27,904	28,011	28,090	28,420
28,420	28,900	28,906	28,950	29,060
29,338	29,392	29,600	29,624	29,630
29,875				

Find the five-number summary and make a boxplot. What distinctive feature of the histogram do these summaries miss? Remember that numerical summaries are not a substitute for looking at the data.

◆ **23.** Figure 5.20 shows boxplots of the incomes of a large sample of people who have a high school diploma but no further education and another large group of people with a bachelor's degree but no higher degree. The data come from a Census Bureau survey, so that they represent all people aged 25 to 64 in the United States. Because there are a few extremely high incomes, the boxplot leaves out the highest 5% in each group. Based on the plot, compare the distributions of income for these two levels of education. Comment on both center and spread.

24. The data for Figure 5.20 include the incomes of 14,959 people whose highest level of education is a bachelor's degree.

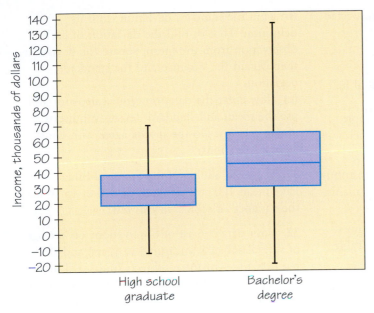

FIGURE 5.20 Boxplots comparing the incomes (in thousands of dollars) of people aged 25 to 64 years who worked full time for two levels of education. Because the highest incomes in any large group are very high indeed, the plot omits the top 5% of incomes in each group.

(a) What is the position of the median in the ordered list of incomes (1 to 14,959)? From the boxplot, about what is the median income of people with a bachelor's degree?

(b) What is the position of the first and third quartiles in the ordered list of incomes for these people? About what are the numerical values of Q_1 and Q_3?

■ **25.** How much oil the wells in a given field will ultimately produce is key information in deciding whether to drill more wells. Here are the estimated total amounts of oil recovered from 64 wells in the Devonian Richmond Dolomite area of the Michigan basin, in thousands of barrels. [J. Marcus Jobe and Hutch Jobe, A statistical approach for additional infill development, *Energy Exploration and Exploitation*, 18 (2000): 89–103.]

21.7	53.2	46.4	42.7	50.4
97.7	103.1	51.9	43.4	69.5
156.5	34.6	37.9	12.9	2.5
31.4	79.5	26.9	18.5	14.7
32.9	196.0	24.9	118.2	82.2
35.1	47.6	54.2	63.1	69.8
57.4	65.6	56.4	49.4	44.9

34.6	92.2	37.0	58.8	21.3
36.6	64.9	14.8	17.6	29.1
61.4	38.6	32.5	12.0	28.3
204.9	44.5	10.3	37.7	33.7
81.1	12.1	20.1	30.5	7.1
10.1	18.0	3.0	2.0	

(a) Graph the distribution and describe its main features.

(b) Find the mean and median of the amounts recovered. Explain how the relationship between the mean and the median reflects the shape of the distribution.

(c) Give the five-number summary and explain briefly how it reflects the shape of the distribution.

■ **26.** Look at the histogram of lengths of words in Shakespeare's plays, Figure 5.17. The heights of the bars tell us what percent of words have each length. The median length is the midpoint, the length with half of all words shorter and half longer. What is the median length of words used by Shakespeare? Similarly, what are the quartiles? Give the five-number summary for Shakespeare's word lengths.

■ **27.** A common criterion for detecting suspected outliers in a set of data is as follows:

1. Find the quartiles Q_1 and Q_3 and the *interquartile range* $IQR = Q_3 - Q_1$. The interquartile range is the spread of the central half of the data.

2. Call an observation an outlier if it falls more than $1.5 \times IQR$ above the third quartile or below the first quartile.

Which states are suspected outliers in the distribution of percent of Hispanics among adult residents, Table 5.1?

■ **28.** According to the $1.5 \times IQR$ rule explained in the previous exercise, which countries in Table 5.3 are suspected outliers? Based on your histogram (Exercise 5), do you agree with the rule's suggestions about which countries are and are not outliers?

Describing Spread: The Standard Deviation

29. Many standard statistical methods are intended for use with distributions that are symmetric and have no outliers. These methods start with the mean and standard deviation, $\bar{x}$ and s. An example of scientific data for which standard methods should work well are Cavendish's measurements of the density of the earth in Exercise 10.

(a) Summarize this data set by giving $\bar{x}$ and s.
(b) Find the median. Is the median quite close to the mean, as we expect it to be for symmetric distributions?

30. The level of various substances in the blood influences our health. Here are measurements of the level of phosphate in the blood of a patient, in milligrams of phosphate per deciliter of blood, made on six consecutive visits to a clinic.

5.6   5.2   4.6   4.9   5.7   6.4

A graph of only six observations gives little information, so we proceed to compute the mean and standard deviation.

(a) Find the mean from its definition. That is, find the sum of the six observations and divide by 6.

(b) Find the standard deviation from its definition. That is, find the deviation of each observation from the mean, square the deviations, then obtain the variance and the standard deviation.

(c) Now enter the data into your calculator and use the mean and standard deviation buttons to obtain $\bar{x}$ and s. Do the results agree with your hand calculations?

31. Scores of adults on the Stanford-Binet IQ test have mean 100 and standard deviation 15. What is the variance of scores on this test?

32. Your data consist of observations on the age of several subjects (measured in years) and the reaction times of these subjects (measured in seconds). In what units are each of the following descriptive statistics measured?

(a) The mean age of the subjects
(b) The standard deviation of the subjects' reaction times
(c) The variance of the subjects' reaction times
(d) The median age of the subjects

33. The mean $\bar{x}$ and standard deviation s measure center and spread but are not a complete description of a distribution. Data sets with different shapes can have the same mean and standard deviation. To demonstrate this fact, use your calculator to find $\bar{x}$ and s for these two small data sets. Then make a stemplot of each and comment on the shape of each distribution.

Data A:	9.14	8.14	8.74	8.77
	9.26	8.10	6.13	3.10
	9.13	7.26	4.74	
Data B:	6.58	5.76	7.71	8.84
	8.47	7.04	5.25	5.56
	7.91	6.89	12.50	

34. The shape of a distribution is a rough guide to whether the mean and standard deviation are a helpful summary of center and spread. For which of these distributions would $\bar{x}$ and s be useful? In each case, give a reason for your decision.

(a) Percents of Hispanics in the states, Figure 5.2.

(b) Iowa Test scores, Figure 5.3.
(c) Annual costs for Massachusetts colleges, Figure 5.4.

35. This exercise requires a calculator with a standard deviation button or statistical software on a computer. The observations

$$1001 \quad 1002 \quad 1003$$

have mean $\bar{x} = 1002$ and standard deviation $s = 1$. Adding a 0 in the center of each number, the next set becomes

$$10{,}001 \quad 10{,}002 \quad 10{,}003$$

The standard deviation remains $s = 1$ as more 0s are added. Use your calculator or computer to calculate the standard deviation of these numbers, adding extra 0s until you get an incorrect answer. How soon did you go wrong? This demonstrates that calculators and computers cannot handle an arbitrary number of digits correctly.

■ **36.** "Conservationists have despaired over destruction of tropical rainforest by logging, clearing, and burning." These words begin a report on a statistical study of the effects of logging in Borneo. [C. H. Cannon, D. R. Peart, and M. Leighton, Tree species diversity in commercially logged Bornean rainforest, *Science*, 281 (1998): 1366–1367.] Researchers compared forest plots that had never been logged (Group 1) with similar plots nearby that had been logged 1 year earlier (Group 2) and 8 years earlier (Group 3). All plots were 0.1 hectare in area. Here are the counts of trees for plots in each group, courtesy of Charles Cannon:

Group 1:	27	22	29	21	19	33
	16	20	24	27	28	19

Group 2:	12	12	15	9	20	18
	17	14	14	2	17	19

Group 3:	18	4	22	15	18
	19	22	12	12	

Give a complete comparison of the three distributions, using both graphs and numerical summaries. To what extent has logging affected the count of trees? The researchers used an analysis based on $\bar{x}$ and s. Explain why this is reasonably well justified.

■ **37.** This is a standard deviation contest. You must choose four numbers from the whole numbers 0 to 10, with repeats allowed.
(a) Choose four numbers that have the smallest possible standard deviation.
(b) Choose four numbers that have the largest possible standard deviation.
(c) Is more than one choice possible in either part (a) or part (b)? Explain.

Normal Distributions and the 68–95–99.7 Rule

38. The length of human pregnancies from conception to birth varies according to a distribution that is approximately normal, with mean 266 days and standard deviation 16 days. Draw a normal curve for this distribution on which the mean and standard deviation are correctly located. (*Hint:* First draw the curve, then mark the axis.)

39. Figure 5.21 shows a smooth curve used to describe a distribution that is not symmetric. The mean and median do not coincide. Which of the points marked is the mean of the distribution, and which is the median? Explain your answer.

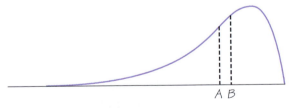

FIGURE 5.21 A curve describing a skewed distribution, for Exercise 39.

40. Sketch a smooth curve that describes a distribution that is symmetric but has two peaks (that is, two strong clusters of observations).

41. Bigger animals tend to carry their young longer before birth. The length of horse pregnancies from conception to birth varies according to a roughly normal distribution, with mean 336 days and standard deviation 3 days. Use the 68–95–99.7 rule to answer the following questions.

(a) Almost all (99.7%) horse pregnancies fall in what range of lengths?

(b) What percent of horse pregnancies are longer than 339 days?

42. Scores on the SAT college entrance test in a recent year were roughly normal, with mean 1026 and standard deviation 209.

(a) What was the range of the middle 68% of SAT scores?

(b) How high must a student score to be in the top 2.5% of SAT scores?

43. What are the quartiles of SAT scores, according to the distribution given in the previous exercise?

44. The Wechsler Adult Intelligence Scale (WAIS) is the most common "IQ test." The scale of scores is set separately for each age group and is approximately normal, with mean 100 and standard deviation 15. People with WAIS scores below 70 are considered mentally retarded when, for example, applying for Social Security disability benefits. By this criterion, what percent of adults are retarded?

45. The yearly rate of return on stock indexes (which combine many individual stocks) is approximately normal. Between 1900 and 2002, U.S. common stocks had a mean yearly return of 8.3%, with a standard deviation of about 20.3%. Take this normal distribution to be the distribution of yearly returns over a long period.

(a) In what range do the middle 95% of all yearly returns lie?

(b) Stocks can go down as well as up. What are the worst 2.5% of annual returns?

46. What is the range of the middle 50% of annual returns on stocks, according to the distribution given in the previous exercise? (*Hint:*

What two numbers mark off the middle 50% of any distribution?)

47. The concentration of the active ingredient in capsules of a prescription painkiller varies according to a normal distribution with $\mu = 10\%$ and $\sigma = 0.2\%$.

(a) What is the median concentration? Explain your answer.

(b) What range of concentrations covers the middle 95% of all the capsules?

(c) What range covers the middle half of all capsules?

48. Answer the following questions for the painkiller in the previous exercise.

(a) What percent of all capsules have a concentration of active ingredient higher than 10.4%?

(b) What percent have a concentration higher than 10.6%?

49. One reason that normal distributions are important is that they describe how the results of an opinion poll would vary if the poll were repeated many times. About 40% of adult Americans say they are afraid to go out at night because of crime. Take many randomly chosen samples of 1050 people. The proportions of people in these samples who stay home for fear of crime will follow the normal distribution with mean 0.4 and standard deviation 0.015. Use this fact and the 68–95–99.7 rule to answer these questions.

(a) In many samples, what percent of samples give results above 0.4? Above 0.43?

(b) In a large number of samples, what range contains the central 95% of proportions of people who stay home because of crime?

■ **50.** You can use the fact that all normal distributions are the same if we measure in standard deviations away from the mean to compare observations from different normal distributions. Scores expressed in standard deviation units are called *standard scores*.

(a) Scores on the ACT college entrance exam in a recent year were roughly normal, with mean 20.8 and standard deviation 4.8. Jermaine scores 27 on the ACT. Express his score in standard deviation units by calculating

$$\text{standard score} = \frac{\text{score} - \text{mean}}{\text{standard deviation}}$$

(b) Scores on the SAT college entrance exam in the same year were roughly normal, with mean 1026 and standard deviation 209. Tonya scores 1318 on the SAT. What is her standard score?
(c) Assuming that the ACT and the SAT measure the same thing, did Jermaine or Tonya have the higher score?

Chapter Review

Different varieties of the tropical flower *Heliconia* are fertilized by different species of hummingbirds. Over time, the lengths of the flowers and the form of the hummingbirds' beaks have evolved to match each other. Here are data on the lengths in millimeters of two varieties of these flowers on the island of Dominica.

H. caribaea Red

41.90	42.01	41.93	43.09	41.47
41.69	39.78	40.57	39.63	42.18
40.66	37.87	39.16	37.40	38.20
38.07	38.10	37.97	38.79	38.23
38.87	37.78	38.01		

H. caribaea Yellow

36.78	37.02	36.52	36.11	36.03
35.45	38.13	37.10	35.17	36.82
36.66	35.68	36.03	34.57	34.63

SOURCE: Thanks to Ethan J. Temeles of Amherst College for providing the data. His work is described in Ethan J. Temeles and W. John Kress, Adaptation in a plant-hummingbird association, *Science*, 300 (2003): 630–633.

Exercises 51 to 55 use these data.

51. Make stemplots of the lengths of each of the two varieties (red and yellow). Briefly describe the overall shape of the two distributions.

52. Find the five-number summaries of the two distributions of flower lengths. Make side-by-side boxplots to give a quick picture that compares the two distributions.

53. The biologists who collected the flower length data compared the two *Heliconia* varieties using statistical methods based on the mean and standard deviation. Find $\bar{x}$ and s for each variety. Based on your stemplots in Exercise 51, which distribution is more suitable for use of $\bar{x}$ and s as summaries? Why?

54. Your stemplot in Exercise 51 suggests that the distribution of lengths of yellow *Heliconia* flowers is roughly normal. Suppose that the distribution is exactly normal. Use the mean and standard deviation you found in Exercise 53 as the μ and σ of the distribution.

(a) What range of lengths covers the middle 50% of yellow flowers?
(b) What range of lengths covers the middle 95% of yellow flowers?

■ **55.** Continue to work with the normal distribution of lengths of yellow flowers from the previous exercise. The shortest red flower was 37.4 millimeters long. Using the 68–95–99.7 rule and the location of the quartiles in normal distributions, what can you say about what percent of yellow flowers that are longer than 37.4 millimeters?

APPLET EXERCISES

To do these exercises, go to www.whfreeman.com/fapp.

1. The *Mean and Median* applet allows you to place observations on a line and see their mean and median visually. Place two observations on the line, by clicking below it. Why does only one arrow appear?

2. In the *Mean and Median* applet, place three observations on the line by clicking below it, two close together near the center of the line and one somewhat to the right of these two. Pull the single rightmost observation out to the right. (Place the cursor on the point, hold down a mouse button, and drag the point.) How does the mean behave? How does the median behave? Explain briefly why each measure acts as it does.

3. In Example 16 we used the fact that SAT scores are close to normal and are adjusted so that the mean is close to 500 and the standard deviation is close to 100. (Actual scores in a particular year have slightly different mean and standard deviation.) Use the *Normal Curve* applet with $\mu = 500$ and $\sigma = 100$ to answer these questions:

(a) What proportion of SAT scores are above 640?
(b) What proportion of SAT scores are between 420 and 640? (If you drag one flag across the other, the applet shows the area between the flags.)

4. Because the resolution of Web browsers is limited, the *Normal Curve* applet can't always get exactly the values you want. Use the applet to come close to exact answers to these questions:

(a) How high must an SAT score be to fall in the top 10% of all scores?
(b) How high must an SAT score be to fall in the top 1% of all scores?

5. The 68–95–99.7 rule for normal distributions is a useful approximation. You can use the *Normal Curve* applet to see how accurate the rule is. Drag one flag across the other so that the applet shows the area under the curve between the two flags.

(a) Place the flags 1 standard deviation on either side of the mean. What is the area between these two values? What does the 68–95–99.7 rule say this area is?
(b) Repeat for locations 2 and 3 standard deviations on either side of the mean. Again compare the 68–95–99.7 rule with the area given by the applet.

✎ WRITING PROJECTS

1. Many social issues involve data and interpreting data. For example, income inequality (roughly speaking, the gap in income between people toward the top of the income scale and people toward the bottom) has increased in the past few decades. A good place to find data is on the Web site of the Census Bureau, www.census.gov. Click on "Income" and look for the latest report on income in the United States. Select a few facts from this detailed collection of income data to describe the extent of income inequality. Write a few paragraphs based on these facts.

2. Let's produce some data and describe them in order to gain insight into chance behavior. The

mathematics of chance is the topic of Chapter 8, but for now we will concentrate on data rather than math. You need two things: a standard six-sided die (raid your Monopoly game) and a thumbtack. Toss the die 180 times and record each outcome (1, 2, 3, 4, 5, or 6). Toss the thumbtack 100 times and record each outcome (point up or point down). Use both graphs and numbers to describe each set of results. Is the die roughly balanced, so that all six outcomes come up about equally? What about the thumbtack: Is point up or point down much more common?

SUGGESTED READINGS

CLEVELAND, WILLIAM S. *The Elements of Graphing Data,* rev. ed., Hobart Press, Summit, N.J., 1994. A careful study of the most effective elementary ways to present data graphically, with much sound advice on improving simple graphs.

MOORE, DAVID S. *The Basic Practice of Statistics,* 3rd ed., Freeman, New York, 2004. This text is a natural next step for more detail on all the material in Part II at about the same mathematical level. The first three chapters provide a more extensive treatment of the material of Chapter 5.

ROSSMAN, ALLAN J., and BETH L. CHANCE. *Workshop Statistics: Discovery with Data,* 2nd ed., Springer, New York, 2001. A different approach to basic data analysis, using hands-on activities. There are several versions, keyed to graphing calculators and to several different software packages.

SUGGESTED WEB SITES

The Web site of the U.S. Census Bureau, www.census.gov, is a prime source of information on many topics. The latest estimates for the populations of the United States and the world are on the home page, updated regularly. Data appear under many headings, not only as numbers but as maps. Try clicking on "American Fact Finder," then on "Maps and Geography," then on "Reference Maps." Enter your zip code, then zoom in on your street. The 1000 tables in the *Statistical Abstract of the United States* are also available. If you need data for a report, this is the place to start. Canadians can find similar help at the Web site of Statistics Canada: www.statcan.ca.

Interested in data about schools, colleges, and students? The National Center for Education Statistics, nces.ed.gov, is the place to look. The "Headlines" entries say what's new. When I last looked, what was new were two international comparisons of what students know (the United States does not do as well as it should) and data about school crime and safety.

Exploring Data: Relationships

A medical study finds that short women are more likely to have heart attacks than women of average height, while tall women have the fewest heart attacks. An insurance group reports that heavier cars have fewer accident deaths per 100,000 vehicles registered than do lighter cars. These and many other statistical studies look at the *relationship between two variables*. To understand such a relationship, we must often examine other variables as well. To conclude that shorter women have higher risk from heart attacks, for example, the researchers had to eliminate the effect of other variables such as weight and exercise habits. Our topic in this chapter is relationships between variables.

To study the relationship between two variables, we measure both variables on the *same individuals*. If we measure both the height and the weight of each of a large group of people, we know which height goes with each weight. These data allow us to study the connection between height and weight. A list of the heights and a separate list of the weights, two sets of single-variable data, do not show the connection between the two variables.

Height and weight are connected: Taller people also tend to be heavier. Neither height nor weight explains or causes the other. They go together in describing bigger or smaller people. Smoking and life expectancy are also connected, and in this case we think that smoking does explain or influence life expectancy: People who smoke more cigarettes per day tend not to live as long as those who smoke fewer. So we call smoking an *explanatory variable* and life expectancy a *response variable*.

Response Variable, Explanatory Variable

A **response variable** measures an outcome or result of a study. An **explanatory variable** is a variable that we think explains or causes changes in the response variables.

6.1 Displaying Relationships: Scatterplots

The most useful graph for displaying the relationship between two numerical variables is a *scatterplot*.

Scatterplot

A **scatterplot** shows the relationship between two numerical variables measured on the same individuals. The values of one variable appear on the horizontal axis, and the values of the other variable appear on the vertical axis. Each individual in the data appears as the point in the plot fixed by the values of both variables for that individual.

Always plot the explanatory variable, if there is one, on the horizontal axis (the *x*-axis) of a scatterplot. As a reminder, we usually call the explanatory variable *x* and the response variable *y*. If there is no explanatory-response distinction, either variable can go on the horizontal axis.

Beer and Blood Alcohol

EXAMPLE 1

How well does the number of beers a student drinks predict his or her blood alcohol content? In a study at Ohio State University, 16 student volunteers drank a randomly assigned number of cans of beer. Thirty minutes later, a police officer measured their blood alcohol content (BAC) in grams of alcohol per deciliter of blood. Here are the data:

Student	1	2	3	4	5	6	7	8
Beers	5	2	9	8	3	7	3	5
BAC	0.10	0.03	0.19	0.12	0.04	0.095	0.07	0.06

Student	9	10	11	12	13	14	15	16
Beers	3	5	4	6	5	7	1	4
BAC	0.02	0.05	0.07	0.10	0.085	0.09	0.01	0.05

FIGURE 6.1 Scatterplot of blood alcohol content (response variable) against the number of beers a student drinks (explanatory variable).

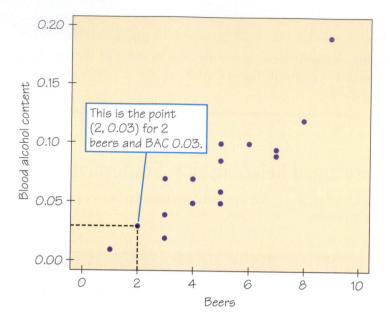

The students were equally divided between men and women and differed in weight and usual drinking habits. Because of this variation, many students don't believe that number of drinks predicts blood alcohol well. What do the data say?

Figure 6.1 is a scatterplot of these data. Because we think that number of beers helps explain BAC, it is the explanatory variable. We plot number of beers on the horizontal axis. One student drank 2 beers and had BAC 0.03. This student's point on the scatterplot is (2, 0.03), above $x = 2$ and to the right of $y = 0.03$. We have marked this point in Figure 6.1. ■

To interpret a scatterplot, apply the usual strategies of data analysis.

Examining a Scatterplot

In any graph of data, look for the **overall pattern** and for striking **deviations** from that pattern.

You can describe the overall pattern of a scatterplot by the **form, direction,** and **strength** of the relationship.

An important kind of deviation is an **outlier,** an individual value that falls outside the overall pattern of the relationship.

The *form* of the relationship in Figure 6.1 is roughly a straight-line pattern. If you look ahead a bit, Figure 6.3 draws a line through the plot to describe the overall pattern. The *direction* of the relationship is clear: as number of beers increases, BAC also increases. We call this a *positive association* between the two variables.

Positive Association, Negative Association

Two variables are **positively associated** when above-average values of one tend to accompany above-average values of the other and below-average values also tend to occur together.

Two variables are **negatively associated** when above-average values of one accompany below-average values of the other, and vice versa.

The *strength* of a relationship describes how closely the points in a scatterplot follow a simple form such as a straight line. Figure 6.1 shows only a small amount of scatter about the straight line, so the relationship is moderately strong. We will soon learn a numerical measure of the strength of a straight-line relationship.

State SAT Scores

EXAMPLE 2

More than a million high school seniors take the SAT college entrance examination each year. We sometimes see the states "rated" by the average SAT scores of their seniors. For example, Illinois students average 1179 on the SAT, which looks better than the 1038 average of Massachusetts students. Rating states by SAT scores makes little sense, because average SAT score is largely explained by what percent of a state's students take the SAT. The scatterplot in Figure 6.2 shows this. Illinois has a high mean score, but only 11% of Illinois seniors take the SAT. In Massachusetts, on the other hand, 82% of seniors take the exam. ■

The *form* of Figure 6.2 is a bit irregular, but there are two distinct *clusters* of states. In one cluster, more than half of high school seniors take the SAT, and the mean scores are low. Fewer than 40% of seniors in states in the other cluster take the SAT—fewer than 20% in most of these states—and these states have higher mean scores. Clusters in a graph suggest that the data describe several distinct kinds of individuals. The two clusters in Figure 6.2 do in fact describe two distinct sets of states. There are two common college entrance examinations, the SAT and the ACT. Each state tends to prefer one or the other. In ACT states (the left cluster in Figure 6.2) most students who take the SAT are applying to selective out-of-state colleges. This select group performs well. In SAT states (the right cluster), many seniors take the SAT, and this broader group has a lower mean score.

The relationship in Figure 6.2 also has a clear *direction:* States in which a higher percent of students take the SAT tend to have lower mean scores. This is true both between the clusters and within each cluster. That is, there is a *negative association* between the two variables.

FIGURE 6.2 Scatterplot of the mean SAT scores in the states (response variable) against the percent of high school seniors who take the SAT (explanatory variable).

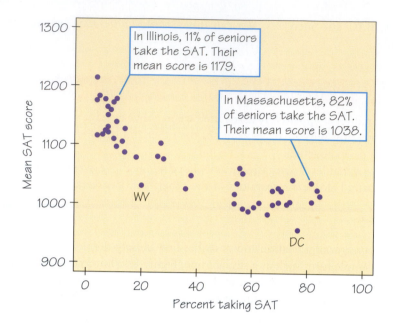

There are no clear *outliers* in Figure 6.2, but each cluster does include a state whose mean SAT score is lower than we would expect from the percent of its students who take the SAT. These points are West Virginia (WV), in the cluster of ACT states, and the District of Columbia (DC)—a city rather than a state—in the cluster of SAT states.

6.2 Regression Lines

If a scatterplot shows a straight-line relationship, we would like to summarize this overall pattern by drawing a line on the scatterplot. A *regression line* summarizes the relationship between two variables, but only in a specific setting: when one of the variables helps explain or predict the other. That is, regression describes a relationship between an explanatory variable and a response variable.

Regression Line

A **regression line** is a straight line that describes how a response variable y changes as an explanatory variable x changes. We often use a regression line to **predict** the value of y for a given value of x.

EXAMPLE 3   Predicting Blood Alcohol

Figure 6.1 shows a straight-line relationship between how many beers a student drinks and his or her blood alcohol content (BAC) 30 minutes later. Figure 6.3

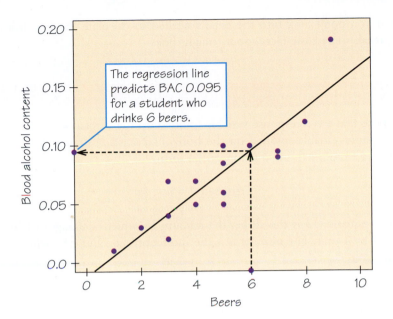

FIGURE 6.3 A regression line for predicting blood alcohol content from the number of beers a student drinks.

repeats this scatterplot, and adds a regression line that we can use to predict BAC for a student based on the number of beers consumed.

Figure 6.3 shows the prediction in graphical form for a student who drinks 6 beers. Start at $x = 6$, go up to the line, then over to the y-axis at the left. We hit the y-axis at BAC 0.095. This is the BAC that corresponds to 6 beers, according to the regression line. (The legal limit for driving in most states is 0.08.) The line represents only the overall pattern of the data, so the BAC of your friends George and Eileen after 6 beers will probably not be exactly 0.095. But because the points for the 16 students in the Ohio State study are not far from the line, we expect the prediction to be reasonably accurate.

It is easier to use the *equation of the line* for prediction. The equation of the line in Figure 6.3 is

$$\text{predicted BAC} = -0.0127 + 0.01796 \times \text{beers}$$

For a student who drinks 6 beers, we have

$$\text{predicted BAC} = -0.0127 + (0.01796)(6) = 0.095$$

You can plot a line from its equation by substituting two values of x, such as $x = 2$ and $x = 8$. Find the corresponding values of y, plot the two points, and draw the line through them. ■

Statistical software and many calculators will give you the equation of a regression line from keyed-in data. You should know how to use a regression line even if you don't look into the details needed to calculate the line from data. First, recall some basic facts about straight lines.

Straight Lines

Suppose that y is a response variable (plotted on the vertical axis) and x is an explanatory variable (plotted on the horizontal axis). A straight line for predicting y from x has an equation of the form

$$\text{predicted } y = a + bx$$

In this equation, b is the **slope**, the amount by which y changes when x increases by 1 unit. The number a is the **intercept**, the value of y when $x = 0$.

The slope of the line in Example 3 is $b = 0.01796$. This says that as we move to the right along the line, predicted BAC goes up by 0.01796 for each additional beer a student drinks. The slope tells us how quickly y changes as we change x, so it is almost always important for understanding the data. The slope is positive ($b > 0$) when there is a positive association between the two variables. It is negative when there is a negative association.

You might think that a big slope (either positive or negative) says that there are big changes in y as x changes and that a small slope means that x has little influence on y. Unfortunately, the size of a slope depends mainly on the units in which we measure the two variables. The slope of the regression in Example 3 is $b = 0.01796$ when we measure BAC y in grams of alcohol per deciliter of blood. That is, when beers consumed increases by 1, BAC increases by 0.01796 grams. There are 1000 milligrams in a gram, so if we measured BAC in milligrams of alcohol, the slope would be 1000 times as large, $b = 17.96$. When beers consumed increases by 1, BAC increases by 17.96 milligrams. *You can't say how important a relationship is by looking at how big the slope is.*

The intercept of the regression line in Example 3 is $a = -0.0127$. This is the predicted value of y when $x = 0$. Although we need the value of the intercept to draw the line, it is statistically meaningful only when x can actually take values close to zero. Even then, you should think of the intercept as describing the line rather than taking it seriously as a prediction. If a student drinks no beers, his or her blood alcohol should be exactly zero. The intercept of the regression line in Example 3 is close to zero, but it is not exactly zero.

6.3 Correlation

A scatterplot displays the form, direction, and strength of the relationship between two numerical variables. Straight-line relations are particularly important

SPOTLIGHT 6.1   Scatterplot Smoothers

Our eyes are good at seeing the overall pattern of a scatterplot. Sometimes the pattern has a simple form, such as a straight line, that we can draw on the plot to summarize the pattern. Clever software for "smoothing" a scatterplot can pick out much more complex overall patterns.

Crash a motorcycle into a wall. The rider, fortunately, is a dummy with an instrument to measure acceleration (change of velocity) mounted in its head. The figure is a scatterplot of the acceleration of the dummy's head against time in thousandths of a second. Acceleration is measured in g's, or multiples of the acceleration due to gravity at the earth's surface. The motorcycle approaches the wall at a constant speed (acceleration near 0). As it hits, the dummy's head snaps forward and decelerates violently (negative acceleration reaching more than 100 g's), then snaps back again (up to 75 g's) and wobbles a bit before coming to rest.

Our eyes can see that the plot has a clear, but complicated, overall pattern. A scatterplot smoother picks out this pattern and draws a line on the plot to display it.

Smoothing a scatterplot. Software draws the smooth curve to describe the overall pattern of the relationship.

because a straight line is a simple pattern that is quite common. We say a straight-line association is strong if the points lie close to a line, and weak if they are widely scattered about a line. This is vague. We need to follow our strategy for data analysis by using a numerical measure along with the graph. *Correlation* is the measure we use.

Correlation

The **correlation** measures the direction and strength of the straight-line relationship between two numerical variables. Correlation is usually written as r.

A correlation r is always a number between -1 and 1. It has the same sign as the slope of a regression line: $r > 0$ for positive association and $r < 0$ for negative association.

Perfect correlation $r = 1$ or $r = -1$ occurs only when all points lie exactly on a straight line. The correlation moves away from 1 or -1 as the straight-line relationship gets weaker. Correlation $r = 0$ indicates no straight-line relationship.

EXAMPLE 4   Scatterplots and Correlation

The scatterplots in Figure 6.4 illustrate how values of r closer to 1 or -1 correspond to stronger straight-line relationships. To make the meaning of r clearer, the standard deviations of both variables in these plots are equal and the horizontal and vertical scales are the same. In general, it is not so easy to guess the value of r from the appearance of a scatterplot. Changing the plotting scales in a scatterplot can alter the appearance of the graph, but it does not change the correlation.

We said that Figure 6.1 shows a moderately strong positive straight-line relationship between how many beers a student drinks and his or her blood alcohol content. The correlation between these variables is $r = 0.894$. Figure 6.2, despite the clusters, also shows a quite strong straight-line relationship between the percent of a state's high school seniors who take the SAT exam and their mean SAT score. The association is negative: Higher percents taking the SAT go with lower mean scores. The correlation is $r = -0.877$. ∎

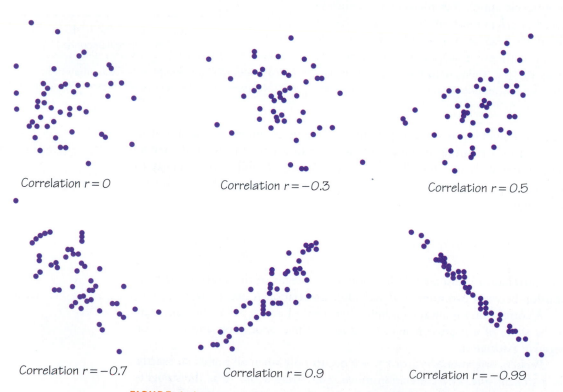

Correlation $r = 0$ Correlation $r = -0.3$ Correlation $r = 0.5$

Correlation $r = -0.7$ Correlation $r = 0.9$ Correlation $r = -0.99$

FIGURE 6.4 How the correlation r measures the direction and strength of a straight-line association.

Here are more facts about the correlation r:

1.   Unlike regression, correlation makes no distinction between explanatory and response variables. It makes no difference which variable you call x and which you call y in interpreting a correlation.

2.   Correlation measures the strength of only a straight-line relationship between two numerical variables. Correlation does not describe curved relationships, no matter how strong they are. Correlation makes no sense for non-numerical variables such as gender and occupation.

3.   The correlation r does not change when we change the units of measurement of x, y, or both. Measuring height in inches rather than centimeters and weight in pounds rather than kilograms does not change the correlation between height and weight. The correlation r itself has no unit of measurement; it is just a number. So correlation is easier to interpret than the slope of a regression line, which changes when the units of x and y change.

4.   Like the mean and standard deviation, the correlation is strongly affected by a few outlying observations. Use r with caution when outliers appear in the scatterplot.

In practice, you will use a calculator or software to find the correlation from keyed-in data. That's fortunate, because using the formula for correlation is quite a bit of work. Nonetheless, all the properties of r come from the formula that defines it. Here is the definition.

Definition of Correlation

Suppose that we have data on variables x and y for n individuals. The means and standard deviations of the two variables are $\bar{x}$ and s_x for the x-values, and $\bar{y}$ and s_y for the y-values. The correlation r between x and y is

$$r = \frac{1}{n-1}\left[\left(\frac{x_1 - \bar{x}}{s_x}\right)\left(\frac{y_1 - \bar{y}}{s_y}\right) + \left(\frac{x_2 - \bar{x}}{s_x}\right)\left(\frac{y_2 - \bar{y}}{s_y}\right) + \cdots + \left(\frac{x_n - \bar{x}}{s_x}\right)\left(\frac{y_n - \bar{y}}{s_y}\right)\right]$$

This formula starts by *standardizing* each observation value. That is, subtract the mean for that variable from the observation and then divide by the standard deviation. Standardizing turns each original data value into "how many standard deviations above or below the mean." This removes the original units and explains why r has no units and doesn't change when we change the units of x or y. The formula says that the correlation is an average of the products of the standardized x and y values for n individuals. Exercise 26 asks you to calculate a correlation step by step from the definition to solidify its meaning.

6.4 Least-Squares Regression

In Example 3, we used the straight line given by the equation

$$\text{predicted BAC} = -0.0127 + 0.01796 \times \text{beers}$$

to predict blood alcohol content from the number of beers consumed. Where did this equation come from? We will now see that the equation is the result of saying what we mean by the *best* line for predicting BAC from beers consumed.

SPOTLIGHT 6.2 Regression in Action

No other statistical method is used as much as regression. Here are some applications of regression that are a bit unusual.

Did the vote counters cheat? Republican Bruce Marks was ahead of Democrat William Stinson when the voting machines were tallied in their Pennsylvania election. But Stinson was ahead after absentee ballots were counted by the Democrats who controlled the election board. A court fight followed. The court called in a statistician, who used regression with data from past elections to predict the counts of absentee ballots based on the results from the voting machines. Marks's lead of 564 votes from the machines predicted that he would get 133 more absentee votes than Stinson. In fact, Stinson got 1025 more absentee votes than Marks. This looks suspicious.

Is regression garbage? No—but garbage can be the setting for regression. The Census Bureau once asked if weighing a neighborhood's garbage would help count its people. So 63 households had their garbage sorted and weighed. It turned out that pounds of plastic in the trash gave the best garbage prediction of the number of people in a neighborhood. Alas, the prediction wasn't good enough to help the Census Bureau.

How much is this injury worth? Jury Verdict Research makes money from regression. The company collects data on more than 20,000 jury verdicts in personal injury lawsuits each year, recording more than 30 variables that describe each case. Multiple regression extends regression to allow more than one explanatory variable to contribute to explaining a response variable. Jury Verdict Research uses multiple regression to predict how much a jury will award in a new case. Lawyers pay for these predictions and use them to negotiate settlements with insurance companies.

Once we say exactly what we mean by the best line, finding that line becomes a mathematical problem. The line in Example 3 is the solution to this problem for the beer and BAC data.

Different people might draw different lines by eye on a scatterplot. This is especially true when the points are widely scattered. We need a way to draw a regression line that doesn't depend on our guess as to where the line should go. No line will pass exactly through all the points, but we want one that is as close as possible. We will use the line to predict y from x, so we want a line that is as close as possible to the points in the *vertical* direction. That's because the prediction errors we make are errors in y, which is the vertical direction in the scatterplot.

The regression line in Example 3 is

$$\text{predicted BAC} = -0.0127 + 0.01796 \times \text{beers}$$

The first student drank 5 beers, so the predicted blood alcohol level is

$$\text{predicted BAC} = -0.0127 + (0.01796)(5) = 0.077$$

The actual BAC for this student was 0.10. The prediction error is

$$\text{error} = \text{observed BAC} - \text{predicted BAC}$$
$$= 0.10 - 0.077 = 0.023$$

Errors are positive if the observed response lies above the line, and negative if the response lies below the line. The "best" regression line makes these prediction errors as small as possible. The most common way to make the errors as a group as small as possible is *least-squares regression*. The line in Example 3 is the least-squares regression line.

Least-Squares Regression Line

The **least-squares regression line** of y on x is the line that makes the sum of the squares of the vertical distances of the data points from the line as small as possible.

The least-squares idea says what we mean by the best-fitting line. How can we find this line from data? Starting with n observations on variables x and y, finding the line that makes the sum of the squares of the vertical errors as small as possible is a mathematical problem. Here is the solution to this problem.

::::: Equation of the Least-Squares Regression Line

We have data on an explanatory variable x and a response variable y for n individuals. From the data, calculate the means $\bar{x}$ and $\bar{y}$ and the standard deviations s_x and s_y of the two variables, as well as their correlation r. The **least-squares regression line** is the line

$$\text{predicted } y = a + bx$$

with **slope**

$$b = r\frac{s_y}{s_x}$$

and **intercept**

$$a = \bar{y} - b\bar{x}$$

This equation gives insight into the behavior of least-squares regression by showing that it is related to the means and standard deviations of the x and y observations and to the correlation between x and y. For example, it is clear that the slope b always has the same sign as the correlation r. In practice, you don't need to calculate the means, standard deviations, and correlation first. Statistical software or your calculator will give the slope b and intercept a of the least-squares line from keyed-in values of the variables x and y.

:::::EXAMPLE 5   Least-Squares Regression of BAC on Number of Beers

Go back to the data in Example 1. Use your calculator to verify that the mean and standard deviation of x, number of beers consumed, are

$$\bar{x} = 4.8125 \qquad \text{and} \qquad s_x = 2.1975$$

The mean and standard deviation of y, blood alcohol content, are

$$\bar{y} = 0.07375 \qquad \text{and} \qquad s_y = 0.04414$$

The correlation between number of beers and BAC is $r = 0.8943$. The least-squares regression line of BAC y on number of beers x has slope

$$b = r\frac{s_y}{s_x} = 0.8943 \times \frac{0.04414}{2.1975}$$
$$= 0.01796$$

and intercept

$$a = \bar{y} - b\bar{x} = 0.07375 - (0.01796)(4.8125)$$
$$= -0.0127$$

The equation of the least-squares line is therefore

$$\text{predicted } y = -0.0127 + 0.01796x$$

just as we claimed earlier.

When doing calculations like this by hand, you may need to carry extra decimal places in the preliminary calculations to get accurate values of the slope and intercept. Using software or a calculator with a regression function eliminates this worry. ■

You now see that correlation and least-squares regression are closely connected. The expression $b = rs_y/s_x$ for the slope says that along the regression line, a change of one standard deviation in x corresponds to a change of r standard deviations in y. When the variables are perfectly correlated ($r = 1$ or $r = -1$), the change in the predicted response is the same (in standard deviation units) as the change in x. Otherwise, because $-1 \leq r \leq 1$, the change in the predicted y is less than the change in x. As the correlation grows less strong, the prediction moves less in response to changes in x.

6.5 Interpreting Correlation and Regression

Correlation and regression are among the most-used statistical methods. Here are a few cautions to keep in mind when you use these methods or see others use them.

Both the correlation r and the least-squares regression line can be strongly influenced by a few outlying points. Always make a scatterplot before doing any calculations. Here is an artificial example that illustrates what can happen.

Beware the Outlier!   EXAMPLE 6

Figure 6.5 shows a scatterplot of data that have a strong positive straight-line relationship. In fact, the correlation is $r = 0.987$, close to the $r = 1$ of a perfect straight line. The line on the plot is the least-squares regression line for predicting y from x. One point is an extreme outlier in both the x and y directions. Let's examine the influence of this outlier.

First, suppose we drop the outlier. The correlation for the 5 remaining points (the cluster at the lower left) is $r = 0.523$. The outlier extends the straight-line pattern and greatly increases the correlation.

FIGURE 6.5 The
outlier increases the
correlation and fixes the
location of the least-
squares line.

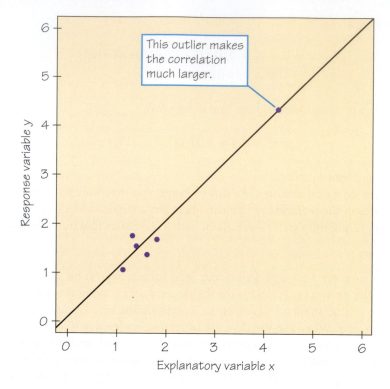

FIGURE 6.5 The
outlier increases the
correlation and fixes the
location of the least-
squares line.

Next, grab the outlier and pull it straight down, as in Figure 6.6. The least-squares line chases the outlier down, pivoting until it has a negative slope. This is the least-squares idea at work: The line stays close to all 6 points, but its location is determined almost entirely by the one outlier. Of course, the correlation is now also negative, $r = -0.796$. Never trust a correlation or a regression line until you have plotted the data. ■

The best way to see how outlying points can influence the correlation and the regression line is to use the *Correlation and Regression* applet. Applet Exercise 1 asks you to animate Example 6 above, watching r change and the regression line move as you pull the outlier down.

Correlation and regression *describe* relationships. *Interpreting* relationships requires more thought. *Often the relationship between two variables is strongly influenced by other variables.* You should always think about the possible effect of other variables before you draw conclusions based on correlation or regression.

EXAMPLE 7   Magic Mozart?

A symphony orchestra once advertised a "Mozart for Minors" program with this statement: "Question: Which students scored 51 points higher in verbal skills and 39 points higher in math? Answer: Students who had experience in music."

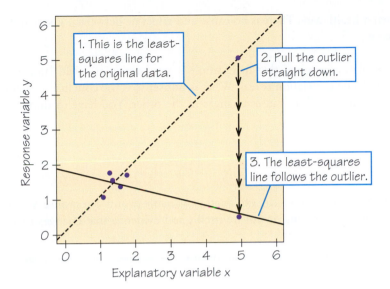

FIGURE 6.6 Moving the outlier changes the correlation and moves the least-squares line.

We could as well answer "Children who played soccer." Why? Children with prosperous and well-educated parents are more likely than poorer children to have experience with music and also to play soccer. They are also likely to attend good schools, get good health care, and be encouraged to study hard. These advantages lead to high test scores. Experience with music and soccer are associated with high scores just because they go along with the other advantages of having prosperous and educated parents. ■

Example 7 brings us to the most important caution about correlation and regression. When we study the relationship between two variables, we often hope to show that changes in the explanatory variable *cause* changes in the response variable. A strong association between two variables is not enough to draw conclusions about cause and effect. Sometimes an observed association really does reflect cause and effect. Drinking more beer does cause an increase in blood alcohol. But in many cases, as in Example 7, a strong association is explained by other variables that influence both *x* and *y*. Here is another example.

Does TV Make You Live Longer?

EXAMPLE 8

Measure the number of television sets per person *x* and the average life expectancy *y* for the world's nations. There is a high positive correlation: Nations with many TV sets have higher life expectancies.

The basic meaning of causation is that by changing *x* we can bring about a change in *y*. Could we lengthen the lives of people in Rwanda by shipping them TV sets? No. Rich nations have more TV sets than poor nations. Rich nations also have longer life expectancies because they offer better nutrition, clean water, and better health care. There is no cause-and-effect tie between TV sets and length of life. ■

Correlations such as that in Example 8 are sometimes called "nonsense correlations." The correlation is real. What is nonsense is the conclusion that changing one of the variables causes changes in the other. You have to think, not just calculate.

Association Does Not Imply Causation

An association between an explanatory variable x and a response variable y, even if it is very strong, is not by itself good evidence that changes in x actually cause changes in y.

Here is a final example in which we use a scatterplot, correlation, and a regression line to understand data.

EXAMPLE 9   Does Fidgeting Keep You Slim?

Some people don't gain weight even when they overeat. Perhaps fidgeting and other "nonexercise activity" (NEA) explains why—the body might spontaneously increase nonexercise activity when fed more. Researchers deliberately overfed 16 healthy young adults for 8 weeks. They measured fat gain (in kilograms) and, as an explanatory variable, increase in energy use (in calories) from activity other than deliberate exercise—fidgeting, daily living, and the like. Here are the data.

NEA increase (cal)	−94	−57	−29	135	143	151	245	355
Fat gain (kg)	4.2	3.0	3.7	2.7	3.2	3.6	2.4	1.3

NEA increase (cal)	392	473	486	535	571	580	620	690
Fat gain (kg)	3.8	1.7	1.6	2.2	1.0	0.4	2.3	1.1

SOURCE: J. A. Levine, N. L. Eberhardt, and M. D. Jensen, Role of nonexercise activity thermogenesis in resistance to fat gain in humans, *Science,* 283 (1999): 212–214.

Figure 6.7 is a scatterplot of these data. The plot shows a moderately strong negative straight-line association with no outliers. The correlation is $r = -0.7786$. People with larger increases in nonexercise activity do indeed gain less fat.

Correlation says that there is a reasonably strong straight-line relationship. A regression line describes this relationship. The least-squares line is

$$\text{predicted fat gain} = 3.505 - (0.00344 \times \text{NEA increase})$$

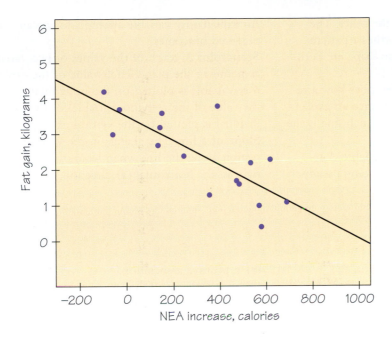

FIGURE 6.7 Scatterplot of fat gain (response variable) versus increase in nonexercise activity (explanatory variable) for 16 volunteers who were deliberately overfed. The line is the least-squares regression line for predicting fat gain.

We have added this line to the plot in Figure 6.7. The slope $b = -0.00344$ says that fat gained goes down by 0.00344 kilogram (or 3.44 grams) for each added calorie of NEA. The intercept, $a = 3.505$ kilograms, is the estimated fat gain if NEA does not change when a person overeats.

These data come from an *experiment* that controlled the subjects' diet and activities. The researchers measured many other variables that might influence fat gain and showed that they don't explain the observed relationship. We can be reasonably confident that NEA actually *causes* subjects to gain less fat. ■

REVIEW VOCABULARY

Correlation A measure of the direction and strength of the straight-line relationship between two numerical variables. Correlations take values between 0 (no straight-line relationship) and ±1 (perfect straight-line relationship).

Intercept of a line The vertical (y) coordinate of the point on the line above 0 on the horizontal (x) axis.

Least-squares regression line A line drawn on a scatterplot that makes the sum of the squares of the vertical distances of the data points from the line as small as possible. The regression line can be used to predict the response variable y for a given value of the explanatory variable x.

Negative association Two variables are negatively associated if above-average values of one tend to go with below-average values of the other. The scatterplot has a northwest-to-southeast pattern, and the correlation and regression slope are both negative.

Outlier An outlier in a scatterplot is a point that lies outside the overall pattern of the other points. Outliers sometimes strongly influence the value of the correlation and the position of the least-squares regression line.

Positive association Two variables are positively associated if above-average values of one tend to go with above-average values of the other. The scatterplot has a southwest-to-northeast pattern, and the correlation and regression slope are both positive.

Regression line Any line that describes how a response variable y changes as we change an explanatory variable x. The most common such line is the least-squares regression line.

Response variable, explanatory variable A response variable measures an outcome of a study.

An explanatory variable attempts to explain the observed outcomes.

Scatterplot A graph of the values of two variables as points in the plane. Each value of the explanatory variable is plotted on the horizontal axis, and the value of the response variable for the same individual is plotted on the vertical axis.

Slope of a line The change in the vertical (y) direction along the line when we move 1 unit to the right in the horizontal (x) direction.

 SKILLS CHECK

1. You have data for many families on the parents' income and the years of education their eldest child completes. When you make a scatterplot, the explanatory variable on the x axis

(a) is parents' income.
(b) is years of education.
(c) doesn't matter.

2. You have data for many families on the parents' income and the years of education their eldest child completes. You expect to see

(a) a positive association.
(b) very little association.
(c) a negative association.

3. Figure 6.8 is a scatterplot of reading test scores against IQ test scores for 15 fifth-grade children. There is one low outlier in the plot. The IQ and reading scores for this child are

(a) IQ = 10, Reading = 124.
(b) IQ = 124, Reading = 72.
(c) IQ = 124, Reading = 10.

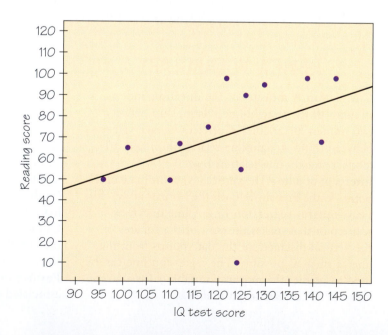

FIGURE 6.8 Scatterplot of the reading test scores of fifth-grade children (response variable) against the children's IQ scores (explanatory variable).

4. The line in Figure 6.8 is a regression line for predicting reading score from IQ score. If another child in this class has IQ score 110, you predict the reading score to be close to

(a) 50.
(b) 60.
(c) 70.

5. The slope of the line in Figure 6.8 is closest to

(a) −1.
(b) 0.
(c) 1.

6. The points on a scatterplot lie close to the line whose equation is $y = 4 - 3x$. The slope of this line is

(a) 4.
(b) 3.
(c) −3.

7. Fred keeps his savings in his mattress. He began with $500 from his mother and adds $100 each year. His total savings y after x years are given by the equation

(a) $y = 500 + 100x$.
(b) $y = 100 + 500x$.
(c) $y = 500 + x$.

8. Starting with a fresh bar of soap, you weigh the bar each day after you take a shower. Then you find the regression line for predicting weight from number of days elapsed. The slope of this line will be

(a) positive.
(b) negative.
(c) can't tell without seeing the data.

9. The amount of water discharged by the Mississippi River has changed over time in roughly a straight-line pattern. A regression line for predicting water discharged (in cubic kilometers) from year is

$$\text{predicted discharge} = -7792 + (4.226 \times \text{year})$$

How much (on the average) does the volume of water increase with each passing year?

(a) −7792 cubic kilometers

(b) 7792 cubic kilometers
(c) 4.226 cubic kilometers

10. According to the regression line in the previous exercise, the predicted Mississippi River discharge in the year 2000 is

(a) −660 cubic kilometers.
(b) 660 cubic kilometers.
(c) 16,244 cubic kilometers.

11. You have data on the body weight x and brain weight y for many species of mammals. Body weight is given in kilograms and brain weight is given in grams. There are 1000 grams in a kilogram. The slope of the regression line for predicting y from x is $b = 1.3$. If brain weight were given in kilograms, the slope would

(a) still be 1.3.
(b) change to 1300.
(c) change to 0.0013.

12. What are all the values that a correlation r can possibly take?

(a) $0 \leq r \leq 1$
(b) $-1 \leq r \leq 1$
(c) $0 \leq r$

13. Starting with a fresh bar of soap, you weigh the bar each day after you shower. The correlation between the weight of the soap and the number of days elapsed will be

(a) positive.
(b) negative.
(c) can't tell without seeing the data.

14. The points on a scatterplot lie very close to the line whose equation is $y = 4 - 3x$. The correlation between x and y is close to

(a) −3.
(b) −1.
(c) 1.

15. If we leave out the low outlier, the correlation for the remaining 14 points in Figure 6.8 is closest to

(a) 0.5.
(b) −0.5.
(c) 0.95.

16. By looking at the equation of the least-squares regression line in Exercise 9, you can see that the correlation is

(a) greater than zero.

(b) less than zero.

(c) can't tell without seeing the data.

17. The correlation between brain weight and body weight in Exercise 11 is $r = 0.86$. If brain weight had been measured in kilograms rather than grams, the correlation would

(a) still be 0.86.

(b) change to 860.

(c) change to 0.00086.

18. High coffee prices give farmers in Indonesia an incentive to cut forest in order to plant more coffee. Here are data on coffee price x (cents per pound) and percent y of deforestation in a national park for five years:

x	29	40	54	55	72
y	0.49	1.59	1.69	1.82	3.10

Use your calculator: What is the correlation between x and y?

(a) About 0.17

(b) About 0.91

(c) About 0.96

19. Look again at the coffee data in the previous exercise. Use your calculator: The equation of the least-squares regression line for predicting y from x is close to

(a) predicted $y = -0.98 + 0.05x$.

(b) predicted $y = 0.05 - 0.98x$.

(c) predicted $y = 0.21 + 0.17x$.

20. There is a strong positive correlation between the number of firefighters at a fire and the amount of damage the fire does. The reason for this is that

(a) more firefighters cause more damage at the fire scene.

(b) bigger fires require more firefighters and also do more damage.

(c) more damage requires more firefighters to clean it up.

EXERCISES   ■ Challenge   ◆ Discussion

Some exercises require use of a calculator (or software) that will find correlation and the slope and intercept of the least-squares regression line from keyed-in data.

1. In each of the following situations, is it more reasonable to simply explore the relationship between the two variables or to view one of the variables as an explanatory variable and the other as a response variable? In the latter case, which is the explanatory variable and which is the response variable?

(a) The amount of time spent studying for a statistics exam and the grade on the exam.

(b) The weight in kilograms and height in centimeters of a person.

(c) Inches of rain in the growing season and the yield of corn in bushels per acre.

(d) A student's scores on the SAT math exam and the SAT verbal exam.

Scatterplots

2. Figure 6.9 shows the calories and salt content (milligrams of sodium) in 17 brands of meat hot dogs. Describe the overall pattern (form, direction, and strength) of these data. In what way is the point marked A unusual?

◆ **3.** Figure 6.10 is a scatterplot of data from the World Bank. The individuals are all the world's nations for which data are available. The explanatory variable is a measure of how rich a country is, the gross domestic product (GDP) per person. GDP is the total value of the goods and services produced in a country, converted into dollars. The response variable is life expectancy at

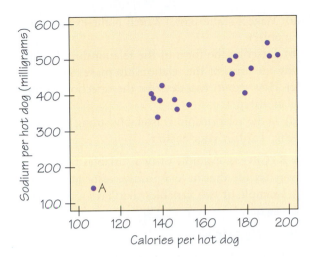

FIGURE 6.9 Scatterplot of sodium content versus calories in 17 brands of meat hot dogs, for Exercise 2.

birth. Three African nations are outliers, with lower life expectancy than usual for their GDP. A full study would ask what special circumstances explain these outliers.

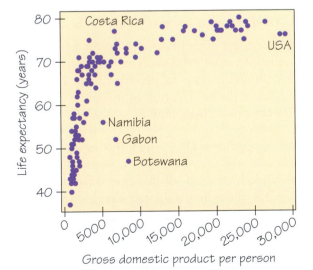

FIGURE 6.10 Scatterplot of the life expectancy of people in many nations against each nation's gross domestic product per person, for Exercise 3.

(a) Describe the direction and form of the relationship. Aside from the outliers, it is moderately strong.
(b) Explain why the direction and form of this relationship make sense.

4. Professor Moore swims 2000 yards regularly in a vain attempt to undo middle age. Here are his times (in minutes) and his pulse rate (in beats per minute) after swimming for 23 sessions in the pool:

Time	34.12	35.72	34.72	34.05	34.13
Pulse	152	124	140	152	146

Time	35.72	36. 17	35.57	35.37	35.57
Pulse	128	136	144	148	144

Time	35.43	36.05	34.85	34.70	34.75
Pulse	136	124	148	144	140

Time	33.93	34.60	34.00	34.35	35.62
Pulse	156	136	148	148	132

Time	35.68	35.28	35.97
Pulse	124	132	139

(a) Make a scatterplot. (Which is the explanatory variable?)
(b) Is the association between these variables positive or negative? Explain why you expect the relationship to have this direction.
(c) Describe the form and strength of the relationship.

5. Table 5.2 (page 188) gives the city and highway gas mileages for 21 two-seater cars. Omit the Honda Insight, which doesn't belong with the remaining 20 gasoline-powered cars.

(a) Make a scatterplot, taking city mileage as the explanatory variable.
(b) Describe in words the form, direction, and strength of the relationship between highway mileage and city mileage.

6. How fast do icicles grow? Here are data for one set of conditions: no wind, temperature −11°C, and water flowing over the icicle at 12 milligrams per second.

Time (minutes)	10	20	30	40	50
Length (centimeters)	0.6	1.8	2.9	4.0	5.0

Time (minutes)	60	70	80	90	100
Length (centimeters)	6.1	7.9	10.1	10.9	12.7

Time (minutes)	110	120	130	140	150
Length (centimeters)	14.4	16.6	18.1	19.9	21.0

Time (minutes)	160	170	180
Length (centimeters)	23.4	24.7	27.8

SOURCE: N. Maeno et al., Growth rates of icicles, *Journal of Glaciology*, 40 (1994): 319–326.

Make a scatterplot. (Which is the explanatory variable?) Describe in words the direction, form, and strength of the relationship.

◆ **7.** How does the fuel consumption of a car change as its speed increases? Here are data for a British Ford Escort. Speed is measured in kilometers per hour, and fuel consumption is measured in liters of gasoline used per 100 kilometers traveled.

Speed (km/h)	10	20	30	40	50
Fuel (liters/100 km)	21.00	13.00	10.00	8.00	7.00

Speed (km/h)	60	70	80	90	100
Fuel (liters/100 km)	5.90	6.30	6.95	7.57	8.27

Speed (km/h)	110	120	130	140	150
Fuel (liters/100 km)	9.03	9.87	10.79	11.77	12.83

SOURCE: T. N. Lam, Estimating fuel consumption from engine size, *Journal of Transportation Engineering*, 111 (1985): 339–357.

(a) Make a scatterplot. (Which is the explanatory variable?)

(b) Describe the form of the relationship. Explain why the form of the relationship makes sense.

(c) How would you describe the direction of this relationship?

(d) Is the relationship reasonably strong or quite weak? Explain your answer.

◆ **8.** Give an example of two variables from everyday life that have a positive association. Give an example of two variables that have a negative association.

Regression Lines

9. Figure 6.9 shows the calories and salt content (milligrams of sodium) in 17 brands of meat hot dogs. If we ignore the outlying point marked A, a regression line for predicting sodium from calories passes close to these two observations:

$$calories = 139, \ sodium = 386 \ mg$$
$$calories = 191, \ sodium = 506 \ mg$$

Use this fact to estimate the slope of this regression line. [*Hint:* Remember that the slope of a line is the "rise" (vertical change) divided by the "run" (horizontal change) for any two points on the line.]

10. Exercise 4 gives Professor Moore's times (in minutes) to swim 2000 yards and his pulse rate after swimming (in beats per minute) for 23 sessions in the pool. A regression line for predicting pulse from time is

$$pulse = 479.9 - (9.695 \times time)$$

(a) What is the slope of this line? What does the slope say about how pulse rate changes with swim time?

(b) The next day's time is 34.30 minutes. Predict Professor Moore's pulse rate. In fact, his pulse rate was 152. How accurate is your prediction?

11. Researchers studying acid rain measured the acidity of precipitation in a Colorado wilderness area for 150 consecutive weeks. Acidity is measured by pH. Lower pH values show higher acidity. The acid rain researchers observed a straight-line pattern over time. They reported that the regression line

$$\text{predicted pH} = 5.43 - (0.0053 \times \text{weeks})$$

fit the data well. [W. M. Lewis and M. C. Grant, Acid precipitation in the western United States, *Science*, 207 (1980): 176–177.]

(a) Draw a graph of this line. Explain in plain language what the line says about how pH was changing over time.
(b) According to the regression line, what was the pH at the beginning of the study (weeks = 1)? At the end (weeks = 150)?
(c) What is the slope of the regression line? Explain clearly what this slope says about the rate of change in pH.

12. Figure 6.2 plots the mean score of each state's high school seniors on the SAT against the percent of seniors who take the exam. In addition to two clusters, the plot shows an overall rough straight-line pattern. The least-squares regression line for predicting SAT score from percent taking is

$$\text{predicted SAT} = 1150 - (2.02 \times \text{percent taking})$$

(a) What is the slope of this line? What does the sign of the slope (positive or negative) tell us about the relationship between these variables?
(b) In New York State, 82% of high school seniors took the SAT. Predict their average score. (The actual average score in New York was 1006.)

13. If women always married men who were two years older than they are, what would be the slope of the regression line for predicting husband's age from wife's age? (*Hint:* Draw a scatterplot for several ages.)

14. Suppose that the slope of the regression line of weight on height for a group of young men is $b = 1.1$ when we measure height x in centimeters and weight y in kilograms. That is, when height increases by 1 centimeter, weight increases by 1.1 kilograms. There are 1000 grams in a kilogram. If

we measured weight in grams, what would be the slope?

Correlation

15. Find the correlation between the city and highway gas mileages for the 20 gasoline-powered two-seater cars in Table 5.2. (That is, omit the Honda Insight.) Explain why the value of r matches the scatterplot that you made in Exercise 5.

16. Exercise 4 gives data on the time to swim 2000 yards and the pulse rate after swimming for a middle-aged professor.

(a) Use a calculator to find the correlation r. Explain from looking at the scatterplot why this value of r is reasonable.
(b) Suppose that the times had been recorded in seconds. For example, the time 34.12 minutes would be 2047 seconds. How would the value of r change?

17. Find the correlation between city and highway mileage for all 21 two-seater cars in Table 5.2, including the Honda Insight. Compare your r with the value you found in Exercise 15. Explain why adding the Insight changes r in this direction.

18. Find the correlation between time and icicle length for the data in Exercise 6. Explain why the value of r matches the scatterplot that you made in Exercise 6.

19. Exercise 7 gives data on gas used versus speed for a small car. Make a scatterplot, if you did not do so in Exercise 7. Calculate the correlation. Explain why r is close to 0 despite a strong relationship between speed and gas use.

20. The length of the icicle in Exercise 6 is measured in centimeters. There are 2.54 centimeters in an inch. If length were measured in inches, how would the correlation you found in Exercise 18 change?

21. If women always married men who were two years older than they are, what would be the correlation between the ages of husband and wife? (*Hint:* Draw a scatterplot for several ages.)

22. Measurements on many people show that the correlation

(a) between the heights of fathers and the heights of their adult sons is about _____.
(b) between the heights of husbands and the heights of their wives is about _____.
(c) between the heights of women at age 4 and their heights at age 18 is about _____.

The answers (in scrambled order) are $r = 0.2$, $r = 0.5$, and $r = 0.8$. Match the answers to the statements and explain your choice.

23. For each of the following pairs of variables, would you expect a substantial negative correlation, a substantial positive correlation, or a small correlation?

(a) The age of secondhand cars and their prices.
(b) The weight of new cars and their gas mileages in miles per gallon.
(c) The heights and the weights of adult men.
(d) The heights and the IQ scores of adult men.

◆ **24.** Each of the following statements contains a blunder. Explain what is wrong in each case.

(a) "There is a high correlation between the gender of American workers and their income."
(b) "We found a high correlation ($r = 1.09$) between students' ratings of faculty teaching and ratings made by other faculty members."
(c) "The correlation between age and income was found to be $r = 0.53$ years."

■ **25.** Mutual-fund reports often give correlations to describe how the prices of different investments are related. You look at the correlations between three Fidelity funds and the Standard & Poor's 500 Stock Index, which describes stocks of large U.S. companies. The three funds are Dividend Growth (stocks of large U.S. companies), Small Cap Stock (stocks of small U.S. companies), and Emerging Markets (stocks in developing countries). For 2003, the three correlations are $r = 0.35$, $r = 0.81$, and $r = 0.98$.

(a) Which correlation goes with each fund? Explain your answer.

(b) The correlations of the three funds with the index are all positive. Does this tell you that stocks went up in 2003? Explain your answer.

■ **26.** *Archaeopteryx* is an extinct beast having feathers like a bird but teeth and a long bony tail like a reptile. Only six fossil specimens are known. If the specimens belong to the same species and differ in size because some are younger than others, there should be a straight-line relationship between the lengths of a pair of bones from all individuals. An outlier from this relationship would suggest a different species. Here are data on the lengths in centimeters of the femur (a leg bone) and the humerus (a bone in the upper arm) for the five specimens that preserve both bones.

Femur length x	38	56	59	64	74
Humerus length y	41	63	70	72	84

SOURCE: M. A. Houck et al., Allometric scaling in the earliest fossil bird, *Archaeopteryx lithographica*, *Science*, 247 (1990): 195–198.

(a) Make a scatterplot. Do you think that all five specimens come from the same species?
(b) Find the correlation r step by step. *Step 1:* Find the mean $\bar{x}$ and standard deviation s_x of the five femur lengths. Find the mean $\bar{y}$ and the standard deviation s_y of the five humerus lengths. (Use your calculator.) *Step 2:* Find the standardized values $(x - \bar{x})/s_x$ of each of the five femur lengths, and do the same for the five humerus lengths. *Step 3:* Substitute your numbers from steps 1 and 2 into the formula for r.
(c) Now enter the x and y data into your calculator and use the calculator's correlation function to find r. Check that you get the same result as in part (b).

Least-Squares Regression

27. In Exercise 5 you made a scatterplot of city and highway gas mileage for the 20 gasoline-powered two-seater cars in Table 5.2 (page 188).

(a) What is the equation of the least-squares regression line for predicting highway mileage from city mileage?

(b) Another two-seater car gets 18 mpg in the city. Predict its highway mileage.

(c) Based on the scatterplot you made in Exercise 5, do you expect the prediction in part (b) to be quite accurate? Why?

28. In Exercise 6 you made a scatterplot of the length of an icicle and the number of minutes water has been flowing over the icicle.

(a) What is the equation of the least-squares regression line for predicting icicle length from time?

(b) Use your regression line to predict the length of the icicle after 75 minutes.

29. Redo your scatterplot of highway mileage against city mileage from Exercise 5. Add your regression line from Exercise 27 to the plot. Be sure to show how you were able to plot the line starting with its equation. Finally, use the "up-and-over" method illustrated in Figure 6.3 to show the predicted highway mileage of a car that gets 18 mpg in the city.

30. Redo your scatterplot of icicle length against time from Exercise 6. Add your regression line from Exercise 28 to the plot. Be sure to show how you were able to plot the line starting with its equation. Finally, use the "up-and-over" method illustrated in Figure 6.3 to show the predicted length of the icicle after 75 minutes.

31. Exercise 7 gives data on gas used versus speed for a small car. Make a scatterplot, if you did not do so in Exercise 7. The least-squares regression line for these data is

$$\text{predicted fuel} = 11.058 - 0.0147 \times \text{speed}$$

Draw this line on your scatterplot. What are the predicted and observed fuel consumption values for speeds of 10, 70, and 150 kilometers per hour (km/h)? *You can fit a regression line to any set of two-variable data. The line is of little use if the plot does not show a straight-line pattern.*

32. The length of the icicle in Exercise 6 is measured in centimeters. There are 2.54

centimeters in an inch. If length were measured in inches, how would the slope of the regression line you found in Exercise 28 change?

33. The mean height of American women in their early twenties is about 64.5 inches and the standard deviation is about 2.5 inches. The mean height of men the same age is about 68.5 inches, with standard deviation about 2.7 inches. If the correlation between the heights of husbands and wives is about $r = 0.5$, what is the equation of the regression line of the husband's height on the wife's height in young couples? Predict the height of the husband of a woman who is 67 inches tall.

34. In Professor Friedman's economics course, the correlation between the students' total scores before the final examination and their final examination scores is $r = 0.6$. The pre-exam totals for all students in the course have mean 280 and standard deviation 30. The final exam scores have mean 75 and standard deviation 8. Professor Friedman has lost Julie's final exam but knows that her total before the exam was 300. He decides to predict her final exam score from her pre-exam total.

(a) What is the slope of the least-squares regression line of final exam scores on pre-exam total scores in this course? What is the intercept?

(b) Use the regression line to predict Julie's final exam score.

■ **35.** Use the equation for the least-squares regression line to show that this line always passes through the point $(\bar{x}, \bar{y})$. That is, set $x = \bar{x}$ and show that the line predicts that $y = \bar{y}$.

■ **36.** Exercise 6 gives data on the growth of an icicle.

(a) Find the mean and standard deviation of the times and icicle lengths. Find the correlation between the two variables. Use these five numbers to find the equation of the regression line for predicting length from time. Verify that your result agrees with that in Exercise 28.

(b) Use the same five numbers to find the equation of the regression line for predicting from an icicle's length the time it has been growing. Use your line to predict the time that an icicle 15 centimeters long has been growing. *There is just one correlation between two variables, but there are two different least-squares lines, depending on which you choose as the response variable.*

■ **37.** Fidelity Investments, like other large mutual fund companies, offers many "sector funds" that concentrate their investments in narrow segments of the stock market. These funds often rise or fall by much more than the market as a whole. Here are the percent returns for 23 Fidelity "Select Portfolios" funds for the years 2002 (when stocks fell) and 2003 (when stocks went up).

2002 return	2003 return	2002 return	2003 return	2002 return	2003 return
−17.1	23.9	−0.7	36.9	−37.8	59.4
−6.7	14.1	−5.6	27.5	−11.5	22.9
−21.1	41.8	−26.9	26.1	−0.7	36.9
−12.8	43.9	−42.0	62.7	64.3	32.1
−18.9	31.1	−47.8	68.1	−9.6	28.7
−7.7	32.3	−50.5	71.9	−11.7	29.5
−17.2	36.5	−49.5	57.0	−2.3	19.1
−11.4	30.6	−23.4	35.0		

Do a careful statistical analysis of these data using both graphs and whatever numerical measures you think are appropriate. Make a side-by-side comparison of the distributions of returns in 2002 and 2003 and also describe the relationship between the returns of the same funds in these two years. What are your most important findings? (The outlier is Fidelity Gold Fund.)

Interpreting Correlation and Regression

38. Make a scatterplot of the following data:

x	1	2	3	4	10	10
y	1	3	3	5	1	11

Use your calculator to show that the correlation is about 0.5. What feature of the data is responsible for

reducing the correlation to this value despite a strong straight-line association between x and y in most of the observations?

39. Table 6.1 presents four sets of data prepared by the statistician Frank Anscombe to illustrate the dangers of calculating without first plotting the data.

(a) Without making scatterplots, find the correlation and the least-squares regression line for all four data sets. What do you notice? Use the regression line to predict y for $x = 10$.

(b) Make a scatterplot for each of the data sets and add the regression line to each plot.

(c) In which of the four cases would you be willing to use the regression line to describe the dependence of y on x? Explain your answer in each case.

◆ **40.** Children who watch many hours of television get lower grades in school on the average than those who watch less TV. Explain clearly why this fact does not show that watching TV *causes* poor grades. In particular, suggest some other characteristics of households where children watch lots of TV that may contribute to poor grades.

◆ **41.** People who use artificial sweeteners in place of sugar tend to be heavier than people who use sugar. Does this mean that artificial sweeteners cause weight gain? Give a more plausible explanation for this association.

◆ **42.** "Based on an examination of twenty-two companies that announced large layoffs during 1994, Downs found a strong (.31) correlation between the size of the layoffs and the compensation of the CEOs." [K. Phillips, *Wealth and Democracy*, Broadway Books, New York, 2002, p. 151.] This correlation is probably explained by a third variable, the size of the company as measured by its number of employees. Explain why this is true.

■ **43.** An article entitled "The Health and Wealth of Nations" says: "The positive correlation between health and income per capita is one of the best-known relations in international

TABLE 6.1	Four Data Sets for Exploring Correlation and Regression

Data Set A

x	10	8	13	9	11	14	6	4	12	7	5
y	8.04	6.95	7.58	8.81	8.33	9.96	7.24	4.26	10.84	4.82	5.68

Data Set B

x	10	8	13	9	11	14	6	4	12	7	5
y	9.14	8.14	8.74	8.77	9.26	8.10	6.13	3.10	9.13	7.26	4.74

Data Set C

x	10	8	13	9	11	14	6	4	12	7	5
y	7.46	6.77	12.74	7.11	7.81	8.84	6.08	5.39	8.15	6.42	5.73

Data Set D

x	8	8	8	8	8	8	8	8	8	8	19
y	6.58	5.76	7.71	8.84	8.47	7.04	5.25	5.56	7.91	6.89	12.50

SOURCE: Frank J. Anscombe, Graphs in statistical analysis, *The American Statistician*, 27 (1973): 17–21.

development. This correlation is commonly thought to reflect a causal link running from income to health. . . . Recently, however, another intriguing possibility has emerged: that the health-income correlation is partly explained by a causal link running the other way–from health to income." [D. E. Bloom and D. Canning, The health and wealth of nations, *Science*, 287 (2000): 1207–1208.]

Explain how higher income in a nation can cause better health. Then explain how better health can cause higher national income. There is no simple way to determine the direction of the link.

■ **44.** The effect of an outside variable can be surprising when individuals are divided into groups. In recent years, the mean SAT score of all high school seniors has increased. But the mean SAT score has decreased for students at each level of high school grades (A, B, C, and so on). Explain how grade inflation in high school can

account for this pattern. *A relationship that holds for each group within a population need not hold for the population as a whole. In fact, the relationship can even change direction.*

Chapter Review

45. How well does the income of a college student's parents predict how much the student will borrow to pay for college? We have data on parents' income and college debt for a sample of 1200 recent college graduates. What are the explanatory and response variables? Do you expect a positive or negative association between these variables? Why?

46. A study of reading ability in schoolchildren chose 60 fifth-grade children at random from a school. The researchers had the children's scores on an IQ test and on a test of reading ability. Figure 6.11 plots reading test score (response) against IQ score (explanatory).

FIGURE 6.11 IQ and reading test scores for 60 fifth-grade children, for Exercise 46.

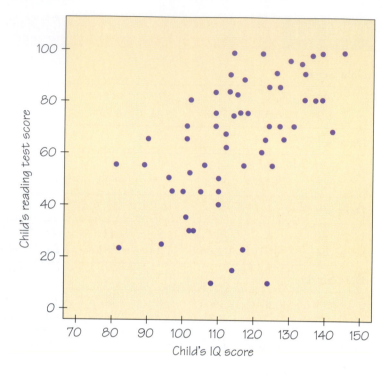

(a) Explain why we should expect a positive association between IQ and reading score for children in the same grade. Does the scatterplot show a positive association?

(b) A group of four points appear to be outliers. In what way do these children's IQ and reading scores deviate from the overall pattern?

(c) Ignoring the outliers, is the form of the association between IQ and reading scores roughly a straight line? Is it very strong? Explain your answers.

47. A student wonders if tall women tend to date taller men than do short women. She measures herself, her dormitory roommate, and the women in the adjoining rooms; then she measures the next man each woman dates. Here are the data (heights in inches):

Women (x)	66	64	66	65	70	65
Men (y)	72	68	70	68	71	65

(a) Make a scatterplot of these data, with the women's heights as the explanatory variable. Based on the scatterplot, do you expect the correlation to be positive or negative? Near ± 1 or not?

(b) Find the correlation r between the heights of the men and women.

48. In the previous exercise, you found the correlation r between the heights in inches of several college women and the heights in inches of the next man each woman dates.

(a) How would r change if all the men were 3 inches shorter than the heights given in the table?

(b) How would r change if heights were measured in centimeters rather than inches? (There are 2.54 centimeters in an inch.)

49. The equation of the least-squares regression line for predicting men's heights from women's heights for the data in Exercise 47 is

predicted male height = 24 + 0.68 × female height

(a) What is the slope of this line? Explain in simple language what the numerical value of the slope tells us about the heights of the men these women date.

(b) Another woman is 67 inches tall. Use the regression line to predict the height of the next man she dates.

 ## APPLET EXERCISES

To do these exercises, go to www.whfreeman.com/fapp.

1. In the *Correlation and Regression* applet, imitate Figure 6.5. That is, click to locate five points at the lower left of the scatterplot, then click "Show least-squares line."

(a) What is the correlation r for these five points? If necessary, move points with the mouse to get a value near $r = 0.5$, as in Example 6.

(b) Now add an outlier at the upper right that lies exactly on the line. What is the correlation r for the six points?

(c) Use the mouse to drag the outlier down and then to the left. Watch the least-squares line follow this one point. How negative can you make the correlation r?

2. You are going to use the *Correlation and Regression* applet to make different scatterplots with 10 points that have correlation close to 0.7. *Many patterns can have the same correlation. Always plot your data before you trust a correlation.*

(a) Stop after adding the first two points. What is the value of the correlation? Why does it have this value no matter where the two points are located?

(b) Make a lower-left to upper-right pattern of 10 points with correlation about $r = 0.7$. (You can drag points up or down to adjust r after you have 10 points.) Make a rough sketch of your scatterplot.

(c) Make another scatterplot with nine points in a vertical stack at the left of the plot. Add one point far to the right and move it until the correlation is close to 0.7. Make a rough sketch of your scatterplot.

(d) Make yet another scatterplot with 10 points in a curved pattern that starts at the lower left, rises to the right, then falls again at the far right. Adjust the points up or down until you have a quite smooth curve with correlation close to 0.7. Make a rough sketch of this scatterplot also.

3. It isn't easy to guess the position of the least-squares line by eye. Use the *Correlation and Regression* applet to compare a line you draw with the least-squares line. Click on the scatterplot to create a group of 15 to 20 points from lower left to upper right with a clear, positive straight-line pattern (with correlation around 0.7). Click the "Draw line" button and use the mouse to draw a line through the middle of the cloud of points from lower left to upper right. Note the "thermometer" that appears above the plot. The red portion is the sum of the squared vertical distances from the points in the plot to the least-squares line. The green portion is the "extra" sum of squares for your line—it shows by how much your line misses the smallest possible sum of squares.

(a) You drew a line by eye through the middle of the pattern. Yet the right-hand part of the bar is probably almost entirely green. What does that tell you?

(b) Now click the "Show least-squares line" box. Is the slope of the least-squares line smaller (the new line is less steep) or larger (line is steeper) than that of your line? If you repeat this exercise several times, you will consistently get the same result. *The least-squares line minimizes the vertical distances of the points from the line. It is not the line through the "middle" of the cloud of points.* This is one reason why it is hard to draw a good regression line by eye.

WRITING PROJECTS

1. Choose two variables that you think have a roughly straight-line relationship. Gather data on these variables and do a statistical analysis: Make a scatterplot, find the correlation, find the regression line (use a statistical calculator or software), and draw the line on your plot. Then write a short report on your work. Some examples of suitable pairs of variables are:

(a) The height and arm span of a group of people
(b) The height and walking stride length of a group of people
(c) The price per ounce and bottle size in ounces for several brands of shampoo and several bottle sizes for each brand

2. We often describe our emotional reaction to social rejection as "pain." A clever study asked whether social rejection causes activity in areas of the brain that are known to be activated by physical pain. If it does, we really do experience social and physical pain in similar ways. Subjects were first included and then deliberately excluded from a social activity while increases in blood flow in their brains were measured. After each activity, the subjects filled out questionnaires that assessed how excluded they felt.

Here are data for 13 subjects. [N. I. Eisenberger, M. D. Lieberman, and K. D. Williams, Does rejection hurt? An fMRI study of social exclusion, *Science,* 302 (2003): 290–292.] The explanatory variable is "social distress" measured by each sub-

ject's questionnaire score after exclusion relative to the score after inclusion. (So values greater than 1 show the degree of distress caused by exclusion.) The response variable is activity in the anterior cingulate cortex, a region of the brain that is activated by physical pain.

Subject	Social distress	Brain activity
1	1.26	−0.055
2	1.85	−0.040
3	1.10	−0.026
4	2.50	−0.017
5	2.17	−0.017
6	2.67	0.017
7	2.01	0.021
8	2.18	0.025
9	2.58	0.027
10	2.75	0.033
11	2.75	0.064
12	3.33	0.077
13	3.65	0.124

Use graphs and numerical measures to describe the relationship between social distress and brain activity. Write a short report of your findings. Do the data suggest that brain activity in the "pain" region really is directly related to the distress from social exclusion?

SUGGESTED READINGS

CLEVELAND, WILLIAM S. *The Elements of Graphing Data*, rev. ed., Hobart Press, Summit, N.J., 1994. A careful study of the most effective elementary ways to present data graphically, with much sound advice on improving graphs such as scatterplots.

MOORE, DAVID S. *The Basic Practice of Statistics*, 3rd ed., Freeman, New York, 2004. Chapters 4 and 5 of this text provide a more extensive treatment of the material of Chapter 6, at about the same mathematical level.

SUGGESTED WEB SITES

The Web sites suggested in Chapter 5 as sources for data provide data to investigate relationships as well. How is the number of medical doctors per 100,000 people in the states related to how rich the state is? To infant mortality in the state? To the cost of medical care? You can study these and many other relationships using data from the Census Web site www.census.gov.

The *Journal of Statistics Education*, www.amstat.org/publications/jse/, is an electronic journal sponsored by the American Statistical Association that presents articles on teaching statistics that often include interesting data and examples. Look, for example, at "Exploring Relationships in Body Dimensions" by Grete Heinz and Louis J. Peterson, in Volume 11, Number 2. (Click on "Archive" to see a list of past issues.) Here you will find information about measuring body dimensions, actual data from 247 men and 260 women, and some examples of both distributions and relationships.

Data for Decisions

Numerical data are raw material for the growth of knowledge. Statistics provides tools that help data produce knowledge rather than confusion. These tools include methods for producing data as well as for interpreting already available data.

Chapters 5 and 6 explored the art of *data analysis*. They showed how to use graphs and numbers to uncover the nature of a set of data. We can distinguish two different attitudes toward data. We may ask, "What do we see in the data?" Or we may ask, "What answer do the data give to this specific question?" For example, we may want data to answer the question "What car colors are most popular among young adults?" or "Do recent college graduates with higher college debt tend to also have higher incomes?" Data that answer such questions are a product of intelligent effort, just like hybrid tomatoes and DVD players. In this chapter, we first address the issue of *how to produce data* that can be trusted to answer specific questions. You will learn to recognize good and bad methods of producing data. Understanding how to produce trustworthy data is the first—and the most important—step toward the ability to judge whether conclusions based on data are reliable or not.

Statistical techniques for producing data open the door to a further advance in data analysis, *formal statistical inference*, which answers specific questions with a known degree of confidence. Statistical inference uses the descriptive tools we have already met, in combination with new kinds of reasoning. We will look briefly at some of this reasoning in the final part of this chapter.

7.1 Sampling

A political scientist wants to know what percent of college-age adults consider themselves to be conservatives. An automaker hires a market research firm to learn what percent of adults aged 18 to 35 recall seeing television advertisements for a new sport utility vehicle. Government economists inquire about average household income. In all these cases, we want to gather information about a large group of individuals. Time, cost, and inconvenience forbid contacting every individual. So we gather information about only part of the group in order to draw conclusions about the whole.

Population and Sample

The **population** in a statistical study is the entire group of individuals about which we want information.

A **sample** is a part of the population from which we actually collect information used to draw conclusions about the whole.

We often draw conclusions about a whole on the basis of a sample. Everyone has sipped a spoonful of soup and judged the entire bowl on the basis of that taste. But a bowl of soup is homogeneous, so that the taste of a single spoonful represents the whole. Choosing a representative sample from a large and varied population is not so easy. The first step in a proper *sample survey* is to say carefully just what population we want to describe. The second step is to say exactly what we want to measure. These preliminary steps can be complicated, as the following example illustrates.

The Current Population Survey

EXAMPLE 1

The monthly unemployment rate comes from the government's Current Population Survey (CPS), a sample of about 60,000 households each month (see Figure 7.1). To measure unemployment, we must first specify the population we want to describe. Which age groups will we include? Will we include illegal aliens or people in prisons? The CPS defines its population as all U.S. residents (whether citizens or not) 16 years of age and over who are civilians and are not in an institution such as a prison. The civilian unemployment rate announced in the news refers to this specific population.

The second question is harder: What does it mean to be "unemployed"? Someone who is not looking for work—for example, a full-time student—should not be called unemployed just because she is not working for pay. If you are chosen for the CPS sample, the interviewer first asks whether you are available to work and whether you actually looked for work in the past four weeks. If not, you are neither employed nor unemployed—you are not in the labor force.

FIGURE 7.1 The Web
page of the Current
Population Survey.

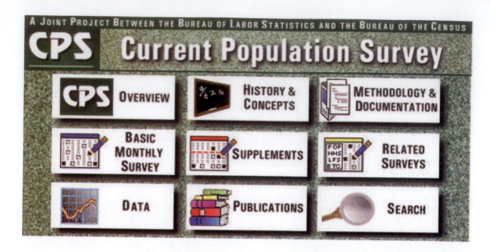

If you are in the labor force, the interviewer goes on to ask about employment. Any work for pay or in your own business the week of the survey counts you as employed. So does at least 15 hours of unpaid work in a family business. You are also employed if you have a job but didn't work because of vacation, being on strike, or for some other good reason. An unemployment rate of 4.7% means that 4.7% of the sample was unemployed, using the exact CPS definitions of both "labor force" and "unemployed." ∎

7.2 Bad Sampling Methods

How can we choose a sample that is truly representative of the population? The easiest—but not the best—way to select a sample is to choose individuals close at hand. If we are interested in finding out how many people have jobs, for example, we might go to a shopping mall and ask people passing by if they are employed. A sample selected by taking the members of the population who are easiest to reach is called a **convenience sample.** Convenience samples often produce unrepresentative data.

EXAMPLE 2   Convenience Samples

Taking a sample of mall shoppers is fast and cheap. But people at shopping malls tend to be more prosperous than typical Americans. They are also more likely to be teenagers or retired. What is more, when we decide which people to question, we will tend to choose well-dressed, respectable people and we will tend to avoid poorly dressed, unfriendly, or tough-looking individuals. In short, our shopping mall interviews will not contact a sample that is representative of the entire population. ∎

Our shopping mall sample will almost surely overrepresent middle-class and retired people and underrepresent the poor. This will happen every time we take such a sample. That is, it is a systematic error caused by a bad sampling method, not just bad luck on one sample. This is *bias*: The outcomes of mall surveys will repeatedly miss the truth about the population in the same ways.

Bias

The design of a statistical study is **biased** if it systematically favors certain outcomes.

Online Polls

EXAMPLE 3

The American Family Association (AFA) is a conservative group that claims to stand for "traditional family values." It regularly posts online poll questions on its Web site—just click on a response to take part. Because the respondents are people who visit this site, the poll results always support AFA's positions. Well, almost always. In 2004, AFA's online poll asked about the heated issue of allowing same-sex marriage. Soon, email lists and social-network sites favored mostly by young liberals pointed to the AFA poll. Almost 850,000 people responded, and 60% of them favored legalization of same-sex marriage. AFA claimed that homosexual rights groups had skewed its poll. ▪

Online polls are now everywhere—some sites will even provide help in conducting your own online poll. As the AFA poll illustrates, you can't trust the results. People who take the trouble to respond to an open invitation are not representative of the entire adult population. That's true of regular visitors to AFA's site, of the activists who made a special effort to vote in the marriage poll, and of the people who bother to respond to write-in, call-in, or online polls in general. Polls like these are examples of *voluntary response sampling*.

Voluntary Response Sample

A **voluntary response sample** consists of people who choose themselves by responding to a general appeal. Voluntary response samples are biased because people with strong opinions are most likely to respond.

7.3 Simple Random Samples

In a voluntary response sample, people choose whether to respond. In a convenience sample, the interviewer makes the choice. In both cases, personal choice produces bias. The statistician's remedy is to allow impersonal chance to choose the sample. A sample chosen by chance allows neither favoritism by the sampler nor self-selection by respondents. Choosing a sample by chance avoids bias by giving all individuals an equal chance to be chosen. Rich and poor, young and old, black and white—all have the same chance to be in the sample.

The simplest way to use chance to select a sample is to place names (the population) in a hat and draw out a handful (the sample). This is the idea of *simple random sampling*.

Simple Random Sample

A **simple random sample (SRS)** of size n consists of n individuals from the population chosen in such a way that every set of n individuals has an equal chance to be the sample actually selected.

Picturing drawing names from a hat helps us understand what an SRS is. The same picture helps us see that an SRS is a better method of choosing samples than convenience or voluntary response sampling because it doesn't favor any part of the population. But writing names on slips of paper and drawing them from a hat is slow and inconvenient. That's especially true if, as in the Current Population Survey, we must draw a sample of size 60,000. We can speed up the process by using a *table of random digits*. In practice, samplers use computers to do the work, but we can do it by hand for small samples.

Random Digits

A **table of random digits** is a list of the digits 0, 1, 2, 3, 4, 5, 6, 7, 8, 9 with these two properties:

1. Each entry in the table is equally likely to be any of the 10 digits 0 through 9.
2. The entries are independent of one another. That is, knowledge of one part of the table gives no information about any other part.

Table 7.1 is a table of random digits. The digits in the table appear in groups of five to make the table easier to read, and the rows are numbered so we can refer to them, but the groups and row numbers are just for convenience. The entire table is one long string of randomly chosen digits. There are two steps in using the random-digit table to choose a simple random sample.

Step 1. Label Give each member of the population a numerical label of the *same length*. Up to 100 items can be labeled with two digits: 01, 02, . . . , 99, 00. Up to 1000 items can be labeled with three digits, and so on.

Step 2. Table To choose a simple random sample, read from Table 7.1 successive groups of digits of the length you used as labels. Your sample contains the individuals whose labels you find in the table. This gives all individuals the same chance because all labels of the same length have the same chance of being found in the table. For example, any pair of digits in the table is equally likely to be any of the 100 possible labels 01, 02, . . . , 99, 00. Ignore any group of digits that was not used as a label or that duplicates a label already in the sample.

Sampling Spring Break Resorts

EXAMPLE 4

A campus newspaper plans a major article on spring break destinations. The authors intend to call four randomly chosen resorts at each destination to ask about

SPOTLIGHT 7.1 Is It Really Random?

Are the random digits in Table 7.1 really random? Not a chance. They were produced by a computer program. A computer program implements an algorithm that does exactly what you tell it to do. Give the program the same input and it will produce exactly the same "random" digits. You can get quite respectable random digits by calculating π = 3.14159265358979 . . . to more and more decimal places. Go to www.cecm.sfu.ca/pi/pi.html and you will see these digits stream by. You get the same digits on every visit, of course. Clever people have devised algorithms that produce output that *looks* like random digits. These are called "pseudo-random numbers," and that's what Table 7.1 contains. Pseudo-random numbers work fine for statistical randomizing, but they have hidden nonrandom patterns that can mess up more refined uses.

For purists, the RAND Corporation long ago published a book titled *One Million Random Digits*. The book lists 1 million digits that were produced by a very elaborate physical randomization and really are random. An employee of RAND once told me that this is not the most boring book that RAND has ever published.

Cryptologists and computer scientists would like an endless supply of really random digits. Really random digits must come from nature, not from a computer program. Radioactive decay is really random, and so is the "thermal noise" in an amplifier, which you can hear as a soft whoosh if you turn up the volume with no music playing. Alas, extracting random digits from these really random sources requires various human devices, and these often impose subtle patterns. As of now, in fact, pseudo-random numbers from the best algorithms actually look more random than numbers refined from the randomness in nature by some human apparatus. "Easy to say, hard to do" applies to making random digits as well as to many other human aspirations.

TABLE 7.1	Random Digits						
101	19223	95034	05756	28713	96409	12531	42544	82853
102	73676	47150	99400	01927	27754	42648	82425	36290
103	45467	71709	77558	00095	32863	29485	82226	90056
104	52711	38889	93074	60227	40011	85848	48767	52573
105	95592	94007	69971	91481	60779	53791	17297	59335
106	68417	35013	15529	72765	85089	57067	50211	47487
107	82739	57890	20807	47511	81676	55300	94383	14893
108	60940	72024	17868	24943	61790	90656	87964	18883
109	36009	19365	15412	39638	85453	46816	83485	41979
110	38448	48789	18338	24697	39364	42006	76688	08708
111	81486	69487	60513	09297	00412	71238	27649	39950
112	59636	88804	04634	71197	19352	73089	84898	45785
113	62568	70206	40325	03699	71080	22553	11486	11776
114	45149	32992	75730	66280	03819	56202	02938	70915
115	61041	77684	94322	24709	73698	14526	31893	32592
116	14459	26056	31424	80371	65103	62253	50490	61181
117	38167	98532	62183	70632	23417	26185	41448	75532
118	73190	32533	04470	29669	84407	90785	65956	86382
119	95857	07118	87664	92099	58806	66979	98624	84826
120	35476	55972	39421	65850	04266	35435	43742	11937
121	71487	09984	29077	14863	61683	47052	62224	51025
122	13873	81598	95052	90908	73592	75186	87136	95761
123	54580	81507	27102	56027	55892	33063	41842	81868
124	71035	09001	43367	49497	72719	96758	27611	91596
125	96746	12149	37823	71868	18442	35119	62103	39244

their attitudes toward groups of students as guests. Here are the resorts listed in one city:

01	Aloha Kai	08	Captiva	15	Palm Tree	22	Sea Shell
02	Anchor Down	09	Casa del Mar	16	Radisson	23	Silver Beach
03	Banana Bay	10	Coconuts	17	Ramada	24	Sunset Beach
04	Banyan Tree	11	Diplomat	18	Sandpiper	25	Tradewinds
05	Beach Castle	12	Holiday Inn	19	Sea Castle	26	Tropical Breeze
06	Best Western	13	Lime Tree	20	Sea Club	27	Tropical Shores
07	Cabana	14	Outrigger	21	Sea Grape	28	Veranda

Step 1. Label   Give each resort a numerical label. Because two digits are needed to label the 28 resorts, all labels will have two digits. We have added labels 01 to 28 in the list of resorts. Always say how you labeled the members of the

126	96927	19931	36809	74192	77567	88741	48409	41903
127	43909	99477	25330	64359	40085	16925	85117	36071
128	15689	14227	06565	14374	13352	49367	81982	87209
129	36759	58984	68288	22913	18638	54303	00795	08727
130	69051	64817	87174	09517	84534	06489	87201	97245
131	05007	16632	81194	14873	04197	85576	45195	96565
132	68732	55259	84292	08796	43165	93739	31685	97150
133	45740	41807	65561	33302	07051	93623	18132	09547
134	27816	78416	18329	21337	35213	37741	04312	68508
135	66925	55658	39100	78458	11206	19876	87151	31260
136	08421	44753	77377	28744	75592	08563	79140	92454
137	53645	66812	61421	47836	12609	15373	98481	14592
138	66831	68908	40772	21558	47781	33586	79177	06928
139	55588	99404	70708	41098	43563	56934	48394	51719
140	12975	13258	13048	45144	72321	81940	00360	02428
141	96767	35964	23822	96012	94591	65194	50842	53372
142	72829	50232	97892	63408	77919	44575	24870	04178
143	88565	42628	17797	49376	61762	16953	88604	12724
144	62964	88145	83083	69453	46109	59505	69680	00900
145	19687	12633	57857	95806	09931	02150	43163	58636
146	37609	59057	66967	83401	60705	02384	90597	93600
147	54973	86278	88737	74351	47500	84552	19909	67181
148	00694	05977	19664	65441	20903	62371	22725	53340
149	71546	05233	53946	68743	72460	27601	45403	88692
150	07511	88915	41267	16853	84569	79367	32337	03316

population. To sample from the 1240 resorts in a major vacation area, you would label the resorts 0001, 0002, . . . , 1239, 1240.

Step 2. Table   Now go to Table 7.1 and read two-digit groups until you have chosen four resorts. Starting at line 130 (any line will do), we find

69051   64817   87174   09517   84534   06489   87201   97245

Because our labels are two digits long, we read successive two-digit groups from the table. Ignore groups not used as labels, like the initial 69. Also ignore any repeated labels, like the second 17 in this row, because we can't choose the same resort twice. Our sample contains the resorts labeled 05, 16, 17, and 20. These are Beach Castle, Radisson, Ramada, and Sea Club. Most statistical software will select an SRS for you, eliminating the need for Table 7.1. The *Simple Random Sample* applet is a convenient way to automate this task (see Applet Exercise 1). ■

Online polls and mall interviews produce samples. We can't trust results from these samples because they are chosen in ways that invite bias. We have more confidence in results from an SRS because it uses impersonal chance to avoid bias. The first question to ask about any sample is whether it was chosen at random. Opinion polls and other sample surveys carried out by people who know what they are doing use random sampling. Most national sample surveys use sampling schemes more complex than an SRS. They may, for example, dial the last four digits of a telephone number at random separately within each exchange (the area code and first three digits). The national sample is pieced together from many smaller samples. The big idea remains the deliberate use of chance to choose the sample. Because simple random sampling is the essential principle behind all random sampling and because it is also the main building block for more complex samples, we will focus our study on simple random sampling.

EXAMPLE 5   Soda or Not?

A Gallup poll on the American diet asked subjects about their attitudes toward various foods. The press release mentioned "the increasing proportion of Americans who say they try to avoid 'soda or pop' (51%, up from 41% in 2002)." Can we trust that 51%? Ask first how Gallup selected its sample. Later in the press release we read this: "These results are based on telephone interviews with a randomly selected national sample of 1005 adults, aged 18 and older, conducted July 8–11, 2004."

This is a good start toward gaining our confidence. Gallup tells us what population it has in mind (people at least 18 years old who live anywhere in the United States). We know that the sample from this population was of size 1005 and, most important, that it was chosen at random. There is more to say, and we will say it later in this chapter, but we have at least heard the comforting words "randomly selected." ■

7.4 Cautions About Sample Surveys

Random sampling eliminates bias in the choice of a sample from a list of the population. Sample surveys of large human populations, however, require more than a good sampling design.

To begin, we need an accurate and complete list of the population. Because such a list is rarely available, most samples suffer from some degree of *undercoverage*. A sample survey of households, for example, will miss not only homeless people but also prison inmates and students in dormitories. An opinion poll conducted by telephone will miss the 6% of American households without residential phones. The results of national sample surveys therefore have some bias if the people not covered—who most often are poor people—differ from the rest of the population.

A more serious source of bias in most sample surveys is *nonresponse,* which occurs when a selected individual cannot be contacted or refuses to cooperate. Nonresponse to sample surveys often reaches 50% or more, even with careful planning and several callbacks. Because nonresponse is higher in urban areas, most sample surveys substitute other people in the same area to avoid favoring rural areas in the final sample. If the people contacted differ from those who are rarely at home or who refuse to answer questions, some bias remains.

Undercoverage and Nonresponse

Undercoverage occurs when some groups in the population are left out of the process of choosing the sample.

Nonresponse occurs when an individual chosen for the sample can't be contacted or refuses to participate.

How Bad Is Nonresponse?

EXAMPLE 6

The Current Population Survey (CPS) has the lowest nonresponse rate of any poll we know: Only about 4% of the households in the CPS sample refuse to take part and another 3% or 4% can't be contacted. People are more likely to respond to a government survey such as the CPS, and the CPS contacts its sample in person before doing later interviews by phone.

What about polls done by the media and by market research and opinion polling firms? We don't know their rates of nonresponse because they won't say. That itself is a bad sign. The Pew Research Center imitated a careful telephone survey and published the results: Out of 2879 households called, 1658 were never at home, refused, or would not finish the interview. That's a nonresponse rate of 58%. ■

When people do respond, we can't rely on them to always tell the truth. People know that they should take the trouble to vote, for example, so many who didn't vote in the last election will tell a pollster that they did. Finally, the *wording of questions* strongly influences the answers given to a sample survey. Confusing or leading questions can introduce strong bias, and even minor changes in wording can change a survey's outcome. Here are some examples.

EXAMPLE 7   Watch That Wording

How do Americans feel about government help for the poor? Only 13% think we are spending too much on "assistance to the poor," but 44% think we are spending too much on "welfare." How do the Scots feel about the movement to become independent from England? Well, 51% would vote for "independence for Scotland," but only 34% support "an independent Scotland separate from the United Kingdom." It seems that "assistance to the poor" and "independence" are nice, hopeful words. "Welfare" and "separate" are negative words. ■

The statistical design of sample surveys is a science, but this science is only part of the art of sampling. Because of nonresponse, false responses, and the difficulty of posing clear and neutral questions, you should hesitate to fully trust reports about complicated issues based on surveys of large human populations. Insist on knowing the exact questions asked, the rate of nonresponse, and the date and method of the survey before you trust a poll result.

7.5 Experiments

Sample surveys gather information on part of the population in order to draw conclusions about the whole. When the goal is to describe a population, statistical sampling is the right tool to use.

Suppose, however, that we want to study the response to a stimulus, to see how one variable affects another when we change existing conditions. Will a new mathematics curriculum improve the scores of sixth-graders on a standard test of mathematics achievement? Will taking small amounts of aspirin daily reduce the risk of suffering a heart attack? Does a mother's smoking during pregnancy reduce the IQ of her child? Studies that simply *observe and describe* are ineffective tools for answering these questions. *Experiments* give us clearer answers.

Observation Versus Experiment

An **observational study**, such as a sample survey, observes individuals and measures variables of interest but does not attempt to influence the responses. The purpose of an observational study is to describe some group or situation.

An **experiment**, on the other hand, deliberately imposes some *treatment* on individuals in order to observe their responses. The purpose of an experiment is to study whether the treatment causes a change in the response.

Experiments are the preferred method for examining the effect of one variable on another. By imposing the specific treatment of interest and controlling other influences, we can pin down cause and effect. A sample survey may show that two variables are related, but it cannot demonstrate that one causes the other. Statistics has something to say about how to arrange experiments, just as it suggests methods for sampling.

An Uncontrolled Experiment

EXAMPLE 8

A college regularly offers a review course to prepare candidates for the Graduate Management Admission Test (GMAT) required by most graduate business schools. This year, it offered only an online version of the course. The average GMAT score of students in the online course was 10% higher than the long-time average for those who took the classroom review course. Is the online course more effective?

This experiment has a very simple design. A group of subjects (the students) were exposed to a treatment (the online course), and the outcome (GMAT scores) was observed. Here is the design:

Online course → Observe GMAT scores

or, in general form

Treatment → Observe response ▪

Most laboratory experiments use a design like that in the example: Apply a treatment and measure the response. In the controlled environment of the laboratory, simple designs often work well. But field experiments and experiments with human subjects are exposed to more variable conditions and deal with more variable subjects. It isn't possible to control outside factors that can influence the outcome. With greater variability comes a greater need for statistical design.

A closer look at the GMAT review course showed that the students in the online review course were quite different from the students who in past years took the classroom course. In particular, they were older and more likely to be employed. An online course appeals to these mature people, but we can't compare their performance with that of the undergraduates who previously dominated the course. The online course might even be less effective than the classroom version. The effect of online versus in-class instruction is hopelessly mixed up with influences lurking in the background. Figure 7.2 shows the mixed-up influences in picture form. We say that student age and background is *confounded* with whatever effect the change to online instruction may have.

Confounding

Variables, whether part of a study or not, are said to be **confounded** when their effects on the outcome cannot be distinguished from each other.

FIGURE 7.2

FIGURE 7.2
Confounding. We can't
distinguish the effects of
the treatment from those
of other influences.

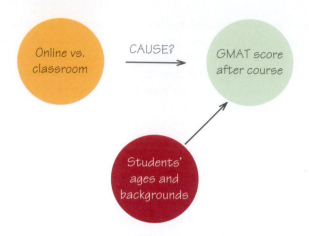

The remedy for confounding is to do a *comparative experiment* in which some
students are taught in the classroom and other similar students take the course
online. The first group is called a **control group.** Most well-designed experiments
compare two or more treatments. Of course, comparison alone isn't enough to
produce results we can trust. If the treatments are given to groups that differ
markedly when the experiment begins, bias will result. For example, if we allow
students to elect online or classroom instruction, older employed students are
likely to sign up for the online course. Personal choice will bias our results in the
same way that volunteers bias the results of call-in opinion polls. The solution
to the problem of bias is the same for experiments and for samples: Use impersonal chance to select the groups.

EXAMPLE 9   A Randomized Comparative Experiment

The college decides to compare the progress of 25 on-campus students taught in
the classroom with that of 25 students taught the same material online. Select
the students who will be taught online by taking a simple random sample of size
25 from the 50 available subjects. The remaining 25 students form the control
group. They will receive classroom instruction. The result is a **randomized comparative experiment** with two groups. Figure 7.3 outlines the design in graphical form.

The selection procedure is exactly the same as it is for sampling: label and
table. Step 1: *Label* the 50 students 01 to 50. Step 2: Go to the *table* of random
digits and read successive two-digit groups. The first 25 labels encountered select

FIGURE 7.3 The
design of a randomized
comparative experiment
to compare online and
classroom instruction.

the online group. As usual, ignore repeated labels and groups of digits not used as labels. For example, if you begin at line 125 in Table 7.1, the first five students chosen are those labeled 21, 49, 37, 18, and 44. The *Simple Random Sample* applet makes it particularly easy to choose treatment groups at random. ∎

The GMAT experiment is *comparative* because it compares two treatments (the two instructional settings). It is *randomized* because the subjects are assigned to the treatments by chance. Randomization creates groups that are similar to each other before we start the experiment. Comparison means that possible confounding variables act on both groups at once. The only difference between the groups is the online versus in-class setting. So if we see a difference in performance, it must be due to the different setting. That is the basic logic of randomized comparative experiments. This logic shows why experiments can give good evidence that the different treatments really *caused* different outcomes.

There is a fine point: The performance of the two groups will differ even if the treatments are identical, just because the individuals assigned at random to the groups differ. It is only differences *larger than would plausibly occur just by chance* that show the effects of the treatments. The laws of chance allow statisticians to say how big an effect is *statistically significant*.

Statistical Significance

An observed effect so large that it would rarely occur by chance is called **statistically significant**.

Here is another example of a randomized comparative experiment, this time comparing three treatments.

Conserving Energy

EXAMPLE 10

Many utility companies have introduced programs to encourage energy conservation among their customers. An electric company considers placing electronic indicators in households to show what the cost would be if the electricity use at that moment continued for a month. Will indicators reduce electricity use? Would cheaper methods work almost as well? The company decides to design an experiment.

One cheaper approach is to give customers a chart and information about monitoring their electricity use. The experiment compares these two approaches (indicator, chart) and also a control. The control group of customers receives information about energy conservation but no help in monitoring electricity use. The outcome is measured by total electricity used in a year. The company finds 60 single-family residences in the same city willing to participate, so it assigns 20 residences at random to each of the three treatments. Figure 7.4 outlines the design.

FIGURE 7.4 The design of a randomized comparative experiment to compare three ways of encouraging households to conserve electricity.

To carry out the random assignment, label the 60 households 01 to 60. Enter Table 7.1 to select an SRS of 20 to receive the indicators. Continue in Table 7.1, selecting 20 more to receive charts. The remaining 20 form the control group. ■

7.6 Thinking About Experiments

Randomized comparative experiments are common tools of industrial and academic research. They are also widely used in medical research. For example, federal regulations require that the safety and effectiveness of new drugs be demonstrated by randomized comparative experiments. Let's look at a medical experiment.

EXAMPLE 11   **St. John's Wort for Depression?**

Although prescription drugs must pass the test of randomized comparative experiments before being sold, herbs and other "natural remedies" are exempt. Because these treatments are so popular, some are now being studied more carefully. Fans of natural remedies often use extracts of the herb St. John's wort to treat depression. Is the herb safe? Does it work? The *Journal of the American Medical Association* reported a "randomized, double-blind, placebo-controlled clinical trial" in which 200 patients with major depression were assigned at random to take either herb extract or a dummy pill that looked and tasted the same. Results: The herb is safe, but "[i]n this study, St. John's wort was not effective for treatment of major depression." ■

If you read accounts of medical studies, you will often meet language like "randomized, double-blind, placebo-controlled clinical trial." A clinical trial is a medical experiment with actual patients as subjects. "Randomized" and "controlled" tell us that this was a randomized comparative experiment (that's good). A "placebo" is a fake treatment, the dummy pill in this study. Here we meet a new idea, the importance of the **placebo effect**, a special kind of confounding. The placebo effect is the tendency of patients to respond favorably to any treatment, even a placebo. If depressed patients given St. John's wort are compared with patients who receive no treatment, the first group gets the benefit of both the herb and the placebo effect. Any beneficial effect that St. John's wort may have is confounded with the placebo effect. To prevent confounding, it is important that some treatment be given to all subjects in any medical experiment.

The depression study was a **double-blind experiment:** Neither the subjects nor the experimenters who worked with them knew which treatment any subject received. Subjects might react differently if they knew they were getting "only a placebo." Knowing that a particular subject was getting "only a placebo" could also influence the health workers who interviewed and examined the subjects. So both subjects and workers were kept "blind." Only the study's statistician knew which treatment each subject received.

The difference between the St. John's wort and placebo groups was *not statistically significant*—that is, it was no larger than would be expected when we divide 200 depressed patients at random into two groups and do nothing else. Larger numbers of subjects would give more precise results. It's unlikely that there is exactly *no* difference between St. John's wort and a placebo. If the clinical trial had used 2000 patients rather than 200, it might have picked up a small effect (in either direction). The researchers thought that 200 patients was enough to pick up any effect large enough to be medically important.

The logic of experimentation, the statistical design of experiments, and the laws that govern chance behavior combine to give compelling evidence of cause and effect. Only experimentation can produce fully convincing evidence of causation.

Smoking and Health

EXAMPLE 12

By way of contrast, consider the statistical evidence linking cigarette smoking to lung cancer. We can't assign groups of people to smoke or not, so a direct experiment isn't possible. The most careful studies have selected samples of smokers and nonsmokers, then followed them for many years, eventually recording the cause of death. These are called **prospective studies** because they follow the subjects forward in time. Prospective studies are comparative, but they are not experiments because the subjects themselves choose whether or not to smoke. A large prospective study of British doctors found that the death rate from lung cancer among cigarette smokers was 20 times that among nonsmokers. Another study of American men aged 40 to 79 found that the lung cancer death rate was 11 times higher among smokers than among nonsmokers. These and many other observational studies show a strong connection between smoking and lung cancer. ■

The connection between smoking and lung cancer is statistically significant. That is, it is far stronger than would occur by chance. We can be confident that something other than chance links smoking to cancer. But observation of samples cannot tell us *what* factors other than chance are at work. Perhaps there is something in the genetic makeup of some people that predisposes them both to nicotine addiction and to lung cancer. In that case, we would observe a strong link even if smoking itself had no effect on the lungs.

The statistical evidence that points to cigarette smoking as a cause of lung cancer is about as strong as nonexperimental evidence can be. First, the connection has been observed in many studies in many countries. This eliminates factors peculiar to one group of people or to one specific study design. Second, there is a *dose-response relationship*: People who smoke more are more likely to get lung cancer than those who smoke less, and quitting cigarettes reduces the cancer risk. Third, specific ways in which smoking could cause cancer have been identified—cigarette smoke contains tars that have been shown by experiment to cause tumors in animals. Finally, no plausible alternative explanation is available. For example, the genetic hypothesis cannot explain the increase in lung cancer among women that occurred as more and more women became smokers. Lung cancer, which has long been the leading cause of cancer deaths in men, has now passed breast cancer as the most fatal cancer for women. This evidence is convincing, but it is not quite as strong as the conclusive statistical evidence we get from randomized comparative experiments.

Despite their attractions, experiments like samples can have weaknesses. The most common weakness is a *lack of realism* that makes it hard to say exactly how far the results of an experiment apply.

EXAMPLE 13   Is the Experiment Realistic?

Clinical trials give medical treatments to actual patients with the condition that the treatments are supposed to help. Many experiments are less realistic. A psychologist studying the effects of stress on teamwork observes teams of students carrying out tasks in a psychology laboratory under different conditions. The students know it's "just an experiment" and that the stress will only last an hour. Do the conclusions of such experiments apply to real-life stress? An engineer uses a small pilot production process in a laboratory to find the choices of pressure and temperature that maximize yield from a complex chemical reaction. Do the results apply to a full-scale manufacturing plant?

These are not statistical questions. The psychologist and the engineer must use their understanding of psychology and engineering to judge how far their results apply. The statistical design enables us to trust the results for the students and the pilot process but not to generalize the conclusions to other settings. ■

7.7 Inference: From Sample to Population

A market research firm interviews a random sample of 2500 adults. Result: 66% find shopping for clothes frustrating and time consuming. That's the truth about the 2500 people in the sample. What is the truth about the almost 220 million American adults who make up the population? Because the sample was chosen at random, it's reasonable to think that these 2500 people represent the entire population fairly well. So the market researchers turn the *fact* that 66% of the *sample* find shopping frustrating into an *estimate* that about 66% of *all adults* feel this way. That's a basic move in statistics: Use a fact about a sample to estimate the truth about the whole population. We call this **statistical inference** because we infer conclusions about the wider population from data on selected individuals. If the selected individuals were chosen at random, we think that they fairly represent the population and inference makes sense. If we have data only from a convenience sample or a voluntary response sample, the data don't represent the population and we can't use them for inference. *Statistical inference works only if the data come from a random sample or randomized comparative experiment.* That's why this chapter starts with producing data before moving on to inference from the data to a larger population.

To think about inference, we must keep straight whether a number describes a sample or a population. Here is the vocabulary we use.

Parameters and Statistics

A **parameter** is a number that describes the **population.** A parameter is a fixed number, but in practice we do not know its value.

A **statistic** is a number that describes a **sample.** The value of a statistic is known when we have taken a sample, but it can change from sample to sample. We often use a statistic to estimate an unknown parameter.

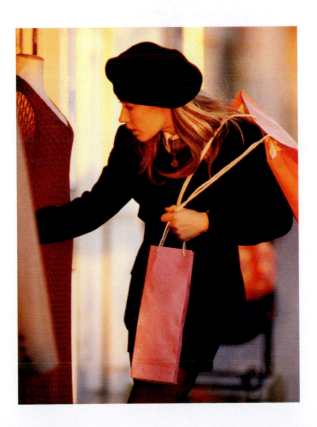

EXAMPLE 14   Do You Find Shopping Frustrating?

Sample surveys show that fewer people enjoy shopping than in the past. A survey by the market research firm Yankelovich Clancy Shulman asked a nationwide random sample of 2500 adults if they agreed or disagreed that "I like buying new clothes, but shopping is often frustrating and time consuming." Of the respondents, 1650 said they agreed. The proportion of the sample who agree is

$$\hat{p} = \frac{1650}{2500} = 0.66 = 66\%$$

The number $\hat{p} = 0.66$ is a *statistic*. The corresponding *parameter* is the proportion (call it p) of all adult U.S. residents who would have said "Agree" if asked the same question. We don't know the value of the parameter p, so we use the statistic $\hat{p}$ to estimate it. ■

If Yankelovich took a second random sample of 2500 adults, the new sample would have different people in it. It is almost certain that there would not be exactly 1650 positive responses. That is, the value of the statistic $\hat{p}$ will vary from sample to sample. If the variation when we take repeat samples from the same population is too great, we can't trust the results of any one sample. We are saved by the second great advantage of random samples. The first advantage is that choosing at random eliminates favoritism. That is, random sampling attacks bias. The second advantage is that if we take lots of random samples of the same size from the same population, the variation from sample to sample will follow a predictable pattern.

All of statistical inference is based on one idea: to see how trustworthy a procedure is, ask what would happen if we repeated it many times. So we must ask, "What would happen if we took many samples?" Here's how to answer that question:

▶ Take a large number of random samples from the same population.
▶ Calculate the sample proportion $\hat{p}$ for each sample.
▶ Make a histogram of the values of $\hat{p}$.
▶ Examine the distribution displayed in the histogram for shape, center, and spread, as well as outliers or other deviations.

In practice it is too expensive to take many samples from a large population such as all adult U.S. residents. But we can program a computer to imitate drawing many samples at random from a population that we specify. This is called *simulation*. Here's what happens when we do this.

What Happens in Many Samples?

EXAMPLE 15

Figure 7.5 illustrates the process of choosing many samples and finding the sample proportion $\hat{p}$ for each one. Follow the flow of the figure from the population at the left, to choosing an SRS and finding the $\hat{p}$ for this sample, to collecting together the $\hat{p}$'s from many samples. The histogram at the right of the figure shows the distribution of the values of $\hat{p}$ from 1000 separate SRSs of size 100 drawn from a population with $p = 0.6$.

Of course, Yankelovich interviewed 2500 people, not just 100. Figure 7.6 is parallel to Figure 7.5. It shows the process of choosing 1000 SRSs, each of size 2500, from a population in which the true proportion is $p = 0.6$. The 1000 values of $\hat{p}$ from these samples form the histogram at the right of the figure. Figures 7.5 and 7.6 are drawn on the same scale. Comparing them shows what happens when we increase the size of our samples from 100 to 2500. These histograms display the *sampling distribution* of the statistic $\hat{p}$ for two sample sizes. ■

Sampling Distribution

The **sampling distribution** of a statistic is the distribution of values taken by the statistic in all possible samples of the same size from the same population.

Strictly speaking, the sampling distribution is the ideal pattern that would emerge if we looked at all possible samples of the same size from our population. A distribution obtained from a fixed number of trials, like the 1000 trials in these figures, is only an approximation to the sampling distribution. Probability

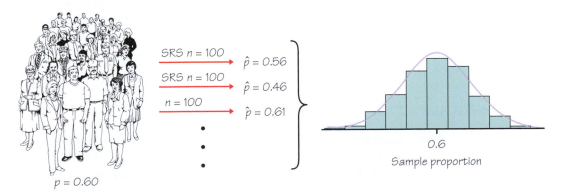

FIGURE 7.5 Draw 1000 SRSs of size 100 from a population with proportion $p = 0.60$ of successes. The histogram shows the distribution of the 1000 sample proportions $\hat{p}$.

FIGURE 7.6 Draw 1000 SRSs of size 2500 from the same population as in Figure 7.5. The histogram shows the distribution of the 1000 sample proportions $\hat{p}$, using the same scale as Figure 7.5. The statistic from the larger sample is less variable.

theory, the mathematics of chance behavior, can sometimes describe sampling distributions exactly. Chapter 8 will introduce you to basic probability theory. The interpretation of a sampling distribution is the same, however, whether we obtain it by simulation or by the mathematics of probability.

We can use the tools of data analysis to describe any distribution. Let's apply those tools to Figures 7.5 and 7.6.

▶ **Shape:** The histograms look normal. The normal curves drawn through the histograms describe the overall shape quite well.

▶ **Center:** In both cases, the values of the sample proportion $\hat{p}$ vary from sample to sample, but the values are centered at 0.6. Recall that $p = 0.6$ is the true population parameter. Some samples have a $\hat{p}$ less than 0.6 and some greater, but there is no tendency to be always low or always high. That is, $\hat{p}$ has **no bias** as an estimator of p. This is true for both large and small samples. (Want the details? The mean of the 1000 values of $\hat{p}$ is 0.598 for samples of size 100 and 0.6002 for samples of size 2500. The median value of $\hat{p}$ is exactly 0.6 for samples of both sizes.)

▶ **Spread:** The values of $\hat{p}$ from samples of size 2500 are much less spread out than the values from samples of size 100. In fact, the standard deviations are 0.051 for Figure 7.5 and 0.0097, or about 0.01, for Figure 7.6.

Although these results describe just two sets of simulations, they reflect facts that are true whenever we use random sampling. We can turn to probability theory to learn the mathematical facts that lie behind the simulations. Here they are.

We'll use the word "success" for whatever we are counting, such as "Agree" responses in the shopping survey.

Sampling Distribution of a Sample Proportion

Choose an SRS of size n from a large population that contains population proportion p of successes. Let $\hat{p}$ be the **sample proportion** of successes,

$$\hat{p} = \frac{\text{count of successes in the sample}}{n}$$

Then:

▶ **Shape:** For large sample sizes, the sampling distribution of $\hat{p}$ is **approximately normal.**

▶ **Center:** The **mean** of the sampling distribution is p.

▶ **Spread:** The **standard deviation** of the sampling distribution is

$$\sqrt{\frac{p(1-p)}{n}}$$

Figure 7.7 summarizes these facts in a form that reminds us that a sampling distribution describes the results of lots of samples from the same population.

Do You Find Shopping Frustrating?

EXAMPLE 16

Suppose that 60% of all adults find shopping for clothes frustrating and time consuming. Take a simple random sample of 2500 adults. That's exactly the setting of the second simulation in Example 15. Now we can apply mathematics to learn how the sample proportion $\hat{p}$ would behave if we took many samples. The population proportion is $p = 0.6$. So the distribution of $\hat{p}$ in many samples

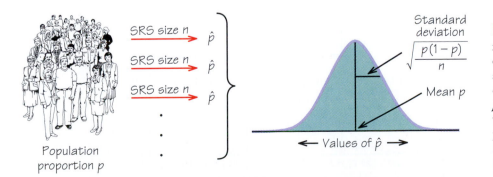

FIGURE 7.7 Repeat many times the process of selecting an SRS of size n from a population of which the proportion p are successes. The values of the sample proportion of successes $\hat{p}$ have this normal sampling distribution.

▶ is close to normal;

▶ has mean 0.6;

▶ has standard deviation

$$\sqrt{\frac{p(1-p)}{n}} = \sqrt{\frac{(0.6)(0.4)}{2500}} = 0.0098$$

The mean 0.6 and standard deviation 0.0098 from the mathematics are very close to the mean 0.6002 and standard deviation 0.0097 we observed in our simulation. If the simulation used more than 1000 trials, the results would be yet closer to the mathematical truth. ■

7.8 Confidence Intervals

The sampling distribution shows why we can trust the results of a large random sample: Almost all such samples give results that are close to the truth about the population.

EXAMPLE 17   **The 68–95–99.7 Rule Again**

In Example 16, the population parameter, the proportion of adults who find shopping frustrating, is $p = 0.6$. If we take SRSs of size 2500, the sample proportions $\hat{p}$ follow the normal distribution with mean 0.6 and standard deviation about 0.01. So the 95 part of the 68–95–99.7 rule says that 95% of all samples give a $\hat{p}$ within ±0.02 (that's 2 standard deviations) of the truth about the population. That is, 95% of all samples have $\hat{p}$ between 0.58 and 0.62. Figure 7.8 illustrates this use of the 68–95–99.7 rule. ■

FIGURE 7.8 The sampling distribution of $\hat{p}$ for Example 16. By the 68–95–99.7 rule, 95% of all samples have a sample proportion $\hat{p}$ within ±0.02 of the true population proportion $p = 0.6$.

Sampling distribution of $\hat{p}$, with mean 0.6 and standard deviation 0.01.

Two standard deviations = 0.02

95% of all samples give a result within this range

0.58   Mean 0.6   0.62

We can repeat this reasoning for any value of the parameter p and the sample size n. It is always true that 95% of all samples give a sample proportion $\hat{p}$ within 2 standard deviations of the population proportion p. That is to say that 95% of all samples catch p in the interval extending 2 standard deviations on either side of $\hat{p}$. That's the interval

$$\hat{p} \pm 2\sqrt{\frac{p(1-p)}{n}}$$

This formula tells us how close the unknown parameter p lies to the observed statistic $\hat{p}$ in 95% of all samples. There is one catch: We can't calculate the interval from the data because the standard deviation involves the population proportion p, and in practice we don't know p. In Examples 16 and 17, we applied the formula for $p = 0.6$, but this may not be the true p for the actual population of all American adults.

What to do? The standard deviation of the statistic $\hat{p}$ does depend on the parameter p, but it doesn't change a lot when p changes. Go back to Example 16 and redo the calculation for other values of p. Here's the result:

Value of p	0.4	0.5	0.6	0.7	0.8
Standard deviation	0.0098	0.01	0.0098	0.0092	0.008

The standard deviations are all 0.01 when rounded to two places. You see that if we guess a value of p reasonably close to the true value, the standard deviation found from the guessed value will be about right. We know that when we take a large random sample, the statistic $\hat{p}$ is almost always close to the parameter p. So we will use $\hat{p}$ as the guessed value of the unknown p. Now we have an interval that we can calculate from the sample data. We call it a *confidence interval*.

Confidence Interval

A **95% confidence interval** is an interval obtained from the sample data by a method that in 95% of all samples will produce an interval containing the true population parameter.

Choose an SRS of size n from a large population that contains an unknown proportion p of successes. A **95% confidence interval for p** is

$$\hat{p} \pm 2\sqrt{\frac{\hat{p}(1-\hat{p})}{n}}$$

Here $\hat{p}$ is the proportion of successes in the sample and $2\sqrt{\hat{p}(1-\hat{p})/n}$ is the **margin of error**.

This recipe is only approximately correct but is quite accurate when the sample size n is large.

This interval is only approximately correct for two reasons. The sampling distribution of the sample proportion $\hat{p}$ isn't exactly normal. And we don't get the standard deviation of $\hat{p}$ exactly right because we used $\hat{p}$ in place of the unknown p. Both of these difficulties go away as the sample size n gets larger. Our method works well enough for many practical uses. More important, it shows how we get a confidence interval from the sampling distribution of a statistic. That's the reasoning behind any confidence interval.

EXAMPLE 18   Risky Behavior in the Age of AIDS

How common is behavior that puts people at risk for AIDS? The National AIDS Behavioral Surveys interviewed a random sample of 2673 adult heterosexuals. Of these people, 170 had had more than one sexual partner in the past year. The sample proportion who admit to multiple partners is

$$\hat{p} = \frac{170}{2673} = 0.0636$$

A 95% confidence interval for the proportion p of all adult heterosexuals with multiple partners is therefore

$$\hat{p} \pm 2\sqrt{\frac{\hat{p}(1 - \hat{p})}{n}} = 0.0636 \pm 2\sqrt{\frac{(0.0636)(0.9364)}{2673}}$$

$$= 0.0636 \pm 0.0094, \text{ or } 6.36\% \pm 0.94\%$$

$$= 0.0542 \text{ to } 0.0730, \text{ or } 5.42\% \text{ to } 7.30\%$$

A report of these calculations might say, "The study found that 6.36% of heterosexuals had more than one sexual partner. The margin of error for this result is 0.94%." ■

We got the interval in Example 18 by using a recipe that catches the true unknown population proportion in 95% of all samples. The shorthand for this is: We are **95% confident** that the true proportion of heterosexuals with multiple partners lies between 5.42% and 7.30%. The margin of error refers to the spread needed to capture the true p in 95% of all samples. The truth lies outside the margin of error in 5% of all samples.

Figure 7.9 lays out the meaning of "95% confidence." The vertical line is the true value of the population proportion p. The normal curve at the top of the figure is the sampling distribution of the sample statistic $\hat{p}$, which is centered at the true p. The 95% confidence intervals from 25 SRSs appear below, one after the other. The central dots are the values of $\hat{p}$, the centers of the intervals. The arrows on either side span the confidence interval. In the long run, 95% of the intervals will cover the true p and 5% will miss. Of the 25 intervals in Figure 7.9, 24 hit and 1 misses. (Remember that the sampling distribution describes what happens in a very large number of samples—we don't expect exactly 95% of 25

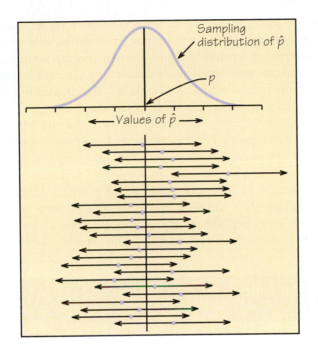

FIGURE 7.9 Twenty-five samples from the same population give these 95% confidence intervals. In the long run, 95% of all such intervals cover the true population proportion, marked by the vertical line.

intervals to capture the true parameter.) The *Confidence Interval* applet animates Figure 7.9. You can use the applet to watch confidence intervals from one sample after another capture or fail to capture the true parameter.

The length of a confidence interval depends on the size n of the sample. Larger samples give shorter intervals because of the $\sqrt{n}$ in the denominator of the margin of error. But the interval does *not* depend on the size of the population. This is true as long as the population is much larger than the sample. The confidence interval in Example 18 works for a sample of 2673 from a city with 100,000 adults as well as for a sample of 2673 from a nation of 220 million. What matters is how many people we interview, not what percent of the population we contact.

The length of a confidence interval also depends on how confident we want to be that the interval does capture the true parameter value. It is common to use 95% confidence, but you can ask for higher or lower confidence if you want. Our 95% confidence interval was based on the middle 95% of a normal distribution. A 99% confidence interval requires the middle 99% of the distribution and so is wider (has a larger margin of error).

Understanding the News

EXAMPLE 19

Here's what the TV news announcer says: "A new Gallup poll on American eating habits finds that 51% of adults are trying to avoid drinking soda. The margin of error for the poll was 3 percentage points." Plus or minus 3 percent starting at 51% is 48% to 54%. People who know no statistics often think that the truth about the entire population lies in that range.

This is what Gallup actually said: "For results based on this sample, one can say with 95% confidence that the maximum error attributable to sampling and other random effects is 3 percentage points. In addition to sampling error, question wording and practical difficulties in conducting surveys can introduce error or bias into the findings of public opinion polls." That is, Gallup tells us that the margin of error only works for 95% of all its samples. "95% confidence" is shorthand for that. The news report left out the "95% confidence." In fact, *almost all margins of error in the news are for 95% confidence*. If you don't see the confidence level in a report, it's usually safe to assume 95%. ∎

Gallup's mention of "question wording and practical difficulties" takes us back to our cautions about sample surveys. *The margin of error ignores nonresponse and other practical difficulties.* The margin of error in a confidence interval comes from the sampling distribution of the statistic. The sampling distribution describes

SPOTLIGHT 7.2   How Accurate Is Our Poll?

Responsible polling organizations tell the public something about the accuracy of their poll results. The margin of error is basic, and so is a warning that there are other sources of error not covered by the announced margin of error. Here are the statements that accompanied poll results from three sources, found on their Web sites.

The Gallup Poll. "For results based on this sample, one can say with 95 percent confidence that the maximum error attributable to sampling and other random effects is plus or minus 3 percentage points. In addition to sampling error, question wording and practical difficulties in conducting surveys can introduce error or bias into the findings of public opinion polls."

The Harris Poll. "In theory, with a probability sample of this size, one can say with 95 percent certainty that the results have a statistical precision of plus or minus 3 percentage points of what they would be if the entire adult population had been polled with complete accuracy.

Unfortunately, there are several other possible sources of error in all polls or surveys that are probably more serious than theoretical calculations of sampling error. They include refusals to be interviewed (nonresponse), question wording and question order, interviewer bias, weighting by demographic control data and screening (e.g., for likely voters). It is difficult or impossible to quantify the errors that may result from these factors."

The *New York Times*. "In theory, in 19 cases out of 20 the results based on such samples will differ by no more than three percentage points in either direction from what would have been obtained by seeking out all American adults. For smaller subgroups, the margin of sampling error is larger. In addition to sampling error, the practical difficulties of conducting any survey of public opinion may introduce other sources of error into the poll. Variation in the wording and order of questions, for example, may lead to somewhat different results."

the variation of the statistic due to chance in repeated random samples. This random variation is the *only* source of error covered by the margin of error. Real-life samples also suffer from undercoverage and nonresponse. Errors from these practical difficulties are usually more serious than random sampling error. The actual error in sample surveys may be much larger than the announced margin of error. Worse, we can't say how much larger. Statistical conclusions are approximations to a complicated truth, not mathematical results that are simply true.

Measuring Risky Behavior

EXAMPLE 20

What about the National AIDS Behavioral Surveys? The interviews were carried out by telephone. This is acceptable for surveys of the general population, because about 94% of American households have telephones. However, some groups at high risk for AIDS, such as intravenous drug users, often don't live in settled households and are underrepresented in the sample. About 30% of the people reached refused to cooperate. A nonresponse rate of 30% is not unusual in large sample surveys, but it may cause some bias if those who refuse differ systematically from those who cooperate. The survey used statistical methods that adjust for unequal response rates in different groups. Finally, some respondents may not have told the truth when asked about their sexual behavior. The survey team tried hard to make respondents feel comfortable. For example, Hispanic women were interviewed by Hispanic women, and Spanish speakers were interviewed by Spanish speakers with the same regional accent (Cuban, Mexican, or Puerto Rican). Nonetheless, the survey report says that some bias is probably present:

> It is more likely that the present figures are underestimates; some respondents may underreport their numbers of sexual partners and intravenous drug use because of embarrassment and fear of reprisal, or they may forget or not know details of their own or of their partner's HIV risk and their antibody testing history. [Joseph H. Catania et al., Prevalence of AIDS-related risk factors and condom use in the United States, *Science*, 258 (1992): 1104.]

Reading the report of a large study like the National AIDS Behavioral Surveys reminds us that statistics in practice involves much more than recipes for confidence intervals. ■

REVIEW VOCABULARY

Bias A systematic error that tends to cause the observations to deviate in the same direction from the truth about the population whenever a sample or experiment is repeated.

95% confidence interval An interval computed from a sample by a method that captures the unknown parameter in 95% of all possible samples. When we calculate the interval for a single sample,

we are **95% confident** that the interval captures the unknown parameter.

Confounding Two variables are confounded when their effects on the outcome of a study cannot be distinguished from each other.

Control group A group of experimental subjects that is given a standard treatment or no treatment (such as a placebo).

Convenience sample A sample that consists of the individuals who are most easily available, such as people passing by in the street. A convenience sample is usually biased.

Double-blind experiment An experiment in which neither the experimental subjects nor the persons who interact with them know which treatment each subject received.

Experiment A study in which treatments are applied to people, animals, or things in order to observe the effect of the treatments.

Margin of error As announced by most national polls, the margin of error says how close to the truth about the population the sample result would fall in 95% of all samples drawn by the method used to draw this one sample.

Nonresponse Some individuals chosen for a sample cannot be contacted or refuse to participate.

Observational study A study (such as a sample survey) that observes individuals and measures variables of interest but does not attempt to influence the responses.

Parameter A number that describes the population. In statistical inference, the goal is often to estimate an unknown parameter or make a decision about its value.

Placebo effect The effect of a dummy treatment (such as an inert pill in a medical experiment) on the response of subjects.

Population The entire group of people or things about which we want information.

Prospective study An observational study that follows two or more groups of subjects forward in time.

Randomized comparative experiment An experiment to compare two or more treatments in which people, animals, or things are assigned to treatments by chance.

Sample A part of the population that is actually observed and used to draw conclusions, or inferences, about the entire population.

Sample proportion The proportion $\hat{p}$ of the members of a sample having some characteristic (such as agreeing with an opinion poll question). The sample proportion from a simple random sample is used to estimate the corresponding proportion p in the population from which the sample was drawn.

Sampling distribution The distribution of values taken by a statistic when all possible random samples of the same size are drawn from the same population. The sampling distributions of sample proportions are approximately normal.

Simple random sample (SRS) A sample chosen by chance, so that every possible sample of the same size has an equal chance to be the one selected.

Statistic A number that describes a sample. A statistic can be calculated from the sample data alone; it does not involve any unknown parameters of the population.

Statistical inference Methods for drawing conclusions about an entire population on the basis of data from a sample. Confidence intervals are one type of inference method.

Statistical significance An observed effect is statistically significant if it is so large that it is unlikely to occur just by chance in the absence of a real effect in the population from which the data were drawn.

Table of random digits A table whose entries are the digits 0, 1, 2, 3, 4, 5, 6, 7, 8, 9 in a completely random order. That is, each entry is equally likely to be any of the 10 digits and no entry gives information about any other entry.

Undercoverage The process of choosing a sample may systematically leave out some groups in the population, such as households without telephones.

Voluntary response sample A sample of people who choose themselves by responding to a general invitation to give their opinions. Such a sample is usually strongly biased.

✔ SKILLS CHECK

1. An opinion poll contacts 1161 adults and asks them, "Which political party do you think has better ideas for leading the country in the twenty-first century?" In all, 696 of the 1161 say, "The Democrats." The sample in this setting is

(a) all 220 million adults in the United States.
(b) the 1161 people interviewed.
(c) the 696 people who chose the Democrats.

2. A committee on community relations in a college town plans to survey local businesses about the importance of students as customers. From telephone book listings, the committee chooses 150 businesses at random. Of these, 73 return the questionnaire mailed by the committee. The population for this study is

(a) all businesses in the college town.
(b) the 150 businesses chosen.
(c) the 73 businesses that returned the questionnaire.

3. The sample in the setting of the previous exercise is

(a) all businesses in the college town.
(b) the 150 businesses chosen.
(c) the 73 businesses that returned the questionnaire.

4. You can find the Excite poll online at poll.excite.com. You simply click on a response to become part of the sample. The poll question for July 2, 2004, was, "Do you support or oppose state laws allowing illegal immigrants to have driver's licenses?" In all, 10,282 people responded, with 8138 (79%) saying they were opposed. You can conclude that

(a) American adults strongly oppose such laws.
(b) the poll uses voluntary response, so the results tell us little about the population of all adults.
(c) the sample is too small to draw any conclusion.

5. You must choose an SRS of 10 of the 440 retail outlets in New York that sell your company's products. How would you label this population in order to use Table 7.1?

(a) 001, 002, 003, . . . , 439, 440
(b) 000, 001, 002, . . . , 439, 440
(c) 1, 2, . . . , 439, 440

6. You are using the table of random digits to choose a simple random sample of 6 students from a class of 30 students. You label the students 01 to 30 in alphabetical order. Go to line 133 of Table 7.1. Your sample contains the students labeled

(a) 45, 74, 04, 18, 07, 65.
(b) 04, 18, 07, 13, 02, 07.
(c) 04, 18, 07, 13, 02, 05.

7. You want to choose an SRS of 5 of the 7200 salaried employees of a corporation. You label the employees 0001 to 7200 in alphabetical order. Using line 111 of Table 7.1, your sample contains the employees labeled

(a) 6694, 5130, 0041, 2712, 3827.
(b) 6694, 0513, 0929, 7004, 1271.
(c) 8148, 6694, 8760, 5130, 9297.

8. A sample of households in a community is selected at random from the telephone directory. In this community, 4% of households have no telephone and another 35% have unlisted telephone numbers. The sample will certainly suffer from

(a) nonresponse.
(b) undercoverage.
(c) false responses.

9. A study of cell phones and the risk of brain cancer looked at a group of 469 people who have brain cancer. The investigators matched each cancer patient with a person of the same sex, age, and race who did not have brain cancer, then asked about use of cell phones. This is

(a) an observational study.
(b) a uncontrolled experiment.
(c) a randomized comparative experiment.

10. A clinical trial compares an antidepression medicine with a placebo for relief of chronic headaches. There are 36 headache patients available to serve as subjects. To choose 18 patients to receive the medicine, you would

(a) assign labels 01 to 36 and use Table 7.1 to choose 18.

(b) assign labels 01 to 18 because only 18 need be chosen.

(c) assign the first 18 who signed up to get the medicine.

11. The Bureau of Labor Statistics announces that last month it interviewed all members of the labor force in a sample of 60,000 households; **4.9%** of the people interviewed were unemployed. The boldface number is a

(a) margin of error.

(b) parameter.

(c) statistic.

12. A study of voting chose 663 registered voters at random shortly after an election. Of these, 72% said they had voted in the election. Election records show that only **56%** of registered voters voted in the election. The boldface number is a

(a) margin of error.

(b) parameter.

(c) statistic.

13. A study of religious practices among college students interviewed a sample of 127 students; 107 of the students said that they prayed at least once in a while. The sample proportion who said they pray is about

(a) $\hat{p} = 0.084$.

(b) $\hat{p} = 0.84$.

(c) $\hat{p} = 84$.

14. A random sample of 1318 Internet users was asked where they would go for information about health or medicine; 606 said that they would use the Internet. The proportion of the sample who would use the Internet is about

(a) $\hat{p} = 0.68$.

(b) $\hat{p} = 0.46$.

(c) $\hat{p} = 46$.

15. An opinion poll asks a simple random sample of 1000 adults how they view the state of the economy. Suppose that 35% of all adults would say "good" if they were asked. In repeated samples, the sample proportion $\hat{p}$ who say "good" would follow a normal distribution with mean

(a) 35.

(b) 3.0.

(c) 0.35.

16. The standard deviation of the distribution of the sample proportion in the previous exercise is about

(a) 0.00023.

(b) 0.015.

(c) 0.03.

17. The margin of error when we use the result of Exercise 13 to estimate what percent of all college students pray is about

(a) 0.0325.

(b) 0.065.

(c) 0.002.

18. If you use the result of Exercise 14 to estimate the proportion of all Internet users who would turn to the net for health information, your 95% confidence interval is about

(a) 0.653 to 0.707.

(b) 0.446 to 0.474.

(c) 0.433 to 0.487.

19. In January 2000, a Gallup poll found that 69% of Americans were "satisfied with the way things are going in the United States" at the beginning of the new century. The poll's margin of error was 3%. This means that

(a) the poll used a method that gets an answer within 3% of the truth about the population 95% of the time.

(b) we can be sure that the percent of all adults who feel this way is between 66% and 72%.

(c) if Gallup takes another poll using the same method, the results of the second poll will lie between 66% and 72%.

20. The sample survey in Exercise 13 actually called 150 students, but 23 of the students refused to say whether they pray. This nonresponse could cause the survey result to be in error. The error due to nonresponse

(a) is in addition to the margin of error found in Exercise 17.
(b) is included in the margin of error found in Exercise 17.
(c) can be ignored because it isn't random.

EXERCISES   ■ Challenge   ◆ Discussion

Sampling

1. A Gallup poll asked, "Do you think the U.S. should take the leading role in world affairs, take a major role but not the leading role, take a minor role, or take no role at all in world affairs?" Gallup's report said, "Results are based on telephone interviews with 1,002 national adults, aged 18 and older, conducted Feb. 9–12, 2004." What is the population for this sample survey? What is the sample size?

2. The American Community Survey (ACS) will replace the census "long form" starting with the 2010 census. The main part of the ACS contacts households at 250,000 addresses by mail each month, with follow-up by phone and in person if there is no response. Each household answers questions about its housing, economic, and social status. What is the population for the ACS?

Bad Sampling Methods

◆ **3.** You see a woman student standing in front of the student center, now and then stopping other students to ask them questions. She says that she is collecting student opinions for a class assignment. Explain why this sampling method is almost certainly biased.

◆ **4.** A member of Congress is interested in whether her constituents favor a proposed gun-control bill. Her staff reports that letters on the bill have been received from 361 constituents and that 323 of these oppose the bill. What is the population of interest? What is the sample? Is this sample likely to represent the population well? Explain your answer.

◆ **5.** Highway planners made a main street in a college town one-way. Local businesses were against the change. The local newspaper invited readers to call a telephone number to record their comments. The next day, the paper reported:

> *Readers overwhelmingly prefer two-way traffic flow to one-way streets. By nearly a 7-1 margin, callers to the newspaper's Express Yourself opinion line on Wednesday complained about the one-way streets that have been in place since May. Of the 98 comments received, all but 14 said no to one-way.*

(a) What population do you think the newspaper wants information about?
(b) Is the proportion of this population who favor one-way streets almost certainly larger or smaller than the proportion 14/98 in the sample? Why?

◆ **6.** Your college wants to gather student opinion about parking for students on campus. It isn't practical to contact all students.

(a) Give an example of a way to choose a sample of students that is poor practice because it depends on voluntary response.
(b) Give another example of a bad way to choose a sample that doesn't use voluntary response.

Simple Random Samples

7. You are planning a report on apartment living in a college town. You decide to select three apartment complexes at random for in-depth interviews with residents. Use Table 7.1, starting at line 117, to select a simple random sample of three of the following apartment complexes.

Ashley Oaks	Country View	Mayfair Village
Bay Pointe	Country Villa	Nobb Hill
Beau Jardin	Crestview	Pemberly
Bluffs	Del-Lynn	Courts
Brandon Place	Fairington	Peppermill
Briarwood	Fairway	Pheasant Run
Brownstone	Knolls	Richfield
Burberry	Fowler	Sagamore Ridge
Cambridge	Franklin Park	Salem
Chauncey	Georgetown	Courthouse
Village	Greenacres	Village Manor
Country	Lahr House	Waterford Court
Squire		Williamsburg

8. You want to ask a sample of college students the question, "How much do you trust information about health that you find on the Internet—a great deal, somewhat, not much, or not at all?" You try out this and other questions on a pilot group of 6 students chosen from your class. The class members are

Anderson	Eckstein	Johnson	Puri
Arroyo	Fernandez	Kim	Richards
Batista	Fullmer	Molina	Rodriguez
Bell	Gandhi	Morgan	Samuels
Burke	Garcia	Nguyen	Shen
Calloway	Glauser	Palmiero	Velasco
Delluci	Helling	Percival	Washburn
Drasin	Husain	Prince	Zhao

Use Table 7.1, starting at line 105, to choose a simple random sample of six of these students.

9. There are approximately 371 active telephone area codes covering Canada, the United States, and some Caribbean areas. (More are created regularly.) You want to choose an SRS of 25 of these area codes for a study of available telephone numbers.

(a) How would you label the area codes in order to use Table 7.1?

(b) Use Table 7.1, starting at line 125, to choose the first 5 members of this sample.

10. Each March, the Current Population Survey is expanded to gather a wider variety of information.

On the Bureau of Labor Statistics Web site, you can find data from this survey on 14,959 people aged 25 to 64 whose highest level of education is a bachelor's degree. Think of these people as a population.

(a) In order to select an SRS of these people, how would you assign labels?

(b) Use Table 7.1, starting at line 137, to choose the first three members of the SRS.

◆ **11.** In using Table 7.1 repeatedly to choose samples, you should not always begin at the same place, such as line 101. Why not?

■ **12.** Which of the following statements are true of a table of random digits and which are false? Explain your answers.

(a) There are exactly four 0's in each row of 40 digits.

(b) Each pair of digits has chance 1/100 of being 00.

(c) The digits 0000 can never appear as a group because this pattern is not random.

■ **13.** The last stage of the Current Population Survey uses a *systematic random sample*. An example will illustrate the idea of a systematic sample. Suppose that we must choose 4 rooms out of the 100 rooms in a dormitory. Because 100/4 = 25, we can think of the list of 100 rooms as 4 lists of 25 rooms each. Choose 1 of the first 25 rooms at random, using Table 7.1. The sample will contain this room and the rooms 25, 50, and 75 places down the list from it. If 13 is chosen, for example, then the systematic random sample consists of the rooms numbered 13, 38, 63, and 88.

(a) Use Table 7.1 to choose a systematic random sample of 5 rooms from a list of 200. Enter the table at line 120.

(b) Your sample gives every room the same chance to be chosen. Explain why. Yet this systematic sample is not a simple random sample. Explain why.

■ **14.** At a party there are 30 students over age 21 and 20 students under age 21. You choose at random 3 of those over 21 and separately choose at random 2 of those under 21 to interview about attitudes toward alcohol. You have given every student at the party the same chance to be interviewed: What is that chance? Why is your sample not an SRS?

Cautions About Sample Surveys

◆ **15.** An opinion poll calls 1800 randomly chosen residential telephone numbers, then asks to speak with an adult member of the household. The interviewer asks, "How many movies have you watched in a movie theater in the past 12 months?"

(a) What population do you think the poll has in mind?

(b) In all, 1131 people respond. What is the rate (percent) of nonresponse?

(c) Many responses to this question are likely to be inaccurate. Why?

◆ **16.** A common form of nonresponse in telephone surveys is "ring-no-answer." That is, a call is made to an active number but no one answers. The Italian National Statistical Institute looked at nonresponse to a government survey of households in Italy during the periods January 1 to Easter and July 1 to August 31. All calls were made between 7:00 and 10:00 P.M., but 21.4% gave "ring-no-answer" in one period versus 41.5% "ring-no-answer" in the other period. Which period do you think had the higher rate of no answers? Why? Explain why a high rate of nonresponse makes sample results less reliable.

◆ **17.** The wording of questions can strongly influence the results of a sample survey. You are writing an opinion poll question about a proposed amendment to the Constitution. You can ask if people are in favor of "changing the Constitution" or "adding to the Constitution" by approving the amendment. One of these choices of wording will produce a much higher percent in favor. Which one? Why?

Experiments

◆ **18.** An educational software company wants to compare the effectiveness of its computer animation for teaching cell biology with that of a textbook presentation. The company tests the biological knowledge of each of a group of first-year college students, then divides them into two groups. One group uses the animation, and the other studies the text. The company retests all the students and compares the increase in understanding of cell biology in the two groups. Is this an experiment? Why or why not? What are the explanatory and response variables?

◆ **19.** Could the magnetic fields from power lines cause leukemia in children? Investigators spent five years and $5 million comparing 638 children who had leukemia and 620 who did not. They went into the homes and actually measured the magnetic fields in the children's bedrooms, in other rooms, and at the front door. They recorded facts about nearby power lines for the family home and also for the mother's residence when she was pregnant. Result: No evidence of more than a chance connection between magnetic fields and childhood leukemia. Explain carefully why this study is *not* an experiment.

◆ **20.** A typical hour of prime-time television shows three to five violent acts. Linking family interviews and police records shows a clear association between time spent watching TV as a child and later aggressive behavior.

(a) Explain why this is an observational study rather than an experiment. What are the explanatory and response variables?

(b) Suggest several variables describing a child's home life that may be confounded with how much TV he or she watches. Explain why confounding makes it difficult to conclude that more TV *causes* more aggressive behavior.

◆ **21.** The Nurses' Health Study has interviewed a sample of more than 100,000 female registered nurses every two years since 1976. Beginning in 1980, the study asked questions about diet, including alcohol consumption. The researchers concluded that "light-to-moderate drinkers had a significantly lower risk of death" than either nondrinkers or heavy drinkers.

(a) Is the Nurses' Health Study an observational study or an experiment? Why?

(b) What does "significant" mean in a statistical report?

(c) Suggest some confounding variables that might explain why moderate drinkers have lower death rates than nondrinkers. (The study adjusted for these variables.)

22. You can use your computer to make telephone calls over the Internet. How would cost affect the behavior of users of this service? You will offer the service to all 200 rooms in a college dormitory. Some rooms will pay a low flat rate. Others will pay higher rates at peak periods and very low rates off-peak. You are interested in the amount and time of use and in the effect on the congestion of the network. Outline the design of an experiment to study the effect of rate structure.

23. Will classroom programs explaining the health advantages of drinking water rather than sugary sodas reduce obesity among children aged 7 to 11 years? Because children are already in school classrooms, we must randomize classes rather than individual children. An experiment assigned 15 classes to receive the program and another 14 to form a control group. After 12 months, obesity had increased in the control group and remained steady in the treatment group. Outline the design of the experiment, label the available classes, and use Table 7.1, beginning at line 103, to carry out the random assignment.

◆ 24. A college allows students to choose either classroom or self-paced instruction in a basic mathematics course. The college wants to compare the effectiveness of self-paced and regular instruction. Someone proposes giving the same final exam to all students in both versions of the course and comparing the average score of those who took the self-paced option with the average score of students in regular sections.

(a) Explain why confounding makes the results of that study worthless.

(b) Given 30 students who are willing to use either regular or self-paced instruction, outline an experimental design to compare the two methods

of instruction. Then use Table 7.1, starting at line 108, to carry out the randomization.

25. Will people spend less on health care if their health insurance requires them to pay some part of the cost themselves? An experiment on this issue asked if the percent of medical costs that are paid by health insurance has an effect either on the amount of medical care that people use or on their health. The treatments were four insurance plans. Each plan paid all medical costs above a ceiling. Below the ceiling, the plans paid 100%, 75%, 50%, or 0% of costs incurred. Outline the design of a randomized comparative experiment suitable for this study.

26. "Chronic tension-type headaches" are headaches that occur almost daily for at least six months. Can antidepressant medications or stress management training reduce the number and severity of these headaches? Are both together more effective than either alone? Investigators compared four treatments: antidepressant alone, placebo alone, antidepressant plus stress management, and placebo plus stress management. Outline the design of the experiment. The headache sufferers named below have agreed to participate in the study. Use Table 7.1 at line 130 to randomly choose the subjects to receive the first treatment. (The complete random assignment to four treatments is a bit time consuming. Applet Exercise 3 shows how the *Simple Random Sample* applet does the work for you.)

Acosta	Farouk	Liang	Solomon
Asihiro	Fleming	Maldonado	Trujillo
Bennett	George	Marsden	Tullock
Bikalis	Han	Montoya	Valasco
Chen	Howard	O'Brian	Vaughn
Clemente	Hruska	Ogle	Wei
Duncan	Imrani	Padilla	Wilder
Durr	James	Plochman	Willis
Edwards	Kaplan	Rosen	Zhang

27. Stores advertise price reductions to attract customers. What type of price cut is most

attractive? Market researchers prepared ads for athletic shoes announcing different levels of discounts (20%, 40%, or 60%). The student subjects who read the ads were also given "inside information" about the fraction of shoes on sale (50% or 100%). Each subject then rated the attractiveness of the sale on a scale of 1 to 7.

(a) Each treatment in this experiment combines values of two explanatory variables, discount level and fraction on sale. List the treatments. How many treatments are there?

(b) Outline a randomized comparative experiment using 60 student subjects. Use Table 7.1 at line 123 to choose the subjects for the first treatment.

28. Healthcare providers are giving more attention to relieving the pain of cancer patients. An article in the journal *Cancer* surveyed a number of studies and concluded that controlled-release (CR) morphine tablets, which release the painkiller gradually over time, are more effective than giving standard morphine when the patient needs it. The "methods" section of the article begins: "Only those published studies that were controlled (i.e., randomized, double-blind, and comparative), repeated-dose studies with CR morphine tablets in cancer pain patients were considered for this review." Explain the terms in parentheses to someone who knows nothing about medical trials.

29. Eye cataracts are responsible for over 40% of blindness around the world. Can drinking tea regularly slow the growth of cataracts? We can't experiment on people, so we use rats as subjects. Researchers injected 18 young rats with a substance that causes cataracts. Half the rats also received tea extract; the other half got a placebo. The response variable was the growth of cataracts over the next six weeks. Yes, the tea extract did slow cataract growth.

(a) Outline the design of this experiment.

(b) Use Table 7.1, starting at line 128, to assign rats to treatments.

■ **30.** The rats in the previous exercise were labeled 01 to 18 in order to use the table of random digits. Unknown to the researchers, the 6 rats labeled 01 to 06 have a genetic defect that favors cataracts. If we simply put rats 01 to 09 in the tea group, the experiment would be biased against tea. We can observe how random selection works to reduce bias by keeping track of how many of these 6 rats are assigned to the tea group. Carry out the random assignment of 9 rats to the tea group 20 times, keeping track of how many of rats 01 to 06 are in the tea group each time. Make a histogram of the count of rats 01 to 06 assigned to tea. What is the average number in your 20 tries?

Thinking About Experiments

◆ **31.** People who eat lots of fruits and vegetables have lower rates of colon cancer than those who eat little of these foods. Fruits and vegetables are rich in antioxidants such as vitamins A, C, and E. Will taking antioxidants help prevent colon cancer? A clinical trial studied this question with 864 people who were at risk for colon cancer. The subjects were divided into four groups: daily beta-carotene, daily vitamins C and E, all three vitamins every day, and daily placebo. After four years, the researchers were surprised to find no significant difference in colon cancer among the groups.

(a) Outline the design of the experiment. Use your judgment in choosing the group sizes.

(b) Assign labels to the 864 subjects and use Table 7.1, starting at line 118, to choose the first five subjects for the beta-carotene group.

(c) The study was double-blind. What does this mean?

(d) What does "no significant difference" mean in describing the outcome of the study?

(e) Suggest some characteristics of the kind of people who eat lots of fruits and vegetables that might explain lower rates of colon cancer. The experiment suggests that these variables, rather than the antioxidants, may be responsible for the observed benefits of fruits and vegetables.

◆ **32.** The financial aid office of a university asks a sample of students about their employment and earnings. The report says that "for academic year earnings, a statistically significant difference was found between the sexes, with men earning more on the average. No significant difference was found between the earnings of black and white students." Explain both of these conclusions, for the effects of sex and of race on average earnings, in language understandable to someone who knows no statistics.

◆ **33.** Do those high center brake lights, required on all cars sold in the United States since 1986, really reduce rear-end collisions? Randomized comparative experiments with fleets of rental and business cars, done before the lights were required, showed that the third brake light reduced rear-end collisions by as much as 50%. Alas, requiring the third light in all cars led to only a 5% drop. Explain why the experiment did not realistically imitate conditions after the lights were required.

◆ **34.** A psychologist studies how much people disclose about themselves to other people met at a party. He arranges for student subjects to be introduced to new people. The subjects are both female and male and both black and white. The results show that "there were no significant race effects, but self-disclosure was significantly higher among females than among males." Explain what this means in language understandable to someone who knows no statistics. Do not use the word *significance* in your answer.

◆ **35.** Give an example of a question about college students, their behavior, or their opinions that would best be answered by

(a) a sample survey.
(b) an experiment.

Inference: From Sample to Population

36. An opinion poll uses random digit dialing equipment to dial 2000 randomly chosen residential telephone numbers. Of these, **621** are unlisted numbers. This isn't surprising, because **35%** of all residential numbers are unlisted. Is each of the boldface numbers a parameter or a statistic?

37. The Tennessee STAR experiment randomly assigned children to regular or small classes during their first four years of school. When these children reached high school, **40.2%** of blacks from small classes took the ACT or SAT college entrance exams. Only **31.7%** of blacks from regular classes took one of these exams. Is each of the boldface numbers a parameter or a statistic?

38. The College Alcohol Study interviewed an SRS of 14,941 college students about their drinking habits. Suppose that half of all college students "drink to get drunk" at least once in a while. That is, $p = 0.5$.

(a) What are the mean and standard deviation of the proportion $\hat{p}$ of the sample who drink to get drunk?
(b) In what range of values do the proportions $\hat{p}$ from 95% of all samples fall?
(c) In what range of values do the proportions $\hat{p}$ from 99.7% of all samples fall?

39. Harley-Davidson motorcycles make up 14% of all the motorcycles registered in the United States. You plan to interview an SRS of 500 motorcycle owners.

(a) What is the approximate distribution of the proportion of your sample who own Harleys?
(b) In 95% of all samples like this one, the proportion of the sample who own Harleys will fall between _____ and _____. What are the missing numbers?

40. Exercise 38 asks what values the sample proportion $\hat{p}$ is likely to take when the population proportion is $p = 0.5$ and the sample size is $n = 14{,}941$. What range covers the middle 95% of values of $\hat{p}$ when $p = 0.5$ and $n = 1000$? When $n = 4000$? When $n = 16{,}000$? What general fact about the behavior of $\hat{p}$ do your results illustrate?

■ **41.** You can use a table of random digits to *simulate* sampling from a population. Suppose that 60% of the population bought a lottery ticket in the last 12 months. We will simulate the behavior of random samples of size 40 from this population.

(a) Let each digit in the table stand for one person in this population. Digits 0 to 5 stand for people who bought a lottery ticket, and 6 to 9 stand for people who did not. Why does looking at one digit from Table 7.1 simulate drawing one person at random from a population with 60% "yes"?

(b) Each row in Table 7.1 contains 40 digits. So the first 10 rows represent the results of 10 samples. How many digits between 0 and 5 does the top row contain? What is the percent of "yes" responses in this sample? How many of your 10 samples overestimated the population truth 60%? How many underestimated it? You could program a computer to continue this process, say 1000 times, to produce a pattern like that in Figure 7.5.

Confidence Intervals

42. A random sample of students who took the SAT college entrance examination twice found that 427 of the respondents had paid for coaching courses and that the remaining 2733 had not. Give a 95% confidence interval for the proportion of coaching among students who retake the SAT.

43. A Gallup poll asked a random sample of 1785 adults, "Did you, yourself, happen to attend church or synagogue in the last 7 days?" Of the respondents, 750 said "yes." Give a 95% confidence interval for the proportion of all adults who claim that they attended church or synagogue during the week preceding the poll. (The proportion who actually attended is no doubt lower—some people say "yes" if they usually attend, often attend, or sometimes attend.)

44. The *New York Times* and CBS News conducted a nationwide survey of 1048 randomly selected 13- to 17-year-olds. Of these teenagers, 692 had a television in their room.

(a) Give a 95% confidence interval for the proportion of all teens who have a TV set in their room.

(b) The news article says, "In theory, in 19 cases out of 20, the survey results will differ by no more than three percentage points in either direction from what would have been obtained by seeking out all American teenagers." Explain how your results agree with this statement.

◆ **45.** A telephone survey of 880 randomly selected drivers asked, "Recalling the last 10 traffic lights you drove through, how many of them were red when you entered the intersections?" Of the 880 respondents, 171 admitted that at least one light had been red.

(a) Give a 95% confidence interval for the proportion of all drivers who ran one or more of the last 10 red lights they met.

(b) A practical problem with this survey is that people may not give truthful answers. What is the likely direction of the bias: Do you think more or fewer than 171 of the 880 respondents really ran a red light? Why?

◆ **46.** The Harris poll asked a sample of 1009 adults which causes of death they thought would become more common in the future. Topping the list was gun violence: 70% of the sample thought deaths from guns would increase.

(a) How many of the 1009 people interviewed thought deaths from gun violence would increase?

(b) Harris says that the margin of error for this poll is plus or minus 3 percentage points. Explain to someone who knows no statistics what "margin of error plus or minus 3 percentage points" means.

(c) Give a 95% confidence interval for this survey. Does your margin of error agree with the 3 percentage points announced by Harris?

◆ **47.** Here is the language used by the Harris poll to explain the accuracy of its results: "In theory, with a sample of this size, one can say with 95 percent certainty that the results have a statistical precision of plus or minus 3 percentage points of what they would be if the entire adult population had been polled with complete accuracy." Explain clearly what Harris means by "95 percent certainty."

◆ **48.** A news article reports that in a recent Gallup poll, 78% of the sample of 1108 adults said they believe there is a heaven. Only 60% said they believe there is a hell. The news article ends, "The poll's margin of sampling error was plus or minus four percentage points." Can we be certain that between 56% and 64% of all adults believe there is a hell? Explain your answer

◆ **49.** A survey of Internet users found that males outnumbered females by nearly 2 to 1. This was a surprise, because earlier surveys had put the ratio of men to women closer to 9 to 1. Later in the article we find that surveys were sent to 13,000 organizations and that 1468 of these responded. The survey report claims that "the margin of error is 2.8 percent, with 95 percent confidence."

(a) What was the *response rate* for this survey? (The response rate is the percent of the planned sample that responded.)

(b) Do you think that the small margin of error is a good measure of the accuracy of the survey's results? Explain your answer.

50. A recent Gallup poll found that 68% of adult Americans favor teaching creationism along with evolution in public schools. The Gallup press release says:

> *For results based on samples of this size, one can say with 95 percent confidence that the maximum error attributable to sampling and other random effects is plus or minus 3 percentage points.*

Give one example of a source of error in the poll result that is *not* included in this margin of error.

■ **51.** The Internal Revenue Service plans to examine an SRS of individual income tax returns from each state that were filed electronically. One variable of interest is the proportion of returns that were filed by a tax practitioner rather than by an individual taxpayer. The total number of e-filed tax returns in a state varies from 4.9 million in California to 97,000 in Vermont.

(a) Will the margin of error for estimating the proportion change from state to state if an SRS of 1000 e-filed returns is selected in each state? Explain your answer.

(b) Will the margin of error change from state to state if an SRS of 1% of all e-filed returns is selected in each state? Explain your answer.

■ **52.** Exercise 46 describes a Harris poll that interviewed 1009 people. Suppose you want a margin of error half as large as the one you found in that exercise. How many people must you plan to interview?

■ **53.** Though opinion polls usually make 95% confidence statements, some sample surveys use other confidence levels. The monthly unemployment rate, for example, is based on the Current Population Survey of about 60,000 households. The margin of error in the unemployment rate is announced as about ±0.15% with 90% confidence. Is the margin of error for 90% confidence larger or smaller than the margin of error for 95% confidence? Why? (*Hint:* Look at Figure 7.8 again.)

Chapter Review

54. Ability to grow in shade may help pines found in the dry forests of Arizona resist drought. How well do these pines grow in shade? Investigators planted pine seedlings in a greenhouse in either full light or light reduced to 5% of normal by shade cloth. At the end of the study, they dried the young trees and weighed them.

(a) Explain why this study is an experiment.

(b) What are the individuals, the treatments, and the response variable in this experiment?

(c) You have 200 pine seedlings available. Outline the design you would use for this experiment.

55. Many medical trials randomly assign patients to either an active treatment or a placebo. These trials are always double-blind. Sometimes the patients can tell whether or not they are getting the active treatment. This defeats the purpose of blinding. Reports of medical research usually ignore this problem. Investigators looked at a random sample of 97 articles reporting on placebo-controlled randomized trials in the top five general medical journals. Only 7 of the 97 discussed the success of blinding. Give a 95% confidence interval for the proportion of all such articles that discuss the success of blinding. [Dean Fergusson et al., Turning a blind eye: The success of blinding reported in a random sample of randomised, placebo-controlled trials, *British Medical Journal*, 328 (2004): 432–436.]

56. Cindy wants to ask a sample of students at her college, "Do you think that Social Security will still be paying benefits when you retire?" She obtains the college email addresses of the 2654 students.

(a) How would you label the addresses in order to choose a simple random sample of 100 students?

(b) Use Table 7.1, starting at line 103, to choose the first three labels in the sample.

(c) Cindy sends her question by email to the 100 addresses in her sample. Although she has chosen an SRS, a serious practical difficulty may make it hard to draw clear conclusions from her sample. What practical difficulty do you expect Cindy to encounter?

■ **57.** Suppose that exactly 10% of all articles in major medical journals that describe placebo-controlled randomized trials discuss the success of blinding. That is, the proportion of "successes" in the population is $p = 0.1$. What is the approximate probability that fewer than 7% of an SRS of 97 articles from this population discuss the success of blinding?

 ## APPLET EXERCISES

To do these exercises, go to www.whfreeman.com/fapp.

1. Use the *Simple Random Sample* applet to choose the sample of spring break resorts in Example 4. Assign labels 01 to 28 by entering 28 in the "Population 1 to" box and clicking "Reset." Then enter 4 in the "Select a sample of size" box and click "Sample." Which resorts from the list in Example 4 make up your sample? Click "Reset" and choose another sample. Which resorts did you choose this time? You see that random sampling gives different samples each time—what matters is that all resorts have the same chance to be chosen.

2. The *Simple Random Sample* applet is handy when the population or sample is large. (The applet will handle population sizes up to 500.) Skills Check 5 asks you to choose an SRS of 10 out of 440 retail outlets. Use the applet to do this and report your result.

3. You can use the *Simple Random Sample* applet to choose treatment groups at random for a randomized comparative experiment. Exercise 26 asks you to choose the subjects to get the first

treatment in an experiment that compares four treatments.

(a) Use the applet to choose an SRS of 9 out of 36 to receive the first treatment. Which subjects make up this group?

(b) The applet allows you to randomly assign subjects to more than two groups. After you choose the first group, the "Population hopper" contains the 27 subjects that were not chosen, in scrambled order. Click "Sample" again to choose 9 of these remaining subjects to receive the second treatment. Do this once more to choose the third group. The 9 subjects that remain in the "Population hopper" form the fourth group. Which of the 36 subjects will receive each of the four treatments?

4. You can use the *Probability* applet to speed up and improve Exercise 41. You have a population in which 60% of the individuals approve of legal gambling. You want to take many samples from this population to observe how the sample proportion that approves of gambling varies from sample to sample. Set the "Probability of heads" in the applet to 0.6 and the number of tosses to 40. This simulates an SRS of size 40 from a large population. Each head in the sample is a person who approves of legal gambling, and each tail is a person who disapproves. By alternating between "Toss" and "Reset" you can take many samples quickly.

(a) Take 50 samples, recording the proportion that approves of gambling in each sample. (The applet gives this proportion at the top left of its display.) Make a histogram of the 50 sample proportions.

(b) Another population contains only 20% who approve of legal gambling. Take 50 samples of size 40 from this population, record the number in each sample that approves, and make a histogram of the 50 sample proportions. How do the centers of your two histograms reflect the differing truths about the two populations?

5. The idea of an 80% confidence interval is that the interval captures the true parameter value in 80% of all samples. That's not high enough confidence for practical use, but 80% hits and 20% misses make it easy to see how a confidence interval behaves in repeated samples from the same population. Go to the *Confidence Interval* applet.

(a) Set the confidence level to 80%. Click "Sample" to choose an SRS and calculate the confidence interval. Do this 10 times to simulate 10 SRSs with their 10 confidence intervals. How many of the 10 intervals captured the true mean μ? How many missed?

(b) You see that we can't predict whether the next sample will hit or miss. The confidence level, however, tells us what percent will hit in the long run. Reset the applet and click "Sample 50" to get the confidence intervals from 50 SRSs. How many hit? Keep clicking "Sample 50" and record the percent of hits among 100, 200, 300, 400, and 500 SRSs. Even 500 samples is not truly "the long run," but we expect the percent of hits in 500 samples to be fairly close to the confidence level, 80%.

✎ WRITING PROJECTS

1. Go to the Web site of the Gallup Organization (www.gallup.com) and click on "Gallup Poll." Users must pay to read most of Gallup's reports, but the "Daily Discourse" is a Gallup press release that you can read freely. (The site design changes regularly, so look for a free report.) Newspapers publish short articles based on press releases like this one. Write a news article about two paragraphs long based on the press release.

2. Write your own examples of bad sample survey questions.

(a) Write two questions on the same topic, designed to draw opposite responses. Explain how the wording of each question is biased.

(b) Write a question that is confusing, so that it is hard to answer.

3. Choose an issue of current interest to students at your school. Prepare a short questionnaire (no more than five questions) to determine opinions on this issue. Choose a sample of about 25 students, administer your questionnaire, and write a brief description of your findings. Also write a short discussion of your experiences in designing and carrying out the survey.

Although 25 students are too few for you to be statistically confident of your results, this project centers on the practical work of a survey. You must first identify a population; if it is not possible to reach a wider student population, use students enrolled in this course. Did the subjects find your questions clear? Did you write the questions so that it was easy to tabulate the responses? At the end, did you wish you had asked different questions?

SUGGESTED READINGS

DE VEAUX, RICHARD D., and PAUL F. VELLEMAN. *Intro Stats*, Addison Wesley, Boston, 2004. Another text at the level just above ours. Like this chapter, it uses estimating a population proportion to introduce confidence intervals (Chapter 19).

MOORE, DAVID S. *The Basic Practice of Statistics*, 3rd ed., Freeman, New York, 2004. Chapters 7 and 8 discuss samples and experiments. Chapter 13 presents the reasoning of confidence intervals, and Chapter 18 presents confidence intervals for a population proportion.

SUGGESTED WEB SITES

The National Council on Public Polls, www.ncpp.org, has a statement on "20 Questions a Journalist Should Ask About a Poll" that makes interesting reading. The explanations expand our cautions about sample surveys in practice. You can find similar information on the Web site of the American Association for Public Opinion Research, www.aapor.org. Click the "For Journalists" box and take a look at the "Guide to Best Practices" and "Survey Practices that AAPOR Condemns" listings.

The single most important sample survey in the United States is probably the government's monthly Current Population Survey (CPS), carried out by the Census Bureau on behalf of the Bureau of Labor Statistics (BLS). The CPS Web site, www.bls.census.gov/cps/, contains an overwhelming amount of information. You might click on "Publications," then on "Chronological List" and look at the latest monthly statement of the BLS Commissioner to Congress about employment and unemployment.

Probability: The Mathematics of Chance

Have you ever wondered how gambling, which is a recreation or an addiction for individuals, can be a business for the casino? A business requires predictable revenue from the service it offers, even when the service is a game of chance. Individual gamblers may win or lose. They can never say whether a day at the casino will turn a profit or a loss. But the casino isn't gambling. Casinos are consistently profitable, and state governments make money both from running lotteries and from selling licenses for other forms of gambling.

It is a remarkable fact that the aggregate result of many thousands of chance outcomes can be known with near certainty. The casino need not load the dice, mark the cards, or alter the roulette wheel. It knows that in the long run each dollar bet will yield its five cents or so of revenue. It is therefore good business to concentrate on free floor shows or inexpensive bus fares to increase the flow of dollars bet. The flow of profit will follow.

Gambling houses are not alone in profiting from the fact that a chance outcome many times repeated is firmly predictable. For example, although a life insurance company does not know *which* of its policyholders will die next year, it can predict quite accurately *how many* will die. It sets its premiums according to this knowledge, just as the casino sets its jackpots. Statisticians also rely on the regular behavior of chance: A 95% confidence interval works 95% of the time because, in the long run, chance behavior is predictable.

Randomness

A phenomenon is said to be **random** if individual outcomes are uncertain but the long-term pattern of many individual outcomes is predictable.

To a statistician, "random" does not mean "haphazard." Randomness is a kind of order, an order that emerges only in the long run, over many repetitions. Many phenomena, both natural and of human design, are random. The hair colors of children, the spread of epidemics, and the decay of radioactive substances are examples of natural randomness. Indeed, quantum mechanics asserts that at the subatomic level the natural world is inherently random.

Games of chance are examples of randomness deliberately produced by human effort. Casino dice are carefully machined, and their drilled holes are filled with material equal in density to the plastic body. This guarantees that the side with six spots has the same weight as the opposite side, which has only one spot. Thus, each side is equally likely to land upward. All the odds and payoffs of dice games rest on this carefully planned randomness. Random sampling and randomized comparative experiments are also examples of planned randomness, although they use tables of random digits rather than dice and cards. The reasoning of statistical inference rests on asking, "How often would this method give a correct answer if I used it very many times?" Probability theory, the mathematical description of randomness, is the basis for gambling, insurance, much of modern science, and statistical inference. Probability is the topic of this chapter.

8.1 Probability Models and Rules

Toss a coin, or choose an SRS. The result can't be predicted in advance, because the result will vary when you toss the coin or choose the sample repeatedly. But

FIGURE 8.1 The proportion of tosses of a coin that give a head varies as we make more tosses. Eventually, however, the proportion approaches 0.5, the probability of a head. This figure shows the results of two trials of 5000 tosses each.

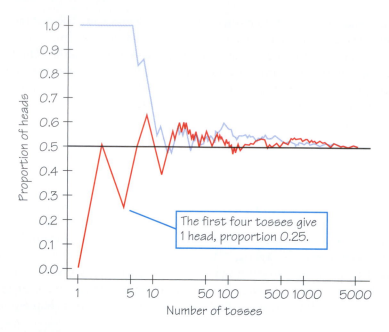

The first four tosses give 1 head, proportion 0.25.

there is nonetheless a regular pattern in the results, a pattern that emerges clearly only after many repetitions. This remarkable fact is the basis for the idea of probability.

EXAMPLE 1   Coin Tossing

When you toss a coin, there are only two possible outcomes, heads or tails. Figure 8.1 shows the results of tossing a coin 5000 times twice. For each number of tosses from 1 to 5000, we have plotted the proportion of those tosses that gave a head. Trial *A* (red line) begins tail, head, tail, tail. You can see that the proportion of heads for Trial *A* starts at 0 on the first toss, rises to 0.5 when the second toss gives a head, then falls to 0.33 and 0.25 as we get two more tails. Trial *B* (blue line), on the other hand, starts with five straight heads, so the proportion of heads is 1 until the sixth toss.

The proportion of tosses that produce heads is quite variable at first. Trial *A* starts low and Trial *B* starts high. As we make more and more tosses, however, the proportions of heads for both trials get close to 0.5 and stay there. If we made yet a third trial at tossing the coin a great many times, the proportion of heads would again settle down to 0.5 in the long run. We say that 0.5 is the *probability* of a head. The probability 0.5 appears as a horizontal line on the graph. ■

Probability

The **probability** of any outcome of a random phenomenon is the proportion of times the outcome would occur in a very long series of repetitions.

The *Probability* applet animates Figure 8.1. It allows you to choose the probability of a head and simulate any number of tosses of a coin with that probability. Try it. You will see that the proportion of heads gradually settles down close to the probability. Equally important, you will also see that the proportion in a small or moderate number of tosses can be far from the probability. *Probability describes only what happens in the long run.* Random phenomena are irregular and unpredictable in the short run.

We might suspect that a coin has probability 0.5 of coming up heads just because the coin has two sides. As Exercises 1 and 2 illustrate, such suspicions are not always correct. The idea of probability is *empirical*. That is, it is based on observation rather than theorizing. Probability describes what happens in very many trials, and we must actually observe many trials to pin down a probability.

Gamblers have known for centuries that the fall of coins, cards, and dice displays clear patterns in the long run. The idea of probability rests on the observed fact that the average result of many thousands of chance outcomes can be known with near certainty. But a definition of probability as "long-run proportion" is vague. Who can say what "the long run" is? We can always toss the coin another 1000 times. Instead, we give a mathematical description of *how probabilities behave,* based on our understanding of long-run proportions. To see how to proceed, think first about a very simple random phenomenon, tossing a coin once. When we toss a coin, we cannot know the outcome in advance. What do we know? We are willing to say that the outcome will be either heads or tails. We believe that each of these outcomes has probability 1/2. This description of coin tossing has two parts:

▶   A list of possible outcomes
▶   A probability for each outcome

This description is the basis for all probability models. Here is the vocabulary we use.

Sample Space, Event, Probability Model

The **sample space** S of a random phenomenon is the set of all possible outcomes.

An **event** is any outcome or any set of outcomes of a random phenomenon. That is, an event is a subset of the sample space.

A **probability model** is a mathematical description of a random phenomenon consisting of two parts: a sample space S and a way of assigning probabilities to events.

The sample space S can be very simple or very complex. When we toss a coin once, there are only two outcomes, heads and tails. So the sample space is $S = \{H, T\}$. If we draw a random sample of 1000 U.S. residents age 18 and over, as opinion polls often do, the sample space contains all possible choices of 1000 of the more than 220 million adults in the country. This S is extremely large. Each member of S is a possible opinion poll sample, which explains the term *sample space*.

EXAMPLE 2   Rolling Dice

Rolling two dice is a common way to lose money in casinos. There are 36 possible outcomes when we roll two dice and record the up faces in order (first die, second die). Figure 8.2 displays these outcomes. They make up the sample space S. "Roll a 5" is an event, call it A, that contains four of these 36 outcomes:

$$A = \left\{ \; \blacksquare \; , \; \blacksquare \; , \; \blacksquare \; , \; \blacksquare \; \right\}$$

If the dice are carefully made, experience shows that each of the 36 outcomes in Figure 8.2 comes up equally often. So a reasonable probability model assigns probability 1/36 to each outcome.

In craps and other games, all that matters is the *sum* of the spots on the up faces. Let's change the random outcomes we are interested in: Roll two dice and count the spots on the up faces. Now there are only 11 possible outcomes, from a sum of 2 for rolling a double 1 through 3, 4, 5, and on up to 12 for rolling a double 6. The sample space is now

$$S = \{2, 3, 4, 5, 6, 7, 8, 9, 10, 11, 12\}$$

Comparing this S with Figure 8.2 reminds us that we can change S by changing the detailed description of the random phenomenon we are describing. The outcomes in this new sample space are *not* equally likely, because there are six ways to roll a 7 and only one way to roll a 2 or a 12. ■

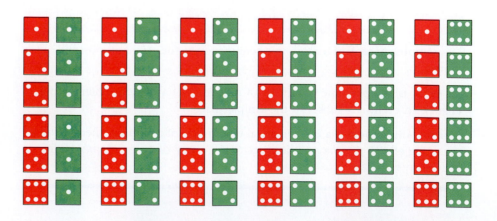

FIGURE 8.2 The 36 possible outcomes for rolling two dice, for Example 2.

There are many ways to assign probabilities, so it is convenient to start with some general rules that any assignment of probabilities to outcomes must obey. These facts follow from the idea of probability as "the long-run proportion of repetitions on which an event occurs."

1. **Any probability is a number between 0 and 1.** Any proportion is a number between 0 and 1, so any probability is also a number between 0 and 1. An event with probability 0 never occurs, and an event with probability 1 occurs on every trial. An event with probability 0.5 occurs in half the trials in the long run.

2. **All possible outcomes together must have probability 1.** Because some outcome must occur on every trial, the sum of the probabilities for all possible outcomes must be exactly 1.

3. **If two events have no outcomes in common, the probability that one or the other occurs is the sum of their individual probabilities.** If one event occurs in 40% of all trials, a different event occurs in 25% of all trials, and the two can never occur together, then one or the other occurs on 65% of all trials because 40% + 25% = 65%.

4. **The probability that an event does not occur is 1 minus the probability that the event does occur.** If an event occurs in (say) 70% of all trials, it fails to occur in the other 30%. The probability that an event occurs and the probability that it does not occur always add to 100%, or 1.

We can use mathematical notation to state Facts 1 to 4 more concisely. We use capital letters near the beginning of the alphabet to denote events. If A is any event, we write its probability as $P(A)$. Here are our probability facts in formal language. As you apply these rules, remember that they are just another form of intuitively true facts about long-run proportions.

Probability Rules

Rule 1. The probability $P(A)$ of any event A satisfies $0 \leq P(A) \leq 1$.

Rule 2. If S is the sample space in a probability model, then $P(S) = 1$.

Rule 3. Two events A and B are **disjoint** if they have no outcomes in common and so can never occur together. If A and B are disjoint,

$$P(A \text{ or } B) = P(A) + P(B)$$

This is the **addition rule for disjoint events**.

Rule 4. The **complement** of any event A is the event that A does not occur, written as A^c. The **complement rule** states that

$$P(A^c) = 1 - P(A)$$

SPOTLIGHT 8.1   Probability and Risk

High exposures to asbestos are dangerous. Low exposures, such as that experienced by teachers and students in schools where asbestos is present in the insulation around pipes, are not very risky. The probability that a teacher who works for 30 years in a school with typical asbestos levels will get cancer from the asbestos is around 15/1,000,000. The probability of dying in a car accident during a lifetime of driving is about 15,000/1,000,000. That is, driving regularly is 1000 times more risky than teaching in a school where asbestos is present.

Risk does not stop us from driving. Yet the much smaller risk from asbestos launched massive cleanup campaigns and a federal requirement that every school inspect for asbestos and make the findings public. Why do we ignore the probabilities when we think about risk? Why do we take asbestos so much more seriously than driving? Why do we worry more about improbable threats from tornadoes

and mad cow disease than about heart attacks? It appears that psychology rather than probability governs our judgment of risk. We feel safer when a risk seems under our control than when we cannot control it. We are in control (or so we imagine) when we are driving, but we can't control the risk from asbestos or tornadoes.

Our reaction to risk also reflects the weakness of our intuition about how probability operates. It is hard to grasp the meaning of very small probabilities. Probabilities of 15 per million and 15,000 per million are both so small that our intuition cannot distinguish between them. Psychologists have shown that we generally overestimate very small risks and underestimate higher risks. Despite our bias toward things we think we can control, we should at least look at the probabilities when we choose what risks to take.

EXAMPLE 3   Probabilities for Rolling Dice

Figure 8.2 displays the 36 possible outcomes of rolling two dice. For casino dice it is reasonable to assign the same probability to each of the 36 outcomes in Figure 8.2. Because all 36 outcomes together must have probability 1 (Rule 2), each outcome must have probability 1/36.

What is the probability of rolling a 5? Because the event "roll a 5" contains the four outcomes displayed in Example 2, the addition rule (Rule 3) says that its probability is

$$P(\text{roll a 5}) = P\left(\boxed{\cdot}\ \boxed{\because}\right) + P\left(\boxed{\because}\ \boxed{\cdot\cdot}\right) + P\left(\boxed{\cdot\cdot}\ \boxed{\because}\right) + P\left(\boxed{\because}\ \boxed{\cdot}\right)$$

$$= \frac{1}{36} + \frac{1}{36} + \frac{1}{36} + \frac{1}{36}$$

$$= \frac{4}{36} = 0.111$$

What is the probability of rolling anything other than a 5? By the complement rule (Rule 4),

$$P(\text{roll does not give a 5}) = 1 - P(\text{roll a 5})$$
$$= 1 - 0.111 = 0.889$$

What about the probability of rolling a 7? Figure 8.2 shows that there are six outcomes for which the sum of the spots is 7. The probability is 6/36, or about 0.167. Continue in this way to get the full probability model (sample space and assignment of probabilities) for rolling two dice and summing the spots on the up faces. Here it is:

Outcome	2	3	4	5	6	7	8	9	10	11	12
Probability	1/36	2/36	3/36	4/36	5/36	6/36	5/36	4/36	3/36	2/36	1/36

This model assigns probabilities to individual outcomes. To find the probability of an event, just add the probabilities of the outcomes that make up the event. For example:

$$P(\text{outcome is odd}) = P(3) + P(5) + P(7) + P(9) + P(11)$$
$$= \frac{2}{36} + \frac{4}{36} + \frac{6}{36} + \frac{4}{36} + \frac{2}{36}$$
$$= \frac{18}{36} = \frac{1}{2} \ \blacksquare$$

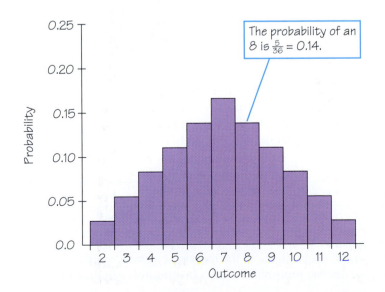

The probability of an 8 is $\frac{5}{36} = 0.14$.

FIGURE 8.3
Probability histogram showing the probability model for rolling two balanced dice and counting the spots on the up faces.

When the outcomes for a probability model are numbers, we can use a histogram to display the assignment of probabilities to the outcomes. Figure 8.3 is a **probability histogram** of the probability model in Example 3. The height of each bar shows the probability of the outcome at its base. Because the heights are probabilities, they add to 1. Think of Figure 8.3 as an idealized picture of the results of very many rolls of a die. As an idealized picture, it is perfectly symmetric.

Example 3 illustrates one way to assign probabilities to events: Assign a probability to every individual outcome, then add these probabilities to find the probability of any event. This idea works well when there are only a finite (fixed and limited) number of outcomes.

8.2 Discrete Probability Models

We will work with two kinds of probability models. The first kind is illustrated by Example 3. It is called a *discrete* probability model.

Discrete Probability Model

A probability model with a finite sample space is called **discrete.** To assign probabilities in a discrete model, list the probability of all the individual outcomes. These probabilities must be numbers between 0 and 1 and must have sum 1.

The probability of any event is the sum of the probabilities of the outcomes making up the event.

EXAMPLE 4   Benford's Law

Faked numbers in tax returns, invoices, or expense account claims often display patterns that aren't present in legitimate records. Some patterns, like too many round numbers, are obvious and easily avoided by a clever crook. Others are more subtle. It is a striking fact that the first digits of numbers in legitimate records often follow a model known as Benford's law. Here it is (note that a first digit can't be 0):

First digit	1	2	3	4	5	6	7	8	9
Probability	0.301	0.176	0.125	0.097	0.079	0.067	0.058	0.051	0.046

Check that the probabilities of the outcomes sum to exactly 1. This is therefore a legitimate discrete probability model. Investigators can detect fraud by comparing the first digits in records such as invoices paid by a business with these probabilities.

Consider the events

$$A = \{\text{first digit is 1}\}$$
$$B = \{\text{first digit is 6 or greater}\}$$

From the table of probabilities,

$$P(A) = P(1) = 0.301$$
$$P(B) = P(6) + P(7) + P(8) + P(9)$$
$$= 0.067 + 0.058 + 0.051 + 0.046 = 0.222$$

Note that $P(B)$ is not the same as the probability that a first digit is strictly greater than 6. The probability $P(6)$ that a first digit is 6 is included in "6 or greater" but not in "greater than 6."

The probability that a first digit is anything other than 1 is, by the complement rule,

$$P(A^c) = 1 - P(A)$$
$$= 1 - 0.301 = 0.699$$

The events A and B are disjoint, so the probability that a first digit is either 1 or 6 or greater is, by the addition rule,

$$P(A \text{ or } B) = P(A) + P(B)$$
$$= 0.301 + 0.222 = 0.523$$

Be careful to apply the addition rule only to disjoint events. Check that the probability of the event C that a first digit is odd is

$$P(C) = P(1) + P(3) + P(5) + P(7) + P(9) = 0.609$$

The probability

$$P(B \text{ or } C) = P(1) + P(3) + P(5) + P(6) + P(7) + P(8) + P(9) = 0.727$$

is *not* the sum of $P(B)$ and $P(C)$, because events B and C are not disjoint. Outcomes 7 and 9 are common to both events. ∎

8.3 Equally Likely Outcomes

A simple random sample gives all possible samples an equal chance to be chosen. Rolling two casino dice gives all 36 outcomes the same probability. When randomness is the product of human design, it is often the case that the outcomes in the sample space are all equally likely. Rules 1 and 2 force the assignment of probabilities in this case.

Equally Likely Outcomes

If a random phenomenon has k possible outcomes, all **equally likely**, then each individual outcome has probability $1/k$. The probability of any event A is

$$P(A) = \frac{\text{count of outcomes in } A}{\text{count of outcomes in } S}$$
$$= \frac{\text{count of outcomes in } A}{k}$$

EXAMPLE 5   Equally Likely First Digits?

You might think that first digits are distributed "at random" among the digits 1 to 9. The 9 possible outcomes would then be equally likely. The sample space for a single digit is

$$S = \{1, 2, 3, 4, 5, 6, 7, 8, 9\}$$

so the probability for each of the 9 outcomes is 1/9. That is, the probability model is

First digit	1	2	3	4	5	6	7	8	9
Probability	1/9	1/9	1/9	1/9	1/9	1/9	1/9	1/9	1/9

The probability of the event B that a randomly chosen first digit is 6 or greater is

$$P(B) = P(6) + P(7) + P(8) + P(9)$$
$$= \frac{1}{9} + \frac{1}{9} + \frac{1}{9} + \frac{1}{9} = \frac{4}{9} = 0.444$$

Compare this with the Benford's law probability in Example 4. A crook who fakes data by using "random" digits will end up with too many first digits 6 or greater and too few 1s and 2s. Figure 8.4 displays probability histograms that compare the probability model for random digits with the model given by Benford's law. ■

When outcomes are equally likely, we find probabilities by counting outcomes. The study of counting methods is called **combinatorics**. We will start with the multiplication method that we called the *fundamental principle of counting* in Chapter 2 (page 41).

(a)

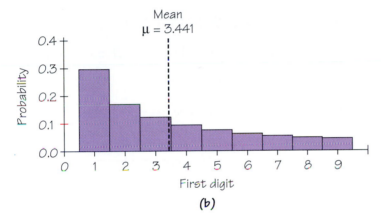

(b)

FIGURE 8.4 Probability histograms of two models for first digits in numerical records. (a) Equally likely digits. (b) Digits follow Benford's law. The vertical lines mark the means of the two models.

Code Words

EXAMPLE 6

A computer system assigns log-in identification codes to users by choosing three letters at random. All three-letter codes are therefore equally likely. What is the probability that the code assigned to you has no x in it?

First count the total number of code words. There are 26 letters that can occur in each position in the word. Any of the 26 letters in the first position can be combined with any of the 26 letters in the second position to give 26×26 choices. (This is true because the order of the letters matters, so that ab and ba are different choices.) Any of the 26 letters can then follow in the third position. The number of different codes is

$$26 \times 26 \times 26 = 17{,}576$$

Now count the number of code words that have no x. These codes are made up of the other 25 letters. So there are

$$25 \times 25 \times 25 = 15{,}625$$

such codes. The probability that your code has no x is therefore

$$P(\text{no } x) = \frac{\text{number of codes with no } x}{\text{number of codes}}$$

$$= \frac{15{,}625}{17{,}576} = 0.889$$

Suppose that the computer is programmed to avoid repeated letters in the identification codes. Any of the 26 letters can still appear in the first position. But only the 25 remaining letters are allowed in the second position, so that there are 26×25 choices for the first two letters in the code. Any of these choices leaves 24 letters for the third position. The number of different codes without repeated letters is

$$26 \times 25 \times 24 = 15{,}600$$

Codes with no x are allowed one fewer choice in each position. There are

$$25 \times 24 \times 23 = 13{,}800$$

such codes. The probability that your code has no x is then

$$P(\text{no } x) = \frac{\text{number of codes with no } x}{\text{number of codes}}$$

$$= \frac{13{,}800}{15{,}600} = 0.885$$

Eliminating repeats slightly lowers your chance of avoiding an x. ∎

The example makes use of two rules about counting that we often apply in finding probabilities.

Counting Arrangements of Distinct Items

Rule A.   Suppose we have a collection of n distinct items. We want to arrange k of these items in order, and the same item can appear several times in the arrangement. The number of possible arrangements is

$$n \times n \times \cdots \times n = n^k$$

Rule B.   Suppose we have a collection of n distinct items. We want to arrange k of these items in order, and any item can appear no more than once in the arrangement. The number of possible arrangements is

$$n \times (n - 1) \times \cdots \times (n - k + 1)$$

In Example 6, n (the number of letters available) is first 26, then 25, and k (the number of letters to be arranged to make a code) is 3. It is easier to think your way through the counting than to memorize the recipes.

How Many Arrangements?

EXAMPLE 7

Seven teams of students are waiting to present reports on group projects. In how many orders can the teams speak?

Because each team speaks once, there are no repeats. This is the case described by Counting Rule B, with n and k both equal to 7. To think through the problem, proceed like this: Any of the 7 teams can speak first; then any of the 6 that remain can speak second; and so on. The number of arrangements is therefore

$$7 \times 6 \times 5 \times 4 \times 3 \times 2 \times 1 = 5040$$

(This number is often called 7!, read "seven factorial.") ■

8.4 Continuous Probability Models

When we use the table of random digits to select a digit between 0 and 9, the discrete probability model assigns probability 1/10 to each of the 10 possible outcomes. Suppose that we want to choose a number at random between 0 and 1, allowing *any* number between 0 and 1 as the outcome. Software random-number generators will do this. You can visualize such a random number by thinking of a spinner (Figure 8.5) that turns freely on its axis and slowly comes to a stop. The pointer can come to rest anywhere on a circle that is marked from 0 to 1. The sample space is now an entire interval of numbers:

$$S = \{\text{all numbers } x \text{ such that } 0 \leq x \leq 1\}$$

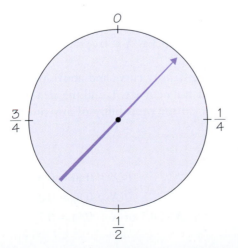

FIGURE 8.5 This spinner chooses a number between 0 and 1 at random. That is, it is equally likely to stop at any point on the circle.

How can we assign probabilities to events such as {$0.3 \leq x \leq 0.7$}? As in the case of selecting a random digit, we would like all possible outcomes to be equally likely. But we cannot assign probabilities to each individual value of x and then sum, because there are infinitely many possible values. Instead we use a second way of assigning probabilities directly to events—as *areas under a curve*. The curve must have total area 1 underneath it, corresponding to total probability 1. We call such curves *density curves*.

Density Curve

A **density curve** is a curve that

▶   is always on or above the horizontal axis and
▶   has area exactly 1 underneath it.

A **continuous probability model** assigns probabilities as areas under a density curve. The area under the curve and above any range of values is the probability of an outcome in that range.

EXAMPLE 8   A Continuous Uniform Model

The random-number generator will spread its output uniformly across the entire interval from 0 to 1 as we allow it to generate a long sequence of numbers. The results of many trials are represented by the density curve of a *uniform probability model*. This density curve appears in red in Figure 8.6. It has height 1 over the interval from 0 to 1, and height 0 everywhere else. The area under the density curve is 1, the area of a square with base 1 and height 1. The probability of any event is the area under the density curve and above the event in question.

As Figure 8.6a illustrates, the probability that the random-number generator produces a number X between 0.3 and 0.7 is

$$P(0.3 \leq X \leq 0.7) = 0.4$$

because the area under the density curve and above the interval from 0.3 to 0.7 is 0.4. The height of the density curve is 1, and the area of a rectangle is the product of height and length, so the probability of any interval of outcomes is just the length of the interval.

Similarly,

$$P(X < 0.5) = 0.5$$
$$P(X > 0.8) = 0.2$$
$$P(X < 0.5 \text{ or } X > 0.8) = 0.7$$

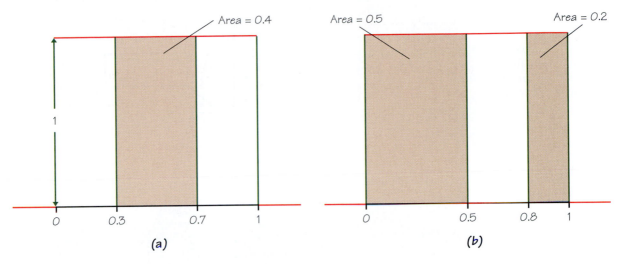

FIGURE 8.6 Assigning probabilities for generating a random number between 0 and 1, for Example 8. The probability of any interval of numbers is the area above the interval and under the density curve. (a) The probability of an outcome between 0.3 and 0.7. (b) The probability of an outcome less than 0.5 or greater than 0.8.

The last event consists of two nonoverlapping intervals, so the total area above the event is found by adding two areas, as illustrated by Figure 8.6b. This assignment of probabilities obeys all of our rules for probability. ▪

The probability model for a continuous random variable assigns probabilities to intervals of outcomes rather than to individual outcomes. In fact, *all continuous probability models assign probability 0 to every individual outcome*. Only intervals of values have positive probability. To see that this is true, consider a specific outcome such as $P(X = 0.8)$ in Example 8. The probability of any interval is the same as its length. The point 0.8 has no length, so its probability is 0.

The density curves that are most familiar to us are the normal curves. Because any density curve describes an assignment of probabilities, *normal distributions are continuous probability models*. Let's redo an example from Chapter 7, now using the language of probability.

Areas Under a Normal Curve Are Probabilities   EXAMPLE 9

Suppose that 60% of adults find shopping for clothes time consuming and frustrating. All adults form a population, with population proportion $p = 0.6$. Interview an SRS of 2500 people from this population and find the proportion $\hat{p}$ of the sample who say that shopping is frustrating. We know that if we take many

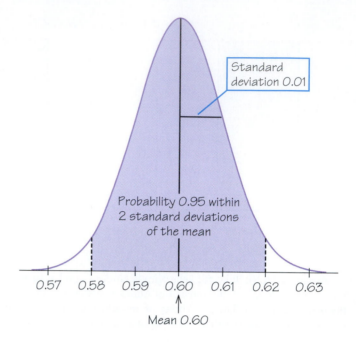

Standard
deviation 0.01

Probability 0.95 within
2 standard deviations
of the mean

0.57   0.58   0.59   0.60   0.61   0.62   0.63

Mean 0.60

such samples, the statistic $\hat{p}$ will vary from sample to sample according to a normal distribution with

$$\text{mean} = p = 0.6$$

$$\text{standard deviation} = \sqrt{\frac{p(1-p)}{n}}$$

$$= \sqrt{\frac{(0.6)(0.4)}{2500}} = 0.01 \quad \text{(approximately)}$$

The 68–95–99.7 rule now gives *probabilities* for the value of $\hat{p}$ from a single SRS. The probability is 0.95 that $\hat{p}$ lies between 0.58 and 0.62. Figure 8.7 shows this probability as an area under the normal density curve.

All that is new is the language of probability. Probability 0.95 is shorthand for "95% of the time in a very large number of samples." ■

8.5 The Mean and Standard Deviation of a Probability Model

Suppose you are offered this choice of bets, each costing the same: Bet *A* pays $10 if you win and you have probability 1/2 of winning, while bet *B* pays $10,000 and offers probability 1/10 of winning. You would very likely choose *B* even though *A* offers a better chance to win, because *B* pays much more if you win. It would be foolish to decide which bet to make just on the basis of the probability of winning. How much you can win is also important. When a random phenomenon has numerical outcomes, we are concerned with their amounts as well as with their probabilities.

What will be the average payoff of our two bets in many plays? Recall that the probabilities are the long-run proportions of plays in which each outcome occurs. Bet A produces $10 half the time in the long run and nothing half the time. So the average payoff should be

$$\left(\$10 \times \frac{1}{2} \right) + \left(\$0 \times \frac{1}{2} \right) = \$5$$

Bet B, on the other hand, pays out $10,000 on 1/10 of all bets in the long run. Bet B's average payoff is

$$\left(\$10{,}000 \times \frac{1}{10} \right) + \left(\$0 \times \frac{9}{10} \right) = \$1000$$

If you can place many bets, you should certainly choose B. Here is a general definition of the kind of "average outcome" we used to compare the two bets.

Mean of a Discrete Probability Model

Suppose that the possible outcomes $x_1, x_2, \ldots, x_k$ in a sample space S are numbers and that p_j is the probability of outcome x_j. The **mean** μ of this probability model is

$$\mu = x_1 p_1 + x_2 p_2 + \cdots + x_k p_k$$

In Chapter 5, we met the mean $\bar{x}$, the average of n observations that we actually have in hand. The mean μ, on the other hand, describes the probability model rather than any one collection of observations. You can think of μ as a theoretical mean that gives the average outcome we expect in the long run. You will sometimes see the mean of a probability model called the *expected value*. This isn't a very helpful name, because we don't necessarily expect the outcome to be close to the mean.

First Digits

EXAMPLE 10

If first digits in a set of records appear "at random," the probability model for the first digit is as in Example 5:

First digit	1	2	3	4	5	6	7	8	9
Probability	1/9	1/9	1/9	1/9	1/9	1/9	1/9	1/9	1/9

The mean of this model is

$$\mu = 1\frac{1}{9} + 2\frac{1}{9} + 3\frac{1}{9} + 4\frac{1}{9} + 5\frac{1}{9} + 6\frac{1}{9} + 7\frac{1}{9} + 8\frac{1}{9} + 9\frac{1}{9}$$

$$= 45 \times \frac{1}{9} = 5$$

If, on the other hand, the records obey Benford's law, the distribution of the first digit is

First digit	1	2	3	4	5	6	7	8	9
Probability	0.301	0.176	0.125	0.097	0.079	0.067	0.058	0.051	0.046

The mean is

$$\mu = (1)(0.301) + (2)(0.176) + (3)(0.125) + (4)(0.097) + (5)(0.079) +$$
$$(6)(0.067) + (7)(0.058) + (8)(0.051) + (9)(0.046)$$
$$= 3.441$$

The means reflect the greater probability of smaller first digits under Benford's law. We have marked the means on the probability histograms in Figure 8.4. Because the histogram for random digits is symmetric, the mean lies at the center of symmetry. We can't locate the mean of the right-skewed Benford's law model by eye; calculation is needed. ▪

What about continuous probability models? Think of the area under a density curve as cut out of solid material. The mean μ is the point at which the shape would balance. Figure 8.8 illustrates this interpretation of the mean. The mean lies at the center of symmetric density curves such as the uniform density in Figure 8.6 and the normal curve in Figure 8.7. Exact calculation of the mean of a distribution with a skewed density curve requires advanced mathematics. The idea that the mean is the balance point of the probabilities applies to dis-

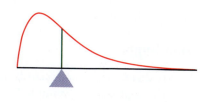

FIGURE 8.8 The mean of a continuous probability model is the point at which the density curve would balance.

crete models as well, but in the discrete case we have a formula that gives us this point.

The mean μ is an average outcome in two senses. The definition for discrete models says that it is the average of the possible outcomes not weighted equally but weighted by their probabilities. More likely outcomes get more weight in the average. An important fact of probability, the *law of large numbers*, says that μ is the average outcome in another sense as well.

Law of Large Numbers

Observe any random phenomenon having numerical outcomes with finite mean μ. According to the **law of large numbers**, as the random phenomenon is repeated a large number of times,

▶ the proportion of trials on which each outcome occurs gets closer and closer to the probability of that outcome, and

▶ the mean $\bar{x}$ of the observed values gets closer and closer to μ.

These facts can be stated more precisely and then proved mathematically. The law of large numbers brings the idea of probability to a natural completion. We first observed that some phenomena are random in the sense of showing long-run regularity. Then we used the idea of long-run proportions to motivate the basic laws of probability. Those laws are mathematical idealizations that can be used without interpreting probability as proportion in many trials. Now the law of large numbers tells us that in many trials the proportion of trials on which an outcome occurs will always approach its probability.

The law of large numbers also explains why gambling can be a business. The winnings (or losses) of a gambler on a few plays are uncertain—that's why gambling is exciting. It is only *in the long run* that the mean outcome is predictable. The house plays tens of thousands of times. So the house, unlike individual gamblers, can count on the long-run regularity described by the law of large numbers. The average winnings of the house on tens of thousands of plays will be very close to the mean of the distribution of winnings. Needless to say, gambling games have mean outcomes that guarantee the house a profit.

We know that the simplest description of a distribution of data requires both a measure of center and a measure of spread. The same is true for probability models. The *mean* is the average value for both a set of data and a discrete probability model. All the observations are weighted equally in finding the mean $\bar{x}$ for data, but the values are weighted by their probabilities in finding the mean μ of a probability model. The measure of spread that goes with the mean is the standard deviation. For data, the standard deviation s is the square root of the average squared deviation of the observations from their mean. We apply exactly the same idea to probability models, using probabilities as weights in the average. Here is the definition.

> ## Standard Deviation of a Discrete Probability Model
>
> Suppose that the possible outcomes $x_1, x_2, \ldots, x_k$ in a sample space S are numbers, and that p_j is the probability of outcome x_j. The **variance** σ^2 of this probability model is
>
> $$\sigma^2 = (x_1 - \mu)^2 p_1 + (x_2 - \mu)^2 p_2 + \cdots + (x_k - \mu)^2 p_k$$
>
> The **standard deviation** σ is the square root of the variance.

EXAMPLE 11   First Digits

If the first digits in a set of records obey Benford's law, the discrete probability model is

First digit	1	2	3	4	5	6	7	8	9
Probability	0.301	0.176	0.125	0.097	0.079	0.067	0.058	0.051	0.046

We saw in Example 10 that the mean is $\mu = 3.441$. To find the variance,

$$
\begin{aligned}
\sigma^2 &= (x_1 - \mu)^2 p_1 + (x_2 - \mu)^2 p_2 + \cdots + (x_k - \mu)^2 p_k \\
&= (1 - 3.441)^2 (0.301) + (2 - 3.441)^2 (0.176) + \cdots + (9 - 3.441)^2 (0.046) \\
&= 1.7935 + 0.3655 + \cdots + 1.4215 \\
&= 6.061
\end{aligned}
$$

The standard deviation is $\sigma = \sqrt{6.061} = 2.46$. You can follow the same pattern to show that the standard deviation of the equally likely model is 2.58. The Benford's law model is less spread out than the equally likely model. ■

Finding the variance and standard deviation of a continuous probability model usually requires advanced mathematics. Chapter 5 told us the answer in one important case: The standard deviation of a normal curve is the distance from the center (the mean) to the change-of-curvature points on either side.

8.6 The Central Limit Theorem

The key to finding a confidence interval that estimates a population proportion (Chapter 7) was the fact that the sampling distribution of a population proportion is close to normal when the sample is large. This fact is an application of

SPOTLIGHT 8.2   Cancer Clusters

Distressing events bring out the desire to pinpoint a cause. Yet sometimes misfortune is just the play of chance. Consider the recurring worry about cancer clusters. A cancer cluster will quite likely appear in a neighborhood near you. You may find, for example, your local newspaper full of worry about a "larger-than-expected" number of cancer cases among people who have been students or teachers at a large elementary school.

Exactly what is "larger than expected"? Cancer is common. More than 14 million people in the United States have cancer or have had cancer in the past. The disease accounts for 23% of all deaths. So it is not surprising that cancer cases sometimes occur in clusters in the same neighborhood. There are bound to be clusters *somewhere*, simply by chance. Think of tossing a large number of marbles at random onto a bed. Some of them will cluster together just by chance, and some spots on the bed will be bare, also just by chance. How many marbles do we expect to see in a square foot of

the bed? We could construct a probability model for "tossing at random" and find a mean number from that model. That's the number we would expect to see. Now someone worried about "marble clusters" looks at the bed, singles out a spot where the marbles lie close together, and says, "Aha! There are more marbles than expected here." But the count in a spot singled out exactly because marbles were clustered there can't be compared with the mean count in a random area.

When a cancer cluster occurs in *our* neighborhood, we naturally tend to suspect the worst and look for someone to blame. State authorities get several thousand calls a year from people worried about "too much cancer" in their area. Some clusters do have specific causes, such as a toxic substance in the local water supply. This is most likely to be true when the cluster involves cases of a specific and relatively rare form of cancer. But most cancer clusters are simply the result of chance.

one of the most important results of probability theory, the *central limit theorem*. This theorem says that the distribution of any random phenomenon tends to be normal if we average it over a large number of independent repetitions. The central limit theorem allows us to analyze and predict the results of chance phenomena when we average over many observations.

Central Limit Theorem

Draw an SRS of size n from any large population with mean μ and finite standard deviation σ. Then

▶ The mean of the sampling distribution of $\overline{x}$ is μ.
▶ The standard deviation of the sampling distribution of $\overline{x}$ is $\sigma/\sqrt{n}$.
▶ The **central limit theorem** says that the sampling distribution of $\overline{x}$ is approximately normal when the sample size n is large.

The first two parts of this statement can be proved from the definitions of the mean and the standard deviation. They are true for any sample size n. The central limit theorem is a much deeper result. Pay attention to the fact that the standard deviation of a mean decreases as the number of observations n increases. Together with the central limit theorem, this makes exact two general statements that help us understand a wide variety of random phenomena:

Averages are less variable than individual observations.

Averages are more normal than individual observations.

The *Central Limit Theorem* applet allows you to watch the central limit theorem in action: It starts with a distribution that is strongly skewed, not at all normal. As you increase the size of the sample, the distribution of the mean $\bar{x}$ gets closer and closer to the normal shape.

EXAMPLE 12   Heights of Young Women

The distribution of heights of young women is approximately normal, with mean 64.5 inches and standard deviation 2.5 inches. This normal distribution describes the population of young women. It is also the probability model for choosing one woman at random from this population and measuring her height. For example, the 68–95–99.7 rule says that the probability is 0.95 that a randomly chosen woman is between 59.5 and 69.5 inches tall.

Now choose an SRS of 25 young women at random and take the mean $\bar{x}$ of their heights. The mean $\bar{x}$ varies in repeated samples—the pattern of variation is the sampling distribution of $\bar{x}$. The sampling distribution has the same center $\mu = 64.5$ inches as the population of young women. In statistical terms, the sample mean $\bar{x}$ has *no bias* as an estimator of the population mean μ. If we take many samples, $\bar{x}$ will sometimes be smaller than μ and sometimes larger, but it has no systematic tendency to be too small or too large.

The standard deviation of the sampling distribution of $\bar{x}$ is

$$\frac{\sigma}{\sqrt{n}} = \frac{2.5}{\sqrt{25}} = \frac{2.5}{5} = 0.5 \text{ inch}$$

The standard deviation σ describes the variation when we measure many individual women. The standard deviation $\sigma/\sqrt{n}$ of the distribution of $\bar{x}$ describes the variation in the average heights of samples of women when we take many samples. The average height is less variable than individual heights.

Figure 8.9 compares the two distributions: Both are normal and both have the same mean, but the average height of 25 randomly chosen women is much less spread out. For example, the 68–95–99.7 rule says that 95% of all averages $\bar{x}$ lie between 63.5 and 65.5 inches because 2 standard deviations of $\bar{x}$ make 1 inch. This 2-inch range is just one-fifth as wide as the 10-inch range that catches the middle 95% of heights for individual women. ■

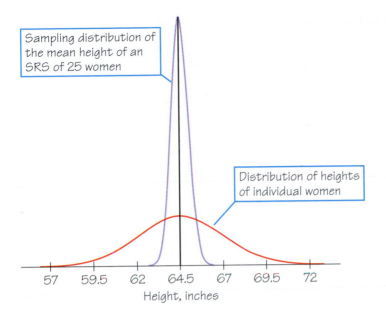

Sampling distribution of the mean height of an SRS of 25 women

Distribution of heights of individual women

| 57 | 59.5 | 62 | 64.5 | 67 | 69.5 | 72 |

Height, inches

FIGURE 8.9 The sampling distribution of the average height of an SRS of 25 women has the same center as the distribution of individual heights but is much less spread out.

The central limit theorem says that in large samples the sample mean $\bar{x}$ is approximately normal. In Figure 8.9, we show a normal curve for $\bar{x}$ even though sample size 25 is not very large. Is that acceptable? How large a sample is needed for the central limit theorem to work depends on how far from a normal curve the model we start with is. The closer to normality we start, the quicker the distribution of the sample mean becomes very normal. In fact, if individual observations follow a normal curve, the sampling distribution of $\bar{x}$ is exactly normal for any sample size. So Figure 8.9 is accurate. The central limit theorem is a striking result because as n gets large it works for *any* model we may start with, no matter how far from normal. Here is an example that starts very far from normal.

Red or Black in Roulette

EXAMPLE 13

An American roulette wheel has 38 slots, of which 18 are black, 18 are red, and 2 are green. The dealer spins the wheel and spins a small ball in the opposite direction within the wheel (see Figure 8.10). Gamblers bet on where the ball will come to rest. One of the simplest wagers chooses red or black. A bet of $1 on red pays off an additional $1 if the ball lands in a red slot. Otherwise, the player loses his $1. When gamblers bet on red or black, the two green slots belong to the house.

Lou bets on red. He wins if the ball stops in one of the 18 red slots. He loses if it lands in one of the 20 slots that are black or green. Because casino roulette wheels are carefully balanced so that all slots are equally likely, the probability model is

FIGURE 8.10 A gambler may win or lose at roulette, but in the long run the casino always wins.

Outcome	Win $1	Lose $1
Probability	18/38	20/38

The mean outcome of a single bet on red is

$$\mu = (1)\left(\frac{18}{38}\right) + (-1)\left(\frac{20}{38}\right)$$

$$= -\frac{2}{38} = -0.053 \text{ (a loss of 5.3 cents)}$$

The law of large numbers says that the mean μ is the average outcome of a very large number of individual bets. In the long run, gamblers will lose (and the casino will win) an average of 5.3 cents per bet. We can similarly find the standard deviation for a single bet:

$$\sigma^2 = (1 - (-0.053))^2 \frac{18}{38} + (-1 - (-0.053))^2 \frac{20}{38}$$

$$= (1.053)^2 \frac{18}{38} + (-0.947)^2 \frac{20}{38}$$

$$= 0.9972$$

$$\sigma = \sqrt{0.9972} = 0.9986 \blacksquare$$

Lou certainly starts far from any normal curve. The probability model for each bet is discrete, with just two possible outcomes. Yet the central limit theorem says that the average outcome of many bets follows a normal curve. Lou is a habitual gambler who places fifty $1 bets on red almost every night. Because we know the probability model for a bet on red, we can simulate Lou's experience over many nights at the roulette wheel. The histogram in Figure 8.11 shows Lou's average winnings for 1000 nights. As the central limit theorem says, the distribution looks normal.

EXAMPLE 14   Lou Gets Entertainment

The normal curve in Figure 8.11 comes from the central limit theorem and the values of the mean μ and standard deviation σ in Example 13. It has

$$\text{mean} = \mu = -0.053$$

$$\text{standard deviation} = \frac{\sigma}{\sqrt{n}} = \frac{0.9986}{\sqrt{50}} = 0.141$$

Apply the 99.7 part of the 68–95–99.7 rule: Almost all average nightly winnings will fall within 3 standard deviations of the mean, that is, between

$$-0.053 - (3)(0.141) = -0.476$$

and

$$-0.053 + (3)(0.141) = 0.370$$

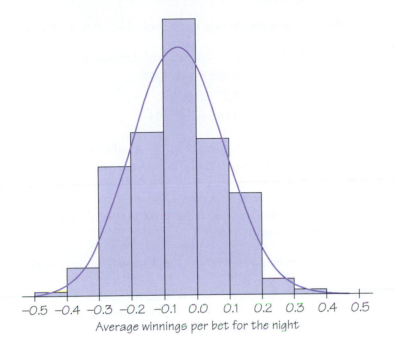

FIGURE 8.11 A gambler's winnings in a night of 50 bets on red or black in roulette vary from night to night. Here is the distribution for 1000 nights. It is approximately normal.

Average winnings per bet for the night

Lou's total winnings after 50 bets of $1 each will then fall between

$$(50)(-0.476) = -23.80$$

and

$$(50)(0.370) = 18.50$$

Lou may win as much as $18.50 or lose as much as $23.80. Gambling is exciting because the outcome, even after an evening of bets, is uncertain. It is possible to walk away a winner. It's all a matter of luck. ■

The casino, however, is in a different position. It doesn't want excitement, just a steady income.

The Casino Has a Business   EXAMPLE 15

The casino bets with all its customers—perhaps 100,000 individual bets on black or red in a week. The central limit theorem guarantees that the distribution of average customer winnings on 100,000 bets is very close to normal. The mean is still the mean outcome for one bet, -0.053, a loss of 5.3 cents per dollar bet. The standard deviation is much smaller when we average over 100,000 bets. It is

$$\frac{\sigma}{\sqrt{n}} = \frac{0.9986}{\sqrt{100,000}} = 0.003$$

Here is what the spread in the average result looks like after 100,000 bets:

$$\text{Spread} = \text{mean} \pm 3 \text{ standard deviations}$$
$$= -0.053 \pm (3)(0.003)$$
$$= -0.053 \pm 0.009$$
$$= -0.062 \text{ to } -0.044$$

Because the casino covers so many bets, the standard deviation of the average winnings per bet becomes very small. And because the mean is negative, almost all outcomes will be negative. The gamblers' losses and the casino's winnings are almost certain to average between 4.4 and 6.2 cents for every dollar bet.

The gamblers who collectively place those 100,000 bets will lose money. The probable range of their losses is

$$(100,000)(-0.062) = -6200 \text{ to } (100,000)(-0.044) = -4400$$

The gamblers are almost certain to lose—and the casino is almost certain to take in—between $4400 and $6200 on those 100,000 bets. What's more, the range of average outcomes continues to narrow as more bets are made. That is how a casino can make a business out of gambling. ■

In Chapter 7, we based a confidence interval for a population proportion p on the fact that the sampling distribution of a sample proportion $\hat{p}$ is close to normal for large samples. The central limit theorem applies to means. How can we apply it to proportions? By seeing that *a proportion is really a mean*. This is our final example of the central limit theorem.

EXAMPLE 16   The Sampling Distribution of a Proportion

We have an SRS of size n from a population that contains proportion p of "successes." (As usual, a success is shorthand for whatever we are counting, such as "Agree" answers to an opinion poll question.) Here's how to write the sample proportion of successes as a sample mean.

Look at the result for the ith individual in our sample. Make the result numerical by writing

$$x_i = 1 \text{ if the result is a success}$$
$$x_i = 0 \text{ if the result is a failure}$$

The count of successes is the count of 1's among the x_i. So

$$\hat{p} = \frac{\text{count of successes}}{n}$$
$$= \frac{x_1 + x_2 + \cdots + x_n}{n} = \bar{x}$$

There you have it: $\hat{p}$ is a mean, so the central limit theorem says that the sampling distribution of $\hat{p}$ is close to normal when the sample size n is large. ■

Because $\hat{p}$ is the mean of the x_i, we can find the mean and standard deviation of $\hat{p}$ from the mean and standard deviation of one observation x_i. Each observation has probability p of being a success, so the probability model for one observation is

Outcome	Success, $x_i = 1$	Failure, $x_i = 0$
Probability	p	$1 - p$

The mean of x_i is therefore

$$\mu = (1)(p) + (0)(1 - p) = p$$

In the same way, after a bit more algebra, you can discover that the variance of one observation x_i is

$$\sigma^2 = (1 - p)^2 p + (0 - p)^2(1 - p) = p(1 - p)$$

The standard deviation of $\hat{p}$ is the standard deviation of the mean of n observations,

$$\frac{\sigma}{\sqrt{n}} = \frac{\sqrt{p(1 - p)}}{\sqrt{n}} = \sqrt{\frac{p(1 - p)}{n}}$$

That's the fact we used in Chapter 7.

Examples 12 to 16 illustrate the importance of the central limit theorem and also the reason for the importance of normal distributions. We can often replace tricky calculations about a probability model by simpler calculations for a normal distribution, courtesy of the central limit theorem.

REVIEW VOCABULARY

Addition rule If two events are disjoint, the probability that one or the other occurs is the sum of their individual probabilities.

Central limit theorem The average of many independent random outcomes is approximately normally distributed. When we average n independent repetitions of the same random phenomenon, the resulting distribution of outcomes has mean equal to the mean outcome of a single trial and standard deviation proportional to $1/\sqrt{n}$.

Combinatorics The branch of mathematics that counts arrangements of objects.

Complement rule The probability that an event does *not* occur is always 1 minus the probability that it does occur.

Continuous probability model A probability model that assigns probabilities to events as areas under a density curve.

Density curve A curve that is always on or above the horizontal axis and has area exactly 1 underneath it. A density curve describes a continuous probability model.

Discrete probability model A probability model that assigns probabilities to each of a finite number of possible outcomes.

Disjoint events Events that have no outcomes in common.

Event Any collection of possible outcomes of a random phenomenon. An event is a subset of the sample space.

Law of large numbers As a random phenomenon is repeated many times, the mean $\bar{x}$ of the observed outcomes approaches the mean μ of the probability model.

Mean of a probability model The average outcome of a random phenomenon with numerical values. When possible values $x_1, x_2, \ldots, x_k$ have probabilities $p_1, p_2, \ldots, p_k$, the mean is the average of the outcomes weighted by their probabilities, $\mu = x_1p_1 + x_2p_2 + \cdots + x_kp_k$.

Probability A number between 0 and 1 that gives the long-run proportion of repetitions of a random phenomenon on which an event will occur.

Probability histogram A histogram that displays a discrete probability model when the outcomes are numerical. The height of each bar is the probability of the outcome or group of outcomes at the base of the bar.

Probability model A sample space S together with an assignment of probabilities to events. The two main types of probability models are *discrete* and *continuous*.

Random phenomenon A phenomenon is random if it is uncertain what the next outcome will be but each outcome nonetheless tends to occur in a fixed proportion of a very long sequence of repetitions. These long-run proportions are the probabilities of the outcomes.

Sample space A list of all possible outcomes of a random phenomenon.

Sampling distribution The distribution of values taken by a statistic when many random samples are drawn under the same circumstances. A sampling distribution consists of an assignment of probabilities to the possible values of a statistic.

Standard deviation of a probability model A measure of the variability of a probability model. When the possible values $x_1, x_2, \ldots, x_k$ have probabilities $p_1, p_2, \ldots, p_k$, the **variance** is the average (weighted by probabilities) of the squared deviations from the mean, $\sigma^2 = (x_1 - \mu)^2p_1 + (x_2 - \mu)^2p_2 + \cdots + (x_k - \mu)^2p_k$. The **standard deviation** σ is the square root of the variance.

✔ SKILLS CHECK

1. You read in a book on poker that the probability of being dealt three of a kind in a five-card poker hand is 1/50. What does this mean?

(a) If you deal thousands of poker hands, the fraction of them that contain three of a kind will be very close to 1/50.

(b) If you deal 50 poker hands, exactly one of them will contain three of a kind.

(c) If you deal 10,000 poker hands, exactly 200 of them will contain three of a kind.

2. A basketball player shoots 8 free throws during a game. The sample space for counting the number she makes is

(a) S = any number between 0 and 1.

(b) S = whole numbers 0 to 8.

(c) S = all sequences of 8 hits or misses, like HMMHHHMH.

Here is the probability model for the blood type of a randomly chosen person in the United States:

Blood type	O	A	B	AB
Probability	0.45	0.40	0.11	?

Exercises 3 to 5 use this information.

3. The probability that a randomly chosen American has type AB blood is

(a) any number between 0 and 1.
(b) 0.04.
(c) 0.4.

4. Maria has type B blood. She can safely receive blood transfusions from people with blood types O and B. What is the probability that a randomly chosen American can donate blood to Maria?

(a) 0.11
(b) 0.44
(c) 0.56

5. What is the probability that a randomly chosen American does not have type O blood?

(a) 0.55
(b) 0.45
(c) 0.04

6. Figure 8.2 shows the 36 possible outcomes for rolling two dice. These outcomes are equally likely. A "soft 4" is a roll of 1 on one die and 3 on the other. What is the probability of rolling a soft 4?

(a) 1/36
(b) 2/36
(c) 3/36

7. In a table of random digits such as Table 7.1, each digit is equally likely to be any of 0, 1, 2, 3, 4, 5, 6, 7, 8, or 9. What is the probability that a digit in the table is a 0?

(a) 1/9
(b) 1/10
(c) 9/10

8. In a table of random digits such as Table 7.1, each digit is equally likely to be any of 0, 1, 2, 3, 4, 5, 6, 7, 8, or 9. What is the probability that a digit in the table is 7 or greater?

(a) 7/10
(b) 4/10
(c) 3/10

9. Toward the end of a game of Scrabble, you hold the letters D, O, G, and Q. In how many orders can you arrange these 4 letters?

(a) 4
(b) (4)(3)(2)(1) = 24
(c) (4)(4)(4)(4) = 256

10. Toward the end of a game of Scrabble, you hold the letters D, O, G, and Q. In how many ways can you choose 3 of these 4 letters and arrange them in order?

(a) (4)(3)(2) = 24
(b) (3)(2)(1) = 6
(c) (3)(3)(3) = 27

11. A bridge deck contains 52 cards, 13 from each of the four suits: clubs, diamonds, hearts, and spades. You deal 4 cards from a well-shuffled deck, so that you are equally likely to deal any 4 cards. What is the probability that you deal 4 spades?

(a) 1/4, because 1/4 of the cards are spades.
(b) (13)(12)(11)(10)/(52)(51)(50)(49) = 0.0026
(c) (13)(12)(11)(10)/(52)(52)(52)(52) = 0.0023

12. You deal 4 cards as in the previous exercise. What is the probability that you deal no spades?

(a) 1 − 0.0026 = 0.9974 by the complement rule
(b) 1 − 0.0023 = 0.9977 by the complement rule
(c) (39)(38)(37)(36)/(52)(51)(50)(49) = 0.3038

13. Figure 8.6 shows the density curve of a continuous probability model for choosing a number at random between 0 and 1. The probability that the number chosen is less than or equal to 0.4 is

(a) 0.4.
(b) 0.5.
(c) 0.6.

14. Figure 5.3 (page 185) shows that the normal distribution with mean $\mu = 6.8$ and standard deviation $\sigma = 1.6$ is a good description of the Iowa Test vocabulary scores of seventh-grade students in Gary, Indiana. The probability that a randomly chosen student has a score higher than 8.4 is

(a) 0.68.
(b) 0.32.
(c) 0.16.

15. "Pick 3" games in many state lotteries work like this: You choose a three-digit number, like 456 or 999. The lottery chooses a winning number at random, so that each of the 1000 three-digit numbers has probability 0.001 of winning. You win $500 if your number is chosen; otherwise you lose the $1 cost of a ticket. What is the mean outcome of one bet?

(a) $5
(b) $0.50
(c) −$0.499

16. Annual returns on the more than 5000 common stocks available to investors vary a lot. In a recent year, the mean return was 8.3% and the standard deviation of returns was 28.5%. The law of large numbers says that

(a) you can get an average return higher than the mean 8.3% by investing in a large number of stocks.
(b) as you invest in more and more stocks chosen at random, your average return on these stocks gets ever closer to 8.3%.
(c) if you invest in a large number of stocks chosen at random, your average return will have approximately a normal distribution.

17. Figure 8.6 shows the density curve of a continuous probability model for choosing a number at random between 0 and 1. The mean of this model is

(a) 0.5 because the curve is symmetric.
(b) 1 because there is area 1 under the curve.
(c) can't tell—this requires advanced mathematics.

18. Scores on the SAT college entrance test in a recent year were roughly normal, with mean 1026 and standard deviation 209. You choose an SRS of 100 students and average their SAT scores. If you do this many times, the mean of the average scores you get will be

(a) 1026.
(b) 209.
(c) $1026/\sqrt{100} = 10.26$.

19. Scores on the SAT college entrance test in a recent year were roughly normal, with mean 1026 and standard deviation 209. You choose an SRS of 100 students and average their SAT scores. If you do this many times, the standard deviation of the average scores you get will be

(a) 209.
(b) $100/\sqrt{209} = 6.92$.
(c) $209/\sqrt{100} = 20.9$.

20. The number of hours a light bulb burns before failing varies from bulb to bulb. The distribution of burnout times is strongly skewed to the right. The central limit theorem says that

(a) as we look at more and more bulbs, their average burnout time gets ever closer to the mean μ for all bulbs of this type.
(b) the average burnout time of a large number of bulbs has a distribution of the same shape (strongly skewed) as the distribution for individual bulbs.
(c) the average burnout time of a large number of bulbs has a distribution that is close to normal.

EXERCISES   ■ Challenge   ◆ Discussion

Probability Models and Rules

1. Hold a penny upright on its edge under your forefinger on a hard surface, then snap it with your other forefinger so that it spins for some time before falling. Based on 50 spins, estimate the probability of heads.

2. Toss a thumbtack on a hard surface 100 times. How many times did it land with the point up? What is the approximate probability of landing point up?

3. The table of random digits (Table 7.1) was produced by a random mechanism that gives each digit probability 0.1 of being a 0. What proportion of the first five lines in the table are 0's? This proportion is an estimate of the true probability, which in this case is known to be 0.1.

4. Probability is a measure of how likely an event is to occur. Match one of the probabilities that follow with each statement about an event. (The probability is usually a much more exact measure of likelihood than is the verbal statement.)

$$0, 0.01, 0.3, 0.6, 0.99, 1$$

(a) This event is impossible. It can never occur.
(b) This event is certain. It will occur on every trial of the random phenomenon.
(c) This event is very unlikely, but it will occur once in a while in a long sequence of trials.
(d) This event will occur more often than not.

In each of Exercises 5 to 8, describe a reasonable sample space for the random phenomena mentioned. In some cases, you must use judgment to choose a reasonable S.

5. Toss a coin 10 times.

(a) Count the number of heads observed.
(b) Calculate the percent of heads among the outcomes.
(c) Record whether or not at least five heads occurred.

6. A randomly chosen subject arrives for a study of exercise and fitness.

(a) The subject is either female or male.
(b) After 10 minutes on an exercise bicycle, you ask the subject to rate his or her effort on the Rate of Perceived Exertion (RPE) scale. RPE ranges in whole-number steps from 6 (no exertion at all) to 20 (maximal exertion).
(c) You also measure the subject's maximum heart rate (beats per minute).

7. A basketball player shoots four free throws.

(a) You record the sequence of hits and misses.
(b) You record the number of shots she makes.

8. Choose a student at random from a large class and ask these questions.

(a) Are you right-handed or left-handed?
(b) What is your height in inches?
(c) How many minutes did you study last night?

9. A majority of email messages are now "spam." Choose a spam email message at random. Here is the probability model for its topic:

Topic	Adult	Financial	Health
Probability	0.145	0.162	0.073

Topic	Leisure	Products	Scams
Probability	0.078	0.210	0.142

(a) What is the probability that a spam email does not concern one of these topics?
(b) Corinne is particularly annoyed by spam offering "adult" content (that is, pornography) and scams. What is the probability that a randomly chosen spam email falls into one or the other of these categories?

10. Choose a young adult (age 25 to 34 years) at random. The probability is 0.12 that the person chosen did not complete high school, 0.31 that the person has a high school diploma but no further education, and 0.29 that the person has at least a bachelor's degree.

(a) What must be the probability that a randomly chosen young adult has some education beyond high school but does not have a bachelor's degree?
(b) What is the probability that a randomly chosen young adult has at least a high school education?

■ **11.** The addition rule $P(A \text{ or } B) = P(A) + P(B)$ is not always true. Give an example of real-world events A and B for which this rule is *not* true.

Discrete Probability Models

12. Choose a new car or light truck at random and note its color. Here are the probabilities of the most popular colors:

Color	Silver	White	Black
Probability	0.201	0.184	0.116

Color	Gray	Dark blue	Light brown
Probability	0.115	0.088	0.085

(a) What is the probability that the car you choose has any color other than the six listed?
(b) What is the probability that a randomly chosen car is either silver or white?

13. North Carolina State University posts the grade distributions for its courses online. Students in Statistics 101 in the fall 2003 semester received 21% A's, 43% B's, 30% C's, 5% D's, and 1% F's. Choose a Statistics 101 student random. Here is the probability model for the student's grade on a four-point scale (with A = 4).

Grade	0	1	2	3	4
Probability	0.01	0.05	0.30	0.43	0.21

(a) Make a probability histogram for this model.
(b) What is the probability that the student got a grade of B or better?

14. How do rented housing units differ from units occupied by their owners? Here are probability models for the number of rooms for owner-occupied units and renter-occupied units, according to the Census Bureau:

Rooms	1	2	3	4	5
Owned	0.000	0.001	0.014	0.099	0.238
Rented	0.011	0.027	0.229	0.348	0.224

Rooms	6	7	8	9	10
Owned	0.266	0.178	0.107	0.050	0.047
Rented	0.105	0.035	0.012	0.004	0.005

Make probability histograms of these two models, using the same scales. What are the most important differences between the models for owner-occupied and rented housing units?

15. In each of the following situations, state whether or not the given assignment of probabilities to individual outcomes is legitimate, that is, satisfies the rules of probability. If not, give specific reasons for your answer.
(a) Roll a die and record the count of spots on the up face: $P(1) = 0$, $P(2) = 1/6$, $P(3) = 1/3$, $P(4) = 1/3$, $P(5) = 1/6$, $P(6) = 0$.
(b) Choose a college student at random and record sex and enrollment status: P(female full-time) = 0.56, P(female part-time) = 0.24, P(male full-time) = 0.44, P(male part-time) = 0.17.
(c) Deal a card from a shuffled deck: P(clubs) = 12/52, P(diamonds) = 12/52, P(hearts) = 12/52, P(spades) = 16/52.

16. What is the probability that a housing unit has five or more rooms? Use the models in Exercise 14 to answer this question for both owner-occupied and rented units.

■ 17. Nonstandard dice can produce interesting distributions of outcomes. You have two balanced, six-sided dice. One is a standard die, with faces having 1, 2, 3, 4, 5, and 6 spots. The other die has three faces with 0 spots and three faces with 6 spots. Find the probability model for the total number of spots on the up faces when you roll these two dice. (Start with a display like Figure 8.2 for the outcomes of rolling the two dice. To get equally likely outcomes, suppose that the three 0 faces on the second die are marked 0a, 0b, and 0c and that the three 6 faces are marked 6a, 6b, and 6c.)

■ 18. Role-playing games like Dungeons & Dragons use many different types of dice. Suppose that a balanced four-sided die has faces marked 1, 2, 3, 4. The intelligence of a character is determined by rolling this die twice and adding 1 to the sum of the spots. Give a probability model for the character's intelligence. (Start with a display like Figure 8.2 for the outcomes of the two rolls of the die. These outcomes are equally likely.) What is the probability that the character has intelligence 7 or higher?

Equally Likely Outcomes

19. A party host gives a door prize to one guest chosen at random. There are 48 men and 42 women at the party. What is the probability that the prize goes to a woman?

20. Abby, Deborah, Julie, Sam, and Roberto work in a firm's public relations office. Their employer must choose two of them to attend a conference in Paris. To avoid unfairness, the choice will be made by drawing two names from a hat. (This is a simple random sample of size 2.)
(a) Write down all possible choices of two of the five names. This is the sample space.
(b) The random drawing makes all choices equally likely. What is the probability of each choice?
(c) What is the probability that Julie is chosen?
(d) What is the probability that neither of the two men (Sam and Roberto) is chosen?

21. You toss a balanced coin 10 times. Each toss gives either a head or a tail. Write down the sequence of results, such as HTTTHHTHHH.
(a) How many possible outcomes are there?
(b) What is the probability of getting all heads or all tails?

22. The Korean language, unlike Chinese and Japanese, uses an alphabet. The Korean alphabet has 10 vowels and 14 consonants. A computer assigns three-letter log-in codes in Korean at random, with repeated letters allowed.
(a) How many different codes are there?
(b) What is the probability that a code contains only consonants?

23. A computer assigns three-character log-in IDs that may contain the digits 0 to 9 as well as the letters *a* to *z*, with repeats allowed. What is the probability that your ID contains no *x*? What is the probability that your ID contains no digits?

24. What is the probability that a code in Korean assigned by the computer in Exercise 22 has the form consonant-vowel-consonant?

25. A monkey at a keyboard presses three keys and hits the letters *a*, *g*, and *s* in random order. How many possible three-letter "words" can the monkey type using only these letters? Which of these are meaningful English words? What is the probability that the word the monkey typed is meaningful?

■ **26.** A computer system allows user IDs that are three, four, or five characters long. Repeated characters are allowed. If only the letters of the alphabet can be used, how many possible IDs are there in all? What is the probability that an ID chosen at random from all possible IDs has no repeated letters?

■ **27.** A computer system allows user IDs that are three, four, or five characters long. Repeated characters are allowed. The first character must be a letter, but the other characters can be either letters or numbers. How many possible IDs are there in all? What is the probability that an ID chosen at random from all possible IDs has no numbers?

■ **28.** An ancient Korean drinking game involves a 14-sided die. The players roll the die in turn and must submit to whatever humiliation is written on the up face: something like "Keep still when tickled on face." Six of the 14 faces are squares. The other eight faces are triangles. Each of the squares is equally likely. Each of the triangles is also equally likely, but the triangle probability differs from the square probability.
(a) If you know that the probability of getting a square is 0.72, what is the probability model for the 14 possible outcomes?
(b) Give a different correct probability model for the 14 possible outcomes.

Continuous Probability Models

29. Generate two random numbers between 0 and 1 and take their sum. The sum can take any value between 0 and 2. The density curve is the triangle shown in Figure 8.12.

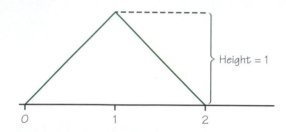

Height = 1

FIGURE 8.12 The density curve for the sum of two random numbers, for Exercise 29.

(a) Verify by geometry that the area under this curve is 1.

(b) What is the probability that the sum is less than 1? (Sketch the density curve, shade the area that represents the probability, then find that area. Do this for part (c) also.)

(c) What is the probability that the sum is less than 0.5?

30. Many random-number generators allow users to specify the range of the random numbers to be produced. Suppose that you specify that the range is to be all numbers between 0 and 2. The density curve for the outcome has constant height between 0 and 2, and height 0 elsewhere.

(a) What is the height of the density curve between 0 and 2? Draw a graph of the density curve.

(b) Use your graph from part (a) and the fact that probability is area under the curve to find the probability of getting an outcome less than or equal to 1.

(c) Find the probability of an outcome between 0.5 and 1.3.

31. Make a sketch of the *unit square,* the square with sides from 0 to 1 in both the *x* (horizontal) and *y* (vertical) directions. If you choose a point in this square at random, the probability that the

point is in any region within the square is equal to the area of that region. What is the probability that the *y*-coordinate of the point is less than the *x*-coordinate? (Shade this region on your sketch and use geometry to find the probability.)

The Mean and Standard Deviation of a Probability Model

32. Linda is a sales associate at a large auto dealership. She expects to earn $400 for each vehicle she sells. Linda motivates herself by using probability estimates of her sales. For a sunny Saturday in April, she estimates her sales as follows:

Vehicles sold	0	1	2	3
Probability	0.3	0.4	0.2	0.1

Give the probability model for Linda's earnings. What are her mean earnings?

33. Exercise 13 gives a probability model for the grade of a randomly chosen student in Statistics 101 at North Carolina State University, using the 4-point scale. What is the mean grade in this course? What is the standard deviation of the grades?

34. In Exercise 18, you gave a probability model for the intelligence of a character in a role-playing game. What is the mean intelligence for these characters?

35. Exercise 14 gives probability models for the number of rooms in owner-occupied and rented housing units. Find the mean number of rooms for each type of housing. Make probability histograms for the two models and mark the mean on each histogram. You see that the means describe an important difference between the two models: Owner-occupied units tend to have more rooms.

36. Typographical and spelling errors can be either "nonword errors" or "word errors." A nonword error is not a real word, as when "the" is typed as "teh." A word error is a real word, but not the right word, as when "lose" is typed as "loose." When undergraduates are asked to write a 250-word essay (without spell-checking), the number of nonword errors has this probability model:

Errors	0	1	2	3	4
Probability	0.1	0.2	0.3	0.3	0.1

The number of word errors has this model:

Errors	0	1	2	3
Probability	0.4	0.3	0.2	0.1

What are the mean numbers of nonword errors and word errors in an essay? How does the difference between the means describe the difference between the two models?

37. What is the mean for the continuous probability model in Exercise 29? What is the mean for the probability model in Exercise 30? Explain how you found these means.

■ **38.** The idea of insurance is that we all face risks that are unlikely but carry high cost. Think of a fire destroying your home. Insurance spreads the risk: We all pay a small amount, and the insurance policy pays a large amount to those few of us whose homes burn down. An insurance company looks at the records for millions of homeowners and sees that the mean loss from fire in a year is $\mu = \$250$ per person. (Most of us have no loss, but a few lose their homes. The $250 is the average loss.) The company plans to sell fire insurance for $250 plus enough to cover its costs and profit. Explain clearly why it would be unwise to sell only 12 policies. Then explain why selling thousands of such policies is a safe business.

◆ **39.** You roll dice balanced so that all six sides are equally likely to land facing upward.

(a) What is the mean number of spots observed in rolling a single die?

(b) Example 3 gives a probability model for rolling two balanced dice. What is the mean number of spots observed?

(c) Explain why your result in part (a) leads you to expect the result you found in part (b).

40. An American roulette wheel has 38 slots numbered 0, 00, and 1 to 36. The ball is equally likely to come to rest in any of these slots when the wheel is spun. The slot numbers are laid out on a board on which gamblers place their bets. One column of numbers on the board contains multiples of 3—that is, 3, 6, 9, . . . , 36. Joe places a $1 column bet that pays out $3 if any of these numbers comes up.

(a) What is the probability model for the outcome of one bet, taking into account the $1 cost of a bet?

(b) What are the mean and standard deviation for this model?

(c) Joe plays roulette every day for years. What does the law of large numbers tell us about his results?

41. You have two balanced, six-sided dice. The first has 1, 3, 4, 5, 6, and 8 spots on its six faces. The second die has 1, 2, 2, 3, 3, and 4 spots on its faces.

(a) What is the mean number of spots on the up face when you roll the first die?

(b) What is the mean number of spots on the up face when you roll the second die?

(c) The sum of these means is the mean number of spots when you roll both dice. What is this mean? How does it compare with the mean when you roll two standard dice, found in Exercise 39?

■ **42.** A state lottery "Pick 3" game offers a choice of several bets. You choose a three-digit number and bet $1. The lottery commission announces the winning three-digit number, chosen at random, at the end of each day. The "box" pays $82.33 if the number you chose has the same digits as the winning number, in any order. Otherwise, you lose your dollar. Find the mean winnings for a bet on the box, taking into account that you paid $1 to play. (Assume that you chose a number having three distinct digits.)

■ **43.** According to the current mortality table adopted by state insurance regulators, a randomly chosen 25-year-old man has these probabilities of dying during the next five years:

Age at death	25	26	27
Probability	0.00039	0.00044	0.00051

Age at death	28	29
Probability	0.00057	0.00060

(a) What is the probability that the man does not die in the next five years?

(b) An online insurance site offers a term insurance policy that will pay $100,000 if a 25-year-old man dies within the next five years. The cost is $175 per year. So the insurance company will take in $875 from this policy if the man does not die within five years. If he does die, the company must pay $100,000. Its loss depends on how many premiums were paid, as follows:

Age at death	25	26	27
Loss	$99,825	$99,650	$99,475

Age at death	28	29
Loss	$99,300	$99,125

What is the insurance company's mean cash intake from such policies?

■ **44.** Here is a simple way to create a probability model that has specified mean μ and standard deviation σ: There are only two outcomes, $\mu - \sigma$ and $\mu + \sigma$, each with probability 0.5. Use the definition of the mean and variance for probability models to show that this model does have mean μ and standard deviation σ.

The Central Limit Theorem

45. Newly manufactured automobile radiators may have small leaks. Most have no leaks, but some have one, two, or more. The number of leaks in radiators made by one supplier has mean 0.15 and standard deviation 0.4. The distribution of the number of leaks cannot be normal because only whole-number counts are possible. The supplier ships 400 radiators per day to an auto assembly plant. Take $\bar{x}$ to be the mean number of leaks in these 400 radiators. Over several years of daily shipments, what range of values will contain the middle 95% of the many $\bar{x}$'s?

46. The scores of twelfth-grade students on the NAEP year 2000 mathematics test have a distribution that is approximately normal, with mean $\mu = 300$ and standard deviation $\sigma = 35$.

(a) Choose one twelfth-grader at random. What is the probability that his or her score is higher than 300? Higher than 335?

(b) Now choose an SRS of four twelfth-graders. What is the probability that their mean score is higher than 300? Higher than 335?

◆ **47.** Antonio measures the alcohol content of whiskey for his Chemistry 101 lab. He actually measures the mass of 5 milliliters of whiskey—a chemical calculation then finds the percent alcohol from the mass. The standard deviation of students' measurements of mass is $\sigma = 10$ milligrams (mg). Antonio repeats the measurement three times and records the mean $\bar{x}$ of his three measurements.

(a) What is the standard deviation of Antonio's mean result?

(b) How many times must Antonio repeat the measurement to reduce the standard deviation of $\bar{x}$ to 5 mg? Explain to someone who knows no statistics the advantage of reporting the average of several measurements rather than the result of a single measurement.

48. In Exercise 40 you found the mean and standard deviation of the outcome of a column bet in roulette. The central limit theorem says that the average outcome of a large number of bets has a distribution that is close to normal.

(a) What is the spread (mean ± 3 standard deviations) of a gambler's average winnings after 100 bets?

(b) What is the spread of a gambler's average winnings after 1000 bets?

49. Averages of several measurements are less variable than individual measurements. The true mass of the whiskey sample in Exercise 47 is 4.6 grams, or 4600 milligrams (mg). Antonio's measurements have the normal distribution with mean 4600 mg and standard deviation 10 mg. In this case, the mean of his three measurements also has a normal distribution.

(a) Sketch on the same graph the two normal curves, for individual measurements and for means of three measurements. Figure 8.9 is an example of this kind of graph.

(b) What range of values covers the middle 95% of Antonio's measurements?

(c) What range of values covers the middle 95% of averages of three measurements?

50. Exercise 18 gives the probability model for the intelligence assigned by chance to a character in a role-playing game. You found the mean intelligence of such characters in Exercise 34. Jermaine plays this character often. What range covers (approximately) the middle 68% of average intelligence scores for 100 of Jermaine's games?

51. The scores of high school seniors on the ACT college entrance examination in 2003 were roughly normal with mean $\mu = 20.8$ and standard deviation $\sigma = 4.8$.

(a) What is the approximate probability that a single student randomly chosen from all those taking the test scores 25.6 or higher?

(b) Now take an SRS of nine students who took the test. What are the mean and standard deviation of the sample mean score $\bar{x}$ of these nine students?

(c) What is the approximate probability that the mean score $\bar{x}$ of these nine students is 25.6 or higher?

■ 52. Although cities encourage carpooling to reduce traffic congestion, most vehicles carry only one person. For example, 70% of vehicles on the roads in the Minneapolis–St. Paul metropolitan area are occupied by just the driver. You choose 84 vehicles at random.

(a) What are the mean and standard deviation of the proportion of vehicles in your sample that carry only one person?

(b) What is the probability that more than 60% of the vehicles in your sample carry only one person? (Use the central limit theorem.)

Chapter Review

53. License plates in Florida have the form A12BCD. That is, a letter followed by two digits followed by three more letters.

(a) How many possible different license plates are there?

(b) Jerry would like a plate that ends in AAA. How many such plates are there?

(c) If license plates are issued at random from all possible plates, what is the probability that Jerry will get a plate that ends in AAA?

54. After you tell Jerry the probability you calculated in the previous exercise, he realizes that he's unlikely to get a plate ending in AAA. So he asks you, "What's the probability I will get a plate in which all four letters are from my name?" These letters are J, E, R, and Y.

(a) Suppose Jerry insists that the letters appear in order, so that his plate reads JnmERY, where "n" and "m" stand for any number. What is the probability?

(b) Suppose Jerry allows his letters to appear in the plate in any order and also allows repeats. What is now the answer to Jerry's question?

55. Choose a person age 19 to 25 years at random and ask, "In the past seven days, how many times did you go to an exercise or fitness center or work out?"

Based on a large sample survey, here is a discrete probability model for the answer you will get:

Days	0	1	2	3
Probability	0.68	0.05	0.07	0.08
Days	4	5	6	7
Probability	0.05	0.04	0.01	0.02

(a) What is the probability that the person you choose worked out either two or three days in the past seven?

(b) What is the probability that the person you choose worked out at least one day in the past seven?

56. What is the mean number of days that randomly chosen 19- to 25-year-olds worked out in the past seven days? (Use the information in the previous exercise.) If you interview many people in this age group, what does the law of large numbers say about the average number of days these people work out?

■ 57. Use the information in Exercise 55 and your result from Exercise 56 to answer these questions.

(a) What is the standard deviation of the number of days in the past seven that a randomly chosen 19- to 25-year-old has worked out?

(b) You interview 100 randomly chosen 19- to 25-year olds. You ask each how many days in the past seven they have worked out and you calculate the average number of days. According to the central limit theorem, there is probability 0.95 that your average will fall between what two values?

 # APPLET EXERCISES

To do these exercises, go to
www.whfreeman.com/fapp.

1. When we toss a coin, experience shows that the probability (long-term proportion) of a head is close to 1/2. Suppose now that we toss the coin repeatedly until we get a head. What is the probability that the first head comes up in an odd number of tosses (1, 3, 5, and so on)? Use the *Probability* applet to estimate this probability. Set the probability of heads to 0.5. Toss coins one at a time until the first head appears. Do this 50 times (click "Reset" after each trial). What is your estimate of the probability that the first head appears on an odd toss?

2. The table of random digits (Table 7.1) was produced by a random mechanism that gives each digit probability 0.1 of being a 0.

(a) What proportion of the digits in the first row of Table 7.1 are 0's? This proportion is an estimate, based on 40 repetitions, of the true probability, which in this case is known to be 0.1.

(b) The *Probability* applet can imitate random digits. Set the probability of heads in the applet to 0.1. Check "Show true probability" to show this value on the graph. A head stands for a 0 in the random digit table and a tail stands for any other digit. Simulate 200 digits (40 at a time–don't click "Reset"). If you kept going forever, presumably you would get 10% heads. What was the percent of heads in your 200 tosses?

3. The basketball player Shaquille O'Neal makes about half of his free throws over an entire season. Use the *Probability* applet to simulate 100 free throws shot independently by a player who has probability 0.5 of making each shot. (Toss 40, 40, and 20 without clicking "Reset.")

(a) What percent of the 100 shots did he hit?

(b) Examine the sequence of hits and misses after each click on "Toss" and keep track of the longest run of shots made and the longest run of shots missed. How long were the longest runs in the 100 shots taken? (Sequences of random outcomes often show runs longer than our intuition thinks likely.)

✎ WRITING PROJECTS

1. Most American states and Canadian provinces operate lotteries. How much is bet in a year in your state or provincial lottery? What percent of the money bet is returned to the bettors in the form of prizes? What percent of the money bet is used by the state to pay lottery expenses? What percent is net revenue to the state? For what purposes does the state use lottery revenue? Write a short essay that answers these questions. The lottery's Web site is the place to look for facts.

2. Psychologists have shown that our intuitive understanding of chance behavior is rather poor. Amos Tversky (1937–1996) was a leader in the study of how we make decisions in the face of uncertainty. In its obituary of Tversky, the *New York Times* cited the following example:

Tversky asked subjects to choose between two public health programs that affect 600 people. One has probability 1/2 of saving all 600 and probability 1/2 that all 600 will die. The other is guaranteed to save exactly 400 of the 600 people. Most people chose the second program. He then offered a different choice. One program has probability 1/2 of saving all 600 and probability 1/2 of losing all 600, while the other will definitely lose exactly 200 lives. Most people chose the first program.

Discuss this example. What is the difference between the two choices offered? What is the mean number of people saved by the two options in each choice? What do the reactions of most subjects to these choices show about how people make decisions?

SUGGESTED READINGS

COMAP. *Principles and Practices of Mathematics*, Springer, New York, 1997. Chapter 4 of this team-authored text presents combinatorics, and Chapter 8 discusses probability. This is a good choice for going beyond *For All Practical Purposes* in these areas.

MOSTELLER, FREDERICK, ROBERT E. K. ROURKE, and GEORGE B. THOMAS. *Probability with Statistical Applications*, Addison-Wesley, Reading, Mass., 1970. A rich treatment of basic probability that requires only high school algebra but is somewhat sophisticated. Although out of print, this book is a classic that nonetheless deserves mention.

SUGGESTED WEB SITES

"Buffon's needle" is a probability problem first stated in 1777 by Count Buffon: If you drop a needle on a sheet of lined paper, what is the probability that the needle crosses one of the lines? In the simplest case, the length of the needle is the same as the distance between the lines. Some fairly advanced math shows that the answer is $2/\pi$, or about 0.637. A number of Web sites simulate dropping a needle many times in order to estimate this probability. One good simulation is at the Web site of Charles Stanton,

www.math.csusb.edu/faculty/stanton/probstat/buffon. html. (If this site has disappeared, type "Buffon's needle" into Google. Be sure to choose a site that actually pictures the needle's position.)

You may be interested in the debate over legalized gambling. For the case against, visit the National Coalition Against Legalized Gambling at www.ncalg.org. For the defense by the casino industry, visit the American Gaming Association at www.americangaming.org.

Voting and Social Choice

The application of mathematics to the study of human beings—their behavior, values, interactions, conflicts, and methods of making decisions—is generally considered to be a recent revolution. Yet the study of voting and social choice, which is very much the root of this revolution, goes back several centuries.

We begin in Chapter 9 with the question of how a group of individuals, each with his or her own set of values, selects one outcome from a list of possibilities. While majority rule is a good system for deciding an election with just two candidates, it turns out that there is no perfect way of deciding an election in which there are three or more candidates.

Group decision-making is often a strategic encounter, and citizens need to be aware of the difficulties that can arise when some participants have an incentive to manipulate the outcome. It is this issue that we turn to in Chapter 10.

In Chapter 11 we consider decision-making bodies in which the individual voters or parties do not have equal power. In particular, we look at weighted voting systems in which a voter's power need not be proportional to the number of votes that he or she is entitled to cast.

In Chapter 12 we analyze not only how the Electoral College influences resource-allocation in a campaign but also how polls and the positioning of candidates on a left–right continuum affect the strategies of candidates and the choices of voters. ■

Social Choice: The Impossible Dream

The basic question of *social choice*, of how groups can best arrive at decisions, has occupied social philosophers and political scientists for centuries. One primary example of a social-choice problem is the selection of a "good" voting system. Indeed, voting is a subject that lies at the very heart of representative government and participatory democracy.

Social-choice theory attempts to address the problem of finding good procedures that will turn individual preferences for different candidates into a single choice by the whole group. An example of such a choice would be the selection of a *winner* of an election. The goal is to find such procedures that will result in an outcome that "reflects the will of the people."

This search for good voting systems, as we shall see, is plagued by a variety of counterintuitive results and disturbing outcomes. In fact, it turns out that one can prove (mathematically) that no one will ever find a completely satisfactory voting system for three or more candidates.

The elections with which we are most familiar often involve only two candidates, and we will begin our discussion of voting systems with this two-candidate case.

There are, however, real-world situations in which elections must be held to choose a single winner from among three or more candidates, as in the presidential election of 2004 in which George W. Bush, John Kerry, and Ralph Nader were the candidates.

There are several methods that can be used to elect a single candidate from a choice of three or more, and we will investigate some in this chapter. Most

In the 2004 election, President George W. Bush was challenged by Massachusetts senator John Kerry. President Bush received a slim majority of the vote. The election results have been dogged by charges of irregularities in the important swing state of Ohio. *(Left: Reuters/Corbis; right: John Gress/Reuters/Corbis.)*

of these methods use a ballot in which a voter provides a rank ordering of the candidates (without ties) that indicates the order in which he or she prefers them.

⠿⠿⠿⠿ Definition of Preference List Ballot

A ballot consisting of such a rank ordering of candidates (which we often picture as a vertical list with the most preferred candidate on top and the least preferred on the bottom) is called a **preference list ballot** because it is a statement of the preferences of one of the individuals who is voting.

Preference list ballots allow voters to make a much clearer statement of their preferences than do ballots allowing a single vote. Preference list ballots are already used in a wide range of applications, such as rating football teams and scoring track meets.

Although we do not allow ties in a preference list ballot, most voting rules of interest will, in some elections, result in a tie for the win among two or more of the candidates. In the real world, the number of voters is often so large that ties seldom occur. Nevertheless, to avoid excessive annoyances in the theory we develop, and to simplify what we do in this chapter, we make the following assumption throughout:

⠿⠿⠿⠿ The Number of Voters Assumption

Throughout this chapter, we assume that the number of voters is odd.

9.1 Majority Rule and Condorcet's Method

When a choice is being made between two candidates, the first type of voting system to suggest itself is **majority rule:** Each voter indicates a preference for one of the two candidates and the one with the most votes wins. With two candidates, there is no real distinction between a ballot that indicates a voter's choice for one of the two candidates and what we have called a preference list ballot. The point is that we can, for example, identify a choice for A (however indicated) with the list that has A over B and a choice B with the list that has B over A.

Majority rule has at least three desirable properties:

1. All voters are treated equally. That is, if any two voters were to exchange (marked) ballots before submitting them, the outcome of the election would be the same.

2. Both candidates are treated equally. That is, if a new election were held and every voter were to reverse his or her vote, then the outcome of the previous election would be reversed as well.

3. It is **monotone.** If a new election were held and a single voter were to change his or her ballot from being a vote for the loser of the previous election to being a vote for the winner of the previous election, and everyone else voted exactly as before, then the outcome of the new election would be the same as the outcome of the previous election.

It is easy to devise voting systems for two candidates in which these fail, but each such voting system quickly reveals its undesirability. For example, condition 1 is not satisfied by a *dictatorship* (in which all ballots except that of the dictator are ignored); condition 2 is not satisfied by *imposed rule* (in which candidate X wins regardless of who votes for whom); and condition 3 is not satisfied by *minority rule* (in which the candidate with the fewest votes wins).

But maybe there are voting systems in the two-candidate case that are superior to majority rule in the sense of satisfying the three properties just listed *and* some other properties that we might also wish to have satisfied. This, however, turns out not to be the case. In 1952, Kenneth May proved the following:

May's Theorem

Among all two-candidate voting systems that never result in a tie, majority rule is the *only* one that treats all voters equally, treats both candidates equally, and is monotone.

This is an important and elegant result. Thus, mathematical reasoning spares us the trouble of searching for a better voting system for two candidates.

But what if there are three or more candidates? Perhaps we can design a voting system for this situation that, in some way, builds on the success of majority rule in the two-candidate case. In point of fact, there does exist a voting system that arises from precisely this hope, and it is known today as **Condorcet's method.**

Our description of Condorcet's method begins with the observation that if we have a sequence of preference list ballots, then—for each pair of candidates—we can determine who the winner would have been had the election involved only these two in a one-on-one contest using majority rule.

To illustrate this notion of a one-on-one contest, consider the following preference list ballots:

Rank	Number of Voters (3)		
First	*A*	*B*	*C*
Second	*B*	*C*	*A*
Third	*C*	*A*	*B*

In this election, candidate *A* would defeat candidate *B* in a one-on-one contest (two votes to one), while *B* would, in turn, defeat *C* in a one-on-one contest, again by a score of 2 to 1. We'll return to this example in a moment, but we now have at hand all we need to describe Condorcet's voting system for three or more candidates.

Description of Condorcet's Method

With the voting system known as Condorcet's method, a candidate is a winner precisely when he or she would, on the basis of the ballots cast, defeat every other candidate in a one-on-one contest using majority rule.

Historically, the voting system we are calling Condorcet's method dates back at least to Ramon Llull in the thirteenth century (see Spotlight 9.1). It was rediscovered and popularized in the eighteenth century by the Marquis de Condorcet (1743–1794).

EXAMPLE 1   Condorcet's Method

Suppose we have four candidates (*GB, AG, RN,* and *PB,* with these initials chosen for a soon-to-be-revealed reason) and the following sequence of preference list ballots, where the heading of "6" indicates that 6 of the 15 voters hold the ballot with *GB* over *PB* over *AG* over *RN,* the heading of "5" indicates that 5 of the 15 voters hold the ballot with *AG* over *RN* over *GB* over *PB,* and so on.

Rank	Number of Voters (15)			
	6	5	3	1
First	GB	AG	RN	PB
Second	AG	RN	AG	GB
Third	PB	GB	GB	AG
Fourth	RN	PB	PB	RN

We claim that *AG* is the winner in this election if we use Condorcet's method. Let's check the one-on-one scores:

AG versus *GB*: *AG* is over *GB* on $5 + 3 = 8$ of the ballots, while the reverse is true on $6 + 1 = 7$ of the ballots. Thus, *AG* defeats *GB* by a score of 8 to 7.

AG versus *RN*: *AG* is over *RN* on $6 + 5 + 1 = 12$ of the ballots, while the reverse is true on 3 of the ballots. Thus, *AG* defeats *RN* by a score of 12 to 3.

AG versus *PB*: *AG* is over *PB* on $6 + 5 + 3 = 14$ of the ballots, while the reverse is true on 1 of the ballots. Thus, *AG* defeats *PB* by a score of 14 to 1.

This shows that *AG* is the winner using Condorcet's method. ■

Like majority rule, Condorcet's method satisfies some very desirable properties, as we'll see later in this section. But it also has a tragic flaw, and this flaw is called **Condorcet's voting paradox.**

Condorcet's Voting Paradox

With three or more candidates, there are elections in which Condorcet's method yields *no* winners. In particular, the following ballots (often called the "Condorcet voting paradox ballots") constitute an election in which Condorcet's method yields no winner.

Rank	Number of Voters (3)		
First	A	B	C
Second	B	C	A
Third	C	A	B

The Condorcet voting paradox ballots given above are the same ones we used earlier in illustrating the notion of a one-on-one contest. We pointed out then that *A* defeats *B* one on one and *B* defeats *C* one on one. The additional observation needed is that we also have *C* defeating *A* one on one. Thus *A* cannot be a winner using Condorcet's method (it loses to *C*), *B* cannot be a winner (it loses to *A*), and *C* cannot be a winner (it loses to *B*). We will revisit Condorcet's voting paradox in section 9.3.

SPOTLIGHT 9.1   The Historical Record

The following letter was written by Friedrich Pukelsheim of the University of Augsburg, Germany. He is imagining what Ramon Llull (1232–1316) might say if he were alive today.

Dear Editors:

It is my distinct pleasure to respond "from the beyond" to your kind invitation to set the historical record straight. I was born in 1232 on the Island of Mallorca in the Mediterranean Sea, which in your times is known as a popular tourist place. In my days it was a strong political center of that part of the world, with a population that was a mix of Christians, Jews, and Muslims. It was my dream to persuade people of the virtues of Christian belief by relying, not on force, but on reason.

Unfortunately, people did not find it easy to follow my arguments, so I was more than pleased to discover some down-to-earth applications, including an election system. My idea was to oppose every pair of candidates, one on one, and ask the electors whom of the two they would prefer—very much like a medieval jousting tournament. But how to combine the results from all the duels into a winner of the election? I first proposed electing the candidate who won the most duels, then later suggested a system of successive eliminations.

I wrote three papers on the topic, the second of which I "smuggled" into my novel *Blanquerna* in 1283. More than a century after my death, in 1428, the young German scholar Nicolaus Cusanus (1401–1464) journeyed to Paris to read my works in libraries there. He even copied out the third of my electoral writings, which I had completed on 1 July 1299 in Paris, and his manuscript is the only copy handed down to your days. Reading my papers, Cusanus was inspired to invent his own electoral system. Did he not understand mine, or just find it inadequate? Who knows?

While I had been concerned with electing church officials, Cusanus sought a system to elect the Holy Roman Emperor. In his system, each elector assigns each candidate a rank score, with the lowest candidate getting a score of 1, the second lowest a score of 2, and the best candidate the highest score possible, that is, 10 when there are 10 candidates. The scores are totaled for each candidate and the candidate with the highest score wins. If you are a soccer player or a hockey player, you will have a good sense for one difference between our systems: Whereas I count victories, Cusanus adds up goals. Cusanus applauded himself for having invented an absolutely ingenious and novel electoral system.

Notice that, because of our assumption that the number of voters is odd, Condorcet's method yields either no winner or a unique winner (see Exercise 5).

It is tempting at this point to suggest modifying Condorcet's method as we have presented it by declaring all the candidates to be tied for the win if there is no candidate who defeats each of the others one on one. The drawback to this modification is that a number of the upcoming desirable properties possessed by Condorcet's method then evaporate. We'll explore this in the exercises.

Also, I advocated open voting, whereas Cusanus favored a secret ballot. He was concerned that voters might sell their votes, or that the candidates might pressure the voters. Well, that certainly happened all of the time in elections for worldly authorities! But for election to clerical office, I thought it good enough if electors take an oath to vote for the most worthy candidate and submit themselves to the social control that comes with an open election.

Cusanus was famous in his times, as I was in mine, but fame indeed is transitory. Sure enough, my electoral system was reinvented by the Marquis de Condorcet (1743–1794), and Cusanus's system was proposed afresh by the Chevalier de Borda (1733–1799)—neither of whom, I am sure, wasted a thought on the possibility that "their" systems might already be on record. But, as my works had fallen into oblivion as had those of Cusanus, neither Condorcet nor Borda should be blamed for failing to acknowledge our priority.

My first electoral paper—actually the one that is longest and most detailed, written around 1280—was rediscovered only in 2000, filed away in the Vatican Library. How would you feel if your work attracts fresh attention after more than 700 years? Actually, I am utterly pleased that mine has resurfaced at last! The text was excavated by a mathematician interested in voting systems, Friedrich Pukelsheim of the University of Augsburg, Germany. Since the text is handwritten in Latin, handling it became an interdisciplinary project that brought together exports on medieval manuscripts, Church Latin and theology, and even computer scientists. As a result, my electoral writings are now on the Internet (in the original and in translations into English and German) at www.uni-augsburg.de/llull/.

Looking back on my lack of success in preaching peace among Christians, Jews, and Muslims, and all the writing and copying by hand of my works, I hope you can appreciate how highly I value the printed book (such as this one) and, even more, instant communication worldwide over the Internet. May that ease of communication help facilitate the religious peace that I so dearly sought.

Yours truly,
Ramon Llull (1232–1316)
Left Choir Chapel
San Francisco Cathedral
Palma de Mallorca

9.2 Other Voting Systems for Three or More Candidates

With three or more candidates, we find no shortage of additional procedures that suggest themselves and that seem to represent perfectly reasonable ways to choose a winner. Closer inspection, however, reveals shortcomings with all of these. We illustrate this with a consideration of several well-known procedures. Additional procedures (and additional shortcomings) can be found in the exercises.

Plurality Voting and the Condorcet Winner Criterion

In **plurality voting,** only first-place votes are considered. Thus, while we will consider plurality voting in the context of preference list ballots, a ballot here might just as well be a single vote for a single candidate. The candidate with the most votes wins, even though this may be considerably fewer than one-half the total votes cast. This is perhaps the most common system in use today. It is how the voters in Florida chose George W. Bush over Al Gore, Ralph Nader, and Patrick Buchanan in the presidential election of 2000.

EXAMPLE 2   Plurality Voting and the 2000 Presidential Election

On the evening of December 12, 2000, Al Gore conceded the presidential election of 2000 to George W. Bush, thus bringing to a close one of the most remarkable elections in modern times. The outcome, ultimately decided in the electoral college, came down to which of Bush or Gore would carry Florida. With more than six million votes cast in Florida, the ultimate margin of victory for George W. Bush was only a few hundred votes.

There is little doubt that if the 2000 presidential election had pitted Al Gore solely against any one of the other three candidates, then Gore would have won both the election in Florida and the presidency. The point is that while most of the Buchanan supporters would have voted for Bush, the far more numerous Nader supporters would have gone largely for Gore. In fact, the illustration of Condorcet's method that we gave in Example 1 is a simplified version of this Florida election (with *GB* standing for George Bush, *AG* for Al Gore, *PB* for Patrick Buchanan, and *RN* for Ralph Nader).

Thus, although plurality voting led to Bush's winning the 2000 election in Florida (and hence the presidency), Gore was, in this example, what is called a **Condorcet winner:** He would have won the election if Condorcet's method had been used. ■

Governor George W. Bush and Vice President Al Gore debate the issues before the 2000 election, possibly the most controversial election in U.S. history. Gore was the Condorcet winner of the election, but Bush eked out a victory that relied on the rules of the Electoral College. Many voters were suddenly put on notice that the U.S. Constitution makes the election of the president indirect—and not a pure expression of the majority's choice. *(Reuters/Corbis.)*

Definition of Condorcet Winner Criterion (CWC)

A voting system is said to satisfy the **Condorcet winner criterion** (CWC) provided that, for every possible sequence of preference list ballots, either (1) there is no Condorcet winner (as is often the case) or (2) the voting system produces exactly the same winner for this election as does Condorcet's method.

The CWC is certainly a property that one would like to see satisfied. We record plurality voting's failure in this respect with the following.

Theorem: The Failure of the CWC with Plurality Voting

The presidential election of 2000—and, in particular, the Florida vote in that election—shows that plurality voting fails to satisfy the Condorcet winner criterion.

Perhaps a more fundamental drawback of plurality voting is the extent to which the ballots provide no opportunity for a voter to express any preferences except for naming his or her top choice. No use is made, for example, of the fact that a candidate may be no one's first choice but everyone's close second choice.

Finally, there is yet another shortcoming of plurality voting. It is subject to what is called **manipulability:** There are elections in which it is to a voter's advantage to submit a ballot that misrepresents his or her true preferences. For example, in the presidential election of 2000, many voters who ranked Ralph Nader or Patrick Buchanan over George W. Bush and Al Gore chose to vote for Bush or Gore rather than to "throw away" their vote on a candidate they felt had no chance. Condorcet's method, it turns out, is not manipulable, and this is one of its most desirable properties. We'll explore this further in the next chapter.

The Borda Count and Independence of Irrelevant Alternatives

In many elections that use preference list ballots, the goal is to arrive at a final group rank ordering of all the contestants that best expresses the desires of the electorate. The purpose is not only to determine the winner—say, the class valedictorian—but also to arrive at who finished second, third, and so on, as in the case of one's rank in the senior class. In other applications, such as an election to a hall of fame, the first few finishers each receive the award, while the remaining nominees are also-rans.

One common mechanism for achieving this objective is to assign points to each voter's rankings and then to sum these for all voters to obtain the total points for each candidate. If there are 10 candidates, for example, then we could assign 10 points to each first-place vote for a given candidate, 9 points for each second-place vote, 8 for each third-place vote, and so forth. The candidate with the highest total number of points is the winner. Subsequent positions are assigned to those with the next-highest tallies.

> ## Description of Rank Methods and the Borda Count
>
> A *rank method* of voting assigns points in a nonincreasing manner to the ordered candidates on each voter's preference list ballot and then sums these points to arrive at a group's final ranking. The special case in which there are n candidates with each first-place vote worth $n - 1$ points, each second-place vote worth $n - 2$ points, and so on down to each last-place vote worth zero points is known as the **Borda count.** The actual point totals are referred to as a candidate's **Borda score.**

The Borda count is named after Jean-Charles de Borda (1733–1799). He was a contemporary of Condorcet.

Rank methods other than the Borda count are common. For example, a track meet can be thought of as an "election" in which each event is a "voter" and each of the schools competing is a "candidate." If the order of finish in the 100-meter dash is school A, school B, school C, school D, then points are often awarded to each school as follows: 5 points for first place, 3 for second place, 2 for third place, and 1 for fourth place.

Sports polls often use point assignments that qualify as rank methods according to our definition. The following example provides an illustration of this.

EXAMPLE 3   Rank Methods and a Thoroughbred Poll

A thoroughbred poll, conducted by the National Thoroughbred Racing Association, was reported as shown in the clipping on the next page.

An interesting question is whether or not this is a ranking system. For example, if it is, what are the candidates, and how many of them are there? In fact, this *can* be regarded as a ranking system, but the number of candidates is not 10. That is, although 10 horses appeared on each ballot, at least one ballot included the horse Voodoo Dancer, as evidenced by her receiving 1 point, whereas other ballots definitely did not.

To regard this as a ranking system, we must consider the set of candidates to be the entire set of eligible horses. We must also interpret each ballot as listing all horses other than that voter's top 10 in, say, alphabetical order and coming below that voter's top 10. The point assignments are then as in the newspaper clipping, except that we also assign 0 points for an 11th-place vote, 0 points for a 12th-place vote, and so on. This is why, in the definition of rank method, we said "assigns points in a *nonincreasing* manner" instead of "assigns points in a *decreasing* manner."

Thoroughbred poll
NTRA

NEW YORK—The thoroughbred poll conducted by the National Thoroughbred Racing Association. Rankings based on the votes of sports and thoroughbred racing media representatives on a 10-9-8-7-6-5-4-3-2-1 basis, with first-place votes in parentheses.

	Pts
1. Kona Gold (12)	199
2. Lido Palace (8)	194
3. Albert The Great	141
4. Tiznow	138
5. Flute (1)	119
6. With Anticipatn	78
7. Skimming	76
8. Include (1)	58
9. Officer	35
10. Guided Tour	33

Others receiving votes: Aptitude 19, Macho Uno 18, England's Legend 16, Silvano 12, Xtra Heat 12, Captain Steve 10, Came Home 7, Bienamado 6, Bet On Sunshine 6, Caller One 5, Touch Tone 5, Broken Vow 4, Tranquility Lake 4, E Dubai 3, Fleet Renee 3, Hap 3, El Corredor 2, Say Florida Sandy 2, Dixie Dot Com 1, Voodoo Dancer 1.

We can use this poll to illustrate how total points are arrived at with a ranking method. Let's see how Kona Gold might have earned the 199 points it is reported to have in the poll. We know that it had 12 first-place votes, and each first-place vote was worth 10 points. We also know that the total number of ballots was 22 (because they give the number of first-place votes that each horse received and $12 + 8 + 1 + 1 = 22$).

Thus, one possibility is that Kona Gold received:

12 first-place votes (at 10 points each)

4 second-place votes (at 9 points each)

3 third-place votes (at 8 points each)

2 fourth-place votes (at 7 points each)

1 sixth-place vote (at 5 points)

The total would then be:

$$120 + 36 + 24 + 14 + 5 = 199 \ \blacksquare$$

There is an easy way to calculate the Borda score of a candidate. You can count the number of occurrences of other candidate names that are below this candidate's name. For example, consider the following ballots:

Rank	Number of Voters (5)					Points
First	A	A	A	B	B	2
Second	B	B	B	C	C	1
Third	C	C	C	A	A	0

Because there are three candidates, each first-place vote is $n - 1$, or $3 - 1 = 2$, each second-place vote is $n - 2 = 1$, and each third-place vote is $n - 3 = 0$. If we were to calculate the Borda score of candidate B algebraically, we would say that B has two first-place votes, worth 2 points each (a total of 4 points), and three second-place votes, worth 1 point each (a total of 3 or more points). Thus, the Borda score of candidate B is $4 + 3 = 7$.

But instead of calculating this Borda score algebraically, we can mentally re-place each occurrence of a letter below B by a box, □, and simply count the boxes.

Rank	Number of Voters (5)				
First	A	A	A	B	B
Second	B	B	B	□	□
Third	□	□	□	□	□

Notice that there are seven boxes, giving us the correct value of 7 as the Borda score for candidate B. Of course, you don't actually have to draw any boxes. We are just emphasizing the fact that, in the counting process, it is "spaces" that we are counting, without regard to which letter occurs in the space. A quick glance at the original ballots (without the boxes) reveals that the Borda score of candidate A is 6 and the Borda score of candidate C is 2.

The Borda count certainly seems to be a reasonable way to choose a winner from among several candidates (or to arrive at a group ranking of the candidates). It also has its shortcomings, however, one of which is the failure of a property known as **independence of irrelevant alternatives.**

To describe this property, suppose that an election yields one candidate (call it A) as a winner and another candidate (call it B) as a nonwinner. Suppose that a new election is now held and that, although some of the voters may have changed their preference list ballots, no one who had previously ranked A and B changed his or her ballot to rank B over A now.

If this new election were to yield B as a winner, the new outcome would seem strange, especially because none of the relative individual preferences for A over B had changed in B's favor. The ballot changes responsible for the new outcome involve candidates *other than A or B*. One could argue that these other candidates ought to be irrelevant to the question of whether A is more desirable than B or B is more desirable than A.

Definition of Independence of Irrelevant Alternatives (IIA)

A voting system is said to satisfy independence of irrelevant alternatives (IIA) if it is impossible for a candidate B to move from nonwinner status to winner status unless at least one voter reverses the order in which he or she had B and the winning candidate ranked.

Condorcet's method satisfies IIA. That is, if we have a sequence of preference list ballots that yield A as a Condorcet winner and B as a nonwinner, then A defeats every other candidate, and B in particular, in a one-on-one contest according to these ballots. If no voter reverses the order in which he or she ranked A and B, then A will still defeat B one on one, and thus B remains a nonwinner.

The following illustration shows that the Borda count, on the other hand, fails to satisfy independence of irrelevant alternatives. Suppose the initial five ballots are as follows:

Rank	Number of Voters (5)				
First	A	A	A	C	C
Second	B	B	B	B	B
Third	C	C	C	A	A

Our counting procedure shows that the Borda scores are as follows:

Borda score of A is 6
Borda score of B is 5
Borda score of C is 4

The winner is A (with 6 points), and B is a nonwinner (with 5 points). But now suppose that the two voters on the right change their ballots by moving C down between A and B. The ballots then become

Rank	Number of Voters (5)				
First	A	A	A	B	B
Second	B	B	B	C	C
Third	C	C	C	A	A

Our counting procedure shows that the Borda scores are as follows:

Borda score of A is 6
Borda score of B is 7
Borda score of C is 2

The Borda count therefore now yields B as the winner (with 7 points). Thus, B has gone from being a nonwinner to being a winner, even though no one changed his or her mind about whether B is preferred to A, or vice versa. This establishes the following:

Theorem: The Failure of IIA with the Borda Count

The Borda count fails to satisfy independence of irrelevant alternatives.

Sequential Pairwise Voting and the Pareto Condition

In our voting-theoretic context, an **agenda** will be understood to be a listing (in some order) of the candidates. This listing is not to be confused with any of the preference list ballots, and, to avoid confusion, we will present agendas as horizontal lists and continue to present preference list ballots vertically.

Description of Sequential Pairwise Voting

Sequential pairwise voting starts with an agenda and pits the first candidate against the second in a one-on-one contest. The winner then moves on to confront the third candidate in the list, one on one. Losers are deleted. This process continues throughout the entire agenda, and the one remaining at the end wins.

For a given sequence of individual preference list ballots, the particular agenda chosen can greatly affect the outcome of the election, as we'll show in the next chapter. Nevertheless, we will see later in this chapter that sequential pairwise voting arises naturally in the legislative process. Notice also that because of our assumption that the number of voters is odd, there is always a unique winner with sequential pairwise voting.

EXAMPLE 4   Sequential Pairwise Voting

Assume we have four candidates and that the agenda is A, B, C, D. Consider the following sequence of three preference list ballots:

Rank	Number of Voters (3)		
First	A	C	B
Second	B	A	D
Third	D	B	C
Fourth	C	D	A

The first one-on-one pits A against B, and A wins by a score of 2 to 1 (meaning that two of the voters—the two on the left—prefer A to B, and one of the voters prefers B to A). Thus, B is eliminated and A moves on to confront C. Because C wins this one on one (by a score of 2 to 1), A is eliminated. Finally, C takes on D, and D wins by a score of 2 to 1. Thus, D is the winner. ■

There is something very troubling about the outcome of the preceding example, especially if you are candidate B. *Everyone* prefers B to D!

Theorem: The Failure of the Pareto Condition with Sequential Pairwise Voting

Sequential pairwise voting fails to satisfy what is called the **Pareto condition,** which says that if everyone prefers one candidate (in this case, B) to another candidate (D), then this latter candidate (D) should not be among the winners of the election.

The Pareto condition is named after Vilfredo Pareto (1848–1923), an Italian economist.

Runoff Systems and Monotonicity

The voting system known as the **Hare system,** which was introduced by Thomas Hare in 1861, is also known by names such as the "single transferable vote system." In 1862, John Stuart Mill described the Hare system as being "among the greatest improvements yet made in the theory and practice of government." Today, the system is used to elect public officials in Australia, Malta, the Republic of Ireland, and Northern Ireland.

Description of the Hare System

The Hare system proceeds to arrive at a winner by repeatedly deleting candidates that are "least preferred" in the sense of being at the top of the fewest ballots. If a single candidate remains after all others have been eliminated, it alone is the winner. If two or more candidates remain and all of these remaining candidates would be eliminated in the next round (because they all have the same number of first-place votes), then these candidates are declared to be tied for the win.

EXAMPLE 5   The Hare System

Suppose we have the following sequence of preference list ballots, where, as before, the heading of "5" indicates that 5 of the 13 voters hold the ballot with A over B over C, the heading of "4" indicates that 4 of the 13 voters hold the ballot with C over B over A, and so forth.

Rank	Number of Voters (13)			
	5	4	3	1
First	A	C	B	B
Second	B	B	C	A
Third	C	A	A	C

Candidates B and C have only 4 first-place votes (while A has 5). Thus, B and C are eliminated in the first round, and A wins the election. ■

In the preceding example, suppose that the voter in the last column moves candidate A up on his list. Let's look at the new election. Notice that, even though A won the last election, the only change we are making in ballots for the new election is one that is favorable to A. The ballots for the new election are as follows:

Rank	Number of Voters (13)			
	5	4	3	1
First	A	C	B	A
Second	B	B	C	B
Third	C	A	A	C

If we apply the Hare system again, only B is eliminated in round one, as it has 3 first-place votes to 4 for C and 6 for A. Thus, after this round, the ballots are as follows:

Rank	Number of Voters (13)			
	5	4	3	1
First	A	C	D	A
Second	C	A	A	C

We now have A on top of 6 lists and C on top of 7 lists. Thus, at stage two, A (our previous winner!) is eliminated and C is the winner of this new election.

Clearly, this is once again quite counterintuitive. Alternative A won the original election, the only change in ballots made was one favorable to A (and no one else), and then A lost the next election.

Theorem: Failure of Monotonicity with the Hare System

Example 5 shows that the Hare system does not satisfy **monotonicity,** which, with three or more candidates, says that if a candidate is a winner, and a new election is held in which the only ballot change made is for some voter to move the former winning candidate higher on his or her ballot, then the original winner should remain a winner.

The fact that the Hare system does not satisfy monotonicity is considered by many—and with good reason—to be a glaring defect. A 17-voter example in which only a single candidate is eliminated in the first round can also be used to show that the Hare system does not satisfy monotonicity—see Exercise 26. For an even more glaring version of this defect, one in which alternative A goes from winning to losing because voters move A from last place on their ballots to first place on their ballots, see Exercise 27.

In spite of these drawbacks, the Hare system is used in important ways today. For example, it is essentially the method that was used to choose Sydney, Australia, as the site of the 2000 Summer Olympics. Beijing would have been the plurality winner, but after the elimination of Istanbul, Berlin, and Manchester (in that order), Sydney defeated Beijing by a vote of 45 to 43. Four years later, Beijing finally did prevail; it will be the site of the 2008 Summer Olympics.

There are other runoff systems, some more frequently used than the Hare system. One such example is the following.

Description of the Plurality Runoff Method

Plurality runoff is the voting system in which there is a runoff (that is, a new election using the same ballots) between the two candidates receiving the most first-place votes. If there are ties, then the runoff is among either those tied for the most first-place votes, or the lone candidate with the most first-place votes along with those tied for the second-most first-place votes (and plurality voting is used).

Alas, the plurality runoff method also is not monotone. Exercise 23 asks you to verify this by making use of the following ballots:

Rank	Number of Voters (13)				
	4	3	3	2	1
First	A	B	C	D	E
Second	B	A	A	B	D
Third	C	C	B	C	C
Fourth	D	D	D	A	B
Fifth	E	E	E	E	A

9.3 Insurmountable Difficulties: Arrow's Impossibility Theorem

All of the voting systems for three or more candidates that we have discussed turn out to be flawed in one way or another. You may well ask at this point why we don't simply present *one* voting method for the three-candidate case that has all the desirable properties we want to have satisfied. That is, after all, exactly what we did for the two-candidate case (with majority rule filling the bill, and being the only one to do so by May's theorem).

The answer to this question is extremely important. The difficulties in the three-candidate case are not in any way tied to a few particular systems that we present in a text such as this (or that we choose to use in the real world). The fact is, there are difficulties that will be present *regardless* of what voting system is used, and this applies even to voting systems not yet discovered.

Nothing in the remarkable body of work produced by Nobel laureate Kenneth J. Arrow of Stanford University is as well known or widely acclaimed as the result known as **Arrow's impossibility theorem** (see Spotlight 9.2).

Arrow's Impossibility Theorem

With three or more candidates and any number of voters, there does not exist—and there never will exist—a voting system that always produces a winner, satisfies the Pareto condition and independence of irrelevant alternatives, and is not a dictatorship.

Arrow's theorem isn't obvious, and we won't be saying anything about the proof. But we can state and prove a much weaker result of some interest in its own right. This version is taken from the 1995 text *Mathematics and Politics,* cited in the Suggested Readings, and replaces Arrow's assumption of the Pareto condition and nondictatorship by the Condorcet winner criterion.

Theorem: A Weak Version of Arrow's Impossibility Theorem

With three or more candidates and an odd number of voters, there does not exist—and there never will exist—a voting system that satisfies both the Condorcet winner criterion and independence of irrelevant alternatives and that always produces at least one winner in every election.

To see why this is true, we'll handle only the case of exactly three voters. Our plan will be to assume that we have some kind of hypothetical voting

SPOTLIGHT 9.2   Kenneth J. Arrow

For centuries, mathematicians have been in search of a prefect voting system. Finally, in 1951, economist Kenneth Arrow proved that finding an absolutely fair and decisive voting system is impossible. Arrow is the Joan Kenney Professor of Economics, as well as a professor of operations research, at Stanford University. In 1972, he received the Nobel Memorial Prize in Economic Science for his outstanding work in the theory of general economic equilibrium. His numerous other honors include the 1986 von Neumann Theory Prize for his fundamental contributions to the decision sciences. He has served as president of the American Economic Association, the Institute of Management Sciences, and other organizations. Dr. Arrow talks about the process by which he developed his famous impossibility theorem and his ideas on the laws that govern voting systems:

Kenneth Arrow

My first interest was in the theory of corporations. In a firm with many owners, how do the owners agree when they have different opinions, for example, about the prospects of the company? I was thinking of stockholders. In the course of this, I realized that there was a paradox involved—that majority voting can lead to cycles. I then dropped that discussion because I was frustrated by it.

I happened to be working with The RAND Corporation one summer about a year or two later. They were very interested in applying concepts of rationality, particularly of game theory, to military and diplomatic affairs. That summer, I felt not like an economist but instead like a general social scientist or a mathematically oriented social scientist. There was tremendous interest in game theory, which was then new.

Some there asked me, "What does it mean in terms of national interest?" I said, "Oh, that's a very simple matter," and he said, "Well, why don't you write us a little memorandum on the subject." Trying to write that memorandum led to a sharper formulation of the social-choice question, and I realized that I had been thinking of it earlier in that other context.

I think that society must choose among a number of alternative policies. These policies may be thought of as quite comprehensive, covering a number of aspects: foreign policy, budgetary policy, or whatever. Now, each individual member of the society has a preference, or a set of preferences, over these alternatives. I guess that you can say one alternative is better than another. And these individual preferences have a property I call rationality or consistency, or more specifically,

(continued on page 358)

SPOTLIGHT 9.2 (*continued*)

what is technically known as transitivity: if I prefer *a* to *b*, and *b* to *c*, then I prefer *a* to *c*.

Imagine that society has to make these choices among a set. Each individual has a preference ordering, a ranking of these alternatives. But we really want society, in some sense, to give a ranking of these alternatives. Well, you can always produce a ranking, but you would like it to have some properties. One is that, of course, it be responsive in some sense to the individual rankings. Another is that when you finish, you end up with a real ranking, that is, something that satisfies these consistency, or transitivity, properties. And a third condition is that when choosing between a number of alternatives, all I should take into account are the preferences of the individuals among those alternatives. If certain things are possible and some are impossible, I shouldn't ask individuals whether they care about the impossible alternatives, only the possible ones.

It turns out that if you impose the conditions I just stated, there is no method of putting together the individual preferences that satisfies all of them.

The whole idea of the axiomatic method was very much in the air among anybody who studied mathematics, particularly among those who studied the foundations of mathematics. The idea is that if you want to find out something, to find the properties, you say, "What would I like it to be?" [You do this] instead of trying to investigate special cases. And I was really accustomed to this approach. Of course, the actual process did involve trial and error.

But I went in with the idea that there was some method of handling this problem. I started out with some examples. I had already discovered that these led to some problems. The next thing that was reasonable was to

write down a condition that I could outlaw. Then I constructed another example, another method that seemed to meet that problem, and something else didn't seem very right about it. Then I had to postulate that we have some other property. I found I was having difficulty satisfying all of these properties that I thought were desirable, and it occurred to me that they couldn't be satisfied.

After having formulated three or four conditions of this kind, I kept on experimenting. And lo and behold, no matter what I did, there was nothing that would satisfy these axioms. So after a few days of this, I began to get the idea that maybe there was another kind of theorem here, namely, that there was no voting method that would satisfy all the conditions that I regarded as rational and reasonable. It was at this point that I set out to prove it. And it actually turned out to be a matter of only a few days' work.

It should be made clear that my impossibility theorem is really a theorem [showing that] the contradictions are possible, not that they are necessary. What I claim is that given any voting procedure, there will be some possible set of preference orders for individuals that will lead to a contradiction of one of these axioms.

But you say, "Well, okay, since we can't get perfection, let's at least try to find a method that works well most of the time." Then when you do have a problem, you don't notice it as much. So my theorem is not a completely destructive or negative feature any more than the second law of thermodynamics means that people don't work on improving the efficiency of engines. We're told you'll never get 100% efficient engines. That's a fact—and a law. It doesn't mean you wouldn't like to go from 40% to 50%.

system that satisfies both the Condorcet winner criterion and independence of irrelevant alternatives and to show that, when confronted by the Condorcet voting paradox ballots, it produces *no* winner.

The argument really comes in three separate, but extremely similar, pieces— one for each of the three candidates. Piece 1 argues that A can't be among the winners, piece 2 that B can't be among the winners, and piece 3 that C can't be among the winners. We'll do piece 1 and leave the others for you. The sequence of ballots that we are considering is the following, which we have already seen has no Condorcet winner.

Rank	Number of Voters (3)		
First	A	B	C
Second	B	C	A
Third	C	A	B

Our starting point, however, will be to ask what our hypothetical voting rule must do when confronted by a slightly different sequence of ballots.

Rank	Number of Voters (3)		
First	A	C	C
Second	B	B	A
Third	C	A	B

Here, C is clearly a Condorcet winner, and thus it must be the unique winner of the election contested under our hypothetical voting rule. Therefore, C is a winner and A is a nonwinner (for *this* sequence of ballots).

However, because our hypothetical voting rule satisfies independence of irrelevant alternatives, we know that A will remain a nonwinner as long as no one reverses his or her ordering of A and C. But to arrive at the voting paradox ballots we can move B (the candidate that is irrelevant to A and C) up one slot in the second voter's list.

Thus, because of IIA, we know that A is a nonwinner when our voting rule is confronted by the voting paradox ballots. This is one-third of the argument. As we mentioned before, similar arguments (see Exercise 31) show that B and C are also nonwinners when our voting rule is confronted by the voting paradox ballots.

9.4 A Better Approach? Approval Voting

Elections in which there are only two candidates present no problem. Majority rule is, as we have seen, an eminently successful voting system in both theory and practice. If there are three or more candidates, however, the situation changes quite dramatically. While several voting systems suggest themselves (plurality, the

Borda count, sequential pairwise voting, and the Hare system), each fails to satisfy one or more desired properties (the Condorcet winner criterion, independence of irrelevant alternatives, the Pareto condition, and monotonicity). Manipulability is an ever-present problem, as we'll see in the next chapter. Moreover, when all is said and done, Arrow's impossibility theorem says that any search for an ideal voting system of the kind we have discussed is doomed to failure.

Where does this leave us? More than intellectual issues are at stake here: More than 550,000 elected officials serve in approximately 80,000 governments in the United States. Whether it is a small academic department voting on the best senior thesis or a democratic country electing a new leader, multicandidate elections will be contested in one way or another. If there is no perfect voting system—and perhaps not even a best voting system (whatever that may mean; that is, best in what way?)—what can we do?

Perhaps the answer is that different situations lend themselves to different voting systems, and what is required is a judicious blend of common sense with an awareness of what the mathematical theory has to say. For example, while both the Hare system and the Borda count are subject to manipulability, it seems easier to manipulate the latter. Thus, people may tend to vote more sincerely, rather than strategically, if the Hare system is used instead of the Borda count. This may be a consideration when choosing a voting system for a faculty governance system, for example.

For national political elections, there are also practical considerations. The kind of ballot we are considering (a preference list ballot) is certainly more complicated than the ballots we now employ, and preference list ballots cannot be used with existing voting machines. There is, however, a voting system that avoids the practical difficulties caused by the type of ballot being used that has much else to commend it. It is called **approval voting.**

Description of Approval Voting

Under approval voting, each voter is allowed to give one vote to as many of the candidates as he or she finds acceptable. No limit is set on the number of candidates for whom an individual can vote. Voters show disapproval of other candidates simply by not voting for them. The winner under approval voting is the candidate who receives the largest number of approval votes. This approach is also appropriate in situations where more than one candidate can win, for example, in electing new members to an exclusive society such as the National Academy of Sciences or the Baseball Hall of Fame.

Approval voting was proposed independently by several analysts in the 1970s. Probably the best-known official elected by approval voting today is the secretary general of the United Nations. In the 1980s, several academic and professional societies initiated the use of approval voting. Examples include the Institute of Electrical and Electronics Engineers (IEEE), with about 400,000 members, and the National Academy of Sciences. In Eastern Europe and some former Soviet republics, approval voting has been used in the form wherein one disapproves of (instead of approving of) as many candidates as one wishes.

Is approval voting the perfect voting system? Certainly not. For example, the type of ballot used limits the extent to which voter preferences can be expressed. However, it is certainly a voting system with much potential, and the reader wishing to explore it in more detail can start with Brams and Fishburn's 1983 monograph, listed in the Suggested Readings.

REVIEW VOCABULARY

Agenda An ordering of the candidates to be considered. Often used in sequential pairwise voting.

Approval voting A method of electing one or more candidates from a field of several in which each voter submits a ballot that indicates which candidates he or she approves of. Winning is determined by the total number of approvals a candidate obtains.

Arrow's impossibility theorem Kenneth J. Arrow's discovery that any voting system can give undesirable outcomes.

Borda count A voting system for elections with several candidates in which points are assigned to voters' preferences and these points are summed for each candidate to determine a winner. The actual point totals are referred to as a candidate's **Borda score.**

Condorcet's method A voting system for elections with several candidates in which a candidate is a winner precisely when he or she would, on the basis of the ballots cast, defeat every other candidate in a one-on-one contest.

Condorcet winner A Condorcet winner in an election is a candidate who, based on the ballots, would have defeated every other candidate in a one-on-one contest.

Condorcet winner criterion (CWC) A voting system satisfies the Condorcet winner criterion if, for every election in which there is a Condorcet winner, it wins the election when that voting system is used.

Condorcet's voting paradox The observation that there are elections in which Condorcet's method yields no winner.

Hare system A voting system for elections with several candidates in which candidates are successively eliminated in an order based on the number of first-place votes.

Independence of irrelevant alternatives (IIA) A voting system satisfies independence of irrelevant alternatives if the only way a candidate (call him A) can go from losing one election to being among the winners of a new election (with the same set of candidates and voters) is for at least one voter to reverse his or her ranking of A and the previous winner.

Manipulability A voting system is subject to manipulability (or is manipulable) if there are elections in which it is to a voter's advantage to submit a ballot that misrepresents his or her true preferences.

Majority rule A voting system for elections with two candidates (and an odd number of voters) in which the candidate preferred by more than half the voters is the winner.

May's theorem Kenneth May's discovery that, for two alternatives and an odd number of voters, majority rule is the only voting system satisfying three natural properties.

Monotonicity A voting system satisfies monotonicity provided that ballot changes favorable to one candidate (and not favorable to any other candidate) can never hurt that candidate.

Pareto condition A voting system satisfies the Pareto condition provided that every voter's ranking of one candidate higher than another precludes the possibility of this latter candidate winning.

Plurality runoff A voting system for elections with several candidates in which, assuming there are no ties, there is a runoff between the two candidates receiving the most first-place votes.

Plurality voting A voting system for elections with several candidates in which the candidate with the most first-place votes wins.

Preference list ballot A ballot that ranks the candidates from most preferred to least preferred, with no ties.

Sequential pairwise voting A voting system for elections with several candidates in which one starts with an agenda and pits the candidates against each other in one-on-one contests (based on preference list ballots), with losers being eliminated as one moves along the agenda.

✔ SKILLS CHECK

1. A preference list ballot

(a) indicates only a voter's top choice.

(b) is a rank ordering of the candidates, with no ties.

(c) will often have ties.

2. To say that a voting system treats all voters equally means that

(a) they all submit some kind of ballot.

(b) if any two voters exchange ballots, the election outcome changes.

(c) if any two voters exchange ballots, the election outcome is unchanged.

3. To say that a voting system for two candidates treats both candidates equally means that

(a) each wins if he or she receives all the votes.

(b) if all voters reverse their ballots, the election outcome changes.

(c) if any two voters exchange ballots, the election outcome is unchanged.

4. A two-candidate voting system is monotone if

(a) there are no highs or lows.

(b) a switch in a ballot from being a vote for the loser to being a vote for the winner doesn't change the election outcome.

(c) it treats both candidates the same.

5. May's theorem says that, with an odd number of voters, among all two-candidate voting systems that never result in a tie, majority rule is the only one that

(a) treats both candidates equally.

(b) treats both candidates equally and all voters equally.

(c) treats both candidates equally and all voters equally and is monotone.

6. Which of the following does not satisfy exactly two of the conditions in May's theorem?

(a) A dictatorship

(b) Imposed rule

(c) Minority rule

(d) None of the above

7. The winner with Condorcet's method is the candidate that

(a) has the most first-place votes.

(b) defeats every other candidate in a one-on-one contest.

(c) has the highest Borda score.

8. The flaw in Condorcet's method is that it

(a) fails to satisfy the Pareto condition.

(b) sometimes produces no winner at all.

(c) is not monotone.

(d) fails to satisfy independence of irrelevant alternatives.

9. Condorcet's voting paradox refers to the fact that

(a) people vote even though an individual vote virtually never affects the outcome of an election.
(b) the statement "This statement is false" can be neither true nor false.
(c) there are elections in which there is no winner using Condorcet's method.

10. With plurality voting, the winner is

(a) the candidate with the most first-place votes.
(b) the candidate who defeats every other candidate in a one-on-one contest.
(c) the candidate with the fewest last-place votes.

11. George W. Bush's defeat of Al Gore in the state of Florida in the 2000 presidential election shows that

(a) plurality voting does not satisfy the Condorcet winner criterion.
(b) majority rule is not monotone.
(c) the Borda count does not satisfy independence of irrelevant alternatives.

12. With the Borda count, the election winner is

(a) the candidate with the most first-place votes.
(b) the candidate who defeats every other candidate in a one-on-one contest.
(c) the candidate with the highest Borda score.

13. Instead of assigning points and doing arithmetic, the Borda score of a candidate can be found by

(a) scanning the ballots and counting the number of occurrences of other candidates below that one.
(b) counting the number of first-place votes and multiplying by 4.
(c) counting the number of candidates that it defeats one on one.

14. Independence of irrelevant alternatives says that a nonwinner can never switch to being a winner unless at least one voter changes his or her ballot in a way that

(a) results in an election with fewer candidates.
(b) reverses the order in which this nonwinner and the winner were ranked.
(c) favors the winning candidate and no other.

15. The Borda count fails to satisfy

(a) monotonicity.
(b) the Pareto condition.
(c) independence of irrelevant alternatives.

16. Sequential pairwise voting is the voting system in which

(a) one-on-one contests take place according to an ordering of the candidates called an "agenda."
(b) the winner is the candidate who wins the most pairwise one-on-one contests.
(c) voters line up two abreast at the polling booths.

17. Sequential pairwise voting fails to satisfy

(a) monotonicity.
(b) the Pareto condition.
(c) the Condorcet winner criterion.

18. Both the Hare system and the plurality runoff method are defective in that

(a) they sometimes yield no winner at all.
(b) they are not monotone.
(c) they fail to satisfy the Pareto condition.

19. Arrow's theorem says that with three or more candidates and any number of voters, there is no voting system that

(a) is not a dictatorship.
(b) satisfies independence of irrelevant alternatives and is not a dictatorship.
(c) satisfies the Pareto condition and independence of irrelevant alternatives, and is not a dictatorship.
(d) always produces a winner, satisfies the Pareto condition and independence of irrelevant alternatives, and is not a dictatorship.

20. The weak version of Arrow's theorem asserts that, with three or more candidates and an odd number of voters, there is no voting system that

(a) satisfies the Condorcet winner criterion.
(b) satisfies the Condorcet winner criterion and independence of irrelevant alternatives.
(c) satisfies the Condorcet winner criterion and independence of irrelevant alternatives, and always produces at least one winner in every election.

EXERCISES   ■ Challenge   ◆ Discussion

Majority Rule and Condorcet's Method

1. In a few sentences, explain why minority rule (the voting procedure for two alternatives that is described on page 341) satisfies conditions (1) and (2) on page 341, but not (3).

2. In a few sentences, explain why imposed rule (the voting procedure for two alternatives that is described on page 341) satisfies conditions (1) and (3) on page 341, but not (2).

3. In a few sentences, explain why a dictatorship (the voting procedure for two alternatives that is described on page 341) satisfies conditions (2) and (3) on page 341, but not (1).

4. Find (or invent) a voting rule for two alternatives that satisfies

(a) condition (1) on page 341, but neither (2) nor (3).
(b) condition (2) on page 341, but neither (1) nor (3).
(c) condition (3) on page 341, but neither (1) nor (2).

5. In a sentence or two, explain why it's impossible, with an odd number of voters, to have two distinct candidates win the same election using Condorcet's method.

6. Construct a real-world example (perhaps involving yourself and two friends) where the individual preference lists for three alternatives are as in the voting paradox of Condorcet.

Other Voting Systems for Three or More Candidates

7. Plurality voting is illustrated by the 1980 U.S. Senate race in New York among Alfonse D'Amato (*D*, a conservative), Elizabeth Holtzman (*H*, a liberal), and Jacob Javits (*J*, also a liberal). Reasonable estimates (based largely on exit polls) suggest that voters ranked the candidates according to the following table:

22%	23%	15%	29%	7%	4%
D	*D*	*H*	*H*	*J*	*J*
H	*J*	*D*	*J*	*H*	*D*
J	*H*	*J*	*D*	*D*	*H*

(a) Is there a Condorcet winner?
(b) Who won using plurality voting?

8. (Everyone wins.) Consider the following set of preference lists:

Rank	Number of Voters (9)						
	3	1	1	1	1	1	1
First	*A*	*A*	*B*	*B*	*C*	*C*	*D*
Second	*D*	*B*	*C*	*C*	*B*	*D*	*C*
Third	*B*	*C*	*D*	*A*	*D*	*B*	*B*
Fourth	*C*	*D*	*A*	*D*	*A*	*A*	*A*

Note that the first list is held by three voters, not just one. Calculate the winner using

(a) plurality voting.
(b) the Borda count.
(c) the Hare system.
(d) sequential pairwise voting with the agenda *A, B, C, D.*

9. Consider the following set of preference lists:

Rank	Number of Voters (7)				
	2	2	1	1	1
First	*C*	*D*	*C*	*B*	*A*
Second	*A*	*A*	*D*	*D*	*D*
Third	*B*	*C*	*A*	*A*	*B*
Fourth	*D*	*B*	*B*	*C*	*C*

Calculate the winner using

(a) plurality voting
(b) the Borda count.
(c) the Hare system.
(d) sequential pairwise voting with the agenda *B, D, C, A.*

10. Consider the following set of preference lists:

Rank						

	Number of Voters (8)					
Rank	2	2	1	1	1	1
First	A	E	A	B	C	D
Second	B	B	D	E	E	E
Third	C	D	C	C	D	A
Fourth	D	C	B	D	A	B
Fifth	E	A	E	A	B	C

Calculate the winner using

(a) plurality voting.
(b) the Borda count.
(c) the Hare system.
(d) sequential pairwise voting with the agenda B, D, C, A, E.

11. Consider the following set of preference lists:

	Number of Voters (5)				
Rank	1	1	1	1	1
First	A	B	C	D	E
Second	B	C	B	C	D
Third	E	A	E	A	C
Fourth	D	D	D	E	A
Fifth	C	E	A	B	B

Calculate the winner using

(a) plurality voting.
(b) the Borda count.
(c) the Hare system.
(d) sequential pairwise voting with the agenda A, B, C, D, E.

12. Consider the following set of preference lists:

	Number of Voters (7)				
Rank	2	2	1	1	1
First	A	B	A	C	D
Second	D	D	B	B	B
Third	C	A	D	D	A
Fourth	B	C	C	A	C

Calculate the winner using

(a) plurality voting.
(b) the Borda count.
(c) the Hare system.
(d) sequential pairwise voting with the agenda B, D, C, A.

13. Consider the following set of preference lists:

	Number of Voters (7)				
Rank	2	2	1	1	1
First	C	E	C	D	A
Second	E	B	A	E	E
Third	D	D	D	A	C
Fourth	A	C	E	C	D
Fifth	B	A	B	B	B

Calculate the winner using

(a) plurality voting.
(b) the Borda count.
(c) the Hare system.
(d) sequential pairwise voting with the agenda A, B, C, D, E.

14. Consider the following set of preference lists:

	Number of Voters (7)						
Rank	1	1	1	1	1	1	1
First	C	D	C	B	E	D	C
Second	A	A	E	D	D	E	A
Third	E	E	D	A	A	A	E
Fourth	B	C	A	E	C	B	B
Fifth	D	B	B	C	B	C	D

Calculate the winner using

(a) plurality voting.
(b) the Borda count.
(c) sequential pairwise voting with the agenda A, B, C, D, E.
(d) the Hare system.

15. An interesting variant of the Hare system was proposed by the psychologist Clyde Coombs. It operates exactly as does the Hare system, but instead of deleting alternatives with the fewest first-place votes, it deletes those with the most last-place votes.

(a) Use the Coombs procedure to find the winner if the ballots are as in Exercise 14.

(b) Show that for two voters and three alternatives, it is possible to have ballots that result in one candidate winning if the Coombs procedure is used and a tie between the other two if the Hare system is used.

◆ **16.** In a few sentences, explain why Condorcet's rule satisfies

(a) the Pareto condition.
(b) monotonicity.

◆ **17.** In a few sentences, explain why plurality voting satisfies

(a) the Pareto condition.
(b) monotonicity.

◆ **18.** In a few sentences, explain why the Borda count satisfies

(a) the Pareto condition.
(b) monotonicity.

◆ **19.** In a few sentences, explain why sequential pairwise voting satisfies

(a) the Condorcet winner criterion.
(b) monotonicity.

◆ **20.** In a few sentences, explain why the Hare system satisfies the Pareto condition.

◆ **21.** In a few sentences, explain why the plurality runoff method satisfies the Pareto condition.

■ **22.** Use the following ballots to show that the plurality runoff method does not satisfy the Condorcet winner criterion:

Rank	Number of Voters (5)		
	2	2	1
First	A	B	C
Second	C	C	B
Third	B	A	A

■ **23.** Use the following ballots to show that the plurality runoff method does not satisfy monotonicity:

Rank	Number of Voters (13)				
	4	3	3	2	1
First	A	B	C	D	E
Second	B	A	A	B	D
Third	C	C	B	C	C
Fourth	D	D	D	A	B
Fifth	E	E	E	E	A

24. Consider the following two elections among candidates A, B, and C:

Rank	Number of Voters (4)			
	1	1	1	1
First	A	A	B	C
Second	B	B	C	B
Third	C	C	A	A

Rank	Number of Voters (4)			
	1	1	1	1
First	A	A	B	B
Second	B	B	C	C
Third	C	C	A	A

(a) Use these two elections to show that plurality voting does not satisfy independence of irrelevant alternatives.

(b) Use these two elections to show that the Hare system does not satisfy independence of irrelevant alternative.

■ **25.** Construct ballots for the alternatives A, B, and C to show that the Borda count does not satisfy the Condorcet winner criterion.

26. Show that the nonmonotonicity of the Hare system can also be demonstrated by the following 17-voter, 4-alternative election. (In a number of recent books, this example is used to show the nonmonotonicity of the Hare system. The easier 13-voter, 3-alternative example given in the text was pointed out to us by Matt Gendron, an undergraduate at Union College.)

	Number of Voters (17)			
Rank	7	5	4	1
First	A	C	B	D
Second	D	A	C	B
Third	B	B	D	A
Fourth	C	D	A	C

27. The following example illustrates how badly the Hare system can fail to satisfy monotonicity. Consider the following sequence of preference lists:

	Number of Voters (21)			
Rank	7	6	5	3
First	A	B	C	D
Second	B	A	B	C
Third	C	C	A	B
Fourth	D	D	D	A

(a) Show that A is the unique winner if the Hare system is used.
(b) Find the winner using the Hare system in the new election wherein the three voters on the right all move A from last place on their preference lists to first place on their preference lists.

◆ **28.** In a few sentences, explain why, with an odd number of voters,

(a) sequential pairwise voting always yields a unique winner.
(b) we can never have exactly two winners with the Hare system.

◆ **29.** In a few sentences, explain why the plurality runoff method can never elect a candidate ranked last on a majority of ballots, assuming there are no ties for first or second place in the voting.

30. Produce ballots showing that plurality voting can, in fact, elect a candidate ranked last on a majority of the ballots.

Insurmountable Difficulties: Arrow's Impossibility Theorem

■ **31.** Complete the proof of the version of Arrow's theorem from the text by showing that neither B nor C can be a winner in the situation described. (Your argument will be almost word for word the same as the proofs in the text.)

A Better Approach? Approval Voting

32. Ten board members vote by approval voting on eight candidates for new positions on their board as indicated in the following table. An X indicates an approval vote. For example, Voter 1, in the first column, approves of candidates A, D, E, F, and G, and disapproves of B, C, and H.

	Voters									
Candidate	1	2	3	4	5	6	7	8	9	10
A	X	X	X			X	X	X		X
B		X	X	X	X	X	X	X	X	
C			X					X		
D	X	X	X	X	X		X	X	X	X
E	X		X		X		X		X	
F	X		X	X	X	X	X	X		X
G	X	X	X	X	X			X		
H		X		X		X		X		X

(a) Which candidate is chosen for the board if just one of them is to be elected?
(b) Which candidates are chosen if the top four are selected?
(c) Which candidates are elected if 80% approval is necessary and at most four are elected?
(d) Which candidates are elected if 60% approval is necessary and at most four are elected?

33. The 45 members of a school's football team vote on three nominees, A, B, and C, by approval voting for the award of "most improved player" as indicated in the following table. An X indicates an approval vote.

Nominee	Number of Voters (45)							
	7	8	9	9	6	3	1	2
A	X			X	X			X
B		X		X			X	X
C			X		X	X		X

(a) Which nominee is selected for the award?

(b) Which nominee gets announced as runner-up for the award?

(c) Note that two of the players "abstained," that is, approved of none of the nominees. Note also that one person approved of all three of the nominees. What would be the difference in the outcome if one were to "abstain" or "approve of everyone"?

WRITING PROJECTS

1. In the 2000 presidential election in Florida, the final results were as follows:

Candidates	Number of Votes	Percentage of Votes
Bush	2,911,872	49
Gore	2,910,942	49
Nader	97,419	2
Buchanan	17,472	0

Making reasonable assumptions about voters' preference schedules, give a one-page discussion of how the election might have turned out under the different voting methods discussed in this chapter.

2. Frequently in presidential campaigns, the winner of the first few primaries is given front-runner status that can lead to the nomination of his or her party. Moreover, there are often several candidates running in early primaries such as New Hampshire. In one page, consider a recent election and discuss how the nominating process might have proceeded through the campaign if approval voting had been used to decide primary winners.

SUGGESTED READINGS

BLACK, DUNCAN. *The Theory of Committees and Elections,* Kluwer, Dordrecht, The Netherlands, 1986. The historical highlights and developments of voting methods in the nineteenth and twentieth centuries are traced in this economist's volume.

BRAMS, STEVEN J., and PETER C. FISHBURN. *Approval Voting,* Birkhäuser, Boston, 1983. This volume is a research-level work on developments in the recently popular (but rediscovered) method now called approval voting. The first chapter, however, is an elementary exposition of this voting method and its uses.

NURMI, HANNU. *Comparing Voting Systems,* Reidel, Dordrecht, The Netherlands, 1987. This monograph provides an excellent treatment, at a somewhat more technical level, of the topics dealt with in this chapter.

SAARI, DONALD G. *Chaotic Elections! A Mathematician Looks at Voting,* American Mathematical Society, Providence, R.I., 2001. This expository book begins with the 2000 presidential election and discusses a number of paradoxical results in voting.

TAYLOR, ALAN D. *Mathematics and Politics: Strategy, Voting, Power, and Proof,* Springer-Verlag, New York, 1995. Chapters 5 and 10 give an expanded treatment of the topics considered here, with proofs included. This book is also intended for nonmajors.

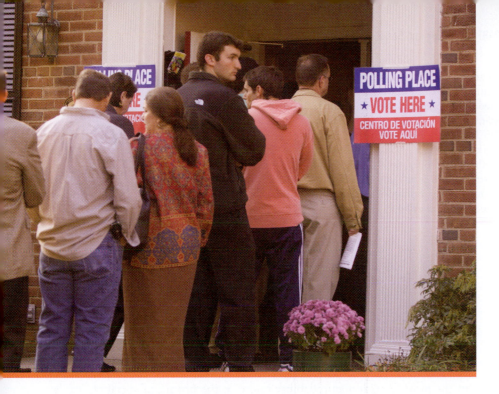

CHAPTER
10

IN THIS CHAPTER WE
COVER . . .

10.1 Majority Rule and
Condorcet's Method

10.2 Other Voting
Systems for Three or
More Candidates

10.3 Impossibility

10.4 The Chair's
Paradox

The Manipulability of Voting Systems

P eople know almost by instinct that, sometimes, you can achieve the election
result you prefer by submitting a ballot that misrepresents your actual preferences. This type of strategic voting is called **manipulation,** and a ballot that
misrepresents a voter's true preferences is referred to as an **insincere** or **disingenuous ballot.**

All three of these terms—manipulation, insincere, disingenuous—are widely used
in the social-choice literature, but in daily life we use these terms pejoratively—
they aren't exactly warm praise. In point of fact, your choice to manipulate a voting system is typically no more inherently evil than your submission of a sealed bid
for a lamp at an auction at a price considerably below its actual worth. "Strategy-
proof"—a term with considerably less negative content—is sometimes used in place
of "nonmanipulable," but "nonmanipulable" is more common and we'll stick
with it here.

Historical references to the manipulability of voting systems include a comment by the nineteenth-century mathematician C. L. Dodgson (1832–1898)—
better known by the pseudonym Lewis Carroll, under which he wrote *Alice's Ad-
ventures in Wonderland* (1865)—that voters have a tendency to "adopt a principle
of voting which makes it more of a game of skill than a true test of the wishes
of the electors," and that it would be "better for elections to be decided according to the wishes of the majority than of those who have the most skill at the
game."

Charles L. Dodgson was a mathematical lecturer at Oxford University. Dodgson, who used the pen name Lewis Carroll, wrote on mathematical topics and even manipulability. But he achieved greater fame for his satirical works. In the *Alice* books, he refers to the mathematical operations as Ambition, Distraction, Uglification, and Derision, and his characters play nonsensical, easily manipulated games. *(Bettmann/Corbis.)*

But the most famous manipulability quote in the history of social choice is Jean Charles de Borda's reply to a colleague who had pointed out to him how easily the Borda count can be manipulated. "My scheme," Borda replied, "is only intended for honest men!"

Let's look at an example to illustrate how the Borda count can be manipulated.

EXAMPLE 1   Manipulating the Borda Count

Suppose there are two voters and four candidates, and suppose the true preferences of the voters are reflected in the following ballots:

Voter 1	Voter 2
A	*B*
B	*C*
C	*A*
D	*D*

Using the Borda count with point values 3, 2, 1, 0 (or by counting the number of occurrences of other candidates below the one in question, as described in Section 9.2), we see that the Borda scores of the four candidates are as follows:

Borda score of *A* is 4

Borda score of *B* is 5

Borda score of *C* is 3

Borda score of *D* is 0

Thus, Candidate B wins this election. Voter 1, however, would have preferred to see Candidate A–his top choice, according to his true preferences–win this election rather than Candidate B, his second choice.

Assume that Voter 1 had known that Voter 2 planned to submit the ballot that he cast above. Could Voter 1 have secured a victory for Candidate A by submitting a disingenuous ballot?

The answer here, as we'll show, turns out to be yes. The intuition is fairly transparent: Voter 1 wants to pretend that B is not her second choice, but her last choice. Let's see if this is enough to bring about the desired switch in winner from B to A. The new ballots and Borda scores are as follows:

Voter 1	Voter 2
A	B
C	C
D	A
B	D

Borda score of A is 4

Borda score of B is 3

Borda score of C is 4

Borda score of D is 1

Close, but not quite what we wanted: Candidates A and C now tie for the win, and we wanted the winner to be just Candidate A. But a moment's inspection reveals that Voter 1 can achieve this if, in addition to plunging Candidate B to the bottom of her ballot, she also flip-flops C and D. That is, the desired ballots (and Borda scores) that yield Candidate A as the sole winner are as follows:

Voter 1	Voter 2
A	B
D	C
C	A
B	D

Borda score of A is 4

Borda score of B is 3

Borda score of C is 3

Borda score of D is 2 ■

In presenting an example of a voting system's susceptibility to manipulation, we will typically present two elections—the original one ("Election 1") in which we assume all ballots are sincere, and the one that contains a disingenuous ballot from some voter ("Election 2"). For example, if we collect the pieces of what we just did, this instance of manipulation of the Borda count could be succinctly presented as follows.

Election 1				Election 2		
Rank	Number of Voters (2)			Rank	Number of Voters (2)	
First	*A*	*B*		First	*A*	*B*
Second	*B*	*C*		Second	*D*	*C*
Third	*C*	*A*		Third	*C*	*A*
Fourth	*D*	*D*		Fourth	*B*	*D*

There are two aspects of manipulation taking place in this example that deserve comment.

First, there is only one voter (the voter on the left, in this example) changing his or her ballot—we call this a **unilateral change** in ballot. An example involving a unilateral change of ballot is sometimes referred to as an instance of "single-voter manipulation" to distinguish it from a situation wherein a group of voters, acting in concert, can change their ballots so that all of them prefer the new winner to the original winner. We'll see examples of group manipulation in Section 10.2.

Second, the original election produced a single winner, as did the new election held after we finished constructing Voter 1's disingenuous ballot. Thus, because we know each voter's sincere preference ranking for the candidates, we also know exactly which of the two election outcomes each voter will prefer. Ties, on the other hand, present a problem. For example, if a voter has sincere preferences that rank *A* over *B* over *C* over *D*, then it's not at all obvious whether this voter will prefer an election outcome that ties *A* and *D* to an election outcome that ties *B* and *C* or vice-versa.

A voting system is manipulable if there is at least one scenario in which some voter can achieve a more preferred election outcome by unilaterally changing his or her ballot. The precise definition follows.

Definition of Manipulability

A voting system is said to be **manipulable** if there exist two sequences of preference list ballots and a voter (call the voter Bob) such that

1. Neither election results in a tie.
2. The only ballot change is by Bob.
3. Bob prefers—assuming that his ballot in the first election represents his true preferences—the outcome of the second election to that of the first election.

In this chapter, as in Chapter 9, we begin with majority rule in the two-candidate case and Condorcet's method in the case of three or more candidates. Condorcet's method again shines, leaving the voting paradox as its only blemish. We then move on to revisit the other voting systems introduced in Chapter 9 that apply to elections with three or more candidates, and we show that each succumbs to some form of manipulation. In fact, there is a striking impossibility result that arises here known as the *Gibbard–Satterthwaite manipulability theorem*. It is related to—indeed, some would say equivalent to—Arrow's impossibility theorem. We conclude with a treatment of a striking first cousin of manipulability known as the "chair's paradox."

10.1 Majority Rule and Condorcet's Method

Throughout this section, we assume that the number of voters is odd. In Section 9.1 we pointed out that with two candidates, majority rule has three very desirable properties: It treats all voters equally, it treats both candidates equally, and it is *monotone*, meaning that a single voter's change in ballot from a vote for the loser to a vote for the winner has no effect on the election outcome. More strikingly, May's theorem told us that among all voting systems in the two-candidate case that never result in a tie, majority rule is the *only* one satisfying these three properties.

But let's consider for a moment what monotonicity is saying in this two-candidate case for voting systems that never yield ties. It says that if you rank A over B on your ballot, and the election winner is B, then the election winner will remain B if you switch to a ballot with B over A. But there are only two possible choices for a ballot in this two-candidate case: B over A and A over B. Monotonicity is thus saying that if you rank A over B, then no unilateral change in your ballot can make the outcome A. This is simply the assertion that you can't manipulate the voting system!

Thus, in the two-candidate case, nonmanipulability and monotonicity are exactly the same thing. This allows us to restate **May's theorem** from Section 9.1 with "monotonicity" replaced by "nonmanipulability."

> ### May's Theorem for Manipulability
>
> Among all two-candidate voting systems that never result in a tie, majority rule is the only one that treats all voters equally, treats both candidates equally, and is nonmanipulable.

There are examples of two-candidate voting systems that are manipulable, even though they treat all voters equally and both candidates equally. For example, the voting system that declares the winner to be the alternative with the fewest first-place votes is manipulable, as is the one that declares the winner to be whichever alternative has an odd number of first-place votes. Exercises 1

Paper ballots are still used in elections in many states. A lingering controversy from the 2004 elections is over the use of electronic ballots, which do not leave physical evidence and thus may not be possible to recount in disputes about the plurality or majority. *(Jonathan Nourok/ PhotoEdit.)*

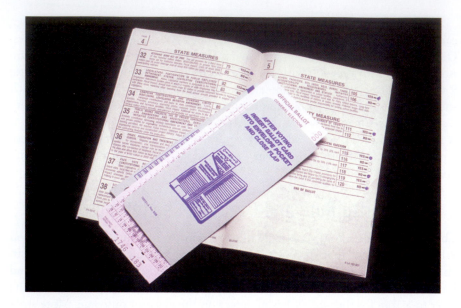

and 2 ask you to provide an example of voter manipulation for each of these systems.

Turning to the case of three or more candidates, we begin with Condorcet's method, as we did in Chapter 9. Condorcet's method is based on majority rule, and, as we've just seen, majority rule is nonmanipulable. So the following result, as pleasing as it is, comes as no surprise.

Theorem: The Nonmanipulability of Condorcet's Method

Condorcet's method is nonmanipulable in the sense that a voter can never unilaterally change an election result from one candidate to another candidate that he or she prefers.

Let's see why Condorcet's method is nonmanipulable. Suppose that we have an election in which you, as one of the voters, prefer Candidate A to Candidate B, but B wins using Condorcet's method. We'll show that any attempt that you might make to manipulate the election so that A becomes the winner is doomed to failure.

Because Candidate B is the winner using Condorcet's method, we know that B defeats every other candidate in a one-on-one contest based on the ballots cast. In particular B defeats A in a one-on-one contest, even with your original ballot that has A over B. This means that more than half of the other voters ranked B

over *A*, so, regardless of how you change your ballot, *B* will *still* defeat *A* in a one-on-one contest. While this need not ensure that *B* remains a winner with Condorcet's method, it certainly guarantees that *A* isn't. Hence, you cannot unilaterally cause *A* to be a winner using Condorcet's method, and so your attempt at manipulation will have failed.

We had to take a little care in stating the above theorem because of the fact that there are elections, as we've seen, in which there is no winner using Condorcet's method. With three voters and three candidates, it is possible for a voter (the one on the left in the example below) to unilaterally change an election from being one that yields his or her second choice (Candidate *C* in the example below) as the sole winner to one in which there is no winner at all, as the following example shows:

Election 1				Election 2			
Rank	Number of Voters (3)			Rank	Number of Voters (3)		
First	*A*	*B*	*C*	First	*A*	*B*	*C*
Second	*C*	*C*	*A*	Second	*B*	*C*	*A*
Third	*B*	*A*	*B*	Third	*C*	*A*	*B*

A voter's ability to unilaterally bring about this kind of change in an election, however, is not something that falls within the scope of our formal definition of manipulation. Nevertheless, one could argue that there are situations in which you might well prefer having an election with no outcome at all to having an election in which some candidate other than your top choice emerges as the sole winner.

But let us now move on to voting systems with three or more candidates that, unlike Condorcet's method, always produce at least one winner. As you might expect from the results in Chapter 9, these voting systems are not, in terms of nonmanipulability, as perfect as one might hope for.

10.2 Other Voting Systems for Three or More Candidates

Manipulability and the Borda Count

Example 1 showed how a single voter can manipulate an election in which the Borda count is being used. But Example 1 involved four candidates. Is there a simpler example involving only three candidates?

The answer turns out to be no, provided that we continue to interpret the notion of a "more preferred election outcome" to be a switch from a single winner to another single winner (as opposed to a switch creating or breaking a tie). This negative answer is formalized in the following theorem.

Theorem: The Nonmanipulability of the Borda Count with Exactly Three Candidates

With exactly three candidates, the Borda count cannot be manipulated in the sense of a voter unilaterally changing an election outcome from one single winner to another single winner that he or she prefers according to that voter's ballot in the first election, which we take to be sincere preferences.

Let's see why this is true. Suppose the candidates are A, B, and C, and that you prefer A to B, but B is the election winner using the Borda count. We'll show that any attempt you make to manipulate the election by changing your ballot so that A emerges as the winner (using the Borda count) is doomed to failure.

Because you prefer A to B, your sincere ballot can be one of only three possibilities, corresponding to whether C is ranked first, second, or third. We'll consider each case in turn.

Case 1. Your sincere ballot is A over B over C.   No ballot change on your part can increase A's Borda score, and you can only decrease B's Borda score by at most 1. Thus, at best, you can make a unilateral change that results in A and B having the same Borda score, whereas successful manipulation on your part requires that A have a strictly higher Borda score than B after your ballot change.

Case 2. Your sincere ballot is C over A over B.   No ballot change on your part can decrease B's Borda score, and you can only increase A's Borda score by at most 1. Thus, at best, you can make a unilateral change that results in A and B having the same Borda score, whereas successful manipulation on your part requires that A have a strictly higher Borda score than B after your ballot change.

Case 3. Your sincere ballot is A over C over B.   No ballot change on your part can increase A's Borda score or decrease B's Borda score. Thus, after your ballot change B will still have a higher Borda score than A, so your attempt at manipulation has failed in this case also.

So with three candidates, the Borda count is nonmanipulable. With more than three candidates, the Borda count does not fare as well, regardless of how many voters there are.

Theorem: The Manipulability of the Borda Count with Four or More Candidates

With four or more candidates (and two or more voters), the Borda count can be manipulated in the sense that there exists an election in which a voter can unilaterally change the election outcome from one single winner to another single winner that he or she prefers according to that voter's ballot in the first election, which we take to be sincere preferences.

We've already established part of this theorem, as we've seen an example of manipulation of the Borda count in the case of four candidates and two voters. This is really half the battle, as we can readily modify that example to serve in any case in which the number of voters is even as follows.

1. Any candidates in addition to A, B, C, and D can be placed below those four on every ballot.

2. The rest of the voters can be paired off with the members of each pair holding ballots that rank the candidates in exactly opposite orders (thus "canceling each other out" in terms of the Borda scores).

The following example illustrates this method of generalizing our earlier instance of manipulation of the Borda count to the case of five candidates and six voters.

Manipulating the Borda Count with Five Candidates and Six Voters   EXAMPLE 2

Consider the following two elections:

Election 1						Election 2					
A	B	A	E	A	E	A	B	A	E	A	E
B	C	B	D	B	D	D	C	B	D	B	D
C	A	C	C	C	C	C	A	C	C	C	C
D	D	D	B	D	B	B	D	D	B	D	B
E	E	E	A	E	A	E	E	E	A	E	A

The ballots of the first two voters (in both elections) are the same as in Example 1 (the manipulation of the Borda count with two voters and four candidates), with the new candidate E placed at the bottom of both ballots. The last four voters contribute exactly 8 to the Borda score of each candidate, and so, taken together, they have no effect on who is the winner of the election. This is what we meant by canceling each other out.

In the first election, as in Example 1, Candidate B wins. But if we take these ballots to represent true preferences, the voter on the far left prefers A to B. Moreover, that voter can achieve this better outcome—Candidate A—by submitting the disingenuous ballot that he or she cast in Election 2. ■

To handle the case where the number of voters is odd, we need a three-voter, four-candidate example of manipulation of the Borda count to start with. Exercise 9 provides this. We can then modify this example to work for any odd number of voters by again adding pairs of ballots that cancel each other out exactly as we did before. Exercises 10 and 11 fill in some of the details needed for this part of the argument, and ask you to provide the necessary explanations and calculations.

Manipulability and Runoff Systems

Both the plurality runoff rule and the Hare system are manipulable. But rather than give the whole story away, we'll just present the sequences of sincere ballots in each case. Exercises 16 and 17 ask you to figure out how the left-most voter in each case can secure a more preferred outcome by a unilateral change of ballot.

Election 1 for the Hare System				
A	*B*	*C*	*C*	*D*
B	*A*	*B*	*B*	*B*
C	*C*	*A*	*A*	*C*
D	*D*	*D*	*D*	*A*

Election 1 for the Plurality Runoff Rule				
A	*A*	*C*	*C*	*B*
B	*B*	*A*	*A*	*C*
C	*C*	*B*	*B*	*A*

Sequential Pairwise Voting and Agenda Manipulability

Sequential pairwise voting can also be manipulated by a single voter, even in the case of three voters and three candidates. Exercise 18 takes you through a particular example. But there is another aspect of manipulability that arises with this particular voting system that is of even more interest, and this is something called **agenda manipulation.**

William H. Riker, in his book *The Art of Political Manipulation*, spoke of the possibility that "those in control of procedures can manipulate the agenda by, for example, restricting alternatives [candidates] or by arranging the order in which they are brought up." The following example provides a striking illustration of this with sequential pairwise voting.

EXAMPLE 3   ## Agenda Manipulation of Sequential Pairwise Voting

Suppose we have four candidates and three voters who we know will be submitting the following preference list ballots:

Rank	Number of Voters (3)		
First	*A*	*C*	*B*
Second	*B*	*A*	*D*
Third	*D*	*B*	*C*
Fourth	*C*	*D*	*A*

Now suppose that we have agenda-setting power in the sense that we get to choose the order in which the one-on-one contests will take place. Remarkably, we can arrange for the winner to be whichever of the four candidates we want!

The intuition behind finding an agenda that will yield a certain candidate as the winner arises from the observation that candidates who appear later in the agenda are favored over candidates who appear early in the agenda. For example, if we want A to win, we place A last and look for which candidates would, in fact, defeat A one on one. Here, only C defeats A, and so we want to arrange for C to be eliminated along the way. But B defeats C one on one, so if we choose the agenda $BCDA$, we have that C is eliminated by B in the first round, then D is eliminated by B in the second round, and finally B is eliminated by A in the third round, leaving A as the winner. Exercise 19 asks you to find the three other agendas that will, in turn, yield B, C, and D as the winner. ◼

Plurality Voting and Group Manipulability

In the real world, all other voting systems pale in comparison to plurality voting in terms of the significance of the role played by disingenuous voting. "Throwing away your vote"—as some accuse Nader voters in Florida of doing in the 2000 presidential election—represents a choice, conscious or otherwise, to forgo obtaining a more desired outcome through strategic considerations.

Ironically, plurality voting, like Condorcet's method, is nonmanipulable according to the formal definition given on page 346. However, a *group* of voters, acting together, can change an election outcome into something they *all* prefer. We'll record this observation in the following, and then explain why it's true.

Theorem: The Group Manipulability of Plurality Voting

Plurality voting cannot be manipulated by a single individual. However, it is **group manipulable** in the sense that there are elections in which a group of voters can change their ballots so that the new winner is preferred to the old winner by everyone in the group, assuming that the original ballots represent the true preferences of each voter in the group.

First of all, let's see why no individual can manipulate plurality voting. Suppose that you prefer A to B, but B is the winner with plurality voting. Then B has at least one more first-place vote than A. Now, because you prefer A to B, we know that B is not on top of your sincere ballot, so no ballot change that you make can subtract from B's number of first-place votes. Moreover, by moving A to the top of your ballot, you only increase A's number of first-place votes by 1. Thus, the best you can do with a unilateral change in ballots is to move A into a tie with B.

To see that plurality voting is group manipulable we only have to look at any real-world election in which a third-party candidate acted as the "spoiler." As we've said, Ralph Nader was exactly this in the state of Florida in the 2000 presidential election. Another example occurs in Exercise 20.

At this point, we've seen that several of our familiar voting systems for three or more candidates—the Borda count, runoff systems, sequential pairwise voting—can be manipulated. Can't we do better than this in attempting to improve on Condorcet's method? We turn to this question next.

The Green Party holds its convention. Ralph Nader ran for the presidency as a Green in the 2000 election. By doing so, he brought up many questions of social choice—some would say deliberately. Was Nader a spoiler candidate? Were Nader supporters casting sincere votes for him? Were other voters who liked his positions hedging their bets and voting insincerely if they chose another candidate? *(Mark Leffingwell/AFP/Getty.)*

10.3 Impossibility

Condorcet's method, as we've seen, has a number of very desirable properties, including the following four:

1. Elections never result in ties.
2. It satisfies the Pareto condition.
3. It is nonmanipulable.
4. It is not a dictatorship.

Unfortunately, Condorcet's voting paradox on page 343 shows that there are elections in which Condorcet's method produces no winner at all.

Can we find a voting system that satisfies all four of these properties and that, unlike Condorcet's method, always yields a winner? Several possibilities suggest themselves. For example, to avoid ties, we could modify any of our usual methods by agreeing to use a fixed ordering of the candidates to break any ties that occur. Or we could extend Condorcet's method by making the winner be the candidate with the best "win-loss record" in one-on-one contests (a method called *Copeland's rule*).

Alas, any such attempt is doomed. In the early 1970s, Allan Gibbard and Mark Satterthwaite independently proved the following remarkable result.

The Gibbard–Satterthwaite Theorem

With three or more candidates and any number of voters, there does not exist—and there never will exist—a voting system that always produces a winner, never has ties, satisfies the Pareto condition, is nonmanipulable, and is not a dictatorship.

The **Gibbard–Satterthwaite theorem** (often called the GS theorem for short) is a deep result that is related in important ways to Arrow's impossibility theorem. In particular, you shouldn't find it at all obvious, and we won't be saying anything about the proof. But we can state and prove a much weaker result that is of some interest in its own right.

Theorem: A Weak Version of the GS Theorem

Any voting system for three candidates that agrees with Condorcet's method whenever there is a Condorcet winner—and that additionally produces a unique winner when confronted by the ballots in the Condorcet voting paradox—is manipulable.

Let's see why this is true. With the Condorcet voting paradox, the winner is either A or B or C. For the moment, we'll assume it is C (and leave the other two cases to you—see Exercise 22). Consider the following two elections:

Election 1				Election 2			
Rank	**Number of Voters (3)**			**Rank**	**Number of Voters (3)**		
First	A	B	C	First	B	B	C
Second	B	C	A	Second	A	C	A
Third	C	A	B	Third	C	A	B

In Election 1, the winner is C (our assumption in this case) and in Election 2, the winner is B (because we are assuming that our voting system agrees with Condorcet's method when there is a Condorcet winner, as B is here). Notice that the voter on the left, by a unilateral change in ballot, has improved the election outcome from his or her third choice to being his or her second choice. This is what that voter set out to do and is the desired instance of manipulation.

But the nonintuitive nature of voting and manipulation does not end here. It also turns out that sometimes "more is less" when it comes to "voting power." We illustrate this with the so-called chair's paradox.

10.4 The Chair's Paradox

We conclude this chapter with an aspect of manipulability that is so counterintuitive that it is referred to as the **chair's paradox**. The situation is as follows. Suppose we have three candidates—*A*, *B*, and *C*—and three voters whom we'll call, for simplicity, the chair, you, and me.

Now the chair prefers *A* to *B* to *C*. You prefer *B* to *C* to *A*. I prefer *C* to *A* to *B*. Thus, if we were to cast sincere preference list ballots, they'd be precisely the Condorcet voting paradox ballots. But that's not what we're going to do.

We're going to assume that each of the three of us gets to vote for one of the candidates. Votes are tallied as follows: If any candidate gets at least two of the three votes, he or she wins. But if each gets one vote, then whichever candidate the chair voted for wins. Thus, the chair has what might be called **tie-breaking power**. In particular, the chair clearly has more power (whatever that means) than you or I.

The goal now is to analyze the situation and to determine how each of us will vote if we're rational in the sense of having a willingness to vote strategically (meaning to manipulate the system) if it's in our own best interest. This is really a game-theoretic analysis, and it's useful to borrow a couple of pieces of game-theoretic terminology.

A choice of which candidate to vote for will be called a **strategy.** So each of us has three strategies at our disposal: Vote for *A*, vote for *B*, and vote for *C*. It will be useful to have our preferences displayed as if they were ballots. But remember, these are just our preferences, not our ballots.

Chair	You	Me
A	*B*	*C*
B	*C*	*A*
C	*A*	*B*

Our first observation is that if we're all rational and acting in our own self-interest, none of us will vote for our least-preferred candidate. The point is that voting for either a first or second choice **weakly dominates** the strategies of voting for a third choice in the sense that the former choices always yield outcomes that are either the same as, or better than, the latter.

But now the chair's strategy of voting for *A* weakly dominates his strategy of voting for *B*. That is, if we both vote for *C*, the outcome is *C* regardless of how he votes, but otherwise he does strictly better by voting for *A* rather than *B*. Hence, assuming the chair is rational, we know the chair will, in fact, vote for his top choice, candidate *A*.

Now, given that we know what the chair will do, the claim is that my strategy of voting for *C* weakly dominates my strategy of voting for *A*. That is, if you vote for *B*, the outcome is *A* regardless of whether I vote for *C* or *A*. On the

A woman casts her ballot on election day, the most important day in the American civic ritual of political campaigns and elections. Although she is acting as a responsible citizen, she may also contribute to some remarkable and contradictory results: Condorcet's voting paradox and the Gibbard–Satterthwaite theorem warn us that some elections don't produce a winner! *(Fred Prouser/Reuters/Corbis.)*

other hand, if you vote for C, then I can secure my best outcome C by voting for C. Assuming I'm rational, we know that I will, like the chair, vote for my top choice, which is candidate C.

But let's see where these decisions leave you. You know that the chair is voting for A and I'm voting for C. So if you vote for B, then the outcome is A—your last choice. However, if you vote for C along with me, then the outcome is C, your second choice. There is no way that you can secure your top choice B as the winner. So if you're rational, then you'll vote for C also, and candidate C will win the election.

So why is this paradoxical? Well, the chair had the most power, but the eventual winner of the election was his least-preferred candidate! He would have been better off handing over the tie-breaking power to either you or me.

The chair's paradox represents only one of manipulability's first cousins, some of which involve not only the fields of mathematics and political science but psychology as well. I recall a third-grade penmanship contest in which each of us had a writing sample taped to the blackboard, and the teacher, Mrs. Levy, announced that we'd get to vote for the one we thought best, with the proviso that the voter couldn't vote for his or her own paper. She also announced that if two or more were tied, we'd have a runoff among those.

I remember being torn as to which of three to vote for, all of which I thought were very good and considerably better than the rest, including my own. When the votes were counted, these three were, in fact, tied for the win, with my writing sample alone in fourth, and only one vote out of the tie.

After announcing the results, Mrs. Levy went on to say that the runoff would involve not three of us, but four, as she had decided also to vote, and she was voting for me! I don't remember the final tally, or what Mrs. Levy then said to the class, or what my three classmates, all plenty smart enough to realize what had just happened, later said to me. But I do remember sitting back and smiling—absolutely sure of the outcome—as soon as she had announced her intention to vote for me.

REVIEW VOCABULARY

Agenda manipulation The ability to control who wins an election with sequential pairwise voting by a choice of the agenda—that is, a choice of the order in which the one-on-one contest will be held.

Chair's paradox The fact that with three voters and three candidates, the voter with tie-breaking power (the "chair") can, if all three voters act rationally in their own self-interest, end up with her or his least-preferred candidate as the election winner.

Disingenuous ballot Any ballot that does not represent a voter's true preferences. Also called an *insincere ballot.*

Gibbard–Satterthwaite (GS) manipulability theorem Alan Gibbard and Mark Satterthwaite's independent discovery that every voting system for three or more alternatives and any number of voters that satisfies the Pareto condition, always produces a unique winner, and is not a dictatorship can be manipulated.

Group manipulability A voting system is group manipulable if there exists at least one election in which a group of voters can change their ballots (with the ballots of voters not in the group left unchanged) in such a way that they all prefer the winner of the new election to the winner of the old election, assuming that the original ballots represent the true preferences of these voters.

Manipulation A voting system is manipulable if there exists at least one election in which a voter can change his or her ballot (with the ballots of all other voters left unchanged) in such a way that he or she prefers the winner of the new election to the winner of the old election, assuming that the original ballots represent the true preferences of the voters.

May's theorem for manipulability Kenneth May's discovery that for two candidates and an odd number of voters, majority rule is the only voting system that treats both candidates equally, treats all voters equally, and is nonmanipulable.

Strategy In the chair's paradox, a choice of which candidate to vote for is called a strategy. This is a special case of the use of the term in general game-theoretic situations.

Tie-breaking power That aspect of the voting rule used in the chair's paradox that says the winner will be whichever candidate the chair votes for if there is a tie (which only happens if each candidate gets exactly one vote).

Unilateral change A change (in ballot) by a voter while every other voter keeps her or his ballot exactly as it was.

Weak-dominance One strategy (for example, a choice of whom to vote for) weakly dominates another if it yields an outcome that is at least as good, and sometimes better, than the other.

✔ SKILLS CHECK

1. A ballot that misrepresents a voter's true preferences is referred to as

(a) an insincere ballot.

(b) a disingenuous ballot.

(c) either an insincere ballot or a disingenuous ballot.

2. The quote "My scheme is intended only for honest men!" is from

(a) Condorcet.
(b) Hare.
(c) Borda.
(d) May.

3. A "unilateral change in ballot" refers to the fact that

(a) only one candidate's position is being altered.
(b) no communication is taking place.
(c) only one voter is changing his or her ballot.

4. If a voter has sincere preferences of *A* over *B* over *C* over *D*, then

(a) she will prefer a tie between *A* and *D* to a tie between *B* and *C*.
(b) she will prefer a tie between *B* and *C* to a tie between *A* and *D*.
(c) it's not at all clear which tie–*AD* or *BC*–she will prefer.

5. Suppose that two elections show that a voting system is manipulable. Then

(a) neither election results in a tie.
(b) the winners are the same in both elections.
(c) every voter has changed his or her ballot.

6. In the two-candidate case, manipulation is equivalent to

(a) the Pareto condition.
(b) independence of irrelevant alternatives.
(c) monotonicity.
(d) the equal treatment of voters and candidates.

7. May's theorem for manipulability says that, with an odd number of voters, among all voting systems for two candidates that never result in a tie, majority rule is the only one that is nonmanipulable and

(a) monotone.
(b) not a dictatorship.
(c) treats both candidates equally and all voters equally.
(d) satisfies the Pareto condition.

8. Condorcet's method

(a) can be manipulated but always produces a winner.
(b) is nonmanipulable but sometimes produces no winner.
(c) sometimes results in a tie, so manipulability is hard to assess.

9. The Borda count is nonmanipulable in the special case in which

(a) there are only three candidates.
(b) there are only two voters.
(c) all but one of the voters is honest.

10. With the Borda count, two ballots "cancel each other out" if

(a) they are identical.
(b) each is arrived at by turning the other one upside down.
(c) other voters also hold these same ballots.

11. A six-voter example of manipulation with the Borda count can be modified to yield a 10-voter example by

(a) adding four ballots that are identical to each other.
(b) adding four ballots that are identical to Voter 1's ballot.
(c) adding two pairs of ballots, with the ballots in each pair canceling each other out.

12. With any voting system that satisfies the Pareto condition, an *n*-voter example of manipulation with *k* candidates can be modified to yield an *n*-voter example with $k + j$ candidates by

(a) placing the additional *j* candidates at the bottom of each ballot (in any order whatsoever).
(b) adding ballots that cancel each other out.
(c) inserting the additional *j* candidates in ballots randomly.

13. Of the Hare system and the plurality runoff method,

(a) only the Hare system is manipulable.
(b) only plurality runoff is manipulable.
(c) both are manipulable.

14. Sequential pairwise voting is susceptible to a kind of manipulation called

(a) group manipulation.
(b) quasimanipulation.
(c) agenda manipulation.

15. Plurality voting is susceptible to a kind of manipulation called

(a) group manipulation.
(b) quasimanipulation.
(c) agenda manipulation.

16. Plurality voting

(a) cannot be manipulated by a single voter.
(b) can be manipulated by a single voter.
(c) is subject to agenda manipulation.

17. The Gibbard–Satterthwaite theorem says that with three or more candidates and any number of voters, there is no voting system that

(a) is not a dictatorship.
(b) is nonmanipulable and is not a dictatorship.
(c) satisfies the Pareto condition, is nonmanipulable, and is not a dictatorship.

(d) always yields a unique winner, satisfies the Pareto condition, is nonmanipulable, and is not a dictatorship.

18. The weak version of the Gibbard–Satterthwaite theorem asserts that if we have a voting system that agrees with Condorcet's method whenever there is a Condorcet winner, and that additionally produces a unique winner when confronted by the ballots in the Condorcet voting paradox, then

(a) the system is a dictatorship.
(b) the system is manipulable.
(c) the system is majority rule.

19. The voters' preferences in the paradox of the chair are

(a) precisely the Condorcet voting paradox ballots.
(b) all the same.
(c) dictated by the chair.

20. The chair's paradox is paradoxical because

(a) all three voters have tie-breaking power.
(b) the chair has the most power, but fares the worst.
(c) no voter gets his or her first choice.

EXERCISES   ■ Challenge   ◆ Discussion

Majority Rule and Condorcet's Method

1. Consider the voting system for two candidates (*A* and *B*) and three voters in which the candidate with the *fewest* first-place votes wins. Produce two elections that show this voting system is manipulable.

2. Consider the voting system for two candidates (*A* and *B*) and three voters in which the candidate receiving an odd number of first-place votes wins. Produce two elections that show this voting system is manipulable.

3. Consider the voting system for two candidates (*A* and *B*) and three voters in which the candidate receiving an even number of first-place votes wins. Produce two elections that show this voting system is manipulable.

4. There are at least two voting systems for two candidates (*A* and *B*) and three voters that are

nonmanipulable and that treat all voters the same (meaning that if two voters were to exchange ballots, then the election outcome would be unchanged).

(a) What does May's theorem tell us about such a voting system?
(b) In one sentence, give an example of such a voting system (that is, produce the rule that determines which of the two candidates (*A* or *B*) wins an election).
(c) In one sentence, give another example that is different from the example you gave in part (b) in that it produces a different winner for at least one election.

5. There are at least three voting systems for two candidates (*A* and *B*) and three voters that are nonmanipulable and that treat both candidates the same (meaning that if all three voters change their ballots, then the election outcome also changes).

(a) What does May's theorem tell us about such a voting system?

(b) In one sentence, give an example of such a voting system (that is, produce the rule that determines which of the two candidates wins an election).

(c) In one sentence, give two other examples that are different from the example you gave in part (b) in that they produce a different winner for at least one election.

6. Alfonse D'Amato (D) won the 1980 U.S. Senate race in New York by defeating Elizabeth Holtzman (H) and Jacob Javits (J). Reasonable estimates (based largely on exit polls) suggest that voters ranked the candidates according to the following table:

22%	23%	15%	29%	7%	4%
D	D	H	H	J	J
H	J	D	J	H	D
J	H	J	D	D	H

Who would have won if Condorcet's method (instead of plurality voting) had been used?

Other Voting Systems for Three or More Candidates

7. Consider the following election with 4 candidates and 2 voters:

B	A
C	D
A	C
D	B

Show that if the Borda count is being used, the voter on the left can manipulate the outcome (assuming the above ballot represents his true preferences).

8. Example 2 showed that the Borda count is manipulable if there are five candidates and six voters. Mimic what was done there in order to construct an example with 7 candidates and 8 voters.

9. Use the following election to illustrate the manipulability of the Borda count with 3 voters and 4 candidates:

A	B	B
B	A	A
C	C	C
D	D	D

10. Show that the Borda count is manipulable if there are 4 candidates and 5 voters. (*Hint:* Start with the ballots in the previous exercise, and then add two ballots that cancel each other out.)

11. Building on the idea in the previous exercise, show that the Borda count is manipulable if there are 6 candidates and 9 voters.

12. Assume the following ballots give the true preferences of the voters and that the Borda count is being used. Show that at least one of the voters can improve the election outcome from her point of view by a unilateral change in her ballot.

B	D	C	B
C	C	A	A
D	A	B	C
A	B	D	D

13. There is a modified version of Condorcet's method called the *weak Condorcet rule:* A candidate is among the winners precisely if he would defeat or tie every other candidate in a one-on-one contest. Notice that with an odd number of voters, the weak Condorcet rule is identical to Condorcet's method. Use the following ballots to show that the weak Condorcet rule is manipulable:

A	C	B	D
B	A	D	C
C	B	C	A
D	D	A	B

14. *Copeland's rule* is a voting system that, like Condorcet's method, looks at one-on-one contests. Copeland's rule, however, takes as the election winner the candidate with the best "win-loss record." Use the following ballots to show that Copeland's rule is manipulable:

A	C	A	D
B	E	E	B
C	D	D	E
D	B	C	C
E	A	B	A

15. *Coombs's rule* is the voting system that operates like the Hare system, except that instead of deleting candidates with the *fewest* first-place votes one after another, it deletes candidates with the *most* last-place votes one after another. Use the following ballots to show that Coombs's rule is manipulable:

A	B	B	A	A
B	C	C	C	C
C	A	A	B	B

16. Use the following election to show that the Hare system is manipulable:

A	B	C	C	D
B	A	B	B	B
C	C	A	A	C
D	D	D	D	A

17. Use the following election to show that the plurality runoff rule is manipulable:

A	A	C	C	B
B	B	A	A	C
C	C	B	B	A

18. Use the following election to show that sequential pairwise voting is manipulable. (Assume the agenda is *ABC*.)

A	B	C
B	C	A
C	A	B

19. Given the following ballots:

A	C	B
B	A	D
D	B	C
C	D	A

mimic what was done in Example 3 to find

(a) an agenda for which *B* is the winner using sequential pairwise voting.

(b) an agenda for which *C* is the winner using sequential pairwise voting.

(c) an agenda for which *D* is the winner using sequential pairwise voting.

◆ **20.** Suppose that we have a voting system that satisfies unanimity: If every voter ranks the same candidate first, then that candidate is the unique winner. In a few sentences, explain why it is that, if the system fails to satisfy the Pareto condition, it can be manipulated by some group.

21. Use the ballots in Exercise 6 to show that the plurality rule is group manipulable.

Impossibility

■ **22.** Complete the proof of the weak version of the Gibbard–Satterthwaite theorem by handling the case where

(a) the winner with the voting paradox ballots is *A*.

(b) the winner with the voting paradox ballots is *B*.

The Gibbard–Satterthwaite theorem says that the following four properties of voting systems can never be simultaneously satisfied:

(1) Elections always have unique winners.

(2) It satisfies the Pareto condition.

(3) It is nonmanipulable.

(4) It is not a dictatorship.

23. Which of the four properties are satisfied by a dictatorship?

24. Which of the four properties are satisfied by an "antidictatorship," where the election winner is whichever candidate Voter 1 ranks *last* on his or her ballot?

25. Which of the four properties are satisfied if we use the plurality rule with Voter 1's ballot used to break any ties that occur?

The Chair's Paradox

Consider the ballots from the chair's paradox:

Chair	You	Me
A	B	C
B	C	A
C	A	B

Assume that we know that the chair will vote for A, but that we don't know anything about how I will vote.

◆ **26.** In a sentence or two, explain why your strategy to vote for B does not weakly dominate your strategy of voting for C.

◆ **27.** In a sentence or two, explain why your strategy to vote for C does not weakly dominate your strategy of voting for B.

 WRITING PROJECT

In a paragraph or two, explain why Condorcet's method is not group manipulable.

SUGGESTED READINGS

MOULIN, HERVÉ. *The Strategy of Social Choice*, North Holland, New York, 1983. Manipulability from an economist's point of view.

RIKER, WILLIAM. *The Art of Political Manipulation*, Yale University Press, New Haven and London, 1986. Manipulability from a political scientist's point of view.

TAYLOR, ALAN. *Social Choice and the Mathematics of Manipulation*, Cambridge University Press, Cambridge, U.K., 2005. Manipulability from a mathematician's point of view.

Weighted Voting Systems

Voting is often used to decide yes or no questions. Legislatures vote on bills, stockholders vote on resolutions presented by the board of directors of a corporation, and juries vote to acquit or convict a defendant. In this chapter, we shall concentrate on situations where there are just two alternatives, such as "yes" or "no." The theorem of Kenneth May quoted in the previous chapter says that majority rule is the only system with the following properties:

1. All voters are treated equally.
2. Both alternatives are treated equally.
3. If you vote "no," and "yes" wins, then "yes" would still win if you switched your vote to "yes," provided that no other voters switched their votes.
4. A tie cannot occur unless there is an even number of voters.

There are many situations in which one or more of these properties are not valid. For example, in a criminal trial the jury is required to reach a unanimous decision on a motion to convict (or on a motion to acquit); thus, if there is one "no" vote, the motion is not adopted. In this case, the alternatives are not treated equally. Here's another familiar example. Stockholders are allowed one vote per share that they own. If I own 10,000 shares and you own 100, then this voting system does not treat us equally.

Some systems where the voters appear to be unequal in power actually have all of the properties required by May's theorem. Any student of politics will attest that not all legislators are equally powerful (think of the speaker of the U.S.

House of Representatives versus a freshman member, or the prime minister versus a backbencher in Parliament). Nevertheless, the voting system actually treats the legislators equally: Each has one vote. Our interest is in the voting system itself and not in the influence that some voters might acquire as a result of experience or accomplishment.

Voting systems that treat participants unequally are often used when the participants are indeed unequal. For example, the Council of Ministers of the European Union accords more power to states such as France, which have large populations, than it does to smaller states, such as Austria. Rather than giving the larger states more representatives, as in the U.S. House of Representatives, the Council of Ministers gives the ministers from the larger states more votes.

We shall find two measures of voting power that apply when voters are not treated equally or alternatives are not treated equally, or both: the Shapley–Shubik power index, and the Banzhaf power index. The Banzhaf power index is an accurate measure of power when there is no spectrum of opinion. For example, if each voter decides which way to vote by tossing a coin, the Banzhaf power index will indicate each voter's share of power. The Shapley–Shubik index is appropriate in a process where measures are crafted so as to attract enough votes to win.

One type of voting system in which the voters or the alternatives may be treated unequally is a **weighted voting system.** Each participant has a specified number of votes, called his or her **weight.** If my voting weight is more than yours, then I might have more power than you to influence the outcome, and certainly I won't have less. (We will see that voters with different numbers of votes may actually have equal power.) In any voting system, there must be a criterion for deciding whether "yes" or "no" has won. In a weighted voting system, this is done by specifying a number called the **quota.** If the sum of the weights of all the voters who favor a motion is equal to the quota, or exceeds it, then "yes" wins. Otherwise, "no" wins. The quota must be greater than half of the total weight of all the voters, to avoid situations where contradictory motions can pass, and it cannot be greater than the total weight, or no motion would ever pass.

The European Union's Council of Ministers uses weighted voting, but in the United States, it is unusual for a legislative body to use a weighted voting system (see Spotlight 11.3). It cannot be said that there is no weighted voting in the United States, because the Electoral College, which elects the president, functions as a weighted voting system in which the voters are the states. See Spotlight 11.1.

Notation for Weighted Voting Systems

To describe a weighted voting system, you must specify the voting weights $w_1, w_2, \ldots, w_n$ of the participants, and the quota, q. The following notation is a shorthand way of making these specifications:

$$[q : w_1, w_2, \ldots, w_n]$$

SPOTLIGHT 11.1   The Electoral College

In a U.S. presidential election, the voters in each state don't actually cast their votes for the candidates. They vote for electors to represent them in the Electoral College. The number of electors allotted to a state is equal to the size of its congressional delegation, so a state with one congressional district gets 3 electors: one for its representative, and one for each of its two senators. A state with 25 representatives would get 27 electors. The District of Columbia, while not a state, is entitled by the 23rd Amendment to the U.S. Constitution to send 3 electors to the College.

All states except two select their electors in a statewide contest. Thus, all of a state's electors are committed to vote for the presidential candidate favored by a plurality of the voters in the state. For example, the candidate who gets a plurality in California receives all 55 of the state's electoral votes. In Maine and Nebraska, there is a different procedure. Two electors (corresponding to the senators) are chosen statewide, and the electors corresponding to the representatives are chosen by congressional district. Nebraska has three congressional districts. It is possible that one or two of the districts might favor one ticket, while the state as a whole might favor another.

Effectively, the Electoral College functions as a weighted voting systems in which there are 56 participants: the 50 states, the District of Columbia, three Nebraska congressional districts, and two Maine congressional districts. The weights range from 1 for individual congressional districts to 55, and the quota, 270, is a simple majority of the 538 electors.

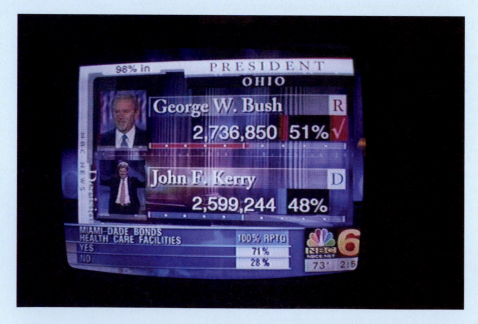

Electoral votes cast in Ohio, election day 2004. *(Thomas Dworzak/Magnum Photos.)*

The weighted voting system [51 : 40, 60] describes a voting system in which there are two voters, with voting weights 40 and 60, and the quota is 51.

A Dictator

EXAMPLE 1

Suppose there is one voter, *D*, who has all of the power. A motion will pass if and only if *D* is in favor, and it doesn't matter how the other participants vote. Most weighted voting systems that we will consider do not have a **dictator,** but if there is one, his or her voting weight must be equal to or more than the quota. The system [51 : 40, 60] has a dictator because the weight-60 voter can pass any motion that she wants. ◼

Dummy Voters

EXAMPLE 2

A voting system may include some participants—called **dummy** voters—whose votes don't count. For example, the U.S. Congress has a nonvoting delegate who represents the District of Columbia. If a voting system has a dictator, all of the participants except the dictator are dummy voters. In the voting system [8 : 5, 3, 1], the weight-1 voter is a dummy, because a motion will pass only if it has the support of the weight-5 and weight-3 voters, and then the additional 1 vote is not needed. For another example, consider [51 : 26, 26, 26, 22]. The voter with weight 22 is not needed when two of the other voters combine to support a motion; they have enough weight to pass the motion without her. If she joins forces with just one of the other voters, their total weight, 48, is not enough to win. Thus, the weight-22 voter is a dummy. ◼

Three More Three-Voter Systems

EXAMPLE 3

By adjusting the quota, the distribution of power in a weighted voting system can be altered. We have seen that the weight-1 voter in [8 : 5, 3, 1] is a dummy, but by increasing the quota to 9 we obtain a system in which the power is equally distributed—unanimous support is required to pass a motion in [9 : 5, 3, 1]. The weight-1 voter is also not a dummy in [6 : 5, 3, 1] because he can join the weight-5 voter to pass a motion, even if the weight-3 voter opposes. Finally, consider [51 : 49, 48, 3]. Although it looks as if the weight-3 voter will have relatively little power, and may even be a dummy, in fact she has the same voting power as the other two voters. Any two of the three voters in this system can pass a measure. ◼

Veto Power

EXAMPLE 4

A voter whose vote is necessary to pass any motion is said to have **veto power.** For example, in the system [6 : 5, 3, 1], the weight-5 voter has veto power because the other two voters do not have enough combined weight to pass a motion. A dictator always has veto power, and it is possible for more than one voter to have veto power as well. In a criminal trial, each juror has veto power. In the system [8 : 5, 3, 1] the voters with weights 5 and 3 each have veto power. ◼

The first widely accepted numerical index for assessing power in voting systems was the **Shapley–Shubik power index**, developed in 1954 by a mathematician, Lloyd S. Shapley, and an economist, Martin Shubik. A particular voter's power as measured by this index is proportional to the number of different permutations (or orderings) of the voters in which he or she has the potential to cast the pivotal vote—the vote that first turns from losing to winning.

The **Banzhaf power index** was introduced in 1965 by John F. Banzhaf III, a law professor who is also well-known as the founder of the antismoking organization ASH (Action on Smoking and Health). The Banzhaf index is the one most often cited in court rulings, perhaps because Banzhaf brought several cases to court and continues to file *amicus curiae* briefs when courts evaluate weighted voting systems. A voter's Banzhaf index is the number of different possible voting **combinations** in which he or she casts a swing vote—a vote in favor of a motion that is necessary for the motion to pass, or a vote against a motion that is essential for its defeat.

Lloyd S. Shapley

John F. Banzhaf III

Martin Shubik

The voters in the system [6 : 5, 3, 1] are not equally powerful—the weight-5 voter has veto power and the other two don't—and yet none of the voters are dummies. We can't compare power by comparing the voting weights because the weight-3 voter has the same voting power as the weight-1 voter. Together, they can stop the weight-5 voter from passing a motion, and either one can combine with the weight-5 voter to pass a motion. A **power index** gives a way to measure the share of power that each participant in a voting system (weighted or otherwise) has. Spotlight 11.2 is a brief history of power indices.

11.1 The Shapley–Shubik Power Index

When an election looms, politicians focus on "moderate voters." These are people who could be convinced to favor one side or the other. Moderate voters can make elected officials pay attention while voters who have an extreme commitment to one side or the other are ignored. However, moderate voters achieve their influence as a result of their political position, and we are primarily interested in the power that voters acquire as a result of the system itself. For example, France has 29 votes in the European Union (EU) Council, and Austria has 10. A motion before the Council, in which both countries were moderate voters, would be more likely to be written so as to acquire France's vote rather than Austria's, because France has so many more votes.

In United States presidential elections, politicians color the states that are likely to vote Republican in the Electoral College red, and the Democratic ones blue. The ones that could go either way are the "battleground states," and the campaigns put most of their investment in these states.

In 1954, Lloyd Shapley and Martin Shubik devised a way to gauge the share of decision-making power of each participant in a voting system. A voter's share of power is called his or her *Shapley–Shubik power index*. The index is defined in terms of *permutations*.

Permutation

A **permutation** of voters is an ordering of all of the voters in a voting system.

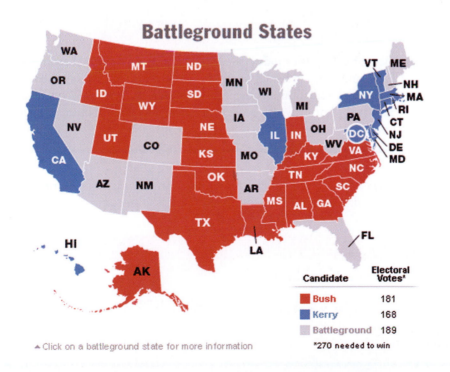

Battleground States

Candidate	Electoral Votes*
Bush	181
Kerry	168
Battleground	189

*270 needed to win

▲ Click on a battleground state for more information

Politicians color the states that are strongly Republican red, and the states that are strongly Democratic blue. We have colored the battleground states—which could go either way—gray. (© 2004 Time Inc. All rights reserved.)

Voters are ordered in accordance with their commitment to an issue, starting with those who are most favorably inclined and ending with those who are most determined to oppose. For example, suppose that the issue is animal rights. Here the spectrum might range from a voter who would outlaw the sale of cow's milk to one who would legalize cockfighting. If an animal rights bill is being drafted, it must be written so as to receive enough votes to meet the quota.

Pivotal Voter

The first voter in a permutation who, when joined by those coming before him or her, would have enough voting weight to win is the **pivotal voter** in the permutation. Each permutation has exactly one pivotal voter.

If the issue is taxation instead of animal rights, the spectrum of opinion will probably be completely different. Voters who have moderate positions on animal rights may or may not be at the extremes when the subject is taxes. Each issue being debated corresponds to some permutation—and the pivotal voter on one issue may well not be pivotal on another issue.

A successful tax cut bill must be drafted so as to secure the support of the pivotal voter of the taxation permutation, an animal welfare bill must be drafted so that the the pivotal voter of the animal rights permutation will support it, and so on.

The Shapley–Shubik Power Index

The **Shapley–Shubik power index** of each voter is computed by counting the number of permutations in which he or she is pivotal, then dividing it by the total number of permutations. Thus, if we consider each permutation to belong to the voter who is pivotal, each voter's Shapley–Shubik index is his or her share of the permutations.

If there are n voters, the number of permutations is called the **factorial** of n and is denoted $n!$. There is a simple formula for $n!$:

Factorial

For a positive whole number n,

$$n! = n \times (n - 1) \times (n - 2) \times \cdots \times 2 \times 1$$

and $0! = 1$.

For example,

$$1! = 1$$
$$2! = 2 \times 1 = 2$$
$$3! = 3 \times 2 \times 1 = 6$$
$$4! = 4 \times 3 \times 2 \times 1 = 24$$

To continue this list, observe that $n! = n \times (n - 1)!$ for $n \geq 1$. Thus

$$5! = 5 \times 4! = 5 \times 24 = 120$$
$$6! = 6 \times 5! = 6 \times 120 = 720$$
$$7! = 7 \times 6! = 7 \times 720 = 5040$$
$$8! = 8 \times 7! = 8 \times 5040 = 40,320$$

and so on. You can imagine that $n!$ increases dramatically as n increases—an instance of the combinatorial explosion. You probably don't want to calculate 100!. It is a 158-digit number.

To justify the formula, suppose that we are listing all of the permutations. There are n voters who could be first; when the first voter is selected, there are $n - 1$ remaining voters who could be in second position, then $n - 2$ who could be third, and so on. When it is time to select for the last position, there is one voter left. By the fundamental principle of counting (see Chapter 1), the number of permutations is the product of the numbers of choices that we have had at each stage.

The Shapley–Shubik Power Index of a Three-Voter System

EXAMPLE 5

Let us calculate the Shapley–Shubik power index of the voting system $[6:5,3,1]$. We will name the participants A, B, and C, and consider their $3! = 6$ permutations. Table 11.1 displays all six permutations. Next to each permutation, the total weights of the first voter, the first two voters, and all three voters are shown in sequence. The first number in the sequence that equals or exceeds the quota is underlined, and the corresponding pivotal voter's symbol is circled. We see that A is pivotal in four permutations, while B and C are each pivotal in one. Hence the Shapley–Shubik index of A is $\frac{4}{6}$, and B and C each have Shapley–Shubik indices of $\frac{1}{6}$. ∎

The Corporation with Four Shareholders

EXAMPLE 6

A corporation has four shareholders, A, B, C, and D, with 40, 30, 20, and 10 shares, respectively. The corporation uses the weighted voting system

$$[51 : 40, 30, 20, 10]$$

The $4! = 24$ permutations of the shareholders are shown in Table 11.2. In 10 of the permutations, A is the pivotal voter; B and C are each pivotal voters in 6;

TABLE 11.1   Permutations and Pivotal Voters for the Three-Person Committee

Permutations			Weights		
A	(B)	C	5	8	9
A	(C)	B	5	6	9
B	(A)	C	3	8	9
B	C	(A)	3	4	9
C	(A)	B	1	6	9
C	B	(A)	1	4	9

TABLE 11.2   Permutations and Pivotal Voters for the Four-Person Corporation

Permutations				Weights				Pivot			
A	(B)	C	D	40	70	90	100		B		
A	(B)	D	C	40	70	80	100		B		
A	(C)	B	D	40	60	90	100			C	
A	(C)	D	B	40	60	70	100			C	
A	D	(B)	C	40	50	80	100		B		
A	D	(C)	B	40	50	70	100			C	
B	(A)	C	D	30	70	90	100	A			
B	(A)	D	C	30	70	80	100	A			
B	C	(A)	D	30	50	90	100	A			
B	C	(D)	A	30	50	60	100				D
B	D	(A)	C	30	40	80	100	A			
B	D	(C)	A	30	40	60	100			C	
C	(A)	B	D	20	60	90	100	A			
C	(A)	D	B	20	60	70	100	A			
C	B	(A)	D	20	50	90	100	A			
C	B	(D)	A	20	50	60	100				D
C	D	(A)	B	20	30	70	100	A			
C	D	(B)	A	20	50	60	100		B		
D	A	(B)	C	10	50	80	100		B		
D	A	(C)	B	10	50	70	100			C	
D	B	(A)	C	10	40	80	100	A			
D	B	(C)	A	10	40	60	100			C	
D	C	(A)	B	10	30	70	100	A			
D	C	(B)	A	10	30	60	100		B		

and D is the pivotal voter in 2 permutations. Therefore, the Shapley–Shubik power index for this weighted voting system is

$$\left(\frac{10}{24}, \frac{6}{24}, \frac{6}{24}, \frac{2}{24} \right) \quad \blacksquare$$

How to Compute the Shapley–Shubik Power Index

For voting systems with no more than four voters, the Shapley–Shubik power index may be calculated by making a list of all the voting permutations and identifying the pivotal voter in each.

Listing all of the permutations is the brute force way of calculating the Shapley–Shubik power index. As we saw in connection with the traveling salesperson problem in Chapter 2, brute force methods can be impossible to carry out, due to the combinatorial explosion. If we were to study a 10-voter system by the brute force method, we would have to list 10! permutations—that's more than $3\frac{1}{2}$ million. If we made the list using the same size page and same typeface as this text, we could probably get about 350 permutations on each page. It would take 10,000 of these pages to list all of the permutations. If we used a computer, each permutation would occupy 10 bytes, and $3\frac{1}{2}$ million of them would take 35 megabytes, a manageable size. But even a computer would not be able to handle a 100-voter system because the number of permutations is a 158-digit number. Even by using all the memory devices—chips, disk drives, magnetic tape, flash memory, and so on—in the world, you would not be able to save a file containing all of the permutations.

If all the voters have the same voting weight, you don't need to make a list of all the permutations, because each has the same share of power. If there are 100 voters, each with 1 vote, the Shapley–Shubik index of each is $\frac{1}{100}$. If all but one or two of the voters have equal power, we can still calculate the Shapley–Shubik power index of each without making a list of permutations. We use two principles to do this:

▶ Voters with the same voting weight have the same Shapley–Shubik power index.

▶ The sum of the Shapley–Shubik power indices of all the voters is 1.

A Seven-Person Committee

EXAMPLE 7

The chairperson of a committee has 3 votes, and there are six other members with 1 vote. The quota for passing a measure is a simple majority, 5 of the 9 votes. In our notation, this voting system is [5 : 3, 1, 1, 1, 1, 1, 1].

Each ordinary member has the same power index. Our strategy is to compute the index of the chair, and then divide the share of power that the chair does *not* have equally among the ordinary members.

There are 7! = 5040 permutations to consider. We will group them by the position occupied by the chairperson. Thus, *CMMMMMM*, in which the chairperson is first, is the first group. Counting from the left, we see that the votes are accumulated in the sequence 3, 4, $\underline{5}$, 6, 7, 8, 9. In these permutations (there are 6! of them), the chairperson is not the pivot; an ordinary member is. In the second group, $MCMMMMM$, the votes accumulate in the sequence 1, 4, $\underline{5}$, 6, 7, 8, 9, and again, the chairperson is not the pivot.

The chairperson is the pivot in the next three groups of permutations, $MMCMMMM$, $MMMCMMM$, and $MMMMCMM$, with vote accumulations 1, 2, $\underline{5}$, 6, 7, 8, 9; 1, 2, 3, $\underline{6}$, 7, 8, 9; and 1, 2, 3, 4, $\underline{7}$, 8, 9, respectively. In the final two groups, $MMMMMCM$ and $MMMMMMC$, with vote accumulations 1, 2, 3, 4, $\underline{5}$, 8, 9 and 1, 2, 3, 4, $\underline{5}$, 6, 9, an ordinary member, not the chairperson, is the pivot again.

Each of the 7 groups of permutations is of the same size, 6!, because the 6 ordinary members can appear in any order in each. The chairperson is the pivot in three groups, for a total of $\frac{3}{7}$ of the total number of permutations. His Shapley–Shubik power index is therefore $\frac{3}{7}$. The remaining $\frac{4}{7}$ of the voting power is shared equally by the 6 ordinary members. Therefore each has $\frac{4}{7} \div 6 = \frac{2}{21}$ of the power.

The Shapley–Shubik power index of this weighted voting system is therefore

$$\left(\frac{3}{7}, \frac{2}{21}, \frac{2}{21}, \frac{2}{21}, \frac{2}{21}, \frac{2}{21}, \frac{2}{21} \right)$$

Because $\frac{3}{7} \div \frac{2}{21} = 4\frac{1}{2}$, the Shapley–Shubik model indicates that the chairperson is $4\frac{1}{2}$ times as powerful as an ordinary member, although his voting power is only 3 times as much. ∎

EXAMPLE 8   A Committee with Two Co-Chairs

A committee has 7 members: two co-chairs who each have 3 votes, and five other members with 1 vote each. The quota is 7, and thus the weighted voting system is [7 : 3, 3, 1, 1, 1, 1, 1]. Our objective is to determine the Shapley–Shubik power index of each member of the committee. Our strategy this time is to determine the voting power of A, a weight-1 voter. A is pivotal when the voters coming before him in the permutation have a combined weight of exactly 6. There are two ways to meet this condition:

▶ $C_1 C_2 A X_1 X_2 X_3 X_4$, where C_1 and C_2 represent co-chairs, and $X_1, \ldots, X_4$ represent members who are not co-chairs and are not A.

▶ $Y_1 Y_2 Y_3 Y_4 A Y_5 Y_6$, where one of $Y_1, \ldots Y_4$ is a co-chair, one of Y_5, Y_6 is a co-chair, and the remaining Y's are weight-1 voters.

For the first type of permutation, there are 2 ways to order the co-chairs (if the co-chairs are P and Q, then C_1 could be P or Q and C_2 would be the remaining co-chair), and there are 4! ways of ordering the other 4 members. There are thus $2 \times 4! = 48$ permutations of the first type.

To count the permutations of the second type, let us start with Y_5 and Y_6. There are 2 ways to choose the co-chair and 4 ways to choose the ordinary member for these positions in the permutation. Once these are chosen, there are 2 ways to put them in order and 4! ways to put the remaining co-chair and three ordinary members in order as Y_1, Y_2, Y_3, and Y_4. Thus, the number of permutations of the second type is $2 \times 4 \times 2 \times 4! = 384$.

In all, there are $48 + 384 = 432$ permutations in which A is pivotal. The Shapley–Shubik index of A is therefore $\frac{432}{7!} = \frac{3}{35}$. The other weight-1 voters have the same Shapley–Shubik index, so the combined share of power of the five weight-1 voters is $5 \times \frac{3}{35} = \frac{3}{7}$. The two co-chairs have the remainder of the power, so each has $(1 - \frac{3}{7}) \div 2 = \frac{2}{7}$ of the power. The Shapley–Shubik index of the system is therefore

$$\left(\frac{2}{7}, \frac{2}{7}, \frac{3}{35}, \frac{3}{35}, \frac{3}{35}, \frac{3}{35}, \frac{3}{35} \right) \blacksquare$$

The Permutation in the 2004 Election

EXAMPLE 9

In the 2004 election, the Electoral College reelected the Bush–Cheney ticket. Spotlight 11.1 explains how the Electoral College operates. Although there are 538 electors in the college, all states except Maine and Nebraska select their electors in a statewide general election, so even if the popular vote in a state was close, all of the electors from that state will vote for the same ticket. There are actually 56 independent votes in the Electoral College.[1]

Each of the 56 voters in the Electoral College is selected by and represents an electorate. Some, such as Nebraska's third district, were heavily in favor of the Bush–Cheney ticket (by more than 3 to 1); others, such as Iowa, New Mexico, and Wisconsin, were almost equally split between the Bush–Cheney and Kerry–Edwards tickets; and still others, such as the District of Columbia, were strongly in the Kerry–Edwards camp (almost 10 to 1). Table 11.3 lists the 56 voters in the Electoral College, ordered by their margin in favor of the Bush–Cheney ticket. The voting weight of each is shown, and a running total of electoral votes gives the total weight of each voter and all who came before it in the table. In listing the states and other voters in this order, we have recorded a permutation of the Electoral College participants. The pivotal voter is the one that brings the running total over the quota (270). If you recall the news about the 2004 election, you will not be surprised to see which voter is pivotal—Ohio. $\blacksquare$

---

[1]Strictly speaking, the votes in Maine and Nebraska are not really independent—see Exercise 39 at the end of this chapter.

TABLE 11.3	The Permutation Resulting from the General Election for President of the United States in 2004					
Voter*	Weight	Bush's Margin	Running Total	Voter*	Weight	Bush's Margin	Running Total
NE Dist 3	1	3.158	1	CO	9	1.112	222
UT	5	2.695	6	FL	27	1.107	249
WY	3	2.370	9	NV	5	1.055	254
ID	4	2.257	13	OH	20	1.051	274
NE	2	2.023	15	IA	7	1.018	281
OK	7	1.904	22	NM	5	1.016	286
ND	3	1.771	25	WI	10	0.992	296
NE Dist 1	1	1.752	26	NH	4	0.973	300
AK	3	1.721	29	PA	21	0.955	321
KS	6	1.705	35	MI	17	0.934	338
AL	9	1.697	44	MN	10	0.932	348
TX	34	1.599	78	OR	7	0.924	355
NE Dist 2	1	1.574	79	ME Dist 2	1	0.888	356
SD	3	1.558	82	NJ	15	0.879	371
MT	3	1.531	85	WA	11	0.864	382
IN	11	1.528	96	DE	3	0.858	385
MS	6	1.506	102	HI	4	0.839	389
KE	8	1.500	110	ME	2	0.836	391
SC	8	1.421	118	CA	55	0.818	446
GA	15	1.405	133	IL	21	0.815	467
LA	9	1.345	142	CT	7	0.810	474
TN	11	1.338	153	ME Dist 1	1	0.783	475
WV	5	1.296	158	MD	10	0.778	485
NC	15	1.288	173	VT	3	0.701	488
AZ	10	1.235	183	NY	31	0.700	519
AR	6	1.221	189	RI	4	0.653	523
VA	13	1.191	202	MA	12	0.595	535
MO	11	1.159	213	DC	3	0.103	538

*The voters are ordered by decreasing margin for the Bush–Cheney ticket. This margin is the number of popular votes cast for Bush–Cheney divided by the votes cast for Kerry–Edwards.

11.2 The Banzhaf Power Index

While the Shapley–Shubik power index is based on a count of permutations in which a voter is pivotal, the Banzhaf index is based on a count of coalitions in which a voter is *critical*. A **coalition** is a set of voters who are prepared to vote for, or to oppose, a motion. A **winning coalition** favors a motion, and has enough votes to pass it. A **blocking coalition** opposes a measure, and has the votes to defeat it. For example, in a dictatorship, a coalition in favor of a motion is a winning coalition if and only if the dictator belongs to it. Similarly, a coalition opposing a measure is a blocking coalition if and only if it includes the dictator.

In a winning or blocking coalition, there may be some voters whose votes are necessary to win. If any one of these voters should switch to the other side, the coalition would not have the votes it needs to have its way: It would become a **losing coalition.** These voters are called **critical voters** in the coalition.

Let's consider the presidential election of 2004 from this viewpoint. The Bush–Cheney ticket won this election with 286 electoral votes—16 more than the quota. The critical voters would be those belonging to the coalition that favored the Bush–Cheney ticket and had a voting weight of more than 16: Florida, Ohio, and Texas. If a voter with voting weight less than 16 were to leave the coalition, the Bush–Cheney ticket would still win, but with a smaller margin. Such a voter would not be critical.

By contrast, the election of 2000 was much closer, with only 271 votes for the Bush–Cheney ticket. With only one vote more than the quota, every voter in the winning coalition, except the congressional districts in Nebraska, was critical. (The Maine congressional districts voted for the Gore–Lieberman ticket and thus did not belong to the winning coalition.)

Critical Voters

EXAMPLE 10

Consider a committee of three members, *A*, *B*, and *C*. The chairperson of the committee, *A*, has two votes, while *B* and *C* each have one. The quota is three, and this voting system is

$$[3: 2, 1, 1]$$

The coalition $\{A, B, C\}$ is a winning coalition because it has all four votes. Suppose that *A* decides to leave the coalition. We can indicate this situation schematically as follows:

A	*B*	*C*	Votes	Outcome
2	1	1	4	Pass
↓				
0	1	1	2	Fail

By changing her vote, *A* has changed the outcome. In this coalition, *A* is a critical voter.

Now let's go back to the original coalition and see what happens if B changes his vote.

A	B	C	Votes	Outcome
2	1	1	4	Pass
	$\downarrow$			
2	0	1	3	Pass

This time, the outcome doesn't change, so B is not a critical voter in this coalition. Because C has the same power as B, he is also not a critical voter in the coalition. ■

EXAMPLE 11   Winning and Blocking

In the committee with members A, B, and C, and voting system [3: 2, 1, 1], A and B have formed a coalition to vote in favor of measure X and to oppose another measure, Y. Member C is voting against X and for Y. Because the coalition $\{A, B\}$ has 3 votes (2 for A, and 1 for B), it is a winning coalition for X and a blocking coalition for Y.

Measure X					Measure Y				
A	B	C	Votes	Outcome	A	B	C	Votes	Outcome
2	1	0	3	Pass	0	0	1	1	Fail

When voting for X, both A and B are critical voters. However, it takes only 2 votes to block a measure. If A leaves the coalition and joins C in voting for Y, then Y will be approved. Therefore A is a critical voter in the coalition to block motion Y. Voter B is not critical in the blocking coalition, because if he decides to switch his vote and support Y, it will not change the outcome. A can block motion Y by herself. ■

The Banzhaf power index was developed in 1965 by an attorney, John F. Banzhaf III, in an analysis of weighted voting that appeared in the *Rutgers Law Review*. Mr. Banzhaf's article was entitled "Weighted Voting Doesn't Work." See Spotlight 11.3.

The Banzhaf Power Index

A voter's **Banzhaf power index** is the number of distinct winning or blocking coalitions in which his or her vote is critical.

SPOTLIGHT 11.3   A Mathematical Quagmire

A county legislature in the United States is usually called a Board of Supervisors. Unlike state legislators, who represent districts that are carefully drawn to be equal in population, supervisors in some counties represent towns within the county. Because the towns differ in population, weighted voting is used to compensate for the resulting inequity.

If each supervisor's voting weight is proportional to the population of the town he or she represents, there will be situations in which one or more supervisors on a board are dummy voters, even if no supervisor is dictator. In a 1965 law review article, John F. Banzhaf III found that three of the six supervisors of Nassau County, New York, were dummies. The article inspired legal action against several elected bodies that employ weighted voting systems.

The first legal challenge to weighted voting was to invalidate the voting system of the Board of Supervisors of Washington County, New York. In its decision, the New York State Court of Appeals provided a way to fix a weighted voting system: Each supervisor's Banzhaf power index, rather than his or her voting weight, should be proportional to the population of the district that he or she represents. The court predicted that its remedy would lead to a "mathematical quagmire."

Five lawsuits, filed over a period of 25 years, challenged weighted voting in the Nassau County Board of Supervisors. These cases proved to be the mathematical quagmire that the appeals court had feared. The courts attempted to force Nassau County to comply with the Washington County decision. Although the county made a sincere attempt to do so, every voting system that it devised faced a new legal challenge. With conflicting expert testimony, the U.S. District Court finally ruled in 1993 that weighted voting was inherently unfair.

Banzhaf's law review article, which initially drew attention to weighted voting in Nassau County, was aptly titled "Weighted Voting Doesn't Work."

Nevertheless, tradition is hard to change. Many boards of supervisors of counties, particularly in the State of New York, still use weighted voting, and legal challenges to the practice, even after the Nassau County decision, have not always been successful.

To determine the Banzhaf power index of a voter A, we must count all possible winning and blocking coalitions of which A is a member and casts a critical vote. The weight of a winning coalition must be q or more, where q is the quota. A blocking coalition must be large enough to deny the "yes" voters the q votes they need to win. If the total weight of all the voters is n, then the weight of the blocking coalition has to be more than $n - q$. Assuming that all weights are integers, this means that the weight of a blocking coalition must be at least $n - q + 1$.

To determine which voters are critical in a given winning or blocking coalition, the following principle is useful.

Extra Votes Principle

A winning coalition with total weight w has $w - q$ **extra votes**. A blocking coalition with total weight w has $w - (n - q + 1)$ extra votes. The critical voters are those whose weight is more than the coalition's extra votes. These are the voters that the coalition can't afford to lose.

We can readily identify the critical voters in any coalition by comparing each voter's weight with the number of extra votes that the coalition has.

Calculating the Banzhaf Power Index

To calculate the Banzhaf power index of a given voting system:

1. Make a list of the winning and blocking coalitions.
2. Use the **extra-votes principle** to identify the critical voters in each coalition.

A voter's Banzhaf power index is then the number of coalitions in which he or she appears as a critical voter.

EXAMPLE 12   Calculating the Banzhaf Index

We will calculate the Banzhaf index for the committee with voting system $[3: 2, 1, 1]$.

The winning coalitions are all those whose weights sum to 3 or 4, and we will start by making a list of them:

Winning Coalition	Weight	Extra Votes	Critical Votes A	B	C
$\{A, B\}$	3	0	1	1	0
$\{A, C\}$	3	0	1	0	1
$\{A, B, C\}$	4	1	1	0	0
		Totals	3	1	1

All members of the coalitions with 0 extra votes are critical voters. Because A is the only voter with more than 1 vote, she is the only critical voter in the coalition that has 1 extra vote. We have thus found that A is a critical voter in three winning coalitions, while B and C are each critical voters in one winning coalition.

Blocking coalitions have total weights of 2, 3, or 4. Here is a list of the blocking coalitions:

Blocking Coalition	Weight	Extra Votes	Critical Votes		
			A	B	C
{A}	2	0	1	0	0
{B, C}	2	0	0	1	1
{A, B}	3	1	1	0	0
{A, C}	3	1	1	0	0
{A, B, C}	4	2	0	0	0
		Totals	3	1	1

Again, all voters in the coalitions with 0 extra votes are critical. In the blocking coalitions with 1 extra vote, only A is critical. The 4-vote blocking coalition {A, B, C} has 2 extra votes. Because no voter has more than 2 votes, there are no critical voters in {A, B, C}, considered as a blocking coalition. Voter A is critical in three blocking coalitions, while B and C are each critical in one. Adding up winning and blocking critical votes, we find that the Banzhaf index of A is 6, while B and C each have a Banzhaf index of 2. We will say that the Banzhaf index of this system is (6, 2, 2).

The Banzhaf index provides a comparison of the voting power of the participants in a voting system. Thus, A, with a Banzhaf index of 6, is three times as powerful as B or C. To determine the way voting power is distributed, we can add the numbers of critical voters for all three voters together to get $6 + 2 + 2 = 10$ critical votes in all. Thus, A has $\frac{6}{10} = 60\%$ of the voting power, while B and C each have 20%. By comparison, the Shapley–Shubik model gives $\frac{2}{3}$ of the power to A, while B and C each have $\frac{1}{6}$. ■

Consider the following three voting systems.

System I: [2 : 1, 1, 1]

System II: [3 : 2, 1, 1]

System III: [3 : 1, 1, 1]

We have studied system II and found that its Banzhaf power index is (6, 2, 2). Although power is distributed equally in systems I and III, and both have the same Shapley–Shubik power index, these systems have different Banzhaf power indices. System III requires a unanimous vote to pass a measure. There is only one winning coalition, in which each of the three voters is critical. Each voter is also critical in exactly one blocking coalition, in which he or she stands alone against the other two voters. The Banzhaf power index for system III is therefore (2, 2, 2). In system I, coalitions with total weight 2 or 3 can either block or win. The coalition with weight 3 has one extra vote, and no voter has more than one vote, so it has no critical voters. All voters are critical in the coalitions of weight 2. Each coalition is counted twice—once as a winning coalition, when both members vote "yes," and once as a blocking coalition, when both voters vote "no." Thus, the two weight-2 coalitions that include voter A, $\{A, B\}$ and $\{A, C\}$, give A a total of 4 critical votes. The Banzhaf index of system I is thus $(4, 4, 4)$.

The voters in system I have greater Banzhaf power indices than in system III. From a practical point of view, this means that in system I, an individual voter has more chances to influence the outcome if system I is used rather than system III.

In the examples that we have discussed so far, each participant has been a critical voter in exactly as many winning coalitions as blocking coalitions. This is not a coincidence, as we will now see.

Winning/Blocking Duality

The number of winning coalitions in which a given voter is critical is equal to the number of blocking coalitions in which the same voter is critical.

To understand why winning/blocking duality works, consider a voter, A, who is critical in a winning coalition, C. The voters who are not in the coalition C are voting "no" but do not have enough votes to block. However, if A changes her vote to "no," then C will be a losing coalition because A was a critical voter. When joined by A, the "no" voters form a blocking coalition, with A as a critical voter. Thus, there is a one-to-one correspondence between winning coalitions in which A is a critical voter and blocking coalitions with A as a critical voter.

By the winning/blocking duality principle, we can determine a voter's Banzhaf power index by doubling the number of winning coalitions in which he or she is a critical voter—this will account for the blocking coalitions.

EXAMPLE 13   The Corporation with Four Shareholders

The corporation with four shareholders (see Example 6) uses the weighted voting system

$$[51 : 40, 30, 20, 10]$$

Table 11.4 displays a list of all the winning coalitions of shareholders and the number of extra votes that each has. The four columns at the right are marked to indicate the critical voters in each coalition. By doubling the critical votes shown in the table, we arrive at the Banzhaf index of the corporation: $(10, 6, 6, 2)$. In this model, A has

$$\frac{10}{10 + 6 + 6 + 2} \qquad \text{or approximately 42\%}$$

TABLE 11.4	Winning Coalitions in the Four-Stockholder Corporation							

Coalition	Weight	Extra Votes	A	B	C	D
$\{A, B, C, D\}$	100	49				
$\{A, B, C\}$	90	39	1			
$\{A, B, D\}$	80	29	1	1		
$\{A, C, D\}$	70	19	1		1	
$\{A, B\}$	70	19	1	1		
$\{B, C, D\}$	60	9		1	1	1
$\{A, C\}$	60	9	1		1	
Critical votes			5	3	3	1

of the voting power, while B and C each have 25% (even though B has more shares than C). Shareholder D has the remaining 8% of the voting power, according to the Banzhaf model. In this case, power is distributed exactly as it was by the Shapley–Shubik model. ■

How to Count Combinations

A **voting combination** is a record of how the voters cast their votes for or against a given proposition. For example, if there are three voters, A, B, and C, and A and C voted "yes" while B voted "no," we might record the voting combination as "Yes, No, Yes." A briefer notation is to visualize voting combinations as **binary numbers.**

A whole number N is represented in binary form as a sequence of binary digits, or **bits,** which can be 0 or 1. This sequence expresses the way that N can be expressed as a sum of powers of 2. Thus, if

$$N = 2^k + 2^p + 2^q + \cdots + 2^z$$

where k is the largest exponent, then the binary representation of N is

$$(N)_2 = b_k b_{k-1} \cdots b_1 b_0$$

where the bits $b_k, b_p, b_q, \ldots, b_z$ corresponding to the exponents in the sum are equal to 1, and all other bits are 0. For this and many other purposes, I recommend memorizing the first few powers of 2, as in the following table.

n	0	1	2	3	4	5	6	7	8	9	10
2^n	1	2	4	8	16	32	64	128	256	512	1024

EXAMPLE 14   Expressing the Number 49 in Binary Notation

The largest power of 2 less than 49 is $2^5 = 32$. Subtract 32 from 49 to get 17. The largest power of 2 less than 17 is $16 = 2^4$. Subtract 16 from 17 to get 1, which is a power of 2 ($1 = 2^0$). Thus $49 = 2^5 + 2^4 + 2^0$, and hence the nonzero bits of $(49)_2$ are b_5, b_4, and b_0, while $b_3 = b_2 = b_1 = 0$. Listing the bits in order, $(49)_2 = 110001$. ∎

Now suppose that we have n voters. A sequence of n bits can represent a voting combination, where each voter is associated to a particular bit, which is 1 if the voter approves and 0 if the voter disapproves. A sequence of n bits also gives the binary representation of a number between 0 (the sequence with n 0's) and $2^n - 1$, which, as a binary number, is a sequence of n 1's. It follows that a set of n voters can have 2^n different voting combinations.

The number of voting combinations with n voters and exactly k "yes" votes is denoted C_k^n. For example, there is only one combination, $000\cdots0$ where no one votes "yes," so $C_0^n = 1$. You can show that by the same reasoning, $C_n^n = 1$. There are n combinations with exactly one "yes" vote:

$$100\cdots0, \ 010\cdots0, \ \ldots, \ 000\cdots1,$$

and hence $C_1^n = n$.

If each voter in a combination with k "yes" votes and $n - k$ "no" votes were to switch his or her vote to the opposite side, there would be $n - k$ "yes" votes and k "no" votes. Thus, the number of combinations of n voters with k "yes" votes is equal to the number of combinations of n voters with $n - k$ "yes" votes. In symbols, we can state this fact as follows:

Duality Formula for Combinations

$$C_k^n = C_{n-k}^n$$

Another formula that is useful is the **addition formula.** Suppose that there are $n + 1$ voters, one of whom is called Z. There are C_k^n combinations in which Z votes "no" and k other voters vote "yes," and C_{k-1}^n combinations in which Z and $k - 1$ of the other voters all vote "yes." This encompasses all C_k^{n+1} voting combinations of the $n + 1$ voters in which there are exactly k "yes" votes. Thus we have the following:

Addition Formula

$$C_{k-1}^n + C_k^n = C_k^{n+1}$$

The addition formula enables us to calculate the numbers C_k^n. Starting with $C_0^0 = C_0^1 = C_1^1 = 1$, we obtain $C_1^2 = C_0^1 + C_1^1 = 1 + 1 = 2$. Continuing, it is convenient to display the results in triangular form:

$$
\begin{array}{c}
1 \\
1\ 1 \\
1\ 2\ 1 \\
1\ 3\ 3\ 1 \\
1\ 4\ 6\ 4\ 1 \\
1\ 5\ 10\ 10\ 5\ 1 \\
1\ 6\ 15\ 20\ 15\ 6\ 1
\end{array}
$$

The entries on the left and right edges of the triangle are all equal to 1, and each entry in the interior of the triangle is obtained by adding the two entries just above it. This triangle is called **Pascal's triangle** in honor of Blaise Pascal (1623–1662), the French mathematician and philosopher credited with discovering it. The number C_k^n can be found by counting down to the nth row, remembering that the 1 on top is the 0th row, and then finding the kth entry, counting from the left, where the count starts with 0.

Pascal's triangle is an intriguing pattern, but it is only useful to calculate C_k^n when n is relatively small. The following expression gives a way to calculate C_k^n in more general situations:

To use the combination formula, cancel before multiplying.

Combination Formula

$$
C_k^n = \frac{n!}{k!(n-k)!}
$$

Calculate C_4^{40}

EXAMPLE 15

From the combination formula, $C_4^{40} = \dfrac{40!}{4!\ 36!}$. Notice that $40! = 40 \times 39 \times 38 \times 37 \times 36!$. Thus we can cancel 36! and obtain

$$
C_4^{40} = \frac{40 \times 39 \times 38 \times 37}{4 \times 3 \times 2 \times 1} = 91{,}390 \ \blacksquare
$$

To verify the combination formula, let $D_k^n = \dfrac{n!}{k!(n-k)!}$. It's our job to show that $D_k^n = C_k^n$. Recalling that $0! = 1$, we have $D_0^n = \frac{n!}{0!n!} = 1$ and $D_n^n = \frac{n!}{n!0!} = 1$. Also, the numbers D_k^n obey the addition formula:

$$
D_{k-1}^n + D_k^n = D_k^{n+1}
$$

To see this, we have to add the fractions $D_{k-1}^n = \frac{n!}{(k-1)!(n-k+1)!}$ and $D_k^n = \frac{n!}{k!(n-k)!}$.
Because $k! = k \times (k-1)!$ and $(n-k+1)! = (n-k+1) \times (n-k)!$, the least common denominator is $k!(n-k+1)!$. Therefore

$$
\begin{aligned}
D_{k-1}^n + D_k^n &= \frac{k \times n!}{k!(n-k+1)!} + \frac{(n-k+1) \times n!}{k!(n-k+1)!} \\
&= \frac{k \times n! + (n-k+1) \times n!}{k!(n-k+1)!} = \frac{(n+1) \times n!}{k!(n+1-k)!} = D_k^{n+1}
\end{aligned}
$$

It follows that if we arrange the numbers D_k^n in a triangle, as we did C_k^n, we will again get Pascal's triangle, because the left and right edges are filled with 1's, and each interior entry is equal to the sum of the two entries above it. We thus conclude that $C_k^n = D_k^n$, and hence the combination formula holds.

Efficient counting methods make it possible to compute the Banzhaf power index of large weighted voting systems. The method of counting combinations applies to systems in which most of the voters have the same weight, as in the seven-person committee that we considered in Example 7.

EXAMPLE 16   The Banzhaf Index of the Seven-Person Committee

The chairperson of this committee has 3 votes. Each of the six other members has 1 vote. The quota is 5, so we are considering the voting system

$$[5 : 3, 1, 1, 1, 1, 1, 1]$$

The chairperson, whom we will call C, is a critical voter in any winning coalition with no more than 2 extra votes. To achieve the quota, C's coalition must include at least two weight-1 voters. If there are five or more weight-1 voters in the coalition, then C's vote will not be needed: C will not be a critical voter. The number of coalitions with two weight-1 voters is C_2^6, because we are counting the voting combinations of the six weight-1 voters with 2 "yes" votes. Similarly, there are C_3^6 coalitions consisting of C and three weight-1 voters, and C_4^6 coalitions with C and four weight-1 voters. Referring to Pascal's triangle, displayed on the previous page, there are

$$C_2^6 + C_3^6 + C_4^6 = 15 + 20 + 15 = 50$$

winning coalitions. Counting an equal number of blocking coalitions, the Banzhaf power index of C is 100.

When we calculated the Shapley–Shubik power index, we only had to consider the chairperson. The other members' indices could then be determined because the Shapley–Shubik indices of all the members add up to 1. Because there is no fixed sum of the Banzhaf power indices of all the participants, we have to calculate the indices of the weight-1 voters separately. These voters do have the same voting power, so we only have to consider one of them, whom we will call

M. By the extra votes principle, M is a critical voter in a winning coalition only if this coalition has exactly 5 votes. There are two ways to assemble such a winning coalition:

▶ A three-member coalition consisting of M, C, and one of the other five weight-1 members. There are $C_1^5 = 5$ of these coalitions, because we are considering voting combinations of the 5 weight-1 voters other than M in which there is 1 "yes" and 4 "no" votes.

▶ A five-member coalition consisting of M and 4 other weight-1 voters. There are $C_4^5 = 5$ of these coalitions—here we are counting the voting combinations of the 5 weight-1 voters other than M in which there are 4 "yes" votes and 1 "no."

Adding, we find that M is a critical voter in 10 winning coalitions. Doubling this to account for the blocking coalitions, the Banzhaf power index of each weight-1 voter is 20.

To summarize, the Banzhaf power index of this voting system is

$$(100, 20, 20, 20, 20, 20, 20).$$

The total number of critical votes is $100 + 6 \times 20 = 220$. Thus, according to the Banzhaf model, C has $\frac{100}{220}$, or about 45%, of the power in the committee, and each weight-1 voter has $\frac{1}{11}$, or about 9.1%, of the power. This is in pretty close agreement with the Shapley–Shubik model, where we found that C had $\frac{3}{7}$, or about 43%, of the power, while each weight-1 voter had $\frac{2}{21}$, or approximately 9.5%, of the power. ■

In Example 8, we determined the Shapley–Shubik power index of the voting system $[7 : 3, 3, 1, 1, 1, 1, 1]$ (the committee with two co-chairs). In the following example, we will determine the Banzhaf power index of that committee.

The Committee with Two Co-Chairs

EXAMPLE 17

To determine the voting power of each voter in the system $[7 : 3, 3, 1, 1, 1, 1, 1]$ by the Banzhaf model, let's start with a weight-1 voter, M. He will be a critical voter in a winning coalition if and only if the votes of the other members in the coalition add up to exactly 6. There are two ways to achieve this total:

▶ The two co-chairs, and no other weight-1 voters, could join with M. There is exactly one such coalition.

▶ One of the two co-chairs, and 3 of the other 4 weight-1 voters, could join with M. There are $C_1^2 \times C_3^4 = 8$ such coalitions.

Thus, M is a critical voter in 9 winning coalitions; doubling this, we find that his Banzhaf power index is 18.

Now we must determine the Banzhaf power index of a weight-3 voter, A. She will be a critical voter in a winning coalition in which the other members have a combined total of 4, 5, or 6 votes.

▶ If she is joined by the other co-chair, the coalition would need 1, 2, or 3 of the 5 weight-1 members. The number of coalitions of this sort is

$$C_1^5 + C_2^5 + C_3^5 = 5 + 10 + 10 = 25$$

▶ If the other co-chair is opposed, she could be joined by 4 or all 5 of the weight-1 members. The number of such coalitions is $C_4^5 + C_5^5 = 6$.

It follows that A is a critical voter in 31 winning coalitions; her Banzhaf power index is 62. The Banzhaf power index of this committee is $(62, 62, 18, 18, 18, 18, 18)$. ■

The total number of critical votes in the committee with two co-chairs is $(2 \times 62) + (5 \times 18) = 214$. Thus, each co-chair has $\frac{62}{214}$, or approximately 29.0%, of the power, and each weight-1 member has $\frac{18}{214}$, or about 8.4%, of the power, by the Banzhaf model. Recall that, according to the Shapley–Shubik model, the co-chairs each had $\frac{2}{7}$ (about 28.6%) of the power and the weight-1 members had $\frac{3}{35}$ (about 8.6%). The agreement between the two models is, as in the other examples that we have considered, pretty close.

There are situations in which the difference in the distribution of power given by the models is significant. In Spotlight 11.4, the Banzhaf and Shapley–Shubik power indices of the United States Electoral College are compared. While the differences may seem small, by the Shapley–Shubik model, California has about 11.0% of the voting power in the college, while by the Banzhaf model, California has 11.4% of the power. When the stakes are high, this difference is significant. The following example presents a situation in which the models give dramatically different results.

EXAMPLE 18   The Big Shareholder

B holds 100,000 shares of stock in a corporation. There is a total of one million shares of stock, and the remaining stock is held by 9000 shareholders, each of whom has 100 shares. A weighted voting system, in which each shareholder's voting weight is equal to the number of shares that he or she owns, is used.

The Shapley–Shubik index of this system is determined by the same strategy that we used in Example 7 (the seven-person committee). This time, the permutations of the stockholders are divided into 9001 groups, depending upon the location of B. Each group has the same number of permutations (9000! of them, to be precise), and B is pivotal when she appears in the 4002nd through the 5001st position. If she is 4002nd, then there are $4001 \times 100 = 400,100$ shares preceding her, and her 100,000 shares bring the total to a bare majority of 500,100 shares. If there are more than 5000 shares ahead of B, the 5001st, a small shareholder,

SPOTLIGHT 11.4 The Electoral College: Presidential Elections of 2004 and 2008

The following table displays the Shapley–Shubik (SSPI) and Banzhaf (BPI) power indices of the voters in the Electoral College, as compared with the voter's weight as a percent of 538 (PCT), the total weight of all of the voters. It shows that for the most part, both measures of power agree closely with the actual share of power that a participant in the college has by virtue of its voting weight. There is an exception, though. California, whose voting weight is slightly more than 20% of the quota, has more than its share of power by either measure.

Voter	Weight	PCT (%)	SSPI (%)	BPI (%)
CA	55	10.22	11.04	11.41
TX	34	6.32	6.50	6.39
NY	31	5.76	5.89	5.79
FL	27	5.02	5.09	5.01
IL, PA	21	3.90	3.91	3.87
OH	20	3.72	3.72	3.68
MI	17	3.16	3.14	3.12
GA, NC, NJ	15	2.79	2.76	2.74
VA	13	2.42	2.38	2.37
MA	12	2.23	2.20	2.19
IN, MO, TN, WA	11	2.04	2.01	2.01
AZ, MD, MN, WI	10	1.86	1.82	1.82
AL, CO, LA	9	1.67	1.64	1.64
KY, SC	8	1.49	1.45	1.46
CT, IA, OK, OR	7	1.30	1.27	1.27
AR, KS, MS	6	1.12	1.09	1.09
NM, NV, UT, WV	5	0.93	0.90	0.91
HI, ID, NH, RI	4	0.74	0.72	0.73
AK, DE, DC, MT, ND, SD, VT, WY	3	0.56	0.54	0.55
ME, NE	2	0.37	0.36	0.36
Congressional districts (5 in all)	1	0.19	0.18	0.18

would be the pivot. Thus, B is pivotal in 1000 of the 9001 groups, and her Shapley–Shubik power index is $\frac{1000}{9001}$, or about 11.1%. The 9000 small shareholders have equal shares of the remaining power: The index of each is

$$\left(1 - \frac{1000}{9001}\right) \div 9000$$

which works out to be 0.0099%.

The Banzhaf power index is much harder to calculate, and it involves numbers of critical votes that are unimaginably large. However, we can assess B's share of power according to the Banzhaf model by the following consideration. By this model, each voter's share of the power is proportional to the probability that he or she would be a critical voter in a winning or blocking coalition if every voter tossed a coin and voted "yes" for heads, "no" for tails.

The shareholder B will be critical if between 4001 and 5000 small shareholders vote "yes." (If she votes "yes," she will be a critical voter in a winning coalition with weight between 500,100 and 600,000; if she votes "no," she will be a critical voter in a blocking coalition with between 500,000 and 599,900 shares.) Imagine that you are betting on 9000 people tossing coins. On average, there will be 4500 heads; you win if the number of heads is between 4001 and 5000. You could play this game for the rest of your life and never lose! If you remember about normal distributions (see Chapter 5), the number of heads in this experiment would have mean 4500, and the standard deviation would be approximately 50. Thus, 68% of the time there would be between 4450 and 4550 heads; 95% of the time there would be between 4400 and 4600 heads; 99.7% of the time there would be between 4350 and 4650 heads; and so on. It follows that B is a critical voter in almost all situations.

Now consider a shareholder S who owns 100 shares. He will be critical if the owners of exactly 500,000 other shares vote "yes" (this could be B and 4000 small shareholders, or 5000 small shareholders without B). Imagine again that you are betting on 9000 coin tosses, with B tossing first. You win if either the first toss is heads, and *exactly* 4000 of the following tosses are heads, or if the first toss is tails, and *exactly* 5000 of the following tosses are heads. Otherwise you lose. Because 4000 and 5000 are both far from the average number of heads, 4500, this is a game S could play for the rest of his life and *never* win! It follows that by the Banzhaf model, the small shareholders have virtually no power; the big shareholder possesses almost all of the power. ■

11.3 Comparing Voting Systems

In many cases, different weighted voting systems turn out to have the same winning coalitions. For example, a dictatorship is no different if the dictator's weight is exactly equal to the quota or if it is much more. The dictator will have the same Banzhaf power index (2^n if there are n voters, including the dictator), and the same Shapley–Shubik power index (1) by virtue of being a dictator. To compare voting systems—which may be specified with weights or in some other way—we need to know what the winning coalitions are.

If there are just two voters, A and B, the empty coalition, { }, is surely a losing coalition, and $\{A, B\}$ is a winning coalition. There are only three distinct voting systems in this case: In the first, unanimous consent is required to pass a measure, so the only winning coalition is $\{A, B\}$. In the second, A is a dictator, and $\{A\}$ is also a winning coalition. In the third, B is a dictator, and the winning coalitions are $\{B\}$ and $\{A, B\}$. Although there is an unlimited number of ways to assign weights and a quota to a two-voter system, there are only three ways that the power can be distributed: A as dictator, B as dictator, or consensus rule.

Equivalent Voting Systems

Two voting systems are **equivalent** if there is a way for all of the voters of the first system to exchange places with the voters of the second system and preserve all winning coalitions.

The weighted voting systems [50 : 49, 1] and [4 : 3, 3], involving pairs of voters A, B, and C, D, respectively, are equivalent because in each system, unanimous support is required to pass a measure. We could have A exchange places with C, and B exchange places with D.

Now consider two voting systems [2 : 2, 1] and [5 : 3, 6] involving the same pair of voters, A and B. In the first, [2 : 2, 1], A is a dictator, while in the second, [5 : 3, 6], B dictates. By having A and B exchange places with each other, we see that the two systems are equivalent. "Equivalent" does not mean "the same." Voter A would tell you that the system where he is the dictator is not the same as the system where B is the dictator. The systems are equivalent because each has a dictator.

Every two-voter system is equivalent either to a system with a dictator or to one that requires consensus. As the number of voters increases, the number of different types of voting systems increases.

Minimal Winning Coalitions

A **minimal winning coalition** is a winning coalition in which each voter is a critical voter.

In a dictatorship, every coalition that includes the dictator is a winning coalition, but the only *minimal* winning coalition is the one that includes the dictator and no other voters.

Minimal Winning Coalitions: A Three-Voter System

EXAMPLE 19

The three-member committee from Example 5 uses the voting system [6 : 5, 3, 1]. Let's refer to its members as A, B, and C in order of decreasing weight. There are three winning coalitions. One, {A, B}, has weight 8, more than the quota, but it is minimal because both voters are critical. Another, {A, C}, with weight 6, is also minimal. The third winning coalition, {A, B, C}, is not minimal because only A is a critical voter. ■

EXAMPLE 20   The Four-Shareholder Corporation

Table 11.4 lists the five winning coalitions in the corporation with the voting system [51 : 40, 30, 20, 10]. In each coalition, the critical voters have been identified. The minimal ones are those in which each voter is marked as critical: $\{A, B\}$, $\{A, C\}$, and $\{B, C, D\}$. These minimal winning coalitions are displayed in Figure 11.1. ■

A voting system can be described completely by specifying its minimal winning coalitions. If you want to make up a new voting system, instead of specifying weights and a quota, you could make a list of the minimal winning coalitions. You would have to be careful that your list satisfies the following three requirements:

1. Your list can't be empty. You have to name at least one coalition—otherwise, there would be no way to approve a motion.
2. You can't have one minimal winning coalition that contains another one—otherwise, the larger coalition wouldn't be minimal.
3. Every pair of coalitions in the list has to overlap—otherwise, two opposing motions could pass.

In the four-shareholder corporation (see Figure 11.1), you can see that these requirements are satisfied. Now let's construct some voting systems.

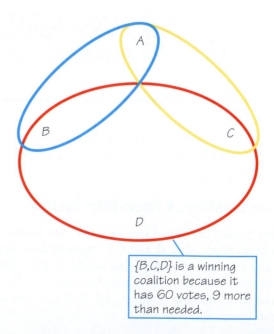

FIGURE 11.1 Each oval surrounds a minimal winning coalition for the four-shareholder corporation.

{B,C,D} is a winning coalition because it has 60 votes, 9 more than needed.

EXAMPLE 21

Three-Voter Systems

We would like to make a list of all voting systems that have three participants, *A*, *B*, and *C*. To keep the size of the list manageable, we will insist that no two voting systems on the list be equivalent.

To start, suppose that {*A*} is a minimal winning coalition. Requirement 3 tells us that every other minimal winning coalition must overlap with {*A*}, but the only way that could happen would be if *A* also belonged to the other coalition. In this case, requirement 2 would be violated. Thus, {*A*} can be the only minimal winning coalition. This is the voting system where *A* is dictator. Systems where *B* or *C* is dictator are not listed because they are equivalent to this one.

Now suppose that there is no dictator. Every minimal winning coalition must contain either two or all three voters. Let's consider the case in which {*A*, *B*, *C*} is a minimal winning coalition. It is the only winning coalition, because any other winning coalition would have to be entirely contained in this coalition, which requirement 2 doesn't allow. In this voting system, a unanimous vote is required to pass a measure. We will call this system *consensus rule.*

Finally, let's suppose that there is a two-voter minimal winning coalition, say {*A*, *B*}. If it is the only minimal winning coalition, then a measure will pass if *A* and *B* both vote "yes" and the vote of *C* does not matter: In other words, *C* is a dummy, and *A* and *B* make all the arrangements. We will call this system the *clique.* Of course, the clique could be {*A*, *C*} or {*B*, *C*}, but these systems are equivalent to the one where {*A*, *B*} is the clique.

There could be two two-voter minimal winning coalitions, say {*A*, *B*} and {*A*, *C*}. Neither coalition contains the other, and there is an overlap, so all of the requirements are satisfied. In this system, *A* has veto power. We have encountered this system in Example 10, and we will call it the *chair veto.* There are two other voting systems equivalent to this one, where *B* or *C* is chair.

It is possible that all three two-member coalitions are minimal winning coalitions. Because there are only three voters, any two distinct two-member coalitions will overlap, so the requirements are still satisfied. This system is called *majority rule.* ∎

Table 11.5 lists all five of these three-voter systems. Each system can be presented as a weighted voting system, and suitable weights are given in the table. If we want to make a similar list of all types of four-voter systems, we can start by making each three-voter system into a four-voter system. This is done by putting a fourth voter, *D*, into the system, without including him in any of the minimal winning coalitions. This makes *D* a dummy. You may be interested to know that there are an additional nine 4-voter systems that don't have any dummies. Try to list as many of these systems as you can.

EXAMPLE 22

The Scholarship Committee

A university offers scholarships on the basis of either academic excellence or financial need. Each application for a scholarship is reviewed by two professors,

TABLE 11.5	Voting Systems with Three Participants		
System	Minimal Winning Coalitions	Weights	Banzhaf Index
Dictator	$\{A\}$	$[3: 3, 1, 1]$	$(8, 0, 0)$
Clique	$\{A, B\}$	$[4: 2, 2, 1]$	$(4, 4, 0)$
Majority	$\{A, B\}, \{A, C\}, \{B, C\}$	$[2: 1, 1, 1]$	$(4, 4, 4)$
Chair veto	$\{A, B\}, \{A, C\}$	$[3: 2, 1, 1]$	$(6, 2, 2)$
Consensus	$\{A, B, C\}$	$[3: 1, 1, 1]$	$(2, 2, 2)$

who rate the student academically, and two financial aid officers, who rate the applicant's need. If both professors or both financial aid officers recommend the applicant for a scholarship, the Dean of Admissions decides whether or not to award a scholarship. Is it possible to assign weights to the professors, the financial aid officers, and the dean to reflect this decision-making system?

To answer this question, let's focus on the minimal winning coalitions. The participants are the two professors, A and B; the financial aid officers E and F; and the dean D. A scholarship will be offered if approved by the professors and the dean, or by the financial aid officers and the dean. Thus, the minimal winning coalitions are

$$\{A, B, D\} \quad \text{and} \quad \{D, E, F\}$$

(see Figure 11.2).

Consider the following two winning coalitions: In the first, all except the financial aid officer F favors an award; while in the second, only the professor B dissents.

$$C_1 = \{A, B, D, E\} \quad \text{and} \quad C_2 = \{A, D, E, F\}$$

In C_1, we notice that A is a critical voter and E isn't, while in C_2 the tables are turned because E is critical while A is not. If this were a weighted voting system, then in any winning coalition, the critical voters would all have greater weight than those who are not critical. Thus A would have to have both more weight than E (because of the situation in C_1) and less weight than E (because of C_2), which is impossible. ∎

FIGURE 11.2 The Scholarship Committee: Minimal winning coalitions.

Although the scholarship committee is not equivalent to any weighted voting system, it is possible to determine the Shapley–Shubik and Banzhaf power indices of each participant.

Power Indices of the Scholarship Committee

EXAMPLE 23

The dean has veto power. Therefore, she will be the pivot in any permutation where she appears last. If she is second-to-last in a permutation, she will still be the pivot, because either both professors or both financial aid officers must come before her. In the middle position, she will be pivotal if and only if both professors or both aid officers come first. Adding this up, we have $2 \times 4! = 48$ permutations in which the dean is in fourth or fifth position. There are four permutations of the form Prof, Prof, Dean, Aid, Aid because the professors and the aid officers can be in either order, and another four of the form Aid, Aid, Dean, Prof, Prof. The dean is not the pivot when she is first or second, because at least three people have to approve a scholarship. We conclude that the dean is pivotal in $48 + 4 + 4 = 56$ permutations in all. Her Shapley–Shubik power index is therefore $\frac{56}{5!} = \frac{7}{15}$. Each of the other participants is equally powerful, and they share the remaining $\frac{8}{15}$ of the power. Thus each professor and each aid officer has a Shapley–Shubik power index of $\frac{2}{15}$.

To compute the Banzhaf index, let's list all the winning coalitions: There are 7 of them.

$$\{A, B, D\}, \{E, F, D\}, \{B, E, F, D\}, \{A, E, F, D\},$$
$$\{A, B, F, D\}, \{A, B, E, D\}, \text{ and } \{A, B, E, F, D\}$$

The dean, who has veto power, is a critical voter in each winning coalition and in 7 blocking coalitions, so her Banzhaf power index is 14. Professor A is critical in each winning coalition that includes B but not both aid officers: There are 3 of these. He is also critical in 3 blocking coalitions, so his Banzhaf power index is 6. The remaining participants, B, E, and F, have the same power, so the Banzhaf power index of the scholarship committee is

$$(14, 6, 6, 6, 6) \blacksquare$$

REVIEW VOCABULARY

Addition formula
$$C^n_{k-1} + C^n_k = C^{n+1}_k$$
Banzhaf power index A count of the winning or blocking coalitions in which a voter is a critical member. This is a measure of the actual voting power of that voter.

Bit A binary digit: 0 or 1.
Binary number The expression of a number in base-2 notation. If b_n denotes the n bit to the left of the binary point (n is 0 or negative for bits to the right of the binary point; this only occurs in representing numbers that are not whole), then the

number being represented is the sum of the terms $b_n \times 2^{n-1}$. The binary number 11001101 has $b_1 = 1$, $b_2 = 0$, $b_3 = b_4 = 1$, $b_5 = b_6 = 0$, and $b_7 = b_8 = 1$. Hence it can be converted to decimal form as
$$2^0 + 2^2 + 2^3 + 2^6 + 2^7 = 205$$
Blocking coalition A coalition in opposition to a measure that can prevent the measure from passing.

C_k^n The number of voting combinations in a voting system with n voters, in which k voters say "yes" and $n - k$ voters say "no." This number, referred to as "n-choose-k," is given by the formula
$$C_k^n = \frac{n!}{k! \times (n - k)!}$$
Coalition The set of participants in a voting system who favor, or who oppose, a given motion. A coalition may be empty (if, for example, the voting body unanimously favors a motion, the opposing coalition is empty), it may contain some but not all voters, or it may consist of all the voters.

Critical voter A member of a winning coalition whose vote is essential for the coalition to win, or a member of a blocking coalition whose vote is essential for the coalition to block.

Dictator A participant in a voting system who can pass any issue even if all other voters oppose it and block any issue even if all other voters approve it.

Duality formula $C_k^n = C_{n-k}^n$

Dummy A participant who has no power in a voting system. A dummy is never a critical voter in any winning or blocking coaliton and is never the pivotal voter in any permutation.

Equivalent voting systems Two voting systems are equivalent if there is a way for all the voters of the first system to exchange places with the voters of the second system and preserve all winning coalitions.

Extra votes The number of votes that a winning coalition has in excess of the quota.

Extra-votes principle The critical voters in the coalition are those whose weights are more than the extra votes of the coalition. For example, if a coalition has 12 votes and the quota is 9, there are 3 extra votes. The critical voters in the coalition are those with more than 3 votes.

Factorial The number of permutations of n voters (or n distinct objects) is called *n-factorial*, or in symbols, $n!$. Because the empty coalition can be ordered in only one way, $0! = 1$. When n is a positive whole number, $n!$ is equal to the product of all the integers from 1 up to n. If n has more than one digit, then $n!$ is a pretty big number: 10! is more than three million, and 1000! has 2568 digits.

Losing coalition A coalition that does not have the voting power to get its way.

Minimal winning coalition A winning coalition that will become losing if any member defects. Each member is a critical voter.

Pascal's triangle A triangular pattern of integers, in which each entry on the left and right edges is 1, and each interior entry is equal to the sum of the two entries above it. The entry that is located k units from the left edge, on the row n units below the vertex, is C_k^n.

Permutation A specific ordering from first to last of the elements of a set; for example, an ordering of the participants in a voting system.

Pivotal voter The first voter in a permutation who, with his or her predecessors in the permutation, will form a winning coalition. Each permutation has one and only one pivotal voter.

Power index A numerical measure of an individual voter's ability to influence a decision, the individual's voting power.

Quota The minimum number of votes necessary to pass a measure in a weighted voting system.

Shapley–Shubik power index A numerical measure of power for participants in a voting system. A participant's Shapley–Shubik index is the number of permutations of the voters in which he or she is the pivotal voter, divided by the number of permutations ($n!$ if there are n participants).

Veto power A voter has veto power if no issue can pass without his or her vote. A voter with veto power is a one-person blocking coalition.

Voting combination A list of voters indicating the vote of each on an issue. There is a total of 2^n

combinations in an n-element set, and C_k^n combinations with k "yes" votes and $n - k$ "no" votes.
Weight The number of votes assigned to a voter in a weighted voting system, or the total number of votes of all voters in a coalition.
Weighted voting system A voting system in which each participant is assigned a voting weight (different participants may have different voting weights). A quota is specified, and if the sum of

the voting weights of the voters supporting a motion is at least equal to that quota, the motion is approved. The notation
$$[q : w_1, w_2, \ldots w_n]$$
is used to denote a system in which there are n voters, with voting weights $w_1, w_2, \ldots w_n$; and the quota is q.
Winning coalition A set of participants in a voting system who can pass a measure by voting for it.

✔ SKILLS CHECK

1. What would be the quota for a voting system that has a total of 20 voters and uses a simple majority quota?

(a) 10
(b) 11
(c) 20

2. For the weighted voting system [65 : 60, 30, 10],

(a) the weight-60 voter is a dictator.
(b) the weight-30 voter has veto power.
(c) the weight-10 voter is not a dummy.

3. Two daughters each hold six votes, and a son has the remaining two votes for a trust fund. The quota for passing a measure is 8. Which statement is true?

(a) The son is a dummy voter.
(b) The son is not a dummy voter but has less power than a daughter.
(c) The three children have equal power.

4. Which participants in the weighted voting system [10 : 4, 4, 3, 2] have veto power?

(a) None of the participants
(b) Only the weight-4 voters
(c) Each participant

5. What is the value of C_2^6?

(a) 12
(b) 15
(c) 32

6. For the weighted voting system [6: 4, 3, 2, 1], find the Banzhaf power index for the voter with three votes.

(a) 3
(b) 6
(c) 14

7. Calculate the Shapley–Shubik power index for the three-vote voter in the weighted voting system [6; 4, 3, 2, 1].

(a) $\frac{1}{4}$

(b) $\frac{5}{24}$

(c) $\frac{1}{12}$

8. A blocking coalition can always

(a) defeat a motion.
(b) pass a motion.
(c) force a reevaluation of a motion.

9. If a winning coalition is minimal, then the number of extra votes

(a) is zero.
(b) is less than the number of votes held by any member of the winning coalition.
(c) is less than the number of votes held by the losing coalition.

10. The best description for a voter who always wins is

(a) a voter with veto power
(b) a dictator.
(c) a pivotal voter.

11. A critical voter in a coalition is a voter

(a) who has the most votes.
(b) who has fewer votes than the number of extra votes of the coalition.
(c) whose defection changes the coalition from a winning coalition to a losing coalition.

12. The weight of a voter is

(a) the number of votes assigned to the voter.
(b) the number of times the voter is pivotal.
(c) the number of times the voter is part of a winning coalition.

13. How large is the number 8! (eight factorial)?
(a) More than a million
(b) Less than a million but more than 10,000
(c) Less than 10,000

14. In how many different ways can six voters be ordered from first to last?

(a) 30
(b) 64
(c) More than 500

15. In how many ways can six voters respond to a "yes–no" question?

(a) 12
(b) 36
(c) 64

16. What number is missing from this row of Pascal's triangle?

$$1 \ 6 \ 15 \ \underline{\quad} \ 15 \ 6 \ 1$$

(a) 20
(b) 25
(c) 30

17. A pivotal voter

(a) casts the deciding vote in a permutation if each voter votes in turn.
(b) is a dictator.
(c) can defeat any motion by voting "no."

18. A voter with veto power

(a) always wins.
(b) has the greatest voting weight.
(c) is pivotal in every permutation where he or she is the last voter.

19. $C_9^{12} =$
(a) C_3^{12}
(b) 220
(c) Both of the above

20. In the weighted voting system [7 : 3, 2, 2, 2] the number of voting combinations with exactly the quota to win is

(a) 2
(b) 3
(c) 4

EXERCISES   ■ Challenge   ◆ Discussion

How Weighted Voting Works

◆ **1.** In the United States Senate, each of the 100 senators has one vote, and the vice president of the United States can vote also if it is necessary to break a tie.

(a) A simple majority is needed to pass a bill. What constitutes a winning coalition, and what constitutes a blocking coalition?
(b) One seat in the Senate is vacant, so that only 99 senators can vote. What constitutes a winning

coalition, and what constitutes a blocking coalition for passing a bill?
(c) To ratify a treaty, a two-thirds majority is required. What constitutes a winning coalition, and what constitutes a blocking coalition?

◆ **2.** Is it possible to have a weighted voting system in which more votes are required to block a measure than to pass a measure?

3. In the weighted voting system [9 : 5, 4, 3]:
(a) Is there a dictator?

(b) Which voters, if any, have veto power?

(c) Is any voter a dummy?

4. If there are four voters, with voting weights of 30, 29, 28, and 13, and if one, and only one, of the voters has veto power,

(a) which voter has veto power?

(b) find the quota.

(c) is any voter a dummy?

Shapley–Shubik Power Index

5. Can a dummy be pivotal in any permutation? Explain why or why not.

6. A jury requires a unanimous vote to convict or to acquit. Give a quick way to determine the pivotal voter in any permutation of the jury's members.

7. For the weighted voting system [51 : 30, 25, 24, 21]:

(a) List all permutations in which the weight-30 voter is pivotal.

(b) List all permutations in which the weight-25 voter is pivotal.

(c) Calculate the Shapley–Shubik index.

8. In the voting system [7 : 3, 2, 2, 2, 2, 2]:

(a) Describe the set of permutations in which the weight-3 voter is pivotal.

(b) How many of these permutations are there?

(c) Use the answer that you have given in part (b) to determine the Shapley–Shubik index of the system.

9. How would the Shapley–Shubik index in Exercise 7 change if the quota were increased to

(a) 52?

(b) 55?

(c) 58?

10. Refer to the permutation of the 2004 presidential election (see Table 11.3). The Republican Bush–Cheney ticket carried Nevada (NV), 414,939 to 393,372. Which state would be the pivot if at the last minute their opponents, Kerry–Edwards, had broadcast an ad that convinced 1000 voters to switch from Bush–

Cheney to Kerry–Edwards? Assume that no votes are changed outside of Nevada.

Banzhaf Power Index

11. (a) List the 16 possible combinations of how four voters, *A*, *B*, *C*, and *D*, can vote either "yes" (Y) or "no" (N) on an issue.

(b) List the 16 subsets of the set {*A*, *B*, *C*, *D*}.

(c) How do the lists in parts (a) and (b) correspond to each other?

(d) In how many of the combinations in part (a) is the vote

(i) 4 Y to 0 N?

(ii) 3 Y to 1 N?

(iii) 2 Y to 2 N?

12. List all winning coalitions in the weighted voting system, [51 : 30, 25, 24, 21].

(a) Identify the critical voters in each coalition.

(b) Determine the Banzhaf power index of each participant.

(c) List all of the blocking coalitions in which the weight-30 voter is critical, and match each with its dual winning coalition in which the same voter is critical.

13. The system in Exercise 12 is modified by increasing the quota to

(a) 52.

(b) 55.

(c) 58.

(d) 73.

(e) 76.

(f) 79.

(g) 82.

Calculate the Banzhaf index in each case. (*Hint:* Increasing the quota will reduce the number of extra votes in each of the original coalitions. When the number of extra votes becomes negative, the coalition is losing and you can cross it off the list. As the number of extra votes decreases, a member of a coalition who was originally not a critical voter will become a critical voter.)

14. Determine the number of extra votes for each winning coalition, and calculate the Banzhaf index for each of the following weighted voting systems.

(a) [51: 52, 48]
(b) [3: 2, 2, 1]
(c) [8: 5, 4, 3]
(d) [51: 45, 43, 8, 4]
(e) [51: 45, 43, 6, 6]

15. Express the following decimal numbers as binary:

(a) 585
(b) 1365
(c) 2005

16. Calculate the following:

(a) C_3^7
(b) C_{100}^{50}
(c) C_2^{15}
(d) C_{13}^{15}

17. Calculate the following:

(a) C_3^6
(b) C_2^{100}
(c) C_{98}^{100}
(d) C_5^9

18. The various weighted voting systems used by the Board of Supervisors of Nassau County, New York turned out to be the mathematical quagmire described in Spotlight 11.3. Before the county's weighted voting was declared unconstitutional by a federal district court in 1993, it was changed several times. The weights in use since 1958 were as follows:

Year	Quota	Weights					
		H_1	H_2	N	B	G	L
1958	16	9	9	7	3	1	1
1964	58	31	31	21	28	2	2
1970	63	31	31	21	28	2	2
1976	71	35	35	23	32	2	3
1982	65	30	28	15	22	6	7

Here H_1 is the presiding supervisor, always from the community of Hempstead; H_2 is the second

supervisor from Hempstead; and N, B, G, and L are the supervisors from the remaining districts: North Hempstead, Oyster Bay, Glen Cove, and Long Beach.

(a) ◆ From 1970 on, more than a simple majority was required to pass any measure. Give an argument in favor of this policy from the viewpoint of a supervisor who would benefit from it, and an argument against the policy from the viewpoint of a supervisor who would lose some power.
(b) In which years were some supervisors dummy voters?
(c) Suppose that the two Hempstead supervisors always vote together. In which years are some of the supervisors dummy voters?
(d) Assume that the two Hempstead supervisors always agree, so that the board is in effect a five-voter system. Determine the Banzhaf index of this system in each year.
(e) In 1982, a special supermajority of 72 votes was needed to pass measures that ordinarily require a two-thirds majority. If the two Hempstead supervisors vote together, what is the Banzhaf index of the resulting five-voter system?
(f) ◆ Table 11.6 gives the 1980 census for each municipality, the number of votes assigned to each supervisor, and the Banzhaf index for each supervisor in 1982. Do you think the voting scheme was fair?

Comparing Voting Systems

19. Consider a four-person voting system with voters A, B, C, and D. The winning coalitions are

$$\{A, B, C, D\}, \{A, B, C\}, \{A, B, D\},$$
$$\{A, C, D\}, \text{ and } \{A, B\}$$

(a) List the minimal winning coalitions.
◆ (b) Show that {A} is a minimal blocking coalition. Are there other minimal blocking coalitions?
(c) Determine the Banzhaf power index for this voting system.
(d) Find an equivalent weighted voting system.
(e) Calculate the Shapley–Shubik index of this system.

TABLE 11.6	Nassau County Board of Supervisors, 1982			
Supervisor From	Population	Number of Votes	Banzhaf Power Index	
Quota			65	72
Hempstead (presiding)	738,517	30	30	26
Hempstead		28	26	22
North Hempstead	218,624	15	18	18
Oyster Bay	305,750	22	22	18
Glen Cove	24,618	6	2	2
Long Beach	43,073	7	6	6
Totals	1,320,582	108	104	92

◆ **20.** A five-member committee has the following voting system. The chairperson can pass or block any motion that she supports or opposes, provided that at least one other member is on her side. Show that this voting system is equivalent to the weighted voting system [4: 3, 1, 1, 1, 1].

21. Calculate the Banzhaf index for the weighted voting system in Exercise 20.

22. Find weighted voting systems that are equivalent to

(a) a committee of three faculty members and the dean. To pass a measure, at least two faculty members and the dean must vote "yes."
(b) a committee of three faculty members, the dean, and the provost. To pass a measure, two faculty, the dean, and the provost must vote "yes."

◆ **23.** A four-member faculty committee and a three-member administration committee vote separately on each issue. The measure passes if it receives the support of a majority of each of the committees. Show that this system is not equivalent to a weighted voting system.

24. Calculate the Banzhaf index of the voting system in Exercise 23. Who is more powerful according to the Banzhaf model, a faculty member or an administrator?

25. Determine the Shapley–Shubik index of the system in Exercise 23. Who is more powerful according to the Shapley–Shubik model, a faculty member or an administrator?

◆ **26.** Explain why a voting system in which no voter has veto power must have at least three minimal winning coalitions.

◆ **27.** How many *distinct* (nonequivalent) voting systems with four voters can you find? Systems that have dummies don't count. The challenge is to find all nine.

28. A corporation has four shareholders and a total of 100 shares. The quota for passing a measure is the votes of shareholders owning 51 or more shares. The number of shares owned are as follows:

A	48 shares
B	23 shares
C	22 shares
D	7 shares

There is also an investor, E, who is interested in buying shares but does not own any shares at present. Sales of fractional shares are not permitted.

(a) List the winning coalitions and compute the number of extra votes for each. Make a separate list of the losing coalitions, and compute the number of votes that would be needed to make the coalition winning.

(b) How many shares can A sell to B without causing any of the winning coalitions listed in part (a) to lose or any of the losing coalitions in part (a) to win?

(c) How many shares can A sell to D without changing the sets of winning or losing coalitions?

(d) How many shares can A sell to E without changing the winning coalitions? Since E is now a dummy, he must remain a dummy after the trade.

(e) How many shares can D sell—without changing the set of winning coalitions—to $A, B, C,$ or E? Again, it is conceivable that D would be able to sell more to one stockholder than to another.

(f) How many shares can D sell to $A, B, C,$ or E without becoming a dummy?

(g) How many shares can B sell to C without changing the set of winning coalitions?

29. Which of the following voting systems is equivalent to the voting system in use by the corporation in Exercise 28?

(a) [3: 1, 1, 1, 1]
(b) [3: 2, 1, 1, 1]
(c) [5: 3, 1, 1, 1]
(d) [5: 3, 2, 1, 1]

30. Determine the Banzhaf and Shapley–Shubik power indices for the corporation in Exercise 28.

31. A nine-member committee has a chairperson and eight ordinary members. A motion can pass if and only if it has the support of the chairperson and at least two other members, or if it has the support of all eight ordinary members.

(a) Find an equivalent weighted voting system.
(b) Determine the Banzhaf power index.
(c) Determine the Shapley–Shubik power index.
(d) Compare the results of parts (b) and (c): Do the power indices agree on how power is shared in this committee?

32. The New York City Board of Estimate consists of the mayor, the comptroller, the city council president, and the presidents of each of the five boroughs. It employed a voting system in which the city officials each had 2 votes and the borough presidents each had 1; the quota to pass a measure was 6. This voting system was declared unconstitutional by the U.S. Supreme Court in 1989 (*Morris* v. *Board of Estimate*).

(a) Describe the minimal winning coalitions.
(b) Determine the Banzhaf power index.

33. Here is a proposed weighted voting system for the New York City Board of Estimate that is based on the populations of the boroughs (see Exercise 32):

[71: 35, 35, 35, 11.3, 7.3, 9.6, 6.0, 1.8]

Find a simpler system of weights that yields an equivalent voting system.

34. The United Nations Security Council has 5 permanent members—China, France, Russia, the United Kingdom, and the United States—and 10 other members that serve two-year terms. To resolve a dispute not involving a member of the Security Council, 9 votes, including the votes of each of the permanent members, are required. (Thus, each permanent member has veto power.)

◆ (a) Show that this voting system is equivalent to the weighted voting system in which each permanent member has 7 votes, each ordinary member has 1 vote, and the quota is 39.

(b) Compute the Banzhaf index for the Security Council.

◆ (c) The Security Council originally had 5 permanent members and 6 members who served two-year terms. Each permanent member had veto power, and 6 votes were required to resolve an issue. Devise an equivalent weighted voting system and compute its Banzhaf index. Do you think that the addition of 4 more nonpermanent members involved a significant loss of power by the permanent members?

35. Find the minimal winning coalitions of the weighted voting system [7 : 3, 3, 3, 1, 1, 1] and determine the Banzhaf index.

36. A new weight-1 voter joins the system of Exercise 35. Again, describe the minimal winning coalitions and determine the Banzhaf power index. Does the presence of this new voter increase or decrease the share of power of each weight-1 voter?

37. Compute the Shapley–Shubik power index for the systems in Exercises 35 and 36. How does the addition of the new voter affect the power of the other three weight-1 voters?

38. An alumni committee consists of 3 rich alumni and 12 recent graduates. To pass a measure, a majority, including at least 2 of the rich alumni, must approve. Is this equivalent to a weighted voting system? If so, find the weights and a quota; if not, explain why not.

39. Show that if a state uses the district system to choose its electors in a two-candidate presidential election, as Maine and Nebraska do, then some electoral permutations are impossible. Give an example of an impossible permutation. How would this affect the calculation of the Shapley–Shubik power index?

✎ WRITING PROJECTS

1. The most important weighted voting system in the United States is the Electoral College (see Spotlight 11.1). Three alternate methods to elect the president of the United States have been proposed:

▶ *Direct election.* The Electoral College would be abolished, and the candidate receiving a plurality of the votes would be elected. Most versions of this system include a runoff election or a vote in the House of Representatives in cases where no candidate receives more than 40% of the vote.

▶ *District system.* This system could be adopted by individual states without amending the Constitution or passing a federal law. It is now in use by two states, Maine and Nebraska. In each congressional district, and in the District of Columbia, the candidate receiving the plurality would select one elector. Furthermore, in each state, including the District of Columbia, the candidate receiving the plurality would receive two electors. In effect, the unit rule would be retained for the District of Columbia and for states with a single congressional district. Larger states might have electors representing both parties.

▶ *Proportional system.* Each state and the District of Columbia would have fractional electoral votes assigned to each candidate in proportion to the number of popular votes received. Under this system, Governor Bush, who received 4,567,429 popular votes out of 10,965,822 cast in California in 2000, would have received

$$\frac{4,567,429}{10,965,822} \times 54 \text{ electoral votes}$$

$$= 22.4918 \text{ electoral votes}$$

Vice President Gore would have received

$$\frac{5,861,203}{10,965,822} \times 54 \text{ electoral votes}$$

$$= 28.8629 \text{ electoral votes}$$

The Green Party received 418,707 votes in California and would be entitled to 2.06188 electoral votes. There were four other parties that received between them less than one electoral vote. Obviously, no actual electors would be chosen.

Should the present electoral college, operating under the unit rule, be replaced by one of these systems? Reference: *The Presidential Election Game*, by Steven Brams, which contains useful references to Senate hearings on electoral college reform.

2. Write an essay on weighted voting in the Council of Ministers of the European Community. Compute the Banzhaf and Shapley–Shubik indices for the system as it was in 1958. In later years, the number of member nations increased significantly, and you may want to use the power index calculator, available on the Web at www.whfreeman.com/fapp. If they differ significantly in their allocation of power, which index represents the true balance of power best?

3. California has 10.22% of the votes in Electoral College, but according to Spotlight 11.4 that state has more than 11% of the power in the Electoral College, as measured by either of our power indices. Discuss the appropriateness of each power index as a measure of voting power in the Electoral College. Is the disproportionate power of California in the Electoral College a problem that the United States should address? Assume that California has acquired additional congressional seats as a result of migration. Calculate the Banzhaf index when California has 65, 75, and 100 electors. In each case take the electoral votes that are to be awarded to California from other states.

What would happen if all states, except California, adopted the district system for choosing electors? See Writing Project 1 for a discussion of the district system.

SUGGESTED READINGS

BRAMS, STEVEN J. *Game Theory and Politics*, 2nd Ed., Dover Publications, New York, 2004.

Iannucci v. *Board of Supervisors of Washington County* 20 N.Y. 2d 244, 251, 229 N.E. 2d 195, 198, 282 N.Y.S. 2d 502, 507 (1967). This code will help a law librarian find this case for you. It opened a "mathematical quagmire."

FELSENTHAL, DAN S., and MOSHÉ MACHOVER, *The Measurement of Voting Power: Theory and Practice, Problems and Paradoxes,* Edward Elgar, Cheltenham, U.K., 1998. This book has a detailed analysis of the Council of Ministers of the European Community, and a thorough treatment of the power indices mentioned in this chapter.

TAYLOR, ALAN D. *Mathematics and Politics: Strategy, Voting Power, and Proof,* Springer-Verlag, New York, 1995. Chapter 4 covers weighted voting systems and their analysis using the Shapley–Shubik and Banzhaf power indices. It has no mathematical prerequisites, but it does include carefully written logical arguments that must be carefully read.

Electing the President

Electing the president of the United States has been a tricky business since the founding of the republic. In 1824, none of the four presidential candidates received a majority of votes in the **Electoral College,** which is a body that elects the president, currently made up of 538 members. (We will say more about this institution later.) Consequently, the election went to the House of Representatives, as mandated by the Twelfth Amendment of the U.S. Constitution, wherein John Quincy Adams defeated Andrew Jackson. Although Jackson had won the most popular votes, Henry Clay threw his support to Adams in the Electoral College. Clay was subsequently appointed secretary of state by Adams in what was called the "corrupt bargain."

There were two more instances in the nineteenth century when the popular-vote winner lost the presidential election. In 1876, Rutherford B. Hayes defeated Samuel J. Tilden. In 1888, Benjamin Harrison defeated Grover Cleveland (although Cleveland got his revenge four years later when he beat Harrison). But it was the recurrence of this divided outcome in the 2000 presidential election that provoked the most controversy.

In the 2000 election, Al Gore received 537,000 more popular votes (0.5%) than did George W. Bush, but Bush won the electoral-vote tally by 4 votes. The outcome of this election turned on who would win Florida. Thirty-six days after the election, the Supreme Court, in a 5-4 decision, blocked further vote recounts in disputed Florida counties. By winning in Florida by a razor-thin margin of 537 votes (less than 0.01% of those cast), George W. Bush won the presidency. Although analysts thought there might be another divided outcome in 2004, George

W. Bush just edged out John Kerry in electoral votes (286 to 252) but beat him substantially in popular votes (by 3.3 million, or 2.8%).

We will not give further details of the 2000 election here but will focus instead on how presidential elections, in general, can be modeled as games, and what insights mathematics can provide about the strategic aspects of campaigning and elections. (In Chapter 15, we provide a more systematic introduction to game theory and discuss a variety of other applications of the theory.) There are several phases in the presidential election process. Because the rules of play change from phase to phase, the optimal strategies candidates pursue in their quest for the presidency change, too.

The first phase begins when Democratic and Republican candidates seek their party's nomination for president by running in state primaries. New Hampshire is the state that kicks off the primary season in February of a presidential election year, although it is preceded by caucuses in Iowa, where people gather in different locations throughout the state to discuss the candidates and then vote for party delegates that represent them. The strategic question in this phase is how to position oneself to gather momentum and thereby beat the competition in a set of contests spread over several months. But the process usually boils down to a heated race in the first few weeks following the New Hampshire primaries.

Because winning in the presidential primaries almost always assures a candidate of the party's nomination in its national convention in July or August, the convention is usually just a rubber stamp for the winner in the primary phase. The last exception to this rule occurred in the Democratic Party convention in 1968. After the incumbent president, Lyndon B. Johnson, withdrew from the race following the New Hampshire primary (which he had won!), Vice President Hubert H. Humphrey won his party's presidential nomination without running in the primaries.

There has not been a national party convention that has taken more than one ballot, or round of voting, to decide its presidential nominee since Adlai E. Stevenson defeated Estes Kefauver for the Democratic Party nomination in three ballots in 1952. Conventions are usually staged with consensus-promoting hoopla to unite the different factions of a party, especially if they have been sharply divided in the primaries, in preparation for the final phase of the election.

This occurs with the general election in the fall, which typically involves only two serious contenders—the nominees of the Democratic and Republican parties—but sometimes includes significant other candidates. One example is Ross Perot, who ran with no party affiliation in 1992 and garnered 19% of the popular vote. Although no minor-party candidate has ever won a presidential election, some have affected which of the two major-party nominees did win, as we will see. In the general election, the role of the Electoral College becomes paramount, as the 2000 election vividly demonstrated.

What can mathematics tell us about presidential elections? First, it can clarify what are better and worse campaign strategies in each phase. In addition, it can shed light on the likely effect that different election reforms would have on both campaign strategies and election outcomes. In fact, we will analyze two prominent reform proposals and indicate how they might have changed the outcome in the 2000 election: (1) the use of approval voting and (2) the abolition of the Electoral College and its replacement by direct popular-vote election of the president. Throughout we will highlight general results that emerge from the mathematical analysis.

We begin by looking at how candidates position themselves in presidential primaries to win their party's nomination. The principal tool of analysis is **spatial models,** which we will describe in the next section and apply to both two-candidate and multicandidate elections.

12.1 Spatial Models for Two-Candidate Elections

While two-candidate contests are most common in the general election, sometimes the nomination race in the Democratic Party or Republican Party also comes down to a contest between just two contenders. As a case in point, Gerald Ford faced one major opponent, Ronald Reagan, in the 1976 Republican race. This race was not decided until the Republican national convention in August, when Ford edged out Reagan in a close vote.

To model such elections, we assume that voters respond to the positions that candidates take on issues. This is not to say that other factors, such as personality, ethnicity, religion, and race, have no effect on election outcomes but rather that issues take precedence in a voter's decision.

How can the position of a candidate on issues be represented? We start by assuming that there is a single overriding issue on which the candidates must take a definite stand, or a set of issues related to a single underlying dimension, such as degree of governmental intervention in the economy. We assume that the attitudes of voters on this issue or dimension can be represented along a left–right continuum, ranging from very liberal on the left (much intervention) to very conservative on the right (little intervention).

To derive conclusions about the behavior of voters from their attitudes and the positions candidates take in a campaign, some assumption is necessary about how voters decide for whom to vote. More important than the attitudes of *individual* voters, however, are the *numbers* of voters who have particular attitudes along the left–right continuum.

Unimodal Distribution

A **voter distribution** can be represented by a curve that gives the number (or percentage) of voters who have attitudes at each point along a horizontal axis. The greater the vertical height of the curve, the more voters have attitudes at that point. Figure 12.1a shows one such distribution. Because this distribution has one peak, or **mode,** it is **unimodal.** For simplicity, we picture the distribution is continuous, although in fact, because the number of voters is finite, there cannot be voters at all points along the continuum.

More important than the mode, from the viewpoint of the candidates, is the *median M* of a distribution.

Median

The **median *M*** of a voter distribution is the point on the horizontal axis where half the voters have attitudes that lie to the left and half to the right.

The notion of the median of a voter distribution is closely related to the notion of a median of a data sample given in Chapter 5.

The Figure 12.1a distribution is **symmetric**—the curve to the left of *M* is a mirror image of the curve to the right. Thus, the same numbers of voters have attitudes that are equal distances to the left and to the right of *M*.

Although the *attitudes* of voters are a fixed quantity in the calculations of the candidates, the *decisions* of voters will depend on the positions that the candidates take. Assuming the candidates know the distribution of voter attitudes, what positions are optimal for them?

Assume candidates *A* (red) and *B* (blue) take the positions along the left–right axis shown in Figure 12.1a, where candidate *A* is to the left of *M* and candidate *B* is to the right. Assume that all voters vote for the candidate with attitudes closer to their own, and that all voters vote (we will consider modifications of this assumption later in the exercises). Then *A* will certainly attract all the voters to the left of his position, and *B* all the voters to the right of her position. If both candidates are an equal distance from *M*, as shown in Figure 12.1a, they will split any votes in the middle, with those to the left of *M* going to *A* and those to the right going to *B*.

Can either candidate do better by changing his or her position? If *B*'s position remains fixed, *A* could move alongside *B*, just to her left, and capture all

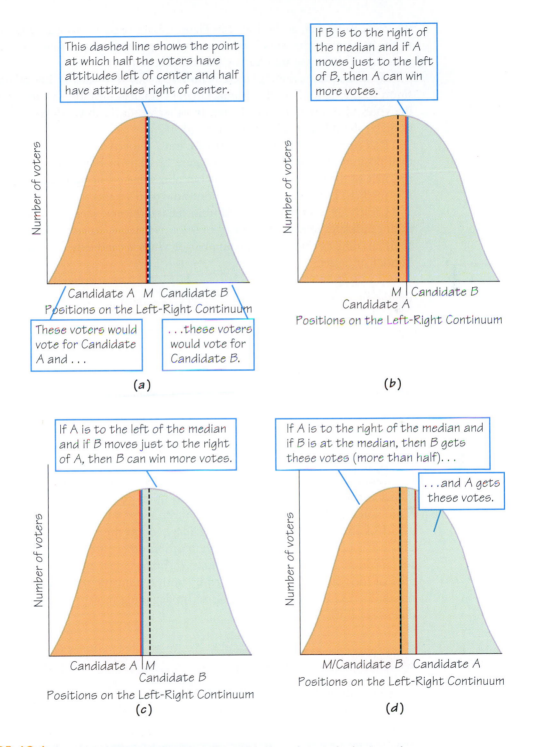

This dashed line shows the point at which half the voters have attitudes left of center and half have attitudes right of center.

These voters would vote for Candidate A and . . .

. . .these voters would vote for Candidate B.

Candidate A M Candidate B
Positions on the Left-Right Continuum

(a)

If B is to the right of the median and if A moves just to the left of B, then A can win more votes.

M Candidate B
Candidate A
Positions on the Left-Right Continuum

(b)

If A is to the left of the median and if B moves just to the right of A, then B can win more votes.

Candidate A M
Candidate B
Positions on the Left-Right Continuum

(c)

If A is to the right of the median and if B is at the median, then B gets these votes (more than half). . .

. . .and A gets these votes.

M/Candidate B Candidate A
Positions on the Left-Right Continuum

(d)

FIGURE 12.1 Unimodal distribution. The median M is the point on the horizontal axis that divides the area under the distribution curve—measuring the number of voters—exactly in half.

the votes to B's left, as illustrated in Figure 12.1b. Because A would have moved to the right of M, he would, by changing his position in this manner, receive a majority of the votes and thereby win the election.

By analogous reasoning, there is no reason for B to stick to her original position to the right of M. By approaching A's original position, B can capture all the votes to A's right (Figure 12.1c). In other words, both candidates, acting rationally, should approach each other and M.

If A were to move rightward past M, but B moved leftward only as far as M, their positions would be as shown in Figure 12.1d. Now B would receive not only the 50% of the votes to the left of M but also some of the votes that lie between B's (median) position and A's position (now to B's right).

Clearly, A loses by crossing M from the left. Hence, there is an incentive for both candidates to move toward M but not overstep it. In fact, taking a position at M maximizes the minimum number of votes a candidate can guarantee for himself or herself.

:::::: Maximin

A position is **maximin** for a candidate if there is no other position that can *guarantee* a better outcome—more votes for that candidate—whatever position the other candidate adopts.

If both candidates choose M, voters will be indifferent to the choice between them on the basis of their positions alone and would presumably make their choice on other grounds.

Taking a position at M, however, guarantees A at least 50% of the total vote *no matter what B does*. Moreover, there is no other position that can guarantee A more votes, and likewise for B.

M is also *stable*, because if one candidate adopts this position, the other candidate has no incentive to choose any position other than M. Thus, M is both the maximin position for each candidate (it offers a guarantee of a minimum of 50% of the votes) and, if M is chosen by both candidates, then these choices are in equilibrium (a candidate does worse by departing from it if the other candidate stays at it).

:::::: Equilibrium

A pair of positions is in **equilibrium** if, once chosen by both candidates, neither candidate has an incentive to depart from it unilaterally (i.e., by himself or herself).

More formally, we have the median-voter theorem.

Median-Voter Theorem

Median-voter theorem: In a two-candidate election with an odd number of voters, M is the unique equilibrium position.

We have already shown that if both candidates choose M, these choices are in equilibrium. Is there another equilibrium position or positions? There are two possibilities: (1) It is the same position for both candidates, which we call a *common position*, or (2) it is two distinct positions, one taken by each candidate. If it were a common position, suppose it is to the left of M (an analogous argument works if it is to the right). Then one candidate can always do better by moving rightward but staying to the left of M. This contradicts the supposition that the common position is in equilibrium. Now suppose the equilibrium were two distinct positions. Then one candidate can always do better by moving alongside the other candidate but staying closer to M. This contradicts the supposition that these two positions are in equilibrium. Thus, in both cases, one candidate would have an incentive to depart from his or her position—holding the position of the other candidate fixed—so a nonmedian position or positions cannot be in equilibrium. Therefore, M is the only equilibrium position.

Bimodal Distribution: Median and Mean Different

The median-voter theorem is applicable *whatever* the distribution of the electorate's attitudes. Consider the distribution in Figure 12.2a, which is **bimodal** (two peaks) and is not symmetric. Applying the logic of the previous analysis, M is once again the maximin and equilibrium position of two candidates, even though the bulk of voters are concentrated at the two modes.

We next compare the M with the mean, which may be quite different:

Mean of a Voter Distribution

The **Mean** $\bar{l}$ of a voter distribution is

$$\bar{l} = \frac{1}{n} \sum_{i=1}^{k} n_i l_i$$

where

k = number of different positions i that voters take on the continuum

n_i = number of voters at position i

l_i = location of position i on the continuum

$$n = \sum_{i=1}^{k} n_i = n_1 + n_2 + \cdots + n_k = \text{total number of voters}$$

FIGURE 12.2 Bimodal distribution in which the median and mean are different. The distribution is skewed to the left; the median M is to the right of the mean $\bar{l}$.

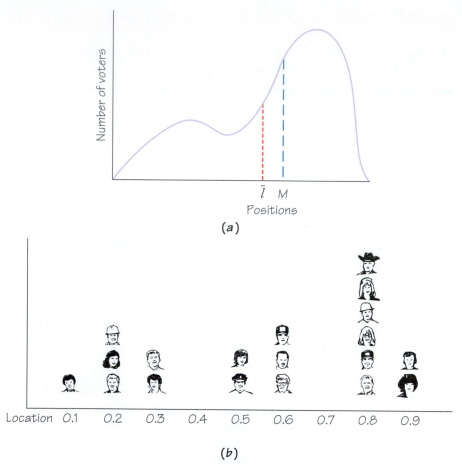

(a)

(b)

The symbol Σ (sigma) is the *summation sign*. It signifies that all subscripted terms to its right (e.g., in the definition of n), beginning with the subscript 1 and continuing to the subscript k, are summed and a total obtained.

The notion of a mean is the same as that for data samples in Chapter 5, except that here we are calculating a *weighted* average: The location l_i of each position is weighted by the number of voters, n_i, at that position. Thus, the mean can be thought of as the position of a typical voter—that is, the expected position of a voter drawn randomly from the set of all voters.

The mean $\bar{l}$ need not coincide with the median M. As an illustration of this point, consider the following discrete distribution of $n = 19$ voters at $k = 7$ different positions over the interval between 0 and 1, or [0, 1], which is illustrated in Figure 12.2b.

Position, i	1	2	3	4	5	6	7
Location (l_i) of position i	0.1	0.2	0.3	0.5	0.6	0.8	0.9
Number of voters (n_i) at position i	1	3	2	2	3	6	2

Whereas M is 0.6 because 8 voters lie to the left and 8 voters lie to the right,

$$\bar{l} = \left(\frac{1}{19}\right)[1(0.1) + 3(0.2) + 2(0.3) + 2(0.5) + 3(0.6) + 6(0.8) + 2(0.9)] = 0.56$$

Taking a position at 0.56 against an opponent who takes a position at 0.6, a candidate would lose the election by 11 to 8 votes.

The distributions of Figures 12.2a and 12.2b are **skewed** to the left–the area under the curve, or the number of voters, is less concentrated to the left of M than to the right. This necessarily pushes the M to the right of the mean $\bar{l}$. The lesson we derived from these figures is that it may *not* be rational for a candidate to take a position at $\bar{l}$ if the distribution is skewed, either to the right or to the left.

Bimodal Distribution: Median and Mean Same

A sufficient condition for M and $\bar{l}$ to coincide is that the distribution be symmetric, but this condition is not necessary: M and $\bar{l}$ can coincide if a distribution is asymmetric, as illustrated in Figure 12.3a. When M and $\bar{l}$ coincide, a candidate need not take a different position to ensure victory–or at least prevent defeat if his or her opponent adopts the same position. However, as Figure 12.3 demonstrates, the noncoincidence of M and $\bar{l}$ is not necessarily related to the lack of symmetry in a distribution: Half the voters may still lie to the left, and half to the right, of $M/\bar{l}$ if the distribution is asymmetric.

What can we say about equilibrium positions if there is an even number of voters? For example, consider the following *discrete distribution* of $n = 26$ voters at $k = 8$ different positions over the interval [0, 1], which is illustrated in Figure 12.3b.

Position, i	1	2	3	4	5	6	7	8
Location (l_i) of position i	0	0.2	0.3	0.4	0.5	0.7	0.8	0.9
Number of voters (n_i) at position i	2	3	4	4	2	3	7	1

▒ Discrete Distribution of Voters

A **discrete distribution of voters** is one in which voters are located at only certain positions along the left–right continuum.

We begin by calculating the mean $\bar{l}$:

$$\bar{l} = \left(\frac{1}{26}\right)[2(0) + 3(0.2) + 4(0.3) + 4(0.4) + 2(0.5)$$
$$+ 3(0.7) + 7(0.8) + 1(0.9)] = 0.5$$

FIGURE 12.3 Bimodal distribution in which median and mean are the same. It is not necessary for a distribution to be symmetric for the median M and mean $\bar{l}$ to coincide.

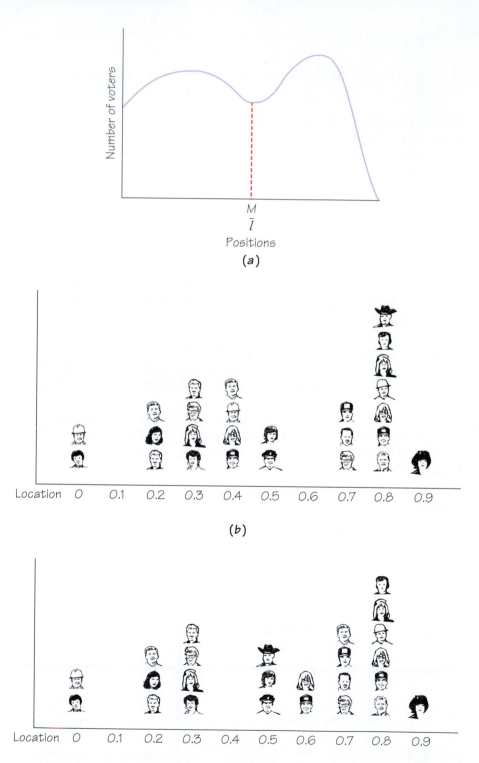

(a)

(b)

(c)

The median M is not 0.5. For an even number of voters, as in this example, M is the average of the two middle positions. The two middle voters are the 13th and 14th voters when they are lined up in the order of their positions from left to right. The 13th voter is at position 0.4, and the 14th voter is at position 0.5, so M is 0.45.

Note that if both candidates position themselves at the median M, their positions will be in equilibrium, as we showed earlier. But it is not the unique equilibrium pair. In this example, there are many other pairs of equilibrium positions. For instance, the same reasoning shows that any pair of positions between 0.4 and 0.5 will be in equilibrium. Moreover, it is easy to show that the position 0.4 for one candidate, and the position 0.5 for the other candidate, are in equilibrium. The candidate at 0.4 will get the support of the 13 voters at or to his left, and the candidate at 0.5 will get the support of the 13 voters at or to her right. If either candidate takes a different position, either to the left of 0.4 or to the right of 0.5, he or she will not be assured of 13 votes.

In general, if the number of voters is even, and the two middle voters adopt different positions, then if the candidates adopt those positions *or any pair of positions in between*, they will be in equilibrium.

It is possible that, for either an odd or even number of voters, there may not be a median position such that half the voters lie to the left and half to the right of this position. As a case in point, consider the following discrete distribution of $n = 25$ voters at $k = 8$ different positions over the interval $[0, 1]$, which is illustrated in Figure 12.3c.

Position, i	1	2	3	4	5	6	7	8
Location (l_i) of position i	0	0.2	0.3	0.5	0.6	0.7	0.8	0.9
Number of voters (n_i) at position i	2	3	4	3	2	4	6	1

We begin by calculating the mean:

$$\bar{l} = \left(\frac{1}{25}\right)[2(0) + 3(0.2) + 4(0.3) + 3(0.5) + 2(0.6)$$
$$+ 4(0.7) + 6(0.8) + 1(0.9)] = 0.52$$

At 0.6, 12 voters lie to the left of 0.6 and 11 voters lie to the right. At 0.5 and 0.7, the imbalances on the left and the right are even more lopsided. Moreover, there is no position, including the mean $\bar{l} = 0.52$, such that exactly half the voters lie to the left and half to the right.

In the absence of such a median position, is there an equilibrium? It is easy to show that 0.6 is indeed the equilibrium for two candidates. Somewhat more difficult to show is that if a distribution is discrete and there is no median position, there is still a unique position for both candidates that is in equilibrium. We call this the **extended median** because it extends the median-voter theorem to the discrete case, in which there may be no median position.

Given the stability of the median or the extended median in a two-candidate, single-issue election, is it any wonder that candidates who want to win try to avoid extreme positions? As shown in Figures 12.2a and 12.3a, even when the greatest concentration of voters does not lie at M but instead at a mode (the mode to the right of the median in both these figures), a candidate would be foolish to adopt this modal position. For although the right-leaning voters would be very pleased, the candidate's opponent could win the votes of a majority by sidling up to this position but staying just to the left.

Voters on the far left may not be particularly pleased to see both candidates situate themselves at M, which is nearer the right mode in Figure 12.2a. But in a two-candidate race, they would have nobody else to turn to. Of course, if left-leaning voters felt sufficiently alienated by both candidates, they might decide not to vote at all, which has implications we explore further later.

EXAMPLE 1   Location of Department Stores

There is a rather different application of the foregoing analysis to business, which in fact was the first substantive area to which spatial modeling was applied. Consider two competitive retail businesses, such as department stores, that consider locating their stores somewhere along the main street that runs through a city. Assume that, because transportation is costly, people will buy at the department store closer to them. Then the analysis says that no matter how the population is distributed along or near the main street, the best location is the median.

Thus, if the city's population is symmetrically distributed—that is, not skewed toward one end or the other of the main street—then this location will, of course, be at the center of the main street. Indeed, clusters of similar stores are frequently bunched together near the center of many main streets, although these stores may not be particularly convenient to people who live far from the city's center. Consequently, their location seems not in the public interest. Wouldn't it be better to have some stores near one end of the main street and some near the other, so no people are discriminated against? In an election, by contrast, not every voter can so easily be satisfied if only one candidate is to be elected, so the median seems the most attractive location in this context. ■

12.2 Spatial Models for Multicandidate Elections

Primary elections, in which candidates seek the nomination of one of the major parties, tend to attract more than two candidates. In presidential primaries, in particular, many candidates are likely to jump into the fray, especially in the states that go early in the season, if the incumbent president or vice president is not running.

Under what conditions is entry into a multicandidate race attractive? If no positions offer a potential candidate any possibility of success, then it will not be rational for him or her to enter the primary in the first place. Therefore, the rationality of entering a race, and the rationality of the positions he or she might take once there, are really two aspects of the same decision.

Suppose that two candidates have already entered a primary, and they both take positions at M. Is there any room for a third candidate?

Entry of a Third Candidate in a Two-Candidate Race

EXAMPLE 2

Look at Figure 12.4, where A and B are both at M and therefore split the vote. Now if a third candidate, C, enters and takes a position on either side of M (say, to the right), the area under the distribution to C's right may encompass less than $\frac{1}{3}$ of the total area and still enable C to win a plurality of votes.

To show why this is so, consider the portion of the electorate's vote that A/B will receive and the portion that C will receive. If C's area (yellow) is greater than half of A/B's area (blue), C will win more votes than A or B, because C's area includes not only the votes to the right of his or her position but also some votes to the left. More precisely, C will attract voters up to the point midway between his or her position on the horizontal axis and that of A/B; A and B will split the votes to the left of this midway point. Because C picks up some votes to the left of his or her position, less than $\frac{1}{3}$ of the electorate may lie to the right and still enable C to win a plurality of more than $\frac{1}{3}$ of the total vote.

By similar reasoning, it is possible to show that a fourth candidate, D, could take a position to the left of A/B and further chip away at the total of the two centrists. Indeed, D could beat candidate C, as well as A and B, by moving closer to A/B from the left than C moves from the right.

Clearly, M has little appeal, and in fact is quite vulnerable, to a third or fourth candidate contemplating a run against two centrists. Indeed, it is not dif-

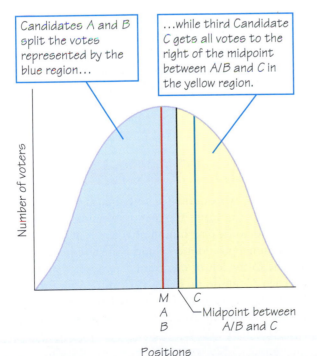

Candidates A and B split the votes represented by the blue region...

...while third Candidate C gets all votes to the right of the midpoint between A/B and C in the yellow region.

Number of voters

M
A
B

C

Midpoint between A/B and C

Positions

FIGURE 12.4
Unimodal distribution with three candidates. Candidate C can take a position with less than $\frac{1}{3}$ of the voters to his or her right and still win if candidates A and B at the median M and mean $\bar{I}$ split the remainder of the vote.

ficult to show that *whatever* positions two candidates adopt—the same or different—at least one of these candidates will be vulnerable to a third candidate. ∎

This is not to say, however, that a third candidate *C* will necessarily win against *both A* and *B*. There are both obstacles and opportunities for *C*, which are summarized in Figure 12.5 (the reasoning behind these is explored in Exercises 18 and 19).

Theodore Roosevelt
(Hulton Getty/Liaison.)

The **1/3-separation obstacle** says that if there is little room in the middle, a third candidate may beat one candidate but in so doing will hand the election to the other. This occurred in the 1912 presidential election, when Theodore Roosevelt ran as the Progressive ("Bull Moose") Party candidate after losing the Republican nomination to William Howard Taft. (Roosevelt had previously been president but had lost favor with his party after sitting out one term.) In the general election, Roosevelt received 27% of the popular vote and Taft, 24%. Both candidates were handily defeated by the Democratic candidate Woodrow Wilson, who got 41% of the popular vote. There was a fourth candidate in this race, socialist Eugene V. Debs, but he received only 6% and was never a serious threat to Wilson on the left. Wilson was also the overwhelming winner in the Electoral College.

The **2/3-separation opportunity** says that if there is a wide separation between *A* and *B*, there may be enough room in the middle for *C* to win, but this event has never occurred in a U.S. presidential election. In fact, the 1912 election is the only election in which even one major-party candidate has been defeated by a third-party candidate in the popular vote.

The stability of the two-party system in the United States may be partially explained by the fact that the two major parties, anticipating the possible entry of a third-party candidate, deliberately position themselves far enough away from the median to discourage entry on the left or right—but not so far away as to make entry in the middle advantageous. The following theoretical result gives some insight into how this can be done (the reasoning behind it is explored in Exercise 20):

Optimal Entry

The optimal entry of two candidates, anticipating a third entrant. Assume *A* and *B* are the first candidates to enter an election and anticipate the later entry of *C*. Assume that the distribution of voters is uniform (rectangular) over [0, 1]. Then the optimal positions of *A* and *B* are to enter at $\frac{1}{4}$ and $\frac{3}{4}$, whereby $\frac{1}{4}$ of the voters lie to the left of one candidate (say, *A*) and $\frac{1}{4}$ to the right of the other (*B*). Then *C* can do no better than win 25% of the vote: He or she will be indifferent among entering just to the left of *A*, just to the right of *B*, or at any position in between. At none of these positions will *C* win.

The 1/3-separation obstacle. If A and B are distinct positions that are equidistant from the median of a symmetric distribution and separated from each other by no more than $\frac{1}{3}$ of the area under the curve (so that no more than $\frac{1}{3}$ of the voters lie between A and B), C can take no position that will displace both A and B and enable C to win (see below).

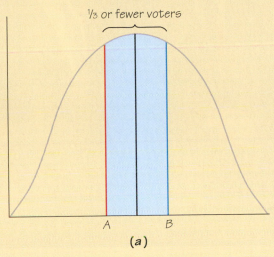

⅓ or fewer voters

A B

(a)

The 2/3-separation opportunity. If A and B are distinct positions that are equidistant from the median of a symmetric unimodal distribution and separated from each other by at least $\frac{2}{3}$ of the area under the curve (so that at least $\frac{2}{3}$ of the voters lies between A and B), C can defeat both A and B by taking a position at M (exactly between them, as shown below).

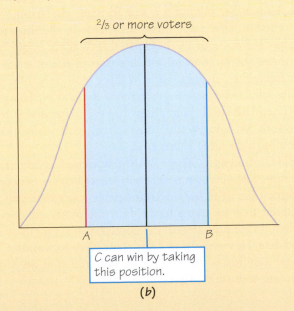

²⁄₃ or more voters

A B

C can win by taking this position.

(b)

FIGURE 12.5 The obstacles and opportunities for a third candidate, C, to enter a race.

445

A presidential debate in 2004 between John Kerry and George W. Bush. (*Brian Snyder/Reuters/ Corbis.*)

Presidential politics in the United States seems to be a reflection of both the median-voter theorem and optimal entry. For example, the median-voter theorem seems to have been operative in 1968, when the Democratic and Republican nominees, Hubert H. Humphrey and Richard M. Nixon, both presented themselves as centrists. This made them vulnerable to the third-party candidate that year, George Wallace—not in the sense that Wallace could win, but rather that he could throw the election in one direction or the other or even into the House of Representatives. In fact, while Wallace won only 14% of the popular vote in 1968, he attracted mostly supporters of Richard N. Nixon on the right, who barely defeated Hubert H. Humphrey that year. Nixon won by less than 1% in the popular vote. Without Wallace in the race, polls show that Nixon's victory would have been far more substantial.

In 1992, Bill Clinton and George Bush were viewed as quite far apart on the left–right spectrum. Ross Perot was generally viewed to be between Clinton on the left and Bush on the right, leaving considerable room in the middle that Perot could better exploit than by trying to displace one of the major-party nominees on the left or right. In winning 19% of the popular vote, Perot drew almost equally from each candidate. However, he did not come close to winning, which the optimal entry of the candidates makes difficult. (Clinton won decisively, with 43% of the popular vote to Bush's 38%.)

12.3 Winnowing the Field

Up to now, we have looked at the spatial game that candidates play as they vie to position themselves optimally in two-candidate and multicandidate races so as to (1) maximize their vote totals or (2) deter new candidates from entering. We will continue to assume that candidates take positions along a left–right spectrum, but now we consider the game from the point of view of the voters. More specifically, we ask the following question in a multicandidate race: When one candidate drops out, perhaps because of performing below expectations in an early primary, to whom will the dropout's supporters shift their votes?

Beating expectations is the name of the game in the early primaries. Thus, when Senator Edmund S. Muskie from Maine ran for the Democratic Party nomination in 1972, he was expected to do well in the neighboring state of New Hampshire. To "win" in this first primary, he had to exceed these expectations, whereas other candidates, who were not expected to do so well, could afford more mediocre performances. As in a horse race, to which primary elections are often compared, the contenders are handicapped; that is, they must beat expectations about their performances if they are to gain momentum.

Momentum, or what George H. W. Bush (the father) called the "Big Mo," can start a *bandwagon*, or a presumption that a candidate will win. The resulting **bandwagon effect** induces voters to vote for the presumed winner, independent of his or her merit.

Assume three candidates take positions, from left to right, as follows: A–B–C. Clearly, if A or C drops out, their supporters mostly likely will switch to B, giving the centrist a boost. But what if B is the first to drop out? Then it is unclear whether A, C, or neither will benefit—it depends on the number of B's supporters that prefer A next, C next, or neither (and hence may not vote at all). In any event, with B out of the race, the winner must be one of the candidates on the extremes.

The possibilities become more interesting when there are four candidates arrayed from left to right as follows: A–B–C–D. If one of the extremists, A or D, drops out, then one of the two centrists, B or C, will benefit. But what if a centrist, say C, drops out? Does this benefit one of the extremists, or does the other centrist (B) benefit? At first glance, one might think that, with only one centrist remaining, he or she will surely benefit.

This will not be the case, however, if most C supporters prefer D to B, which is certainly possible. Then the extremist D will benefit, which will be most upsetting to A's supporters. Conceivably, A's supporters might encourage A to withdraw so they can throw their support to B, whom they definitely prefer to D.

Does this sound implausible? Think back to the 2000 election, in which our four hypothetical candidates are replaced by the following ordering from left to right: Nader–Gore–Bush–Buchanan. (Recall that Ralph Nader was the Green Party candidate and Pat Buchanan the Reform Party candidate.) Just before the election, the polls were showing that Buchanan was not much of a threat to Bush, but Nader–who ended up with 2.7% of the popular vote nationwide (Buchanan got only 0.4%)–was definitely a threat to Gore. Despite pleas from some of his supporters, Nader refused to withdraw and, consequently, gave Bush a victory in Florida and a few other close states.

This 2000 scenario is not the same as the previous four-candidate hypothetical scenario, in which we argued that the extremist on the right, D, might win if one of the centrists, C, dropped out. In the 2000 scenario, the extremist on the left, Nader, could have dropped out to "save" the centrist closest to him, Gore.

Unfortunately for Gore, Nader not only refused to make this sacrifice but contended afterward that Gore's loss was due to Gore's own poor performance, not Nader's presence in the race. We will return to this issue when we discuss the effects of approval voting and the abolition of the Electoral College as possible remedies to the so-called **spoiler problem,** caused by a candidate who cannot win but "spoils" the election for a candidate who otherwise would win.

12.4 What Drives Candidates Out?

So far we have considered the possibility that candidates drop out but not why they do so. Presumably, they do so because of their poor performance in the polls or early primaries, where performance depends in part on expectations of how well they will do. But expectations change over time, and this change in turn affects how voters perceive a race–in particular, who is ahead and who is behind. On this basis, voters choose voting strategies that are likely to benefit their favorites.

Polls make public the standing of candidates in a race, as do the returns from presidential primaries. To be sure, the electorate is different in each primary state, whereas a poll is a sample from the entire electorate. While polls and primaries both provide a glimpse of the "state of the electorate," here we will focus on the change of polls over time. But our results are just as applicable to primaries, whose winnowing-out effects are evident in almost every election in which an incumbent is not running or reelection.

Suppose the election procedure is **plurality voting,** in which each voter votes for one candidate and the candidate with the most votes wins. Assume that voters rank candidates. For example, $A\ B\ C$ indicates that a voter prefers A to B to C. Before a poll, we assume that each voter votes **sincerely**–for his or her favorite candidate–because, in the absence of poll information, there is no reason to do otherwise.

EXAMPLE 3   Poll for Three Candidates

Nine voters, who can be divided into three classes, have the following preferences for candidates A, B, and C:

I. 4: $A\ C\ B$

II. 3: $B\ C\ A$

III. 2: $C\ A\ B$

Because the four voters in class I prefer A to C to B, the poll would indicate A to have 4 votes (44%), whereas B and C would have 3 and 2 votes (33% and 22%), respectively, if voters vote sincerely.

After the poll, we make the following assumption:

Poll Assumption

Voters adjust, if necessary, their sincere voting strategies to differentiate between the top two candidates as revealed in the poll, voting for the one they prefer.

Since class I and class II voters chose one of the top two candidates, only the two class III voters, who voted for C, would change their votes. Because they prefer A to B, it would be in their interest to vote insincerely for A (instead of C), thereby distinguishing A from the other top candidate, B, by giving A their votes. This would result in A's winning with 6 votes, B's getting 3 votes, and C's getting no votes.

Paradoxically, it is C who is the Condorcet winner. ■

Condorcet Winner

A **Condorcet winner** is a candidate who can defeat each of the other candidates in pairwise contests.

C is preferred to A by class II and III voters (5 to 4) and to B by class I and III voters (6 to 3). Hence, if there were a series of pairwise contests (in any order), C would win. Yet the poll not only does not make C victorious but instead magnifies A's plurality victory (4 votes) by inducing C's supporters, thinking that their candidate is out of the running, to throw their support to A, giving A a $\frac{2}{3}$-majority victory (6 out of the 9 votes).

This example can be generalized to yield the following:

Condorcet Winner Unsuccessful

Given the poll assumption, a Condorcet winner will always lose if he or she is not one of the top two candidates identified by the poll.

Why this is true is considered in Exercise 28. Here we note that even if *C* were given serious consideration in a tight race, *A* would still win with a plurality of votes.

One might think that *C*'s problem is due solely to the fact that the poll assumption puts him or her out of the running by presuming that his or her supporters "jump ship"—desert *C* for the apparently more viable candidates, *A* and *B*. But, surprisingly, even when a Condorcet winner is on top before a poll that distinguishes the top three candidates (instead of the top two), the Condorcet winner may be hurt by the poll after strategic adjustments are made by the voters, as we next illustrate.

EXAMPLE 4   Poll for Four Candidates

Add a fourth candidate, *D*, to the three in the preceding example, and assume that there are 12 voters with the following preferences for the four candidates:

$$\text{I. } 3: A\ C\ B\ D$$
$$\text{II. } 3: B\ C\ A\ D$$
$$\text{III. } 4: C\ A\ B\ D$$
$$\text{IV. } 2: D\ A\ B\ C$$

After the poll establishes that *A*, *B*, and *C* are the top three candidates, the two class IV voters would be motivated to switch to their second choice, *A*. *A* would thereby increase his or her total from 3 to 5 votes—and win after the poll. Yet *C* is again the Condorcet winner. In staying the same at 4 votes, *C* is hurt, relative to *A*, by the poll. In fact, *C* would lose to *A* after the poll (because *D*'s supporters would continue to vote for *A*), even though *C* was the winner before the poll. ■

Thus, the poll assumption, which induces strategic adjustments that favor the top two candidates, may hurt the Condorcet winner when a larger number of candidates (possibly including the Condorcet winner) are considered to be contenders. However, when one of the top two contenders distinguished by the poll assumption is the Condorcet winner, we have the following result:

Condorcet Winner Successful

Given the poll assumption, a Condorcet winner will always win if he or she is one of the top two candidates identified by the poll.

Why this is true is considered in Exercise 29. Here we note that a Condorcet winner need not be the winner in the poll but instead can place second and be successful. If a Condorcet winner places second, and the poll has the effect of turning him or her into a majority winner, then it is proper to say that the poll is instrumental in electing this candidate.

We conclude that a poll may either hurt or help a Condorcet winner. If the poll assumption is modified to distinguish more than two candidates, a Condorcet winner may be hurt *even if he or she is among those distinguished by the poll,* as the previous example showed.

12.5 Election Reform: Approval Voting

The furor caused by the divided outcome in the 2000 presidential election, in which George W. Bush won the electoral vote and Al Gore won the popular vote, has spurred efforts for reform of the election system. But except for calls for abolition of the Electoral College, whose effects we will analyze in the next section, most of the discussion has centered on making balloting more accurate and reliable and eliminating election irregularities, especially those that discriminate among different classes of voters.

Unfortunately, such reforms ignore a fundamental problem that plagues *multicandidate elections*—elections with three or more candidates—namely, that the candidate who wins under plurality voting may not be a Condorcet winner. Indeed, there will be no Condorcet winner if each candidate can be beaten by at least one other candidate. In such a situation, who is the rightful winner?

Chapter 9 presents alternatives to plurality voting. Most of these alternatives allow a voter to rank candidates from best to worst; analysts have investigated their ability to elect a Condorcet winner if one exists. Here, however, we will examine in depth a simple election reform, approval voting, that does not require voters to rank candidates.

Approval Voting

Under **approval voting,** voters can vote for as many candidates as they like or find acceptable. Each candidate approved of receives 1 vote, and the candidate with the most approval votes wins the election.

What if approval voting had been used in the 2000 presidential election? Nader supporters, knowing that voting for Nader would be only a protest vote because Nader had no chance of winning, might also have voted for Gore. In fact, because polls show that Gore was the second choice of most Nader voters, Gore almost certainly would have won in Florida and possibly a few other close states if there had been approval voting.

To be sure, Bush would have benefited from the approval votes of Buchanan supporters, but the number of these votes would not have come close to matching the number of votes Gore would have received from Nader supporters. There is therefore little doubt that Gore was the Condorcet winner in this election, because polls show that he could have defeated Bush (with help from Nader voters) as well as each of his less popular opponents in pairwise contests.

Arguments for an election reform like approval voting, however, should not be based on the outcome of only one election. Moreover, even in this one election, we cannot be entirely sure that Gore would have won under approval voting, because the nature of the campaign almost surely would have changed if there had been approval voting.

For example, John McCain, the Republican senator from Arizona who defeated George Bush in the New Hampshire primary but ultimately lost the Republican nomination to him, might have run as an independent candidate if there had been approval voting. As a centrist, he would have been attractive to both Democrats and Republicans and, conceivably, could have won under approval voting. But even if McCain had not run, it is likely that the candidates would have pitched their campaign appeals somewhat differently to try to attract as much approval as possible, especially from Nader and Buchanan supporters.

Although we cannot make precise predictions of the effects of approval voting in a presidential election like that of 2000, we can say what voters will do in certain *types* of situations:

Voting Only for a Second Choice

In a three-candidate election under approval voting, it is never rational for a voter to vote only for a second choice. If a voter finds a second choice acceptable, he or she should also vote for a first choice.

This is certainly not true under plurality voting. If you were a Nader supporter *and* found Gore also acceptable as a second choice, you should have voted for Gore rather than Nader if (1) you thought Gore could win but Nader could not and (2) electing an acceptable candidate was important to you. (Certainly some Nader supporters switched to Gore for these reasons.) By comparison, because you lose nothing by voting for *both* Nader and Gore under approval voting—and may gain by doing so (if Nader cannot win, you at least help Gore)—you should never vote for just your second choice (Gore in this example).

Why voting only for a second choice is never rational is considered in Exercise 35. In effect, a voter whose favorite candidate in a three-candidate race seems to be out of the running can have his or her cake and eat it, too—by casting a *sincere* vote for a favorite candidate and a *strategic* vote for a second choice (to try to prevent a worst choice from winning). Roughly speaking, **strategic voting** (for instance, by Nader supporters for Gore in the 2000 presidential election) is voting that is not sincere but nevertheless has a strategic purpose—namely, to elect an acceptable candidate if one's first choice is not viable.

Another general result about approval voting that is helpful to know uses the concept of dichotomous preferences:

Dichotomous Preferences

A voter has **dichotomous preferences** if he or she divides the set of candidates into two subsets—a preferred subset and a nonpreferred subset—and is indifferent among all candidates in each subset.

In other words, a dichotomous voter sees the world in two colors, white and black, and there is nothing in between. True, most of us see grays, but it is useful to analyze the dichotomous case first. In this case, it is not hard for voters to make a rational choice by choosing a **dominant strategy,** which is a strategy that is at least as good as, and sometimes better than, any other strategy for that person. With this definition, we can show a general condition under which Condorcet winners will be elected under approval voting:

Effect of Dichotomous Preferences

A Condorcet winner will always be elected under approval voting if all voters have dichotomous preferences and choose their dominant strategies.

A dichotomous voter's dominant strategy is to vote for all candidates in his or her preferred subset and no others. This strategy is dominant because the preferred candidates are assumed all to be equally good, so a voter has no reason to distinguish among them. Furthermore, the voter has no reason to vote for a nonpreferred candidate or candidates because they are all equally bad and voting for any one of them could help that candidate win. As an illustration of the effect of dichotomous preferences, consider the following example.

Dichotomous Preferences

EXAMPLE 5

For each class of voters, the preferred subset of candidates is enclosed in the first set of parentheses, the nonpreferred subset in the second set of parentheses. Thus, the four class I voters prefer A and B, between whom they are indifferent, to C and D, between whom they are also indifferent:

I. 4: $(A\ B)\ (C\ D)$
II. 3: $(C)\ (A\ B\ D)$
III. 2: $(B\ C\ D)\ (A)$

Assuming that each class of voters chooses its dominant strategy, B wins with 6 votes to 5 votes for C and 4 votes for A.

In pairwise contests, notice that B is preferred to A by the two class III voters (class I and II voters are indifferent between these two candidates), so B would defeat A by 2 to 0 votes. (We assume that indifferent voters express no preference.) Because B is preferred to C by the four class I voters, and C is preferred to B by the three class II voters (class III voters are indifferent between these two candidates), B would defeat C by 4 to 3 votes. Thus, B, the approval-vote winner, is also the Condorcet winner, which must always be the case when voter preferences are dichotomous and voters vote for all their approved candidates. ■

Insofar as voters in the 2000 presidential election thought equally well (or badly) of Bush and Buchanan on the one hand, and Gore and Nader on the other, they would have preferences like those of class I voters in the previous example. Of course, most voters probably made finer distinctions, in which case there is no guarantee that approval voting will elect a Condorcet winner.

12.6 The Electoral College

As we have noted, the Electoral College had a decisive effect in the 2000 presidential election. In winning the popular vote in Florida by the slimmest of margins, George Bush captured all 25 of Florida's electoral votes, which gave him a majority in the Electoral College. This won him the presidency even though he lost the popular vote.

What is the justification for the Electoral College? Its original purpose was to place the selection of a president in the hands of a body that, while its members would be chosen by the people, would be sufficiently removed from them that it could make more deliberative choices. As for its composition, each state gets 2 electoral votes for its two senators (total for all states: 100). In addition, a state receives 1 electoral vote for each of its representatives in the House of Representatives, whose numbers are based on population (see Chapter 11) and range from 1 representative for the seven smallest states to 53 representatives for the largest state, California. The House has a total of 435 representatives. The District of Columbia, like the smallest states, is given 3 electoral votes. Altogether, there are 538 electoral votes, and a candidate needs 270 to win. In 2000, George W. Bush got 271 electoral votes.

Although there is nothing in the U.S. Constitution mandating that the popular-vote winner in a state receive all its electoral votes, this has been the tradition almost from the founding of the republic. Only in Maine and Nebraska can the electoral votes be split among candidates, depending on who wins each of the two congressional districts in Maine and the three congressional districts in Nebraska. Because the statewide winner receives the two senatorial electoral votes, the closest split possible in these two states is 3-1 in Maine and 3-2 in Ne-

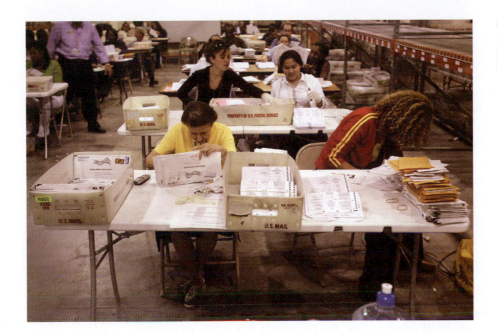

Checking ballots in Florida after the 2000 presidential election. (*Joe Raedle/Getty Images.*)

braska. In the actual election, Gore won all of Maine's 4 electoral votes, and Bush won all of Nebraska's 5 electoral votes, so winner-take-all prevailed in all 50 states and the District of Columbia.

Effectively, then, the presidency is decided by 51 players: members of the Electoral College from each of the 50 states and the District of Columbia, who almost always cast their votes as blocs for a single candidate. The voting weights of states, which depend in part on their populations, are related to their voting power (see Chapter 11).

Although the percentage of voting power of a state closely tracks its percentage of electoral votes, this is not the full story. More important is the power of *individual* voters in each state, based on their ability to be pivotal in their states and their states, in turn, to be pivotal in the Electoral College. Amazingly, individual voters in California are about three times as powerful as individual voters in the smallest states, despite the fact that the smallest states (with only one representative) get a 200% (2/1) boost from having 2 "senatorial" electoral votes— besides the 1 electoral vote they are entitled to on the basis of their populations— whereas California receives less than a 4% (2/53) boost.

But there is more to the story than just the size of states. Because California was never considered a close state in the 2000 presidential election (polls indicated that it would almost surely go for Al Gore), it received relatively little attention from both candidates. The real battle was fought in the so-called battleground states, or *toss-up states,* where the outcome was expected to be close (as it certainly was in Florida!). These states received the bulk of the candidates' time, money, and other resources.

Instead of viewing the Electoral College as a 105-million-person game in 2000, in which the voters are the players and their power is a function of the size of the states in which they vote, we view it as a game between the two major-party candidates. We will develop two different models: the first in which the candidates seek to maximize their expected popular vote, and the second in which they seek to maximize their expected electoral vote, in the toss-up states that determine the outcome in a close race.

Common to both models is the assumption that the probability, p_i, that a voter in toss-up state i votes for the Democratic candidate is

$$p_i = d_i/(d_i + r_i)$$

or the proportion of campaign resources that the Democratic candidate (d_i), compared with the Republican (r_i), spends in state i. The probability that a voter in state i votes for the Republican candidate will be the probability $1 - p_i = r_i/(d_i + r_i)$. Thus, we ignore the effects of other candidates in the race and assume that either the Democratic or the Republican candidate will win with certainty: $p_i + (1 - p_i) = 1$. This assumption is plausible in light of the fact that no third-party candidate has ever won the presidency and rarely any states.

Expected Popular Vote

The **expected popular vote** (**EPV**) of the Democratic candidate in toss-up states, EPV_D, is the number of voters, n_i, in toss-up state i, multiplied by the probability, p_i, that a voter in toss-up state i votes for the Democrat, summed across all toss-up states:

$$EPV_D = \sum_{i=1}^{t} n_i p_i$$

where t is the number of toss-up states.

EPV_D bears some similarity to the expression for the mean ($\bar{l}$) discussed earlier. Whereas $\bar{l}$ is an average weighted by the proportions, n_i/n, EPV_D is an average weighted by the probabilities p_i.

The candidates seek strategies for optimally allocating their resources to each toss-up state. Recall that d_i for the Democrat and r_i for the Republican are the resources each candidate allocates to state i. For the Democrat, if d_i changes, this affects the value of p_i, the probability that a voter votes for him or her, which in turn affects the value of EPV_D. Thus, the Democrat seeks a strategy d_i that will make EPV_D as large as possible.

Proportional Rule

The strategy of the Democrat who maximizes his or her EPV_D (indicated by the asterisk), given that the Republican also chooses a maximizing strategy, is

$$d_i^* = (n_i/N)D$$

where $N = \sum_{i=1}^{t} n_i$, the total number of voters in the toss-up states, and $D = \sum_{i=1}^{t} d_i$, the sum of the Democrat's expenditures across all states.

In words, the Democrat should allocate his or her resources in proportion to the size of each state (n_i/N) if the Republican behaves similarly by following a strategy of $r_i^* = (n_i/N)R$, where $R = \sum_{i=1}^{t} r_i$.

To show that d_i^* maximizes EPV_D when the Republican chooses r_i^* requires calculus, but we can readily illustrate why departures from d_i^* by the Democrat will cost him or her popular votes.

Departures from a Popular-Vote Maximizing Strategy

EXAMPLE 6

Suppose there are three toss-up states with 2, 3, and 4 electoral votes and the candidates, because they accepted public financing for the election, are limited to spending the same total of $63 million (M). If the Republican follows his or her optimal strategy of spending in the proportion 2:3:4 ($14M:$21M:$28M), but the Democrat, ignoring the smallest state, spends in the proportion of 0:3:4 ($0M:$27M:$36M), the Republican will receive on average

$$EPV_R = 2[14/14] + 3[21/(21 + 27)] + 4[28/(28 + 36)] = 5.06 \text{ votes}$$

or 56% of the 9 votes in the three states.

If the Republican anticipates that the Democrat will spend nothing in the smallest state, the Republican can do even better. By spending only a minuscule amount in the smallest state, the Republican can almost match the Democrat in the other two states and win an average of about 5.5 votes (2 votes from the smallest state and $7/2 = 3.5$ votes from the other two states), or 61%. ∎

Let's assume that the goal of the candidates is not to maximize EPV but, instead, their *expected electoral vote (EEV)*, which is an entirely different quantity that we define below. To illustrate the difference, a candidate who wants to maximize EEV might think of throwing all of his or her resources into the 11 largest states—and ignoring the 39 other states and the District of Columbia if all states are toss-up states—because the 11 largest states have a majority of electoral votes (271 to be exact). Moreover, the candidate need not win "big" in these states. Winning them by small margins will work just fine, because the candidate will still win *all* their electoral votes and thereby the election.

But this strategy has a problem. An opponent can readily defeat it by spending very small amounts in all the other states, which will defeat the candidate if he or she spends nothing in these states. In addition, by using his or her leftover funds to outspend the candidate in, say, one or two big states, the opponent will end up winning more electoral votes. However, there is a counterstrategy to this strategy, and indeed to every other pure strategy (no randomization—see Chapter 15) one can think of in a winner-take-all system like that of the Electoral College.

To prevent being exploited if an opponent anticipates one's strategy and selects a best counterstrategy against it, each candidate should try to keep secret exactly what he or she intends to do. The best way to keep a secret is to randomize one's choices, using mixed strategies. But mixed strategies are difficult to calculate in a system as complicated as that of the Electoral College, so we make simplifying assumptions. First, however, we need a definition.

Expected Electoral Vote

The **expected electoral vote (*EEV*)** of the Democratic candidate in toss-up states, EEV_D, is

$$EEV_D = \sum_{i=1}^{t} v_i P_i$$

where v_i is the number of electoral votes of toss-up state i, and P_i is the probability that the Democrat wins *more than* 50% of the popular votes in this state, which would give the Democrat *all* that state's electoral votes, v_i.

To determine P_i, we need to count the number of ways that a candidate can win a majority of electoral votes in each state i and then compute the probabilities that each of these ways occurs. These probabilities, in turn, are used to determine EEV_D. EEV_R can be defined similarly.

EXAMPLE 7   Computing the Democrat's Expected Electoral Vote

Consider our earlier example of three states, *A*, *B*, and *C*, with 2, 3, and 4 electoral votes. For simplicity, assume here that the number of electoral votes of each state is equal to the number of voters in that state.

To calculate P_i for each state i, we must determine the probabilities that a majority of voters in states *A*, *B*, and *C* vote Democratic. (We will ignore the possibility of ties in states *A* and *C*, which have an even number of voters.) To obtain these probabilities, we multiply the probabilities that individual voters in each state, who are assumed to act independently of each other, vote Democratic or Republican (based on the resources the two candidates allocate to each state).

For state A to vote Democratic, for example, both voters in this state must vote Democratic, so $P_A = (p_A)(p_A) = (pp_A)^2$. For state B to vote Democratic, either two of the three voters (in 3 possible ways) or all three voters must vote Democratic, so $P_B = 3[(p_B)^2(1 - p_B)] + (p_B)^3$. For state C to vote Democratic, either three of the four voters (in 4 possible ways) or all four voters must vote Democratic, so $P_C = 4[(p_C)^3(1 - p_C)] + (p_C)^4$. Substituting these probabilities into the formula for EEV_D, we obtain

$$EEV_D = v_A P_A + v_B P_B + v_C P_C$$
$$= 2[(p_A)^2] + 3[3(p_B)^2(1 - p_B) + (p_B)^3] + 4[4(p_C)^3(1 - p_C) + (p_C)^4] \blacksquare$$

We indicated earlier that the strategy of the Democrat that maximizes EEV_D, given that the Republican adopts a similar strategy, is mixed, involving randomizing his or her choice of states in which to allocate resources. Because this randomization is difficult to determine, we simplify the task by assuming that $d_i = r_i$ in each toss-up state i, and necessarily $D = R$ across all toss-up states.

This assumption is defensible if the candidates have the same total amount to spend in all the states ($D = R$). If they perceive the value of each toss-up state to be the same, which is reasonable, they will allocate equal resources to each. But how much should these amounts be? It is possible to show the following:

The 3/2's Rule

The strategies of the Democratic and the Republican candidates that maximize their *EEV*s are

$$d_i^* = \left(\frac{v_i \sqrt{n_i}}{S} \right) D \qquad r_i^* = \left(\frac{v_i \sqrt{n_i}}{S} \right) R$$

where

$$S = \sum_{i=1}^{t} v_i \sqrt{n_i}$$

In words, the candidates should allocate their resources in proportion to the number of electoral votes of each state (v_i) multiplied by the square root of its size (n_i). The allocations, d_i^* and r_i^*, will be the same if the candidates spend equally in each toss-up state ($d_i = r_i$), as previously assumed.

Because the number of electoral votes in each state i (v_i) is approximately proportional to the number of voters in each state (or their populations), n_i, the maximizing strategies of the candidates can be approximated by

$$d_i^* = \left(\frac{v_i^{3/2}}{T}\right) D \qquad r_i^* = \left(\frac{v_i^{3/2}}{T}\right) R$$

where

$$T = \sum_{i=1}^{t} v_i^{3/2}$$

Thus, if the candidates allocate the same amount of resources to each of the toss-up states, the 3/2's rule is that they should spend approximately in proportion to the 3/2's power of the electoral votes of these states to maximize EEV.

EXAMPLE 8   Applying the 3/2's Rule

Assume states A, B, and C have, respectively, 9, 16, and 25 voters, and these are also their numbers of electoral votes. If each of these states is a toss-up state, the 3/2's rule says that the candidates should allocate their resources in the proportions 27:64:125, because the 3/2's powers of their electoral votes are their numbers of voters multiplied by the square root of these numbers. Thus for state A, $9^{3/2} = 9\sqrt{9} = (9)(3) = 27$.

To illustrate the difference between the proportional rule (which maximizes EPV) and the 3/2's rule (which maximizes EEV), assume both candidates can spend 100 units of resources. Then the proportional rule says that states A, B, and C should get resources in approximately the amounts 18, 32, and 50, whereas the 3/2's rule says these states should get resources in the amounts 13, 30, and 58.

Clearly, the smallest state gets less and the largest state gets more under the 3/2's rule, whereas the middle states stay about the same. In the actual Electoral College, the voters in California, when it is a toss-up state, are about three times as attractive per capita as voters in the smallest states. Following the 3/2's rule, therefore, the candidates should allocate about three times as much per voter to California as to a small toss-up state with only three electoral votes.

In fact, presidential candidates greatly overspend in the largest toss-up states, well out of proportion to their size. This large-state bias is far out of line with the democratic principle of "one person, one vote." For Californians, compared to small-state voters, this principle should read "one person, three votes." ∎

To be sure, it is not the Electoral College itself that creates this bias but, rather, its winner-take-all feature. If this feature were abolished and the electoral votes of a state were split according to the popular votes received by the candidates—insofar as possible—then the large-state bias would disappear.

Departures from an Electoral-Vote Maximizing Strategy

EXAMPLE 9

We illustrated earlier how a candidate's departure from the popular-vote maximizing strategy—the proportional rule—lowers that candidate's expected popular vote, given that the candidate's opponent adheres to this rule. Because of the loss candidates suffer when they depart from the proportional rule, it is an equilibrium strategy for both. This is also true of *some* departures from the 3/2's rule: If a candidate's departure from this rule is "small," he or she will lower his or her expected electoral vote, given the candidate's opponent sticks to that rule.

Suppose, for example, that the Republican follows the 3/2's rule in allocating resources to states A, B, and C with 9, 16, and 25 voters/electoral votes, respectively. If each candidate has 100 units of resources to spend, we showed in the previous example that this translates into allocating approximately 13, 30, and 58 units to states A, B, and C, respectively.

Now if the Democrat deviates slightly from the 3/2's rule, and allocates 14, 30, and 57 units to these states (more to A, less to C), he or she increases the chances of winning in A and decreases the chances of winning in C. It can be shown that this deviation from the 3/2's rule hurts the Democrat, because even though his or her chances go up more in A than they go down in C (because A is smaller than C), C has almost three times as many electoral votes. On balance, this deviation lowers the Democrat's expected electoral vote.

Now suppose the Democrat makes a "large" deviation from the 3/2's rule, ignoring state B entirely and throwing all his or her resources into state A (9 electoral votes) and state C (25 electoral votes). If he or she wins in these two states, this would give the Democrat 34 of the 50 electoral votes, which is more than enough to win.

If the Republican adheres to the 3/2's rule, he or she will put approximately 13% of his or her resources into state A and 58% into state C. By following the 3/2's rule in these two states and ignoring state B, the Democrat will put approximately 18% into state A and approximately 82% into state C. This translates into *each voter* in states A and C supporting the Democrat with probability 0.58, which means that the Democrat will most certainly win in both these states, giving him or her an expected electoral vote of almost 34. The Republican *will* certainly win in state B, because the Democrat spends nothing in this state, but this is small consolation if the Republican loses in the two other states and receives an expected electoral vote of somewhat more than 16. ∎

This example illustrates why the 3/2's rule is a local maximum but not a global maximum.

Local Maximum and Global Maximum

A **local maximum** is a maximizing strategy from which small deviations are nonoptimal but large deviations may be optimal. A **global maximum** is a maximizing strategy from which *all* deviations (small or large) are nonoptimal. The proportional rule is a global maximum (and equilibrium) for candidates whose goal is to maximize their expected popular vote, whereas the 3/2's rule is only a local maximum for candidates whose goal is to maximize their expected electoral vote.

If the goal of candidates is to maximize their expected electoral vote, there is no *determinate* maximizing strategy—randomizing one's choices is necessary to prevent exploitation by an opponent. How this randomization is done for some simple games is analyzed in Chapter 15.

In the case of the Electoral College, we illustrated how a radical departure from the 3/2's rule by a candidate, who ignores some state or states entirely, may be the best response to an opponent who follows this rule. But then there is a best response to this best response, and so on, so no determinate strategy is invulnerable.

However, insofar as the candidates view the toss-up states in similar terms, the 3/2's rule offers a good rule of thumb as to how much to spend in each, as a function of its size, to maximize their expected electoral vote. But we must remember that it is only a local maximum. Hence, unlike the proportional rule, which is robust against all other popular-vote maximizing strategies, the 3/2's rule is vulnerable to radically different strategies, such as those in which candidates concentrate their efforts on relatively few states.

Usually those who try such strategies, however, take big risks. In 1964, the Republican presidential candidate, Barry Goldwater, said that he would like to "saw off the Eastern seaboard" (Goldwater was from Arizona), but in the end he went, in his memorable phrase, "shooting where the ducks [voters] are." In doing so, however, he appeared not so much to want to win as to present voters with "a choice, not an echo," by taking relatively extreme (conservative) positions. Is it any wonder, then, that Goldwater lost in a huge landslide to his Democratic opponent, Lyndon Johnson?

12.7 Is There a Better Way to Elect a President?

The quest for the presidency is the greatest spectacle in American politics. While there is nothing to match its excitement, especially when the race is close, the quieter gamelike features of a presidential campaign are no less consequential.

We have emphasized these features in this chapter, showing how mathematics can be used to analyze optimal positions in two-candidate and multicandidate races, and how polls and presidential primaries may affect who stays in and who drops out of the race. Sometimes, as we have seen, Condorcet candidates, who in pairwise contests can beat every other candidate, may not survive. And, as was dramatically illustrated in the 2000 presidential election, the popular-vote winner may not win in the Electoral College.

Some people think that approval voting would better enable voters, especially in the early presidential primaries—which typically draw many candidates (if an incumbent is not running for reelection)—to express their preferences. However, other election procedures, including those discussed in Chapter 9, possess features that may make them desirable as election reforms.

All these procedures would probably be of more help to centrist candidates, who not only better represent the entire electorate than extremist candidates but who also are more likely to be a party's strongest contender in the general election. In the past 40 years, the biggest losers in presidential elections, Republican Barry Goldwater in 1964 and Democrat George McGovern in 1972, came from the right and left extremes, respectively, of their parties.

Taking the choice of a president out of the hands of voters and putting it in the hands of members of the Electoral College may no longer be justified. The Electoral College, with its winner-take-all feature, creates a large-state bias, as we showed. In the 2000 presidential election, a few hundred voters in one large toss-up state, Florida, determined the outcome.

If one thinks that the votes of *all* voters, wherever they reside, should count equally, then direct popular-vote election of a president would best accomplish this goal. Short of abolishing the Electoral College, however, which would require a constitutional amendment, electoral votes could be allocated proportionally in each state to approximate this goal.

If approval voting were used, then it would be approval votes rather than the single votes of each voter that would determine the allocation of electoral votes to the candidates. Because the general election in recent years has drawn major third- and fourth-party candidates, approval voting, or one of the other voting procedures discussed in Chapter 9, seems worthy of consideration if one wants to reduce the role of spoilers.

In summary, mathematics illuminates strategic aspects of campaigning and voting in presidential elections not apparent to the naked eye. It also points the way to possible reforms that may ameliorate some of the problems that plague our current system.

REVIEW VOCABULARY

Approval voting Allows voters to vote for as many candidates as they like or find acceptable. Each candidate approved of receives one vote, and the candidate with the most approval votes wins.
Bandwagon effect Voting for a candidate not on the basis of merit but, instead, because of the expectation that he or she will win.

Condorcet winner A candidate who can defeat each of the other candidates in pairwise contests.
Dichotomous preferences Held by voters who divide the set of candidates into two subsets—a preferred subset and a nonpreferred subset—and are indifferent among all candidates in each subset.

Discrete distribution of voters A distribution in which voters are located at only certain positions along the left–right continuum.

Dominant strategy A strategy that is at least as good as, and sometimes better than, any other strategy.

Electoral College A body of 538 electors that selects a U.S. president.

Equilibrium position A position is in equilibrium if no candidate has an incentive to depart from it unilaterally.

Expected electoral vote (*EEV*) In toss-up states, the number of electoral votes of each toss-up state, multiplied by the probability that the Democratic (or Republican) candidate wins more than 50% of the popular votes in that state, summed across all toss-up states.

Expected popular vote (*EPV*) In toss-up states, the number of voters in each toss-up state, multiplied by the probability that that voter votes for the Democratic (or Republican) candidate, summed across all toss-up states.

Extended median The equilibrium position of two candidates when there is no median.

Global maximum A maximizing strategy from which *all* deviations (small or large) are nonoptimal.

Local maximum A maximizing strategy from which small deviations are nonoptimal but large deviations may be optimal.

Maximin position A position is maximin for a candidate if there is no other position that can guarantee a better outcome—more votes—whatever position another candidate adopts.

Mean ($\bar{I}$) A weighted average, wherein the positions of voters are weighted by the fraction of voters at that position.

Median (M) The point on the horizontal axis of a voter distribution where half the voters have attitudes that lie to the left and half to the right.

Median-voter theorem In a two-candidate election with an odd number of voters, the median is the unique equilibrium position.

Mode A peak of a distribution. A distribution is **unimodal** if it has one peak, and **bimodal** if it has two peaks.

1/3-separation obstacle An obstacle for the entry of a third candidate created if two previous entrants are sufficiently close together.

Poll assumption Voters adjust their sincere voting strategies, if necessary, to differentiate between the top two candidates—as revealed in the poll—by voting for the one they prefer.

Plurality voting Allows voters to vote for one candidate, and the candidate with the most votes wins.

Proportional rule Presidential candidates allocate their resources to toss-up states according to their size. This allocation rule maximizes the expected popular vote of a candidate, given that his or her opponent adheres to it. It is a global maximum.

Sincere voting Voting for a favorite candidate, whatever his or her chances are of winning.

Spatial models The representation of candidate positions along a left–right continuum in order to determine the equilibrium or optimal positions of the candidates.

Spoiler problem Caused by a candidate who cannot win but "spoils" the election for a candidate who otherwise would win.

Strategic voting Voting that is not sincere but nevertheless has a strategic purpose—namely, to elect an acceptable candidate if one's first choice is not viable.

3/2's rule Presidential candidates allocate their resources to toss-up states according to the 3/2's power of their electoral votes. This allocation rule maximizes the expected electoral vote of a candidate, given that his or her opponent adheres to it. It is a local maximum.

2/3-separation opportunity An opportunity for the entry of a third candidate created if two previous entrants are sufficiently far apart.

Voter distribution Gives the number (or percentage) of voters who have attitudes at points along the left–right continuum, which can be represented by a curve. The distribution is **symmetric** if the curve to the left of the median is a mirror image of the curve to the right. It is **skewed** if the area under the curve is concentrated more on one side of the median than the other.

✔ SKILLS CHECK

1. In a two-candidate election, suppose the attitudes of the voters are distributed symmetrically with median M. Of the two candidates A and B, A is positioned far to the left of M and B is positioned just to the right of M. How will the voters choose between candidates A and B?

(a) A will receive a majority of the votes.
(b) B will receive a majority of the votes.
(c) A and B will both receive exactly one-half of the votes.

2. In a three-candidate election, suppose the attitudes of the voters are distributed symmetrically with median M. Of the three candidates A, B, and C, A is positioned far to the left of M, B is positioned just to the right of M, and C is positioned at M. How will the voters choose among these three candidates?

(a) A will receive the most votes.
(b) B will receive the most votes.
(c) C will receive the most votes.

3. In a two-candidate election, which of the following positions is an optimal position for both candidates A and B?

(a) A and B just to the left and right of M
(b) A and B far to the left and right of M
(c) A and B both at M

4. In a three-candidate election, if candidates A and B are positioned at M, what is an election-winning position for candidate C?

(a) Just to the right of M
(b) Far to the right of M
(c) There is no election-winning position for candidate C.

5. In a three-candidate election, if candidates A and B are positioned just to the left and just to the right of M, what is an election-winning position for candidate C?

(a) At M
(b) Far to the right of M

(c) There is no election-winning position for candidate C.

6. In a four-candidate election, if candidates are aligned in order A–B–C–D, who benefits if D drops out of the race?

(a) Candidate A
(b) Candidate B
(c) Candidate C

7. When is a maximin position not an equilibrium position?

(a) When it is not the median
(b) When it is the median
(c) When it is the mean

8. Why is it desirable that two candidates take median positions but undesirable that two department stores locate themselves at the center of a main street?

(a) Only one candidate can win, and a median choice is not too far away from anybody.
(b) Two department stores at different ends of a main street are closer to most consumers than one in the center.
(c) Both (a) and (b)

9. In a three-candidate election, suppose 12 voters can be divided into three classes according to their preferences: Five voters prefer (in order) A, B, C; 4 voters prefer C, A, B; 3 voters prefer B, C, A. To elect one of their top 2 candidates, which group of voters will not vote sincerely (for their first choice)?

(a) The 5 voters
(b) The 4 voters
(c) The 3 voters

10. Making the poll assumption, the Condorcet winner will win

(a) precisely when he or she is not the top candidate identified by the poll.
(b) precisely when he or she is one of the top two candidates identified by the poll.
(c) regardless of his or her rank in the poll.

11. In an election with a large number of candidates, approval voting benefits

(a) candidates at the extreme left and right.
(b) candidates at or near M.
(c) only candidates precisely at M.

12. In a four-candidate approval voting election with 12 voters, if 5 voters approve of A and B, 4 voters approve of B and C, and 3 voters approve of A and D, who wins the election?

(a) Candidate A
(b) Candidate B
(c) Candidate C

13. In Exercise 12, assume that the 5 voters who approve of A and B actually prefer A. Would they have an incentive to vote strategically?

(a) Yes
(b) No
(c) Some would and some wouldn't.

14. In the election of the president using the Electoral College, voters in smaller swing states have

(a) less power than voters in large swing states.
(b) more power than voters in large swing states.
(c) the same power as voters in large swing states.

15. In the election of the president, if the Democrats believe that the Republicans will not allocate resources in a way that maximizes their EEV, then

(a) they can successfully counter by allocating their resources according to the method that maximizes EEV.
(b) they can successfully counter by not allocating their resources according to the method that maximizes EEV.
(c) they cannot successfully counter.

16. Suppose three voting blocs, A, B, and C, control 16, 25, and 36 votes, respectively, and 39 votes are required to win. If each bloc is a toss-up, for every \$1 that is allocated to lobby the voters in bloc A, how much should be allocated to lobby the voters in bloc C?

(a) About 2 times as much
(b) About 3 times as much
(c) About 20 times as much

17. In Exercise 16, suppose your opponent spends all his money lobbying blocs A and B and ignores bloc C. What would be a good strategy to use in response?

(a) Lobby just A
(b) Lobby just B
(c) Lobby A and C

18. Voting for president using the Electoral College is

(a) biased in favor of the largest states.
(b) biased in favor of the smallest states.
(c) not biased.

19. In Exercise 18, is your answer correct if the only swing states are medium-size?

(a) Yes
(b) No, it's biased in favor of medium states.
(c) No, it's biased against medium states.

20. Is it possible that a candidate for president could win the popular vote and yet lose the Electoral College vote?

(a) Yes, and it occurs on occasion.
(b) Yes, but it has never occurred in the past 100 years.
(c) No.

EXERCISES   ■ Challenge   ◆ Discussion

Spatial Models for Two-Candidate Elections

1. Why do M and $\bar{I}$ not coincide if a distribution is skewed?

2. Show that 0.6 is the equilibrium in Figure 12.3b.

■ **3.** Prove that if a distribution is discrete and there is no median position, there is always an *extended median*. (*Hint:* Show that there is always one position at which a majority of voters lies neither to the left nor to the right, and neither candidate would have an incentive to depart from this position.)

4. Assume that the one voter at 0.1 in Figure 12.2b decides not to vote because he is "too far away" from the two candidates who take the median position at 0.6. Would either candidate depart from $M = 0.6$ to try to do better if he or she knew that this voter had decided not to vote—but he or she would vote for the closer of two candidates less than a distance of 0.5 away? What if the candidates knew that the three voters at 0.2 had also decided not to vote—but they would vote for the closer of two candidates less than a distance of 0.4 away?

5. If you are a far-left or a far-right voter, are you helping your cause when you announce, like the voters in Exercise 4, that you will not support candidates who are too far away?

6. Consider the two most extreme voters at 0 in Figure 12.3b (those who are farthest from the extended median of 0.6). Would their nonvoting change the extended median? How about, as well, the nonvoting of the somewhat less extreme voter at 0.9? Show when, if at all, M or the extended median will change as fewer and fewer extreme voters decide not to vote in this example?

◆ 7. In Figure 12.2a, $\bar{l}$ is not in equilibrium—one candidate would do better if he or she moved from 0.56 to 0.6. But is 0.6 really a better reflection of the views of the electorate than 0.56?

8. Define an outcome to be in equilibrium if, given that one candidate chooses it, the other candidate cannot do better than take the same position. Show that this definition is equivalent to the text's definition of being in equilibrium.

■ 9. Consider a trimodal distribution (three peaks). When will taking a position at the middle peak be in equilibrium? Is it possible that one of the other peaks can ever be in equilibrium? If so, give a discrete-distribution example.

10. Define A's position in a two-candidate race to be *opposition-optimal* if, given that the position of B is fixed, it maximizes A's vote total. Show that A's

opposition-optimal position must be adjacent to B's position and closer to M, except when B is at the median. (Roughly speaking, being "adjacent" means being a very small distance away.)

◆ 11. Assume the population along a main street is uniformly distributed over [0, 1], so there are equal numbers of people located at all equally spaced intervals from $M/\bar{l}$. (This makes the distribution rectangular, or "flat.") It has been argued that the "social optimum" for the location of two stores are at the points $\frac{1}{4}$ and $\frac{3}{4}$, because then no person would have to travel more than $\frac{1}{4}$ of the length of the street to buy at one store. Is this desirable if the population is not uniformly distributed?

◆ 12. What is a social optimum in an election if only one candidate is to be elected? How about five candidates to a city council? Is it better that the city council members' positions all be centered around 0.5, or should they be more spread out?

◆ 13. Which is better for consumers: (a) to minimize the maximum distance they must travel to a store; or (b) to foster price competition, which would presumably be encouraged if two stores are located at $M = \frac{1}{2}$?

■ 14. Assume a city comprises three equal-sized districts, each of which elects a candidate to the city council. The mayor is elected by the entire city. Show with an example that the median or extended median for the mayor need not be the median or extended median for any of the three city council districts. Does this explain why mayors and city council members often disagree?

■ 15. In Exercise 14, must the median or extended median for the mayor be between the leftmost and rightmost medians or extended medians of the three districts? How about the mean $\bar{l}$?

Spatial Models for Multicandidate Elections

16. Assume that A and B take the *same* non-median position. What position should C take to maximize his or her vote total? Is C's position always a winning one?

■ **17.** Assume that *A* and *B* take *different* positions, with one possibly being at *M*. What position should *C* take to maximize his or her vote total? Is *C*'s position always a winning one?

18. Is there a 1/3-separation obstacle if the distribution is not symmetric but no more than $\frac{1}{6}$ of the area under the curve separates *A* (on the left) from *M*, and no more than $\frac{1}{6}$ of the area separates *B* (on the right) from *M*? What if these $\frac{1}{6}$-or-less areas on the left and the right are not the same?

19. Is there a 2/3-separation opportunity if the distribution is not unimodal but at least $\frac{1}{3}$ of the area under the curve separates *A* (on the left) from *M*, and at least $\frac{1}{3}$ separates *B* (on the right) from *M*? (*Hint:* Start by assuming that the distribution is uniform between *A* and *B*—and hence not unimodal—and that exactly $\frac{2}{3}$ of the voters lie between *A* and *B*. Can *C* always win by taking a position at *M*? If not, is there a distribution that affords *C* this opportunity?)

20. Show that *C* cannot win under the conditions for the optimal entry of two candidates, anticipating a third entrant (page 442). (*Hint:* Indicate which candidate will win when *C* enters to the left of *A*, to the right of *B*, or in between.)

21. It is known that *A*, *B*, and *C* will enter an election in that order, with *A* announcing his position first, then *B*, and finally *C*. If the distribution is uniform over [0, 1], what position should each candidate take to maximize his or her vote total, anticipating—in the case of *A* and *B*—the entry of future candidates? [*Hint:* Start by assuming that *A* takes a position at $\frac{1}{4}$. Is *B*'s position at $\frac{3}{4}$ optimal, anticipating the entry of *C*? Or can *B* do better at some other position (perhaps by influencing *C*'s choice of a maximizing position)?]

■ **22.** If *A* and *B* are equidistant from the median of a symmetric distribution and separated from each other by exactly $\frac{1}{2}$ of the area under the curve, under what conditions is this separation an obstacle and under what conditions is it an opportunity? (*Hint:* Start by constructing examples

of symmetric distributions in which *C* would either win or lose by taking a position at *M*.)

■ **23.** What are the vote-maximizing positions for four candidates to take if it is known that they will enter in the order *A*, *B*, *C*, *D*?

Winnowing the Field

24. Assume that the four candidates in the 2000 presidential election can be arrayed from left to right as follows: Nader–Gore–Bush–Buchanan. Suppose a poll reveals Gore at 48%, Bush at 47%, Nader at 3%, and Buchanan at 2%. Would Bush be well advised to offer Buchanan a cabinet position to drop out of the race (as Adams offered Clay the secretary-of-state post after the 1824 election)? What if Bush knew that, after Buchanan dropped out, only half of Buchanan's supporters would switch to him, with most of the remainder not voting, except for a few who would switch to Gore?

25. Assuming the same poll results as in Exercise 24, now suppose that Gore offered the same deal to Nader, knowing that only one-third of Nader supporters would switch to him and the rest would not vote. However, suppose Gore also thought that if Nader dropped out, so would Buchanan, and all Buchanan supporters would vote for Bush. Should Gore set off this train of events?

◆ **26.** Is there any evidence that the four presidential candidates in 2000 might have contemplated "deals" of the kind indicated in Exercises 24 and 25? If you cannot find any evidence, do you think this is because the candidates found such ploys unethical or because they thought they might be found out if they tried to engage in them?

◆ **27.** One tactic that was considered by Nader supporters who thought that their votes for Nader might kill Gore's chances in some states was to swap votes: In close states that Gore might lose if Nader supporters stuck with their candidate, these supporters would switch to Gore if Gore supporters in less contested states, where Gore would almost surely win, would switch to Nader. Thereby the popular-vote totals for the two candidates would not change overall, but Gore would be able to win in the close states he might otherwise lose. Is this a sensible way of dealing with problems created by the Electoral College, which puts a premium on winning in large states?

What Drives Candidates Out?

28. Show why the Condorcet-winner-unsuccessful result (page 449) is true. Is it true that if the poll assumption was modified to differentiate the top three (rather than the top two) candidates from the rest, and the Condorcet winner was not among the top three, that he or she would still lose?

◆ **29.** Show why the Condorcet-winner-successful result (page 450) is true. Is it proper that the candidate who comes in second in the poll should win after the results of the poll are announced? Why?

◆ **30.** In Example 4 (page 450), after class IV voters switch from D to A, the vote totals for the top three candidates are A–5, B–3, and C–4. Now assume a second poll is taken, differentiating A and C, the top two contenders, from B. If B supporters switch at this point to their second choice, which candidate will win? Do you consider this a desirable outcome?

◆ **31.** Assume there are four classes of voters that rank four candidates as follows:

> I. 4: $A\,D\,B\,C$
> II. 3: $B\,D\,A\,C$
> III. 2: $C\,D\,B\,A$
> IV. 1: $D\,C\,B\,A$

Which candidate is the Condorcet winner? Do you find this result strange in the light of what a poll would tell the voters?

32. Assume there is a poll that differentiates the top two candidates in Exercise 31. Which candidate will win the election after the poll?

◆ **33.** Assume there is a poll that differentiates the top three candidates in Exercise 31. Which candidate will win the election after the poll?

Comment on the different outcomes in this exercise and the previous one.

◆ **34.** Assume there are three classes of voters who rank three candidates as follows:

> I. 4: $A\,B\,C$
> II. 3: $B\,C\,A$
> III. 2: $C\,A\,B$

Show that there is no Condorcet winner. Applying the poll assumption to this example, which candidate will win? Is this fair, given the preferences of class II and III voters, who, together, are a majority?

Election Reform: Approval Voting

■ **35.** Prove the voting-only-for-a-second-choice result (page 452).

36. In a three-candidate election, show that your strategy of voting for your top two choices under approval voting is not always better than voting only for your top choice.

37. Consider a four-candidate election under approval voting. Is there ever a situation in which a voter would vote for a first and a third choice without also voting for a second choice? [*Hint:* Assume a voter ranks the four candidates $A\,B\,C\,D$ and believes that one of two things can happen: The electorate will favor either liberals (say, A and B) or conservatives (say, C and D) but never favor each side equally.]

38. Is there ever a situation under approval voting in which a voter would vote for a worst choice?

39. Is there ever a situation under approval voting in which a voter would *not* vote for a first choice if he or she finds acceptable one or more lower-ranked candidates? (*Note:* This question asks whether the voting-only-for-a-second-choice result can be generalized to more than three candidates.)

■ **40.** Prove the effect-of-dichotomous-preferences result (page 453). (*Hint:* If all voters have dichotomous preferences and vote for all candidates in their preferred subsets, which candidate will get the most approval votes? What does this say about the preferences of voters for the approval-vote winner, compared to their preferences for each of the other candidates?)

41. In the following example, class I and II voters have dichotomous preferences, but the class III voter has *trichotomous preferences* (he or she divides the four candidates into three indifference subsets):

I. 2: $(A\ B)\ (C\ D)$

II. 2: $(C)\ (A\ B\ D)$

III. 1: $(D)\ (C)\ (A\ B)$

Is it rational for the class III voter to vote only for his or her top choice, D? If not, who else should he or she approve of? Which class of voters will be most unhappy if the class III voter does not vote just for D? Can voters in this class, by voting strategically, do anything about their situation?

◆ **42.** Assume the class III voter's preferences change to a different trichotomous ordering:

III. 1: $(D)\ (A\ C)\ (B)$

Suppose, as in Exercise 41, that the class III voter indicates in an initial poll that he or she intends to vote only for D but then, in response to the poll, switches to voting for the candidates in his or her second-choice subset as well. If there is a new poll, based on these results, what will be the outcome? What if there is a third poll, fourth poll, and so on? Do you regard this result as desirable? Why?

43. In Exercise 31, we saw that under plurality voting the Condorcet winner, D, comes in fourth in a poll and, therefore, cannot be helped by subsequent polling, even when the poll distinguishes the top three candidates and voters differentiate among them:

I. 4: $A\ D\ B\ C$

II. 3: $B\ D\ A\ C$

III. 2: $C\ D\ B\ A$

IV. 1: $D\ C\ B\ A$

What are the outcomes under approval voting—both with and without polling—if voters approve of their (i) top-ranked, (ii) two top-ranked, and (iii) three top-ranked candidates initially? [*Note:* In making adjustments to the poll results, assume that voters approve not only of the preferred of their two top-ranked candidates identified by the poll but also of *all* candidates ranked above their preferred candidate. For example, when the poll based on (i) above identifies A and B as the two top-ranked candidates, with 4 and 3 votes, respectively, the class III and class IV voters after the poll will approve not only of their preferred candidate, B, but also of C and D, because they rank the latter two candidates above B.]

◆ **44.** On the basis of your answers to the foregoing problems, do you think approval voting would be beneficial in finding Condorcet winners—either with or without polling—in multicandidate elections?

The Electoral College

45. Assume there are three states with 3, 7, and 9 voters, and that they are all toss-up states. If both the Democratic and Republican candidates choose strategies that maximize their expected popular vote (the proportional rule), and they have the same total resources ($D = R$), what is the expected number of votes that each will receive?

46. Assume the Republican knows in advance what allocations, d_i, to each state i the Democrat will make in Exercise 45. Then the Republican's optimal response can be shown to be

$$r_i = \frac{\sqrt{n_i d_i}}{\sum_{i=1}^{t}\sqrt{n_i d_i}}\,(R + D) - d_i$$

Suppose that the Democrat ignores the smallest state and makes proportional allocations to the two largest states. (For concreteness, assume both candidates have 100 units of resources.) What is the Republican's optimal response? What if the Democrat makes proportional allocations to all three states?

■ **47.** In Exercise 46, show that if the Democrat makes proportional allocations, and the Republican responds optimally according to the formula given there, this formula simplifies to $r_i = (n_i/N)R$, which does not depend on d_i. What does this say about the proportional rule? [*Hint:* If the Republican finds out (say, through a spy) that the Democrat is making proportional allocations, does the Democrat have anything to worry about?]

48. In Exercise 47, if the Democrat has only half the resources of the Republican, would you recommend that he or she behave differently from proportional allocations to maximize his or her expected popular vote? Why? If the Republican allocates his or her resources proportionally to the three states, is there any way the Democrat can allocate his or her resources to win a majority of votes in states with more than half the votes?

49. Instead of maximizing their expected popular vote, assume the candidates in Exercise 45 want to win in states with more than half the votes. Suppose the candidate who allocates more resources to a state wins that state. Is there any state to which a candidate should not consider allocating resources? Should the states that receive allocations receive equal allocations?

50. In Exercise 45, assume you can choose specific voters in each state to whom you can allocate resources. Suppose the candidate who allocates more resources to a voter wins that voter's vote. If your goal is to win the votes of a majority of voters in states that have more than half the votes,

which voters would you target, and how much would you spend on each? (*Hint:* First show which states you would target; then show that these states should receive equal allocations, which in turn should be divided equally among a certain set of voters.)

51. Assume there are three toss-up states, A, B, and C, with, respectively, 2, 3, and 4 voters, which are also the number of electoral votes of each state. In the text, we gave the formulas for the probabilities, P_A, P_B, and P_C, that the Democrat wins a majority of popular votes in each state and, therefore, wins all the electoral votes of that state. Show that the formula for the probability that the Democrat *wins the election* under the Electoral College, PWE_D, is

$$PWE_D = P_A P_B(1 - P_C) + P_A P_C(1 - P_B) + P_B P_C(1 - P_A) + P_A P_B P_C$$

(*Hint:* Winning in any two states is sufficient to win the election.)

■ **52.** Compare the formula for PWE_D with the formula for EEV_D (in the text). Which quantity is it better to maximize? What would be a good resource-allocation strategy for maximizing PWE_D?

53. Is the square root in the formulas for the EEV maximizing strategies of the Democratic and Republican candidates related to the square-root rule for the Electoral College discussed in Chapter 11?

 ## WRITING PROJECTS

1. Do you think polling is useful in helping voters choose the "best" candidate? Or would it be better, as in some countries, to ban the publication of polls before an election? In one to two pages, discuss these questions in light of the theoretical effects polling has when voters react to polls and possibly change their voting strategies. Is there empirical evidence that voters behave in this way?

2. How serious a problem do you think the large-state bias of the Electoral College is? How would you explain the fact that some of the strongest advocates of the Electoral College come from small states? Has the theoretical bias been a reality in the campaign behavior of candidates in recent presidential elections? Discuss in one to two pages.

SUGGESTED READINGS

BRAMS, STEVEN J. *The Presidential Election Game,* Yale University Press, New Haven, Conn., 1978. Focuses on the strategic aspects of presidential elections—from primaries to conventions to general elections—and also includes an analysis of the "game" played between President Richard Nixon and the Supreme Court over the release of the Watergate tapes that led to Nixon's resignation in 1974 (Nixon has been the only president to resign the presidency). Approval voting and direct popular-vote election of a president are recommended as election reforms.

BRAMS, STEVEN J., and PETER C. FISHBURN. *Approval Voting.* Birkhäuser, Boston, Mass., 1983. An in-depth analysis of approval voting, which includes several case studies.

HINICH, MELVIN J., and MICHAEL C. MUNGER. *Analytical Politics,* Cambridge University Press, Cambridge, U.K., 1997. Extends spatial modeling to more than one dimension, analyzes probabilistic voting, and introduces game-theoretic solution concepts relevant to the study of elections.

SAARI, DONALD G. *Chaotic Elections! A Mathematician Looks at Voting,* AMS [American Mathematical Society], Providence, R.I., 2001. Argues that elections—in particular, the 2000 presidential election, but others as well—have chaotic features that can be understood through mathematics. The mathematics used is an unusual kind of geometry that will be accessible to those with some mathematical background.

SHEPSLE, KENNETH A., and MARK S. BONACHEK. *Analyzing Politics: Rationality, Behavior, and Institutions,* Norton, New York, 1997. Rational strategies in voting and elections are a major component of this text, but it also includes sections on collective action and political institutions, such as courts and legislatures. Several case studies illustrate the theory.

SUGGESTED WEB SITES

www.fec.gov Federal Election Commission—About Elections and Voting
www.ifes.org International Foundation for Election Systems

bcn.boulder.co.us government approvalvote center.html
electionmethods.org

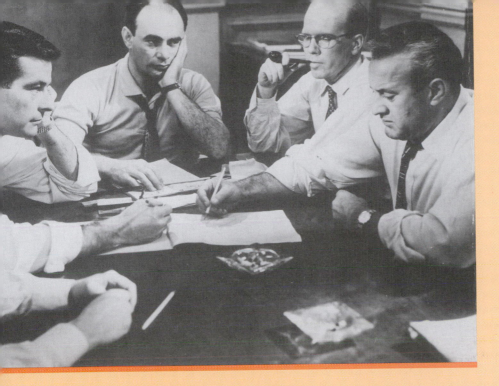

Fairness and Game Theory

The central thrust of the first two chapters in Part IV is the fair division of divisible and indivisible objects. Whereas a cake or a parcel of land is divisible, the representatives who are apportioned to the different states are indivisible. Sometimes, however, seemingly indivisible objects, like a car, can be shared, rendering them divisible. By contrast, the chapter on game theory chapter focuses on what rational players will choose in different strategic situations, which may be highly unfair to some.

Chapter 13 describes fair-decision schemes in which a group of individuals with different values can be assured of each receiving what he or she views as a fair share when dividing objects like cakes or the goods in an estate.

Chapter 14 discusses the apportionment problem, which is to round a set of fractions to whole numbers while preserving their sum; of course, the sum of the original fractions must be a whole number to start. Apportionment problems occur when resources must be allocated in integer quantities—for instance, when legislators allocated seats in the U.S. House of Representatives to the 50 states.

Chapter 15 introduces the mathematical field called game theory, which describes situations involving two or more decision makers having different goals. Game theory provides a collection of models to assist in the analysis of conflict and cooperation as well as strategies for resolution. Interestingly, you will find that the games covered in this chapter provide us with insights into certain social paradoxes that we routinely encounter in our daily lives. ■

Fair Division

When the demands or desires of one party are in conflict with those of another—be it a divorce, a labor–management negotiation, or an international dispute—no one wants to be treated unfairly. And with 1.2 million divorces every year in the United States alone, and crises such as we've seen in the Middle East for decades, it is certainly worth considering how mathematics might help in the search for procedures that can ensure fair and equitable resolutions of such conflicts.

We begin this chapter with one such procedure that was developed in the mid-1990s. This procedure—called **adjusted winner**—allows two parties to settle any dispute involving either issues (as in an international dispute) or objects (as in a divorce or a two-person inheritance) with certain mathematical guarantees of "fairness." Disagreement, it turns out, is both a bad thing and a good thing. On the one hand, disagreement as to how each issue should be resolved typically lies at the heart of a conflict. On the other hand, procedures such as adjusted winner are designed to capitalize on the parties' disagreement as to the importance of each issue, thus allowing each party to end the negotiations thinking it has been met more than halfway.

But adjusted winner is just one of several so-called fair-division procedures that have been developed over the past 60 years. So following our discussion of adjusted winner, we describe a procedure for handling inheritances that was discovered by the Polish mathematician Bronislaw Knaster during World War II. This is followed by a discussion of an extremely basic fair-division procedure—taking turns—and the question of what the optimal strategy is when taking turns choosing objects.

475

Bridging the gap between fair-division procedures with obvious real-world potential, such as divorce and inheritance procedures, and procedures that address fundamental mathematical questions of fairness (as do the procedures treated later in this chapter) is the ancient two-person procedure known as divide-and-choose. An application of this procedure to the Law of the Sea Treaty is described.

Divide-and-choose sets the stage for the mathematical investigations of fair division that have gone on for more than half a century. These investigations have often been phrased within the metaphor of "cake cutting." We present three cake-cutting procedures. The first two of these—found by Steinhaus and Banach–Knaster in the 1940s—yield allocations that are **proportional,** meaning that each player receives what he or she perceives to be at least his or her fair share of the cake. The last one—found by Selfridge–Conway in 1960—yields allocations that are **envy-free** in the sense that each player receives what he or she perceives to be a piece at least tied for largest.

13.1 The Adjusted Winner Procedure

To illustrate the **adjusted winner procedure,** we will consider an application to the multibillion-dollar world of business mergers. It turns out that one of the most elusive ingredients in the success of a merger is what deal-makers call *social issues*—how power, position, sacrifice, and status are allocated between the merging companies and their executives.

As a case in point, let's revisit the 1998 proposed merger between two giant pharmaceutical companies, Glaxo Wellcome and SmithKline Beecham. While most of the details underlying this aborted deal are still unknown to outsiders, the role of social issues is clearly underscored by reports that the companies "saw

(Pascal Plessis/AP Photo.)

nearly 19 billion dollars of stock market value vanish in the clash of two corporate egos."

Exactly what kinds of issues might bring on a "clash of two corporate egos"? While not privy to the details of the Glaxo Wellcome–SmithKline Beecham merger attempt, we can speculate as to their nature. For purposes of illustration, let's assume that the following five social issues were paramount:

1. The name that the combined company would use
2. The location of the headquarters of the combined company
3. The question of who would serve as chairman of the combined company
4. The question of who would serve as CEO of the combined company
5. The question of where the necessary layoffs would come from

Each of these five social issues is known to have been a major factor in other recent proposed mergers. For example, when Chrysler merged with Daimler-Benz in 1998, the issue of the choice of a name for the combined company was described as a "standoff" before both sides finally agreed to DaimlerChrysler.

So, let's assume that these were the five social issues confronting Glaxo Wellcome and SmithKline Beecham, and let's see how the adjusted winner procedure would have suggested a resolution. The starting point—and something that is quite difficult when dealing with issues (as in a negotiation) as opposed to objects (as in a divorce)—is to have each side quantify the importance it attaches to getting its own way on each of the issues.

With the adjusted winner procedure, quantification is done by having each side—independently and simultaneously—spread 100 points over the issues in a way that reflects the relative worth of each issue to that party. In our present example, let's assume that the companies allocated their 100 points as shown in Table 13.1. Adjusted winner is now used to decide which side gets its way on

TABLE 13.1 Applying the Adjusted Winner Procedure to a Merger of Two Companies

	Point Allocations	
Issue	Glaxo Wellcome	SmithKline Beecham
Name	5	10
Headquarters	25	10
Chairman	35	20
CEO	15	35
Layoffs	20	25
Total	100	100

which issues, but the procedure requires that a compromise of sorts may have to be reached on one of the issues.

Here's how the procedure works. Suppose we have two parties and a list of either issues to be resolved in one party's favor or the other's (as in our merger example) or objects to be awarded either to one party or to the other (as in a divorce or a two-person inheritance). To have a single word covering both issues and objects, we will speak of "items." The adjusted winner procedure follows these basic steps:

Basic Steps in the Adjusted Winner Procedure

Step 1. As described earlier, each party distributes 100 points over the items in a way that reflects their relative worth to that party.

Step 2. Each item is initially given to the party that assigned it more points. Each party then assesses how many of his or her own points he or she has received. The party with the fewest points is now given each item on which both parties placed the same number of points.

Step 3. Since the point totals are most likely not equal, let A denote the party with the higher point total and B be the other party. Start transferring items from A to B, in a certain order, until the point totals are equal. The point at which equality is achieved may involve a fractional transfer of one item.

Step 4. The order in which this is done is extremely important and is determined by going through the items in order of increasing **point ratio.**

An item's point ratio is the fraction

$$\frac{A\text{'s point value of the item}}{B\text{'s point value of the item}}$$

where A is the party with the higher point total.

Let's demonstrate the adjusted winner procedure by continuing with our analysis of the proposed merger between Glaxo Wellcome and SmithKline Beecham. Why step 4 is so important will be explained later.

1. Assume that Glaxo Wellcome and SmithKline Beecham have given us the point assignments shown in Table 13.1.

2. Because Glaxo Wellcome has placed more points on headquarters (25) and chairman (35), it is initially "given" these issues, while SmithKline Beecham is initially given name (10), CEO (35), and layoffs (25). Notice that SmithKline Beecham now has $10 + 35 + 25 = 70$ of its points, whereas Glaxo Wellcome only has $25 + 35 = 60$ of its points.

3. We now start transferring issues from SmithKline Beecham to Glaxo Wellcome until the point totals of the two sides are equal. SmithKline Beecham has initially been given three issues (name, CEO, and layoffs), and step 4 will help us decide in what order to start transferring them.

4. Layoffs has point ratio $25/20 = 1.25$, name has point ratio $10/5 = 2.00$, and CEO has point ratio $35/15 = 2.33$. Because layoffs has the lowest point ratio, we start to transfer that item first.

We now see that transferring the entire layoff item (worth 25 to SmithKline Beecham and 20 to Glaxo Wellcome) gives Glaxo Wellcome more points ($60 + 20 = 80$) than SmithKline Beecham has ($70 - 25 = 45$).

Thus, the entire layoff item cannot be transferred. Glaxo Wellcome and SmithKline Beecham will need to compromise on the issue of layoffs. But compromise may not mean meeting each other half way. Our goal is to equalize points between the two companies, and a little algebra will tell us exactly the extent to which Glaxo Wellcome and SmithKline Beecham should get their way on the issue of layoffs. Conceptually, it's easier to think of SmithKline Beecham retaining some fraction x of the issue in question and Glaxo Wellcome receiving the complementary fraction $1 - x$ of that same issue.

Because x is the fraction of the issue that SmithKline Beecham retains, the number of points SmithKline Beecham gets from this issue is x times 25. The fraction that Glaxo Wellcome gets is $1 - x$, so the number of points it gets from this issue is $1 - x$ times 20. Thus, if we want a fraction that will make SmithKline Beecham's total points and Glaxo Wellcome's total points equal, then x must satisfy the following equation:

$$10 + 35 + 25x = 25 + 35 + 20(1 - x)$$

We use algebra to solve this equation:

$$45 + 25x = 60 + 20 - 20x$$
$$45 + 25x = 80 - 20x$$
$$45x = 35$$
$$x = \frac{35}{45} = \frac{7}{9}$$

Inserting $\frac{7}{9}$ back into the equation, we see that

$$45 + 25\left(\frac{7}{9}\right) = 60 + 20\left(\frac{2}{9}\right)$$

or approximately 64 points for each side. In rough terms, equality of points is achieved when SmithKline Beecham gets about three-fourths ($7/9 \cong 3/4$) of its way on the issue of layoffs and Glaxo Wellcome gets about one-fourth of its way.

Having seen how the adjusted winner procedure works, we must now ask the following question: Exactly what is it about the allocation produced by this scheme that would make someone want to use it? The answer is given by the following theorem (whose proof can be found in *Fair Division* by Brams and Taylor, listed in the Suggested Readings):

> ### ::::::: Theorem: Properties of the Adjusted Winner Allocation
>
> For two parties, the adjusted winner procedure produces an allocation, based on each player's assignment of 100 points over the items to be divided, that has the following properties:
>
> ▶ The allocation is **equitable:** Both players receive the same number of points.
>
> ▶ The allocation is **envy-free:** Neither player would be happier with what the other received.
>
> ▶ The allocation is **Pareto-optimal:** No other allocation, arrived at by any means, can make one party better off without making the other party worse off.

Economists consider Pareto optimality (also named after Vilfredo Pareto) to be an extremely important property, and the order of transfer in step 4 on page 478 of the adjusted winner procedure is so important because it guarantees that the outcome is Pareto-optimal. The fact that the adjusted winner procedure produces an allocation that is efficient in this sense leads us to hope that it can and will play a future role in real-world dispute resolution.

13.2 The Knaster Inheritance Procedure

The adjusted winner procedure can be applied in the case of an inheritance if there are only two heirs. For *more than two heirs*, there is quite a different scheme, the **Knaster inheritance procedure,** first proposed by Bronislaw Knaster in 1945. It has a drawback, though, in that is requires the heirs to have a large amount of cash at their disposal.

A Four-Person Inheritance

EXAMPLE 1

Suppose (for the moment) that there is just one object—a house—and four heirs—Bob, Carol, Ted, and Alice. Knaster's scheme begins with each heir bidding (simultaneously and independently) on the house. Assume, for example, that the bids are

Bob	Carol	Ted	Alice
$120,000	$200,000	$140,000	$180,000

Carol, being the high bidder, is awarded the house. Her fair share, however, is only one-fourth of the $200,000 she thinks the house is worth, and so she places $150,000 (which is three-fourths of the $200,000 she bid) into a temporary "kitty."

Each of the other heirs now withdraws from the kitty his or her fair share, that is, one-fourth of his or her bid:

$$\begin{array}{ll} \text{Bob withdraws} & \$120,000/4 = \$30,000 \\ \text{Ted withdraws} & \$140,000/4 = \$35,000 \\ \text{Alice withdraws} & \$180,000/4 = \$45,000 \end{array}$$

Thus, from the $150,000 kitty, a total of $30,000 + $35,000 + $45,000 = $110,000 is withdrawn, and each of the four heirs now feels that he or she has the equivalent of one-fourth of the estate. Moreover, there is a $40,000 surplus ($150,000 kitty − $110,000 withdrawn), which is now divided equally among the four heirs (so each receives an additional $10,000). The final settlement is

Bob	Carol	Ted	Alice
$40,000	house − $140,000	$45,000	$55,000

This illustrates Knaster's procedure for the simple case in which there is only one object. But what if our same four heirs have to divide an estate consisting of, say, a house (as before), a cabin, and a boat? The easiest answer is to handle the estate one object at a time (proceeding for each object as we just did for the house). To illustrate, assume that our four heirs submit the following bids:

	Bob	Carol	Ted	Alice
House	$120,000	$200,000	$140,000	$180,000
Cabin	60,000	40,000	90,000	50,000
Boat	30,000	24,000	20,000	20,000

We have already settled the house. Let's handle the cabin the same way. Thus, Ted is awarded the cabin based on his high bid of $90,000. His fair share is one-fourth of this, so he places three-fourths of $90,000 (which is $67,500) into the kitty.

Bob withdraws from the kitty $60,000/4 = $15,000. Carol withdraws $40,000/4 = $10,000, and Alice withdraws $50,000/4 = $12,500. Thus, from the $67,500 kitty, a total of $15,000 + $10,000 + $12,500 = $37,500 is withdrawn.

The surplus left in the kitty is thus $30,000, and this is again split equally ($7500 each) among the four heirs. The final settlement on the cabin is

Bob	Carol	Ted	Alice
$22,500	$17,500	cabin − $60,000	$20,000

If we were now to do the same for the boat (we leave the details to you), the corresponding final settlement would be

Bob	Carol	Ted	Alice
boat − $20,875	$7625	$6625	$6625

Putting the three separate analyses (house, cabin, and boat) together, we get a final settlement of

Bob: boat + ($40,000 + $22,500 − $20,875 = $41,625)
Carol: house + (−$140,000 + $17,500 + $7625 = −$114,875)
Ted: cabin + ($45,000 − $60,000 + $6625 = −$8375)
Alice: $55,000 + $20,000 + $6625 = $81,625. ■

Notice that here, Carol gets the house but must pay $114,875 in cash (and Ted gets the cabin but must put up $8375 in cash). This cash is then disbursed to Bob and Alice. In practice, Carol's having this amount of cash available may be a real problem—the key drawback to Knaster's procedure. Nevertheless, Knaster's procedure shows again that whenever some participants have different evaluations of some objects, there is an allocation in which everyone obtains more than a fair share.

13.3 Taking Turns

For many of us, an early lesson in fair division happens in elementary school with the choosing of sides for a spelling bee or when picking teams on the playground. In terms of importance, these pale in comparison with the issue of property settlement in a divorce. Remarkably, however, the same fair-division procedure—*taking turns*—is often used in both.

Taking turns is fairly self-explanatory. With two parties (and that's all we'll consider here), one party selects an object, then the other party selects one, then the first party again, and so on. But in this context, there are several interesting questions that suggest themselves:

1. How do we decide who chooses first?

2. Because choosing first is often quite an advantage, shouldn't we compensate the other party in some way, perhaps by giving him or her extra choices at the next turn?

3. Should a player always choose the object he or she most favors from those that remain, or are there strategic considerations that players should take into account?

The answer to question 1 is often "toss a coin," but there are other possibilities—for example, the two parties could "bid" for the right to go first, as in an auction. The answer to question 2 is less clear, but we outline a discussion of the issue it raises in Writing Project 2.

Question 3, on the other hand, is remarkably interesting, and it is this one that we want to pursue. Let's look at an easy example. Suppose that Bob and Carol are getting a divorce, and their four main possessions, ranked from best to worst by each, are as follows:

	Bob's Ranking	Carol's Ranking
Best	Pension	House
Second best	House	Investments
Third best	Investments	Pension
Worst	Vehicles	Vehicles

If Carol knows nothing of Bob's preferences, then we can assume that she will choose sincerely—selecting at her turn whichever item she most prefers from those not yet chosen. Now, if Bob is also sincere, and if he chooses first, the items will be allocated as follows:

First turn:	Bob takes the pension.
Second turn:	Carol takes the house.
Third turn:	Bob takes the investments.
Fourth turn:	Carol is left with the vehicles.

Hence, Bob gets his first and third favorites (the pension and the investments). However, if Bob opens by choosing the house—and bypassing the pension for the moment—then the allocation will be as follows:

First turn:	Bob takes the house.
Second turn:	Carol takes the investments.
Third turn:	Bob takes the pension.
Fourth turn:	Carol is left with the vehicles.

Thus, by being insincere, Bob does better—getting his first and second favorites (the pension and the house).

In general, then, what is the optimal strategy for rational players to use, assuming that both know the preferences of the other? The answer is something called the **bottom-up strategy,** discovered by the mathematicians D. A. Kohler and R. Chandrasekaran in 1969. We will illustrate it with an example.

Suppose we have five objects—A, B, C, D, E—and Bob is choosing first. Suppose that Bob and Carol have the following rankings of the objects (called **preference lists** in what follows):

Bob	Carol
A	C
B	E
C	D
D	A
E	B

It will turn out that Bob should open with C (his third choice) followed by Carol's choice of D (skipping over E, for the moment). Bob will then take A, Carol will follow with E, and finally Bob will get B. Bob gets his first, second, and third choices without selecting his first choice first! Where does this strategy come from?

The intuition here is quite easy. Let's make two assumptions about rational players: A rational player will never willingly choose his or her least preferred alternative, and a rational player will avoid wasting a choice on an object that he or she knows will remain available and thus can be chosen later.

With these assumptions as motivation, let's return to the preceding example and think about the mental calculation Bob will go through in deciding what his first choice will be. Bob knows the eventual sequence of choices will fill in all of the following blanks:

Bob: _____   _____   _____
Carol: _____   _____

Now, working mentally from right to left, Bob knows that Carol will not choose B, because it is the bottom thing on her list. Thus, he will get stuck with B, and so he will avoid wasting anything but his last choice on alternative B. Thus, Bob can pencil in alternative B as his last choice:

Bob: _____   _____   __B__
Carol: _____   _____

Bob, placing himself momentarily in Carol's shoes, knows she will reason the same way, and thus he pencils Carol in for the bottom alternative, E, on his list:

Bob: _____   _____   __B__
Carol: _____   __E__

Mentally now, Bob reasons as if alternatives B and E never existed (and the choice sequence had been Bob–Carol–Bob) and continues to pencil in alternatives from right to left, with Bob working from bottom to top on Carol's preference list and Carol working from bottom to top on Bob's preference list. This yields the following sequence of choices mentally penciled in by Bob:

Bob: $\underline{\quad C \quad}$ $\underline{\quad A \quad}$ $\underline{\quad B \quad}$
Carol: $\underline{\quad D \quad}$ $\underline{\quad E \quad}$

Remember, this is just a mental calculation that Bob went through to decide upon the actual choice—in this case, C—with which he will open. Bob has no guarantee that Carol will, in fact, respond with D, so the use of this strategy involves some risk on Bob's part.

This bottom-up strategy can also be viewed as a procedure that a mediator could use to specify a division of several objects between two parties. Given the preference lists of both parties, the mediator could construct a list—exactly as we did for Bob and Carol above—and then offer this to the parties as the suggested allocation. In effect, the mediator is simultaneously playing the role of two rational parties who choose to employ optimal strategies.

13.4 Divide-and-Choose

There are vast mineral resources under the seabed, all of which, one might argue, should be available to both developed and developing countries. In the absence of some kind of agreement, however, what is to prevent the developed countries from mining all of the most promising tracts before the developing countries have reached a technological level where they can begin their own mining operations? Such an agreement went into effect on November 16, 1994, with 159 signatories (including the United States). It was called the **Convention of the Law of the Sea,** and it protects the interests of developing countries by means of the following fair-division procedure.

Whenever a developed country wants to mine a portion of the seabed, that country must propose a division of the portion into two tracts. An international mining company called the Enterprise, funded by the developed countries but representing the interests of the developing countries through the International Seabed Authority, then chooses one of the two tracts to be reserved for later use by the developing countries.

Description of Divide-and-Choose

With **divide-and-choose,** one party divides the object into two parts in any way that he desires, and the other party chooses whichever part she wants.

As a fair-division procedure, the origins of divide-and-choose go back thousands of years. The Hebrew Bible tells the story of Abram (later to be called Abraham) and Lot, who settled a dispute over land via a proposed division by Abram— "If you go north, I will go south; and if you go south, I will go north" (Gen. 13:8–9)—and a choice (of the plain of Jordan) by Lot. Divide-and-choose resurfaced later in Hesiod's book *Theogony*. The Greek gods Prometheus and Zeus had to divide a portion of meat. Prometheus began by placing the meat into two piles, and Zeus selected one.

Actually, a fair-division procedure consists of both rules and strategies, and all we have described so far are the rules of divide-and-choose. But the natural strategies here are quite obvious: The divider makes the two parts equal in his estimation, and the chooser selects whichever piece she feels is more valuable.

Rules and strategies differ from each other in the following sense: A referee could determine if a rule is being followed, even without knowing the preferences of the players. Strategies represent choices of how players follow the rules, given their individual preferences (and any other knowledge or goals they may have).

The strategies on which we focus in our discussion of fair-division procedures are those that require no knowledge of the preferences of the other players and yet provide some kind of minimal degree of satisfaction even in the face of collusion by the other players. For example, the strategies just given for divide-and-choose guarantee each player a piece that he or she would not wish to trade for that received by the other.

There are, to be sure, other strategic considerations that might be relevant. For example, in divide-and-choose, would you rather be the divider or the chooser? The answer, given our assumptions that nothing is known of the preferences of the others, is to be the chooser. However, if you knew the preferences of your opponent (and how much she may value spite), then you might want to be the divider.

As a final comment on strategic considerations, we need only look to the origins of the well-known expression "the lion's share." It comes from one of Aesop's fables, as reported by Todd Lowry in *Archaeology of Economic Ideas* (1987, p. 130):

It seems that a lion, a fox, and an ass participated in a joint hunt. On request, the ass divides the kill into three equal shares and invites the others to choose. Enraged, the lion eats the ass, then asks the fox to make the division. The fox piles all the kill into one great heap except for one tiny morsel. Delighted at this division, the lion asks, "Who has taught you, my very excellent fellow, the art of division?" to which the fox replies, "I learnt it from the ass, by witnessing his fate."

13.5 Cake-Division Procedures: Proportionality

The modern era of fair division in mathematics began in Poland during World War II (see Spotlight 13.1). At this time, Hugo Steinhaus asked what is, in retrospect, the obvious question: What is the "natural" generalization of divide-and-choose to three or more people? The metaphor that has been used in this context, going back at least to the English political theorist James Harrington (1611–1677), is a cake. We picture different players valuing different parts of the cake differently because of concentrations of certain flavors or depth of frosting.

Definition of Cake-Division Procedure

A **cake-division procedure** for n players is a procedure that the players can use to allocate a cake among themselves (no outside arbitrators) so that each player has a strategy that will guarantee that player a piece with which he or she is "satisfied," even in the face of collusion by the others.

As we have seen, divide-and-choose is a cake-division procedure for two players, if by "satisfied" we mean either "thinks his piece is of size or value at least one-half" or "does not want to trade what she received for what anyone else received." These two notions of satisfaction are so important that we define them precisely.

Definition of Proportional and Envy-Free

A cake-division procedure (for n players) will be called *proportional* if each player's strategy guarantees that player a piece of size or value at least $1/n$ of the whole in his or her own estimation. It will be called *envy-free* if, as in the adjusted winner context, each player's strategy guarantees that player a piece he or she considers to be at least tied for largest or most valuable.

It turns out that for $n = 2$, a procedure is envy-free if and only if it is proportional; that is, for $n = 2$, the two notions of fair division are exactly the same. For $n > 2$, however, all we can say is that an envy-free procedure is automatically proportional. For example, if a three-person allocation is not proportional, then one player (call him Bob) thinks that he received less than one-third. Bob then feels that the other two are sharing more than two-thirds between them, and thus that at least one of the two (call her Carol) must have more than one-third. But then Bob will envy Carol, and so the allocation is not envy-free. Because all non-proportional allocations fail to be envy-free, it follows that if an allocation is envy-free, then it must be proportional.

Many procedures that are proportional, however, fail to be envy-free, as we shall soon show. Thus, proportional procedures are fairly easy to come by, but envy-free procedures are fairly hard to come by.

The modern era of cake cutting began with the investigations of the Polish mathematician Hugo Steinhaus during World War II. His research, and that of dozens of others since, involved dealing with two fundamental difficulties. First, allocation schemes that work in the context of two or three players often do not generalize easily to the context of four or more players. Second, procedures that yield envy-free allocations are considerably harder to obtain than procedures that yield proportional allocations.

The mathematics inspired by these two difficulties constitutes a rather elegant corner of the large and important area of fair division. Steinhaus's investigations in the 1940s led to his observation that there is a rather natural extension of divide-and-choose to the case of three players. This is the "lone-divider procedure" described on page 489. Steinhaus's method was generalized to an arbitrary number of players by Harold W. Kuhn of Princeton University in 1967.

Unable to extend his procedure from three to four players, Steinhaus proposed the problem to some Polish colleagues. Two of them, Stefan Banach and Bronislaw Knaster, solved this problem in the mid-1940s by producing the "last-diminisher procedure" described on page 489.

In addition to the procedures devised by Banach, Knaster, and Kuhn, there are other well-known constructive procedures for obtaining a proportional allocation among four or more players. One of these is by A. M. Fink of Iowa State University and appears in Exercise 26.

Another constructive procedure of note, although different in flavor from the others, is the 1961 recasting by Lester E. Dubins and Edwin H. Spanier of the University of California at Berkeley of the last-diminisher method as a "moving-knife procedure" (illustrated in Exercise 28). The trade-off here involves giving up the "discrete" nature of the last-diminisher method in exchange for the conceptual simplicity of the moving knife.

Although the existence of an envy-free allocation (even for four or more players) was known to Steinhaus in the 1940s, the first constructive procedure for producing an envy-free allocation among three players was not found until around 1960. At that time, John L. Selfridge of Northern Illinois University and, later but independently, John H. Conway of Princeton University found the elegant procedure presented on page 490. Although never published by either, the procedure was quickly and widely disseminated by Richard K. Guy of the University of Calgary and others. Eventually it appeared in several treatments of the problem by different authors.

In 1980, a moving-knife procedure for producing an envy-free allocation among three players was found by Walter R. Stromquist of Daniel Wagner Associates. Then, another procedure, capable of being recast as a moving-knife solution of the three-player case, was found by a law professor at the University of Virginia, Saul X. Levmore, and a former student of his, Elizabeth Early Cook.

In 1992, Steven J. Brams, a political scientist at New York University, and Alan D. Taylor, a mathematician at Union College, succeeded in finding a constructive procedure for producing an envy-free allocation among four or more players. In 1994, Brams, Taylor, and William S. Zwicker (also from Union College) found a moving-knife solution to the four-person envy-free problem. No moving-knife procedure is known that will produce an envy-free allocation among five or more players.

The Steinhaus Proportional Procedure for Three Players (Lone Divider)

EXAMPLE 2

Given three players—Bob, Carol, and Ted—we have Bob divide the cake into three pieces, call them X, Y, and Z, each of which he thinks is of size or value exactly one-third. Let's speak of Carol as "approving of a piece" if she thinks it is of size or value at least one-third. Similarly, we will speak of Ted as "approving of a piece" if the same criterion applies. Notice that both Carol and Ted must approve of at least one piece.

(Brand X Pictures/Punchstock.)

If there are distinct pieces—say, X and Y—with Carol approving of X and Ted approving of Y, then we give the third piece, Z, to Bob (and, of course, X to Carol and Y to Ted), and we are done. The problem case is where both Carol and Ted approve of only one piece and it is the *same* piece.

Let's assume that Carol and Ted approve of only one piece, X, and hence (of more importance to us) both *disapprove* of piece Z. Let XY denote the result of putting piece X and piece Y back together to form a single piece. Notice that both Carol and Ted think that XY is at least two-thirds of the cake because both disapprove of Z. Thus, we can give Z to Bob and let Carol and Ted use divide-and-choose on XY. Because half of two-thirds is one-third, both Carol and Ted are guaranteed a proportional share (as is Bob, who approved of all three pieces). ■

The procedure just described, which guarantees proportional shares but is not necessarily envy-free and is sometimes called the **lone-divider method,** was discovered by Hugo Steinhaus around 1944. Unfortunately, it does not extend easily to more than three players. It was left to Steinhaus's students, Stefan Banach and Bronislaw Knaster, to devise a method for more than three players. Picking up where Steinhaus left off (and traveling in quite a different direction), they devised the proportional procedure that today is referred to as the **last-diminisher method.** Like the lone-divider method, it is proportional but not envy-free. We illustrate it for the case of four players (Bob, Carol, Ted, and Alice), and we include both the rules and the strategies that guarantee each player his or her fair share.

The Banach–Knaster Proportional Procedure for Four or More Players (Last Diminisher)

EXAMPLE 3

Bob cuts from the cake a piece that he thinks is of size one-fourth and hands it to Carol. If Carol thinks the piece handed her is larger than one-fourth, she trims it to size one-fourth in her estimation, places the trimmings back on the cake, and passes the diminished piece to Ted. If Carol thinks the piece handed her is of size at most one-fourth, she passes it unaltered to Ted.

Ted now proceeds exactly as did Carol, trimming the piece to size one-fourth if he thinks it is larger than this and passing it (diminished or unaltered) on to Alice. Alice does the same, but, being the last player, simply holds onto the piece momentarily instead of passing it to anyone.

Notice that everyone now thinks the piece is of size at most one-fourth, and the last person to trim it (or Bob, if no one trimmed it) thinks the piece is of size exactly one-fourth. Thus, the procedure now allocates this piece to the last person who trimmed it (and to Bob if no one trimmed it).

Assume for the moment that it was Ted who trimmed the piece last, so he takes this piece and exits the game. Bob, Carol, and Alice all think that at least three-fourths of the cake is left, so they can start the process over with (say) Bob beginning by cutting a piece from what remains that he thinks is one-fourth of the original cake. Carol and Alice are both given a chance to trim it to size one-fourth in their estimation, and again, the last one to trim it takes that piece and exits the game. The two remaining players both think that at least half the cake is left, so they can use divide-and-choose to divide it between themselves and thus be assured of a piece that is of size at least one-fourth in their estimation. ∎

13.6 Cake-Division Procedures: The Problem of Envy

Divide-and-choose has a property that neither of the last two procedures possesses: It can ensure that each player receives a piece of cake he or she considers the largest or tied for the largest. In the case of only two players, this means that each player can get what he or she perceives to be at least half the cake, no matter what the other player does. Thus, divide-and-choose is an envy-free procedure.

Steinhaus's $n = 3$ proportional procedure (the lone-divider method) is not envy-free. For example, consider the case where Carol and Ted both find one piece unacceptable (and this piece is given to Bob). Carol and Ted will not envy each other when one divides and the other chooses, but Bob may think that this is not a 50-50 split. Indeed, if Bob divided the cake initially into what he thought was three equal pieces, an unequal split of the remaining two-thirds of the cake by Carol and Ted means that Bob will prefer the larger of these two pieces to the one-third he got. Consequently, Bob will envy the person who got this larger piece.

Nor is the last-diminisher method envy-free. For example, if Bob initially cuts a piece of cake of size one-fourth, and no one else trims it, then Bob receives this piece and exits the game. If Carol is the one to make the next initial cut, she may well cut a piece from the cake that she thinks is of size one-fourth but that Bob thinks is of size considerably more than one-fourth. But Bob is out of the game. Thus, if Ted and Alice think this piece is of size less than one-fourth, then Carol receives it, and so Bob will envy Carol.

Nevertheless, there do exist cake-division procedures that are envy-free. We present one of these in what follows.

EXAMPLE 4   The Selfridge–Conway Envy-Free Procedure for Three Players

We start with a cake and three people. The point we wish to arrive at is an envy-free allocation of the entire cake among the three people in a finite number of steps. This task may seem formidable, but quite often in mathematics, an important part of solving a problem involves breaking the problem into identifiable parts. In this case, let's call our starting point A and the final point we wish to reach C. Now let's identify an appropriate in-between point B that makes going from A to C–via B–more manageable. Our in-between point B is the following:

> *Point B:* Getting a constructive procedure that gives an envy-free allocation of *part* of the cake.

Can we constructively obtain three pieces of cake, whose union may not be the whole cake, which can be given to the three people so that each thinks he or she received a piece at least tied for largest? This turns out to be quite easy with the solution given by John Selfridge and John Conway. The following process and strategies do the trick:

1. Player 1 cuts the cake into three pieces he considers to be the same size. He hands the three pieces to player 2.

2. Player 2 trims at most one of the three pieces to create at least a two-way tie for largest. Setting the trimmings aside, player 2 hands the three pieces (one of which may have been trimmed) to player 3.

3. Player 3 now chooses, from among the three pieces, one that he considers to be at least tied for largest.

4. Player 2 next chooses, from the two remaining pieces, one that she considers to be at least tied for largest, with the proviso that if she trimmed a piece in step 2, and player 3 did not choose this piece, then she must now choose it.

5. Player 1 receives the remaining piece.

Let's reconsider the five steps of this trimming procedure to assure ourselves that each player experiences no envy. Recall that player 1 cuts the cake into three pieces, and player 2 trims one of these three pieces. Now player 3 chooses, and, as the first to choose, he certainly envies no one. Player 2 created a two-way tie for largest, and at least one of these two pieces is still available after player 3 selects his piece. Hence, player 2 can choose one of the tied pieces she created and will envy no one. Finally, player 1 created a three-way tie for largest and, because of the proviso in step 4, the trimmed piece is not the one left over. Thus, player 1 can choose an untrimmed piece and therefore will envy no one.

So far we have gone from point A to point B: Starting with a cake and three players, we have constructively obtained (in finitely many steps) an envy-free allocation of all of the cake, except the part T that player 2 trimmed from one of the pieces. We will now describe how T can be allocated among the three players in such a way that the resulting allocation of the whole cake is envy-free. (This is the rest of the **Selfridge–Conway envy-free procedure.**)

The key observation for the $n = 3$ case is that player 1 will not envy the player who received the trimmed piece, even if that player were to be given all of T. Recall that player 1 created a three-way tie and received an untrimmed piece. The union of the trimmed piece and the trimmings yields a piece that player 1 considers to be exactly the same size as the one he received. Thus, assume that it is player 3 who received the trimmed piece (it could as well be player 2). Then player 1 will not envy player 3, no matter how T is allocated.

The next step ensures that neither player 2 nor player 3 will envy another player when it comes time to allocate T. Let player 2 cut T into three pieces she considers to be the same size. Let the players choose which of the three pieces they want in the following order: player 3, player 1, player 2.

To see that this yields an envy-free allocation, notice that player 3 envies no one, because he is choosing first. Player 1 does not envy player 2, because he is choosing ahead of her; and player 1 does not envy player 3 because, as pointed out earlier, player 1 will not envy the player who received the trimmed piece. Finally, player 2 envies no one, because she made all three pieces of T the same size.

Hence, for $n = 3$, the Selfridge–Conway procedure will give an envy-free allocation of all the cake except T, followed by an allocation of T that gives an envy-free allocation of all the cake. ■

A naive attempt to generalize to $n = 4$ what we have done for $n = 3$ would proceed as follows: We would begin by having player 1 cut the cake into four pieces he considers to be the same size. Then we would have players 2 and 3 trim some pieces (but how many?) to create ties for the largest. Finally, we would have the players choose from among the pieces–some of which would have been trimmed–in the following order: player 4, player 3, player 2, player 1.

This approach fails because player 1 could be left in a position of envy. To understand how the approach could fail, consider how many pieces player 3 might have to trim to create a sufficient supply of pieces tied for largest so that he is guaranteed to have one available when it is his turn to choose. Player 3 might have to trim one piece to create a two-way tie for largest. Player 2 might need to trim two pieces to create a three-way tie for largest (because if there were only a two-way tie for largest, player 3 might further trim one of these pieces and player 4 might choose the other). This leaves player 1 in a possible position of envy, because we could have a situation where player 2 trims two pieces and player 3 trims a third piece, and player 4 then chooses the only untrimmed piece. If this happens, player 1, by being forced to choose a trimmed piece, will definitely envy player 4.

All is not lost, however, because there are modifications of the Selfridge–Conway procedure that will work for arbitrary n. For more on this, see *Fair Division* in the Suggested Readings.

Although we have used the metaphor of cake cutting throughout our discussion of the problem of envy, the idea of successive trimming is nonetheless applicable to problems of fair division other than parceling out the last crumbs of a cake. The main practical problem in applying the trimming procedure is that many fair-division problems involve goods that cannot be divided up at all, much less trimmed in fine amounts. Such goods are said to be *indivisible*.

It is interesting to recall that when the Allies agreed in 1944 to partition Germany into sectors after World War II (first stage), they initially did not reach agreement about what to do with Berlin. Subsequently, they decided to partition Berlin itself into sectors (second stage), even though this city fell 110 miles within the Soviet sector. Berlin was simply too valuable a "piece" for the Western Allies (Great Britain, France, and the United States) to cede to the Soviets, which suggests how, after a leftover piece is trimmed off, it can be subsequently divided under the trimming procedure.

Yet, what if a large piece like Berlin is not divisible? In the settlement of an estate, this might be the house, which may be worth half the estate to the claimants. In this situation, there may be no alternative but to sell this big item and use the proceeds to make the remaining estate more liquid or, in our terms, "trimmable."

REVIEW VOCABULARY

Adjusted winner procedure A fair-division procedure introduced by Steven Brams and Alan Taylor in 1993. It works only for two players, and begins by having each player independently spread 100 points over the items to be divided so as to reflect the relative worth of each object to that player. The allocation resulting from this procedure is equitable, envy-free, and Pareto-optimal. It requires no cash from either player, but one of the objects may have to be divided or shared by the two players.

Bottom-up strategy A bottom-up strategy is a strategy under an alternating procedure in which sophisticated choices are determined by working backward.

Cake-division procedure A fair-division procedure that uses a cake as a metaphor. Such procedures involve finding allocations of a single object that is finely divisible, as opposed to the situation encountered with either the adjusted winner procedure or Knaster's procedure. In a cake-division procedure, each player has a strategy that will guarantee that player a piece with which he or she is "satisfied," even in the face of collusion by the others.

Convention of the Law of the Sea An agreement based on divide-and-choose that protects the interests of developing countries in mining operations under the sea.

Divide-and-choose A fair-division procedure for dividing an object or several objects between two players. This method produces an allocation that is both proportional and envy-free (the two being equivalent when there are only two players).

Envy-free A fair-division procedure is said to be envy-free if each player has a strategy that can guarantee him or her a share of whatever is being divided that is, in the eyes of that player, at least as large (or at least as desirable) as that received by any other player, no matter what the other players do.

Equitable An allocation (resulting from a fair-division procedure like adjusted winner) is said to be equitable if each player believes he or she received the same fractional part of the total value.

Knaster inheritance procedure A fair-division procedure for any number of parties that begins by having each player (independently) assign a dollar value (a "bid") to the item or items to be divided so as to reflect the absolute worth of each object to that player. The allocation resulting from this procedure leaves each party feeling that he or she received a dollar value at least equal to his or her

fair share (and often more so). It never requires the dividing or sharing of an object, but it may require that the players have a large amount of cash on hand.

Last-diminisher method A cake-division procedure introduced by Stefan Banach and Bronislaw Knaster. It works for any number of players and produces an allocation that is proportional but not, in general, envy-free.

Lone-divider method A cake-division procedure introduced by Hugo Steinhaus. It works only for three players and produces an allocation that is proportional but not, in general, envy-free.

Pareto-optimal When no other allocation, achieved by any means whatsoever, can make any one player better off without making some other player worse off.

Point ratio The fraction in which the numerator is the number of points one party placed on an object and the denominator is the number of points the other party placed on the object.

Preference lists Rankings of the items to be allocated, from best to worst, by each of the participants.

Proportional A fair-division procedure is said to be proportional if each of n players has a strategy that can guarantee that player a share of whatever is being divided that he or she considers to be at least $1/n$ of the whole in size or value.

Selfridge–Conway envy-free procedure A cake-division procedure introduced independently by John Selfridge and John Conway. It works only for three players but produces an allocation that is envy-free (as well as proportional).

Taking turns A fair-division procedure in which two or more parties alternate selecting objects.

✔ SKILLS CHECK

1. The adjusted winner procedure

(a) applies only to two-party disputes or disputes that can be recast as two-party disputes.

(b) applies to either two-party or three-party disputes.

(c) applies to n-party disputes for all n.

2. The starting point with adjusted winner is to have each side—independently and simultaneously—spread 100 points over the issues in such a way that it

(a) reflects the correct monetary values.

(b) reflects the relative worth of each issue to that party.
(c) represents a compromise in the perceived values of the two parties.

3. The "winner" part of the name "adjusted winner" refers to the fact that

(a) each party initially "wins" (that is, is given) each issue on which he or she places more points than the other party.
(b) objects ultimately go to whichever party bid more.
(c) it is impossible for both parties to win with any fair-division scheme.

4. The "adjusted" part of the name "adjusted winner" refers to

(a) the fact that parties may have to change their point assignments.
(b) the transfer of items (or parts thereof) from one party to the other until points are equalized.
(c) the final allocation that makes one party a winner and the other a loser.

5. In transferring items from one party to the other with the adjusted winner procedure,

(a) the order in which items are transferred is extremely important.
(b) only one item will need to be split or shared.
(c) the order of transfer is obtained by looking at so-called point ratios.
(d) all of the above.

6. Chris and Terry must make a fair division of three objects. They assign points to the objects and use the adjusted winner procedure. What does Chris end up with?

Object	Chris	Terry
Boat	30	20
Land	50	60
Car	20	20

(a) Boat and car

(b) Boat and land
(c) Boat, car, and part of the land

7. Chris and Terry use the Knaster inheritance procedure to divide a coin collection. Chris bids $1000 and Terry bids $800. What is the outcome?

(a) Chris gets the coins and pays Terry $200.
(b) Chris gets the coins and pays Terry $450.
(c) Chris gets the coins and pays Terry $500.

8. Four children bid on two objects. Using the Knaster inheritance procedure, what does Adam end up with after the division?

Object	Adam	Beth	Carl	Dietra
House	$80,000	$75,000	$90,000	$60,000
Car	$10,000	$12,000	$13,000	$15,000

(a) House and cash
(b) Car and cash
(c) Cash only

9. With the procedure known as taking turns, the optimal strategy for players is called

(a) the sincere strategy.
(b) the bottom-up strategy.
(c) the top-down strategy.

10. With taking turns, we assume that a rational player will

(a) never willingly choose his or her least preferred item.
(b) avoid wasting a choice on an item that he or she knows will remain available and can be chosen later.
(c) do as in both parts (a) and (b).

11. Two people use the divide-and-choose procedure to divide a field. Suppose Jeff divides and Karen chooses. Which statement is true?

(a) Karen always believes she gets more than her fair share.
(b) Karen can guarantee that she always gets at least her fair share.

(c) Karen can possibly believe she gets less than her fair share.

12. A fair-division procedure is envy-free when each player believes that

(a) every player received an equal portion.
(b) he or she received at least a fair share.
(c) no other player received more than he or she did.

13. Using the Steinhaus procedure for three players (lone divider), what happens if there is a single portion that is the only one approved of by both nondividers?

(a) One of the other portions is given to the divider.
(b) The two nondividers flip a coin to determine who receives the approved portion.
(c) All portions are returned to the cake and a different person serves as the new divider.

14. Using the Steinhaus procedure for three players (lone divider), what happens if the two nondividers approve different portions?

(a) Each nondivider receives a portion that he or she has approved.
(b) The divider receives his or her choice of the portions.
(c) The divider selects portions for each person.

15. Using the Banach–Knaster procedure for three or more players (last diminisher), what happens to the first portion after each person has inspected and possibly trimmed it?

(a) The portion goes to the last person to approve the portion, whether or not it was trimmed.
(b) The portion goes to the last person to trim the portion.
(c) The portion goes to the first person to approve and not trim the portion.

16. Using the Banach–Knaster procedure for three or more players (last diminisher), what does the player who receives the first portion then do?

(a) That player becomes the first person to approve the second portion.

(b) That player becomes the last person to approve the second portion.
(c) That player leaves the game.

17. Using the Banach–Knaster procedure for three or more players (last diminisher), suppose that Scott initially cuts a piece and passes it among the other people, none of whom trims it. What happens next?

(a) Scott gets this piece.
(b) The last person who is handed the piece keeps it.
(c) The piece is returned to the cake and someone else cuts a piece.

18. Using the Banach–Knaster procedure for three or more players (last diminisher), what happens when only two people remain?

(a) One player chooses the scraps and the other chooses the portion.
(b) One player separates the remainders into two portions and the other chooses.
(c) The players flip a coin to decide who takes the remainders.

19. For the Selfridge–Conway envy-free procedure for three players, which of the following statements is true?

(a) Each of the three players has the opportunity to trim the portions if they appear to be unfair.
(b) Each player receives a portion that he or she believes to be exactly one-third of the total.
(c) The first player may believe that the third player received more than a fair share.

20. For the Selfridge–Conway envy-free procedure for three players, which of the following statements is true?

(a) The first player will always receive one of the three portions as originally cut.
(b) The second player will always receive one of the trimmed portions.
(c) The third player will always believe that he or she received exactly a fair share.

EXERCISES   ■ Challenge   ◆ Discussion

The Adjusted Winner Procedure

1. The 1991 divorce of Ivana and Donald Trump was widely covered in the media. The marital assets included a 45-room mansion in Greenwich, Connecticut; the 118-room Mar-a-Lago mansion in Palm Beach, Florida; an apartment in the Trump Plaza; a 50-room Trump Tower triplex; and just over $1 million in cash and jewelry. Assume points are assigned as follows:

Marital asset	Point Allocations	
	Donald's Points	Ivana's Points
Connecticut estate	10	38
Palm Beach mansion	40	20
Trump Plaza apartment	10	30
Trump Tower triplex	38	10
Cash and jewelry	2	2

Use the adjusted winner procedure to determine a fair allocation of the marital assets. (Exercise 1 courtesy of Catherine Duran.)

2. Suppose that Calvin and Hobbes discover a sunken pirate ship and must divide their loot. They assign points to the items as follows:

Object	Calvin's Points	Hobbes's Points
Cannon	10	5
Anchor	10	20
Unopened chest	15	20
Doubloon	11	14
Figurehead	20	30
Sword	15	6
Cannon ball	5	1
Wooden leg	2	1
Flag	10	2
Crow's nest	2	1

Use the adjusted winner procedure to determine a fair allocation of the loot. (Exercise 2 courtesy of Erica DeCarlo.)

3. This exercise illustrates how the adjusted winner procedure can be used to resolve disputes as well as to achieve fair allocations. Suppose Mike and Phil are roommates in college, and they encounter serious conflicts during their first week at school. Their resident adviser decides to use the adjusted winner procedure to resolve the dispute. The issues agreed upon, and the (independently assigned) points, turn out to be the following:

Issue	Mike's Points	Phil's Points
Stereo level	4	22
Smoking rights	10	20
Room party policy	50	25
Cleanliness	6	3
Alcohol use	15	15
Phone time	1	8
Lights-out time	10	2
Visitor policy	4	5

Use the adjusted winner procedure to resolve this dispute. (Exercise 3 courtesy of Erica DeCarlo.)

4. Suppose that a labor union and management are trying to resolve a dispute that involves four issues: the base salary of the workers, the annual salary increase that workers can expect, the benefits package the workers will receive, and the amount of vacation time to which each worker will be entitled. Suppose they use adjusted winner to resolve this dispute, with the following point assignments:

Issue	Labor	Management
Base salary	30	50
Salary increases	20	40
Benefits	35	5
Vacation time	15	5

Use adjusted winner to resolve this dispute.

5. Make up an example involving two people and several *objects* for which the adjusted winner procedure can be used, and then use the adjusted winner procedure to determine a fair division.

6. Make up an example involving two people and several *issues* for which the adjusted winner procedure can be used, and then use the adjusted winner procedure to determine a fair resolution of the dispute.

■ **7.** Suppose we have three items (X, Y, and Z) and three people (Bob, Carol, and Ted). Assume that each of the people spreads 100 points over the items (as in adjusted winner) to indicate the relative worth of each item to that person:

Item	Bob	Carol	Ted
X	40	30	30
Y	50	40	30
Z	10	30	40

For each of the following allocations, indicate

(a) whether or not it is proportional.
(b) whether or not it is envy-free.
(c) whether or not it is equitable.
(d) for the ones that are *not* Pareto-optimal, another allocation that makes one person better off without making anyone else worse off.

Allocation 1: Bob gets Z, Carol gets Y, and Ted gets X. (This is not Pareto-optimal.)

Allocation 2: Bob gets Y, Carol gets Z, and Ted gets X. (This is not Pareto-optimal.)

Allocation 3: Bob gets X, Y, and Z. (This *is* Pareto-optimal; explain why.)

Allocation 4: Bob gets Y, Carol gets X, and Ted gets Z. (This is Pareto-optimal.)

Allocation 5: Bob gets X, Carol gets Y, and Ted gets Z. (This is Pareto-optimal.)

The Knaster Inheritance Procedure

8. If John bids $28,225 and Mary bids $32,100 on their aging parents' old classic car, which they no longer drive, how would you reach a fair division?

9. John and Mary inherit their parents' old house and classic car. John bids $28,225 on the car and $55,900 on the house. Mary bids $32,100 on the car and $59,100 on the house. How should they arrive at a fair division?

10. Can you modify your fair-division procedure in Exercise 9 so that both John and Mary receive one of the two objects while still considering the allocation as fair?

11. Describe a fair division for three heirs, A, B, and C, who inherit a house in the city, a small farm, and a valuable sculpture, and who submit sealed bids (in dollars) on these objects as follows:

	A	B	C
House	145,000	149,999	165,000
Farm	135,000	130,001	128,000
Sculpture	110,000	80,000	127,000

12. Describe a fair division for three children, E, F, and G, who inherit equal shares of their parents' classic car collection and who submit sealed bids (in dollars) on these five cars as follows:

	E	F	G
Duesenberg	18,000	15,000	15,000
Bentley	18,000	24,000	20,000
Ferrari	16,000	12,000	16,500
Pierce-Arrow	14,000	15,000	13,500
Cord	24,000	18,000	22,000

Taking Turns

13. Suppose that Bob and Carol rank a series of objects, from most preferred to least preferred, as follows:

Bob	Carol
Car	Boat
Investments	Investments
CD player	Car
Boat	Washer–dryer
Television	Television
Washer–dryer	CD player

Assume that Bob and Carol use the bottom-up strategy and that Bob gets to choose first. Determine Bob's first choice and the final allocation.

14. Repeat Exercise 13 under the assumption that Carol gets to choose first.

15. Mark and Fred have inherited a number of items from their parents' estate, with no indication of who gets what. They rank the items from most preferred to least preferred as follows:

Mark	Fred
Truck	Boat
Tractor	Tractor
Boat	Car
Car	Truck
Tools	Motorcycle
Motorcycle	Tools

Assume that Mark and Fred use the bottom-up strategy and that Mark gets to choose first. Determine Mark's first choice and the final allocation.

16. Repeat Exercise 15 under the assumption that Fred gets to choose first.

17. Suppose that Donald and Ivana Trump decide to settle their divorce (described in Exercise 1) by taking turns. Assume they both use the bottom-up strategy, and Donald chooses first. Determine Donald's first choice and the final allocation. (Assume that Donald values the Connecticut estate slightly more than he values the Trump Plaza apartment.)

18. Repeat Exercise 17 under the assumption that Ivana gets to choose first.

Divide-and-Choose

♦ **19.** If you and another person are using divide-and-choose to divide something between you, would you rather be the divider or the chooser? (Assume that neither of you knows anything about the preferences of the other.)

♦ **20.** Suppose that Bob is entitled to one-fourth of a cake and Carol is entitled to three-fourths. In a few sentences, explain how divide-and-choose can be used to achieve an allocation in which each

party is guaranteed of receiving at least as much as he or she is entitled to.

21. Suppose that Bob, Carol, and Ted view a cake as having 18 units of value, with each unit of value represented by a small square (as in the accompanying figure). Suppose, however, that the players value various parts of the cake differently (or that Bob views the cake as being perfectly rectangular, whereas Carol and Ted see it as skewed in opposite ways). We represent this pictorially as follows:

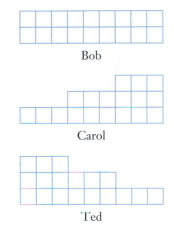

Assume that all cuts that will be made are vertical.

(a) If Bob and Carol use divide-and-choose to divide the cake between them, how large a piece will each receive (assuming they follow the suggested strategies that go with divide-and-choose and that Bob is the divider)?

(b) If Carol and Ted use divide-and-choose to divide the cake between them, how large a piece will each receive (assuming they follow the suggested strategies that go with divide-and-choose and that Carol is the divider)?

22. Assume that Bob and Carol view the cake as in Exercise 21, but assume also that each knows how the other values the cake, and that neither is spiteful. Suppose they are to divide the cake using the *rules*, but not necessarily the *strategies*, of divide-and-choose.

(a) Is Bob better off being the divider or the chooser?

(b) Discuss this in relation to Exercise 19.

Cake-Division Procedures: Proportionality

23. Suppose that players 1, 2, and 3 view a cake as in Exercise 21. Notice that each player views the cake as having 18 square units of area (or value). Assume that each player regards a piece as acceptable if and only if it is at least $18/3 = 6$ square units of area (his or her "fair share"). Assume also that all cuts made correspond to vertical lines.

(a) Provide a total of three drawings to show how each player views a division of the cake by player 1 into three pieces he or she considers to be the same size or value. Label the pieces A, B, and C.

(b) Identify two of these pieces that player 2 finds acceptable and two that player 3 finds acceptable.

(c) Show that a feasible assignment of fair pieces can be achieved by letting the players choose in the following order: player 3, player 2, player 1. Indicate how many square units of value each player thinks he or she received. Is there any other order in which players can choose pieces (in this example) that also results in a feasible assignment?

24. Suppose that players 1, 2, and 3 view a cake as follows:

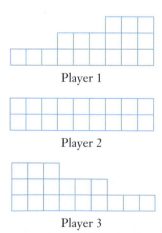

Player 1

Player 2

Player 3

(a) Provide a total of three drawings to show how each player views a division of the cake by player 1 into three pieces he or she considers to be the same size or value. Label the pieces A, B, and C. (We are still assuming that all cuts correspond to vertical lines, so this will require a cut along a vertical center line of some of the squares.)

(b) Show that neither player 2 nor player 3 finds more than one of the three pieces acceptable (with "acceptable" defined as in Exercise 23).

(c) Identify a single piece that player 2 and player 3 agree is *not* acceptable. (There are actually two such pieces; for definiteness, find the one on the right.)

(d) Assume that players 2 and 3 give the piece from part (c) to player 1. Suppose they reassemble the rest and players 2 and 3 divide it between themselves using divide-and-choose (with a single vertical cut). Determine what size piece each of the three players will think he or she received (1) if player 2 divides and player 3 chooses, and (2) if player 3 divides and player 2 chooses.

25. Suppose players 1, 2, and 3 view a cake as in Exercise 24. Illustrate the last-diminisher method (still restricting attention to vertical cuts and, in addition, assuming that the piece potentially being diminished is a piece off the left side of the cake) by following steps (a) through (h) below:

(a) Draw a picture showing the third of the cake (6 squares) that player 1 will slice off the cake.

(b) Determine whether player 2 will pass or further diminish this piece. If he or she would further diminish it, make a new drawing.

(c) Determine whether player 3 will pass or further diminish this piece. If he or she would further diminish it, make a new drawing.

(d) Determine who receives the piece cut off the cake and what size or value he or she thinks it is. (Actually, we knew what size the person receiving this first piece would think it was, assuming he or she followed the prescribed strategy. How did we know this?)

(e) Finish the last-diminisher method using divide-and-choose on what remains, with the lowest-numbered player who remains doing the dividing.

(f) Redo step (e) with the other player doing the dividing.

26. The Banach–Knaster last-diminisher method is not the only well-known cake-division procedure that yields a proportional allocation for any number of players. There is also one due to A. M. Fink (sometimes called the *lone-chooser method*). For three players (Bob, Carol, and Ted) it works as follows:

(i) Bob and Carol divide the cake into two pieces using divide-and-choose.
(ii) Bob now divides the piece he has into three parts that he considers to be the same size. Carol does the same with the piece she has.
(iii) Ted now chooses whichever of Bob's three pieces that he (Ted) thinks is largest, and Ted chooses whichever of Carol's three pieces that he thinks is largest.
(iv) Bob keeps his remaining two pieces, as does Carol.

(a) Explain why Ted thinks he is getting at least one-third of the cake.
(b) Explain why Bob and Carol each think they are receiving at least one-third of the cake.
(c) Explain why, in general this scheme is not envy-free.

■ **27.** In A. M. Fink's procedure (described in Exercise 26), suppose that a fourth person (Alice) comes along after Bob, Carol, and Ted have already divided the cake among themselves so that each of the three thinks he or she has a piece of size at least one-third. Mimic what was done in the three-person case to obtain an allocation among the four that is proportional. (*Hint:* Begin by having Bob, Carol, and Ted divide the pieces they have into a certain number—how many?—of equal parts.)

28. There is a moving-knife version of the Banach–Knaster procedure that is due to Dubins and Spanier. To describe it, we picture the cake as being rectangular, and the procedure beginning with a referee holding a knife along the left edge, as illustrated below.

Assume, for the sake of illustration, that there are four players (Bob, Carol, Ted, and Alice). The referee starts moving the knife from left to right over the cake (keeping it parallel to the position in which it started) until one of the players (assume it is Bob) calls "cut." At this time, a cut is made, the piece to the left of the knife is given to Bob, and he exits the game. The knife starts moving again, and the process continues. The strategies are for each player to call "cut" whenever it would yield him or her a piece of size at least one-fourth.

(a) Explain why this procedure produces an allocation that is proportional.
(b) Explain why the resulting allocation is not, in general, envy-free.
(c) Explain why, if you are not the first player to call "cut," there is a strategy different from the one suggested that is never worse for you, and sometimes better.

Cake-Division Procedures: The Problem of Envy

29. Suppose players 1, 2, and 3 view the cake as in Exercise 24. Illustrate the envy-free procedure for $n = 3$ (yielding an allocation of part of the cake) by following steps (a) through (c) below. Again, restrict attention to vertical cuts.

(a) Provide a total of three drawings to show how each player views a division of the cake by player 1 into three pieces he or she considers to be the same size or value. Label the pieces A, B, and C. (This is the same as Exercise 24a.)
(b) Redraw the picture from player 2's view, and illustrate the trimming of piece A that he or she would do. Label the trimmed piece A' and the actual trimmings T.

(c) Indicate which piece each player would choose (and what he or she thinks its size is) if the players choose in the following order: player 3, player 2, player 1, according to the envy-free procedure.

■ **30.** There is a two-person moving-knife cake-division procedure due to A. K. Austin that leads to each player receiving a piece of cake that he or she considers to be of size exactly one-half. It begins by having one of the two players (Bob) place two knives over the cake, one of which is at the left edge, and the other of which is parallel to the first and placed so that the piece between the knives (*A* in the picture below) is of size exactly one-half in Bob's estimation.

If Carol agrees that this is a 50-50 division, we are done. Otherwise, Bob starts moving both knives to the right—perhaps at different rates—so that the piece between the knives remains of size one-half in his eyes. Carol calls "stop" at the point when she also thinks the piece between the two knives is of size exactly one-half.

(a) If the knife on the right were to reach the right-hand edge, where would the knife on the left be?
(b) Explain why there definitely is a point where Carol thinks the piece between the two knives is of size exactly one-half. (*Hint:* If Carol thinks the piece is too small at the beginning, what will she think of it at the end?)

Knife 1   Knife 2

A   Cake

 WRITING PROJECTS

1. It turns out that there is no way to extend the adjusted winner procedure to three or more players. That is, there are point assignments by three players to three objects so that no allocation satisfies the three desired properties of equability (equal points), envy-freeness, and Pareto optimality. On the other hand, there are separate procedures that will realize any two of the three properties. Thus, trade-offs must be made, and these may depend on the circumstances. In a few paragraphs, discuss the relative importance of the three properties and circumstances that may affect the choice of which two of the three properties one might wish to have satisfied.

2. If we use taking turns to divvy up a collection of objects between two people (Bob and Carol),

then there is an obvious advantage to going first. Assume that we have decided that Bob will, in fact, choose first (say, by the toss of a coin). Let's think about how Carol might be compensated. First of all, if there are only three objects, then the "choice sequence" Bob–Carol–Carol seems to be the only reasonable one. Do you agree? For four objects, however, there are two choice sequences that suggest themselves: Bob–Carol–Carol–Carol and Bob–Carol–Carol–Bob. Do you think that one of these is obviously more fair than the other? What if there are four identical objects? What if both Bob and Carol value object *A* twice as much as *B*, and *B* twice as much as *C*, and *C* twice as much as *D*? What sequences suggest themselves for five objects? For eight objects? In one page or less, discuss these questions. (For more on this, see *The Win–Win Solution* in the Suggested Readings.)

3. One of the most important differences between the three-person and the *n*-person envy-free procedures is that the latter procedure may take more than two stages. And, of course, the more stages there are, the more cuts and trimmings may be necessary. Do you consider this a serious practical problem, or is it mainly a theoretical problem? In one paragraph, explain your reasons.

4. One often hears of the importance of "process" versus "product," the latter referring to *what* is achieved and the former referring to *how* it was achieved. In a couple of sentences, comment on the relevance of this to fair division as illustrated by the following rough paraphrasing of an exchange between two old friends, Ralph Kramden (played by Jackie Gleason) and Ed Norton (played by Art Carney) in the 1950s sitcom *The Honeymooners.*

> *Ralph to Ed* (as the two are sitting alone at the dinner table): I can't believe you did that.
>
> *Ed:* Did what, Ralph?
>
> *Ralph:* There were two potatoes there, and you reached right out and took the big one.
>
> *Ed:* What would you have done, Ralph?
>
> *Ralph:* Why, I'd have taken the little one.
>
> *Ed:* You *got* the little one, Ralph.

SUGGESTED READINGS

BRAMS, S. J., and A. D. TAYLOR. *The Win–Win Solution: Guaranteeing Fair Shares to Everybody,* Norton, New York, 1999. Brams and Taylor do more with adjusted winner, as well as divide-and-choose and taking turns.

BRAMS, S. J., and A. D. TAYLOR. *Fair Division: From Cake-Cutting to Dispute Resolution,* Cambridge University Press, Cambridge, 1996. Brams and Taylor provide a book-length treatment of the kinds of topics introduced in this chapter, as well as divide-and-choose in the political arena, moving-knife procedures for cake cutting, and fairness as it applies to different auction and election procedures.

BRAMS, S. J., and A. D. TAYLOR. An envy-free cake division protocol. *American Mathematical Monthly,* 102 (1995): 9–18. Brams and Taylor describe in detail the finite version of their envy-free procedure for $n = 4$; in addition, they review earlier work on "protocols" (step-by-step procedures) that led up to their constructive solution of the envy-freeness problem for $n > 3$.

ROBERTSON, J., and W. WEBB. *Cake Cutting Algorithms: Be Fair If You Can,* A. K. Peters, Wellesley, Mass., 1998. Robertson and Webb cover a great deal of cake-cutting ground in a text that includes exercises.

STEINHAUS, H. *Mathematical Snapshots,* Oxford University Press, Oxford, 1960. A brief but significant introduction to both fair division and apportionment is provided in this popular book on interesting mathematical topics.

Apportionment

14.1 The Apportionment Problem

The Sports Magnet High School field hockey team played 23 games in the 2004–2005 season. It won 18 games, lost 4, and one game ended in a tie. Table 14.1 expresses these results as percentages.

The coach objects to this table—it looks too complicated. "Just express the percents as whole numbers," she says. We round off the percentages: 78% of the games were won, 17% were lost, and 4% were tied. The coach notices that these numbers add to 99% and changes the winning percentage to 79%. "Now you have 100%!"

::::::: Apportionment

The **apportionment problem** is to round a set of fractions so that their sum is maintained at its original value. The rounding procedure must not be an arbitrary one, but one that can be applied constantly. Any such rounding procedure is called an **apportionment method**.

When rounding percentages so as to preserve the sum of 100%, we face an apportionment problem. The coach used the apportionment method called "make it look good for the team," but a serious apportionment method should be

504

TABLE 14.1	The Sports Magnet High School Field Hockey Team: 2004–2005 Season	
		Percentage
Games won	18	$\frac{18}{23} \times 100\% = 78.26\%$
Games lost	4	$\frac{4}{23} \times 100\% = 17.39\%$
Games tied	1	$\frac{1}{23} \times 100\% = 4.35\%$
Games played	23	100.00%

unbiased, especially when it comes to even *more* critical issues, such as apportioning seats in the U.S. House of Representatives.

The framers of the U.S. Constitution wrote that seats in the House of Representatives "shall be apportioned among the several states within this union according to their respective Numbers. . . ." They may not have realized that the apportionment problem that they set in this clause was of any significance, but it caused trouble right from the start. In 1790, Delaware's population, 55,540, was the least among the 15 states. The total population of the country at the time was 3,615,920, and the House of Representatives was to have 105 members. Thus,

Howard Chandler Christy's "Scene at the Signing of the Constitution of the United States." *(Art Resource, NY.)*

TABLE 14.2	The Congressional Apportionment that George Washington Vetoed		
State	Population	Quota	Apportion- ment
Virginia	630,560	18.310	18
Massachusetts	475,327	13.803	14
Pennsylvania	432,879	12.570	13
North Carolina	353,523	10.266	10
New York	331,589	9.629	10
Maryland	278,514	8.088	8
Connecticut	236,841	6.878	7
South Carolina	206,236	5.989	6
New Jersey	179,570	5.214	5
New Hampshire	141,822	4.118	4
Vermont	85,533	2.484	2
Georgia	70,835	2.057	2
Kentucky	68,705	1.995	2
Rhode Island	68,446	1.988	2
Delaware	55,540	1.613	2
Totals	3,615,920	105	105

the average congressional district should have a population of 3,615,920 ÷ 105 = 34,437. To obtain its fair share of the House seats, we can divide Delaware's population by the average congressional district population. Ideally, Delaware should have had

$$\frac{55,540}{34,437} = 1.613 \text{ congressional districts}$$

Table 14.2 displays the apportionment proposed in a bill that Congress passed and sent to President Washington. Each state's fair share of the 105 congressional districts, or *quota*, was calculated by dividing its population by the average congressional district population of 34,437, as in the case of Delaware, shown above. None of the quotas were whole numbers. The bill gave Delaware two seats in Congress. As you will see in the table, Virginia's quota was 18.310 seats, and Virginia was awarded 18 seats in the House.

Although this apportionment may seem fair enough, President Washington vetoed that bill.[1] Washington came from Virginia, a state that lost something in

[1]This apportionment bill was the first bill in U.S. history to be vetoed.

the apportionment bill. Was he biased in favor of his home state, as the field hockey coach was in favor of her team? We will return to this question.

In the following example, we will see how to set up an apportionment problem.

The High School Mathematics Teacher

EXAMPLE 1

A high school has one mathematics teacher who teaches all geometry, pre-calculus, and calculus classes. She has time to teach a total of five sections. One hundred students are enrolled as follows: 52 for geometry, 33 for pre-calculus, and 15 for calculus. How many sections of each course should be scheduled?

The number of students enrolled in each course is called the *population.* Thus, the population for geometry is 52, the population for pre-calculus is 33, and the population for calculus is 15. There are five sections for the 100 students, so the average section will have $100 \div 5 = 20$ students. We will call this average section size the *standard divisor,* because each quota can be determined by dividing the corresponding population by this number. Table 14.3 displays these calculations.

As shown in the table, the quotas add up to 5. It is tempting to round each to the nearest whole number, as in the right column of the table, but this makes 6 sections in all—too many! The purpose of an apportionment method is to find an equitable way to round a set of numbers like this without increasing or decreasing the original sum. ■

Although many apportionment problems do not involve the House of Representatives, our terminology refers to *states, populations,* and a *house size.* In a course-scheduling problem, the courses to be taught correspond to the states, the numbers of students enrolled in each course correspond to the populations, and the house size is the total number of sections to be scheduled. Thus, in our example, there are three states: Geometry, Pre-calculus, and Calculus, with populations

$$p_1 = 52, \ p_2 = 33, \text{ and } p_3 = 15$$

and the house size is 5.

TABLE 14.3	Calculation of the Quotas for High School Mathematics Courses		
Course	**Population**	**Quota**	**Rounded**
Geometry	52	$52 \div 20 = 2.60$	3
Pre-calculus	33	$33 \div 20 = 1.65$	2
Calculus	15	$15 \div 20 = 0.75$	1
Totals	100	5	6

In the problem of rounding percentages so that their sum is 100, the house size is 100. The categories (such as wins, losses, and ties in the field hockey example) correspond to the states, and the numbers in each category (in the field hockey example, 18 wins, 4 losses, and 1 tie) correspond to the populations of the states.

Divisor and Quota

The quotient of the total population, p, divided by the house size is called the **standard divisor.** If h denotes the house size and s is the standard divisor, then

$$s = \frac{p}{h}$$

In an apportionment problem, the **quota** is the exact share that would be allocated *if a whole number were not required.* To obtain a state's quota, divide its population by the standard divisor. State i, with population p_i, has quota

$$q_i = \frac{p_i}{s}$$

EXAMPLE 2　California's Quota

The Census Bureau recorded the apportionment population[2] of the United States, as of April 1, 2000, to be 281,424,177. There are 435 seats in the House of Representatives; therefore, the standard divisor is

$$s = \frac{281,424,177}{435} = 646,952$$

California's apportionment population was $p_1 = 33,930,798$. Its quota, q_1, is determined by dividing this population by the standard divisor. Thus,

$$q_1 = \frac{p_1}{s} = \frac{33,930,798}{646,952} = 52.447 \text{ seats}$$

California's apportionment, which is required to be a whole number, was set at 53 seats. ■

---

[2] The apportionment population includes the resident population and the overseas population.

An apportionment is given by whole numbers,

$$a_1, a_2, \ldots, a_i, \ldots, a_n$$

with a_i representing the number of representatives given to state i. Ideally, each state's apportionment should be close to its quota. It is unrealistic to expect that for any state, $a_i = q_i$, because a_i has to be a whole number and q_i is unlikely to be a whole number. In choosing an apportionment method, we must decide what we mean by the phrase "a_i should be close to q_i."

Apportionment always involves rounding, and there are many ways to round. "Rounding down" means discarding the fractional part of a number q to obtain a whole number that we will denote $\lfloor q \rfloor$. Thus, $\lfloor 7.00001 \rfloor = 7$, $\lfloor 7 \rfloor = 7$, and $\lfloor 6.99999 \rfloor = 6$. "Rounding up" gives the next whole number, $\lceil q \rceil$. Thus, $\lceil 7.00001 \rceil = 8$, but $\lceil 7 \rceil = 7$.

There are numerous different apportionment methods. Each has flaws, and our goal is to understand how to choose a method that is appropriate for a particular apportionment problem.

14.2 The Hamilton Method

The congressional apportionment bill that was vetoed by President Washington was written by Alexander Hamilton. While it may appear that he simply rounded each quota to the nearest whole number, he was aware that there would be occasions—analogous to the examples of the field hockey team's percentages of wins, losses, and ties not summing to 100%, or the high school teacher receiving an extra class to teach—when the total number of seats apportioned in this way would be either more or less than the statutory house size. He developed an apportionment method that he called the *method of largest fractions*.

Alexander Hamilton's Method

With the **Hamilton method**, state i receives either its **lower quota**—which is its quota rounded down in the notation just introduced, $\lfloor q_i \rfloor$—or its **upper quota**, $\lceil q_i \rceil$, obtained by rounding the quota up. The states that receive their upper quotas are those whose quotas have the largest fractional parts.

Implementing the Hamilton method is a three-step procedure:

1. Calculate each state's quota.
2. Tentatively assign to each state its lower quota of representatives. Each state whose quota is not a whole number loses a fraction of a seat at this

The apportionment method of largest fractions, also known as the Hamilton method, was named for Alexander Hamilton. *(National Portrait Gallery/Art Resource, NY.)*

stage, so the total number of seats assigned at this point is less than the house size *h*. This leaves additional seats to be apportioned.

3.   Allot the remaining seats, one each, to the states whose quotas have the largest fractional parts, until the house is filled.

It is possible that a tie will occur, with the quotas of two states having identical fractional parts, but in practice, this rarely happens when large populations are involved.

EXAMPLE 3   The High School Teacher's Dilemma

Let us use the Hamilton method to determine how many sections of geometry, pre-calculus, and calculus the high school teacher should teach. We have found that the quotas for the three subjects were 2.60, 1.65, and 0.75, respectively (see Table 14.3). The lower quotas are $\lfloor 2.60 \rfloor = 2$, $\lfloor 1.65 \rfloor = 1$, and $\lfloor 0.75 \rfloor = 0$, so we start by assigning two sections of geometry and one section of pre-calculus. With three sections apportioned, we have two more to assign to "fill the house." These two sections go to the subjects with the largest fractions: Calculus has the largest fraction, 0.75, and gets a section; the second section goes to pre-calculus, whose fraction is 0.65. The final apportionment is as follows: geometry, two sections; pre-calculus, two sections; calculus, one section. ∎

The Field Hockey Team

EXAMPLE 4

In Table 14.1, we saw that the percentages of wins, losses, and ties for the field hockey team were 78.26%, 17.39%, and 4.35%, respectively. These are the quotas that we must round so that they sum to 100%. To start, we apportion $\lfloor 78.26 \rfloor =$ 78%, $\lfloor 17.39 \rfloor = 17\%$, and $\lfloor 4.35 \rfloor = 4\%$ to the three categories. These lower quotas add up to 99%. The remaining 1% to be apportioned goes to the losses because their fraction, 0.39, is the largest. The final apportionment is 78% wins, 18% losses, and 4% ties. Coach will veto this apportionment. ∎

Hamilton's Apportionment

EXAMPLE 5

The first congressional apportionment involved 15 states, and the House had 105 seats. According to the 1790 census, the U.S. population was 3,615,920. The standard divisor, $3,615,920 \div 105 = 34,437$, represents the population of the average congressional district.

Table 14.2 displays Alexander Hamilton's proposed apportionment. Each quota shown in the table was calculated by dividing the state's population by this standard divisor. If you are using a calculator to follow the entries in the table, store the standard divisor in the calculator's memory. Then each quota can be figured by entering the individual state's population and dividing by the divisor recalled from memory.

If each state were given its lower quota, only 97 seats would have been apportioned. The remaining 8 seats go to the 8 states whose quotas had the largest fractional parts. ∎

President Washington's veto message stated that the fractions of seats gained by some states in the third step of the Hamilton apportionment were not related to the states' total populations.

The veto prevented the Hamilton method from being used in 1792, but it was adopted by Congress in 1850 and remained in use until 1900. The half-century of experience with the Hamilton method revealed a paradox.

Paradoxes of the Hamilton Method

A *paradox* is a fact that seems obviously false. The first Hamilton apportionment paradox, called the *Alabama paradox,* was discovered in 1881. As part of the reapportionment procedure mandated by the Constitution, the Census Bureau had supplied Congress with a table of congressional apportionments for a range of different house sizes from 275 to 350, based on the 1880 census. The table revealed a strange phenomenon. With a 299-seat house, Alabama's quota was 7.646. The fractional part ranked twentieth of the 38 states, and that was just enough to give Alabama its upper quota of $\lceil 7.646 \rceil = 8$ seats. Illinois and Texas, with quotas of 18.640 and 9.640, respectively, ranked below Alabama and received their lower quotas: 18 for Illinois and 9 for Texas. The next column of the table, corresponding to a house size of 300 seats, showed changes in the apportionments for each of these three states: Alabama now had 7 seats instead of 8, while

Illinois and Texas received increased apportionments of 19 seats and 10 seats, respectively. Alabama had *lost* a seat as a result of an increase in the house size. This happened because the Illinois and Texas quotas increased more than Alabama's quota did. With a 300-seat house, Alabama's quota increased to 7.671, Illinois's to 18.702, and Texas's to 9.672. Because the fractional parts of the quotas for Illinois and Texas were larger, those states were given their upper quotas, while Alabama was left with its lower quota.

The Alabama Paradox

The **Alabama paradox** occurs when a state loses a seat as the result of an increase in the house size.

The Alabama paradox validates President Washington's veto message, issued 90 years before its discovery. If the fractional parts of the quotas, which determine the way the last few seats are apportioned, were related to the populations in any sensible way, the paradox could not have occurred.

EXAMPLE 6   A Mathematics Department Meets the Alabama Paradox

A mathematics department has 30 teaching assistants to cover recitation sections for College Algebra, Calculus I, Calculus II, Calculus III, and Contemporary Mathematics. The enrollments of these courses are given in Table 14.4. The department will use the Hamilton method to apportion the teaching assistants to the five subjects. In this problem, the house size is $h = 30$ (the number of teaching assistants) and the population is $p = 750$. Therefore, the standard divisor is $750 \div 30 = 25$; this represents the average number of students in each recitation section. Each quota shown in the table was determined by dividing the enrollment of the course by this divisor.

TABLE 14.4   Apportioning 30 Teaching Assistants

Course	Enrollment	Quota	Lower Quota	Apportionment
College Algebra	188	7.52	7	7
Calculus I	142	5.68	5 ↑	6
Calculus II	138	5.52	5	5
Calculus III	64	2.56	2 ↑	3
Contemporary Mathematics	218	8.72	8 ↑	9
Totals	750	30.00	27	30

TABLE 14.5	Apportioning 31 Teaching Assistants			
Course	Enrollment	Quota	Lower Quota	Apportion-ment
College Algebra	188	7.771	7 ↑	8
Calculus I	142	5.869	5 ↑	6
Calculus II	138	5.704	5 ↑	6
Calculus III	64	2.645	2	2
Contemporary Mathematics	218	9.011	9	9
Totals	750	31.000	28	31

The lower quotas add up to 27, so the three courses whose quotas have the largest fractional parts, Calculus I and III and Contemporary Mathematics, are entitled to their upper quotas.

After these calculations were finished, the graduate school authorized the department to hire an additional teaching assistant. With 31 teaching assistants, the standard divisor becomes $750 \div 31 = 24.19355$. The new quotas, determined by dividing each population by this new divisor, are shown in Table 14.5. Now the lower quotas add up to 28, so again three additional teaching assistants go to the subjects whose quotas have the largest fractions. The Calculus III fraction, which had been larger than the College Algebra fraction when there were just 30 teaching assistants, has been surpassed. The instructor will be dismayed, because increasing the total number of teaching assistants for the department has caused her course to *lose* one section! ■

The size of the House of Representatives is now fixed at 435 members, so the Alabama paradox cannot occur. A second paradox, called the *population paradox,* is associated with a fixed house size. In Exercise 10 we will consider a situation involving an imaginary country with four states and a 100-seat legislature, apportioned by the Hamilton method. The new census shows substantial increases in population for the three largest states and a slight loss of population for the smallest state. In the new apportionments, the largest and smallest states both gain seats, while the middle two states lose seats. Thus, the smallest state lost population and gained a seat, while the middle two seats gained population and lost seats.

The Population Paradox

The **population paradox** occurs when one state's population increases, and its apportionment decreases, while simultaneously another state's population increases proportionally less, or decreases, and its apportionment increases.

Thomas Jefferson favored a method of apportionment biased in favor of states with large populations. *(National Portrait Gallery/Art Resource, NY.)*

14.3 Divisor Methods

The Jefferson Method

The Constitution requires congressional districts to be drawn so that the population of each is at least 30,000. President Washington could have vetoed the Hamilton apportionment bill because Delaware's population—only 55,540—was too small for the two congressional districts assigned to it. Thomas Jefferson proposed an apportionment method, now called the **Jefferson method,** to replace the Hamilton method.

In any apportionment, the standard divisor, *s,* obtained by dividing the total population by the house size, represents the average district population. In developing his method, Thomas Jefferson focused on the population of the smallest district, *d.* Using *d,* rather than *s,* as the divisor, each state receives an **adjusted quota** that is rounded *down* to obtain its apportionment. If *d* is chosen correctly, the apportionments will add up exactly to the statutory house size.

In effect, the Jefferson method apportions to each state the maximum number of congressional districts of population *d* that will be accommodated by the state's population. Any leftover population is divided among these districts, so that, in effect, most districts will have populations greater than *d.*

Once the divisor *d* is known, the Jefferson apportionment is easy to compute. The apportionment for state *i* is

$$a_i = \left\lfloor \frac{p_i}{d} \right\rfloor$$

The actual divisor that Jefferson used was $d = 33{,}000$. Thus, Virginia's apportionment was

$$a_1 = \left\lfloor \frac{\text{population of Virginia}}{33{,}000} \right\rfloor = \left\lfloor \frac{630{,}560}{33{,}000} \right\rfloor = \lfloor 19.108 \rfloor = 19$$

Therefore, Virginia received 19 seats, rather than 18, which Hamilton's bill would have allocated. Delaware's apportionment was 1 seat instead of 2, as the following calculation shows.

$$a_{15} = \left\lfloor \frac{\text{population of Delaware}}{33{,}000} \right\rfloor = \left\lfloor \frac{55{,}540}{33{,}000} \right\rfloor = \lfloor 1.683 \rfloor = 1$$

The Jefferson method is one of a class of apportionment methods called *divisor methods*.

Divisor Methods

A **divisor method** of apportionment determines each state's apportionment by dividing its population by a common divisor d and rounding the resulting quotient. Divisor methods differ in the rule used to round the quotient.

The divisor must be carefully chosen to achieve the correct house size. In the first congressional apportionment, $d = 30{,}000$ would have resulted in larger apportionments for several states and a house size of 112, while $d = 36{,}000$ would have decreased several apportionments, and the house size would have been 91.

Critical Divisors

To implement the Jefferson method, we must determine the divisor, d. Start by determining the standard divisor, s, and the quota for each state, as we did with the Hamilton method. Each state is assigned, as a **tentative apportionment**, its lower quota. Thus, the tentative apportionment of state i is

$$n_i = \left\lfloor \frac{p_i}{s} \right\rfloor$$

As we noted with the Hamilton method, these tentative apportionments are not enough to fill the house, so the apportionments of some states will have to be increased. To determine which states should receive additional seats, calculate a *critical divisor* for each state.

Critical Divisor: The Jefferson Method

Let n_i be the tentative apportionment of a state, and let p_i be its population. The **critical divisor,** d_i, for that state is

$$d_i = \frac{p_i}{n_i + 1}$$

EXAMPLE 7　Jefferson's Critical Divisors

Referring to Table 14.2, the lower quota for Virginia was $n_1 = \lfloor 18.310 \rfloor = 18$ and the lower quota for Delaware was $n_{15} = \lfloor 1.613 \rfloor = 1$. The populations of the two states were 630,560 and 55,540, respectively, so their critical divisors are

$$d_1 = \frac{630,560}{18 + 1} = 33,187$$

$$d_{15} = \frac{55,540}{1 + 1} = 27,770 \quad \blacksquare$$

The significance of the critical divisor is that if the Jefferson divisor d is equal to d_i, then state i will contain *exactly* $n_i + 1$ districts of population d. Thus, with $d = 33,187$, Virginia's apportionment would be equal to 19. If we take $d = 27,770$, then Delaware would get 2 seats. However, the same divisor must be used for all states, so with this lower value of d, Virginia would get

$$\left\lfloor \frac{630,560}{27,770} \right\rfloor = 22 \text{ seats}$$

Once the critical divisors are determined, the state with the largest critical divisor is entitled to another seat, because when that divisor is used, no other state will receive a changed apportionment—a state receives additional seats only when a divisor smaller than or equal to its critical divisor is used. The total apportionment is thus increased by 1 in this step.

If there remain additional seats to be apportioned, we recompute the critical divisor for the state whose tentative apportionment has increased, and repeat the process. When the house is filled, the critical divisor most recently used is the divisor d, representing the minimum district population. With the Jefferson method (and any other divisor method), the critical divisors determine the priority of a state for receiving additional seats. The state with the largest critical divisor gets the next seat, but after receiving that seat, its critical divisor is recomputed, and usually another state will then be first in line.

The High School Mathematics Teacher

EXAMPLE 8

A mathematics teacher can teach five classes. There are 52 students enrolled in geometry, 33 in pre-calculus, and 15 in calculus. We will use the Jefferson method to determine how many sections of each subject to teach. We have previously determined that the lower quotas for the subjects are 2, 1, and 0, respectively (see Table 14.3). Thus, the critical divisors are

$$\frac{52}{2+1} = 17.333 \text{ for geometry}$$

$$\frac{33}{1+1} = 16.500 \text{ for pre-calculus}$$

$$\frac{15}{0+1} = 15.000 \text{ for calculus}$$

Geometry, with the largest critical divisor, has first priority for a new section, and its tentative apportionment is increased to 3. Its new critical divisor will be $52 \div (3 + 1) = 13$. Now pre-calculus has top priority, and its tentative apportionment is increased to 2. The house is now full, so the final apportionments are 3 sections of geometry and 2 sections of pre-calculus. The minimum section size will be the 16.500, because that is the critical divisor for the subject that was the last to receive an increased apportionment: pre-calculus. In practice, the two sections will have 16 and 17 students. Because the enrollment for calculus is less than the minimum section size, there will be no calculus class. ■

The Field Hockey Team

EXAMPLE 9

The Sports Magnet High School field hockey team had 18 wins, 4 losses, and 1 tie last season. We will use the Jefferson method to express this record as percentages. The house size is 100%, and the total population is the 23 games played. Therefore, the standard divisor is $23 \div 100 = 0.23$. The lower quotas are

$$\left\lfloor \frac{18}{0.23} \right\rfloor = 78\% \text{ wins}$$

$$\left\lfloor \frac{4}{0.23} \right\rfloor = 17\% \text{ losses}$$

$$\left\lfloor \frac{1}{0.23} \right\rfloor = 4\% \text{ ties}$$

and these are the tentative apportionments. The critical divisors are

$$\frac{18}{78+1} = 0.22785 \text{ for wins}$$

$$\frac{4}{17+1} = 0.22222 \text{ for losses}$$

$$\frac{1}{4+1} = 0.20000 \text{ for ties}$$

The category with the largest critical divisor is wins, so we increase its apportionment to 79%. The house is now full; we have the apportionment that Coach wanted: 79% wins, 17% losses, and 4% ties. ∎

These examples demonstrate that different methods may yield different apportionments, because the Hamilton method gave different apportionments in both cases. The Jefferson and Hamilton methods also differed in the first apportionment of congressional seats. We have seen that the critical divisor for Virginia was greater than the critical divisor for Delaware; thus, Virginia had greater priority than Delaware to receive increased apportionment with the Jefferson method, while the opposite was true with the Hamilton method.

With the Jefferson method, no state can receive less than its lower quota as its apportionment—because the lower quota is the initial tentative apportionment—but a state can be apportioned more than its upper quota. This phenomenon occurred for the first time in the apportionment based on the 1820 census, which recorded that New York had a population of 1,368,775. The total population of the United States was found to be 8,969,878, and the house size was 213. Therefore, the standard divisor was 8,969,878 ÷ 213 = 42,112, and New York's quota was

$$q = \frac{1,368,775}{42,112} = 32.503$$

The Hamilton method would have apportioned 33 seats to New York, but the Jefferson method, using the divisor $d = 39,900$, gave New York

$$\left\lfloor \frac{1,368,775}{39,900} \right\rfloor = 34 \text{ seats}$$

An apportionment method is said to satisfy the **quota condition** if in every situation each state's apportionment is equal to either its lower quota or its upper quota. It takes only one example like the 1820 apportionment to show that the Jefferson method does not satisfy the quota condition. In fact, if the house had continued to use the Jefferson method, it would have violated the quota condition in every apportionment since 1850.

With the Hamilton method, each state starts with its lower quota, and some states are given their upper quotas to fill the house. There is no way for a state to receive less than its lower quota or more than its upper quota, so the Hamilton method does satisfy the quota condition. This was obvious to Congress in 1850, so it based its apportionment on the Hamilton method.[3]

The Jefferson method, however, is not troubled by the Alabama and population paradoxes. Consider the Alabama paradox, in which a state loses a seat as a result of an increase in the house size. With any apportionment method, the

---

[3]The origins of the Hamilton method had been forgotten in 1850, and the method was named for Congressman Samuel Vinton, who had rediscovered the method.

apportionments of some states must increase when the size of the house increases. The Jefferson method awards seats in order of critical divisors. When the house size increases, the next seat will go to the state with the largest critical divisor. There is no opportunity for a state to lose a seat.

Congress has never used an apportionment method that satisfies the quota condition and avoids the paradoxes. It would be desirable to use such a method, and in the 1970s, the mathematicians Michel L. Balinski and H. Peyton Young set out to find one. They succeeded in finding a method, which they called the *quota method*, that satisfies the quota condition (as the Hamilton method does) and avoids the Alabama paradox (as the Jefferson method does). However, the population paradox remained. They subsequently proved that the only apportionment methods that are free of the population paradox are the divisor methods. It is known that every divisor method is capable of violating the quota condition, so Balinski and Young have proved an impossibility theorem: *No apportionment method that satisfies the quota condition is free of paradoxes.* This theorem is like Kenneth Arrow's theorem that there is no completely satisfactory way to decide multicandidate elections based on voter preference schedules (see section 9.4).

The Jefferson method favors the larger states. It is not an accident that in all the examples that we considered, the "state" with the largest population fared better with the Jefferson method than it did with the Hamilton method. In an apportionment problem, let s be the standard divisor and let d be the divisor used in the Jefferson method. The apportionment given to state i is then $\lfloor u_i \rfloor$, where $u_i = p_i/d$ is the state's *adjusted quota*. Comparing u_i with the quota, $q_i = p_i/s$, we find that their ratio

$$M = \frac{u_i}{q_i} = \frac{p_i}{d} \div \frac{p_i}{s} = \frac{p_i}{d} \times \frac{s}{p_i} = \frac{s}{d}$$

is the same for each state. Since we know that $s > d$, it follows that $M > 1$. Furthermore, we have $u_i = M \times q_i$. Thus, the apportionment for state i is

$$a_i = \lfloor M \times q_i \rfloor$$

For example, let's consider the congressional apportionment of 1820. The standard divisor was $s = 42,112$, and the Jefferson divisor was $d = 39,900$. Thus, the quotient, M, is

$$M = \frac{s}{d} = \frac{42,112}{39,900} = 1.0554$$

Now suppose that a state has a quota of exactly 19. We obtain an adjusted quota, u_i as follows:

$$u_i = 1.0554 \times 19 = 20.0526$$

In other words, that state gets another seat, violating the quota condition. In fact, every state whose quota is 19 or more will be guaranteed an additional seat with this value of M. If a state has a quota of 38, an identical calculation shows that it will receive two additional seats. On the other hand, consider a small state whose lower quota is 1. To increase its apportionment to 2, its quota must be at least $2/M = 1.89502$. Thus, a state with quota 19.00 is guaranteed an increase, and a state with quota 1.89 is denied an increase.

The Webster Method

The Method of Daniel Webster

The **Webster method** is the divisor method that rounds the quota (adjusted if necessary) to the nearest whole number, rounding up when the fractional part is greater than or equal to $\frac{1}{2}$, and rounding down when the fractional part is less than $\frac{1}{2}$.

The Webster and Jefferson methods are immune to the Alabama and population paradoxes, but neither satisfies the quota condition. However, in many respects, Webster's is a better method than Jefferson's. While the Jefferson method obviously favors the large states, the Webster method is neutral, favoring neither the

Statesman and orator Daniel Webster (1782–1852), who developed a divisor method for apportioning the U.S. House of Representatives. *(National Portrait Gallery/Art Resource, NY.)*

large nor the small states. Furthermore, while the Jefferson method almost always gives the most populous state more than its upper quota, the Webster method would not have violated the quota condition in any of the 22 congressional apportionments that have occurred so far.

To calculate an apportionment by the Webster method, determine the standard divisor, and use it to find the quota for each state. The tentative apportionment, which we will call n_i for state i, is its rounded quota. When all of the tentative apportionments have been determined, calculate their sum. If the sum is equal to the house size, the job is finished: The tentative apportionments become the actual apportionments.

It is possible that the sum of the tentative apportionments will be either more than or less than the house size. In this case, we must reduce the apportionment or increase the apportionment of one or more states. To choose which apportionments to change, we must calculate a critical divisor for each state.

Critical Divisors for the Webster Method

If the tentative apportionments do not fill the house, then the critical divisor for state i, with population p_i and tentative apportionment n_i, is

$$d_i^+ = \frac{p_i}{n_i + \frac{1}{2}}$$

The state with the largest critical divisor is first in line to receive a seat. If the tentative apportionments overfill the house, then the critical divisor for state i is

$$d_i^- = \frac{p_i}{n_i - \frac{1}{2}}$$

The state with the smallest critical divisor is first in line to lose a seat.

When the divisor d_i^+ is used, state i will get the adjusted quota

$$\frac{p_i}{d_i^+} = n_i + \frac{1}{2}$$

which is the smallest number to be rounded up to $n_i + 1$. Thus, the tentative apportionment of state i would increase. None of the states with critical divisors less than d_i^+ would see any change in their tentative apportionments.

Similarly, when the population p_i is divided by d_i^-, the resulting quotient, $n_i - \frac{1}{2}$, is on the borderline for rounding down to $n_i - 1$. If d_i^- is the smallest critical divisor, then state i will be the only state whose tentative apportionment would be affected by the divisor d_i^-.

EXAMPLE 10   Apportioning Classes

Let us return to the case of the mathematics teacher who teaches a total of five classes in geometry, pre-calculus, and calculus. The enrollments are 52 for geometry, 33 for pre-calculus, and 15 for calculus. With a total of 100 students enrolled, and a house size of 5, the standard divisor is 20. The quotas, determined by dividing the enrollments for the three subjects by the standard divisor, are 2.6, 1.65, and 0.75, respectively. The tentative apportionments are 3, 2, and 1; their total, 6, exceeds the house size. We therefore compute the divisors d_i^-.

Subject	Tentative Apportionment	d_i^-
Geometry	3	$52 \div (3 - \frac{1}{2}) = 20.8$
Pre-calculus	2	$33 \div (2 - \frac{1}{2}) = 22$
Calculus	1	$15 \div (1 - \frac{1}{2}) = 30$

The subject with the smallest critical divisor is geometry; it therefore receives the reduced apportionment of 2. The final apportionment is therefore two sections each of geometry and pre-calculus, and one section of calculus. ▨

EXAMPLE 11   The Field Hockey Team

As you may remember, the Sports Magnet High School field hockey team has a fine record for the previous season: 18 wins, 4 losses, and 1 tie. We would like to express the record in the form of whole percentages, and Coach has said she is willing to do this with the Webster method. The quotas are 78.26% for wins, 17.39% for losses, and 4.35% for ties; these are rounded to get the tentative percentages 78%, 17%, and 4%, respectively. The total of these tentative percentages is 99%, less than the "house size" of 100%. We therefore compute the critical divisors d_i^+:

Category	Tentative Percentage	d_i^+
Wins	78	$18 \div (78 + \frac{1}{2}) = 0.2293$
Losses	17	$4 \div (17 + \frac{1}{2}) = 0.2286$
Ties	4	$1 \div (4 + \frac{1}{2}) = 0.2222$

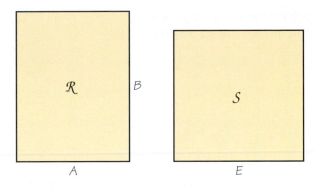

FIGURE 14.1 The edge of the square is the geometric mean of the edges of the rectangle, because the two figures have the same area.

The largest critical divisor belongs to the wins, so it receives the extra 1%. The final apportionment is 79% wins, 17% losses, and 4% ties. ∎

To see why the Webster method is not biased in favor of large states or small states, let s be the standard divisor, which is equal to the average district population. Let d be the divisor that is used in the Webster method (d is equal to the critical divisor belonging to the last state to gain or lose a seat in the apportionment). Finally, let $M = s/d$. In effect, each state's adjusted quota can be obtained by multiplying its quota by M, as we saw in our discussion of the Jefferson method. When $M > 1$ (this happens when the tentative apportionments are not enough to fill the house), the larger states are favored, as they were in the case of the Jefferson method. On the other hand, when $M < 1$, the reverse is true, because a large number multiplied by M will decrease more than a small number would. Thus, the Webster method favors neither large nor small states when the tentative apportionments exactly fill the house; it favors small states when they overfill the house, and it favors large states when they underfill the house. On balance, the Webster method is neutral, because it is equally likely that the house will be overfilled or underfilled by the tentative apportionments.

The Hill–Huntington Method

The **Hill–Huntington method** is a divisor method that has been used to apportion the U.S. House of Representatives since 1940. Like the Jefferson and Webster methods, the apportionment is calculated by rounding the quotas, after adjusting them if necessary. The only difference between the three divisor methods is in the rounding procedure.

The Hill–Huntington rounding procedure is related to the **geometric mean.** Consider the rectangle $\mathcal{R}$ displayed in Figure 14.1. The area of $\mathcal{R}$ is the product of the lengths A and B, or AB. The geometric mean of A and B is defined to be the length E of the edge of a square S with the same area as $\mathcal{R}$. The area of S is E^2, so $E^2 = AB$, and therefore $E = \sqrt{AB}$.

Given a positive number q, let q^* be the geometric mean of $\lfloor q \rfloor$ and $\lceil q \rceil$. The Hill–Huntington rounding of a number q is equal to

$$\lfloor q \rfloor \text{ if } q < q^*$$

and

$$\lceil q \rceil \text{ if } q \geq q^*$$

For example, suppose that $q = 7.485$. Jefferson and Webster would round 7.485 down to 7. Because $\lfloor 7.485 \rfloor = 7$ and $\lceil 7.485 \rceil = 8$, the geometric mean is $q^* = \sqrt{7 \times 8} = 7.48331 \ldots$, which is less than 7.485. Therefore, the Hill–Huntington rounding of 7.485 is $\lceil 7.485 \rceil = 8$.

Hill–Huntington apportionment calculations follow the general plan of the Jefferson and Webster methods. Round each state's quota the Hill–Huntington way to obtain a first tentative apportionment. If the sum of the tentative apportionments is equal to the house size, the job is finished. If not, a list of critical divisors must be constructed, each chosen to be just sufficient to change the corresponding state's apportionment by one seat in the desired direction. If n_i is the number of seats apportioned tentatively to state i, and the total apportionment is too small, the critical divisor for that state is

$$d_i^+ = \frac{p_i}{\sqrt{n_i(n_i + 1)}}$$

where p_i is the state's population. The state with the largest critical divisor gains a seat. If the house is still not full, recompute the critical divisor for the state whose apportionment changed, and repeat the process.

If the total apportionment is too large, the critical divisor for state i is

$$d_i^- = \frac{p_i}{\sqrt{n_i(n_i - 1)}}$$

In this case, the state with the smallest critical divisor loses a seat.

A zero apportionment is impossible with the Hill–Huntington method, because the rounding point for quotas between 0 and 1 is $\sqrt{0 \times 1} = 0$. Any quota less than 1 will be rounded to 1.

EXAMPLE 12   Apportionment of the 1790 Congress, Revisited

The calculations leading to the Hill–Huntington apportionment of the Congress are given in Table 14.6. The rounded quotas add up to 106, more than the house size of 105, so we have to reduce the apportionment of one state. The table shows the critical divisors. The smallest critical divisor is Pennsylvania's, 34,658. Therefore Pennsylvania will lose a seat, while the other apportionments will remain the same. ■

TABLE 14.6	Apportioning the House of Representatives by the Hill–Huntington Method					
State	Population	Quota	q^*	Tentative Apportionment	Critical Divisor	Final Apportionment
Virginia	630,560	18.310	18.493	18	36,046	18
Massachusetts	475,327	13.803	13.491	14	35,234	14
Pennsylvania	432,879	12.570	12.490	13	34,658	12
North Carolina	353,523	10.266	10.488	10	37,265	10
New York	331,589	9.629	9.487	10	34,952	10
Maryland	278,514	8.088	8.485	8	37,217	8
Connecticut	236,841	6.877	6.481	7	36,545	7
South Carolina	206,236	5.989	5.477	6	37,653	6
New Jersey	179,570	5.214	5.477	5	40,153	5
New Hampshire	141,822	4.118	4.472	4	40,940	4
Vermont	85,533	2.484	2.449	3	34,918	3
Georgia	70,835	2.057	2.449	2	50,088	2
Kentucky	68,705	1.995	1.414	2	48,582	2
Rhode Island	68,446	1.988	1.414	2	48,399	2
Delaware	55,540	1.613	1.414	2	39,273	2
Totals	3,615,920	105	—	106	—	105

14.4 Which Divisor Method Is Best?

The three divisor methods that we have considered often yield different results. When this happens, a state that is favored by a method not in use will often argue that the apportionment was unfair. Spotlight 14.1 summarizes a Supreme Court case about this issue. We can try to decide which divisor method is fairest by comparing the apportionments actually produced by the methods.

One Measure of Fairness: Representative Share

Let a_i be the apportionment given to a state whose population is p_i. The quotient a_i/p_i is called the **representative share.** It represents the share of a congressional seat given to each citizen of the state.

In an ideal apportionment, every state would have the same representative share. Because this is impossible, we can measure how close a given apportionment is to being ideal by using representative share as the standard of comparison. We

SPOTLIGHT 14.1   A Legal Challenge to Apportionment

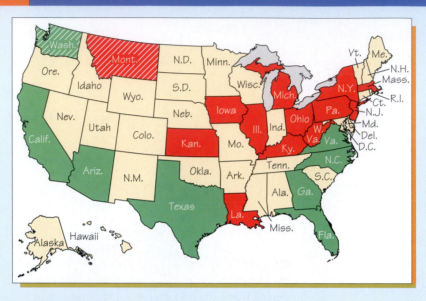

In 1991, the Census Bureau reported the new apportionment for the congressional elections in the years 1992–2000. The states colored red on the above map lost representatives, and those in green gained representatives. New York lost 3 representatives, and Ohio and Pennsylvania lost 2 apiece. Montana, whose apportionment decreased from 2 to 1, sustained the greatest percentage loss, and Montana sued to restore the lost seat. As precedents, Montana referred to the two famous cases, *Baker* v. *Carr* and *Wesberry* v. *Sanders,* in which the U.S. Supreme Court required legislative and congressional district boundaries to be drawn so as to make district populations equal.

Montana argued that the correct apportionment would be the one that met the *Baker* and *Wesberry* criterion of having districts as nearly equal in population as possible, and asked the Court to require the Census Bureau to recompute the apportionments using the Dean method, a divisor method that minimizes differences in district populations. This would

have resulted in the transfer of a congressional seat from Washington to Montana.

The Montana case coincided with another federal lawsuit, *Massachusetts* v. *Mosbacher,* which asked the Court to order the apportionment to be calculated by the Webster method. If Massachusetts had won this suit, it would have gained an additional seat, but Montana would not benefit.

In *U.S. Department of Commerce* v. *Montana,* the Supreme Court unanimously rejected Montana's claim. The opinion of the Court, written by Justice John Paul Stevens, pointed out that *intra*state districts, which were the subject of the *Baker* and *Wesberry* cases, could be equalized in population by drawing district boundaries correctly. Because congressional districts can't cross state lines, some inequity is inevitable in congressional apportionment. The opinion conceded that there were alternatives to the Hill–Huntington method but concluded that the choice of apportionment method was best left to Congress.

are given an apportionment done by some method. The method could be Webster's, Hamilton's, or an old-fashioned smoke-filled room. We would compute the representative share for each state and identify the state that has the largest share and the state that has the smallest. The discrepancy between these two values is a measure of how far the apportionment is from being perfectly equitable. It can be shown that among all conceivable apportionments, the one for which this discrepancy is the least is provided by the Webster method. Thus, while no apportionment can be perfect, the Webster method is the fairest when comparisons are based on representative share.

Inequity in the 77th Congress

EXAMPLE 13

In 1940, Michigan had a population of 5,256,106 (5.256106 million) and was apportioned 17 seats in the House of Representatives. This apportionment gave Michigan a representative share of

$$\frac{17}{5.256106} = 3.234 \text{ seats per million population}$$

Arkansas—with a population of 1,949,387, or 1.949387 million—received 7 seats; so that state had a representative share of $7 \div 1.949387 = 3.591$ seats per million. Arkansas was favored over Michigan by

$$3.591 - 3.234 = 0.357 \text{ seats per million}$$

If a seat had been transferred from Arkansas to Michigan, then the representative share for Michigan would have been $18 \div 5.256106 = 3.425$ seats per million, while Arkansas would have been left with a representative share of $6 \div 1.949387 = 3.078$ seats per million.

Now Michigan, with the larger representative share, has the advantage, but the discrepancy

$$3.425 - 3.078 = 0.347 \text{ seats per million}$$

is less than it was before the transfer was made. In terms of representative share, it would have been more equitable to have given Arkansas 6 seats and Michigan 18 seats. In the Webster apportionment, the seats would have been distributed in this way, because that method minimizes discrepancies in representative share. This shows that the Webster method is better than the Hill–Huntington method, which was used in the 1940 apportionment, *if we view minimizing discrepancies in representative share as being of prime importance.* ■

In an ideal apportionment, each congressional district in the country would have the same population. This is impossible unless we allow congressional districts to overlap state lines, which is forbidden by the Constitution.

▓▓▓▓▓ Another Measure of Fairness: District Population

The **district population** of state i is $p_i \div a_i$. It represents the average population of a congressional district in the state.

The district population of each state can also serve as a standard of comparison of apportionments. In a given apportionment, the district population for each state would be computed. The best apportionment by this standard would be the one for which the worst difference in district populations between states is minimal.

▓▓▓▓EXAMPLE 14   Comparing District Populations

If we consider differences in district population rather than representative share, it was correct to give Michigan 17 seats and Arkansas 7 in the 1940 apportionment. The district population for Michigan was

$$\frac{\text{population of Michigan}}{\text{Michigan's apportionment}} = \frac{5{,}256{,}106 \text{ people}}{17 \text{ districts}}$$

$$= 309{,}183 \text{ people per district}$$

The district population for Arkansas was $1{,}949{,}387 \div 7 = 278{,}484$. The Arkansas average district population was 30,699 less than Michigan's. If Michigan had 18 seats and Arkansas had 6, Michigan would have the lesser district population, 292,006, while Arkansas's would have increased to 324,898. This adjustment in apportionment would have increased the inequity between the two states, because now Arkansas would be worse off than Michigan by 32,892 in district population. ■

For state i, the representative share is a_i/p_i and the district population is p_i/d_i. Thus:

$$\text{representative share} = \frac{1}{\text{district population}}$$

It is surprising that these two ways of evaluating the fairness of an apportionment could disagree, but we have just seen that they can. Those who think it is most important for states to have representative shares as nearly equal as possible would say that Michigan should have 18 seats and Arkansas should have 6, but those who prefer to focus on district population instead of representative share would say that the two states should have 17 and 7 seats, respectively.

A mathematician, Edward V. Huntington, suggested a compromise between the two measures of inequity that we have considered—differences in representative share and differences in district populations. He pointed out that if *relative differences* are compared instead of absolute differences, then either district population and representative share would give identical comparisons of apportionments.

Absolute and Relative Differences

Given two positive numbers A and B, with $A > B$, the **absolute difference** is $A - B$ and the **relative difference** is the quotient $(A - B)/B \times 100\%$.

For any two states, it turns out that the *relative difference* in district populations is equal to the relative difference in representative share (see Exercise 36). Therefore, an apportionment method that minimizes *relative* difference in representative shares will also minimize the relative difference in district populations. The Hill–Huntington method gives the apportionment in which the relative differences in representative shares (or district populations) are as small as possible.

Relative Inequity in the 77th Congress

EXAMPLE 15

Recall that Michigan was given 17 seats in the 77th Congress and had a representative share of 3.234 seats per million. Arkansas had 7 seats and a representative share of 3.591 seats per million. The relative difference was

$$\frac{3.591 - 3.234}{3.234} \times 100\% = 11.04\%$$

Thus, Arkansas was 11.04% better represented in the 77th Congress than Michigan was.

If Michigan had 18 seats and Arkansas had 6, the relative difference in representative shares would be found by subtracting the smaller representative share (Arkansas's) from the larger (Michigan's), and expressing the result as a percentage of the smaller representative share. Thus, the relative difference would have been

$$\frac{3.425 - 3.078}{3.078} \times 100\% = 11.27\%$$

in Michigan's favor. Because the relative inequity was less when Michigan had 17 seats and Arkansas had 7, the Hill–Huntington method gave this apportionment. Spotlight 14.2 explains how the Arkansas–Michigan apportionment in 1940 influenced the decision of Congress to adopt the Hill–Huntington method. ■

SPOTLIGHT 14.2 Mathematics and Politics: A Strange Mixture

Walter F. Willcox

Edward V. Huntington

(Willcox: Department of Manuscripts and University Archives, Cornell University Libraries, Huntington: Courtesy of Harvard University Archives.)

The first American to consider apportionment from a theoretical point of view was Walter Willcox (1861–1964), who strongly advocated the Webster method and had computed the apportionment of 1902. His arguments convinced Congress to use the Webster method again in 1912. In 1911, Joseph Hill, a statistician at the Census Bureau, proposed the Hill–Huntington method–with the strong endorsement of Edward V. Huntington, a mathematics professor at Harvard.

In 1920, the two methods were in competition. There were significant differences in the apportionments determined by the two methods, and the result was Washington gridlock: No apportionment bill passed during the decade, and the 1912 apportionments were retained throughout the 1920s. In preparation for the 1930 census results, the National Academy of Sciences formed a committee to study apportionment. In 1929, the committee endorsed the Hill–Huntington method.

The 1930 census was remarkable in that the apportionments calculated by the Webster method were the same as the Hill–Huntington apportionments. The House was therefore reapportioned, but the method used could be claimed to be either one of the competing methods. The coincidence was almost repeated in the 1940 census, but there was one difference. The Hill–Huntington method gave the last seat to Arkansas, while Webster's method gave it to Michigan (see page 530). At the time, Michigan was a predominantly Republican state, and Arkansas was in the Democratic column. The vote on the apportionment bill split strictly along party lines, with Democrats supporting the Hill–Huntington method and Republicans voting for the Webster method. Because the Democrats had the majority, the Hill–Huntington method became the law.

Because each divisor method can lead to a slightly different apportionment, one way to determine which apportionment method to use is to decide, in a political debate, which type of inequity should be minimized. Challenges to apportionments have followed this approach (see Spotlight 14.1).

An apportionment method could also be chosen by minimizing *bias* in favor of large or small states. We have seen that the Jefferson method favors large states and that the Webster method is neutral. What about the Hill–Huntington method?

Bias in a divisor method is in favor of large states when the quotas are adjusted by using a divisor that is smaller than the standard divisor. If the quotas must be adjusted downward—that is, a divisor larger than the standard divisor is used—small states are favored. Because the rounding point for the Webster method is halfway between whole numbers, it is just as likely for the divisor to be smaller than the standard divisor as it is for it to be larger.

The geometric mean used by the Hill–Huntington method is closer to $\lfloor q \rfloor$ than to $\lceil q \rceil$ (see Exercise 24). This means that a random number q is more likely to be greater than the geometric mean of $\lfloor q \rfloor$ and $\lceil q \rceil$ and thus rounded up to $\lceil q \rceil$ than it is to be less and thus rounded down to $\lfloor q \rfloor$. The difference between the Webster and Hill–Huntington ways of rounding is not significant for relatively large numbers. For example, the geometric mean of 50 and 51 is 50.498. Therefore, a number q between 50 and 51 will be rounded up to 51 by Hill–Huntington if it is larger than 50.498. The Webster method would round q to 51 if $q \geq$ 50.500. The differences are more significant when rounding smaller numbers. Thus, Hill–Huntington rounds all numbers between 0 and 1 up to 1; Webster rounds only the numbers in the range 0.500–1 up to 1. When the Hill–Huntington method is used for apportionment, the sum of the tentative apportionments is more likely to exceed the house size than it is to be less, especially if there are many states with small populations. Therefore, it is more likely that a divisor larger than the standard divisor will be needed. This favors the less populous states.

REVIEW VOCABULARY

$\lfloor q \rfloor$ The result of rounding a number q down; for example, $\lfloor \pi \rfloor = 3$.

$\lceil q \rceil$ The result of rounding a number q up to the next integer; for example, $\lceil \pi \rceil = 4$.

Absolute difference The result of subtracting a smaller number from a larger number.

Adjusted quota The result of dividing a state's quota by a divisor other than the standard divisor. The purpose of adjusting the quotas is to correct a failure of the rounded quotas to sum to the house size.

Alabama paradox A state loses a representative solely because the size of the House is increased. This paradox is possible with the Hamilton method but not with divisor methods.

Apportionment method A systematic way of computing solutions of apportionment problems.

Apportionment problem To round a list of fractions to whole numbers in a way that preserves the sum of the original fractions.

Critical divisor The number closest to the standard divisor that can be used as a divisor of a state's population to obtain a new tentative apportionment for the state. The following table lists formulas for critical divisors for some divisor methods. In the table, p stands for the state's population, and n is its tentative apportionment.

Method	Critical Divisor Causing Tentative Apportionment	
	To Increase	To Decrease
Jefferson	$\dfrac{p}{n + 1}$	Not necessary
Webster	$\dfrac{p}{n + 0.5}$	$\dfrac{p}{n - 0.5}$
Hill-Huntington	$\dfrac{p}{\sqrt{n(n + 1)}}$	$\dfrac{p}{\sqrt{n(n - 1)}}$

District population A state's population divided by its apportionment.

Divisor method One of many apportionment methods in which the apportionments are determined by dividing the population of each state by a common divisor to obtain adjusted quotas. The apportionments are calculated by rounding the adjusted quotas. Divisor methods differ in the way that the rounding of the quotas is carried out. The methods of Jefferson, Webster, and Hill–Huntington are divisor methods.

Geometric mean For positive numbers A and B, the geometric mean is defined to be $\sqrt{A \times B}$.

Hamilton method An apportionment method, advocated by Alexander Hamilton, that assigns to each state either its lower quota or its upper quota. The states that receive their upper quotas are those whose quotas have the largest fractional parts.

Hill–Huntington method A divisor method, named for the statistician Joseph Hill and the mathematician Edward V. Huntington, that minimizes relative differences in both representative shares and district populations.

Jefferson method A divisor method invented by Thomas Jefferson, based on rounding all fractions

down. Thus, if u_i is the adjusted quota of state i, the state's apportionment is $\lfloor u_i \rfloor$.

Lower quota The integer part $\lfloor q_i \rfloor$ of a state's quota q_i.

Population paradox A situation is which state A gains population and loses a congressional seat, while state B loses population (or increases population proportionally less than state A) and gains a seat. This paradox is possible with all apportionment methods *except* divisor methods.

Quota The quotient p/s of a state's population divided by the standard divisor. The quota is the number of seats a state would receive if fractional seats could be awarded.

Quota condition A requirement that an apportionment method should assign to each state either its lower quota or its upper quota in every situation. The Hamilton method satisfies this condition, but none of the divisor methods do.

Relative difference The relative difference between two positive numbers is obtained by subtracting the smaller number from the larger, and expressing the result as a percentage of the smaller number. Thus, the relative difference of 120 and 100 is 20%.

Representative share A state's representative share is the state's apportionment divided by its population. It is intended to represent the amount of influence a citizen of that state would have on his or her representative.

Standard divisor The ratio p/h of the total population p to the house size h. In a congressional apportionment problem, the standard divisor represents the average district population.

Tentative apportionment The result of rounding a state's quota or adjusted quota to obtain a whole number.

Upper quota The result of rounding a state's quota *up* to a whole number. A state whose quota is q has an upper quota equal to $\lceil q \rceil$.

Webster method A divisor method of apportionment invented by Representative Daniel Webster. It is based on rounding fractions the usual way. The Webster method minimizes the differences of representative share between states.

✔ SKILLS CHECK

1. A county is divided into 3 districts with the following populations: Southern, 3600; Western, 3100; Northeastern, 1600. There are 6 seats on the county council to be apportioned. What is the quota for the Southern district?

(a) 0.43
(b) 3
(c) 2.6

2. The Hill–Huntington method of apportionment
(a) is currently used to apportion the House of Representatives.
(b) is not a divisor method.
(c) always rounds up the fractional parts.

3. A county is divided into 3 districts with the following populations: Southern, 3600; Western, 3100; Northeastern, 1600. There are 10 seats on the school board to be apportioned. What is the apportionment for the Northeastern district using the Jefferson method?

(a) 1
(b) 2
(c) 3

4. A county is divided into 3 districts with the following populations: Southern, 3600; Western, 3100; Northeastern, 1600. There are 10 seats on the school board to be apportioned. What is the apportionment for the Southern district using the Webster method?

(a) 4
(b) 5
(c) 6

5. A county is divided into 3 districts with the following populations: Southern, 3600; Western, 3100; Northeastern, 1600. There are 10 seats on the school board to be apportioned. If the Hill–Huntington method is used instead of the Webster method, how do the results differ?

(a) The apportionments are the same.
(b) Southern gains an extra seat and Western loses a seat.
(c) Southern gains an extra seat and Northeastern loses a seat.

6. The geometric mean of 6 and 7 is
(a) 6.5.
(b) less than 6.5.
(c) more than 6.5.

7. The relative difference of 6 and 7 is
(a) 1.
(b) approximately 14.29%.
(c) approximately 16.67%.

8. The absolute difference of 6 and 7 is
(a) 1.
(b) 0.5.
(c) 6.5.

9. The Jefferson method is based on the premise that
(a) each representative's district must meet a threshold population size.
(b) each representative must represent no more than a certain population size.
(c) each representative must represent the same-sized population.

10. The Hamilton method is based on which of the following premises?
(a) The apportionment is determined by rounding each state's quota to the nearest integer.
(b) States whose quotas have the largest fractional parts receive an additional apportionment.
(c) Each apportionment unit represents an identical population size.

11. The Webster method is based on the premise that
(a) each state's apportionment is the integer closest to its adjusted quota.
(b) smaller states should be favored in the apportionment process.
(c) adjusted quotas should not be necessary.

12. The Hill–Huntington method utilizes
(a) a round-up procedure.
(b) the geometric mean.
(c) relative differences.

13. The method used to apportion the U.S. House of Representatives since 1940 is
(a) the Webster method.

(b) the Jefferson method.

(c) the Hill–Huntington method.

14. The population paradox is possible when

(a) the Hill–Huntington method is used.

(b) the Webster method is used.

(c) the Hamilton method is used.

15. If the initial calculations leading to the Hill–Huntington apportionment result in a sum that is too large, what happens next?

(a) A seat is taken from the state with the smallest critical divisor.

(b) The largest apportionment is reduced.

(c) A different method must be used.

16. A state has a population of 1 million, and its tentative apportionment is 9, under the Jefferson method. What is the state's critical divisor?

(a) 10

(b) 100,000

(c) 111,111

17. The Webster method is being used to calculate an apportionment. State A, with a population of 1 million, receives a tentative apportionment of 9 seats. State B has a population of 600,000 and tentatively receives 5 seats, and state C, with a population of 1,500,000, receives 14 seats. If the total number of seats in this tentative apportionment is one more than the house size,

which of these states has the lowest priority for retaining all of its seats?

(a) A

(b) B

(c) C

18. Which apportionment method never gives a state zero seats, no matter how small its population is?

(a) Hill–Huntington

(b) Webster

(c) Hamilton

19. States A, B, and C have populations of 810,000, 1,250,000, and 2,000,000, respectively. Their tentative apportionments by the Hill–Huntington method are 5 seats for A, 8 for B, and 13 for C. The tentative apportionments for all the states add up to 1 less than the house size, and it turns out that an additional seat is given to one of these three states. Which state has the highest priority, and therefore gets the seat?

(a) A

(b) B

(c) C

20. Which divisor method never apportions to a state fewer seats than its lower quota?

(a) Hill–Huntington

(b) Webster

(c) Jefferson

EXERCISES   ■ Challenge   ◆ Discussion

The Apportionment Problem

1. Jane has decided to track her daily expenses, and finds them to be as follows:

Rent	$31
Food	16
Transportation	7
Gym	12
Miscellaneous	5

Express these as percentages. If rounded to whole numbers, do the percentages add up to 100%?

2. A mathematics department uses 20 teaching assistants to aid in its four-semester calculus course. The number of teaching assistants assigned to each level of the course depends on enrollment. Here are the fall enrollments:

Calculus I	500
Calculus II	100
Calculus III	350
Calculus IV	175
Total	1125

How many teaching assistants should be assigned to each level of the course?

3. Help the mathematics department in Exercise 2 reapportion its teaching assistants. Grades have been posted for the previous semester, and some students need to repeat the previous level of the course. Forty-five students move from Calculus II to Calculus I, 41 students move from Calculus III to Calculus II, and 12 students move from Calculus IV to Calculus III.

◆ **4.** Here is a typical apportionment problem. Round the numbers in the sum to integers:

$$8.37 + 10.33 + 12.38 + 5.47 + 3.45 = 40$$

The rounded numbers must add up to 40. How would you approach this?

The Hamilton Method

5. Use the Hamilton method to round each of the following numbers in the sum to a whole number, preserving the total of 10.

$$0.36 + 1.59 + 0.99 + 2.33 + 2.38 + 2.35 = 10$$

6. *The 37th pearl.* Three friends have bought a bag guaranteed to contain 36 high-quality pearls for $14,900 at an auction. Abe contributed $5900, Beth's contribution was $7600, and Charles supplied the remaining $1400. After taking the bag to your house, they pour the 36 pearls from the bag onto the kitchen table.

(a) How many should each friend get if the Hamilton method is used to apportion the pearls according to the size of the contributions?
(b) Charles has noticed the bag isn't empty! Another pearl comes out, and you are asked to recalculate the apportionment.
◆ **(c)** How do you explain the result to Charles?

◆ **7.** Suppose that a country has three states, with populations 254,000, 153,000, and 103,000, respectively. The legislature has 102 seats. Show that if the Hamilton method is used to apportion seats, a tie will result. How would you suggest breaking the tie?

8. A small high school has one mathematics teacher who can teach a total of five sections. The subjects that she teaches, and their enrollments, are as follows: geometry, 52; algebra, 33; calculus, 12. Apportion sections to the subjects using the Hamilton method.

9. Repeat Exercise 8 using the following enrollments: geometry, 77; algebra, 18; calculus, 20.

10. A country has four states, *A*, *B*, *C*, and *D*. Its house of representatives has 100 members, apportioned by the Hamilton method. A new census is taken, and the house is reapportioned. Here are the data:

State	Old Census	New Census
A	5,525,381	5,657,564
B	3,470,152	3,507,464
C	3,864,226	3,885,693
D	201,203	201,049
Totals	13,060,962	13,251,770

(a) Apportion the house using the old census.
(b) Reapportion, using the new census.
◆ **(c)** Explain how this is an example of the population paradox.

11. Abe, Beth, Charles, and David have decided to invest in rare coins. A dealer has offered to sell them a parcel containing 100 identical coins for $10,000. Each person invests all that he or she can afford, but there is not quite enough money, so Charles asks his Aunt Esther to join the group. The coins will be apportioned by the Hamilton method. Here are the amounts contributed:

Abe	$3,619
Beth	1,862
Charles	2,258
David	2,010
Esther	251
Total	$10,000

How should the coins be apportioned among the five contributors? After this is figured out, and the coins are distributed, the dealer mentions that there will be $50 in excise tax! Everyone empties their wallet: Abe finds $16 more, Beth has $2, Charles has $1, and David finds $32. This adds up to $51, so a dollar is returned to Aunt Esther. The apportionment is recalculated and one of the coins changes hands. Who has to give a coin to whom, and how do you explain it?

To see how this situation works out with a different apportionment method, refer to Exercise 27.

12. A small country has four provinces and a sparsely populated desert area. The censuses for this year and last year are given below.

	Censuses	
Province	Last Year	This Year
Ash	93,311	95,450
Beech	35,857	35,926
Chestnut	40,526	40,572
Date	56,856	56,870
The desert	3,450	3,447

Apportion the country's 100-seat legislature by the Hamilton method, using both censuses. Is there a paradox? (See Exercise 28 for a different apportionment method applied to this problem.)

13. A country has five states, with populations 5,576,330, 1,387,342, 3,334,241, 7,512,860, and 310,968. Its house of representatives is apportioned by the Hamilton method.

(a) Calculate the apportionments for house sizes of 82, 83, and 84. Does the Alabama paradox occur?

(b) Repeat the calculations for house sizes of 89, 90, and 91.

Divisor Methods

◆ **14.** Explain why the tentative Webster apportionment of a state with quota q is $\lfloor q + 0.5 \rfloor$.

15. Reapportion the classes in Exercise 9 using the Jefferson method.

16. Reapportion the classes in Exercise 8 using the Webster method.

◆ **17.** The three friends who bought the pearls (see Exercise 6) ask you to suggest a different apportionment method to distribute their purchase. Before answering, determine the apportionments given by each of the divisor methods covered in this text–Jefferson, Webster, and Hill–Huntington–for the 36- and 37-pearl house sizes. Then make your suggestion.

◆ **18.** The three friends have bought a lot of 36 identical diamonds, at a total cost of $36,000; Abe's investment was $15,500, Beth's was $10,500, and Charles's was $10,000. They decided to apportion the diamonds using the Webster method, and they can't make it work out. Can you help?

19. Round the following to whole percentages using the methods of Hamilton, Jefferson, and Webster:

$$87.85\% + 1.26\% + 1.25\% + 1.24\% + 1.23\%$$
$$+ 1.22\% + 1.21\% + 1.20\% + 1.19\% + 1.18\%$$
$$+ 1.17\% = 100\%$$

Do any of these methods violate the quota condition?

20. Round the following percentages to whole numbers, using the methods of Hamilton, Jefferson, and Webster.

$$92.15\% + 1.59\% + 1.58\% + 1.57\% + 1.56\%$$
$$+ 1.55\% = 100\%$$

Do any of these methods violate the quota condition?

21. Find the geometric mean of each pair of numbers.

(a) 0, 1
(b) 1, 2
(c) 2, 3
(d) 3, 4

22. A high school has one math teacher, who can teach five sections. Fifty-six students have enrolled in the algebra class, 28 have signed up for geometry, and 7 students will take calculus. Use the Hill–Huntington method to decide how many sections of each course to schedule.

23. One year later, the high school described in Exercise 22 still has just one math teacher who teaches five sections. The enrollments are algebra, 36; geometry, 61; and calculus, 3. Apportion the classes by the Webster and Hill–Huntington methods. Which apportionment do you think the school principal would prefer?

◆ **24. (a)** Show that for any positive numbers A and B, the geometric mean is less than the arithmetic mean,[4] except when $A = B$; then the two means are equal. (*Hint:* Show that the triangle in Figure 14.2 is a right triangle.)

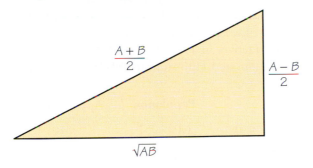

FIGURE 14.2 Is this a right triangle?

(b) Show that for any number $q > 0$, the Hill–Huntington rounding of q is not less than the Webster rounding of q. If the two roundings differ, then the Hill–Huntington rounding of q is equal to $\lceil q \rceil$ and the Webster rounding is equal to $\lfloor q \rfloor$.
(c) Explain why the fact established in part (b) implies that the Hill–Huntington method is more favorable to small states than the Webster method.

■ **25.** Suppose that in 2000, the governor of Utah believed that the population of his state was undercounted. What increase in population would be large enough to entitle Utah to take a seat from California if the apportionment is by the Hill–Huntington method? The data needed for this problem are given in Exercise 35.

[4] The arithmetic mean of A and B is equal to $(A + B)/2$.

■ **26.** Ties can occur when the Hill–Huntington method is used to apportion if two states have identical populations or if the number of states is larger than the house size. Is there any other way that ties can occur with this method?

27. Recalculate the apportionment of the coins in Exercise 11 by the Webster method. Again, after the excise tax is paid, a coin changes hands. Who gives it to whom?

28. Reapportion the legislature of the country in Exercise 12 using the Hill–Huntington method. Compare the results based on the two censuses. Is there a paradox?

29. A country that is governed by a parliamentary democracy has two political parties, the Liberals and the Tories. The number of seats awarded to a party is supposed to be proportional to the number of votes it receives in the election. Suppose the Liberals receive 49% of the vote. If the total number of seats in parliament is 99, how many seats do the Liberals get with the Hamilton method? With the Webster method? With the Jefferson method?

◆ **30.** A country with a parliamentary government has two parties that capture 100% of the vote between them. Each party is awarded seats in proportion to the number of votes received.

(a) Show that the Webster and Hamilton methods will always give the same apportionment in this two-party situation.
(b) Show that the Alabama and population paradoxes cannot occur when the Hamilton method is used to apportion seats between two parties or states.
(c) Show that the Webster method satisfies the quota condition when the seats are apportioned between two parties or states.
■ **(d)** Will the Jefferson and Hill–Huntington methods also yield the same apportionments as the Hamilton method?

31. If the congressional apportionment based on the 1790 census were computed by the Webster method, would the result differ from the

Hill–Huntington method apportionment (see Table 14.6)?

Which Divisor Method Is Best?

32. Determine the relative difference between the numbers 5 and 7.

33. Jim is 72 inches tall and Alice is 65 inches tall. What is the relative difference of their heights?

34. In the 2001 apportionment of Congress, the average congressional district in North Carolina had a population of 620,590. Montana, with a population of 905,316, was the most populous state to receive only one district.

(a) Which state is the more favored in this apportionment?
(b) What is the relative difference in the district sizes?

35. According to the 2000 census, the population of California was 33,930,798 and the population of Utah was 2,236,714. California was apportioned 53 House seats, and Utah received 3 House seats.

(a) Determine the average congressional district sizes for these states.
(b) Determine the absolute and relative differences in these district sizes.
(c) Suppose a seat were transferred from California to Utah, giving Utah 4 seats and California 52. What would now be the absolute and relative differences in district sizes?
(d) Does this evidence indicate that the apportionment following the 2000 census fulfilled the criterion of keeping absolute differences in district populations as small as possible? What about relative differences?

◆ **36.** Let the populations of states A and B be p_A and p_B, respectively. The apportionments will be a_A and a_B. Assuming that district populations for state A are larger than district populations for state B, show that the relative difference in district populations is

$$\frac{p_A a_B - p_B a_A}{p_B a_A} \times 100\%$$

Also show that this expression is equal to the relative difference in representative share. Hence the relative difference in district populations is equal to the relative difference in representative shares.

37. In *Massachusetts* v. *Mosbacher*, Massachusetts contested its 1991 apportionment, claiming a systematic census undercount of Massachusetts residents living abroad. Another issue in the suit was the claim by Massachusetts that the Hill–Huntington method of apportionment is unconstitutional, because it does not reflect the "one person, one vote" principle as well as the Webster method does. Massachusetts sought an additional House seat that had been awarded to Oklahoma. Would Massachusetts have gained a seat from Oklahoma if the Webster method had been used to apportion the House of Representatives in 1991? Use the following populations and Hill–Huntington apportionments:

State	Population	Apportionment
Massachusetts	6,029,051	10
Oklahoma	3,145,585	6

◆ **38.** Here is an apportionment method that should please everyone! Just give each state its upper quota. Of course, the house size would vary.

(a) Show that the house size will be more than the planned house size h.
(b) Do you think California would be enthusiastic about this method, or would that state prefer to give each state its *lower* quota?

◆ **39.** The following apportionment method was invented by Congressman William Lowndes of South Carolina in 1822. Lowndes starts, as Hamilton does, by giving each state its lower quota. But where Hamilton apportions the remaining seats to the states whose quotas have the largest fractional parts—in other words, the states for which the *absolute difference* between q_i and $\lfloor q_i \rfloor$ is greatest—Lowndes gives the extra seats to the states where the *relative difference* between q_i

and $\lfloor q_i \rfloor$ is greatest, raising as many as necessary to their upper quotas to fill the House.

(a) Would this method be more beneficial to states with large populations or small populations, as compared with the Hamilton method?

(b) Does the Lowndes method satisfy the quota condition?

(c) Would there be any trouble with paradoxes with the Lowndes method?

(d) Use the method to apportion the 1790 House of Representatives.

◆ **40.** John Quincy Adams, the sixth president of the United States, proposed that the House of Representatives should be apportioned by a divisor method based on the rounding rule that rounds each fraction up to the next whole number.

(a) Is it likely that the initial tentative apportionment will be final?

■ **(b)** Find a formula for a state's critical divisor in terms of the state's tentative apportionment n_i and its population p_i.

(c) Does the method favor small states or large states?

(d) Is it possible for a state to be apportioned zero seats by using this method?

41. The Marquis de Condorcet, who proposed a criterion for deciding elections (see Chapter 9), also designed a divisor method for apportionment. His rounding rule was to round down numbers whose fractional parts are less than 0.4 and to round up otherwise.

(a) Show that the Condorcet rounding of a number q is $\lfloor q + 0.6 \rfloor$.

(b) Does the method favor large states, small states, or is it neutral?

(c) Find a formula for a state's critical divisor in terms of the state's tentative apportionment n_i and its population p_i.

◆ **42.** The U.S. Constitution requires that each state be apportioned at least one seat in the House of Representatives.

(a) Show that the Hill–Huntington method is consistent with this requirement. What about the methods of Hamilton, Jefferson, and Webster?

(b) Does the Adams method (see Exercise 40) meet the requirement?

■ **(c)** The Dean method is the divisor method that minimizes absolute differences in district population. Using this information, explain why a Dean apportionment will never give any state zero seats, unless the house size is less than the number of states.

■ **43.** Let $q_1, q_2, \ldots, q_n$ be the quotas for n states in an apportionment problem, and let the apportionments assigned by some apportionment method be denoted $a_1, a_2, \ldots, a_n$. The *absolute deviation* for state i is defined to be $|q_i - a_i|$; it is a measure of the amount by which the state's apportionment differs from its quota. The *maximum absolute deviation* is the largest of these numbers. Show that the Hamilton method always gives the least possible maximum absolute deviation.

◆ **44.** The choice of a divisor method for apportioning classes to subjects according to enrollments, as in the senior high school example, depends on what the school principal considers most important.

(a) The principal wants to set a minimum class size. For example, if the minimum class size is 20, and 39 students are signed up for English III, there would be one section, because there are not enough students for two sections with enrollment of at least 20. If there were 40 students, there would be two sections. The minimum class size is adjusted so that as many sections as possible are running. What apportionment method should she use, and what will the minimum class size be?

■ **(b)** The principal prefers to set a maximum class size. For example, if the maximum class size is 33, and 67 students are taking History I, there will be three sections, because there are too many students to fit in two 33-student sections. If there were only 66 taking History I, there would be two sections. The maximum class size is adjusted so that as many sections as possible are running. What apportionment method should she use, and what will the maximum class size be? (*Hint:* This

divisor method is not described in the text but is mentioned in one of the previous exercises.)
(c) The principal wants to treat students as equitably as possible, so that the differences between students' share of teachers vary as little as possible from course to course. What apportionment method should she use now?

(d) The principal wants to minimize relative difference in class size. What divisor method would work best for her?
(e) The principal wants to cancel any class that has an enrollment of just one student. Which apportionment methods should she avoid using?

APPLET EXERCISES

To do these exercises, go to www.whfreeman.com/fapp7e.

A bus company has three lines—*A, B,* and *C*—and a total of 48,000 riders. *A* has 21,700 riders daily, *B* has 17,200, and *C* has 9100. The company has 40 buses to allocate to the three lines. Use the applet *Apportionment* to help you find the standard divisor and determine the allocation of the buses according to the methods of apportionment of Hamilton, Jefferson, Webster, and Hill–Huntington.

WRITING PROJECTS

1. Does the Hill–Huntington method best reflect the intentions of the Founding Fathers, as these intentions were set down in the Constitution and in the debate during the 1787 Constitutional Convention? Good sources of information here include all of the publications listed in the Suggested Readings. This writing project requires that you state your answer to the question and make a case for it.

2. Suppose that in 2000, Congress had reverted to its nineteenth-century habit of increasing the size of the House of Representatives so that no state would have a decrease in the size of its delegation. How many seats would have been added, and which states would have gotten them? (*Warning:* The apportionments of some states might *increase* as a result of this practice.) As the first step of this project, obtain the populations and apportionments for the 50 states from the Census Bureau Web site (www.census.gov/population/www/censusdata/apportionment.html).

3. In 2004, the state of Colorado considered an amendment to its constitution regarding the way

electors representing Colorado in the Electoral College would be selected. Until 2000, Colorado's electors were selected by the president/vice president ticket that received a plurality of the votes. The proposed amendment, which would have taken effect in 2004, apportioned electoral votes to each ticket in proportion to the number of popular votes received.

Write an essay exploring the implications of various apportionment methods that Colorado could have chosen. Among those mentioned in this text, which would be the least desirable, and which do you think would work best? Keep in mind that if a third-party ticket receives some electoral votes, and the election is close, then the third party would have the choice of making a deal with one of the major tickets (for example, a cabinet post for my electoral votes) or withholding the votes, forcing the election into the House of Representatives.

The method proposed in the amendment was as follows: Determine each ticket's quota, and round to the nearest whole number. Tickets with quotas less than 0.5 receive no electors. If the number of electors thus apportioned is less than Colorado is entitled to have, give the remaining electors to the ticket that received the most votes. If the number of electors is more than Colorado is entitled to have, take electors from that ticket that, among all who received electors, received the smallest number of popular votes. If more electors need to be removed, take from the ticket that received the next smallest number of popular votes, and so on.

Is this method the same as any method discussed in this chapter? Work out some examples, using 9 as the number of electors to be selected, and comparing with the Jefferson, Hamilton, Webster, and Hill–Huntington methods. You will want to consider situations where there are just two tickets competing, and where there are more than two. If there were 20 candidates, each with a quota less than 0.5, what would happen?

SUGGESTED READINGS

BALINSKI, M. L., and H. P. YOUNG. *Fair Representation: Meeting the Ideal of One Man, One Vote*, Yale University Press, New Haven, Conn., 1982. In the 1970s, Balinski and Young analyzed apportionment methods in depth. Their approach was to postulate the desirable properties of an apportionment method as axioms and to deduce from the axioms which method is best. This book combines an account of the history of apportionment of the U.S. House of Representatives with the results of their research.

ERNST, LAWRENCE R. Apportionment methods for the House of Representatives and the court challenges, *Management Science,* 40 (1994): 1207–1227. Ernst, who wrote briefs for the government in both the *Montana* and the *Massachusetts* cases, reviews the apportionment problem and the arguments in favor of and against each of the divisor methods. The article includes a summary of the arguments used by both sides in the two court cases.

YOUNG, H. PEYTON. *Equity,* Princeton University Press, Princeton, N.J., 1994. Chapter 3 covers apportionment and focuses on which apportionment method is the most equitable.

SUGGESTED WEB SITE

www.census.gov This site contains a two-page history of apportionment of the Congress.

Game Theory: The Mathematics of Competition

onflict has been prevalent throughout human history. It arises whenever two or more individuals, with different values or goals, compete to try to control the course of events. *Game theory* uses mathematical tools to study situations, called games, involving both conflict and cooperation. Its study was greatly stimulated by the publication in 1944 of the monumental *Theory of Games and Economic Behavior* by John von Neumann and Oskar Morgenstern (see Spotlight 15.1).

The *players* in a game, who may be people, organizations, or even countries, choose from a list of options available to them—that is, courses of action they may take—that are called **strategies.** The strategies chosen by the players lead to *outcomes,* which describe the consequences of their choices. We assume that the players have *preferences* for the outcomes: They like some more than others.

Game theory analyzes the **rational choice** of strategies—that is, how players select strategies to obtain preferred outcomes. Among areas to which game theory has been applied are bargaining tactics in labor–management disputes, resource-allocation decisions in political campaigns, military choices in international crises, and the use of threats by animals in habitat acquisition and protection.

Unlike the subject of *individual* decision making, which researchers in psychology, statistics, and other disciplines study, game theory analyzes situations in which there are at least two players, who may find themselves in conflict because of different goals or objectives. The outcome depends on the choices of *all* the players. In this sense, decision making is *collective,* but this is not to stay that the players necessarily cooperate when they choose strategies. Indeed, many strategy choices are noncooperative, such as those between combatants in warfare or

SPOTLIGHT 15.1   The Early History of Game Theory

John von Neumann

Oskar Morgenstern

As early as the seventeenth century, such outstanding scientists as Christiaan Huygens (1629–1695) and Gottfried W. Leibniz (1646–1716) proposed the creation of a discipline that would make use of the scientific method to study human conflict and interactions. Throughout the nineteenth century, several leading economists created simple mathematical models to analyze particular examples of competitive encounters. The first general mathematical theorem on this subject was proved for games of perfect information by the distinguished logician Ernst Zermelo (1871–1953) in 1912. A game is said to have *perfect information* if at each stage of the play, every player is aware of all past moves (by itself and others) as well as all future choices that are possible. The theorem stated that any finite game with perfect information, such as checkers or chess, has an optimal solution in *pure* strategies; that is, no randomization or secrecy is necessary. This theorem is an example of an *existence theorem:* It demonstrates that there must exist a best way to play such a game, but it does not provide a detailed plan for playing a complex game, like chess, to achieve victory.

The famous mathematician F. E. Émile Borel (1871–1956) introduced the notion of a *mixed*, or randomized, strategy when he investigated optimal strategies in duels around 1920. The fact that every two-person *zero-sum* game must have a solution in optimal mixed strategies that ensure a particular outcome was proved by John von Neumann (1903–1957) in

1928. Von Neumann's result was extended to the existence of equilibrium outcomes in mixed strategies for multiperson games that are either *constant-sum* or *variable-sum* by John F. Nash, Jr. (b. 1928) in 1951.

Modern game theory dates from the publication in 1944 of *Theory of Games and Economic Behavior* by the Hungarian-American mathematician John von Neumann and the Austrian-American economist Oskar Morgenstern (1902–1977). They introduced the first general model and solution concept for multiperson *cooperative games,* which are primarily concerned with coalition formation (by economic cartels, voting blocs, and military alliances) and the resulting distribution of gains or losses. Several other suggestions for a solution to such games have since been proposed. These include the value concept of Lloyd S. Shapley (b. 1923), which relates to fair allocation and serves as well as an index of voting power (see Chapter 11).

The French artist Georges Mathieu designed a medal for the Musée de la Monnaie in Paris in 1971 to honor game theory. It was the seventeenth medal to "commemorate 18 stages in the development of Western consciousness." Game theory also has a mascot, the tiger, arising from the Princeton University tiger and the Russian abbreviation of the term *game theory* (ТЕОРИЯ ИГР, where the underlined letters correspond to the sounds of the English *T, G,* and *R,* respectively).

competitors in sports. In these encounters, the adversaries' objectives may be at cross-purposes: A gain for one means a loss for the other. But in many activities, especially in economics and politics, there may be joint gains that can be realized from cooperation.

Many interactions involve a delicate mix of cooperative and noncooperative behavior. In business, for example, firms in an industry cooperate to gain tax breaks even as they compete for shares in the marketplace.

Game theory has provided important theoretical foundations in economics, starting with microeconomics but now extending to macroeconomics and international economics. It also has been increasingly applied in political science, especially in the study of voting, elections, and international relations. In addition, game theory has contributed major insights in biology, particularly in understanding the evolution of species and conditions under which animals—humans included—fight each other for territory or act altruistically. It has also illuminated certain fields in philosophy, including ethics, the philosophy of religion, and political philosophy, and inspired many experiments in social psychology.

In the next two sections, we present several simple examples of two-person games of **total conflict,** in which what one player wins the other player loses, so cooperation never benefits the players. We distinguish two different kinds of solutions to such games. Then we analyze two well-known games of **partial conflict,** in which the players can benefit from cooperation but may have strong incentives not to cooperate. We next turn to the analysis of a larger three-person voting game, in which we show how to eliminate undesirable strategies in stages. Finally, we offer some general comments on solving games and discuss different applications of game theory.

15.1 Two-Person Total-Conflict Games: Pure Strategies

For some games with two players, determining the best strategies for the players is straightforward. We begin with such a case.

EXAMPLE 1   A Location Game

Two young entrepreneurs, Henry and Lisa, plan to locate a new restaurant at a busy intersection in the nearby mountains. They agree on all aspects of the restaurant except one. Lisa likes low elevations, whereas Henry wants greater heights—the higher, the better. In this one regard, their preferences are diametrically opposed. What is better for Henry is worse for Lisa, and likewise what is good for Lisa is bad for Henry.

The layout for their location problem is shown in Figure 15.1. Observe that three routes, Avenue A, Boulevard B, and County Road C (blue lines), run in an east–west direction, and that three highways, numbered 1, 2, and 3 (red lines), run in a north–south direction. Table 15.1 shows the altitudes at the nine corresponding intersections. The same information is shown in three dimensions in Figure 15.2.

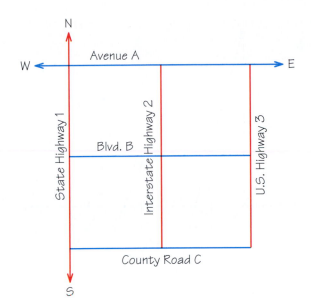

To maximize the number of customers, Henry and Lisa agree that the restaurant should be at a location where one of the east–west routes intersects one of the three highways. But they cannot agree on which intersection, so they decide to turn their decision into the following competitive game: Henry will select one of the three routes, *A*, *B*, or *C*, and Lisa will simultaneously choose one of the three highways, 1, 2, or 3. Because their choices will be made simultaneously, neither one can predict beforehand what the other will do.

Henry, worried that Lisa will choose a low elevation, tries to determine the highest altitude he can guarantee by picking one of the three routes. For each choice of a route, this means considering the worst-case (lowest) elevation on each route. These are the numbers 4, 5, and 2, which are the respective *row minima,* indicated in the right-hand column of Table 15.2. He notes that the highest of these values is 5. By choosing the corresponding route, *B*, Henry can guarantee himself an altitude of at least 5000 feet.

TABLE 15.1	Heights (in thousands of feet) of the Nine Intersections		

Routes	Highways		
	1	2	3
A	10	4	6
B	6	5	9
C	2	3	7

FIGURE 15.2 Three-dimensional road map showing Henry's and Lisa's possible choices (in thousands of feet).

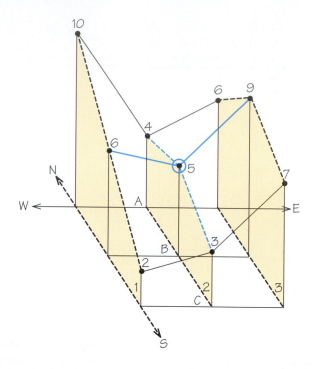

Maximin

The number 5 in the right-hand column is referred to as the **maximin,** which is circled in Table 15.2. It is the maximum value of the minimum numbers in the three rows in the table. The strategy that corresponds to the maximin (for Henry, route *B*) is called his **maximin strategy.**

TABLE 15.2 Heights (in thousands of feet) in Table 15.1, with the Row Minima (maximum circled) and Column Maxima (minimum circled)

	Routes	Lisa Highways			Row Minima
		1	2	3	
Henry	*A*	10	4	6	4
	B	6	5	9	⑤
	C	2	3	7	2
	Column Maxima	10	⑤	9	

Lisa likewise does a worst-case analysis and lists the highest—for her, the worst—elevations for each highway. These numbers, 10, 5, and 9, are the column maxima and are listed in the bottom row of Table 15.2. From Lisa's point of view, the best of these outcomes is 5. If she picks Interstate Highway 2, then she is assured of an elevation of no more than 5000 feet.

Minimax

The number 5 in the bottom row of Table 15.2 is referred to as the **minimax,** which is circled in the table. It is the minimum value of the maximum numbers in the three columns. The strategy that corresponds to the minimax (for Lisa, Highway 2) is called her **minimax strategy.**

To summarize, Henry has a strategy that will ensure the height is 5000 or higher, and Lisa has a strategy that will ensure the height is 5000 or lower. The height of 5000 at the intersection of Route B and Highway 2 is, simultaneously, the lowest value along Boulevard B and the highest along Interstate Highway 2. In other words, the maximin and the minimax are both equal to 5000 for the location game.

Saddlepoint

When a row minimum and a column maximum are the same, the resulting outcome is called a **saddlepoint.**

The reason for the term *saddlepoint* should be clear from the saddle-shaped payoff surface shown in Figure 15.2. The middle point on a horse saddle is simultaneously the lowest point along the spine of the horse and the highest point between the rider's legs. (In Figure 15.2, the rider would be facing leftward or rightward.) In our example, one might also think of the saddlepoint as a mountain pass: As one drives through the pass, the car is at a high point on a highway (in the north–south direction) and at a low point on a route (in the east–west direction).

The resolution of this contest is for Henry to pick B and Lisa to pick 2. This puts them at an elevation of 5000, which is simultaneously the maximin and minimax.

Value

In total-conflict games, the **value** is the best outcome that both players can guarantee. If a game has a saddlepoint, it gives the value (5 in our example): Players can guarantee this outcome by choosing their maximin and minimax strategies.

Total-conflict games without saddlepoints also have a "value," as we shall see later.

There is no need for secrecy in a game with a saddlepoint. Even if Henry were to reveal his choice of B in advance, Lisa would be unable to use this knowledge to exploit him. In fact, both players can use the height information in our example to compute the optimal strategy for their opponent as well as for themselves. In games with saddlepoints, players' worst-case analyses lead to the best *guaranteed* outcome—in the sense that each player can ensure that he or she does not do worse than a certain amount (5 in our example) and may do better (if the opponent deviates from a maximin or minimax strategy). ■

Another well-known game with a saddlepoint is tic-tac-toe. Two players alternately place an ✕ or an ○, respectively, in one of the unoccupied spaces in a 3 × 3 grid with 9 cells. The winner is the first player to have three ✕'s, or three ○'s, in the same row, in the same column, or along a diagonal.

An explicit list of all strategies for either the first- or second-moving player in tic-tac-toe is long and complicated, because it specifies a complete plan for all possible contingencies that can arise. For the first-moving player, for example, a strategy might say "put an ✕ in the middle cell, then an ✕ in the corner if your opponent puts an ○ in a noncorner cell, and so on." While young children initially find this game interesting to play, before long they discover that each player can always prevent the other player from winning by forcing a tie, making the game quite boring. All strategies that force a tie, it turns out, are a saddlepoint in tic-tac-toe.

EXAMPLE 2   The Restricted-Location Game

Assume in our location game that Henry and Lisa are informed by county officials that it is against the law to locate a restaurant on either Boulevard B or Interstate Highway 2. These two choices, which provided our earlier solution, are now forbidden. The resulting location game without these two strategies is given in Table 15.3 (with payoffs again expressed in thousands of feet).

As before, Henry and Lisa can each do a worst-case analysis. Henry is worried about the minimum number in each row, and Lisa is concerned with the

TABLE 15.3   Heights (in thousands of feet) without Boulevard B and Interstate Highway 2

Routes	Highways	
	1	3
A	10	6
C	2	7

TABLE 15.4	Heights (in thousands of feet) in Table 15.3, with Row Minima (maximum circled) and Column Maxima (minimum circled)			
		Lisa Highways		
	Routes	1	3	Row Minima
Henry	A	10	6	⑥
	C	2	7	2
	Column Maxima	10	⑦	

maximum number in each column. These are listed in the right column and bottom row, respectively, in Table 15.4.

Henry sees that his maximin is 6, so he can guarantee a height of 6000 feet or more by choosing route A. Likewise, Lisa observes that her minimax is 7, so she can keep the elevation of the restaurant down to 7000 feet or less by selecting highway 3. There is a gap of $7 - 6 = 1$ between the minimax and maximin. When the maximin is less than the minimax, as in this case, then a game does *not* have a saddlepoint, but it does have a value (described in the next section).

If Henry plays his maximin strategy, route A, and Lisa plays her minimax strategy, highway 3, then the resulting payoff is 6. However, Henry may be motivated to gamble in this case by playing his other strategy, route C. If Lisa sticks to her conservative strategy, highway 3, then the payoff is 7. Henry will have gained one unit (1000 feet), going from 6 to 7.

This is, however, a risky move. If Lisa suspected it, she might counter by selecting highway 1. The payoff would then be 2, the best for Lisa and the worst for Henry. So Henry's gamble to gain 1 unit (6 to 7) by moving has the risk that he might lose 4 units (6 to 2) if Lisa also moves.

But then there is no incentive for Lisa to play her nonminimax strategy (that is, to play highway 1) if she believes Henry, in turn, will move back to his maximin strategy (route A), leading to a payoff of 10. This is worse than 6 from her viewpoint. ■

In two-player games that have saddlepoints, like our original 3 × 3 location game and tic-tac-toe, each player can calculate the maximin and minimax strategies for both players before the game is even played. Once the solution has been determined by either mathematical analysis or practical experience (as was probably true of tic-tac-toe), there may be little interest in actually playing the game.

But this is decidedly not the case for much more complex games, like chess, whose solution has not yet been determined—and is unlikely to be in the foreseeable future. Even though computers are able to beat world champions on occasion, the computer's winning moves will not necessarily be optimal against those of *all* other opponents. Nevertheless, we know that chess, like tic-tac-toe, has a saddlepoint. What we do not know is whether it yields a win for white, a win for black, or a draw.

Unlike chess, many games, like the 2×2 restricted-location game, do not have an outcome that can always be guaranteed. These games, which include poker, involve uncertainty and risk. In such games, one does not want to have one's strategy detected in advance, because this information can be exploited by an opponent. It is no surprise, then, that poker players are told to keep a "poker face," revealing nothing about their likely choices. But this advice is not very helpful in telling the players what actually to do in the game, such as how many cards to ask for in draw poker.

We will show that there are optimal ways to play two-person total-conflict games without a saddlepoint so as not to reveal one's choices. But their solution is by no means as straightforward as that of games with a saddlepoint.

15.2 Two-Person Total-Conflict Games: Mixed Strategies

Probably most competitive games do not have a saddlepoint like the one found in our first location-game example. Rather, as is illustrated in our restricted-location game—in which the maximin and minimax are not the same—players must try to keep secret their strategy choices, lest their opponent use this information to his or her advantage.

In particular, players must take care to conceal the strategy they will select until the encounter actually takes place, when it is too late for the opponent to alter his or her choice. If the game is repeated, a player will want to *vary* his or her strategy in order to surprise the opponent.

In parlor games like poker, players often use the tactic of *bluffing*. This tactic involves a player's sometimes raising the stakes when he has a low hand so that opponents cannot guess whether or not his hand is high or low—and may, therefore, miscalculate whether to stay in or drop out of the game (a player would prefer opponents to stay in when he has a high hand and drop out when he has a low hand). In military engagements, too, secrecy and even deception are often crucial to success.

In many sporting events, a team tries to surprise or mislead the opposition. A pitcher in baseball will not signal the type of pitch he or she intends to throw in advance, varying the type throughout the game to try to keep the batter off balance. In fact, we next consider a confrontation between a pitcher and batter in more detail.

EXAMPLE 3   A Duel Game

Assume that a particular baseball pitcher can throw either a blazing fastball or a slow curve into the strike zone and so has two strategies: *fast* (denoted by *F*) and

The pitcher and the batter use mixed strategies. *(Alan Schein/The Stock Market.)*

curve (*C*). The pitcher faces a batter who attempts to guess, before each pitch is thrown, whether it will be a fastball or a curve, giving the batter two strategies also: guess *F* and guess *C*. Assume that the batter has the following batting averages, which are known by both players.

▶   0.300 if the batter guesses fast (*F*) and the pitcher throws fast (*F*)
▶   0.200 if the batter guesses fast (*F*) and the pitcher throws curve (*C*)
▶   0.100 if the batter guesses curve (*C*) and the pitcher throws fast (*F*)
▶   0.500 if the batter guesses curve (*C*) and the pitcher throws curve (*C*)

A player's batting average is the number of times he hits safely divided by his number of times at bat. If a batter hit safely 3 times out of 10, for example, his average would be 0.300.

This game is summarized in Table 15.5. We see from the right-hand column in the table that the batter's maximin is 0.200, which is realized when he selects his first strategy, *F*. Thus, the batter can "play it safe" by always guessing a fastball, which will result in his batting 0.200, hardly enough for him to remain on the team.

We see from the bottom row of the table that the pitcher's minimax is 0.300, which is obtained when he throws fast (*F*). Note that the batter's maximin of 0.200 is less than the pitcher's minimax of 0.300, so this game does not have a saddlepoint. There is a gap of 0.300 − 0.200 = 0.100 between these two numbers.

Each player would like to play so as to win for himself as much of the 0.100 payoff in the gap as possible. That is, the batter would like to average more than 0.200, whereas the pitcher wants to hold the batter down to less than 0.300. ■

TABLE 15.5	Batting Averages in a Baseball Duel			
		Pitcher		Row Minima (maximum circled)
		F	C	
Batter	F	0.300	0.200	(0.200)
	C	0.100	0.500	0.100
Column Maxima (minimum circled)		(0.300)	0.500	

A Flawed Approach

If the batter and pitcher in our example consider how they might outguess each other, they may reason along the following lines:

1. *Pitcher* (to himself): If I choose strategy F, I hold the batter down to 0.300 (the minimax) or less. However, the batter is likely to guess F because it guarantees him at least 0.200 (his maximin), and it actually provides him with 0.300 against my F pitch. In this case, the batter wins all the 0.100 payoff in the gap.

2. *Batter* (to himself): Because the pitcher will try to surprise me with C by reasoning as in step 1, I should fool him and guess C. I would thus average 0.500, which will show him up for trying to gamble and outguess me!

3. *Pitcher* (to himself): But if the batter is thinking as in step 2—that is, guessing C—I, on second thought, should really throw F. This will lead to an average of only 0.200 for the batter and teach him to not try to outguess me!

This type of cyclical reasoning can go on forever: "I think that he thinks that I think that he thinks. . . ." It provides no resolution to the players' decision problem.

Clearly, there is no pitch, or guess, that is best under all circumstances. Nevertheless, both the pitcher and the batter *can* do better, but not by trying to anticipate each other's choices. The answer to their problem lies in the notion of a *mixed strategy*.

A Better Idea

The play of many total-conflict games requires an element of surprise, which can be realized in practice by making use of a mixed strategy.

Pure and Mixed Strategy

Each of the definite courses of action that a player can choose is called a **pure strategy**. A **mixed strategy** is a strategy in which the course of action is randomly chosen from one of the pure strategies in the following way: Each pure strategy is assigned some probability, indicating the relative frequency with which that strategy will be played. The specific strategy used in any given play of the game can be selected using some appropriate random device.

Note that a pure strategy is a special case of a mixed strategy, with the probability of 1 assigned to just one pure strategy and 0 to all the rest. When a player resorts to a mixed strategy, the resulting outcome of the game is no longer predictable in advance. (For example, if a pitcher throws a curve ball or a fastball with probability 0.5 each, the batter cannot predict which pitch he or she is about to receive.) Rather, the outcome must be described in terms of the probabilistic notion of an *expected value*.

Expected Value *E*

If each of the n payoffs, $s_1, s_2, \ldots, s_n$ will occur with the probabilities $p_1, p_2, \ldots, p_n$, respectively, then the average, or **expected value *E*,** is given by

$$E = p_1 s_1 + p_2 s_2 + \cdots + p_n s_n$$

We assume that the probabilities sum to 1 and that each probability p_i is never negative. That is, we assume that $p_1 + p_2 + \cdots + p_n = 1$, and $p_i \geq 0$ ($i = 1, 2, \ldots, n$).

To see how mixed strategies and expected values are used in the analysis of games, we turn to what is perhaps the simplest of all competitive games without a saddlepoint.

Matching Pennies

EXAMPLE 4

In matching pennies, each of two players simultaneously shows either a head H or a tail T. If the two coins match, with either two heads or two tails, then the first player (player I) receives both coins (a win of 1 for player I). If the coins do not match, that is, if one is an H and the other is a T, then the second player (player II) receives the two coins (a loss of 1 for player I). These wins and losses for player I are shown in Table 15.6.

TABLE 15.6	Wins and Losses for Player I in Matching Pennies	

		Player II	
		H	T
Player I	H	1	−1
	T	−1	1

Payoff Matrix

A **payoff matrix** (illustrated by Table 15.6) is a table whose rows and columns correspond to the strategies of the two players. The numerical entries give the payoffs to player I when these strategies are chosen.

Although the entries in our earlier tables for the location game also gave payoffs, they were not monetary, as here. A game represented by a payoff matrix is called *a game in strategic form.*

The two rows in Table 15.6 correspond to player I's two pure strategies, H and T, and the two columns to player II's two pure strategies, also H and T. The numbers in the table are the corresponding winnings for player I and losses for player II. If two H's or two T's are played, player I wins 1 from player II. When both an H and a T are played, player I pays out 1 to player II.

It is fruitless for one player to attempt to outguess the other in this game. They should instead resort to mixed strategies and use expected values to estimate their likely gains or losses.

The best thing for player I to do is randomly to select H half the time and T half the time. This mixed strategy can be expressed as

$$(p_H, p_T) = (p_1, p_2) = (p, 1 - p) = \left(\frac{1}{2}, \frac{1}{2} \right)$$

Note that the probability p of choosing H, and $(1 - p)$ of choosing T, do indeed sum to 1, as required; in particular, when $p = \frac{1}{2}$, $1 - p = 1 - \frac{1}{2} = \frac{1}{2}$.

This mixture can be realized in practice by the flip of a coin. Player I's resulting expected value is

$$E_H = \frac{1}{2}(1) + \frac{1}{2}(-1) = 0$$

whenever player II plays H (first column of Table 15.6). Whenever player II plays T (second column), player I's resulting expected value is

$$E_T = \frac{1}{2}(-1) + \frac{1}{2}(1) = 0$$

Mixed-Strategy Value

Player I's average outcome of 0 is the **mixed-strategy value** of the game. Unlike the use of this notion in games with a saddlepoint, the value here can be realized only by the use of mixed strategies.

The value of 0 is really an expected value and so must be understood in a statistical sense. That is, in a given play of the game, player I will either win 1 or lose 1. However, his or her expectation over many plays of this game is 0. The optimal mixed strategy for player II is likewise a 50-50 mix of H and T, which also leads to an expectation of 0, making the game fair.

Fair Game

A game is **fair** if its value is 0 and, consequently, it favors neither player when at least one player uses an *optimal* (mixed) strategy—one that guarantees that the resulting payoff is the best that this player can obtain against all possible strategy choices (pure or mixed) by an opponent.

Player II gains nothing by knowing that player I is using the optimal mixed strategy $(\frac{1}{2}, \frac{1}{2})$. However, player I must not reveal to player II whether H or T will be displayed *in any given play* of the game before player II makes his or her own choice of H or T. Even without this information, if player II knew that player I was using a particular *nonoptimal* mixed strategy $(p_1, p_2) = (p, 1 - p)$, where $p \neq \frac{1}{2}$ (that is, not choosing a 50-50 mixture between H and T), then player II could take advantage of this knowledge and increase his or her average winnings over time to something greater than the value of 0. (See Exercise 14.) ■

Nonsymmetrical Matching   EXAMPLE 5

In this game, players I and II can again show either heads H or tails T. When two H's appear, player II pays $5 to player I. When two T's appear, player II pays $1 to player I. When one H and one T are displayed, then player II collects

TABLE 15.7	Payoffs for Player I in a Nonsymmetrical Matching Game	

		Player II	
		H	T
Player I	H	5	−3
	T	−3	1

$3 from player I. Note that although the sum of player I's gains ($5 + $1 = $6) when there are two *H*'s or two *T*'s, and the sum of player II's gains ($3 + $3 = $6) otherwise, are the same, the game is nonsymmetrical.

Nonsymmetrical Game

A **nonsymmetrical** two-person total-conflict game is one in which the row player's gains ($5 and $1 in our example) are different from the column player's gains (always $3), except when there is a tie (not possible in our example, but possible in tic-tac-toe). In the original matching pennies, on the other hand, the payoff for winning is the same for each player, so the game is *symmetrical*.

The game just described is given by the payoff matrix in Table 15.7, which shows the payoffs that player I receives from player II. A worst-case analysis, like that which solved our initial location game, is of little help here. Player I may lose $3 whether he plays *H* or *T*, making his maximum −3. Player II can keep her losses down to $1 by always playing *T* (and thus avoiding the loss of $5 when two *H*'s appear), so player II's minimax is 1. However, if player II chooses *T* and player I knows this, then player I will also play *T* and collect $1 from player II. Can player II do better than lose $1 in each play of the game?

Consider the situation where player I uses a mixed strategy $(p_H, p_T) = (p, 1 − p)$, which involves playing *H* with probability p and playing *T* with probability $1 − p$, where $0 \leq p \leq 1$. Against player II's pure strategy *H*, player I's expected value is

$$E_H = (5)(p) + (−3)(1 − p) = 8p − 3$$

Against player II's pure strategy *T*, player I's expected value is

$$E_T = (−3)(p) + (1)(1 − p) = −4p + 1$$

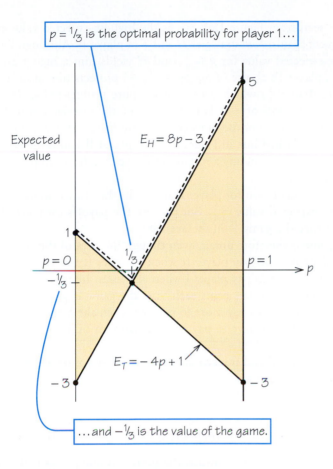

FIGURE 15.3 Solution to the nonsymmetrical matching pennies.

These two linear equations in the variable p are depicted in Figure 15.3. Note that the four points where these two lines intersect the two vertical lines, $p = 0$ and $p = 1$, are the four payoffs appearing in the payoff matrix.

The point at which the lines given by E_H and E_T intersect can be found by setting $E_H = E_T$, yielding

$$8p - 3 = -4p + 1$$
$$12p = 4$$

so $p = \frac{1}{3}$. To the left of $p = \frac{1}{3}$, $E_T > E_H$, and to the right, $E_H > E_T$; at $p = \frac{1}{3}$, $E_H = E_T$. If player I chooses $(p_H, p_T) = (p, 1 - p) = (\frac{1}{3}, \frac{2}{3})$, he can ensure

$$E_H = 8\left(\frac{1}{3}\right) - 3 = E_T = -4\left(\frac{1}{3}\right) + 1 = -\frac{1}{3}$$

regardless of what player II does.

In other words, player I's optimal mixed strategy is to pick H and T with probabilities $\frac{1}{3}$ and $\frac{2}{3}$, respectively, which gives player I an expected value of $-\frac{1}{3}$.

As can be seen from Figure 15.3, $-\frac{1}{3}$ is the highest expected value that player I can guarantee against *both* strategies H and T of player II. Although T yields player I a higher expected value for $p < \frac{1}{3}$, and H yields him a higher expected value for $p > \frac{1}{3}$, player I's choice of $p_H, p_T = (\frac{1}{3}, \frac{2}{3})$ protects him against an expected loss greater than $-\frac{1}{3}$, which neither of his pure strategies does (each may produce a maximum loss of -3). Put another way, the intersection of E_H and E_T at $p = \frac{1}{3}$ is the minimum of the function given by E_T to the left and E_H to the right (shown by the dashed line in Figure 15.3). If player II had more than two strategies, this approach to finding a minimum that puts a floor on player I's expected loss can be extended.

A similar calculation for player II results in the same optimal mixed strategy $(\frac{1}{3}, \frac{2}{3})$ and expected value $-\frac{1}{3}$. But because the payoffs for player II are losses, $-\frac{1}{3}$ means that she gains $\frac{1}{3}$ on the average.

This game is therefore unfair, even though the sum of the amounts ($\$6$) that player I might have to pay player II when he loses is the same as the sum that player II might have to pay player I when she loses. Interestingly, player II, who will win an average of $33\frac{1}{3}$ cents each time the game is played, is favored, even though she may have to pay more to player I when she loses (a maximum of $\$5$) than player I will ever have to pay her (a maximum of $\$3$). ∎

The symmetrical and nonsymmetrical matching games are examples of what are called *zero-sum games.*

Zero-Sum Game

A **zero-sum game** is one in which the payoff to one player is the negative of the corresponding payoff to the other, so the sum of the payoffs to the two players is always zero. These games can be completely described by a payoff matrix, in which the numbers represent the payoffs to player I, while their negatives are the payoffs to player II.

Zero-sum games are total-conflict games in which what one player wins the other loses. But not all total-conflict games are zero-sum—in particular, the sum of the payoffs could be some constant other than zero. Nevertheless, the strategic nature of these latter games is the same as that of zero-sum games: What one player wins, the other player still loses. This was true in our location game, in which Henry's payoff was greater the higher the altitude, and Lisa's greater the lower the altitude.

Scoring in professional chess tournaments usually assigns a payoff of 1 for winning, 0 for losing, and $\frac{1}{2}$ to each player for a tie, making the sum of the payoffs to the two players always 1. Such games, called **constant-sum games,** can readily be converted to zero-sum games. Thus, chess could as well be scored

−1 for a loss, +1 for a win, and 0 for a tie, making the constant 0 in this case. Although constant-sum and zero-sum games have the same strategic nature, constant-sum games are a more general class because the constant need not be zero.

The solution in the symmetrical version of matching pennies illustrated how the mixed strategy of $(\frac{1}{2}, \frac{1}{2})$ guarantees each player the value of 0, but we did not give a *solution technique* for finding optimal mixed strategies. In the nonsymmetrical version of matching pennies, we illustrated a procedure that can be applied to *every* payoff matrix in which each player has only two strategies.

We must use more complex methods, which we will not describe here, to find mixed-strategy solutions when one or both players have more than two strategies. However, one should always check first to see whether a game has a saddlepoint before employing any technique for finding optimal mixed strategies.

In our next example, which is the earlier duel between the pitcher and the batter given by the 2×2 payoff matrix in Table 15.5, we already showed that there is no saddlepoint. Thus, the solution will necessarily be in mixed strategies. We now proceed to find what mix is optimal.

The Duel Game Revisited

EXAMPLE 6

In Table 15.8, we add probabilities, which we explain next, to Table 15.5, where F indicates fastball and C indicates curve ball. The pitcher should use a mixed strategy $(p_1, p_2) = (p_F, p_C) = (p, 1 - p)$. The probabilities p and $1 - p$ (where $0 \le p \le 1$) are indicated below the game matrix and under the corresponding strategies, F and C, for the pitcher. If the pitcher plays a mixed strategy $(p, 1 - p)$ against the two pure strategies, F and C, for the batter, he realizes the respective expected values:

$$E_F = (0.3)p + 0.2(1 - p) = 0.1p + 0.2$$
$$E_C = (0.1)p + 0.5(1 - p) = -0.4p + 0.5$$

As in the nonsymmetrical matching-pennies game, the solution to this game occurs at the intersection of the two lines given by E_F and E_C. Setting the equa-

TABLE 15.8 A Baseball Duel with Probabilities

		Pitcher		
		F	C	
Batter	F	0.300	0.200	q
	C	0.100	0.500	$1 - q$
		p	$1 - p$	

tions of these lines equal to each other yields $p = 0.6$, giving $E_F = E_C = E = 0.260$.

Thus, the pitcher should use his optimal mixed strategy, which selects F with probability $p = \frac{3}{5}$ and C with probability $1 - p = \frac{2}{5}$. This choice will hold the batter down to a batting average of 0.260, which is the value of the game. We stress that 0.260 is an average and must be interpreted in a statistical manner. It says that about one time in four the batter will get a hit, but not what will happen on any particular time at bat.

Assume that the batter uses a mixed strategy $(q_1, q_2) = (q_F, q_C) = (q, 1 - q)$, as indicated to the right of the game matrix in Table 15.8. This mixed strategy, when played against the pitcher's pure strategies, F and C, results in the respective expected values:

$$E_F = (0.3)q + 0.1(1 - q) = 0.2q + 0.1$$
$$E_C = (0.2)q + 0.5(1 - q) = -0.3q + 0.5$$

The intersection of these two lines occurs at the point $q = 0.8$, giving $E_F = E_C = E = 0.260$. The batter's optimal mixed strategy is therefore $(q_F, q_C) = (\frac{4}{5}, \frac{1}{5})$, which gives him the same batting average of 0.260. ∎

We have seen that the outcome of 0.260, which is the value of the game, occurs when either the pitcher selects his optimal mixed pitching strategy $(\frac{3}{5}, \frac{2}{5})$ or the batter selects his optimal mixed guessing strategy $(\frac{4}{5}, \frac{1}{5})$. This particular result holds true for every two-person zero-sum game; it is the fundamental theorem for such games and is known as the *minimax theorem*.

Minimax Theorem

The **minimax theorem** guarantees that there is a unique game value (0.260 in our example), and an optimal strategy for each player, so that either player alone can realize at least this value by playing this strategy, which may be pure or mixed.

15.3 Partial-Conflict Games

The 2×2 matrix games (two players, each with two strategies) presented so far have been total-conflict games: One player's gain was equal to the other player's loss. Although most parlor games, like chess or poker, are games of total conflict, and therefore constant-sum, most real-life games are surely not. (Elections, in which there are usually a clear-cut winner and one or more losers, probably come as close to being games of total conflict as we find in the real world.) We will consider two games of partial conflict, in which the players' preferences are not diametrically opposed, that have often been used to model real-world conflicts.

::::::: Variable-Sum Games

Games of partial conflict are **variable-sum games,** in which the sum of pay-offs to the players at the different outcomes varies.

There is some mutual gain to be realized by both players if they can cooperate in partial-conflict games, but this may be difficult to do in the absence of either good communication or trust. When these elements are lacking, players are less likely to comply with any agreement that is made. *Noncooperative games* are games in which a binding agreement cannot be enforced. Even if communication is allowed in such games, there is no assurance that a player can trust an opponent to choose a particular strategy that he or she promises to select.

In fact, the players' self-interests may lead them to make strategy choices that yield both lower payoffs than they could have achieved by cooperating. Two partial-conflict games illustrate this problem.

Prisoners' Dilemma

EXAMPLE 7 :::::::

Prisoners' Dilemma is a two-person variable-sum game. It provides a simple explanation of the forces at work behind arms races, price wars, and the population problem. In these and other similar situations, the players can do better by cooperating. But there may be no compelling reasons for them to do so unless the players have credible threats of retaliation for not cooperating. The term *Prisoners' Dilemma* was first given to this game by Princeton mathematician Albert W. Tucker (1905–1994) in 1950.

Before defining the formal game, we introduce it through a story, which involves two persons, accused of a crime, who are held incommunicado. Each has two choices: to maintain his or her innocence, or to sign a confession accusing the partner of committing the crime.

Now it is in each suspect's interest to confess and implicate the partner, thereby trying to receive a reduced sentence. Yet if both suspects confess, they ensure a bad outcome—namely, they are both found guilty. What is good for the prisoners as a pair—to deny having committed the crime, leaving the state with insufficient evidence to convict them—is frustrated by their pursuit of their own individual rewards.

Prisoners' Dilemma, as we already noted, has many applications, but we will use it here to model a recurrent problem in international relations: Arms races between antagonistic countries, which earlier included the superpowers but more recently have included such countries as India and Pakistan and Israel and some of its Arab neighbors. Other countries, such as North Korea and Iran, may be antagonistic to more than one other country.

For simplicity, assume there are two nations, Red and Blue. Each can independently select one of two policies:

A: Arm in preparation for a possible war (noncooperation).
D: Disarm, or at least try to negotiate an arms-control agreement (cooperation).

There are four possible outcomes:

(*D, D*):   Red and Blue disarm, which is *next best* for both because, while
             advantageous to each, it also entails certain risks.
(*A, A*):   Red and Blue arm, which is *next worst* for both because they spend
             needlessly on arms and are comparatively no better off than at (*D, D*).
(*A, D*):   Red arms and Blue disarms, which is *best for Red* and *worst for Blue*
             because Red gains a big edge over Blue.
(*D, A*):   Red disarms and Blue arms, which is *worst for Red* and *best for Blue*
             because Blue gains a big edge over Red.

This situation can be modeled by means of the matrix in Table 15.9, which
gives the possible outcomes that can occur. Here, Red's choice involves picking
one of the two rows, whereas Blue's choice involves picking one of the two
columns.

We assume that the players can rank the four outcomes from best to worst,
where 4 = best, 3 = next best, 2 = next worst, and 1 = worst. Thus, the higher
the number, the greater the payoff, but these payoffs are only **ordinal**: They indicate an ordering of outcomes from best to worst but say nothing about the *degree* to which a player prefers one outcome over another. To illustrate, if a player
despises the outcome that he or she ranks 1 but sees little difference among the
outcomes ranked 4, 3, and 2, the "payoff distance" between 4 and 2 will be less
than that between 2 and 1, even though the numerical difference between 4 and
2 is greater.

The ordinal payoffs to the players for choosing their strategies of *A* and *D*
are shown in Table 15.10, where the first number in the pair indicates the payoff to the row player (Red), and the second number the payoff to the column
player (Blue). Thus, for example, the pair (1, 4) in the second row and first column signifies a payoff of 1 (worst outcome) to Red and a payoff of 4 (best outcome) to Blue. This outcome occurs when Red unilaterally disarms while Blue
continues to arm, making Blue, in a sense, the winner and Red the loser.

TABLE 15.9	The Outcomes in an Arms Race, as Modeled by Prisoners' Dilemma

		Blue	
		A	*D*
Red	*A*	Arms race	Favors Red
	D	Favors Blue	Disarmament

TABLE 15.10	Ordinal Payoffs in an Arms Race, as Modeled by Prisoners' Dilemma

		Blue	
		A	*D*
Red	*A*	(2, 2)	(4, 1)
	D	(1, 4)	(3, 3)

Let's examine this strategic situation more closely. Should Red select strategy *A* or *D*? There are two cases to consider, which depend on what Blue does:

▶ If Blue selects *A*: Red will receive a payoff of 2 for *A* and 1 for *D*, so it will choose *A*.

▶ If Blue selects *D*: Red will receive a payoff of 4 for *A* and 3 for *D*, so it will choose *A*.

In both cases, Red's first strategy (*A*) gives it a more desirable outcome than its second strategy (*D*). Consequently, we say that *A* is Red's **dominant strategy,** because it is always advantageous for Red to choose *A* over *D*.

In Prisoners' Dilemma, *A dominates D* for Red, so we presume that a rational Red would choose *A*. A similar argument leads Blue to choose *A* as well—that is, to pursue a policy of arming. Thus, when each nation strives to maximize its own payoffs independently, the pair is driven to the outcome (*A, A*), with payoffs of (2, 2). The better outcome for both, (*D, D*), with payoffs of (3, 3), appears unobtainable when this game is played noncooperatively.

The outcome (*A, A*), which is the product of dominant strategy choices by both players in Prisoners' Dilemma, is a *Nash equilibrium*.

Nash Equilibrium

When no player can benefit by departing unilaterally (by itself) from its strategy associated with an outcome, the strategies of the players constitute a **Nash equilibrium.** Technically, while it is the set of strategies that define the equilibrium, the choice of these strategies leads to an outcome that we shall also refer to as the equilibrium.

Note that in Prisoners' Dilemma, if either player departs from (*A, A*), the payoff for the departing player who switches to *D* drops from 2 to 1 at (*D, A*) and (*A, D*). Not only is there no benefit from departing, but there is actually a loss, with the D player punished with its worst payoff of 1. These losses would presumably deter each nation from moving away from the Nash equilibrium of (A, A), assuming the other nation sticks to A.

Even if both nations agreed in advance jointly to pursue the socially beneficial outcome, (D, D), $(3, 3)$ is unstable. This is because if either nation alone reneges on the agreement and secretly arms, it will benefit, obtaining its best payoff of 4. Consequently, each nation would be tempted to go back on its word and select A. Especially if nations have no great confidence in the trustworthiness of their opponents, they would have good reason to try to protect themselves against the other side's defection from an agreement by arming.

Prisoners' Dilemma

Prisoners' Dilemma is a two-person variable-sum game in which each player has two strategies, cooperate or defect (not cooperate). Defect dominates cooperate for both players, even though the mutual-defection outcome, which is the unique Nash equilibrium in the game, is worse for both players than the mutual-cooperation outcome.

Note that if 4, 3, 2, and 1 in Prisoners' Dilemma were not just ranks but numerical payoffs, their sum would be $2 + 2 = 4$ at the mutual-defection outcome and $3 + 3 = 6$ at the mutual-cooperation outcome. At the other two outcomes, the sum, $1 + 4 = 5$, is still different, illustrating why Prisoners' Dilemma is a variable-sum game.

In real life, of course, people often manage to escape the noncooperative Nash equilibrium in Prisoners' Dilemma. Either the game is played within a larger context, wherein other incentives are at work, such as cultural norms that prescribe cooperation (though this is just another way of saying that defection from (D, D) is not rational, rendering the game not Prisoners' Dilemma), or the game is played on a repeated basis—it is not a one-short affair—so players can induce cooperation by setting a pattern of rewards for cooperation and penalties for noncooperation.

In a repeated game, factors like reputation and trust may play a role. Realizing the mutual advantages of cooperation in costly arms races, players may inch toward the cooperative outcome by slowly phasing down their acquisition of weapons over time, or even destroying them (the United States and Russia have been doing exactly this). They may also initiate other productive measures, such as improving their communication channels, making inspection procedures more reliable, writing agreements that are truly enforceable, or imposing penalties for violators when their violations are detected (as may be possible by reconnaissance or spy satellites).

Prisoners' Dilemma illustrates the intractable nature of certain competitive situations that blend conflict and cooperation. The standoff that results at the

Nash equilibrium of (2, 2) is obviously not as good for the players as that which they could achieve by cooperating–but they risk a good deal if the other player defects.

While saddlepoints are Nash equilibria in total-conflict games, they can never be worse for *both* players than some other outcome (as in partial-conflict games like Prisoners' Dilemma). The reason is that if one player does worse in a total-conflict or zero-sum game, the other player must do better.

The fact that the players must forsake their dominant strategies to achieve the (3, 3) cooperative outcome (see Table 15.10) makes this outcome a difficult one to sustain in one-shot play. On the other hand, assume that the players can threaten each other with a policy of tit-for-tat in repeated play: "I'll cooperate on each round unless you defect, in which case I will defect until you start cooperating again." If these threats are credible, the players may well shun their defect strategies and try to establish a pattern of cooperation in early rounds, thereby fostering the choice of (3, 3) in the future. Alternatively, they may look ahead, in a manner that will be described at the end of this chapter, to try to stabilize (3, 3). ■

Chicken

Let's look at one other two-person game of partial conflict, known as Chicken, that can also lead to troublesome outcomes. Two drivers approach each other at high speed. Each must decide at the last minute whether to swerve to the right or not swerve. Here are the possible consequences of their actions:

1. Neither driver swerves, and the cars collide head-on, which is the worst outcome for both because they are killed (payoff of 1).

2. Both drivers swerve–and each is mildly disgraced for "chickening out"–but they do survive, which is the next-best outcome for both (payoff of 3).

3. One of the drivers swerves and badly loses face, which is his next-worst outcome (payoff of 2), whereas the other does not swerve and is perceived as the winner, which is her best outcome (payoff of 4).

These outcomes and their associated strategies are summarized in Table 15.11.

EXAMPLE 8

TABLE 15.11	Payoffs in a Driver Confrontation, as Modeled by Chicken		
		Driver 2	
		Swerve	Not Swerve
Driver 1	Swerve	(3, 3)	(2, 4)
	Not Swerve	(4, 2)	(1, 1)

If both drivers persist in their attempts to "win" with a payoff of 4 by not swerving, the resulting outcome will be mutual disaster, giving each driver his or her worst payoff of 1. Clearly, it is better for both drivers to back down and each obtain 3 by swerving, but neither wants to be in the position of being intimidated into swerving (payoff of 2) when the other does not (payoff of 4).

Notice that neither player in Chicken has a dominant strategy. His or her better strategy depends on what the other player does: Swerve if the other does not, don't swerve if the other player swerves, making this game's choices highly interdependent, which is characteristic of many games. The Nash equilibria in Chicken, moreover, are (4, 2) and (2, 4), suggesting that the compromise of (3, 3) will not be easy to achieve because both players will have an incentive to deviate in order to try to be the winner.

Chicken

Chicken is a two-person variable-sum game in which each player has two strategies: to swerve to avoid a collision or not to swerve and possibly cause a collision. Neither player has a dominant strategy. The compromise outcome, in which both players swerve, and the disaster outcome, in which both players do not, are not Nash equilibria. The other two outcomes, in which one player swerves and the other does not, are Nash equilibria.

In fact, there is a third Nash equilibrium in Chicken, but it is in mixed strategies, which can be computed only if the payoffs are not ranks, as we have assumed here, but numerical values. Even if the payoffs were numerical, however, it can be shown that this equilibrium is always worse for both players than the cooperative (3, 3) outcome. Moreover, it is implausible that players would sometimes swerve and sometimes not—randomizing according to particular probabilities—in the actual play of this game, compared with either trying to win outright or reaching a compromise.

The two pure-strategy Nash equilibria in Chicken suggest that, insofar as there is a "solution" to this game, it is that one player will succeed when the other caves in to avoid the mutual-disaster outcome. But there are certainly real-life cases in which a major confrontation was defused and a compromise of sorts was achieved in a Chicken-type game. This fact suggests that the one-sided solutions given by the two pure-strategy Nash equilibria may not be the only pure-strategy solutions, especially if the players are farsighted and think about the possible untoward consequences of their actions.

International crises, labor–management disputes, and other conflicts in which escalating demands may end in wars, strikes, and other catastrophic outcomes have been modeled by the game of Chicken (see Spotlight 15.2 for more on game theorists who have analyzed these and other games). But Chicken, like Prisoners' Dilemma, is only one of the 78 essentially different 2×2 ordinal games in which each player can rank the four possible outcomes from best to worst.

SPOTLIGHT 15.2   The 1994 Nobel Prize in Economics

John C. Harsanyi   **John F. Nash, Jr.**   **Reinhard Selten**

The Nobel Memorial Prize in Economics was awarded to three game theorists in 1994, marking the 50th anniversary of the publication of von Neumann and Morgenstern's *Theory of Games and Economic Behavior* (see Spotlight 15.1).

▶ *John C. Harsanyi* (1920–2000) of the University of California, Berkeley, a Hungarian-American who emigrated from Hungary to Australia in 1950 and then to the United States in 1956. He is well known for extending game theory to the study of ethics and showing how societal institutions, each of whose members' satisfaction can be measured against that of others, choose among alternatives. His other major contribution was to give a precise definition to "incomplete information" in games in which players may be thought of as different types, and probabilities are assigned to each type.

▶ *John F. Nash, Jr.* (b. 1928) of Princeton University, an American mathematician

who did path-breaking work on both noncooperative game theory (the Nash equilibrium is named after him) and cooperative game theory, especially on bargaining, in which axioms or assumptions are specified and a unique solution that satisfies these axioms is derived. Nash obtained his results in the early 1950s, when he was only in his 20s, after which he became mentally ill and was unable to work. Fortunately, he has made a remarkable recovery and has now resumed research.

▶ *Reinhard Selten* (b. 1930) of the University of Bonn, a German mathematician who proposed significant refinements in the concept of the Nash equilibrium that help to distinguish those that are most plausible in games (often there are many such equilibria, which creates a selection problem). Selten is also noted for pioneering work on developing game-theoretic models in evolutionary biology.

Chicken and Prisoners' Dilemma, however, are especially disturbing, because the cooperative (3, 3) outcome in each is not a Nash equilibrium. Unlike a constant-sum game, in which the losses of one player are offset by the gains of the other, *both* players can end up doing badly—at (2, 2) in Prisoners' Dilemma and (1, 1) in Chicken—in these variable-sum games. ∎

15.4 Larger Games

We have shown how to compute optimal pure and mixed strategies, and the values ensured by using them, in 2×2 constant-sum games. In 2×2 variable-sum games, we focused on Nash equilibria as a solution concept in Prisoners' Dilemma and Chicken, but we found that this notion of a stable outcome did not justify the choice of cooperative strategies in either of these games.

We turn next to a somewhat larger game, in which there are three players, each of whom can choose among three strategies, which is technically a $3 \times 3 \times 3$ game. In this game, we eliminate certain undesirable strategies, but in stages, to arrive at a Nash equilibrium that seems quite plausible.

If one of the three players has a dominant strategy in the $3 \times 3 \times 3$ game, we suppose this player will choose it, thereby reducing the game to a 3×3 game between the other two players. Of course, if no player has a dominant strategy in a three-person game, it cannot be reduced in this manner to a two-person game.

If this game is not one of total conflict, the minimax theorem, which guarantees players the value in a two-person zero-sum game, is not applicable. Even if the game were zero-sum, the fact that we assume the players in the $3 \times 3 \times 3$ game can only rank outcomes, not assign numerical values to them, means that they cannot calculate optimal mixed strategies in it.

The problem in finding a solution to the reduced 3×3 game is not a lack of Nash equilibria. Rather, there are too many! So the question becomes which, if any, are likely to be selected by the players. Specifically, is one more appealing than the others? The answer is "yes," but it requires extending the idea of dominance, discussed in the previous section, to its successive application in different stages of play.

EXAMPLE 9   The Status-Quo Paradox

The $3 \times 3 \times 3$ game we analyze involves voting, illustrating the applicability of game theory to politics. We will show that the status quo, or existing state of affairs, may be defeated by another alternative, despite its privileged position as the policy in place at the time.

To illustrate this problem, suppose there is a set of three voters, $V = \{X, Y, Z\}$, and a set of three alternatives, $A = \{x, y, z\}$, from which the voters choose. Assume that voter X prefers x to y to z, indicated by *xyz;* voter Y's preference is *yzx;* and voter Z's is *zxy*. These preferences give rise to a *Condorcet voting paradox* (discussed in Chapter 9), because the social ordering, according to majority rule, is *intransitive:* Although a majority (voters X and Z) prefer x to y, and a majority (voters X and Y) prefer y to z, a majority (voters Y and Z) prefer z to x. So there is no **Condorcet winner**—an alternative that would beat all others in separate pairwise contests. Instead, every alternative can be beaten by one other.

Assume that the voting procedure used by the three voters, who choose from among the three alternatives, is the **plurality procedure,** under which the alternative with the most votes wins. If there is a three-way tie (there can never be a two-way tie if there are three voters), we assume that x wins—because it is the status quo—giving X what would appear to be an edge over the other two voters, Y and Z.

To begin, assume that voting is sincere.

Sincere Voting

Under **sincere voting,** every voter votes for his or her most-preferred alternative, based on his or her true preferences, without taking into account what the other voters might do (see Chapter 9).

In this case, X will prevail, because he or she supports the status quo. However, X's apparent advantage disappears if voting is "sophisticated," as demonstrated on the next page.

To see why, first note that X has a dominant strategy of "vote for x": It is never worse and sometimes better than her other two strategies, whatever the other two voters do. Thus, if the other two voters vote for the same alternative, it wins, and X cannot do better than vote sincerely for x, so voting sincerely is never worse. On the other hand, if the other two voters disagree, X's tie-breaking vote (along with her regular vote) for x will be decisive in x's selection, which is X's best outcome.

Given the dominant choice of x on the part of X, Y and Z face the strategy choices shown in Figure 15.4. Y has one, and Z has two, **dominated strategies,** which are never better and sometimes worse than some other strategy, whatever the other two voters do. For example, observe that "vote for x" by Y always leads to his worst alternative, x. The dominated strategies are crossed out in the top matrix in Figure 15.4.

This leaves Y with two *undominated* strategies that are neither dominant nor dominated: "vote for y" and "vote for z." "Vote for y" is better than "vote for z" if Z chooses y (leading to y rather than x), whereas the reverse is the case if Z chooses z (leading to z rather than x). By contrast, Z has a dominant strategy of "vote for z," which leads to outcomes at least as good and sometimes better than his other two strategies.

If voters have complete information about each other's preferences, then they can perceive the situation represented by the top matrix of Figure 15.4. Reason-

FIGURE 15.4

Sophisticated voting, given X chooses "vote for x." The dominated strategies of each voter are crossed out in the first reduction, leaving two (undominated) strategies for Y and one (dominant) strategy for Z. Given these eliminations, Y would then eliminate "vote for y" in the second reduction, making z the sophisticated outcome.

FIRST REDUCTION

SECOND REDUCTION

ing that no player would ever choose a dominated strategy, they would eliminate such strategies from consideration (these have been crossed out in the first reduction).

The elimination of these strategies gives the bottom matrix in Figure 15.4. Then Y, choosing between "vote for y" and "vote for z" in this matrix, would cross out "vote for y" (second reduction), now dominated because that choice would result in x's winning because it is the status quo. Instead, Y would choose "vote for z," which is not Y's sincere strategy. This ensures z's election, which is Z's best outcome but only the next-best outcome for Y. In this manner, z, which is not the first choice of a majority and could in fact be beaten by y in a pairwise contest, becomes the sophisticated outcome.

Sophisticated Voting

The *successive* elimination of dominated strategies by voters (insofar as this is possible), beginning in our example with X's choice of x in favor of y and z, is called **sophisticated voting.**

Sophisticated voting results in a Nash equilibrium, because none of the three players can do better by departing from his or her sophisticated strategy when the other two players choose theirs. This is clearly true for X, because x is her dominant strategy; given X's choice of x, z is dominant for Z; and given these choices by X and Z, z is dominant for Y. These "contingent" dominance relations, in general, make sophisticated strategies a Nash equilibrium.

Observe, however, that there are four other Nash equilibria in this game. First, the choice of each of x, y, or z by all three voters are all Nash equilibria, because no single voter's departure can change the outcome to a different one, much less a better one, for that player. In addition, the choice of x by X, y by Y, and x by Z—resulting in x—is also a Nash equilibrium, because no voter's departure would lead to a better outcome for him or her.

In game-theoretic terms, sophisticated voting produces a different and smaller game in which some formerly undominated strategies in the larger game become dominated in the smaller game. The removal of such strategies, sometimes in several successive stages, in effect enables sophisticated voters to determine what outcomes eventually *will* be chosen by eliminating those outcomes that definitely *will not* be chosen. Voters can thereby ensure that their worst outcomes will not be chosen by successively eliminating dominated strategies, given the presumption that other voters do likewise.

Status-Quo Paradox

How does sophisticated voting affect the status quo? Supporting the status quo is not only *not* helpful but positively harmful: It guarantees that X's worst outcome (z) will be chosen if voting is sophisticated! This situation, in which supporting the apparently favored outcome hurts, is called the **status-quo paradox**.

Clearly, the apparent advantage that the status quo enjoys in the absence of a Condorcet winner disappears if the voters are sophisticated. The strategic situation intervenes and causes them to reassess their sincere strategies in light of the favored position of the status quo. In so doing, they may be led to "gang up" against X, handing X a worse outcome than she would have achieved if she had not supported the status quo but some other alternative.

We stress that Y and Z do not form a coalition against X in the sense of coordinating their strategies and agreeing to act together in a way that can be enforced. Rather, they behave as isolated individuals. At most they could be said to form an "implicit coalition." Such a coalition does not imply communication between its members but simply choices based on their common perceived strategic interests. ■

So far we have used payoff matrices to describe games in strategic form. In these games, the row and column players' choices of strategies led to an outcome from which each player received a payoff. These strategy choices were assumed to be simultaneous, though mentally the players might eliminate them sequentially.

In the next example of a larger game, we start by assuming simultaneous choices and show what outcome would occur. Then we assume that the choices of the players need not be simultaneous; it is possible for one player to move first. We will use a "game tree" to analyze the *sequential choices* players can then make, as occurs when first you move, then I move, and so on, which are called *games in extensive form*. As we will see, the outcome in such a game may be wholly different from what it is in a game with simultaneous choices, which raises the question of which kind of game is the most realistic model of a situation.

EXAMPLE 10   A Truel

A *truel* is like a duel, except that there are three players. Each player can either fire, or not fire, his or her gun at either of the other two players. We assume the goal of each player is, *first*, to survive and, *second*, to survive with as few other players as possible. Each player has one bullet and is a perfect shot; no communication (for example, to pick out a common target) leading to a binding agreement with other players is allowed, making the game noncooperative. We will discuss the answers that simultaneous choices, on the one hand, and sequential choices, on the other, give to what is optimal for the players to do in the truel.

If choices are simultaneous, *at the start of play, each player will fire at one of the other two players, killing that player.*

Why will the players all fire at each other? Because their own survival does not depend an iota on what they do. Since they cannot affect what happens to themselves, but can affect how many others survive (the fewer the better, according to the postulated secondary goal), they should all blaze away at each other. Even if the rules of the play permitted shooting oneself, the primary goal of survival would preclude committing suicide. In fact, the players all have dominant strategies to shoot at each other, because whether or not a player survives—we will discuss shortly the probabilities of doing so—he or she does at least as well shooting an opponent.

The game, and optimal strategies in it, would change if the players (1) were allowed more options, such as to fire in the air and thereby disarm themselves, or (2) did not have to choose simultaneously but, instead, a particular order of play were specified. Thus, if the order of play were *A*, followed by *B* and *C* choosing simultaneously, followed by any player with a bullet remaining choosing, then *A* would fire in the air, and *B* and *C* would subsequently shoot each other. (*A* is no threat to *B* or *C*, so neither of the latter will fire at *A* and waste a bullet; on the other hand, if one of *B* or *C* did not fire immediately at the other, that player would not survive to get in the last shot, so they both fire.) Thus, *A* will be the sole survivor.

In 1992, a modified version of this scenario was played out in late-night television programming among the three major TV broadcasting networks of the time, with ABC's effectively going first with *Nightline*, its well-established news program, and CBS's and NBC's dueling about which host, David Letterman or Jay Leno, to choose for their entertainment shows. Regardless of their ultimate

David Letterman *(Black Star.)*

Jay Leno *(Black Star.)*

Ted Koppel *(Syracuse Newspapers/Dick Blume/The Image Works.)*

choices, ABC "won" when CBS and NBC were forced to divide the entertainment audience. In 2002, ABC, presumably to attract a younger audience than that which watched *Nightline*, attempted unsuccessfully to hire Letterman from CBS. Subsequently, Ted Koppel retired, but *Nightline* continues.

To return to the original game (all choose simultaneously), the players' strategies of all firing have two possible consequences: Either one player survives (even if two players fire at the same person, the third must fire at one of them, leaving only one survivor), or no player survives (if each payer fires at a different person). In either event, there is no guarantee of survival. In fact, if each player has an equal probability of firing at one of the two other players, the probability that any particular player will survive is only 0.25.

The reason is that if the three players are *A, B,* and *C, A* will be killed when *B* fires at him or her, *C* does, or both do. The only circumstance in which *A* will survive is if *B* and *C* fire at each other, which gives *A* one chance in four.

If choices are sequential, no player will fire at any other, so all will survive.

At the start of the truel, all the players are alive, which satisfies their primary goal of survival, though not their secondary goal of surviving with as few others as possible. Now assume that *A* contemplates shooting *B*, thereby reducing the number of survivors, and cannot fire into the air. But looking ahead, *A* knows that by firing first and killing *B*, he or she will be defenseless and be immediately shot by *C*, who will then be the sole survivor.

It is in *A*'s interest, therefore, not to shoot anybody at the start, and the same logic applies to each of the other players. Hence, everybody will survive, which is a happier outcome than when choices are simultaneous, in which case everyone's primary goal is not satisfied—or, quantitatively speaking, satisfied only 25% of the time. ■

While sequential choices produce a "happier" outcome, do they provide a plausible model of a strategic situation that mimics what people might actually think and do in such a situation? We believe that the players in the truel, artifi-cial as this kind of shoot-out may seem, would be motivated to think ahead, given the dire consequences of their actions. Therefore, they would hold their fire, knowing that if one fired first, he or she would be the next target.

In Figure 15.5, we show this logic somewhat more formally with a **game tree**, in which A has three strategies, as indicated by the three branches that sprout from A: not shoot ($\bar{S}$), shoot B ($S \to B$), or shoot C ($S \to C$). The latter two branches, in turn, give survivors C and B, respectively, two strategies: not shoot ($\bar{S}$) or shoot A ($S \to A$).

We assume that the players rank the outcomes as follows, which is consis-tent with their primary and secondary goals: 4 = best (lone survivor), 3 = next best (survivor with one other), 2 = next worst (survivor with two others), and 1 = worst (nonsurvivor). These payoffs are given for ordered triples (A, B, C); thus (3, 3, 1) indicates the next-best payoffs for A and B and the worst payoff for C.

Note that play necessarily terminates when there is only one survivor, as is the case at (1, 1, 4) and (1, 4, 1). To keep the tree simple, we assume that play also terminates when either A initially chooses $\bar{S}$, or B or C subsequently chooses $\bar{S}$, giving outcomes of (2, 2, 2), (3, 3, 1), and (3, 1, 3), respectively. Of course, we could allow the two or three surviving players in the latter cases to make subse-quent choices in an extended game tree, but this example is meant only to illus-trate the analysis of a game tree, not be the definitive statement on truel possi-bilities (more will be explored in the exercises).

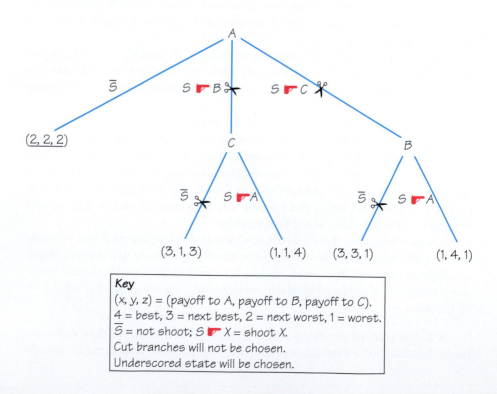

Key
(x, y, z) = (payoff to A, payoff to B, payoff to C).
4 = best, 3 = next best, 2 = next worst, 1 = worst.
$\bar{S}$ = not shoot; S ☛ X = shoot X.
Cut branches will not be chosen.
Underscored state will be chosen.

FIGURE 15.5 A game tree of a truel.

In a game in extensive form represented by a game tree, players work backward, starting the analysis at the bottom of the tree. By "bottom" we mean where play terminates; because this is where the tree branches out, the tree looks upside down in Figure 15.5. The players then work up the tree, using backward induction. **Backward induction** is a reasoning process in which players, working backward from the last possible moves in a game, anticipate each other's rational choices.

To illustrate, because C prefers $(1, 1, 4)$ to $(3, 1, 3)$, we indicate that C would not choose $\overline{S}$ by "cutting" this branch with a scissors. Similarly, B would not choose $\overline{S}$. Thus, if play got to the bottom of the tree, C would shoot A (choose $S \rightarrow A$) if C were the survivor, and B would shoot A (choose $S \rightarrow A$) if B were the survivor, following A's shooting C or B, respectively.

Moving up to the next level, A would know that if he or she chose $S \rightarrow B$, $(1, 1, 4)$ would be the outcome. If he or she chose $S \rightarrow C$, $(1, 4, 1)$ would be the outcome, making one or the other the outcome from the bottom level. Choosing between these two outcomes and $(2, 2, 2)$, A would prefer the latter, so A would cut the two branches, $S \rightarrow B$ and $S \rightarrow C$. Hence, A would choose $\overline{S}$, terminating play with nobody's shooting anybody else.

This, of course, is the conclusion we reached earlier, based on the reasoning that if A shot either B or C, he or she would end up dead, too. Because we could allow each player, like A, to choose among his or her three initial strategies in a $3 \times 3 \times 3$ game, and subsequently make moves and countermoves from the initial state (if feasible), the foregoing analysis applies to all players.

Underlying the completely different answers given by the simultaneous and sequential choices in a truel is a change in the rules of play. If play is sequential, the players do not have to fire simultaneously at the start, as assumed in a $3 \times 3 \times 3$ strategic-form game. Rather, a player who moves first (A in our example)—and then the later players—would not fire, given that play continues until all bullets are expended or nobody chooses to fire.

In the extensive-form analysis, we ask of each player (it need not be A): Given your present situation (all alive), and the situation you anticipate will ensure if you fire first, should you do so? Because each player prefers living to the state he or she would induce by being the first to shoot (certain death), no one shoots. This analysis suggests that truels might be more effective than duels in preventing the outbreak of conflict.

We will not try to develop this argument into a more general model. The main point is that a game tree allows for a look-ahead approach whereby players compare the present state with possible future states—perhaps several steps ahead—to determine which moves to make. These choices, as we have seen, may lead to radically different outcomes compared with those based on simultaneous choices.

15.5 Using Game Theory

Solving Games

Given any payoff matrix, the first thing we ask is whether it is zero-sum (or constant-sum). If so, we check to see whether it has a saddlepoint by determining the minimum number of each row and the maximum number of each column, as we did in several earlier examples. If the maximum of the row minima (maximin) is equal to the minimum of the column maxima (minimax), then the game has a saddlepoint. The resulting value, and the corresponding pure strategies, provide a solution to the game.

This value will appear in the payoff matrix as the smallest number in its row and the largest in its column. In the 3×3 location game in Example 1, this number was 5 (5000 feet).

Like our voting game, dominated strategies can successively be eliminated in the 3×3 location game. Thus, route B dominates route C, and highway 2 dominates highway 3. Having made these eliminations, highway 2 dominates highway 1. Having made this elimination, route B dominates route A. Thus, highway 2 and route B survive the successive eliminations, yielding the saddlepoint of 5. The successive-elimination procedure therefore provides an alternative method for finding the saddlepoint in this 3×3 location game. Unfortunately, it does not work to find the saddlepoint in *all* two-person zero-sum games larger than 2×2.

Recall that instead of eliminating dominated strategies in the 3×3 location game, we eliminated route B and highway 2, which dominated other strategies, to obtain the 2×2 restricted-location game in Table 15.3. In this game, there were no dominated strategies and, hence, no saddlepoint.

If a two-person zero-sum game does not have a saddlepoint—which was the case not only in the restricted-location game but also for matching pennies, the nonsymmetrical matching game, and the baseball duel—the solution will be in mixed strategies. To find the optimal mix in a 2×2 game, we calculate the expected value to a player from choosing its first strategy with probability p and its second with probability $1 - p$, assuming that the other player chooses its first pure strategy (yielding one expected value) and its second pure strategy (yielding another expected value).

Setting these two expected values equal to each other yields a unique value for p that gives the optimal mix, $(p, 1 - p)$, with which the player should choose its first and second strategies. Substituting the numerical solution of p back into either expected-value equation gives the value of the game, which each player can guarantee for itself, whatever strategy its opponent chooses.

Several general algorithms have been developed since 1945 to find mixed-strategy solutions to large constant-sum games. This work has mostly been done in the field of linear programming, using such algorithms as the simplex method of G. B. Dantzig and the more recent method of N. K. Karmarkar (see Chapter 4).

In variable-sum games, we also begin by successively eliminating dominated strategies, if there are any. The outcomes that remain do not depend on the numerical values we attach to them but only on their ranking from best to worst by the players, as illustrated in the three-person voting example.

Care must be taken in interpreting this solution, however. It began with the choice of a dominant strategy by X—and her elimination of her two dominated strategies. Presuming these eliminations, Y and Z were then able to eliminate their own dominated strategies in the first reduction, and Y in turn eliminated a dominated strategy in the second reduction, leading finally to the outcome z, supported by Y and Z.

This solution is a fairly demanding one, because it assumes considerable calculational abilities on the part of the players. Less demanding, of course, is that players simply choose their dominant strategies, as is possible in Prisoners' Dilemma, but of course games may not have such strategies.

In the game of Chicken, for example, neither player has a dominant (or dominated) strategy, so the game cannot be reduced. In such situations, we ascertain what outcomes are Nash equilibria. There are two (in pure strategies) in Chicken, suggesting that the only stable outcomes in this game occur when one player gives in and the other does not. In Prisoners' Dilemma, by comparison, the choice by the players of their dominant strategies singles out the mutual-defection outcome as the unique Nash equilibrium, which is worse for both players than the cooperative outcome.

In both Chicken and Prisoners' Dilemma, there seems no good reason for the choice of the (3, 3) cooperative outcome, at least if each game is played only once, because this outcome is not a Nash equilibrium. However, there is an alternative theory, called the *theory of moves*, that assumes different rules of play and renders the cooperative outcomes in both Prisoners' Dilemma and Chicken stable (at least if the players think ahead).

Theory of Moves

Theory of moves (TOM) is a dynamic theory that describes optimal strategic choices in strategic-form games in which the players, thinking ahead, can make moves and countermoves.

TOM is *dynamic* in the sense that it allows players, after choosing strategies that lead to an outcome in a payoff matrix, to make alternating subsequent moves and countermoves, with row's being able to move vertically by changing his row strategy, and column's being able to move horizontally by changing her column strategy. The reasoning that the players use in deciding whether to move or not move is backward induction, as illustrated earlier in the truel.

We informally illustrate this reasoning in Prisoners' Dilemma, starting from each of the four possible outcomes:

▶ If the play starts at the noncooperative (2, 2) outcome, players are stuck, no matter how far ahead they look, because as soon as one player departs, the other player, enjoying its best outcome at (4, 1) or (1, 4), will not move on. *Result:* The players stay at the noncooperative outcome.

▶ If play starts at the cooperative (3, 3) outcome, then neither player will defect, because if he or she does, the other player will also defect, and both players will end up worse off at (2, 2). Thinking ahead, therefore, neither player will defect. *Result:* The players stay at the cooperative outcome.

▶   If play starts at one of the (4, 1) or (1, 4) win–lose outcomes, the player doing best (4) will know that if he or she does not move to the cooperative (3, 3) outcome, his or her opponent will move to the noncooperative (2, 2) outcome, inflicting on the best-off player a next-worst (2) outcome. Therefore, it is in this player's interest—as well as the worst-off player's interest—that the best-off player act cooperatively and move first to (3, 3), anticipating that if he or she does not, the (2, 2) rather than the (3, 3) outcome will be chosen. *Result:* The best-off player will move to the cooperative outcome, where play will stop.

Thus, TOM does not predict unconditional cooperation in Prisoners' Dilemma but, instead, makes it a function of the starting point of play. As in the truel, a change in rules from simultaneous choices to sequential choices can induce cooperation.

The calculations we have described for Prisoners' Dilemma, which are grounded in backward induction and could be formalized by a game tree, are not, we believe, beyond the ability of most players. Farsighted players *can* escape the dilemma in Prisoners' Dilemma, provided play begins at a state other than the noncooperative one. But we must be careful in interpreting this result: With the change in the rules, the original game changes, so this "solution" to Prisoners' Dilemma is not for the original dilemma. Similar reasoning in Chicken indicates that if play starts at the cooperative (3, 3) outcome, players will stay at this outcome, but the reasoning in this game is somewhat more complicated than in Prisoners' Dilemma.

Practical Applications

The element of surprise, as captured by mixed strategies, is essential in many encounters. For example, mixed strategies are used in various inspection procedures and auditing schemes to deter potential cheaters. By making inspection or auditing choices random, they are rendered unpredictable.

Police and regulatory agencies monitor certain activities to check for faults, errors, or illegal actions. Investigators who conduct surveillance include FBI agents, customs agents, bank auditors, insurance investigators, and quality-control experts. The National Bureau of Standards is responsible for monitoring the accuracy of measuring instruments and for maintaining reliable standards. The Nuclear Regulatory Agency demands an accounting of dangerous nuclear material as part of its safeguards program. The Internal Revenue Service attempts to identify those cheating on taxes.

Military or intelligence services may wish to intercept a weapon hidden among many decoys or plant a secret agent disguised to look like a respectable individual. Because it is prohibitively expensive to check the authenticity of each and every possible item or person, efficient methods must be used to check for violations. Both optimal detection and optimal concealment strategies can be modeled as a game between an inspector trying to increase the probability of detection and a violator trying to evade detection. Since the World Trade Center attack on September 11, 2001, we have seen government agencies take much stronger measures to prevent such evasion.

Some inspection games are constant-sum: The violator "wins" when the evasion is successful and "loses" when it is not. On the other hand, cheating on arms-control agreements may well be variable-sum if both the inspector and the cheater would prefer that no cheating occur to there being cheating and public disclosure of it. The latter could be an embarrassment to both sides, especially if it undermines an arms-control agreement both sides wanted and the cheating is minor.

We alluded earlier to the strategy of bluffing in poker, which is used to try to keep the other player or players guessing about the true nature of one's hand. The optimal probability with which one should bluff can be calculated in a particular situation (see Exercise 17). Besides poker, bluffing is common in many bargaining situations, whereby a player raises the stakes (for instance, labor threatens a strike in labor–management negotiations), even if it may ultimately have to back down if its "hand is called."

Perhaps the greatest value of game theory is the framework it provides for understanding the rational underpinnings of conflict in the world today. As a case in point, a confrontation over the budget between the Democratic President Bill Clinton and the Republican Congress resulted in the shutdown of part of the federal government on two occasions between November 1995 and January 1996. Many government workers were frustrated in not being able to do their jobs, even though they knew they would be paid for not working, not to mention the many citizens who were either greatly hurt or substantially inconvenienced by the shutdown. Viewed as a game of Chicken, in which each side wanted not only to get its way for the moment but also to establish a precedent for the future, this conflict was not so foolish as it might seem at first glance.

The Northern Ireland conflict—settled in principle by a peace agreement in April 1998 after 30 years of fighting and more than 3200 deaths but which has had its ups and downs since—can be viewed in similar terms. As still another example, the constant price wars among the airlines suggest competitors caught up in a Prisoners' Dilemma, in which they all suffer from lower fares but cannot avoid their dominant strategies of not cooperating, perhaps to try to seize a quick advantage or hurt the competition even more (and possibly even eliminate a competitor). This has led to the bankruptcy of such major airlines as United Airlines and US Airways.

To be sure, if the airlines cooperate by colluding on fares, which is definitely not advantageous to consumers, the consumers may be thought of as a collective player whose interests are represented by the government. The government can prosecute the airlines for price fixing, or the consumers themselves can file a class-action suit in a "larger" game. The government has frequently been involved in antitrust suits (for example, against Microsoft) and in setting the rules for auctions of airwaves, in which telecommunication companies—advised by game theorists—have paid billions of dollars for the right to construct cellular phone and other networks.

All in all, game theory offers fundamental insights into conflicts at all levels, especially *seemingly* irrational features which, on second look, are often well conceived and effective.

:::::::::: REVIEW VOCABULARY

Backward induction A reasoning process in which players, working backward from the last possible moves in a game, anticipate each other's rational choices.

Chicken A two-person variable-sum symmetric game in which each player has two strategies: to swerve to avoid a collision, or not to swerve and cause a collision if the opponent has not swerved. Neither player has a dominant strategy; the compromise outcome, in which both players swerve, is not a Nash equilibrium, but the two outcomes in which one player swerves and the other does not are Nash equilibria.

Condorcet winner A candidate that defeats all others in separate pairwise contests.

Constant-sum game A game in which the sum of payoffs to the players at each outcome is a constant, which can be converted to a zero-sum game by an appropriate change in the payoffs to the players that does not alter the strategic nature of the game.

Dominant strategy A strategy that is sometimes better and never worse for a player than every other strategy, whatever strategies the other players choose.

Dominated strategy A strategy that is sometimes worse and never better for a player than some other strategy, whatever strategies the other players choose.

Expected value E If each of the n payoffs, s_1, s_2, . . . , s_n, occurs with respective probabilities p_1, p_2, . . . , p_n, then the expected value E is

$$E = p_1 s_1 + p_2 s_2 + \cdots + p_n s_n$$

where $p_1 + p_2 + \cdots + p_n = 1$ and $p_i \geq 0$ ($i = 1$, 2, . . . , n).

Fair game A zero-sum game is fair when the (expected) value of the game, obtained by using optimal strategies (pure or mixed), is zero.

Game tree A symbolic tree, based on the rules of play in a game, in which the vertices, or nodes, of the tree represent choice points, and the branches represent alternative courses of action that the players can select.

Maximin In a two-person zero-sum game, the largest of the minimum payoffs in each row of a payoff matrix.

Maximin strategy In a two-person zero-sum game, the pure strategy of the row player corresponding to the maximin in a payoff matrix.

Minimax In a two-person zero-sum game, the smallest of the maximum payoffs in each column of a payoff matrix.

Minimax strategy In a two-person zero-sum game, the pure strategy of the column player corresponding to the minimax in a payoff matrix.

Minimax theorem The fundamental theorem for two-person constant-sum games, stating that there always exist optimal pure or mixed strategies that enable the two players to guarantee the value of the game.

Mixed strategy A strategy that involves the random choice of pure strategies, according to particular probabilities. A mixed strategy of a player is optimal if it guarantees the value of the game.

Nash equilibrium Strategies associated with an outcome such that no player can benefit by choosing a different strategy, given that the other players do not depart from their strategies.

Nonsymmetrical game A two-person constant-sum game in which the row player's gains are different from the column player's gains, except when there is a tie.

Ordinal game A game in which the players rank the outcomes from best to worst.

Partial-conflict game A variable-sum game in which both players can benefit by cooperation but may have strong incentives not to cooperate.

Payoff matrix A rectangular array of numbers. In a two-person game, the rows and columns correspond to the strategies of the two players, and the numerical entities give the payoffs to the players when these strategies are selected.

Plurality procedure A voting procedure in which the alternative with the most votes wins.

Prisoners' Dilemma A two-person variable-sum symmetric game in which each player has two strategies, cooperate or defect. Cooperate dominates defect for both players, even though the mutual-defection outcome, which is the unique Nash equilibrium in the game, is worse for both players than the mutual-cooperation outcome.

Pure strategy A course of action a player can choose in a game that does not involve randomized choices.

Rational choice A choice that leads to a preferred outcome.

Saddlepoint In a two-person constant-sum game, the payoff that results when a row minimum and a column maximum are the same, which is the value of the game. The saddlepoint has the shape of a saddle-shaped surface and is also a Nash equilibrium.

Sincere voting Voting for one's most-preferred alternative in a situation.

Sophisticated voting Voting that involves the successive elimination of dominated strategies by voters.

Status-quo paradox The status quo is defeated by another alternative, even if there is no Condorcet winner, when voters are sophisticated.

Strategy One of the courses of action a player can choose in a game; strategies are mixed or pure, depending on whether they are selected in a randomized fashion (mixed) or not (pure).

Theory of moves (TOM) A dynamic theory that describes optimal choices in strategic-form games in which players, thinking ahead, can make moves and countermoves.

Total-conflict game A zero-sum or constant-sum game, in which what one player wins the other player loses.

Value The best outcome that both players can guarantee in a two-person zero-sum game. If there is a saddlepoint, this is the value. Otherwise, it is the expected payoff resulting when the players choose their optimal mixed strategies.

Variable-sum game A game in which the sum of the payoffs to the players at the different outcomes varies.

Zero-sum game A constant-sum game in which the payoff to one player is the negative of the payoff to the other player, so the sum of the payoffs to the players at each outcome is zero.

 # SKILLS CHECK

1. In the following two-person zero-sum game, the payoffs represent gains to row player I and losses to column player II.

$$\begin{bmatrix} 3 & 7 & 2 \\ 8 & 5 & 1 \\ 6 & 9 & 4 \end{bmatrix}$$

What is the maximin strategy for player I?

(a) Play the first row.

(b) Play the second row.

(c) Play the third row.

2. In the following two-person zero-sum game, the payoffs represent gains to row player I and losses to column player II.

$$\begin{bmatrix} 3 & 7 & 2 \\ 8 & 5 & 1 \\ 6 & 9 & 4 \end{bmatrix}$$

What is the minimax for player II?

(a) Play the first column.
(b) Play the second column.
(c) Play the third column.

3. In a two-person zero-sum game, suppose the first player chooses the third row as the maximin strategy and the second player chooses the first column as the minimax strategy. Based on this information, which of the following statements is true?

(a) This game definitely has no saddlepoint.
(b) This game may or may not have a saddlepoint.
(c) This game definitely has a saddlepoint.

Consider the following three two-person zero-sum games, wherein the payoffs represent gains to the row player I and losses to the column player II:

$$\begin{bmatrix} 3 & 6 \\ 5 & 4 \end{bmatrix}$$

$$\begin{bmatrix} -1 & 3 \\ 2 & 0 \end{bmatrix}$$

$$\begin{bmatrix} 6 & 5 & 6 & 5 \\ 1 & 4 & 2 & -1 \\ 8 & 5 & 7 & 5 \\ 0 & 2 & 6 & 2 \end{bmatrix}$$

4. Which of these games has a saddlepoint(s)?

(a) The first game
(b) The second game
(c) The third game

5. In which two games does neither player have a dominant strategy?

(a) The first two games
(b) The last two games
(c) The first and third games

6. How many strategies of player I are dominated in the third game?

(a) One
(b) Two

(c) Three

7. In the game of matching pennies, player I wins a penny if the coins match; player II wins a penny if the coins do not match. Given this information, it can be concluded that the 2×2 matrix that represents this game

(a) has two -1's and two 1's.
(b) has four 1's.
(c) has four -1's.

8. If a game has a saddlepoint, then

(a) it is the value of the game.
(b) it is never the maximin strategy.
(c) it is never the minimax strategy.

9. Which of these games does not have a saddlepoint?

(a) Tic-tac-toe
(b) Chess
(c) Poker

10. A mixed strategy uses randomization

(a) to choose among several pure strategies each time a round of the game is played.
(b) to select a pure strategy for the game.
(c) to anticipate the opponent's next move.

11. A game is fair when

(a) its value is 0.
(b) it requires a mixed strategy.
(c) its payoff is unlimited.

12. A game is symmetrical when

(a) each player has an equal chance to win.
(b) each player receives a winning payoff of the same value.
(c) either player can take the first turn.

13. In the following game of batter-versus-pitcher in baseball, the batter's batting averages are given in the game matrix.

		Pitcher	
		Fastball	Curve
Batter	Fastball	0.400	0.200
	Curve	0.100	0.500

What is the pitcher's optimal strategy?

(a) Throw more curves than fastballs.
(b) Throw more fastballs than curves.
(c) Throw about the same number of curves and fastballs.

14. In the following game of batter-versus-pitcher in baseball, the batter's batting averages are given in the game matrix.

		Pitcher	
		Fastball	Curve
Batter	Fastball	0.400	0.200
	Curve	0.100	0.500

What is the batter's optimal strategy?

(a) Expect more curves than fastballs.
(b) Expect more fastballs than curves.
(c) Expect about the same number of curves and fastballs.

15. Consider the following partial-conflict game, played in a noncooperative manner.

		Player II	
		Choice A	Choice B
Player I	Choice A	(4, 4)	(1, 3)
	Choice B	(3, 1)	(2, 2)

What outcomes constitute a Nash equilibrium?

(a) Only when both players select A
(b) Only when both players select A or both select B
(c) Only when one player selects A and the other selects B

16. Consider the game played between the opposing goalie and a soccer player who, after a penalty, is allowed a free kick. The kicker can elect to kick toward one of the two corners of the net, or else aim for the center of the goal. The goalie can decide to commit in advance (before the kicker's kick) to either one of the sides, or else remain in the center until he sees the direction of the kick. This two-person zero-sum game can be represented as follows, wherein the payoffs are the probability of scoring a goal:

		Goalie		
		Breaks left	Remains center	Breaks right
Kicker	Kicks left	0.5	0.9	0.9
	Kicks center	0.1	0	0.1
	Kicks right	0.9	0.9	0.5

If we assume that decisions between the left or right side are made symmetrically (with equal probabilities), then this game can be represented by a 2×2 matrix, where $0.7 = (\frac{1}{2})(0.5) + (\frac{1}{2})(0.9)$:

		Goalie	
		Remains center	Breaks side
Kicker	Kicks center	0	1
	Kicks side	0.9	0.7

The optimal strategies of the kicker and goalie are to

(a) more often kick center and remain center.
(b) more often kick side and break side.
(c) choose center and side equally often.

17. In the following game, player I has the preferences of the row player in Prisoners' Dilemma, and player II has the preferences of the column player in Chicken.

		Player II	
		Choice A	Choice B
Player I	Choice A	(3, 3)	(1, 4)
	Choice B	(4, 2)	(2, 1)

Does the player with a dominant strategy benefit from having it?

(a) Yes
(b) No

(c) It doesn't make any difference—both players do equally well from choosing their strategies associated with the Nash-equilibrium outcome.

18. Games of partial conflict are examples of

(a) fair games.
(b) zero-sum games.
(c) games in which cooperation yields greater payoffs.

19. A game tree is used to

(a) determine the possible strategies of a player.
(b) anticipate each other's choices through backward induction.
(c) plan a deception strategy.

20. Theory of moves is dynamic in that players

(a) randomly change their strategy.
(b) can consider the opponent's moves to change their strategy plans.
(c) move in alternation.

EXERCISES   ■ Challenge   ◆ Discussion

Total-Conflict Games

Consider the following five two-person zero-sum games, wherein the payoffs represent gains to the row player I and losses to the column player II:

1. $\begin{bmatrix} 6 & 5 \\ 4 & 2 \end{bmatrix}$

2. $\begin{bmatrix} 0 & 3 \\ -5 & 1 \\ 1 & 6 \end{bmatrix}$

3. $\begin{bmatrix} -2 & 3 \\ 1 & -2 \end{bmatrix}$

4. $\begin{bmatrix} 13 & 11 \\ 12 & 14 \\ 10 & 11 \end{bmatrix}$

5. $\begin{bmatrix} -10 & -17 & -30 \\ -15 & -15 & -25 \\ -20 & -20 & -20 \end{bmatrix}$

(a) Which of these games have saddlepoints?
(b) Find the maximin strategy of player I, the minimax strategy of player II, and the value for those games given in part (a).
(c) List dominated strategies in these games that the players should avoid because the resulting payoffs are worse than those for some alternative strategy.

Solve the following three games of batter-versus-pitcher in baseball, wherein the pitcher can throw one of two pitches and the batter can guess either of these two pitches. The batter's batting averages are given in the game matrix.

6.

		Pitcher	
		Fastball	Curve
Batter	Fastball	0.300	0.200
	Curve	0.100	0.400

7.

		Pitcher	
		Fastball	Knuckleball
Batter	Fastball	0.500	0.200
	Knuckleball	0.200	0.300

8.

		Pitcher	
		Blooperball	Knuckleball
Batter	Blooperball	0.400	0.200
	Knuckleball	0.250	0.250

9. A businessperson has the choice of either not cheating on his income tax or cheating and making $1000 if not audited. If caught cheating, he will pay a fine of $2000 in addition to the $1000 he owes. He feels good if he does not cheat and is not audited (worth $100). If he does not cheat and is audited, he evaluates this outcome as −$100 (for the lost day). Viewing the game as a two-person zero-sum game between the businessperson and the tax agency, what are the optimal mixed strategies for each player and the value of the game?

10. When it is third down and short yardage to go for a first down in American football, the quarterback can decide to run the ball or pass it. Similarly, the other team can commit itself to defend more heavily against a run or a pass. This can be modeled as a 2×2 matrix game, wherein the payoffs are the probabilities of obtaining a first down. Find the solution of this game.

		Defense	
		Run	Pass
Offense	Run	0.5	0.8
	Pass	0.7	0.2

11. You have the choice of either parking illegally on the street or parking in the lot and paying $16. Parking illegally is free if the police officer is not patrolling, but you receive a $40 parking ticket if she is. However, you are peeved when you pay to park in the lot on days when the officer does not patrol, and you are willing to assess this outcome as costing $32 ($16 for parking plus $16 for your time, inconvenience, and grief). It seems reasonable to assume that the police officer ranks her preferences in the order (1) giving you a ticket, (2) not patrolling with you parked in the lot, (3) patrolling with you in the lot, and (4) not patrolling with you parked illegally.

(a) Describe this as a matrix game, assuming that you are playing a zero-sum game with the officer.
(b) Solve this matrix game for its optimal mixed strategies and its value.
◆ **(c)** Discuss whether it is reasonable or not to assume that this game is zero-sum.
◆ **(d)** Assuming that you play this parking game each working day of the year, how do you implement an optimal mixed strategy?

12. Describe how a pure strategy for a player in a matrix game can be considered as merely a special case of a mixed strategy.

■ **13.** (a) Describe in detail *one* pure strategy for the player who moves first in the game of tic-tac-toe. This strategy must tell how to respond to all possible moves of the other player. (*Hint:* You may

wish to make use of the symmetry in the 3×3 grid in this game; that is, there are one "center" box, four "corner" boxes, and four "side" boxes.)
(b) Is your strategy optimal in the sense that it will guarantee the first player a tie (and possibly a win) in the game?

14. In the matching-pennies example, consider the case where player I favors heads H over tails T. For example, assume that player I plays H three-fourths of the time and T only one-fourth of the time—a nonoptimal mixed strategy. What should player II do if he knows this?

15. Assume in the nonsymmetrical matching example that player II is using the nonoptimal mixed strategy $(p, 1-p) = (\frac{1}{2}, \frac{1}{2})$; that is, he is playing H and T with the same frequency. What should player I do in this case if she knows this?

16. You plan to manufacture a new product for sale next year, and you can decide to make either a small quantity, in anticipation of a poor economy and few sales, or a large quantity, hoping for brisk sales. Your expected profits are indicated in the following table.

		Economy	
		Poor	Good
Quantity	Small	$500,000	$300,000
	Large	$100,000	$900,000

If you want to avoid risk and believe that the economy is playing an optimal mixed strategy against you in a two-person zero-sum game, then what is your optimal mixed strategy and the resulting expected value? Discuss some alternative ways to go about making your decision.

■ **17.** Consider the following miniature poker game with two players, I and II. Each antes $1. Each player is dealt either a high card H or a low card L, with probability $\frac{1}{2}$. Player I then folds or bets $1. If player I bets, then player II either folds, calls, or raises $1. Finally, if II raises, I either folds or calls.

Most choices by the players are rather obvious, at least to anyone who has played poker: If either player holds *H*, that player always bets or raises if he or she gets the choice. The question remains of how often one should bluff— that is, continue to play (by calling or raising) while holding a low card in the hope that one's opponent also holds a low card.

This poker game can be represented by the following matrix game, wherein the payoffs are the expected winnings for player I (depending upon the random deal) and the dominated strategies have been eliminated.

		Player II (when holding L)		
		Folds	Calls	Raises
Player I (when holding L)	Folds initially	−0.25	0	0.25
	Bets first, folds later	0	0	−0.25
	Bets first, calls later	−0.25	−0.25	0

(a) Are there any strategies in this matrix game that a player should avoid playing?
(b) Solve this game.
(c) Which player is in the more favored position?
(d) Should one ever bluff?

■ **18.** (a) Describe in detail *one* pure strategy for the player who moves second in the game of tic-tac-toe.

(b) Is your strategy in part (a) optimal in the sense that it will guarantee the second player a tie (and possibly a win) in the game?

19. On an overcast morning, deciding whether to carry your umbrella can be viewed as a game between yourself and nature as follows:

		Weather	
		Rain	No rain
You	Carry umbrella	Stay dry	Lug umbrella
	Leave it home	Get wet	Hands free

Let's assume that you are willing to assign the following numerical payoffs to these outcomes, and that you are also willing to make decisions on the basis of expected values (that is, average payoffs):

(Carry umbrella, rain) = −2
(Carry umbrella, no rain) = −1
(Leave it home, rain) = −5
(Leave it home, no rain) = 3

(a) If the weather forecast says there is a 50% chance of rain, should you carry your umbrella or not? What if you believe there is a 75% chance of rain?
(b) If you are conservative and wish to protect against the worst case, what pure strategy should you pick?
(c) If you are rather paranoid and believe that nature will pick an optimal strategy in this two-person zero-sum game, then what strategy should you choose?
(d) Another approach to this decision problem is to assign payoffs to represent what your *regret* will be after you know nature's decision. In this case, each such payoff is the best payoff you could have received under that state of nature, minus the corresponding payoff in the previous table:

		Weather	
		Rain	No rain
You	Carry umbrella	$0 = (-2) - (-2)$	$4 = 3 - (-1)$
	Leave it home	$3 = (-2) - (-5)$	$0 = 3 - 3$

What strategy should you select if you wish to minimize your maximum possible regret?

Partial-Conflict Games

Consider the following three two-person variable-sum games. Discuss the players' possible behavior when these games are played in a noncooperative manner (with no prior communication or agreements). The first payoff is to the row player; the second, to the column player. Are the Nash equilibria in these games sensible? Why or why not?

20.

	Player II	
Player I	(4, 4) (1, 3)	
	(3, 1) (2, 2)	

21. Battle of the sexes:

		She buys a ticket for:	
		Boxing	Ballet
He buys a ticket for:	Boxing	(4, 3)	(2, 2)
	Ballet	(1, 1)	(3, 4)

22.

	Player II	
Player I	(2, 1) (4, 2)	
	(1, 4) (3, 3)	

23.

	Player II	
Player I	(2, 4) (4, 3)	
	(1, 2) (3, 1)	

24.

	Player II	
Player I	(3, 4) (2, 3)	
	(1, 2) (4, 1)	

◆ **25.** Assume that two countries in an arms race assign points to all their own weapons so that the total for each is 1000. Each side can then designate weapons of the *other* side, totaling 100 points, that must be eliminated in the next year, thereby effecting a 10% reduction. Would these countries have any reason to lie about how they value their own weapons? Is this procedure practical as an arms-reduction scheme?

Larger Games

26. For the preferences of the players given in the text—*xyz* for *X*, *yzx* for *Y*, and *zxy* for *Z*—verify that the strategy choices of *x* by *X*, *y* by *Y*, and *x*

by *Z* are a Nash equilibrium. Does this equilibrium seem to you defensible as the social choice by the voters? Under what circumstances might the voters choose these strategies rather than their sophisticated strategies?

■ **27.** Under a voting system called approval voting, a voter can vote for as many alternatives as he or she wishes. (If there are three alternatives, the only undominated strategies of a voter under approval·voting are to vote for his or her best, or two best, choices). If voters *X*, *Y*, and *Z* have paradox-of-voting preferences of *xyz*, *yzx*, and *zxy*, and *X* is the status quo, show that *x* is the sophisticated outcome, obtained by all voters voting for their two best choices.

■ **28.** Show by example that approval voting is not immune to the status-quo paradox.

29. Odd and Even play Low Person Wins, whose rules are as follows:

(a) Odd announces an odd number between 1 and 5 (inclusive).
(b) Independently, Even announces an even number between 2 and 6 (inclusive).
(c) Whoever announces the lower number gets *twice* this number as its payoff.
(d) Whoever announces the higher number gets the *lower* number as its payoff.

What is the Nash equilibrium of this game, based on the successive elimination of dominated strategies? Is there another Nash equilibrium?

■ **30.** Find a two-person zero-sum game with a saddlepoint in which the successive elimination of dominated strategies does *not* lead to the saddlepoint (unlike the 3 × 3 location game; you may restrict yourself to 3 × 3 games).

◆ **31.** Why will the first player to act in a truel shoot in the air (if this option is allowed by the rules)? Is this choice optimal if a second player should succeed in firing in the air at the same time?

32. In a sequential truel with no firing in the air allowed, suppose *A*, who hates *B*, goes first; *B*, who hates *C*, goes second; *C*, who hates *A*, goes third. (If a player fires, he will shoot only his *antagonist*–the player he hates.)

(a) Which player is in the best position, and why?
(b) Does the outcome change if *B* hates *A* rather than *C*?

Answer these questions if each player can take only one turn (if alive); and if, after one round, the game continues (if there is more than one player alive) and players can take more than one turn.

33. If a fourth player is added to the original truel, show that every player will have an incentive to shoot another player (as in a duel).

34. Extend the game tree of the truel in Figure 15.5 to allow the additional possibility that if *A* does not shoot initially, then *B* has the choice of shooting or not shooting *C*. Will *A*, in fact, not shoot initially, and will *B* then shoot *C*?

35. Extend the game tree in Exercise 34 to still another level to allow for the possibility that if *A* does not shoot initially, and *B* shoots *C*, then *A* has the choice of shooting or not shooting *B*. What will happen in this case?

36. Change Exercise 35 to allow for the possibility that if *A* does not shoot initially, and *B* shoots *or does not shoot C*, then *A* has the choice of shooting or not shooting *B*. What will happen in this case?

◆ **37.** What general conclusions would you draw in light of your answers to Exercise 34, 35, and 36?

APPLET EXERCISES

To do these exercises, go to www.whfreeman.com/fapp7e.

What happens to the value of a game if you or your opponent deviates from the optimal strategy?

Can you exploit such deviations by your opponent to your advantage? Explore these possibilities in the *Game Theory* applet.

WRITING PROJECTS

1. Consider a conflict that you, personally, had–with a parent, a boss, a girlfriend or boyfriend, or some other acquaintance–in which each of you had to make a choice without being sure of what the other person would do. What strategies did you seriously consider adopting, and what options do you think the other person considered? What plausible outcomes do you think each set of strategy choices would have led to? How would you rank these outcomes from best to worst, and how do you think the other player would have ranked them? In two to three pages, analyze the resulting game, and state whether you think you and the other person made optimal choices. If not, what upset your or the other person's rationality?

2. It is sometimes argued that game theory does not take account of the (irrational?) emotions of people, such as anger, jealousy, or love. What is your opinion about this question? In one to two pages, give an example, real or hypothetical, that supports your position, paying particular attention to whether the players acted consistently with, or contrarily to, their preferences.

3. In tennis, one player often prefers to play from the baseline while her opponent prefers a serve-and-volley game (likes to come to the net). The baseline player attempts to hit passing shots. This player has a choice of hitting "down the line" or "crosscourt." The net player must often correctly guess in which direction the ball will go to cover the shot. In one to two pages, formulate this situation as a duel game and discuss appropriate strategies for the players.

4. Quentin Tarantino's films *Reservoir Dogs* (1992) and *Pulp Fiction* (1994) both have truels, but the choices that the characters make in each are completely different. Does the truel analysis offer any insight into why? Discuss in one to two pages.

5. In one to two paragraphs, discuss the relationship between the status-quo paradox and the chair's paradox (section 10.4).

SUGGESTED READINGS

AUMANN, ROBERT J., and SERGIU HART, eds. *Handbook of Game Theory with Economic Applications*, Elsevier, Amsterdam, 1992 (vol. 1), 1994 (vol. 2), 2002 (vol. 3). A comprehensive treatment of game theory and its applications, developed in long chapters written by leading experts.

BRAMS, STEVEN J. *Theory of Moves*, Cambridge University Press, New York, 1994. Describes in detail the theory of moves and applies it to a wide variety of conflicts.

DIXIT, AVINASH, and SUSAN SKEATH. *Games of Strategy*, 2nd ed., Norton, New York, 2004. An excellent game-theory text that requires only a minimal mathematical background.

NASAR, SYLVIA. *A Beautiful Mind*, Simon & Shuster, New York, 1998. A biography of John Nash that is also a fascinating account of the early history of game theory. A fictionalized version of this biography was made into a movie in 2001.

OSBORNE, MARTIN J. *An Introduction to Game Theory*. Oxford, New York, 2004. A fine intermediate text with several interesting applications.

SUGGESTED WEB SITES

www.chass.utoronto.ca/~osborne Martin Osborne's home page (game theory)

www.economics.harvard.edu/~aroth/alroth.html Alvin Roth's Game Theory and Experimental Economics Page

www.gametheory.net is the most comprehensive general Web site.

75468 61617

The Digital Revolution

I dentification numbers are ubiquitous in modern society. Unlike Social Security numbers and serial numbers on currency, today's identification numbers are more sophisticated than those introduced before the advent of fast, inexpensive computers. Modern numbers have a built-in "check" to partially ensure that the numbers have been correctly scanned or entered into a computer.

Identification numbers are examples of codes. A code is a group of symbols that represent information. Codes existed thousands of years ago: hieroglyphics, the Greek alphabet, and Roman numerals. Many codes have been developed for a particular application: musical scores, the Morse code, and the "genetic code" used to describe the makeup of DNA. In the past 50 years, codes have been invented for storing, securing, and transmitting information. In this part of the text we will examine some of the ways that information is coded. ■

Identification Numbers

Modern identification numbers serve at least two functions. An identification number should unambiguously identify the person or thing with which it is associated. Less obvious is a "self-checking" aspect of the number.

16.1 Check Digits

Look at the ISBN printed on the back cover of this book. The number 0-7167-5965-9 (0-7167-6901-8 for the paperback version) is a **code**. It distinguishes this book from all others. The last digit 9 is there solely to detect errors that may occur when the ISBN is entered into a computer. Look at the bottom of the airline ticket shown in Figure 16.1. Notice the letters CK (for "check") above the last digit of the stock control number and above the last digit of the document number. These digits are also there for the purpose of error detection. Grocery items, credit cards, overnight mail, magazines, personal checks, traveler's checks, soft-drink cans, automobiles, and many other items you encounter daily have identification numbers that code data and include an extra digit called a **check digit** for error detection. In this chapter, we examine some of the methods used to assign identification numbers and check digits.

Let's begin by considering the U.S. Postal Service money order shown in Figure 16.2. The first 10 digits of the 11-digit number 63024383845 simply identify the money order. The last digit, 5, serves as an **error-detecting** mechanism. Let's see how this mechanism works. The eleventh (last) digit of a Postal Service money order number is the remainder obtained when the sum of the first 10 digits of the number is divided by 9. In our example, the last digit is 5 because $6 + 3 + 0 + 2 + 4 + 3 + 8 + 3 + 8 + 4 = 41$ and the remainder when 41 is divided by 9 is 5.

FIGURE 16.1 Airline ticket with identification number 127881879532 and check digit 1; stock control number 3026164775 and check digit 4.

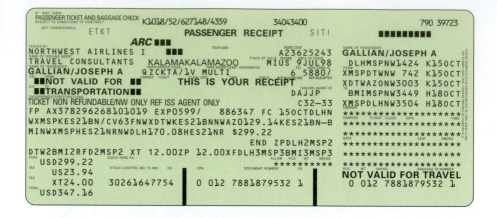

Now suppose that instead of the correct number, the number 63054383845 (an error in the fourth position) were entered into a computer programmed for error detection in money orders. The machine would divide the sum of the first 10 digits of the entered number, 44, by 9 and obtain a remainder of 8. Because the last digit of the entered number is 5 rather than 8, the entered number cannot be correct. This crude method of error detection will not detect the mistake of replacing a 0 with a 9, or vice versa. Nor will it detect the transposition of digits, such as 63204383845 instead of 63024383845 (the digits in positions three and four have been transposed).

American Express traveler's checks, VISA traveler's checks, and Euro banknotes (see Figure 16.3) also use a check digit determined by division by 9. In these cases, the check digit is chosen so that the sum of the digits, including the check digit, is evenly divisible by 9.

EXAMPLE 1   The American Express Travelers Cheque

The American Express Travelers Cheque with the identification number 387505055 has check digit 7 because $3 + 8 + 7 + 5 + 0 + 5 + 0 + 5 + 5 = 38$ and $38 + 7$ is evenly divisible by 9. ■

FIGURE 16.2 Money order with identification number 6302438384 and check digit 5. The check digit is the remainder after dividing the sum of the digits by 9.

FIGURE 16.3 A five-euro note with identification number X1183806538 and check digit 4. To calculate the check digit, the letter (which indicates in which country the note was issued) is converted to a number. The check digit is then chosen so that the sum of all the digits, including the check digit, is evenly divisible by 9. For the note shown here, the letter X is converted to 7.

The scheme used on airline tickets, UPS packages, and Avis and National rental cars assigns the remainder after division by 7 of the number itself as the check digit (see Figure 16.1) rather than dividing the sum of the digits by 7. For example, the check digit for the number 540047 is 4 because $540047 = 7 \times 77149 + 4$. This method will not detect the substitution of 0 for a 7, 1 for an 8, 2 for a 9, or vice versa. However, unlike the Postal Service method, it will detect transpositions of adjacent digits with the exceptions of the pairs 0, 7; 1, 8; and 2, 9. For example, if 5400474 were entered into a computer as 4500474 (the first two digits are transposed), the machine would determine that the check digit should be 3 since $450047 = 7 \times 64292 + 3$. Because the last digit of the entered number is not 3, the error has been detected.

The scheme used on grocery products, the **Universal Product Code (UPC)**, is more sophisticated. Consider the number 0 38000 00127 7 found on the bottom of a box of cornflakes. The first digit identifies a broad category of goods, the next five digits identify the manufacturer, the next five identify the product, and the last is a check digit. Suppose this number were entered into a computer as 0 58000 00127 7 (a mistake in the second position). How would the computer recognize the mistake?

For any UPC number $a_1a_2a_3a_4a_5a_6a_7a_8a_9a_{10}a_{11}a_{12}$, the computer is programmed to carry out the following computation: $3a_1 + a_2 + 3a_3 + a_4 + 3a_5 + a_6 + 3a_7 + a_8 + 3a_9 + a_{10} + 3a_{11} + a_{12}$. If the result doesn't end with a 0, the computer knows the entered number is incorrect.

For the incorrect cornflakes number, we have $3 \cdot 0 + 5 + 3 \cdot 8 + 0 + 3 \cdot 0 + 0 + 3 \cdot 0 + 0 + 3 \cdot 1 + 2 + 3 \cdot 7 + 7 = 62$. Because 62 doesn't end with 0, the error is detected. Notice that had we used the correct digit 3 in the second position instead of 5, the sum would have ended in a 0 as it should. This simple scheme detects *all* single-position errors and about 89% of all other kinds of errors.

Beginning in January 2005, U.S. retailers were required to have software that could read the 12-digit UPC code used in the United States and the 13-digit

European Article Number (EAN) code used in Europe. This change paves the way for the 13-digit EAN to become the worldwide standard. Existing UPC numbers will be converted to EAN numbers by adding an extra 0 at the beginning. The check digit for a 13-digit EAN number $a_1a_2a_3a_4a_5a_6a_7a_8a_9a_{10}a_{11}a_{12}a_{13}$ is selected so that $a_1 + 3a_2 + a_3 + 3a_4 + a_5 + 3a_6 + a_7 + 3a_8 + a_9 + 3a_{10} + a_{11} + 3a_{12} + a_{13}$ ends with 0. Adding an extra 0 in the front of a UPC number does not affect the check digit.

The U.S. banking system uses a variation of the UPC scheme that appends check digits to the numbers assigned to banks. Each bank has an eight-digit routing number $a_1a_2 \cdots a_8$ together with a check digit a_9 so that a_9 is the last digit of $7a_1 + 3a_2 + 9a_3 + 7a_4 + 3a_5 + 9a_6 + 7a_7 + 3a_8$. The numbers 7, 3, and 9 used in this formula are called the **weights**. (The weights for the UPC scheme are 3 and 1.) The weights were carefully chosen so that all single-digit errors and most transposition errors are detected. (The use of different weights in adjacent positions permits the detection of most transposition errors.)

EXAMPLE 2   Bank Identification Number

The First Chicago Bank has the routing number 071000013 on the bottom of all its checks (see Spotlight 16.1). The check digit 3 is the last digit of $7 \cdot 0 + 3 \cdot 7 + 9 \cdot 1 + 7 \cdot 0 + 3 \cdot 0 + 9 \cdot 0 + 7 \cdot 0 + 3 \cdot 1 = 33$. ■

You may wonder if there is any advantage to using three weights as opposed to using two, as is the case for the UPC error-detection scheme. The answer is yes. While both the UPC scheme and the bank scheme detect 100% of single-position errors and the same transposition errors involving adjacent digits, the bank scheme will detect most transposition errors of the form $\cdots abc \cdots \rightarrow \cdots cba \cdots$, whereas the UPC scheme does not detect such errors.

For example, say we look at a number that begins with 241. In the UPC scheme, these digits contribute $3 \cdot 2 + 4 + 3 \cdot 1 = 13$ toward the total calcula-

SPOTLIGHT 16.1   Bank Checks

What do the string of numbers at the bottom of a check represent? Here is the answer.

0710	The bank's Federal Reserve District, office, and state or special collection arrangement
0001	The bank's identification number
3	The check digit
22 633 78	The checking account number
0134	The check number

tion, while the string 142 (the first and third digits are transposed) also contributes $3 \cdot 1 + 4 + 3 \cdot 2 = 13$ toward the total calculation. So the error is not detected. In contrast, using the bank scheme, 241 contributes $7 \cdot 2 + 3 \cdot 4 + 9 \cdot 1 = 35$ toward the total, while 142 contributes $7 \cdot 1 + 3 \cdot 4 + 9 \cdot 2 = 37$ toward the total. Since the total for the correct number ends with 0, the total for the number that had the transposition error would end with the digit 2. Thus, the error is detected.

One of the most efficient error-detection methods is used by all major credit-card companies as well as by many libraries, blood banks, photofinishing companies, and the South Dakota driver's license department. It is called **Codabar**. Say a bank intends to issue a credit card with the identification number 312560019643001. It must then add an extra digit for error detection. This is done as follows. Add the digits in positions 1, 3, 5, 7, 9, 11, 13, and 15 and double the result: $(3 + 2 + 6 + 0 + 9 + 4 + 0 + 1) \times 2 = 50$. Next, count the number of digits in positions 1, 3, 5, 7, 9, 11, 13, and 15 that exceed 4 and add this to the total. For our example, only 6 and 9 exceed 4, so the count is 2 and our running total is 52. Now add in the remaining digits: $52 + (1 + 5 + 0 + 1 + 6 + 3 + 0) = 68$.

The check digit is whatever is needed to bring the final tally to a number that ends with 0. Because $68 + 2 = 70$, the check digit for our example is 2. This digit is appended to the end of the number the bank issues for identification purposes. Errors in input data are detected by applying the same algorithm to the input, including the check digit. If the correct number is entered into a computer, the result will end in 0. If the result doesn't end with 0, a mistake has been made. The credit card shown in Figure 16.4 is reproduced from an ad promoting the Citibank VISA card. Notice that the check digit on the card is not valid because the Codabar algorithm yields

$$(4 + 2 + 0 + 1 + 3 + 5 + 7 + 9) \times 2 + 3$$
$$+ (1 + 8 + 0 + 2 + 4 + 6 + 8) + 0 = 94$$

FIGURE 16.4 VISA card with an invalid Codabar number.

which does not end in 0. This method allows computers to detect 100% of single-position errors and about 98% of other common errors.

Besides detecting errors, the check digit offers partial protection against fraudulent numbers. A person who wanted to create a phony credit card, bank account number, or driver's license number would have to know the appropriate check-digit scheme for the number to go unchallenged by the computer.

Thus far we have not discussed any schemes that detect 100% of single errors and 100% of transposition errors. The **International Standard Book Number (ISBN)** method used throughout the world is one that detects all such errors. (See the copyright page or the back cover of this book.)

A correctly coded 10-digit ISBN $a_1a_2 \cdots a_{10}$ has the property that $10a_1 + 9a_2 + 8a_3 + 7a_4 + 6a_5 + 5a_6 + 4a_7 + 3a_8 + 2a_9 + a_{10}$ is evenly divisible by 11. Consider the ISBN of the book you are now reading: 0-7167-5965-9 (0-7167-6901-8 for the paperback version). The initial digit 0 indicates that the book is published in an English-speaking country (*not* that the book is written in English). The next block of digits—7167—identifies the publisher, W. H. Freeman and Company. The third block—5965 for the hardback version—is assigned by the publisher and identifies the particular book. The last digit 9 (for the hardback version) is the check digit. Let's verify that this number is a legitimate possibility. We must compute $10 \cdot 0 + 9 \cdot 7 + 8 \cdot 1 + 7 \cdot 6 + 6 \cdot 7 + 5 \cdot 5 + 4 \cdot 9 + 3 \cdot 6 + 2 \cdot 5 + 9 = 253$. Because $253 = 11 \cdot 23$, it is evenly divisible by 11, and no error has been detected.

How can we be sure that this method detects 100% of the single-position errors? Well, let's say that a correct number is $a_1a_2a_3a_4a_5a_6a_7a_8a_9a_{10}$ and that a mistake is made in the second position. (The same argument applies equally well in every position.) We may write this incorrect number as $a_1a_2'a_3a_4a_5a_6a_7a_8a_9a_{10}$, where $a_2' \neq a_2$. For this error to go undetected, it must be the case that $10 \cdot a_1 + 9 \cdot a_2' + 8 \cdot a_3 + 7 \cdot a_4 + 6 \cdot a_5 + 5 \cdot a_6 + 4 \cdot a_7 + 3 \cdot a_8 + 2 \cdot a_9 + a_{10}$ is evenly divisible by 11. Then, because both $10a_1 + 9a_2 + 8a_3 + 7a_4 + 6a_5 + 5a_6 + 4a_7 + 3a_8 + 2a_9 + a_{10}$ and $10a_1 + 9a_2' + 8a_3 + 7a_4 + 6a_5 + 5a_6 + 4a_7 + 3a_8 + 2a_9 + a_{10}$ are divisible by 11, so is their difference:

$$(10 \cdot a_1 + 9 \cdot a_2 + 8 \cdot a_3 + \cdots + a_{10})$$
$$- (10 \cdot a_1 + 9 \cdot a_2' + 8 \cdot a_3 + \cdots + a_{10}) = 9 \cdot (a_2 - a_2')$$

Because a_2 and a_2' are distinct digits between 0 and 9, their difference must be one of $\pm 1, \ldots, \pm 9$. Thus, the only possibilities for the number $9 \cdot (a_2 - a_2')$ are $\pm 9, \pm 18, \pm 27, \pm 36, \pm 45, \pm 54, \pm 63, \pm 72, \pm 81$, and none of these is divisible by 11. So, a single-position error cannot go undetected.

To verify that the method detects all adjacent transposition errors, let's suppose that the first two digits are transposed (the same argument applies to all positions). Say the correct number is $a_1a_2a_3 \cdots a_{10}$. As before, for the incorrect number $a_2a_1a_3 \cdots a_{10}$ to go undetected, it must be the case that the difference of the correct number and the incorrect number is evenly divisible by 11. That is,

$$(10a_1 + 9a_2 + 8a_3 + \cdots + a_{10}) - (10a_2 + 9a_1 + 8a_3 + \cdots + a_{10})$$

is evenly divisible by 11. This reduces to $a_1 - a_2$ is divisible by 11. But the only possible differences of two numbers between 0 and 9 are plus or minus the numbers between 0 and 9. Thus $a_1 - a_2 = 0$. But then $a_1 = a_2$ and there is no error.

Since this method, in contrast to the others we have described, detects all single-position errors and all transposition errors, why is it not used more? Well, it does have a drawback. Say the next title published by Freeman is to have 1910 for the third block. (All Freeman books begin with 0-7167-.) What check digit should be assigned? Call it a. Then $10 \cdot 0 + 9 \cdot 7 + 8 \cdot 1 + 7 \cdot 6 + 6 \cdot 7 + 5 \cdot 1 + 4 \cdot 9 + 3 \cdot 1 + 2 \cdot 0 + a = 199 + a$. Because the next integer after 199 that is divisible by 11 is 209, we see that $a = 10$. But appending 10 to the existing 9-digit number would result in an 11-digit number instead of a 10-digit one. This is the only flaw in the ISBN scheme. To avoid this flaw, publishers use an X to represent the check digit 10. As a result, not all ISBNs consist solely of digits (some end with X). Publishers could avoid this inconsistency by simply refraining from using numbers that require an X.

It is possible to detect 100% of all single-digit errors and 100% of transposition errors involving adjacent digits without the drawback of introducing an extra character such as X, but to do so requires a noncommutative multiplication system—that is, a system in which a times b is not always the same as b times a. One situation in which such a system was used is described in Spotlight 16.2.

SPOTLIGHT 16.2 German Banknotes

Before adopting the euro in 2002, Germany was one of the few countries in the world that included a check digit on currency. Each German banknote had a 10-character serial number made up of letters and numbers. To compute the check digit, the letters were converted to numbers, and complicated schemes for weighting the numbers in each position and for calculating the check digit were employed. Interestingly, the method used to calculate the check digit was not commutative (that is, a times b need not be the same as b times a). For example, in this unusual mathematical system, 4 times 5 is 9 but 5 times 4 is 6. Using a noncommutative system of calculation results in greater error-detection capability than is possible with a commutative system. The banknote shown here features the mathematician Carl Gauss. The last digit is the check digit.

German banknote with serial number DZ6768309Y and check digit 9.

To expand the inventory of ISBNs and make them compatible with the EAN numbering scheme for other retail items that will be used worldwide by 2007, publishers have begun to expand the 10-digit ISBN to 13 digits. Except for the check digit, existing ISBNs remain intact with a prefix 978 attached. The prefix 979 will be used on all newly assigned numbers. The check digit for the 13-digit ISBN will be calculated using the EAN method for retail products. The bar code shown on the back cover of this book has the traditional ISBN above the bars and the new ISBN below them.

Many identification numbers use both alphabetic and numerical characters. When a check digit is included, the alphabetic characters are assigned numerical values. The vehicle identification number (VIN) used to identify cars and trucks is one such example, as explained in Spotlight 16.3.

SPOTLIGHT 16.3   The VIN System

Automobiles and trucks are given a vehicle identification number (VIN) by the manufacturer. A typical VIN has 17 alphanumeric characters that code information, such as country where the vehicle was built, manufacturer, make, body style, engine type, plant where the vehicle was built, model year, model, type of restraint, a check digit, and a production sequence number. The check digit is calculated by converting the 26 consecutive letters of the alphabet, respectively, to the numbers 1, 2, 3, 4, 5, 6, 7, 8, 9, 1, 2, 3, 4, 5, 6, 7, 8, 9, 2, 3, 4, 5, 6, 7, 8, 9 (note the skipped digit after the second 9) to obtain a 16-digit number $a_1 a_2 \cdots a_{15} a_{16}$ that is weighted with 8, 7, 6, 5, 4, 3, 2, 10, 9, 8, 7, 6, 5, 4, 3, 2. The check digit is the remainder when the weighted sum $8 \cdot a_1 + 7 \cdot a_2 + \cdots + 3 \cdot a_{15} + 2 \cdot a_{16}$ is divided by 11 unless the remainder is 10, in which case an X is used instead. The check digit is inserted in position 9.

PASSENGER CAR VIN SYSTEM

16.2 The ZIP Code

Identification numbers sometimes **encode** geographic data. The ZIP code, Social Security numbers, and telephone numbers are the foremost examples. In 1963, the U.S. Postal Service numbered every American post office with a five-digit **ZIP code.** The numbers begin with 0's at the points farthest east—00601 for Adjuntas, Puerto Rico—and work up to 9's at the the points farthest west—99950 for Ketchikan, Alaska (see Figure 16.5).

Let's use one of the ZIP codes for Duluth, Minnesota, as an example:

55812

5   The first digit represents one of 10 geographic areas, usually a group of states. The numbers begin at the points farthest east (0) and end at the points farthest west (9).

58   The second two digits, in combination with the first, identify a central mail-distribution point known as a sectional center. The location of a sectional center is based on geography, transportation facilities, and population density. Although just four centers serve the entire state of Utah, there are six of them to take care of New York City.

12   The last two digits indicate the town or local post office. The order is often alphabetical for towns within a delivery area—for example, towns with names beginning with A usually have low numbers. There are many exceptions to this, such as towns that came into existence after the ZIP code scheme was created. In many cases, the largest city in a region will be given the digits 01 and surrounding towns assigned succeeding digits alphabetically, as shown for the Farmville, Virginia, area.

Andersonville	23911	Hampden-Sydney	23943
Boydton	23917	Kenbridge	23944
Buckingham	23921	Keysville	23947
Burkeville	23922	Lunenburg	23952
Charlotte Court House	23923	Meherrin	23954
Chase City	23924	Nottoway	23955
Clarksville	23927	Pamplin	23958
Crewe	23930	Phenix	23959
Cullen	23934	Prospect	23960
Darlington Heights	23935	Red Oak	23964
Dillwyn	23936	Rice	23966
Drakes Branch	23937	Skipwith	23968
Dundas	23938	Victoria	23974
Farmville	23901	Wylliesburg	23976
Green Bay	23942		

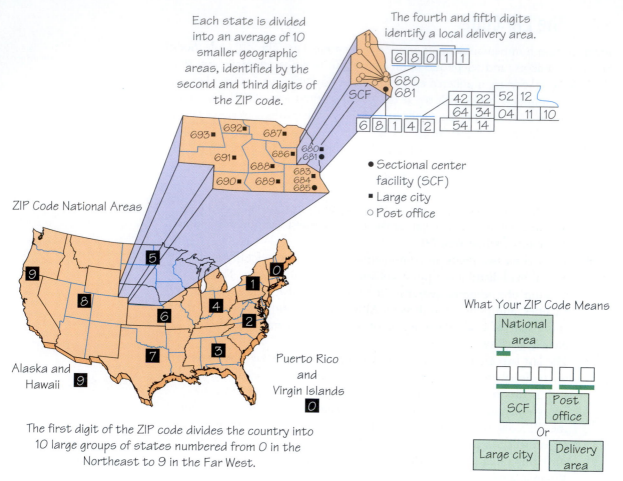

FIGURE 16.5 ZIP code scheme.

In 1983, the U.S. Postal Service added four digits to the ZIP code. When four digits are added after a dash—for example, 68588-1234—the number is called the **ZIP + 4 code.** Mail with ZIP + 4 coding is eligible for cheaper bulk rates, being easier to sort with automated equipment. It's also helpful for businesses that wish to sort the recipients of their mailings by geographic location. The first two numbers of the four-digit suffix represent a delivery sector, which may be several blocks, a group of streets, several office buildings, or a small geographic area. The last two numbers narrow the area further. They might denote one floor of a large office building, a department in a large firm, or a group of post office boxes.

For businesses that receive an enormous volume of mail, the ZIP + 4 code permits automation of in-house mailroom sorting. For example, the first seven digits of all mail sent to the University of Minnesota Duluth, are 55812-24. The school has designated nine pairs of digits for the last two positions to direct the mail to the appropriate dormitory or apartment complex.

16.3 Bar Codes

In modern applications, bar codes and identification numbers go hand in hand. Bar coding is a method for automated data collection. It is a way to transmit information rapidly, accurately, and efficiently to a computer.

Bar Code

A **bar code** is a series of dark and light spaces that represent characters.

To **decode** the information in a bar code, a beam of light is passed over the bars and spaces via a scanning device, such as a handheld wand or a fixed-beam device. The dark bars reflect very little back to the scanner, whereas the light spaces reflect much light. The differences in reflection intensities are detected by the scanner and converted to strings of 0's and 1's that represent specific numbers and letters. Such strings are called a *binary coding* of the numbers and letters.

Binary Code

Any system for representing data with only two symbols is a **binary code.**

ZIP Code Bar Code

The simplest bar code is the **Postnet code** used by the U.S. Postal Service and commonly found on business reply forms (see Figure 16.6). For a ZIP + 4 code there are 52 vertical bars of two possible lengths (long and short). The long bars at the beginning and end are called *guard bars* and together provide a frame for the remaining 50 bars. In blocks of five, the 50 bars within the guard bars represent the ZIP + 4 code and a tenth digit for error correction. Each block of five is composed of exactly two long bars and three short bars, according to the pattern shown below:

Decimal Digit	Bar Code
1	ıııll
2	ıılıl
3	ııllı
4	ılııl
5	ılılı
6	ıllıı
7	lıııl
8	lıılı
9	lılıı
0	llııı

Handheld scanner reading the shipping bar code on a crate. *(Stewart Cohen/Tony Stone Images.)*

FIGURE 16.6 ZIP + 4 bar code.

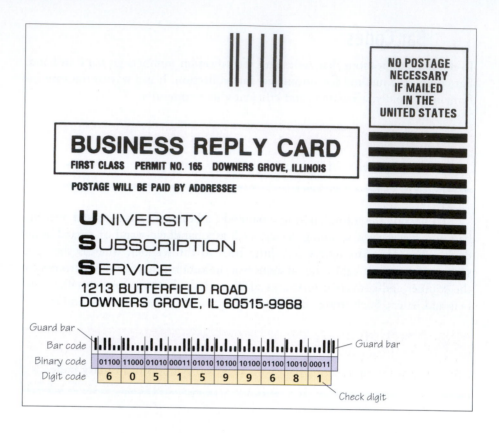

The tenth digit of a Postnet code number is a check digit chosen so that the sum of the nine digits of the ZIP + 4 code and the tenth one is evenly divisible by 10. That is, the check digit C for the ZIP + 4 code $a_1 a_2 \cdots a_9$ is the digit with the property that the sum $a_1 + a_2 + \cdots + a_9 + C$ ends with 0. For example, the ZIP + 4 code 80321-0421 has the check digit 9, because $8 + 0 + 3 + 2 + 1 + 0 + 4 + 2 + 1 = 21$ and $21 + 9 = 30$ ends with 0.

Because each digit is represented by exactly two long bars and three short ones, any error in reading or printing a single bar would result in a block of five with only one long bar or three long bars. In either case, the error is detected. This is the reason behind the choice of five bars to code each digit rather than four bars. With five bars per digit, there are exactly 10 arrangements composed of two long bars and three short bars. Any misreading of a single bar in such a block is therefore recognizable, because it does not match any other of the blocks for the 10 digits. And because the block location of the error is known, the check digit permits the correction of the error. Let's look at an example of an incorrectly printed bar code and see how the error is correctable.

The scanner ignores the guard bars at the beginning and the end and reads the remaining bars in blocks of five, as shown on the next page. (We have inserted dashed dividing lines for readability.)

$$3 \quad 0 \quad 7 \quad 2 \quad 2 \quad ? \quad 9 \quad 0 \quad 1 \quad 7$$

Because the sixth block has only one long bar, it is an incorrect one. To correct the error, the computer linked with the bar-code scanner sums the remaining 9 digits to obtain 31. Because the sum of all 10 digits ends with 0, the correct value for the sixth digit must be 9.

Beginning in 1993, large organizations and businesses that wanted to receive reduced rates for ZIP + 4 bar-coded mail were required to use a 12-digit bar code called the *delivery-point bar code.* This code permits machines to sort a letter into the order in which it will be delivered by the carrier. Mail for the first location on a mail route occurs first, mail for the second location on a route occurs second, and so on.

The 12-digit bar code uses the Postnet bar scheme to code the 12-digit string composed of the 9-digit ZIP + 4 number followed by the last two digits of the street address or box number and a check digit chosen so that the sum of all 12 digits is evenly divisible by 10. For example, a letter addressed to 1738 Maple Street with ZIP + 4 code 55811-2742 would have the Postnet bar code for the digits 558112742384 (38 is from the street address and 4 is the check digit).

The UPC Bar Code

The bar code that we encounter most often is the UPC, which was first used on grocery items in 1973 and has since spread to most retail products. As Figure 16.7 shows, it has other applications as well. The UPC bar code translates a 12-digit number into bars that can be quickly and accurately read by a laser scanner. The number has four components—two five-digit numbers sandwiched between two single digits—as shown in Figure 16.8.

Here is what the four components represent:

5	The first digit identifies the kind of product. For example, a 2 signals random-weight items, such as cheese and meat; a 3 means drug and certain other health-related products; a 4 means products marked for price reduction by the retailer; a 5 signals cents-off coupons (see Figure 16.8).
13000	The next five digits identify the manufacturer. This number is assigned by the Uniform Code Council in Dayton, Ohio.
22020	The next five digits, assigned by the manufacturer to identify the product, can include size, color, or other important information (but not price).
5	The final digit is the check digit. This digit is often not printed, but it is always included in the bar code.

FIGURE 16.7
Entomologist Stephen Buchmann developed a reliable, inexpensive way to track bees using the same technology that supermarkets use to speed up the checkout lines and keep track of inventory. He glued bar-code labels onto the backs of 100 bees and placed a laser scanner above the hive. In the past, researchers marked bees with paint or tags, but monitoring activity required the presence of a human observer. *(Scott Camazine/Sue Trainor.)*

FIGURE 16.8 UPC identification number 5 13000 22020 5. The initial 5 indicates that the number is a manufacturer's coupon. The block 13000 identifies the manufacturer as Heinz. The block 22020 identifies the product. The last digit, 5, is a check digit.

Each digit of the UPC code is represented by a space divided into seven modules of equal width, as illustrated in Figure 16.9. How these seven modules are filled depends on the digit being represented and whether the digit being represented is part of the manufacturer's number or the product number. In every case, there are two light spaces and two dark bars of various thicknesses that alternate. A UPC code has on each end two long bars of one-module thickness separated by a light space of one-module thickness. These two modules are called the *guard bar patterns* (Figure 16.10). The guard bar patterns define the thickness of a single module of each type. They are not part of the identification number. The manufacturer's number and the product number are separated by a center bar pattern consisting of the following five modules: a light space, a (long) dark bar, a light space, a (long) dark bar, and a light space (see Figure 16.10). The center bar pattern is not part of the identification number but merely serves to separate the manufacturer's number and product number. Figure 16.9 shows how the digits 6 and 0 in a manufacturer's number are coded.

Observe the following pattern in Figure 16.9: a light space of one-module thickness, a dark bar of one-module thickness, a light space of one-module thickness, a dark bar of four-module thickness. Symbolically, such a pattern of light spaces and dark bars is represented as 0101111. Here each 0 means a one-module-thickness light space and each 1 means a one-module-thickness dark bar. Figure 16.11 illustrates how the thicknesses of the spaces and bars are translated into a binary code.

Table 16.1 shows the binary code for all digits. Notice that the code for the digits in the product number (the block of five digits on the right side) can be obtained from the code for the digits in the manufacturer's number (the block

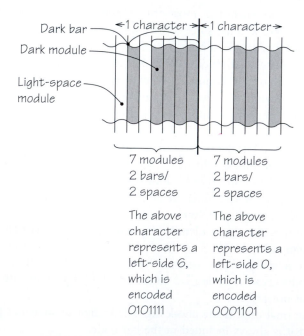

Dark bar

Dark module

Light-space module

←1 character→ ←1 character→

7 modules
2 bars/
2 spaces

7 modules
2 bars/
2 spaces

The above character represents a left-side 6, which is encoded 0101111

The above character represents a left-side 0, which is encoded 0001101

FIGURE 16.9 UPC bar coding for a left-side 6 and left-side 0, part of the manufacturer's number.

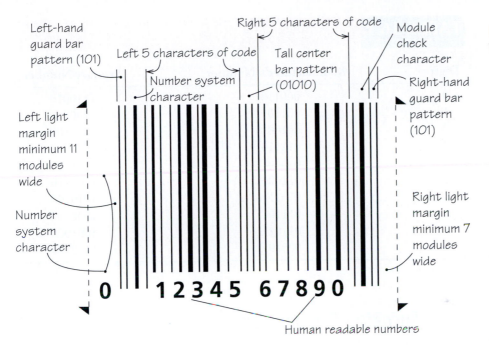

Left-hand guard bar pattern (101)

Left 5 characters of code

Right 5 characters of code

Module check character

Number system character

Tall center bar pattern (01010)

Right-hand guard bar pattern (101)

Left light margin minimum 11 modules wide

Number system character

Right light margin minimum 7 modules wide

0   12345   67890

Human readable numbers

FIGURE 16.10 UPC bar-code format.

of digits on the left side), and vice versa, by replacing each 0 by a 1 and each 1 by a 0. Thus, the code 0111011 for 7 in a manufacturer's number becomes 1000100 in the product number. Also notice that each manufacturer's number has an odd number of 1's, whereas each product number has an even number of 1's. This permits a computer linked with an optical scanner to determine whether the bar code was scanned left to right or right to left. (If the first block of digits has an even number of 1's for each digit, the scanning is being done right to left.) Thus, scanning can be done in either direction without ambiguity.

1010001101010101111

0

6 5 9 2 1 4 3 2 1 0

FIGURE 16.11
Translation of bars and space modules into binary code (see top of bars). The guard pattern defines a single-module thickness for a bar and a space.

TABLE 16.1	Binary UPC Coding	
Digit	Manufacturer's Number	Product Number
0	0001101	1110010
1	0011001	1100110
2	0010011	1101100
3	0111101	1000010
4	0100011	1011100
5	0110001	1001110
6	0101111	1010000
7	0111011	1000100
8	0110111	1001000
9	0001011	1110100

16.4 Encoding Personal Data

Consider this Social Security number: 189-31-9431. What information about the holder can be deduced from the number? Only that the holder obtained it in Pennsylvania (see Spotlight 16.4). Figure 16.12 shows an Illinois driver's license number: M200-7858-1644. What information about the holder can be deduced from this number? This time we can determine the date of birth, sex, and much about the person's name.

These two examples illustrate the extremes in coding personal data. The Social Security number has no personal data encoded in the number. It is entirely determined by the place and time it is issued, not the individual to whom it is assigned. In contrast, in some states the driver's license numbers are entirely determined by personal information about the holders. It is no coincidence that

FIGURE 16.12 Illinois driver's license.

SPOTLIGHT 16.4 Social Security Numbers

The first three digits of Social Security numbers show where the number was applied for. Changes in population have forced some numbers to be moved or assigned out of sequence over the years.

001–003	New Hampshire	387–399	Wisconsin	526–527	Arizona
004–007	Maine	400–407	Kentucky	& 600–601	
008–009	Vermont	408–415	Tennessee	528–529	Utah
010–034	Massachusetts	416–424	Alabama	530	Nevada
035–039	Rhode Island	425–428	Mississippi	531–539	Washington
040–049	Connecticut	& 587–588		540–544	Oregon
050–134	New York	429–432	Arkansas	545–573	California
135–158	New Jersey	433–439	Louisiana	& 602–626	
159–211	Pennsylvania	440–448	Oklahoma	574	Alaska
212–220	Maryland	449–467	Texas	575–576	Hawaii
221–222	Delaware	468–477	Minnesota	577–579	District
223–231	Virginia	478–485	Iowa		of Columbia
232–236	West Virginia	486–500	Missouri	580	Virgin Islands
232, 237–246	North Carolina	501–502	North Dakota	580–584	Puerto Rico
247–251	South Carolina	503–504	South Dakota	& 596–599	
252–260	Georgia	505–508	Nebraska	586	Guam
261–267	Florida	509–515	Kansas	586	American Samoa
& 589–595		516–517	Montana	586	Philippines
268–302	Ohio	518–519	Idaho	700–728	*through July 1,*
303–317	Indiana	520	Wyoming		*1963, reserved for*
318–361	Illinois	521–524	Colorado		*railroad employees*
362–386	Michigan	525 & 585	New Mexico		

Source: Social Security Administration.

the unsophisticated Social Security numbering scheme predates computers. Agencies that have large databases that include personal information such as names, sex, and dates of birth find it convenient to encode these data into identification numbers. Examples of such agencies are the National Archives (where census records are kept), genealogical research centers, the Library of Congress, and state motor vehicle departments.

There are many methods in use to encode personal data such as name, sex, and date of birth. These methods are perhaps most widely used in assigning driver's license numbers in some states. Coding license numbers solely from personal data enables automobile insurers, government entities, and law enforcement agencies to determine the number from the personal data.

Many states encode the surname, first name, middle initial, date of birth, and sex by quite sophisticated schemes (see Figure 16.12).

In one scheme, the first four characters of the license number are obtained by applying the **Soundex Coding System** to the surname as follows.

1. Delete all occurrences of *h* and *w*. (For example, *Schworer* becomes *Scorer* and *Hughgill* becomes *uggill*.)

2. Assign numbers to the remaining letters as follows:

$$a, e, i, o, u, y \rightarrow 0 \qquad l \rightarrow 4$$
$$b, f, p, v \rightarrow 1 \qquad m, n \rightarrow 5$$
$$c, g, j, k, q, s, x, z \rightarrow 2 \qquad r \rightarrow 6$$
$$d, t \rightarrow 3$$

3. If two or more letters with the same numeric value are adjacent, omit all but the first. (For example, *Scorer* becomes *Sorer* and *uggill* becomes *ugil*).

4. Delete the first character of the original name if still present. (*Sorer* becomes *orer*).

5. Delete all occurrences of *a, e, i, o, u,* and *y*.

6. Retain only the first three digits corresponding to the remaining letters; append trailing 0's if fewer than three letters remain; precede the three digits obtained in step 6 with the first letter of the surname.

Figure 16.13 shows three examples.

What is the advantage of this method? It is an error-correcting scheme. Indeed, it is designed so that likely misspellings of a name nevertheless result in the correct coding of the name. For example, frequent misspellings of the name *Erickson* are *Ericksen, Eriksen, Ericson,* and *Ericsen*. Observe that all of these yield the same coding as *Erickson*. If a law enforcement official, a genealogical researcher,

	Step 1		Step 2	
Schworer	$\rightarrow$	Scorer	$\rightarrow$	Scorer
				220606

Step 3		Step 4		Step 5		Step 6	
$\rightarrow$	Sorer	$\rightarrow$	orer	$\rightarrow$	rr	$\rightarrow$	S-660
	20606		0606		66		

	Step 1		Step 2	
Hughgill	$\rightarrow$	uggill	$\rightarrow$	uggill
				022044

Step 3		Step 4		Step 5		Step 6	
$\rightarrow$	ugil	$\rightarrow$	ugil	$\rightarrow$	gl	$\rightarrow$	H-240
	0204		0204		24		

	Step 1		Step 2	
Schmidlapper	$\rightarrow$	Scmidlapper	$\rightarrow$	Scmidlapper
				22503401106

Step 3		Step 4		Step 5		Step 6	
$\rightarrow$	Smidlaper	$\rightarrow$	midlaper	$\rightarrow$	mdlpr	$\rightarrow$	S-534
	250340106		50340106		53416		

FIGURE 16.13 The Soundex Coding System.

a librarian, or an airline reservation agent wanted to pull up the file from a data bank for someone whose name was pronounced "Erickson," the correct spelling isn't essential because the computer searches for records that are coded as E-625 for all spelling variations. The search feature of a Web site where many mathematicians post their research papers uses the Soundex Coding System. This system was designed for the U.S. Census Bureau when much census information was obtained orally.

There are many schemes for encoding the date of birth and the sex in driver's license numbers. For example, the last five digits of Illinois and Florida driver's license numbers capture the year and date of birth as well as the sex. In Illinois, each day of the year is assigned a three-digit number in sequence beginning with 001 for January 1. However, each month is assumed to have 31 days. Thus, March 1 is given the number 063 because both January and February are assumed to have 31 days. These numbers are then used to identify the month and day of birth of male drivers. For females, the scheme is identical except that 600 is added to the number. The last two digits of the year of birth, separated by a dash (probably to obscure the fact that they represent the year of birth), are listed in the fifth and fourth positions from the end of the driver's license number. Thus, a male born on October 13, 1940, would have the last five digits 4-0292 ($292 = 9 \cdot 31 + 13$), whereas a female born on the same day would have 4-0892. The scheme to identify birth date and sex in Florida is the same as in Illinois except that each month is assumed to have 40 days and 500 is added for women. Moreover, a dash occurs between the two digits for the year and the three digits for the day. For example, the five digits 49-585 belong to a woman born on March 5, 1949.

In this chapter, we have investigated how mathematics is used to append a check digit to an identification number for error detection. In the next chapter, we will show how codes consisting of 0's and 1's can be devised so that errors can be corrected.

REVIEW VOCABULARY

Bar code A code that employs bars and spaces to represent information.

Binary code A coding scheme that uses two symbols, usually 0 and 1.

Check digit A digit included in an identification number for the purpose of error detection.

Codabar An error-detection method used by all major credit-card companies, many libraries, blood banks, and others.

Decoding Translating code into data.

Encoding Translating data into code.

Error-detecting code A code in which certain types of errors can be detected.

International Standard Book Number (ISBN) A 10-digit identification number used on books throughout the world that contains a check digit for error detection.

Postnet code The bar code used by the U.S. Postal Service for ZIP codes.

Soundex Coding System An encoding scheme for surnames based on sound.

Universal Product Code (UPC) A bar code and identification number that are used on most retail items. The UPC code detects 100% of all single-digit errors and most other types of errors.

Weights Numbers used in the calculation of check digits.

ZIP code A five-digit code used by the U.S. Postal Service to divide the country into geographic units to speed sorting of the mail. ZIP stands for Zone Improvement Plan.

ZIP + 4 code The nine-digit code used by the U.S. Postal Service to refine ZIP codes into smaller units.

✔ SKILLS CHECK

1. Suppose a U.S. Postal Service money order is numbered 1012065994X, where X indicates that the last digit is obliterated. What is the missing digit?

(a) 1

(b) 8

(c) The missing digit cannot be determined.

2. Suppose an American Express Travelers Cheque is numbered X425036790, where X indicates that the first digit is obliterated. What is the missing digit?

(a) 0

(b) 9

(c) The missing digit cannot be uniquely determined.

3. Is the number 105408970012 a legitimate airline ticket number?

(a) Yes.

(b) No, but if the final digit is changed to a 5, the resulting number 105408970015 is legitimate.

(c) No, but if the final digit is changed to a 6, the resulting number 105408970016 is legitimate.

4. Determine the check digit that should be appended to the UPC code 0-14300-25433.

(a) 1

(b) 9

(c) Another number

5. Determine the check digit that should be appended to the bank routing number 01500085.

(a) 1

(b) 9

(c) Another number

6. Suppose the ISBN 0-1750-3549-0 is incorrectly reported as 0-1750-3540-0. Which of the following statements is true?

(a) This error will not be detected by the check digit.

(b) While this particular error will be detected, the check digit does not detect all single-digit errors in ISBNs.

(c) Any single-digit error in an ISBN is detectable by the check digit.

7. Suppose that the Postnet code is incorrectly reported as 20001-5800-7. You know that only the sixth digit is incorrectly reported. Which of these statements is true?

(a) It is impossible to determine the correct digit for the sixth position.

(b) It is possible to determine the correct digit in this case. However, it is necessary to know which digit was incorrectly reported.

(c) It is possible to determine the correct digit in this case. Moreover, it is not necessary to know which digit was incorrectly reported.

8. When a single incorrect digit is entered, an error-detecting code

(a) will sometimes detect the error.

(b) will always detect the error but may not be able to correct the error.

(c) will always detect and correct the error.

9. The conversion of data to code is called

(a) encoding.

(b) decoding.

(c) Codabar.

10. If the first five digits of a valid U.S. Postal Service money order number are rearranged, will the resulting number have the same check digit as the original number?

(a) Yes, always

(b) Sometimes

(c) Never

11. If the first two digits of a valid airline ticket identification number are exchanged, will the resulting number also be valid?

(a) Yes, always
(b) Sometimes
(c) Never

12. The U.S. banking system uses a system that detects

(a) all transpositions and some single-digit errors.
(b) all single-digit errors and some transpositions.
(c) all single-digit errors and all transpositions.

13. A correctly coded 10-digit ISBN has a weighted sum that is evenly divisible by what integer?

(a) 7
(b) 10
(c) 11

14. A correctly coded 12-digit UPC has a weighted sum that is evenly divisible by what integer?

(a) 7
(b) 10
(c) 11

15. The U.S. Postal Service Postnet code converts the ZIP + 4 code into

(a) a coding of 9 digits using long and short bars.
(b) a coding of 10 digits whose sum is a multiple of 10.
(c) a coding of 10 digits for which the sum of the first 9 digits minus the tenth digit is a multiple of 10.

16. If an error in an identification number is made by transposing the first and third digits, the error is

(a) usually detected by the UPC scheme.
(b) always detected by the bank scheme.
(c) always detected by the ISBN scheme.

17. If a scanner misreads exactly one bar of a Postnet code, the computer will

(a) not always detect the error.
(b) always detect the error but will not always be able to correct it.
(c) always detect the error and correct it.

18. The bank routing number error-detection scheme detects

(a) the same errors as the UPC scheme.
(b) fewer errors than the UPC scheme.
(c) more errors than the UPC scheme.

19. The error-detection scheme used by airlines detects 100% of

(a) single-digit errors.
(b) adjacent-digit transposition errors.
(c) single-digit errors and 100% of adjacent-digit transposition errors not involving the pairs 0, 7; 1, 8; 2, 9.

20. As far as the ability to detect single-position errors and adjacent-digit transposition errors,

(a) the Codabar scheme is superior to the UPC scheme.
(b) the Codabar scheme is superior to the ISBN scheme.
(c) the UPC scheme is superior to the bank scheme.

EXERCISES ■ Challenge ◆ Discussion

Check Digits

1. Determine the check digit for a money order with identification number 3953981640.

2. Determine the check digit for a money order with identification number 7234541780.

3. Determine the check digit for the United Parcel Service (UPS) identification number 873345672.

4. Suppose a money order with the identification number and check digit 21720421168 is erroneously copied as 27750421168. Will the check digit detect the error? Explain your reasoning.

5. Determine the check digit for the airline ticket number 30860422052.

6. Determine the check digit for the Avis rental car with identification number 540047.

7. Determine the check digit for the UPC number 38137009213.

8. Determine the check digit for the UPC number 05074311502.

9. Determine the check digit for the ISBN 0-669-19493.

10. Determine the check digit for the ISBN 0-669-33907.

11. Determine the check digit for the bank routing number 09190204.

12. Determine the check digit for the bank routing number 09100001.

13. Determine the check digit for the American Express Travelers Cheque with identification number 461212023.

14. Determine whether the Master Card number 3541 0232 0033 2270 is valid.

15. Create a check digit for the UPC number 38137009213 using the weights 7, 1, 7, 1, 7, 1, . . . , 7, 1, instead of 3, 1, 3, 1, 3, 1, . . . , 3, 1. Test to see whether this check digit will detect single-digit errors by trying several examples.

16. Create a check digit for the UPC number 38137009213 using the weights 2, 1, 2, 1, 2, 1, . . . , 2, 1, instead of 3, 1, 3, 1, 3, 1, . . . , 3, 1. Is the error caused by replacing the 3 in the first position with an 8 detected? What about the error caused by replacing the 1 in the third position with a 6? Explain why or why not.

17. If the weights 5, 1, 5, 1, 5, 1, . . . , 5, 1 were used for the UPC code, which single-digit errors would go undetected?

18. Exercises 15, 16, and 17 reveal that using the weights 1, 3, or 7 for a particular position detects all errors in that position, whereas using weights 2 or 5 in a position does not detect all errors. Using this observation, make a guess about error-detection capability using weights 9, 4, 6, or 8.

19. Use the Codabar scheme to determine the check digit for the number 300125600196431.

20. Determine the check digit for the VIN JM1GD222J1581570 (see Spotlight 16.3 for a description of the method to be used).

21. For some products, such as soft-drink cans and magazines, an 8-digit UPC number called Version E is used instead of the 12-digit number. The method of calculating the eighth digit, which is the check digit, depends on the value of the seventh digit. The check digit a_8 for a UPC Version E identification number $a_1a_2a_3a_4a_5a_6a_7$, where a_7 is 0, 1, or 2, is chosen so that $a_1 + a_2 + 3a_3 + 3a_4 + a_5 + 3a_6 + a_7 + a_8$ is divisible by 10. Use this fact to determine the check digit for the following Version E numbers:

(a) 0121690
(b) 0274551
(c) 0760022
(d) 0496580

22. The check digit a_8 for a UPC Version E identification number $a_1a_2a_3a_4a_5a_6a_7$, where a_7 is 4, is chosen so that $a_1 + a_2 + 3a_3 + a_4 + 3a_5 + 3a_6 + a_8$ is divisible by 10. Use this fact to determine the check digit for the following Version E numbers:

(a) 0754704
(b) 0774714
(c) 0724444

■ **23.** The ISBN 0-669-03925-4 is the result of a transposition of two adjacent digits not involving the first or last digit. Determine the correct ISBN.

24. Explain why the bank scheme will detect the error 751 $\cdots \rightarrow$ 157 $\cdots$ but the UPC scheme will not.

25. Suppose the check digit a_9 for the bank routing number was chosen to be the last digit of $3a_1 + 7a_2 + a_3 + 3a_4 + 7a_5 + a_6 + 3a_7 + 7a_8$ instead of the way described in this chapter. How would this compare with the actual check digit?

26. Explain why an error caused by transposing the first two digits of a Postal Service money order is not detected by the check-digit scheme. Explain why the same is true for the second and third digits. What about the last two digits?

27. In computing the check digit for German banknotes, the letter Z is counted as having the numerical value 9 (see Spotlight 16.2). Criticize this choice of value. (You do not have to know how the scheme works to answer this question.)

28. Explain why an error caused by transposing any two digits of an American Express Travelers Cheque is not detected by the check-digit scheme.

29. Is there any mathematical reason for a check digit to be at the end of an identification number? Explain your reasoning.

30. Explain why the Postal Service money order check-digit scheme does not detect the mistake of substituting a 0 for a 9, or vice versa.

31. Which digit never appears as a check digit on a Postal Service money order?

32. Which digit never appears as a check digit on an American Express Travelers Cheque?

33. Which digits never appear as a check digit for an airline identification number?

34. Suppose four-digit numbers $a_1a_2a_3a_4$ are assigned a check digit a_5 so that $a_1 + 2a_2 + a_3 + 2a_4 + a_5$ is evenly divisible by 10. Test the number 43216 created in this way to see whether the method detects adjacent-digit transposition errors.

35. Starting with the ISBN 0-7167-4782-0, create three new numbers by transposing any two different digits. (They need not be adjacent.) Are these errors detected by the scheme?

36. Suppose in a UPS number an 8 is mistaken for a 5. Is the error detected? What if a 9 is mistaken for a 2?

37. Give an argument to show that the ISBN error-detection method will detect a transposition error involving the first and third digits. Does the same argument work for the fourth and sixth digits?

38. Suppose the check digit a_{10} of ISBNs were chosen so that $a_1 + 2a_2 + 3a_3 + 4a_4 + 5a_5 +$ $6a_6 + 7a_7 + 8a_8 + 9a_9 + 10a_{10}$ is divisible by 11 instead of the way described in the chapter. How would this compare with the actual check digit?

39. Consider a UPC number in which the digits 7 and 2 appear consecutively (that is, the number has the form $\cdots 72 \cdots$). Will the error caused by transposing these digits (that is, the number is taken as $\cdots 27 \cdots$) be detected? What if the digits 6 and 2 were transposed instead? State the general criterion for the detection of an error of the form $\cdots ab \cdots \rightarrow \cdots ba \cdots$ by the UPC scheme.

40. If the first three digits of a routing number for a checking account are 537 and the 5 and 3 are transposed, will the error be detected? If the first three numbers are 237 and the 2 and 7 are transposed, will the error be detected?

41. State a general criterion for the detection of an error of the form $\cdots abc \cdots \rightarrow \cdots cba \cdots$ for the routing number of a checking account.

42. The state of Utah appends a ninth digit a_9 to an eight-digit driver's license number $a_1a_2 \cdots a_8$ so that $9a_1 + 8a_2 + 7a_3 + 6a_4 + 5a_5 + 4a_6 + 3a_7 + 2a_8 + a_9$ is divisible by 10.

(a) If the first eight digits of a Utah driver's license number are 14910573, what is the ninth digit?
(b) Suppose a legitimate Utah driver's license number 149105767 is miscopied as 149105267. How would you know a mistake was made? Is there any way you could determine the correct number? Suppose you know the error was in the seventh position. Could you correct the mistake?
(c) If a legitimate Utah driver's license number 149105767 were miscopied as 199105767, would you be able to tell a mistake was made? Explain.
(d) Explain why any transposition error involving adjacent digits of a Utah driver's license number would be detected.

43. The Canadian province of Quebec assigns a check digit a_{12} to an 11-digit driver's license number $a_1a_2 \cdots a_{11}$ so that $12a_1 + 11a_2 + 10a_3 + 9a_4 + 8a_5 + 7a_6 + 6a_7 + 5a_8 + 4a_9 + 3a_{10} +$

$2a_{11} + a_{12}$ is divisible by 10. Criticize this method. Describe all single-digit errors that are undetected by this scheme.

44. Speculate on the reason why telephone numbers, Social Security numbers, and serial numbers on most currency do not have check digits.

45. Suppose a company uses a check-digit scheme similar to the UPC scheme, except that instead of using the UPC weights 3, 1, 3, 1, . . . it uses w, 1, w, 1, ... If two of the ID numbers used by the company are 73215674 and 73215661, determine w.

46. If a publishing company has headquarters in both the United States and Germany and publishes the same book in both countries, it is likely that the ISBN for the book will be identical except for the first and last digits (because the first digit for U.S. publications is 0 and the first digit for German publications is 3). If the last digit of the U.S. edition is 1, what is the last digit for the German publication?

Postnet Codes

47. Determine the ZIP + 4 code and check digit for each of the following Postnet bar codes:

(a) |ɪɪlɪ.ɪɪlɪɪlɪlɪlɪlɪɪ.ɪɪlɪɪɪlllɪɪɪlɪɪlɪɪ.ɪlɪll

(b) |ɪlɪlllɪɪɪɪlɪlɪlɪlllɪɪllɪɪɪlɪlɪɪlɪlll

(c) |ɪlɪ.ɪlɪl.ɪɪlllɪllɪɪlɪlɪlɪlɪlɪlllɪɪɪɪlllll

48. Determine the ZIP + 4 code and check digit for each of the following Postnet bar codes:

(a) |ɪɪɪlllɪlɪllɪɪɪlɪɪɪlɪlɪlɪllɪɪɪlllɪlllɪɪɪlɪɪll

(b) |ɪllɪɪllɪɪɪlɪɪɪlɪɪlllɪlɪlllɪlɪlɪɪɪlɪllɪɪɪlɪlɪl

(c) |ɪɪllɪɪɪlɪɪlɪlɪɪllɪɪɪlllɪlɪlɪlɪlɪɪɪlɪɪɪlllɪlɪll

49. In each of the following Postnet bar codes, exactly one mistake occurs (that is, a long bar appears instead of a short one, or vice versa). Determine the correct ZIP code.

(a) |ɪlɪlllɪɪllɪɪlɪlɪlɪlɪlllɪɪɪlɪlɪɪlɪllɪɪll

(b) |ɪlɪɪlɪlɪlɪlɪlɪlɪlɪllɪlɪlɪlɪlɪlɪlɪlɪlllɪl

(c) |ɪlɪlɪɪɪllɪɪɪlɪɪɪlɪllɪlɪlɪlɪɪlɪlllllɪɪɪɪlllɪɪll

50. Below is a 12-digit delivery-point bar code. Determine the ZIP + 4 number, the last two digits of the street address, and the check digit.

|ɪlɪlɪɪlɪlɪlɪɪɪɪɪllɪɪɪllɪlɪllɪɪɪlɪlɪɪlɪlɪlɪɪlɪɪlɪɪɪlll

51. Explain why any two errors in a particular block of five bars in a Postnet code are always detectable. Explain why not all such errors can be corrected.

52. Change 173 into a Postnet code.

53. Form all possible strings consisting of exactly three a's and two b's and arrange the strings in alphabetical order (for example, the first two possibilities are *aaabb* and *aabab*). Do you see any relationship between your list and the Postnet code?

Bar Codes

54. Most recently published books include a bar code on the back cover that has the ISBN above the bars and a 13-digit identification number below the bars. Examine several books with a bar code on the back cover. How does the number below the bar code differ from the UPC code? How is the number below the bar code related to the ISBN? Given the fact that the last digit in the number below the bar code is a check digit, determine how it is calculated.

55. Suppose the first block of a UPC bar code following the guard bar pattern that a scanner reads is 1000100. Is the scanner reading left to right or right to left?

56. The following is an actual identification number and bar code from a roll of wallpaper. What appears to be wrong with them? Speculate on the reason for the apparent violation of the UPC format.

5 011419 194056

Building Regulations: 1985 Class 0
FINE ART WALLCOVERINGS LTD.
HOLMES CHAPEL, CHESHIRE
MADE IN ENGLAND
FABRIQUE EN ANGLETERRE

Encoding Personal Data

57. Judging from the information in Spotlight 16.4, which three states were most likely to have had the smallest populations when Social Security numbers were allocated to the states?

58. What geographical information was used in allotting the first three digits of Social Security numbers to the states?

59. What demographic information was used in allotting the first three digits of Social Security numbers to the states?

60. Which state would you guess will be the next one to be allocated a new block of Social Security numbers?

◆ **61.** The Canadian postal system has assigned each geographic region a six-character code composed of alternating letters and digits, such as P7B5E1 and K7L3N6. Discuss the advantages this scheme has over the five-digit ZIP code used in the United States.

62. Determine the Soundex code for Smith, Schmid, Smyth, and Schmidt.

63. Determine the Soundex code for Skow, Sachs, Lennon, Lloyd, Ehrheart, and Ollenburger.

64. In Florida, the last three digits of the driver's license number of a female with birth month m and birth date b are $40(m - 1) + b + 500$. For both males and females, the fourth and fifth digits from the end give the year of birth. Determine the last five digits of a Florida driver's license number for a female born on July 18, 1942.

65. Explain why an Illinois driver's license number that ends with the last five digits 99817 cannot be valid.

66. Determine the last five digits of an Illinois driver's license number for a male born on June 18, 1942.

67. In Illinois, one obtains the last three digits of the driver's license number for a female by adding 600 to the number for a male with the same birthday. In Florida 500 is added to the number for a male. Why can't Florida use 600?

68. Explain why an Illinois driver's license number that ends with 77061 cannot be valid.

69. Determine the birth date of a person whose Illinois driver's license number ends with 58818.

70. In Florida, the last three digits of the driver's license number of a male with birth month m and birth date b are $40(m - 1) + b$. For both males and females, the fourth and fifth digits from the end give the year of birth. Determine the birth dates of people with numbers whose last five digits are 42218 and 53953.

71. For driver's license numbers issued in New York before September 1992, the last two digits were the year of birth. The three digits preceding the year encoded the sex and the month and day of birth. For a woman with birth month m and birth date b, the three digits were $63m + 2b + 1$ (insert a 0 in front for numbers less than 100). For a man with birth month m and birth date b, the three digits were $63m + 2b$. Determine the birth months, birth dates, and sexes of drivers with the three digits 248 and 601 preceding the year.

72. The state of Washington encodes the last two digits of the year of birth into driver's license numbers (in positions 8 and 9) by subtracting the two-digit number from 100. For example, a person born in 1942 has 58 in positions 8 and 9, whereas a person born in 1971 has 29 in positions 8 and 9. Speculate on the reason for subtracting the birth year from 100.

73. Driver's license number-assignment schemes that use personal data sometimes produce the same number for different people. Speculate about circumstances under which this is more likely to occur.

74. Apply the Soundex code to common ways to misspell your name. Do they give the same code as your name does?

75. Why would the Soundex system of coding last names be a poor method for encoding names in China?

WRITING PROJECTS

1. Prepare a two-page report on coded information in your location. Possibilities for investigation include driver's license numbers in your state, student ID numbers and bar codes at your school, and bar codes used by your school library and city library. Identify the coding schemes and, when possible, determine whether a check digit is employed. Include samples. The Suggested Readings for this chapter contain information that will assist you.

2. Prepare a two-page report on the driver's license coding schemes used by Michigan, Maryland, and Washington (the first two states use the same method). J. Gallian's "Assigning Driver's License Numbers" (see the Suggested Readings) has the information you will need.

3. Prepare a two-page report on the check-digit method used on German currency before 2002. J. Gallian's article "Error Detection Methods" (see the Suggested Readings) has the information you will need.

SUGGESTED READINGS

GALLIAN, J. The mathematics of identification numbers, *College Mathematics Journal,* 22 (1991): 194–202. A survey of check-digit schemes associated with identification numbers.

GALLIAN, J. Assigning driver's license numbers, *Mathematics Magazine,* 64 (1992): 13–22. Discusses various methods used by the states to assign driver's license numbers. Several of these methods include check digits for error detection.

GALLIAN, J. Error detection methods, *ACM Computing Surveys,* 28 (1996): 504–517. A detailed description of many error-detection methods.

GALLIAN, J., and S. WINTERS. Modular arithmetic in the marketplace, *American Mathematical Monthly,* 95 (1988): 548–551. A detailed analysis of the check-digit schemes presented in this chapter. In particular, the error-detection rates for the various schemes are given.

KIRTLAND, J. *Identification Numbers and Check Digit Schemes,* Mathematical Association of America, Washington, D.C., 2001. Provides more examples and exercises for the check-digit schemes discussed in this chapter.

SUGGESTED WEB SITE

www.d.umn.edu/~jgallian/fapp7 This Web site enables users to calculate check digits using the various methods discussed in this chapter.

Information Science

With the enormous volume of email, faxes, Internet traffic, and cellular phone calls, the Digital Revolution has brought about many mathematical challenges. One is how to correct errors in data transmission. Another is how to electronically send and store information economically. A third is how to ensure security of transmitted data. Yet another is how to improve Web search efficiency. In this chapter we illustrate some of the ways in which mathematicians and engineers have responded to these challenges.

17.1 Binary Codes

A system for coding data made up of two states (or symbols) is called a *binary code*. Binary codes are the hidden language of computers. The Postnet code and the Universal Product Code (UPC) bar code are two examples of binary codes. Morse code and Braille are two more. Automatic 35mm cameras send a current across five silver or black squares on the film canister that code the film speed. The Ebert and Roeper "thumbs up/thumbs down" rating of films is a binary code with four messages. Compact disc players, fax machines, digital televisions, modems, cell phones, and space probes represent data as strings of 0's and 1's rather than the usual digits 0 through 9 and letters *A* through *Z*. In this section, we will illustrate one way binary codes can be devised so that errors in the transmission of the code can be corrected.

The idea behind error-correction schemes is simple and one you often use. To illustrate, suppose you are reading the employment section of a newspaper

619

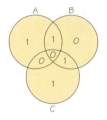

FIGURE 17.1 Diagram for message 1001.

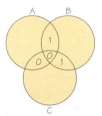

FIGURE 17.2 Diagram for encoded message 1001101.

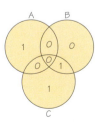

FIGURE 17.3 Diagram for received message 0001101.

and you see the phrase "must have a minimum of *bive* years experience." Instantly you detect an error because *bive* is not a word in the English language. Moreover, you are fairly confident that the intended word is *five*. Why so? Because *five* is a word derived from *bive* by changing a single letter and it makes the phrase understandable. In other phrases, words such as *bike* or *give* might be sensible alternatives to *bive*. Using the extra information provided by the context, we are often able to infer the intended meaning when errors occur.

To demonstrate the way error-correcting schemes work, suppose that NASA sends a spacecraft to land at one of 16 possible landing sites on Mars. The spacecraft orbits Mars while surveying the sites for the most favorable landing conditions. NASA officials have coded the 16 landing sites with four-digit strings of 0's and 1's such as 0000, 0001, 0010, 0100 (see Table 17.1 for the complete list). Once the best site has been selected, NASA will inform the spacecraft where to land by sending the code for the site. However, signals sent through space are subject to interference called *noise*. The noise might cause the spacecraft to interpret the signal as 0001 when the signal actually sent was 1001. Fortunately, over the past 50 years mathematicians and engineers have devised highly sophisticated schemes to build extra information into messages composed of 0's and 1's that often permits one to infer the correct message even though the message may have been received incorrectly (see Spotlights 17.1 and 17.2).

As a simple example, let's assume our message is 1001. We will build extra information into this message with the aid of the diagram in Figure 17.1. Begin by placing the four message digits in the four overlapping regions I, II, III, IV, with the digit in the first position (starting at the left of the sequence) in region I, the digit in the second position in region II, and so on. For regions V, VI, and VII, assign 0 or 1 so that the total number of 1's in each circle is even. See Figure 17.2.

We have now encoded our message 1001 using the diagram as 1001101. Now suppose that this encoded message is received as 0001101 (an error in the first

TABLE 17.1					
Message	→	Code Word	Message	→	Code Word
0000	→	0000000	0110	→	0110010
0001	→	0001011	0101	→	0101110
0010	→	0010111	0011	→	0011100
0100	→	0100101	1110	→	1110100
1000	→	1000110	1101	→	1101000
1100	→	1100011	1011	→	1011010
1010	→	1010001	0111	→	0111001
1001	→	1001101	1111	→	1111111

position). How would we know an error was made? We place each digit from the received message in its appropriate region, as in Figure 17.3.

Noting that in both circles A and B there is an odd number of 1's, we instantly realize that something is wrong, because the intended message had an even number of 1's in each circle. How do we correct the error? Because circles A and B have the wrong parity (parity refers to the oddness or evenness of a number—even integers have **even parity**; odd integers have **odd parity**) and C does not,

---

SPOTLIGHT 17.1 The Ubiquitous Reed–Solomon Codes

One of the mathematical ideas underlying current error-correcting techniques for everything from computer hard-disk drives to CD players was first introduced in 1960 by Irving Reed and Gustave Solomon. Reed–Solomon codes made possible the stunning pictures of the outer planets sent back by the space probes *Voyager 1* and *2*. They make it possible to scratch a compact disc and still enjoy the music.

"When you talk about CD players and digital audio tape and now digital television, and various other digital imaging systems that are coming—all of those Reed–Solomon [codes] are an integral part of the system," says Robert McEliece, a coding theorist at Caltech.

Why? Because digital information consists of 0's and 1's, and a physical device may occasionally confuse the two. *Voyager 2*, for example, was transmitting data at incredibly low power over billions of miles. Error-correcting codes are a kind of safety net, mathematical insurance against the vagaries of an imperfect material world.

In 1960, the theory of error-correcting codes was only about a decade old. Through the 1950s, a number of researchers began experimenting with a variety of error-correcting codes. But the Reed–Solomon paper, McEliece says, "hit the jackpot." "In hindsight it seems obvious," Reed later said. However, he added, "Coding theory was not a subject when we

Irving Reed and Gustave Solomon
At the Jet Propulsion Laboratory in 1989 to monitor the encounter of *Voyager 2* with Neptune. *(Rex Ridenhouse.)*

published the paper." The two authors knew they had a nice result; they didn't know what impact the paper would have.

Four decades later, the impact is clear. The vast array of applications, both current and pending, has settled the questions of the practicality and significance of Reed–Solomon codes. Billions of dollars in modern technology depend on ideas that stem from Reed and Solomon's original work.

Source: Adapted from an article by Barry Cipra, with permission from *SIAM News*, January 1993, p. 1. © by SIAM. All rights reserved.

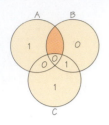

FIGURE 17.4 Circles A and B but not C have wrong parity.

the error is located in the portion of the diagram in circles A and B, but not in circle C; that is, region I (see Figure 17.4). Here we also see the advantage of using only 0's and 1's to encode data. If you have only two possibilities and one of them is incorrect, then the other one must be correct. Because the 0 in region I is incorrect, we know 1 is correct.

This technique can be used to encode all 16 possible binary messages of length 4, as shown in Table 17.1. The encoded messages are called *code words*. The three digits appended to each string of length 4 provide the "extra information" that is sufficient to infer the intended four-digit message as long as the received seven-digit message has at most one error. If a received message has two or more errors, this method will not always yield the correct message.

SPOTLIGHT 17.2 Vera Pless

Vera Pless was born on March 5, 1931, to Russian immigrants on the West Side of Chicago. The neighborhood was intellectually stimulating, and there was a tradition of teaching each other things. At age 12, Pless was taught some calculus by a mathematics graduate student. She accepted a scholarship to attend the University of Chicago at age 15. The program at Chicago emphasized great literature but paid little attention to physics and mathematics. At age 18, with no more than one precalculus course in mathematics, she entered the prestigious graduate program in mathematics at Chicago, where, at that time, there were no women on the mathematics faculty nor even women colloquium speakers. After receiving her master's degree, Pless took a job as a research associate at Northwestern University while pursuing a Ph.D. there. In the midst of writing her thesis, she moved to Boston with her husband and continued to work on her thesis at home. She defended her thesis two weeks before her daughter was born.

Over the next several years, Pless stayed at home to raise her children and taught part time at Boston University. When she decided to work full time, she found that women were not welcome at most colleges and universities. Some

Vera Pless
A leader in the field of coding theory. *(Courtesy of Vera Pless.)*

people told her outright, "I would never hire a woman." Fortunately, there was an Air Force lab in the area that had a group working on error-correcting codes. Although she had never even heard of coding theory, she was hired because of her background in algebra. When the lab discontinued basic research, she took a position as a research associate at MIT. In 1975, she went to the University of Illinois–Chicago, where she remains. Having written more than 100 research papers and a widely used book on coding theory, Pless is a leader in the field.

17.2 Encoding with Parity-Check Sums

Strings of 0's and 1's with extra digits for error correction can be used to send full-text messages. A simple way to do this is to assign a space the string 00000, the letter a the string 00001, b the string 00010, c the string 00100, and so on. Because there are 32 possible binary strings of length 5, the five unassigned strings can be used for special purposes, such as indicating uppercase letters or numerals. For example, we might use the string 11111 to indicate a "shift" from lowercase to uppercase when it precedes the code for a letter (1111100010 represents B). This is like the shift key on a keyboard. Similarly, we could use 11110 to indicate we are "shifting" from letters to numerals. Here 11110 followed by the code for a represents the numeral 0, 11110 followed by the code for b represents the numeral 1, and so on up to 9. Punctuation marks could be handled in the same fashion. However, our diagram method for assigning extra digits does not work for strings with five or more digits. Rather, the messages are encoded by appending extra digits determined by the parity of various sums of certain portions of the messages. We illustrate this method for the 16 messages shown in the left-hand column of Table 17.1. (See also Spotlight 17.3.)

Our goal is to take any binary string $a_1a_2a_3a_4$ and append three check digits $c_1c_2c_3$ so that any single error in any of the seven positions can be corrected. This is done as follows: Choose

$$c_1 = 0 \text{ if } a_1 + a_2 + a_3 \text{ is even}$$
$$c_1 = 1 \text{ if } a_1 + a_2 + a_3 \text{ is odd}$$
$$c_2 = 0 \text{ if } a_1 + a_3 + a_4 \text{ is even}$$
$$c_2 = 1 \text{ if } a_1 + a_3 + a_4 \text{ is odd}$$
$$c_3 = 0 \text{ if } a_2 + a_3 + a_4 \text{ is even}$$
$$c_3 = 0 \text{ if } a_2 + a_3 + a_4 \text{ is odd}$$

The sums $a_1 + a_2 + a_3$, $a_1 + a_3 + a_4$, and $a_2 + a_3 + a_4$ are called **parity-check sums.** They are so named because their function is to guarantee that the sum of various components of the encoded message is even. Indeed, c_1 is defined so that $a_1 + a_2 + a_3 + c_1$ is even. (Recall that this is precisely how the value in region V was defined.) Similarly, c_2 is defined so that $a_1 + a_2 + a_4 + c_2$ is even, and c_3 is defined so that $a_2 + a_3 + a_4 + c_3$ is even.

Let's revisit the message 1001 we considered in Figure 17.1. Then $a_1a_2a_3a_4 = 1001$ and

$$c_1 = 1 \text{ because } 1 + 0 + 0 \text{ is odd}$$
$$c_2 = 0 \text{ because } 1 + 0 + 1 \text{ is even}$$

and

$$c_3 = 1 \text{ because } 0 + 0 + 1 \text{ is odd}$$

So, because $c_1c_2c_3 = 101$, we have $1001 \rightarrow 1001101$.

In the middle of Neil Sloane's office, which is in the center of AT&T Bell Laboratories, which in turn is at the heart of the Information Age, there sits a tidy little pyramid of shiny steel balls stacked up like oranges at a neighborhood grocery. Sloane has been pondering different ways to pile up balls of one kind or another for most of his professional life. Along the way he has become one of the world's leading researchers in the field of sphere packing, a field that has become indispensable to modern communications. Without it, we might not have modems or compact discs or satellite photos of Neptune. "Computers would still exist," says Sloane. "But they wouldn't be able to talk to one another."

To exchange information rapidly and correctly, machines must code it. As it turns out, designing a code is a lot like packing spheres: Both involve cramming things together into the tightest possible arrangement. Sloane, fittingly, is also one of the world's leading coding theorists, not least because he has studied the shiny steel balls on his desk so intently.

Here's how a code might work. Imagine, for example, that you want to transmit a child's drawing that uses every one of the 64 colors found in a jumbo box of Crayola crayons. For transmission, you could code each of those colors as a number—say, the integers from 1 to 64. Then you could divide the image into many small units, or pixels, and assign a code to each one based on the color it contains. The transmission would then be a steady stream of those numbers, one for each pixel.

In digital systems, however, all those numbers would have to be represented as strings of 0's and 1's. Because there are 64 possible combinations of 0's and 1's in a six-digit string, you could handle the entire Crayola palette with 64 different six-digit "code words." For example, 000000 could represent the first

Neil Sloane

At work, wearing his famous "Codemart" T-shirt (952 points in a sphere). *(Courtesy of Neil Sloane/AT&T Labs.)*

color, 000001 the next color, 000010 the next, and so on.

But in a noisy signal, two different code words might look practically the same. A bit of noise, for example, might shift a spike of current to the wrong place, so that 001000 looks like 000100. The receiver might then wrongly color someone's eyes. An efficient way to keep the colors straight in spite of noise is to add four extra digits to the six-digit code words. The receiver, programmed to know the 64 permissible combinations, could now spot any other combination as an error introduced by noise and it would automatically correct the error to the "nearest" permissible color.

In fact, says Sloane, "If any of those ten digits were wrong, you could still figure out what the right crayon was."

Source: Adapted from an article by David Berreby, *Discover*, October 1990.

Now how is the intended message determined from a received encoded message? This process is called **decoding.** Say, for instance, that the message 1000, which has been encoded using parity-check sums as $u = 1000110$, is received as $v = 1010110$ (an error in the third position). We simply compare v with each of the 16 code words (that is, the possible correct messages) in Table 17.1 and decode it as the one that differs from v in the fewest positions. (Put another way, we decode v as the code word that agrees with v in the most positions.) This method works even if the error in the message is one of the check digits rather than one of the digits of the original message string. In a situation when there is more than one code word that differs from v in the fewest positions, we do not decode. To carry out this comparison, it is convenient to define the distance between two strings of equal length.

Distance Between Two Strings

The **distance between two strings** of equal length is the number of positions in which the strings differ.

For example, the distance between $v = 1010110$ and $u = 1000110$ is 1, because they differ in only one position (the third). In contrast, the distance between 1000110 and 0111001 is 7, because they differ in all seven positions. Thus, our decoding procedure is simply to decode any received message v as the code word v' that is "nearest" to v in the sense that among all distances between v and code words, the distance between v and v' is a minimum. If there is more than one possibility for v', we do not decode. Table 17.2 shows the distance between $v = 1010110$ and all 16 code words. From this table, we see that v will be decoded as u, because it differs from u in only one position, whereas it differs from all others in the table in at least two positions. This method is called *nearest-neighbor decoding.*

Nearest-Neighbor Decoding

The **nearest-neighbor decoding** method decodes a received message as the code word that agrees with the message in the most positions.

Assuming that errors occur independently, the nearest-neighbor method decodes each received message as the one it most likely represents.

The scheme we have just described was first proposed in 1948 by Richard Hamming, a mathematician at Bell Laboratories. It is one of a family of codes that are called the Hamming codes.

TABLE 17.2

v	1010110	1010110	1010110	1010110	1010110	1010110	1010110	1010110
Code word	0000000	0001011	0010111	0100101	1000110	1100011	1010001	1001101
Distance	4	5	2	5	1	4	3	4
v	1010110	1010110	1010110	1010110	1010110	1010110	1010110	1010110
Code word	0110010	0101110	0011100	1110100	1101000	1011010	0111001	1111111
Distance	3	4	3	2	5	2	6	3

Strings obtained from all possible messages of a given length of 0's and 1's by appending extra 0's and 1's using parity-check sums, as illustrated earlier, are called *binary linear codes*. The strings with the appended digits are called *code words*.

Binary Linear Code

A **binary linear code** consists of words composed of 0's and 1's obtained from all possible messages of a given length by using parity-check sums to append check digits to the messages. The resulting strings are called **code words.**

Think of a binary linear code as a set of n-digit strings in which each string is composed of two parts: the message part, consisting of the original messages, and the remaining check-digit part.

The longer the messages are, the more check digits are required to correct errors. For example, binary messages consisting of six digits require four check digits to ensure that all messages with one error can be decoded correctly.

Given a binary linear code, how can we tell whether it will correct errors and how many errors it will detect? It is remarkably easy. We examine all the code words to find one that has the fewest number of 1's, excluding the code word consisting entirely of 0's. Call this minimum number of 1's in any nonzero code word the *weight* of the code and denote it by t.

Weight of a Binary Code

The **weight of a binary code** is the minimum number of 1's that occur among all nonzero code words of that code.

If t is odd, the code will correct any $(t - 1)/2$ or fewer errors. If t is even, the code will correct any $(t - 2)/2$ or fewer errors. If we prefer simply to detect errors rather than to correct them (as is often the case in applications), the code will detect any $t - 1$ or fewer errors.

Applying this test to the code in Table 17.1, we see that the weight is 3, so it will correct any $(3 - 1)/2 = 1$ error or it will detect any $3 - 1 = 2$ errors. Be careful here. We must decide *in advance* whether we want our code to correct single errors or detect any two errors. It can do whichever we choose, but not both. If we decide to detect errors, then we will not decode any message that was not among our original list of encoded messages (just as *bive* is not a word in the English language). Instead, we simply note that an error was made and, in most applications, request a retransmission. An example of this occurs when a bar-code reader at the supermarket detects an error and therefore does not emit a sound (in effect, requesting a rescanning). On the other hand, if we decide to correct errors, we will decode any received message as its nearest neighbor.

Here is an example of another binary linear code. Let the set of messages be {000, 001, 010, 100, 110, 101, 011, 111} and append three check digits c_1, c_2, and c_3 using

$$c_1 = 0 \text{ if } a_1 + a_2 + a_3 \text{ is even}$$
$$c_1 = 1 \text{ if } a_1 + a_2 + a_3 \text{ is odd}$$
$$c_2 = 0 \text{ if } a_1 + a_3 \text{ is even}$$
$$c_2 = 1 \text{ if } a_1 + a_3 \text{ is odd}$$
$$c_3 = 0 \text{ if } a_2 + a_3 \text{ is even}$$
$$c_3 = 1 \text{ if } a_2 + a_3 \text{ is odd}$$

For example, if we take $a_1 a_2 a_3$ as 101, we have

$$c_1 = 0 \text{ because } 1 + 0 + 1 \text{ is even}$$
$$c_2 = 0 \text{ because } 1 + 1 \text{ is even}$$
$$c_3 = 1 \text{ because } 0 + 1 \text{ is odd}$$

So we encode 101 by appending 001, that is, $101 \rightarrow 101001$. The entire code is shown in Table 17.3.

Because the minimum number of 1's of any nonzero code word is three, this code will either correct any single error or detect any two errors, whichever we choose.

TABLE 17.3

Message	$\rightarrow$	Code Word	Message	$\rightarrow$	Code Word
000	$\rightarrow$	000000	110	$\rightarrow$	110011
001	$\rightarrow$	001111	101	$\rightarrow$	101001
010	$\rightarrow$	010101	011	$\rightarrow$	011010
100	$\rightarrow$	100110	111	$\rightarrow$	111100

It is natural for you to ask how the method of appending extra digits with parity-check sums enables us to detect or even correct errors. Error detection is obvious. Think of how a computer spell-checker works. If you type *bive* instead of *five*, the spell-checker detects the error because the string *bive* is not on its list of valid words. On the other hand, if you type *give* instead of *five*, the spell-checker will not detect the error because *give* is on its list of valid words.

Our error-detection scheme works the same way, except that if we add extra digits to ensure that our code words differ in many positions—say, t positions—then even as many as $t - 1$ mistakes will not convert one code word into another code word. And if every pair of code words differ from each other in at least three positions, we can correct any single error because the incorrect received word will differ from the correct code word in one position, but it will differ from all others in two or more positions. Thus, in this case, the correct word is the "nearest neighbor." So the role of the parity-check sums is to ensure that code words differ in many positions. For example, consider the code in Table 17.1. The messages 1000 and 1100 differ in only the second position. But the two parity-check sums $a_1 + a_2 + a_3$ and $a_2 + a_3 + a_4$ will guarantee that encoded words for these messages will have different values in positions 5 and 7 as well as in position 2. It is the job of mathematicians to discover the appropriate parity-check sums to correct several errors in long, complicated codes.

Data Compression

Binary linear codes are fixed-length codes. In a fixed-length code, each code word is represented by the same number of digits (or symbols). In contrast, the Morse code (see Figure 17.5), designed for the telegraph, is a **variable-length code,** that is, a code in which the number of symbols for each code word may vary.

Notice that in the Morse code the letters that occur most frequently have the shortest coding, whereas the letters that occur the least frequently have the longest coding. By assigning the code in this manner, telegrams could convey more information per line than would be the case for fixed-length codes or a randomly assigned variable-length coding of the letters. The Morse code is an example of data compression.

A	·—
B	—···
C	—·—·
D	—··
E	·
F	··—·
G	——·
H	····
I	··
J	·———
K	—·—
L	·—··
M	——
N	—·
O	———
P	·——·
Q	——·—
R	·—·
S	···
T	—
U	··—
V	···—
W	·——
X	—··—
Y	—·——
Z	——··

FIGURE 17.5 Morse code.

> ## ⣿ Data Compression
>
> **Data compression** is the process of encoding data so that the most frequently occurring data are represented by the fewest symbols.

Figure 17.6 shows a typical frequency distribution for letters in English-language text material.

Data compression provides a means to reduce the costs of data storage and transmission. A compression algorithm converts data from an easy-to-use format to one optimized for compactness. Conversely, an uncompression algorithm con-

	A	B	C	D	E	F	G	H	I	J	K	L	M
Percentage:	8	1.5	3	4	13	2	1.5	6	6.5	0.5	0.5	3.5	3
	N	O	P	Q	R	S	T	U	V	W	X	Y	Z
Percentage:	7	8	2	0.25	6.5	6	9	3	1	1.5	0.5	2	0.25

FIGURE 17.6 A widely used frequency table for letters in normal English usage.

verts the compressed information back to its original form or approximately its original form. Downloaded files in the ZIP format are an example of a particular kind of data compression. When you "unzip" the file, you return the compressed data to its original state. In some applications, such as data sets that represent images, the original data need only be recaptured in approximate form. In these cases, there are algorithms that result in a great saving of space. Graphics Interchange Format (GIF) encoding returns compressed data to its exact original form, while JPEG encoding and MPEG encoding return data only approximately to its original state.

Let's illustrate the principles of data compression with a simple example. Biologists are able to describe genes by specifying sequences composed of the four letters A, T, G, and C, which represent the four nucleotides adenine, thymine, guanine, and cytosine, respectively. One way to encode a sequence such as AAACAGTAAC in fixed-length binary form would be to encode the letters as

$$A \rightarrow 00 \qquad C \rightarrow 01 \qquad T \rightarrow 10 \qquad G \rightarrow 11$$

The corresponding binary code for the sequence AAACAGTAAC is then

$$00000001001110000001$$

On the other hand, if we knew from experience that A occurs most frequently, C second most frequently, and so on, and that A occurs much more frequently than T and G together, the most efficient binary encoding would be

$$A \rightarrow 0 \qquad C \rightarrow 10 \qquad T \rightarrow 110 \qquad G \rightarrow 111$$

For this encoding scheme, the sequence AAACAGTAAC is encoded as

$$0001001111100010$$

Notice that this binary sequence has 20% fewer digits than our previous sequence, in which each letter was assigned a fixed length of 2 (16 digits versus 20 digits). However, to realize this savings, we have made decoding more difficult. For the binary sequence using the fixed length of two symbols per character, we decode the sequence by taking the digits two at a time in succession and converting them to the corresponding letters. For the compressed coding, we can decode by examining the digits in groups of three.

EXAMPLE 1   Decode 0001001111100010

Consider the compressed binary sequence 0001001111100010. Look at the first three digits: 000. Since our code words have one, two, or three digits and neither 00 nor 000 is a code word, the sequence 000 can represent only the *three* code words 0, 0, and 0. Now look at the next three digits: 100. Again, because neither 1 nor 100 is a code word, the sequence 100 represents the *two* code words 10 and 0. The next three digits, 111, can represent only the code word 111 because the other three code words all contain at least one 0. Next consider the sequence 110. Because neither 1 nor 11 is a code word, the sequence 110 can represent only 110 itself. Continuing in this fashion, we can decode the entire sequence to obtain AAACAGTAAC. ■

The following observation can simplify the decoding process for compressed sequences. Note that 0 occurs only at the end of a code word. Thus, each time you see a 0, it is the end of the code word. Also, because the code words 0, 10, and 110 end in a 0, the only circumstances under which there are three consecutive 1's is when the code word is 111. So, to quickly decode a compressed binary sequence using our coding scheme, insert a comma after every 0 and after every three consecutive 1's. The digits between the commas are code words.

EXAMPLE 2   Code AGAACTAATTGACA and Decode the Result

Recall: A → 0, C → 10, T → 110, and G → 111. So

$$AGAACTAATTGACA \rightarrow 01110010110001101101110100$$

To decode the encoded sequence, we insert commas after every 0 and after every occurrence of 111 and convert to letters:

0,	111,	0,	0,	10,	110,	0,	0,	110,	110,	111,	0,	10,	0
A,	G,	A,	A,	C,	T,	A,	A,	T,	T,	G,	A,	C,	A

■

Delta Encoding

For data sets of numbers that fluctuate little from one number to the next, the method of compression called the *delta function* works well. Consider the following closing prices (rounded to the nearest integer) of the Standard & Poor's index of the stock prices of 500 companies in March 2004.

1156 1149 1151 1155 1157 1147 1141 1124 1107 1121 1104
1111 1124 1122 1110 1095 1094 1091 1109 1108 1122 1127 1126

These numbers use 92 characters in all. To compress this data set using the delta method, we start with the first number and continue by listing only the change from each entry to the next. So our list becomes

$$1156 \ -7 \ 2 \ 4 \ 2 \ -10 \ -6 \ -17 \ -17 \ 14 \ -17 \ 7 \ 13$$
$$-2 \ -12 \ -15 \ -1 \ -3 \ 18 \ -1 \ 14 \ 5 \ -1$$

This time we have used only 49 characters, counting the minus signs, to represent the same data—a savings of almost 47%.

Huffman Coding

The methods we have shown previously are too simple for general use, but in 1951 a graduate student named David Huffman (see Spotlight 17.4) devised a scheme for data compression that became widely used. As was the case for the first scheme we discussed, Huffman coding assigns short code words to those characters with high probabilities of occurring and long code words to those with low probabilities of occurring. A Huffman code is made using a so-called *code tree* by arranging the characters from top to bottom according to increasing probability; it proceeds by combining, at each stage, the two least probable combinations and repeating this process until there is only one combination remaining. To illustrate the method, say we have a data set of six letters that occur with the following probabilities.

A	0.125
B	0.051
C	0.215
D	0.173
E	0.210
F	0.226

Rearranging them in increasing order, we have:

B	0.051
A	0.125
D	0.173
E	0.210
C	0.215
F	0.226

Because B and A are the two least likely to occur, we begin our tree by merging them, adding their probabilities, and rearranging the resulting items in increasing order:

D	0.173
BA	0.176
E	0.210
C	0.215
F	0.226

SPOTLIGHT 17.4   David Huffman

Large networks of IBM computers use it. So do high-definition televisions, modems, and a popular electronic device that takes the brainwork out of programming a videocassette recorder. All these digital wonders rely on the results of a 50-year-old term paper by an MIT graduate student—a data-compression scheme known as Huffman encoding.

In 1951, David Huffman and his classmates in an electrical engineering graduate course on information theory were given the choice of a term paper or a final exam. For the term paper, Huffman's professor had assigned what at first appeared to be a simple problem. Students were asked to find the most efficient method of representing numbers, letters, or other symbols using binary code. Huffman worked on the problem for months, developing a number of approaches, but none that he could prove to be the most efficient. Finally, he despaired of ever reaching a solution and decided to start studying for the final. Just as he was throwing his notes in the garbage, the solution came to him. "It was the most singular moment of my life," Huffman says. "There was the absolute lightning of sudden realization. It was my luck to be there at the right time and also not have my professor discourage me by telling me that other good people had struggled with the problem," he says. When presented with his student's discovery, Huffman recalls, his professor exclaimed: "Is that all there is to it!"

"The Huffman code is one of the fundamental ideas that people in computer science and data communications are using all the time," says Donald Knuth of Stanford University. Although others have used Huffman's code to help make millions of dollars, Huffman's main compensation was

David Huffman
(Matthew Mulbry.)

dispensation from the final exam. He never tried to patent an invention from his work and experienced only a twinge of regret at not having used his creation to make himself rich. "If I had the best of both worlds, I would have had recognition as a scientist, and I would have gotten monetary rewards," he says. "I guess I got one and not the other."

But Huffman received other compensation. A few years ago an acquaintance told him that he had noticed that a reference to the code was spelled with a lowercase *h*. Remarked his friend to Huffman, "David, I guess your name has finally entered the language."

David Huffman died October 7, 1999.

Source: Adapted from an article by Gary Stix, *Scientific American*, September 1991, pp. 54, 58.

This time *D* and *BA* are the two least likely remaining entries so we merge them, add their probabilities, and re-sort from smallest to largest. This gives:

E	0.210
C	0.215
F	0.226
DBA	0.349

Next we combine *E* and *C* and re-sort to get:

F	0.226
DBA	0.349
EC	0.425

Then we combine *F* and *DBA* and re-sort:

EC	0.425
FDBA	0.575

And finally,

ECFDBA	1.000

To assign a binary code word to each letter, we work our way back from the end of the tree to each letter by assigning, at each merging juncture, 0 to the branch with the lower probability, as shown in Figure 17.7. (There is more than one way to draw the tree. Moreover, we can assign the 0's and 1's to the branches in any fashion, but we do it in this specific way for convenience.)

FIGURE 17.7 A
Huffman tree.

The path to each letter determines the code word for that letter. So we have:

A	1111
B	1110
C	01
D	110
E	00
F	10

Notice that the letters that occur least often have the longest codes and the letters that occur most often have the shortest codes. Decoding codes created from a Huffman tree is possible because at each stage there is only one way a particular string could have occurred. Here is an example. Consider the Huffman code created using the code words given in the previous display:

$$1110100001101001111010010$$

How can we determine the corresponding string of letters that has this Huffman code? The method is quite simple. Look at the first two digits. If they correspond to a code word, then decode them as that letter. If not, then look at the next digit. If the three digits correspond to a code word, then decode it as the corresponding letter. If not, then the resulting four digits will be a code word. Looking at our example 1110100001101001111010010 we see that neither 11 nor 111 is a code word but 1110 is the code word for B. So, replacing 1110 with B, we have B100001101001111010010. The next possibility is 10, which is the code word for F, so we have BF0001101001111010010. Next we have 00, which is the code word for E, giving us BFE01101001111010010. Continuing in this way, we obtain BFECFFCACEF. Of course, in practice, coding and decoding are done by computers.

17.3 Cryptography

Thus far, we have discussed ways in which data can be encoded to detect errors or correct errors in transmission. In many situations, there is also a desire for security against unauthorized interpretation of coded data (that is, a desire for secrecy). The process of disguising data is called **encryption. Cryptology** is the study of methods to make and break secret codes.

Historically, encryption was used primarily for military and diplomatic transmissions. Today, encryption is essential for securing electronic transactions of all kinds. Cryptography is what allows you to have a Web site safely receive your credit-card number. Cryptographic schemes prevent hackers from charging calls to your cellular phone. Cryptography is also used for authenticating electronic transactions. In September 1998, history was made when former President Bill Clinton and Ireland's Prime Minister Bertie Ahern used digital signatures to sign an intergovernmental document. Each leader had a unique signing code and a digital certificate that served as a "digital ID," thereby ensuring that the document was approved by them. Although modern encryption schemes are extremely complex, we will illustrate the fundamental concepts involved with a few simple examples.

Among the first known cryptosystems is the so-called **Caesar cipher** used by Julius Caesar to send messages to his troops. To encrypt a message with the method employed by Caesar, we use the following table to replace each letter in the top row with the letter below it.

A B C D E F G H I J K L M N O P Q R S T U V W X Y Z
D E F G H I J K L M N O P Q R S T U V W X Y Z A B C

For example, the message ATTACK AT DAWN is encrypted as DWWDFN DW GDZQ.

To decrypt the message, replace each letter with the letter above it in the table. Obviously, it would not require much effort for someone to "crack" this code.

To describe a more sophisticated scheme for transmitting messages secretly, it is convenient to introduce a special kind of arithmetic used in cryptography. For any positive integers a and n, we define $a \bmod n$ (read: "a modulo n" or just "$a \bmod n$") to be the remainder when a is divided by n. Thus,

$$3 \bmod 2 = 1 \text{ because } 3 = 1 \cdot 2 + 1$$
$$6 \bmod 2 = 0 \text{ because } 6 = 3 \cdot 2 + 0$$
$$5 \bmod 3 = 2 \text{ because } 5 = 1 \cdot 3 + 2$$
$$37 \bmod 10 = 7 \text{ because } 37 = 3 \cdot 10 + 7$$
$$38 \bmod 26 = 12 \text{ because } 38 = 1 \cdot 26 + 12$$
$$342 \bmod 85 = 2 \text{ because } 342 = 4 \cdot 85 + 2$$
$$62 \bmod 85 = 62 \text{ because } 62 = 0 \cdot 85 + 62$$

Arithmetic involving mod n is called **modular arithmetic.** Although this arithmetic may appear unfamiliar, you often unconsciously use it. For example, if it is now September, what month will it be 25 months from now? Of course, you answer "October," but the interesting fact is that you didn't arrive at the answer by starting with September and counting off 25 months. Instead, without even thinking about it, you simply observed that $25 = 2 \cdot 12 + 1$ so that 25 mod $12 = 1$, and you added one month to September. Similarly, if it is now Wednesday, you know that in 23 days it will be Friday. This time, you arrived at your answer by noting that $23 = 3 \cdot 7 + 2$ (that is, 23 mod $7 = 2$), so you added 2 days to Wednesday instead of counting off 23 days. Likewise, if your electricity is off for 26 hours, you know you must advance your clock 2 hours, because 26 mod $12 = 2$. An application of modular arithmetic to genetics is described in Spotlight 17.5.

SPOTLIGHT 17.5 Modeling the Genetic Code

The way that genetic material is composed can be conveniently modeled using modulo 4 arithmetic. A DNA molecule is made up of two long strands in the form of a double helix. Each strand is made up of strings of the four nitrogen bases adenine (A), thymine (T), guanine (G), and cytosine (C). Each base on one strand binds to a complementary base on the other strand. Adenine is always bound to thymine, and guanine is always bound to cytosine. To model this situation, we identify A with 0, T with 2, G with 1, and C with 3. Thus, the DNA segment ACGTAACAGGA and its complement segment TGCATTGTCCT are identified by 03120030110 and 21302212332.

Using modulo 4 arithmetic, $0 + 2 = 2$, $2 + 2 = 0$, $1 + 2 = 3$, and $3 + 2 = 1$, and we see that adding 2 to any of the integers 0, 1, 2, or 3 interchanges 0 and 2 and 1 and 3. So, for any DNA segment $a_1 a_2 \cdots a_n$ represented by strings of 0's, 1's, 2's, and 3's, we see that its complementary segment is represented by $a_1 a_2 \cdots a_n + 22 \cdots 2$, where we add the integers in each component using modulo 4. In particular, $03120030110 + 22222222222 = 21302212332$.

Source: Adapted from *Discrete Mathematics* by S. Washburn, T. Marlow, and C. Ryan, Addison-Wesley, 1999.

With modular arithmetic, we can easily describe the Caesar cipher as follows. Begin by saying that the letter *A* is in position 0, *B* is in position 1, *C* is in position 2, and so on. Then the Caesar cipher replaces the letter in position *i* with the letter in position $(i + 3) \bmod 26$. This formula expresses the fact that the Caesar cipher shifts each letter from *A* through *W* three positions to the right while *X*, *Y*, and *Z* are replaced with *A*, *B*, and *C*, respectively.

Modular arithmetic also provides the basis for a more sophisticated cryptosystem called the **Vigenère cipher.** For this method we first select a **key word,** which can be any word. The letters of the key word are then used to determine the amount of shifting for each letter of our message.

EXAMPLE 3 Vigenère Cipher

We will use the Vigenère system to encrypt the message ATTACK AT DAWN. Choosing the key word MATH, we shift the first letter of the message by 12 because *M* is in position 12; the second letter of the message is shifted by 0 (unchanged) because *A* is in position 0; the third letter of the message is shifted by 19 because *T* is in position 19, and so on. A shift of *j* means that the letter in position *i* is replaced by the letter in position $(i + j) \bmod 26$. When we have used all the letters of the key word, we start over at the beginning. To encrypt ATTACK AT DAWN using the key word MATH, we first note that the letters in the key word MATH are in positions 12, 0, 19, and 7, respectively. So, the *A* in ATTACK is converted to *M* ($0 + 12 = 12$), the first *T* in ATTACK is converted to *T* ($19 + 0 = 19$), the second *T* in ATTACK is converted to *M* (($19 +

TABLE 17.4												
ATTACK AT DAWN	0	19	19	0	2	10	0	19	3	0	22	13
MATHMA TH MATH	12	0	19	7	12	0	19	7	12	0	19	7
MTMHOK TA PAPU	12	19	12	7	14	10	19	0	15	0	15	20

19) mod 26 = 12), and so on. The first two lines of Table 17.4 show the position numbers for the letters of the message and the key word. The third line of the table is obtained from the first two by adding the values in the columns mod 26 and converting the results back to letters. ■

Cryptograms

A form of cryptography that is the basis for popular word puzzles called *cryptograms* can be found in newspapers, in puzzle books, and on Web sites. A **cryptogram** is a sentence (or message) that has been encrypted. Figure 17.6, showing letter frequencies, together with the following statistical information is helpful when trying to solve cryptograms.

Obviously, a word consisting of a single letter must be the word *A* or *I*.

The most common two-letter words in order of frequency are: *of, to, in, it, is, be, as, at, so, we, he, by, or, on, do, if, me, my, up, an, go, no, us, am.*

The most common three-letter words in order of frequency are: *the, and, for, are, but, not, you, all, any, can, had, her, was, one, our, out, day, get.*

The most common four-letter words in order of frequency are: *that, with, have, this, will, your, from, they, know, want, been, good, much, some, time.*

The most commonly used words in the English language in order of frequency are: *the, of, and, to, in, a, is, that, be, it, by, are, for, was, as, he, with, on, his, at, which, but, from, has, this, will, one, have, not, were, or.*

The most common double letters in order of frequency are: *ss, ee, tt, ff, ll, mm, oo.*

Using this information, you can often solve the cryptogram. Since the letter *e* is the most frequent letter, it is best to start by assuming that one of the most frequent letters in the cryptogram is *e*. Once you have the *e* assigned, you can usually find the code for *the*, since it is the most frequent three-letter word. Correctly decoding *the* will determine the word *that*, since *a* is the only vowel that results in a word. This, in turn, identifies the code for the single-letter word *I*, because *A* and *I* are the only two one-letter words. Knowing the correct code for *t*, *h*, *e*, and *i* will give you the code for *r* and *w* provided that the cryptogram contains the words *their* and *with*. Next is the code for *s* from the word *this*, since it is the most likely four-letter word that has the first three letters *thi*. Continuing in this fashion, you can eventually solve the puzzle.

It should come as no surprise that longer cryptograms are easier to solve than shorter ones. Generally speaking, the longer the cryptograms, the more closely they follow predicted patterns. Also, longer cryptograms offer more clues and provide greater opportunities to reveal incorrect choices by yielding incompatible consecutive letters such as *qw, hw,* or *wf.* Using the information given previously and computer software such as that found as www.d.umn.edu/~jgallian/project. petr/cryptogram.html that counts the frequencies of single letters, letter pairs, and letter triples, we can quickly solve the following cryptogram.

P NVLS V GASVE JNVJ WUS GVQ JNPM UVJPWU KPOO APMS RY VUG OPLS WRJ JNS JARS ESVUPUH WB PJM ZASSG: "KS NWOG JNSMS JARJNM JW CS MSOB-SLPGSUJ: JNVJ VOO ESU VAS ZASVJSG SIRVO." P NVLS V GASVE JNVJ WUS GVQ WU JNS ASG NPOOM WB HSWAHPV JNS MWUM WB BWAESA MOVLSM VUG JNS MWUM WB BWAESA MOVLSWKUSAM KPOO CS VCOS JW MPJ GWKU JWHSJNSA VJ V JVCOS WB CAWJNSANWWG. P NVLS V GASVE JNVJ WUS GVQ SLSU JNS MJVJS WB EPMMPMMPYYP, V GSMSAJ MJVJS, MKSOJSAPUH KPJN JNS NSVJ WB PUXRMJPZS VUG WYYASMMPWU, KPOO CS JAVUMBWAESG PUJW VU WVMPM WB BASSGWE VUG XRMJPZS. P NVLS V GASVE JNVJ EQ BWRA ZNPOGASU KPOO WUS GVQ OPLS PU V UVJPWU KNSAS JNSQ KPOO UWJ CS XRGHSG CQ JNS ZWOWA WB JNSPA MFPU CRJ CQ JNS ZWUJSUJ WB JNSPA ZNVAVZJSA. P NVLS V GASVE JWGVQ.

The Web software at www.d.umn.edu/~jgallian/project.petr/cryptogram.html reveals that the most frequent letters in this cryptogram are: *S* 14%, *J* 10%, *V* 9%, *W* 8%; the most common letter pairs are *JN, NS, AS*; and the most common letter triples are *JNS, GAS, ASV,* and *POO*.

Since *S* is the most frequent letter in the cryptogram and *e* is the most frequent letter in English, we suspect that $S = e$. (To distinguish which letters in the cryptogram have been replaced, we use lowercase letters for the ones that have been changed.) This is reinforced by the observation that the most common triple in the cryptogram is *JNS* and the most common three-letter word is *the*. So we try $J = t$, $N = h$, and $S = e$.

We now see from *thVt* that $V = a$, and this gives us from *thPs* that $P = i$. Knowing which letter is *i* gives us $A = r$ from *theiA* and $K = w$ from *Kith*. Moreover, we are able to determine three more letters from the strings *tW, Mtate,* and *haLe*. Mentally trying the unused letters for *W* in *tW, M* in *Mtate,* and *L* in *haLe*, we see that only $W = o$, $M = s$, and $L = v$ make sense. Continuing in this manner we easily solve the puzzle as

I have a dream that one day this nation will rise up and live out the true meaning of its creed: "We hold these truths to be self-evident: that all men are created equal." I have a dream that one day on the red hills of Georgia the sons of former slaves and the sons of former slaveowners will be able to sit down together at a table of brotherhood. I have a dream that one day even the state of Mississippi, a desert state, sweltering with the heat of injustice and oppression, will be transformed into an oasis of freedom and justice. I have a dream that my four children will one day live in a nation where they will not be judged by the color of their skin but by the content of their character. I have a dream today.

The technique of using the frequencies of the letters to break a cryptogram was employed by the Arab mathematician Abu al-Kindi in the ninth century. His treatise entitled "A Manuscript on Deciphering Cryptographic Messages" has survived.

Spotlight 17.6 discusses a famous cryptogram that is the basis of a Sherlock Holmes story.

Encrypting Credit-Card Data on the Web

Suppose that you want to purchase a compact disc from Amazon.com. Should you be concerned that a hacker will intercept your credit-card number during the transaction? As you might expect, your credit-card number is sent to Amazon in encrypted form to protect the data.

To describe one way that this encryption can be done, we need to perform addition of binary strings. We add two binary strings $a_1 a_2 \cdots a_n$ and $b_1 b_2 \cdots b_n$ as follows:

$$
\begin{array}{r}
a_1 a_2 \cdots a_n \\
+ \ b_1 b_2 \cdots b_n \\
\hline
c_1 c_2 \cdots c_n
\end{array}
$$

where $c_i = 0$ if $a_i = b_i$ and $c_i = 1$ if $a_i \neq b_i$. Equivalently, $c_i = (a_i + b_i) \bmod 2$. (Add a_i and b_i in the ordinary way, but replace 2 by 0.)

SPOTLIGHT 17.6   Adventure of the Dancing Men

In Arthur Conan Doyle's story "The Adventure of the Dancing Men," Sherlock Holmes solves a murder mystery by realizing that the series of messages consisting of stick figures of dancing men shown below are cryptograms. Holmes used facts about the frequencies of letters and words in the English language to decipher the code.

EXAMPLE 4   Sum of Binary Strings

$$
\begin{array}{ccc}
11000111 & 00111011 & 10011100 \\
+\ \underline{01110110} & +\ \underline{01100101} & +\ \underline{10011100} \\
10110001 & 01011110 & 00000000
\end{array}
$$

■

We can now explain one way to send credit-card numbers over the Web securely. When you place an order with Amazon, the company sends your computer a randomly generated string of 0's and 1's called a **key.** This key has the same length as the binary string corresponding to your credit-card number, and the two strings are added (think of this process as "locking" the data). The resulting sum is then transmitted to Amazon. Amazon in turn adds the same key to the received string, which then produces the original string corresponding to your credit-card number (adding the key a second time "unlocks" the data).

To illustrate the idea, say you want to send an eight-digit binary string such as $s = 10101100$ to Amazon (actual credit-card numbers have very long strings), and Amazon sends your computer the key $k = 00111101$. Your computer returns the string $s + k = 10101100 + 00111101 = 10010001$ to Amazon, and Amazon adds k to this string to get $10010001 + 00111101 = 10101100$, which is the string representing your credit-card number. If someone intercepts the number $s + k = 10010001$ during transmission, it is of no value without knowing the k. This method works because of the property of binary addition that $a_1 a_2 \cdots a_n + b_1 b_2 \cdots b_n = 00 \cdots 0$ if and only if the two strings are identical. Thus, $(s + k) + k = s + (k + k) = s + 00 \cdots 0 = s$. The method is secure because the key sent by Amazon is randomly generated and used only one time.

You can tell when you are using an encryption scheme on a Web transaction by looking to see if the Web address begins with "https" rather than the customary "http." You will also see a small padlock in the status bar at the bottom of the browser window.

Public Key Cryptography

In the mid-1970s Ronald Rivest, Adi Shamir, and Leonard Adleman devised an ingenious method that permits each person who is to receive a secret message to publicly tell how to scramble messages sent to him or her. Even though the method used to scramble the message is known publicly, only the person for whom it is intended will be able to unscramble the message.

To illustrate their method for transmitting messages secretly, we need the following property of modular arithmetic:

$$(ab) \bmod n = ((a \bmod n)(b \bmod n)) \bmod n$$

This property allows you to replace integers that are greater than or equal to n with integers that are less than n to simplify calculations. You should think of it as saying, "mod before you multiply."

Multiplication Property for Modular Arithmetic

EXAMPLE 5

$$(17 \cdot 23) \bmod 10 = ((17 \bmod 10)(23 \bmod 10)) \bmod 10$$
$$= (7 \cdot 3) \bmod 10 = 21 \bmod 10 = 1$$
$$(22 \cdot 19) \bmod 8 = ((22 \bmod 8)(19 \bmod 8)) \bmod 8$$
$$= (6 \cdot 3) \bmod 8 = 18 \bmod 8 = 2$$
$$(100 \cdot 8) \bmod 85 = ((100 \bmod 85)(8 \bmod 85)) \bmod 85$$
$$= (15 \cdot 8) \bmod 85 = 120 \bmod 85 = 35 \;\blacksquare$$

We now describe the Rivest, Shamir, and Adleman method by way of a simple example. Say we wish to send the message "IBM." We convert the message to digits by replacing A by 1, B by 2, . . . , and Z by 26. So the message IBM becomes 9213. The person to whom the message is to be sent has picked two primes p and q, say, $p = 5$ and $q = 17$. (Recall that a *prime* is an integer greater than 1 whose only divisors are 1 and itself.) The receiver has also picked a number r, such as 3, that has no divisors in common with the least common multiple m of $(p - 1) = 4$ and $(q - 1) = 16$ other than 1, and published $n = pq = 85$ and $r = 3$ in a public directory. To decode our message, the receiver must find a number s so that $r \cdot s = 1 \bmod m$ (this is where knowledge of p and q is necessary). That is, $3 \cdot s = 1 \bmod 16$. This number is 11. (The number s can be found by calculating successive powers of $r \bmod m$. When 1 is reached, the previous power of r is s. In our example, we have $3 \bmod 16 = 3$, $3^2 \bmod 16 = 9$, $3^3 \bmod 16 = 11$, $3^4 \bmod 16 = 1$, so $s = 3^3 \bmod 16 = 11$.)

To send our message to this person, we consult the public directory to find $n = 85$ and $r = 3$, then send the "scrambled" numbers $9^3 \bmod 85$, $2^3 \bmod 85$, and $13^3 \bmod 85$ rather than 9, 2, and 13, and the receiver will unscramble them. Thus, we send:

$$9^3 \bmod 85 = 49$$
$$2^3 \bmod 85 = 8$$
$$13^3 \bmod 85 = 72$$

Now the receiver must take the numbers he or she receives—49, 8, and 72—and convert them back to 9, 2, and 13 by calculating $49^{11} \bmod 85$, $8^{11} \bmod 85$, and $72^{11} \bmod 85$.

The calculation of 49^{11} and 85 can be simplified as follows:[1]

[1]To determine $49^2 \bmod 85$ with a calculator, enter 49×49 to obtain 2401, then divide 2401 by 85 to obtain 28.247058. Finally, enter $2401 - (28 \times 85)$ to obtain 21. Provided that the numbers are not too large, the search engine Google at www.google.com will do modular arithmetic by simply entering the number and the mod value separated by a % sign in the search box. Be careful, however, because entering 49^11%85 yields 0, which is incorrect ($m^r \bmod n$ is never 0 when the greatest common divisor of m and n is 1). Instead, we can use Google to compute smaller powers such as 49^4%85 = 16 and 49^7%85 = 59, then compute $(16 \times 59)\%85 = 9$.

$49 \bmod 85 = 49$

$49^2 \bmod 85 = 2401 \bmod 85 = 21$

$49^4 \bmod 85 = 49^2 \cdot 49^2 \bmod 85 = 21 \cdot 21 \bmod 85 = 441 \bmod 85 = 16$

$49^8 \bmod 85 = 49^4 \cdot 49^4 \bmod 85 = 16 \cdot 16 \bmod 85 = 1$

So,

$$49^{11} \bmod 85 = (49^8 \bmod 85)(49^2 \bmod 85)(49 \bmod 85)$$
$$= (1 \cdot 21 \cdot 49) \bmod 85$$
$$= 1029 \bmod 85$$
$$= 9$$

Thus, the receiver has correctly determined the code for *I*. The calculations for $8^{11} \bmod 85$ and $72^{11} \bmod 85$ are left as exercises. Notice that without knowing how $n = pq$ factors, we cannot find the least common multiple of $p - 1$ and $q - 1$ (in our case, 16), and therefore the *s* that is needed to determine the intended message.

The procedure just described is called the **RSA public key encryption scheme** in honor of Rivest, Shamir, and Adleman, who discovered it. The method is practical and secure because efficient methods exist for finding very large prime numbers (say, about 100 digits long) and for multiplying large numbers, but no one knows an efficient algorithm for factoring large integers (say, about 200 digits long).

The algorithm is summarized below. In practice, the messages are not sent one letter at a time. Rather, the entire message is converted to decimal form, with *A* represented by 01, *B* by 02, . . . , and a space by 00. The message is then broken up into blocks of uniform size and the blocks are sent. See step 2 under Sender below.

Receiver

1. Pick very large primes *p* and *q* and compute $n = pq$.
2. Compute the least common multiple of $p - 1$ and $q - 1$; let's call it *m*.
3. Pick *r* so that it has no divisors in common with *m* other than 1 (any such *r* will do).
4. Find *s* so that $rs = 1 \bmod m$. (To find *s*, simply compute $r^2 \bmod m$, $r^3 \bmod m$, $r^4 \bmod m$, . . . until you reach $r^t \bmod m = 1$. Then $s = r^{t-1} \bmod m$.)
5. Publicly announce *n* and *r*, but keep *p*, *q*, and *s* secret.

Sender

1. Convert the message to a string of digits.

2. Break up the message into uniformly sized blocks of digits, appending 0's in the last block if necessary. Call them $M_1, M_2, \ldots, M_k$. For example, for a string such as 2105092315, we would use $M_1 = 2105$, $M_2 = 0923$, and $M_3 = 1500$.

3. Check to see that the greatest common divisor of each M_i and n is 1. If not, n can be factored and the code is broken. (In practice, the primes p and q are so large that they exceed all M_i, so this step may be omitted.)

4. Calculate and send $R_i = M_i^r \bmod n$.

Receiver

1. For each received message R_i, calculate $R_i^s \bmod n$.

2. Convert the string of digits back to a string of characters.

Let's do another example step by step with $p = 7$, $q = 11$, and the message "HI."

Receiver

1. $n = 77$.
2. The least common multiple m of $7 - 1 = 6$ and $11 - 1 = 10$ is 30.
3. We pick $r = 7$.
4. Since $7^4 = 1 \bmod 30$, we have $s = 7^3 \bmod 30 = 13$.
5. Make public $n = 77$ and $r = 7$.

Sender

1. HI converts to 89.
2. We will send 8 and 9 individually (that is, our blocks have size 1).
3. The greatest common divisor of 8 and 77 is 1, and the greatest common divisor of 9 and 77 is 1, so we can proceed.
4. Send $8^7 \bmod 77 = 57$ and $9^7 \bmod 77 = 37$.

Receiver

1. $57^{13} \bmod 77 = 8$; $37^{13} \bmod 77 = 9$.
2. 89 converts to HI.

This method works because of a basic property of modular arithmetic and the choice of r. As a result of choosing the number m as we described, it has the property that for each positive integer x having no common divisors with n except 1, we have $x^m = 1 \bmod n$. So, in the case of our first example with $n = 85$, $m = 16$, and $r = 3$, for the original message 9 and the received message 49, we have mod 85

$$49^{11} = (9^3)^{11} = 9^{33} = 9 \cdot 9^{32} = 9 \cdot (9^{16})^2 = 9 \cdot 1^2 = 9$$

In 2002, Rivest, Shamir, and Adleman received the Association for Computing Machinery A. M. Turing Award, which is considered to be the "Nobel Prize of Computing," for their seminal contribution to public key cryptography.

17.4 Web Searches and Mathematical Logic

In 2004, the number of Web pages indexed by large Internet search engines such as Google exceeded 8 billion. Spotlight 17.7 shows a visualization of the network structure of the Internet. Computer scientists and mathematicians attempt to manage massive data sets by taking advantage of the associated network structure, which represents the interrelations of the data. The algorithm used by the Google

SPOTLIGHT 17.7 The Network Structure of the Internet

The image below appeared as a portion of the Mathematics Awareness Month April 2004 poster. See http://mathforum.org/mam/04/ for the entire poster.

search engine, for instance, ranks all pages on the Web using these interrelations to determine their relevance to the user's search. Factors such as the frequency, location near the top of the page of key words, font size, and number of links are taken into account. Spotlight 17.8 discusses the "Kevin Bacon" game that illustrates the so-called small-world phenomenon of the interconnectedness of data. In this section we will show how a branch of mathematics called **Boolean logic,** after the nineteenth-century mathematician George Boole (1815–1864), can be used to make search engine queries more efficient.

An *expression* in Boolean logic is simply a statement that is either true or false. For example, the expression, "There are infinitely many prime numbers," is either true or false. For purposes of mathematical reasoning, it is not necessary that we know whether this statement is true or false, but simply that it is one or the other. (It was proved to be true by Euclid more than 2000 years ago.) The expression, "The integer 51 is prime," is an example of an expression that is false, since $51 = 3 \cdot 17$. When combining expressions in logic, we avoid statements that are subject to opinion or various interpretations, such as "Ralph Nader cost Al Gore the 2000 United States presidential election." When we enter a phrase such as "Florida election" as a query to a search engine, the search engine automatically interprets it as the expression, "This Web page contains the phrase 'Florida election'." The search engine then returns the list of Web pages for which this expression is true.

In this section we discuss how complex expressions can be constructed by connecting individual expressions with the *connectives* AND, OR, and NOT. For example, to obtain Web pages containing the phrase "Florida election," but not pages that also contain "Gore" or "Bush," we could formulate the query using the expression "Florida election AND (NOT Gore) AND (NOT Bush)" (Figure 17.8a). The search engine interprets this as "This Web page contains the phrase 'Florida election' AND it is NOT the case that this Web page contains the word 'Gore' AND it is NOT the case that this Web page contains the word 'Bush'" (Figure 17.8b). The parentheses are not necessary but are sometimes useful, as we will see shortly. Most search engines are not *case sensitive*. No distinction is made between uppercase and lowercase letters.

Each search engine has slightly different conventions for formulating queries. Virtually every search engine has a hyperlink on its Web page that explains how to formulate queries. Although we use traditional terminology and notation from logic for our connectives, most popular search engines employ a more user-friendly format for advanced searches such as the one shown in Figure 17.8a. For Google, "Find results with all of the words" is a substitute for our AND connective; "with at least one of the words" plays the role of our OR connective; and "without the words" is the same as our NOT connective. Some search engines use + for our AND connective and − for our NOT connective. Despite the differences in format, the logic is the same.

Our interest here is finding out how we can use Boolean logic to decide whether two different expressions have the same meaning. For example, is the

FIGURE 17.8a

Example of Google
search on Florida
election AND (NOT
Gore) AND (NOT
Bush).

FIGURE 17.8b

Results of Google search
on Florida election AND
(NOT Gore) AND
(NOT Bush).

expression "Florida election AND (NOT Gore) AND (NOT Bush)," equivalent
to "Florida election AND NOT (Gore AND Bush)"? Or is it equivalent to the
expression "Florida election AND NOT (Gore OR Bush)"? To answer these questions, we will now take a closer look at the connectives AND, OR, and NOT.

The NOT connective allows us to take an expression P and create a new expression NOT P, called the *negation* of P. If P is true, then NOT P is false. If P
is false, then NOT P is true. Rather than writing NOT P, we will use the more
standard mathematical notation $\neg P$. The negation relationship can be summarized in the following format, known as a **truth table**:

P	$\neg P$
T	F
F	T

Notice that T and F are used here as shorthand for *true* and *false*, respectively.
The left column of the truth table shows the two possible values of P: T and F.
The right column shows the values of $\neg P$ for each of the corresponding values
of P.

The AND connective allows us to combine two expressions, P and Q, into
a new expression P AND Q called the *conjunction* of P and Q. The new expression is true when both statements P and Q are true and is otherwise false. The
mathematical notation for P AND Q is $P \wedge Q$. (You can remember this by not-

SPOTLIGHT 17.8   Six Degrees of Kevin Bacon

The actor Kevin Bacon is probably better known for a game named after him than for any of the movies he has been in. The game works like this. Every actor who has been in a film with Kevin Bacon has Bacon number 1. Every actor who does not have Bacon number 1, but has been in a film with someone with Bacon number 1, has Bacon number 2. Any actor who does not have Bacon number 1 or 2, but has been in a film with someone who has Bacon number 2, has Bacon number 3, and so on. For example, Nicole Kidman has not been in a film with Bacon but was in *The Interpreter* with Sean Penn, and Sean Penn was in *Mystic River* with Kevin Bacon. So Kidman's Bacon number is 2.

This game was conceived by three college students who first explained it to the public on an MTV show hosted by Jon Stewart. It sometimes goes by the name "Six Degrees of Kevin Bacon" because nearly every actor has a Bacon number that is at most 6. In fact, it is a challenge to think of an actor with a Bacon number exceeding 3. The game is an example of what scientists call the "small-world phenomenon," by which they mean that every person is connected to every other person by a surprisingly short number of links. You can play the Bacon game at www.cs.virginia.edu/oracle/.

The first experiment involving the small-world phenomenon occurred in 1967, when social psychologist Stanley Milgram mailed a series of traceable letters from points in Kansas and Nebraska to "targets" in Boston. The letters could be sent only to someone whom the holders knew on a first-name basis, and who they thought was more likely to know the target than they were themselves. The data revealed a median chain length of about 6.

Mathematicians have their own version of the Bacon game called the "Erdös number," where coauthors of research papers with Paul Erdös have Erdös number 1. The author of this spotlight has Erdös number 3. Ironically, Erdös himself has a Bacon number of 4. In April 2004, a person with Erdös number 4 auctioned on eBay the opportunity for someone to get an Erdös number of 5 by offering to write a joint paper with the highest bidder. The winning bid was $1031.

ing that ∧ has a shape like the first letter of AND.) This relationship can also be summarized in a truth table:

P	Q	$P \wedge Q$
T	T	T
T	F	F
F	T	F
F	F	F

Here, the first two columns are used to show all possible values of P and Q, and the right column shows the value of $P \wedge Q$.

Finally, the OR connective allows us to combine two expressions P and Q into a new expression P OR Q, called the *disjunction* of P and Q, which is true if either P or Q, or both, are true and is otherwise false. The mathematical notation for P OR Q is $P \vee Q$. This relationship is summarized by the truth table

P	Q	$P \vee Q$
T	T	T
T	F	T
F	T	T
F	F	F

A statement involving three expressions P, Q, and R such as $P \wedge Q \wedge R$ appears to be ambiguous. Does this expression mean that P and Q are first combined into a new expression $P \wedge Q$ and this new expression is then combined with R? In other words, should we interpret this expression as $(P \wedge Q) \wedge R$? Perhaps the intention was to connect P with the single expression $Q \wedge R$. In this case, the expression is interpreted as $P \wedge (Q \wedge R)$. Just as with the case for arithmetic statements such as $5 + 3 + 6$, which we can interpret to mean $(5 + 3) + 6$ or $5 + (3 + 6)$, it turns out that both interpretations are the same. For example, consider the three requirements for the office of president of the United States. A candidate for president must be at least 35 years old, must be a natural-born U.S. citizen, and must have lived in the United States for at least 14 years. Let P be the statement "a candidate must be at least 35 years old," let Q be the statement "a candidate must be a natural-born U.S. citizen," and let R be the statement "a candidate must have lived in the United States for at least 14 years."

The expression $(P \wedge Q) \wedge R$ can then be interpreted as "a candidate must be at least 35 years old and a natural-born U.S. citizen and also must have lived in the United States for at least 14 years." The expression $P \wedge (Q \wedge R)$ can be interpreted as "a candidate must be at least 35 years old and also a natural-born U.S. citizen who has lived in the United States for at least 14 years."

Both of these descriptions are effectively the same. Both expressions are true only in the event that each of P, Q, and R is true. One way to verify this is to construct the truth table for expression $(P \wedge Q) \wedge R$ and the truth table for $P \wedge (Q \wedge R)$ and show that they give the same values for every possible value of P, Q, and R. Since the order of operations does not matter in this case, we can simply write $P \wedge Q \wedge R$ without worrying about any possible ambiguity. The same thing is true for $P \vee Q \vee R$. The connectives $\wedge$ and $\vee$ have the associative property. Notice this terminology is consistent with the "associative property" of real-number addition and multiplication: $(a + b) + c = a + (b + c)$ and $(ab)c = a(bc)$.

In some cases, however, the ambiguity is not easy to resolve. For example, consider the expression $P \wedge Q \vee R$, which can be interpreted as either $(P \wedge Q)$

∨ R or as P ∧ (Q ∨ R). Using the statements P, Q, and R as before, (P ∧ Q) ∨ R can be interpreted as "a candidate must be at least 35 years old and a natural-born U.S. citizen, or must have lived in the United States at least 14 years." On the other hand, P ∧ (Q ∨ R) can be interpreted as "a candidate must be at least 35 years old, and be a natural-born citizen or have lived in the United States at least 14 years." These two descriptions are certainly not the same! For example, the first expression excludes Arnold Schwarzenegger as a candidate for president, whereas the second one includes him. One way to see exactly how the two statements differ is to compare the truth table for (P ∧ Q) ∨ R to the truth table for P ∧ (Q ∨ R). The truth table for (P ∧ Q) ∨ R is:

P	Q	R	(P ∧ Q)	(P ∧ Q) ∨ R
T	T	T	T	T
T	T	F	T	T
T	F	T	F	T
T	F	F	F	F
F	T	T	F	T
F	T	F	F	F
F	F	T	F	T
F	F	F	F	F

The truth table for P ∧ (Q ∨ R) is:

P	Q	R	(Q ∨ R)	P ∧ (Q ∨ R)
T	T	T	T	T
T	T	F	T	T
T	F	T	T	T
T	F	F	F	F
F	T	T	T	F
F	T	F	T	F
F	F	T	T	F
F	F	F	F	F

Because the last columns of these two truth tables differ for some values of P, Q, and R, the two expressions are not equivalent. For example, notice that if P is false and Q and R are both true, then (P ∧ Q) is false. Because R is true, however, (P ∧ Q) ∨ R is true. On the other hand, P ∧ (Q ∨ R) is false regardless of whether (Q ∨ R) is true or false, because P is false. To avoid ambiguity, it is often best to use parentheses.

A way to avoid ambiguity without using parentheses is to adopt a convention on the order of operations. For example, in arithmetic, the convention is that multiplication takes precedence over addition. Therefore, $3 + 4 \times 5$ is determined by first evaluating 4×5 and then adding 3. Of course, we could have written $3 + (4 \times 5)$ to avoid the ambiguity altogether. Similarly, in Boolean logic we adopt the convention that $\wedge$ (AND) takes precedence over $\vee$ (OR). Therefore, the expression $P \wedge Q \vee R$, by convention, is to be interpreted as $(P \wedge Q) \vee R$. Furthermore, the convention states that $\neg$ (NOT) takes the highest precedence of all. Thus, $\neg P \wedge \neg Q \wedge R$ is interpreted as $((\neg P) \wedge (\neg Q)) \vee R$.

EXAMPLE 6   Applying Boolean Logic to a Web Search

Let's revisit the Web queries mentioned before. Let P represent the query "Florida election," which corresponds to the expression, "This Web page contains the phrase 'Florida election'." Let Q represent the expression, "This Web page contains the word 'Gore'." Let R represent the expression, "This Web page contains the word 'Bush'." We now translate the query "Florida election AND (NOT Gore) AND (NOT Bush)" as $P \wedge (\neg Q) \wedge (\neg R)$ and write its truth table:

P	Q	R	$\neg Q$	$\neg R$	$P \wedge (\neg Q) \wedge (\neg R)$
T	T	T	F	F	F
T	T	F	F	T	F
T	F	T	T	F	F
T	F	F	T	T	T
F	T	T	F	F	F
F	T	F	F	T	F
F	F	T	T	F	F
F	F	F	T	T	F

For every possible value of P, Q, and R, the truth table gives us the value of our expression. The fourth and fifth columns of the table are not strictly necessary, but they are helpful in determining the values in the last column. As expected, this table tells us that the expression $P \wedge (\neg Q) \wedge (\neg R)$ is true precisely when P is true, Q is false, and R is false. ■

Two expressions are said to be **logically equivalent** if they have the same value, true or false, for each possible assignment of the Boolean variables. To decide whether two expressions are logically equivalent, we construct the truth tables for each one and then check if they have the same values for each of the possible assignments of the Boolean variables. If they do, the expressions are logically equivalent. If they differ for even one case, however, then the expressions are not equivalent.

So, to determine whether the expression, "Florida election AND (NOT Gore) AND (NOT Bush)," is equivalent to the expression, "Florida election AND NOT (Gore AND Bush)," we need only compare their corresponding truth tables.

Applying Boolean Logic to a Web Search

EXAMPLE 7

We first determine the truth table for the expression, "Florida election AND NOT (Gore AND Bush)," which is represented as $P \wedge \neg(Q \wedge R)$. Its truth table is:

P	Q	R	$Q \wedge R$	$\neg(Q \wedge R)$	$P \wedge \neg(Q \wedge R)$
T	T	T	T	F	F
T	T	F	F	T	T
T	F	T	F	T	T
T	F	F	F	T	T
F	T	T	T	F	F
F	T	F	F	T	F
F	F	T	F	T	F
F	F	F	F	T	F

This truth table differs in the last column from the truth table for $P \wedge (\neg Q) \wedge (\neg R)$ in the previous example. For instance, when P is true, Q is true, and R is false, we see that $P \wedge (\neg Q) \wedge (\neg R)$ is false but $P \wedge \neg(Q \wedge R)$ is true. Therefore, we must conclude that $P \wedge (\neg Q) \wedge (\neg R)$ is not logically equivalent to $P \wedge \neg(Q \wedge R)$.

On the other hand, the expression, "Florida election AND NOT (Gore OR Bush)," is represented by $P \wedge \neg(Q \vee R)$. Its truth table is:

P	Q	R	$Q \vee R$	$\neg(Q \vee R)$	$P \wedge \neg(Q \vee R)$
T	T	T	T	F	F
T	T	F	T	F	F
T	F	T	T	F	F
T	F	F	F	T	T
F	T	T	T	F	F
F	T	F	T	F	F
F	F	T	T	F	F
F	F	F	F	F	F

Because the last column of this table agrees with the last column for the expression $P \wedge (\neg Q) \wedge (\neg R)$, we know the expression, "Florida election AND (NOT Gore) AND (NOT Bush)," is equivalent to the expression, "Florida election AND NOT (Gore OR Bush)." ■

We conclude this chapter with a table that illustrates some of the special features of the popular search engine Google.

Google Feature	Enter
Calculator	31 + 4^6
Dictionary	define:ubiquitous
FedEx tracking	4852 9445 1402
Duluth Airport weather	dlh airport
Spell-checker	recieve
Area code map	218
Street map	1522 Triggs Avenue, Duluth, MN
Search Web for pages for words synonymous with wagering	~wagering

REVIEW VOCABULARY

Binary linear code A code consisting of words composed of 0's and 1's obtained by using parity-check sums to append check digits to messages.

Boolean logic Logic attributed to George Boole that uses operations such as $\wedge$, $\vee$, and $\neg$ to connect statements.

Caesar cipher A cryptosystem used by Julius Caesar whereby each letter is shifted the same amount.

Cryptogram is a sentence (or message) that has been encrypted.

Cryptology The study of how to make and break secret codes.

Data compression The process of encoding data so that the most frequently occurring data are represented by the fewest symbols.

Decoding The process of translating received data into code words.

Distance between two strings The distance between two strings of equal length is the number of positions in which they differ.

Encryption The process of encoding data to protect against unauthorized interpretation.

Even parity Even integers are said to have even parity.

Key A string used to encode and decode data.

Key word A word used to determine the amount of shifting for each letter while encoding a message.

Logically equivalent Two expressions are said to be logically equivalent if they have the same values for all possible values of their Boolean variables.

Modular arithmetic Addition and multiplication involving modulo n.

Nearest-neighbor decoding A method that decodes a received message as the code word that agrees with the message in the most positions.

Odd parity Odd integers are said to have odd parity.

Parity-check sums Sums of digits whose parities determine the check digits.

RSA public key encryption scheme A method of encoding that permits each person to announce publicly the means by which secret messages are to be sent to him or her.

Truth table A tabular representation of an expression in which the variables and the intermediate expressions appear in columns and the last column contains the expression being evaluated.

Variable-length code A code in which the number of symbols for each code word may vary.

Vigenère cipher A cryptosystem that utilizes a key word to determine how much each letter is shifted.

Weight of a binary code The minimum number of 1's that occur among all nonzero code words of a code.

✔ SKILLS CHECK

1. If you use the circular diagram method to encode the message 1011, what is the encoded message?

(a) 1011001
(b) 1011010
(c) 1010001

2. Suppose the message 1010010 is encoded using Table 17.1 and is received and decoded using the nearest-neighbor method. What message is recovered?

(a) 1010
(b) 1011
(c) 1110

3. Using the nearest-neighbor method and the code in Table 17.2, the word 1110011 decodes as

(a) 0110010.
(b) 1100011.
(c) 1010001.

4. What is the distance between received words 1011001 and 1000101?

(a) One
(b) Two
(c) Three

5. The weight of the binary linear code {0000000, 0011111, 0101011, 0110100} is

(a) 3.
(b) 4.
(c) 5.

6. If the two messages 0 and 1 are encoded as 000 and 111, respectively, how many errors does the code correct?

(a) Zero
(b) One
(c) Two

7. If every pair of code words differs in at least five positions, then nearest-neighbor decoding can accurately decode words that have

(a) two mistakes.
(b) three mistakes.
(c) four mistakes.

8. If a binary linear code has weight 4, it will

(a) detect up to four errors.
(b) correct up to two errors.
(c) correct any single error.

9. Using the encoding scheme A → 0, B → 10, C → 11, the string 010110 decodes as

(a) ABCB.
(b) ABCA.
(c) ABACA.

10. The sum of the binary string 1011001 and 1001101 is

(a) 0100110.
(b) 0010100.
(c) 1011101.

11. The Caesar cipher encrypts GO HOME NOW as

(a) JR KRPH QRZ.
(b) DL ELJB KLT.
(c) Neither of these.

12. The Caesar cipher decrypts EIEIO as

(a) HLHLR.
(b) BFBFL.
(c) Neither of these.

13. Using the Vigenère cipher and the key word ADAM to decrypt EIEIO, we obtain

(a) ELELR.
(b) EFEFL.
(c) EFEWO.

14. Using modular arithmetic, 3^5 mod 20 is equal to

(a) 3.
(b) 12.
(c) 15.

15. Using the RSA scheme with $n = 91$ and $s = 5$, the message 4 decodes as

(a) 11.
(b) 20
(c) 23.

16. If we use $p = 7$, $q = 17$, and $r = 5$ in the RSA scheme, the value of s is

(a) 25.
(b) 29.
(c) 49.

17. Which messages are the hardest to break?

(a) Messages encrypted with the Vigenère cipher.
(b) Cryptograms.
(c) Messages encrypted with the RSA scheme.

18. Which of the following statements is not a Boolean expression?

(a) Thirteen is a prime number.
(b) Thirteen is an even integer.
(c) Thirteen is an unlucky number.

19. The statement "P OR Q" is true if and only if

(a) either P or Q is true.
(b) either P or Q is true, but not both.
(c) neither P nor Q is false.

20. When the statement "P AND NOT Q" is true, it must be the case that

(a) P is true.
(b) P is false.
(c) either P or Q is false.

EXERCISES   ■ Challenge   ◆ Discussion

Binary Codes

1. Use the diagram method shown in Figures 17.1 and 17.2 to verify the code words in Table 17.1 for the messages 0101, 1011, and 1111.

2. Use the diagram method to decode the received messages 0111011 and 1000101.

3. Find the distance between each of the following pairs of words:

(a) 11011011 and 10100110
(b) 01110100 and 11101100

4. Referring to Table 17.1, use the nearest-neighbor method to decode the received words 0000110 and 1110100.

5. If the code word 0110010 is received as 1001101, how is it decoded using the diagram method?

6. Suppose a received word has the Venn diagram arrangement shown here:

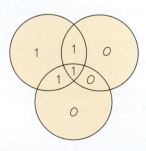

What can we conclude about the received word?

Parity-Check Sums

7. Determine the binary linear code that consists of all possible three-digit messages with three check digits appended using the parity-check sums $a_2 + a_3$, $a_1 + a_3$, and $a_1 + a_2$. (That is, $c_1 = 0$ if $a_2 + a_3$ is even, $c_1 = 1$ if $a_2 + a_3$ is odd, and similarly for c_2 and c_3.)

8. Let C be the code

{0000000, 1110100, 0111010, 0011101,
1001110, 0100111, 1010011, 1101001}

What is the error-correcting capability of C? What is the error-detecting capability of C?

9. Find all code words for binary messages of length 4 by adding three check digits using the parity-check sums $a_2 + a_3 + a_4$, $a_2 + a_4$, and $a_1 + a_2 + a_3$. Will this code correct any single error?

10. Consider the binary linear code

$C = \{00000, 10011, 01010, 11001,$
$00101, 10110, 01111, 11100\}$

Use nearest-neighbor decoding to decode 11101 and 01100. If the received word 11101 has exactly one error, can you determine the intended code word? Explain your reasoning.

11. Construct a binary linear code using all eight possible binary messages of length 3 and appending three check digits using the parity-check sums $a_1 + a_2$, $a_2 + a_3$, and $a_1 + a_3$. Decode each of the received words below by the nearest-neighbor method:

$$001001, 011000, 000110, 100001$$

12. Extend the code words listed in Table 17.1 to eight digits by appending a 0 to words of even weight and a 1 to words of odd weight. What are the error-detecting and error-correcting capabilities of the new code?

13. Extend the code words listed in Table 17.2 to eight digits by appending a 0 to words of even weight and a 1 to words of odd weight. What are the error-detecting and error-correcting capabilities of the new code?

14. Suppose the weight of a binary linear code is 6. How many errors can the code correct? How many errors can the code detect?

15. How many code words are there in a binary linear code that has all possible messages of length 5 with three check digits appended? How many possible received words are there with this code?

■ **16.** Explain why no binary linear code with all three possible message digits together with three check digits can correct all possible errors involving two digits.

■ **17.** A *ternary* code is formed by starting with all possible strings of a fixed length composed of 0's, 1's, and 2's and appending extra digits that are also 0's, 1's, or 2's. Form a ternary code by appending to each message a_1a_2 the check digits c_1c_2 using:

$$c_1 = (a_1 + a_2) \bmod 3$$
$$c_2 = (2a_1 + a_2) \bmod 3$$

■ **18.** Use the ternary code in the preceding exercise and the nearest-neighbor method to decode the received word 1211.

■ **19.** Suppose a ternary code is formed by starting with all possible strings of 0's, 1's, and 2's

of length 4 and appending two extra digits that are also 0's, 1's, and 2's. How many code words are there in this code? How many possible received words are there in this code?

Data Compression

20. Suppose we code a four-symbol genetic set {A, C, T, G} into binary form as follows:

$$A \rightarrow 0 \qquad C \rightarrow 10 \qquad T \rightarrow 110 \qquad G \rightarrow 111$$

Convert the sequence ACAAGTAAC into binary code.

21. Use the code in the previous exercise to determine the sequence of symbols represented by the binary code 001100001111000.

■ **22.** Suppose we code a five-symbol set {*A, B, C, D, E*} into binary form as follows:

$$A \rightarrow 0 \qquad B \rightarrow 10 \qquad C \rightarrow 110$$
$$D \rightarrow 1110 \qquad E \rightarrow 1111$$

Convert the sequence of *AEAADBAABCB* into binary code. Determine the sequence of symbols represented by the binary code 01000110100011111110.

23. Use the code in the previous exercise to convert the sequence *EABAADABB* into binary code. Determine the sequence of letters represented by the binary code 001000110011110111010.

■ **24.** Devise a variable-length binary coding scheme for a six-symbol set {*A, B, C, D, E, F*}. Assume the *A* is the most frequently occurring symbol, *B* is the second most frequently occurring symbol, and so on.

25. Judging from the Morse code, what are the three most frequently occurring consonants in English text material? What is the most frequently occurring vowel?

26. In English, the letter *H* occurs more often than *D*, *G*, *K*, and *W*, but in Morse code, *H* has a longer code than *D*, *G*, *K*, and *W*. Speculate on the reason for this apparent violation of data-compression principles.

27. Explain why the Morse code must include a space after each letter but fixed-length codes do not.

28. Guess the percentage of the occurrences of spaces in typical English text material.

29. Following are the closing values (rounded to the nearest integer) of the NASDAQ stock market between April 1, 2004, and April 15, 2004. Use the delta function method to compress these values. What percentage reduction in characters is there?

2015 2057 2079 2060 2050
2053 2065 2030 2025 2002

30. The following numbers were encoded using delta function encoding. Determine the original numbers.

1207 373 −57 −97 −234 −105 178 −73 275
79 −183 −146 −94 129

31. Decode the binary string 111001000010011101100011010, which has been encoded using the Huffman code given on page 634.

32. Use a Huffman tree code to assign a binary code to the letters that occur with the probabilities

A	0.025
B	0.150
C	0.015
D	0.170
E	0.200
F	0.225
G	0.215

33. Suppose a Huffman tree has been used to create a binary code for the letters A through J, and the results include $B = 111110$, $J = 111111$, and $G = 11110$. If the code has only two code words of length 6 and one of length 5, what can you say about the probability of the occurrence of the letters B, J, and G?

Cryptography

34. For each part below, explain how modular arithmetic can be used to answer the question.

(a) If today is Wednesday, what day of the week will it be in 16 days?

(b) If a clock (with hands) indicates that it is now four o'clock, what will it indicate in 37 hours?

(c) If a military person says it is now 0400, what time would it be in 37 hours? (Instead of A.M. and P.M., military people use 1300 for 1:00 P.M., 1400 for 2:00 P.M., and so on.)

(d) If it is now July 20, what day will it be in 65 days?

(e) If the odometer of an automobile reads 97,000 now, what will it read in 12,000 miles?

35. Use the Caesar cipher to encrypt the message RETREAT. Determine the intended message corresponding to the message DGYDQFH that was encrypted using the Caesar cipher.

36. Suppose you take a message and repeatedly apply the Caesar cipher to it until you return to the original message. How many iterations must be done before this occurs?

37. Using 0, 1, 2, . . . , 25 to label the positions of the letters A, B, C, . . . , Z, suppose we create a cipher by replacing the letter in position i with the letter in position $(i + 8)$ mod 26. How many iterations of this cipher must be done before a message will return to its original state?

38. Use the Vigenère cipher with the key word HELP to encrypt the message PHONE HOME.

39. Given the BEATLES was used as the key word for the Vigenère cipher to encrypt SSLETRY TXOGPW, decrypt the message.

40. Use the Vigenère cipher with the key word CLUE to encrypt the message THE WALRUS WAS PAUL.

41. Name a word that has the cryptogram form C'MM.

42. What is the most likely possibility for the cryptogram word GXFG?

43. Solve the cryptogram X GH G HGK.

44. In 1937, Ernest Vincent Wright wrote a 50,100-word novel called *Gadsby*. Below is an excerpt from the first page. What is remarkable about this passage?

> If youth, throughout all history, had a champion to stand up for it; to show a doubting world that a child can think; and, possibly, do it practically; you wouldn't constantly run across folks today who claim that "a child don't know anything." A child's brain starts functioning at birth; and has, amongst its many infant convolutions, thousands of dormant atoms, into which God has put a mystic possibility for noticing an adult's act, and figuring out its purport.

45. What are two likely possibilities for a cryptogram word of the form LSTWQ'UX?

46. If both of the words IPHV and IPH appear in a cryptogram and you know that none of the letters is *e*, what is the most likely possibility for these words?

47. Following is a phrase from this chapter that has been made into a cryptogram. FROMQFE HSXF MGQBIYX, JQK OQFNSLIX, HSPSEQG EIGIDSXSRLX, ORHIOX, FIGG MNRLIX, QLH XMQFI MYRCIX. Decrypt QLH.

48. Use the Web site www.oneacross.com/cryptograms/ to decrypt the following passage. (If you wish to copy and paste this text rather than type it, you can find it at www.d.umn.edu/~jgallian/fapp7/cryptogram1.html)

JHU, LC V OVWZKW AKCAK, IK YVC CDU EKELYVUK–IK YVC CDU YDCAKYWVUK–IK YVC CDU BVOODI–UBLA ZWDHCE. UBK JWVGK FKC, OLGLCZ VCE EKVE, IBD AUWHZZOKE BKWK, BVGK YDCAKYWVUK LU, NVW VJDGK DHW TDDW TDIKW UD VEE DW EKUWVYU. UBK IDWOE ILOO OLUUOK CDUK, CDW ODCZ WKFKFJKW IBVU IK

AVQ BKWK, JHU LU YVC CKGKW NDWZKU IBVU UBKQ ELE BKWK. LU LA NDW HA UBK OLGLCZ, WVUBKW, UD JK EKELYVUKE BKWK UD UBK HCNLCLABKE IDWX IBLYB UBKQ IBD NDHZBU BKWK BVGK UBHA NVW AD CDJOQ VEGVCYKE. LU LA WVUBKW NDW HA UD JK BKWK EKELYVUKE UD UBK ZWKVU UVAX WKFVLCLCZ JKNDWK HA–UBVU NWDF UBKAK BDCDWKE EKVE IK UVXK LCYWKVAKE EKGDULDC UD UBVU YVHAK NDW IBLYB UBKQ ZVGK UBK OVAU NHOO FKVAHWK DN EKGDULDC–UBVU IK BKWK BLZBOQ WKADOGK UBVU UBKAK EKVE ABVOO CDU BVGK ELKE LC GVLC–UBVU UBLA CVULDC, HCEKW ZDE, ABVOO BVGK V CKI JLWUB DN NWKKEDF–VCE UBVU ZDGKWCFKCU DN UBK TKDTOK, JQ UBK TKDTOK, NDW UBK TKDTOK, ABVOO CDU TKWLAB NWDF UBK KVWUB.

49. Decrypt the following cryptogram of a Homer Simpson quote:

NE, D WNFV QNHK OIZITDJV. VUYYVRDIWWQ MEV "VJKDRE QNHK LNKC YNLVK" UVRMDNJ. D MEDJS DM'U KVIWWQ, KVIWWQ, KVIWWQ ZNNC. (*Hint:* Z is *g*; Q is *y*.)

50. Use the Web site www.oneacross.com/cryptograms/ to solve the following cryptogram. (If you wish to copy and paste this text rather than type it, you can find it at www.d.umn.edu/~jgallian/fapp7/cryptogram2.html)

SB EOTYBS YZRCC XO ZOCM JB RSYLOT GBT R NREDJRC, BT BJZOTLDYO DSGRABUY NTDAO, USCOYY BS R ETOYOSJAOSJ BT DSMDNJAOSJ BG R HTRSM WUTI, OPNOEJ DS NRYOY RTDYDSH DS JZO

CRSM BT SRFRC GBTNOY, BT DS JZO
ADCDJDR, LZOS DS RNJURC YOTFDNO
DS JDAO BG LRT BT EUXCDN MRSHOT;
SBT YZRCC RSI EOTYBS XO YUXWONJ
GBT JZO YRAO BGGOSYO JB XO JLDNO
EUJ DS WOBERTMI BG CDGO BT CDAX;
SBT YZRCC XO NBAEOCCOM DS RSI
NTDADSRC NRYO JB XO R LDJSOYY
RHRDSYJ ZDAYOCG, SBT XO
MOETDFOM BG CDGO, CDXOTJI, BT
ETBEOTJI, LDJZBUJ MUO ETBNOYY BG
CRL; SBT YZRCC ETDFRJO ETBEOTJI XO
JRVOS GBT EUXCDN UYO, LDJZBUJ
WUYJ NBAEOSYRJDBS.

Encrypting Credit-Card Data on the Web

51. Add the following pairs of binary strings:

(a) 10111011 and 01111011
(b) 11101000 and 01110001

52. All binary linear codes have the property that
the sum of two code words is another code word.
Use this fact to determine which of the following
sets cannot be a binary linear code

(a) {0000, 0011, 0111, 0110, 1001, 1010, 1100,
1111}
(b) {0000, 0010, 0111, 0001, 1000, 1010, 1101,
1111}
(c) {0000, 0110, 1011, 1101}

53. Use the RSA scheme with $p = 5$, $q = 17$, and
$r = 3$ to determine the numbers sent for the
message VIP.

54. Use the RSA scheme with $p = 5$, $q = 17$, and
$r = 3$ to decode the received numbers 52 and 72.

55. In the RSA scheme with $p = 5$, $q = 17$, and
$r = 5$, determine the value of s.

56. Why can't we use the RSA scheme with $p = 7$,
$q = 11$, and $r = 3$?

57. Explain why we can't employ the RSA scheme
to send the message "NO" with $p = 7$ and $q = 11$
using blocks of length 2, but we can send it if we
use blocks of length 4.

58. Use the search box at www.google.com to
compute 13^9 mod 77, 13^6 mod 77, and 13^{15} mod
77. (To compute 13^9 mod 77 enter 13^9%77.)

Web Searches and Mathematical Logic

59. Show that $P \vee (P \wedge Q)$ is logically equivalent
to P.

60. Show that $\neg(P \vee Q)$ is logically equivalent to
$\neg P \wedge \neg Q$.

61. Show that $\neg(P \wedge Q)$ is logically equivalent to
$\neg P \vee \neg Q$. This relationship and the one in the
previous exercise are known collectively as *De
Morgan's Laws*.

62. Show that $P \vee (Q \wedge R)$ is logically equivalent
to $(P \vee Q) \wedge (P \vee R)$.

63. Show that $P \wedge (Q \vee R)$ is logically equivalent
to $(P \wedge Q) \vee (P \wedge R)$.

64. A patron at a restaurant tells the waiter to
bring her the chef's recommendation as long as it
has "lots of anchovies or is not spicy and in
addition the portion must be large." The waiter
goes to the kitchen and tells the chef to prepare a
dish that has "lots of anchovies and is also large or
is spicy and is also large." Did the waiter
communicate the patron's wishes correctly to the
chef? Use truth tables to support your answer.

65. The *implication connective* is defined by the
following truth table:

P	Q	$P \to Q$
T	T	T
T	F	F
F	T	T
F	F	T

Use truth tables to show that $P \to Q$ is logically
equivalent to $\neg P \vee Q$.

66. The Minnesota Vikings football coach tells his
team before the last game of the regular season
that if the team wins, they will be in the playoffs.
Use the truth table given in the previous exercise

to verify that if the Vikings lose and are still in the playoffs, the coach made a truthful statement to the team.

67. Using the implication connective and other connectives, variables, and truth tables, determine whether the statement "If it snows, there will be no school" is logically equivalent to the statement "It is not the case that it snows and there is school."

WRITING PROJECTS

1. Prepare a two-page report on applications of modular arithmetic. Explain the calculation of the check digits described in Exercises 7, 9, and 11 with modular arithmetic. Use modular arithmetic to describe the error-detection schemes used in Chapter 16.

2. Prepare a two-page report on the Braille system of coding. The book by Petzold in the Suggested Readings has the information you will need.

3. Prepare a two-page report on the Morse code. The book by Petzold in the Suggested Readings has the information you will need.

SUGGESTED READINGS

DENEEN, L. Secret encryption with public keys. *UMAP Journal,* 8 (1987): 9–29. Describes several ways in which modular arithmetic can be used to code secret messages.

PETZOLD, C. *Code,* Microsoft Press, Redmond, Wash., 1999. The first three chapters of this book provide an excellent explanation of the Morse code and the Braille system of coding.

SUGGESTED WEB SITES

www.d.umn.edu/~jgallian/fapp7 This site implements the nearest-neighbor decoding method for seven-digit binary strings using the code given in Table 17.1.

www.nyise.org/blind/america2.htm This site describes the Braille writing system.

www.freelearning.com/crypto/crypto.cgi This site allows users to convert text into cryptograms.

www.oneacross.com/cryptograms/ This site provides help for solving cryptograms.

www.d.umn.edu/~jgallian/project.petr/cryptogram.html This site will count frequencies of single letters, letter pairs, and letter triples of entered text.

www.google.com/help/features.html This site explains the many features that Google offers.

On Size and Growth

Mathematics is the study of patterns and relationships. It can explain why there are no King Kongs, analyze designs on ancient pottery, and suggest new and beautiful artistic designs. Mathematicians search for and classify numerical, geometric, and even abstract patterns. In these chapters, we follow some of those searches. We concentrate on geometric patterns but find that those lead to numerical considerations, too.

In Chapter 18, Growth and Form, we look at how the size of objects influences their forms. We investigate some big things: King Kong, tall trees, mile-high buildings, mountains. Seeing the underlying principles of scaling will help you to appreciate why objects in the world have the shapes and sizes that they do.

We start with a simple numerical pattern in Chapter 19, Symmetry Patterns, which leads to questions about the proportions of the most esthetically pleasing rectangle and the importance of bilateral symmetry. We expand our notion of symmetry and reveal surprising limitations that even broader notions of symmetry face. We examine the beauty of fractal patterns, ones that resemble themselves at finer and finer scales, in traditional art from Africa and elsewhere.

Chapter 20, Tilings, answers the question of how to arrange objects symmetrically on a surface. What shapes can we use? What patterns can arise if the objects themselves are symmetrical? Or if we allow irregular shapes but demand that they all face the same way? Most curious of all, you can arrange shapes in a pattern that does not repeat but is nevertheless systematic. ■

Growth and Form

Fantasy films have made us familiar with giant creatures, including King Kong, Godzilla, and the auliphants in *The Lord of the Rings.* We also find supergiants in literature, such as the giant of "Jack and the Beanstalk," the Big Friendly Giant in Roald Dahl's *The BFG,* and the Brobdingnagians of *Gulliver's Travels.*

Much as we appreciate those stories, even from an early age we don't really believe in monsters and giants. But could such beings ever exist? What problems would their enormous size cause them? How would they have to adapt to cope? (See Figure 18.1.)

Every species survives by adapting to its environment. In particular, it faces the **problem of scale:** how to adapt and survive at the different sizes from the beginning of life to the final size of a mature adult. For example, consider the giant panda, which ranges from barely 1 pound (lb) at birth to 275 lb in adulthood. A baby panda could be crushed by its mother. An adult panda needs to eat a lot.

For contrast, consider the horse. A newborn foal that weighed as little as a newborn panda would be too small to keep up with the herd and could not survive. An adult horse weighs much more than a panda and has to consume far more food, but the horse can move much more quickly and cover great distances to take advantage of wide-ranging sources of sustenance.

There have been large land mammals (mammoths) and huge sea mammals (the blue whale)—not to mention the dinosaurs. But the tallest humans have been only 9 to 10 feet (ft) tall. The largest mammoth was 16 ft at the shoulder (about twice as tall as an elephant). Even the tallest dinosaur, *Supersaurus,* stood only 40 ft high.

What about supergiants and utterly huge monsters? That they have never existed suggests physical limits to size. In fact, with a few simple principles of geometry, we can show not only that lizards and apes of such size are impossible, but also that no objects or living beings could exist, unchanged in shape, on a vastly different scale, larger or smaller.

18.1 Geometric Similarity

The powerful mathematical idea that we use is *geometric similarity.*

:::::::: Geometric Similarity

Two objects are **geometrically similar** if they have the same shape, regardless of the materials of which they are made.

Similar objects need not be the same size, but measurements of corresponding distances on the two objects must be proportional. For example, when a photo is enlarged, it is enlarged by the same factor in both the horizontal and vertical directions—in fact, in any direction (such as a diagonal). We call this enlargement factor the *linear scaling factor,* and the scaling is called a **dilation.**

FIGURE 18.2 Two geometrically similar photographs. *(image 100/Punchstock.)*

In Figure 18.2, the linear scaling factor is 3; the enlargement is three times as wide and three times as high as the original. In fact, every pair of points goes to a new pair of points three times as far apart as the original ones.

How Area and Volume Scale

The enlargement can be divided into $3 \times 3 = 9$ rectangles, each the size of the original. Hence, the enlargement has $3 \times 3 = 3^2 = 9$ times the area of the original. More generally, if the linear scaling factor is some general number L (not necessarily 3), the resulting enlargement has an area $L \times L = L^2$ ("L squared") times the area of the original.

How Area Scales

The *area* of a scaled-up object goes up with the *square* of the linear scaling factor.

We symbolize the relationship between the area A and the linear scaling factor L by

$$A \propto L^2$$

where the symbol $\propto$ is read as "is proportional to" or "scales as."

What about enlarging three-dimensional objects? If we take a cube and enlarge it by a linear scaling factor of 3, it becomes three times as long, three times as high, and three times as deep as the original (see Figure 18.3).

What about volume? The enlarged cube has three layers, each with $3 \times 3 = 9$ little cubes, each the same size as the original. Thus, the total volume is $3 \times 3 \times 3 = 3^3 = 27$ times as much as the original cube.

How Volume Scales

The *volume* of a scaled-up object goes up with the *cube* of the linear scaling factor.

Denoting the volume by V, we can write

$$V \propto L^3$$

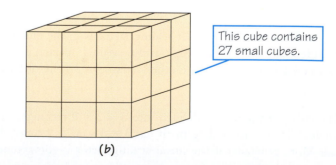

This cube contains 27 small cubes.

FIGURE 18.3 Cube (b) is made by enlarging cube (a) by a factor of 3.

(a)   (b)

Thus, for an object enlarged by a linear scaling factor of L, the enlargement has $L \times L \times L = L^3$ ("L cubed") times the volume of the original. Like the relationship between surface area and L^2, this relationship holds even for irregularly shaped objects, such as science fiction monsters.

You can see, however, that the area of each face (side) of the enlarged cube is $3^2 = 9$ times as large as that of a face of the original cube, just as the area of the photo enlarged by a factor of 3 has nine times the area of the original. The total surface area of the enlarged cube is nine times the total surface area of the original cube.

More generally, for objects of any shape, the total *surface area* of a scaled-up object goes up with the *square* of the linear scaling factor. Thus, the surface area of an object scaled up by a factor of L is L^2 times the surface area of the original. This feature holds true even for irregular shapes.

Before we discuss scaling real three-dimensional objects, you should understand the pitfalls of the language for describing increases and decreases.

The Language of Growth, Enlargement, and Decrease

*The new mission to Mercury, appropriately called Messenger, since Mercury is the winged messenger, is about **ten times less expensive than** the three and a half billion dollar Cassini spacecraft now orbiting Saturn.*

—Bernie Reim, www.northern-stars.com/whats_up.htm

*The 42,000-seat major league park in Houston was a great deal for the Astros. The total cost of $248.5 million . . . was 38 to **108 percent less than** the cost of the three other U.S. retractable ballparks.*

—Ken Conrad and Timothy R. Santi, "Sports Engineering Requires Speed, Skill, Endurance," *Design Intelligence*, 10(4) (April 2004), www.di.net/article.php?article_id=322

The space mission to Mercury, launched in August 2004 and set to last through 2012, was undoubtedly less expensive than the one to Saturn. The baseball park in Houston certainly sounds like a bargain compared to the cost of just a single player, $252 million (for the contract with Alex "A-Rod" Rodriguez in 2001). However, let's explore why the bolded phrases are incorrect and confusing ways to say so. First, we set out usual and correct ways of discussing percentages, increases, and decreases:

Meanings of Percentage

"$x\%$ of A" or "$x\%$ as large as" means $\frac{x}{100} \times A$.

"$x\%$ more than A" means A plus $x\%$ of A, in other words, $(1 + \frac{x}{100}) \times A$. Saying that A has "increased by $x\%$" means the same thing.

"$x\%$ less than A" means A minus $x\%$ of A, in other words, $(1 - \frac{x}{100}) \times A$. Saying that A has "decreased by $x\%$" means the same thing.

Safeco Field in Seattle cost $517 million. To say that Minute Maid (formerly Enron) Field in Houston cost "108% less than" that would mean:

$$\$248{,}500{,}000 \times \left(1 - \frac{108}{100}\right) = \$248{,}500{,}000 \times (1 - 1.08)$$

$$= \$248{,}500{,}000 \times (-0.08)$$

$$= -\$19{,}880{,}000$$

which is clearly nonsense.

The terms *of*, *times*, and *as much as* refer to *multiplication* of the original amount, while the terms *more*, *larger*, and *greater* refer to *adding* to the original amount. For instance, "five times as much" means the same as "four times more than" (the original plus four times as much in addition). Similarly, the relationship of the original amount to the larger amount can be expressed either in multiplicative terms ("one-fifth as much," "20% as much") or in subtractive terms ("four-fifths less than," "80% less than")–but don't mix the two.

People often say "five times more than" when they mean "five times as much." In the quotation about the Mercury mission, the author wrote "ten times less than" when he meant "one-tenth as much." All you can do is be aware of the potential confusion, try to figure out what was meant, and be careful in your own expression. In particular:

Correct Comparisons

Don't use *times* with *more* or with *less*.

Finally, we need to distinguish *percent* from percentage *points*: If support for the president decreased from 60% to 30%, it dropped 30 *percentage points* but decreased 50% (because the drop of 30 percentage points is 50% of the original 60 percentage points).

EXAMPLE 1   What About That Stadium?

Returning to the Houston stadium, how could you state correctly what the author was trying to say?

SOLUTION Its cost, $248.5 million, is about 0.48 times the $517 million cost of the Seattle stadium, so it cost "48% of the Seattle stadium" or "52% less than the Seattle stadium," or–a little less precisely–"half as much."

Caution: The dollar comparison here could be slightly misleading, since it does not take into account that a dollar was worth less in purchasing power in 2000, when the Houston stadium was ready, than in 1999, when the Seattle stadium opened, because of inflation. There would be a greater mismatch of the

TABLE 18.1	Units of the U.S. Customary System

Distance:

1 mile (mi) = 1760 yards (yd) = 5280 feet (ft) = 63,360 inches (in.)
1 yard (yd) = 3 feet (ft) = 36 inches (in.)
1 foot (ft) = 12 inches (in.)

Area:

$$\begin{aligned}
1 \text{ square mile} &= 1 \text{ mi} \times 1 \text{ mi} = 5280 \text{ ft} \times 5280 \text{ ft} \\
&= 27{,}878{,}400 \text{ ft}^2 \approx 28 \times 10^6 \text{ ft}^2 \\
&= 63{,}360 \text{ in.} \times 63{,}360 \text{ in.} \\
&= 4{,}014{,}489{,}600 \text{ in.}^2 \approx 4 \times 10^9 \text{ in.}^2 \\
&= 640 \text{ acres} \\
1 \text{ acre} &= 43{,}560 \text{ ft}^2
\end{aligned}$$

Volume:

$$\begin{aligned}
1 \text{ cubic mile} &= 1 \text{ mi} \times 1 \text{ mi} \times 1 \text{ mi} \\
&= 5280 \text{ ft} \times 5280 \text{ ft} \times 5280 \text{ ft} \\
&= 147{,}197{,}952{,}000 \text{ ft}^3 \\
&\approx 147 \times 10^9 \text{ ft}^3 \\
&= 63{,}360 \text{ in.} \times 63{,}360 \text{ in.} \times 63{,}360 \text{ in.} \\
&\approx 2.5 \times 10^{14} \text{ in.}^3
\end{aligned}$$

1 U.S. gallon (gal) = 4 U.S. quarts (qt) = 231 in.^{3}, exactly

Mass:

1 ton (t) = 2000 pounds (lb)

dollar figures if they spanned a longer period of time. Shortly we will explore the "numerical similarity" of dollars in different years and show how to take inflation into account. ■

18.2 How Much Is That in . . . ?

We give a brief introduction to units in which physical quantities are measured, together with a handy table of *scaling factors* (or *conversion factors*) and examples of how to convert from one system of units to another.

U.S. Customary System

You are probably familiar with units of the *U.S. Customary System* of measurement and their abbreviations. But please examine Table 18.1, and pay close attention to the systematic way of converting from one unit to another and to the expression of approximate numbers in scientific notation. The symbol ≈ means "is approximately equal to."

TABLE 18.2	Units of the Metric System

Distance:

$\qquad$ 1 meter (m) = 100 centimeters (cm)
1 kilometer (km) = 1000 meters (m)
$\qquad\qquad$ = 100,000 centimeters (cm) = 1×10^5 cm

Area:

1 square meter (m^2) = 1 m $\times$ 1 m
$\qquad\qquad$ = 100 cm $\times$ 100 cm = 10,000 (cm^2) = 1×10^4 cm^2
$\qquad$ 1 hectare (ha) = 10,000 m^2

Volume:

$\qquad$ 1 liter (L) = 1000 cm^3 = 0.001 m^3
1 cubic meter (m^3) = 1 m $\times$ 1 m $\times$ 1 m
$\qquad\qquad$ = 100 cm $\times$ 100 cm $\times$ 100 cm
$\qquad\qquad$ = 1,000,000 cm^3 = 1×10^6 cm^3 (or cc)

Mass:

1 kilogram (kg) = 1000 grams (g)

Metric System

Except for the United States, the world uses the metric system in science, industry, and commerce. It was proposed by Gabriel Mouton, a vicar in Lyons, in 1670, and was adopted in France in 1795. The fundamental unit of length, the *meter* (m), was originally defined as one ten-millionth of the distance from the North Pole to the equator, as measured on the meridian through Paris. The length is now defined as the distance that light travels in a vacuum in 1/299,792,458 second (s). The second, in turn, is defined as the time that it takes an atom of the metal cesium to vibrate 9,192,631,770 times.

All other units of length, area, and volume are *defined* in terms of the meter. For example, a centimeter (cm) is a hundredth of a meter.

Mass is quantity of matter. The metric unit of mass, the *kilogram* (kg), is defined as the mass of a platinum–iridium standard kept in Paris. Since you can't determine the mass of a sack of potatoes by comparing it to that, we measure the mass indirectly by seeing how much force gravity exerts on it—that is, we weigh it, on a scale conveniently calibrated in pounds or kilograms. However, a mass of 1 kg would "weigh" (register on the scale) only one-sixth as much on the Moon.

Table 18.2 lists the units of the metric system.

Converting Between Systems

What are the scaling factors between the U.S. Customary System and the metric system? Since 1959, the fundamental units of the U.S. Customary System, the

TABLE 18.3	Conversions Between the U.S. Customary System and the Metric System

Distance:

1 in. = 2.54 cm, exactly
 1 ft = 12 in. = 12 × 2.54 cm = 30.48 cm = 0.3048 m, exactly
1 yd = 0.9144 m, exactly
1 mi = 5280 ft = 5280 × 30.48 cm
     = 160,934.4 cm, exactly ≈ 1.61 km
1 cm ≈ 0.393701 in. ≈ 0.4 in.
 1 m ≈ 39.37 in. ≈ 3.281 ft
1 km ≈ 0.621 mi

Area:

1 hectare (ha) ≈ 2.47 acres

Volume:

1 cubic meter (m^3) = 1000 liters ≈ 264.2 U.S. gallons ≈ 35.31 ft^3
1 liter (L) = 1000 cm^3 ≈ 1.057 U.S. quarts (qt) ≈ 0.2642 U.S. gallons

Mass:

1 lb = 0.45359237 kg, exactly
1 kg ≈ 2.205 lb

yard (yd, for length) and the pound (for mass), have been *defined* in terms of metric units, so that we have

$$1 \text{ yd} = 0.9144 \text{ m, exactly}$$
$$1 \text{ lb} = 0.45359237 \text{ kg, exactly}$$

Table 18.3 illustrates the conversion factors. We explain how to convert a measurement in one unit to the corresponding measurement in a different unit.

What's That in Feet?

EXAMPLE 2

An international student tells her American student friends that she is 160 cm tall. They ask how much that is in feet and inches.

We approach this common conversion by using the scaling factor that 1 cm = $\frac{1}{2.54}$ inch (in.):

$$160 \text{ cm} = 160 \times 1 \text{ cm} \approx 160 \text{ cm} \times \frac{1 \text{ in.}}{2.54 \text{ cm}}$$

$$\approx 63.0 \text{ in.} = 63.0 \text{ in.} \times \frac{1 \text{ ft}}{12 \text{ in.}} \approx \frac{63.0}{12} \text{ ft} \approx 5.25 \text{ ft}$$

However, because we normally give height in feet and a whole number of inches, the height is

$$63.0 \text{ in.} = 5 \times (12 \text{ in.}) + 3.0 \text{ in.} = 5 \text{ ft} + 3.0 \text{ in.} \approx 5 \text{ ft } 3 \text{ in.}$$

Another way to approach the problem is by means of a proportion:

$$\frac{\text{height in in.}}{\text{height in cm}} = \frac{\text{length of 1 in. in inches}}{\text{length of 1 in. in cm}} = \frac{1 \text{ in.}}{2.54 \text{ cm}}$$

so that

$$\text{height in in.} = \text{height in cm} \times \frac{1 \text{ in.}}{2.54 \text{ cm}}$$

$$= 160 \text{ cm} \times \frac{1 \text{ in.}}{2.54 \text{ cm}} \approx 63.0 \text{ in.} \ \blacksquare$$

18.3 Scaling a Mountain

Gravity exerts an enormous effect on the size and shape that objects and beings can assume. **Weight** (force under gravity) is relative. We consider only Earth's gravity. For us, weight is the reading at sea level on a scale *calibrated in pounds of mass*. When we talk about strength of a material such as steel or wood, however, you need to realize that the pressure at the bottom of a mass on Earth would not be the same on the Moon, where gravity is one-sixth as strong. A boulder might have enough mass that its weight (force) would crush you on Earth, but you might easily toss off its weight on the Moon.

Suppose that the two cubes in Figure 18.3 are made of steel and that the first is 1 ft on a side and the second is 3 ft on a side. For each, its bottom face supports the weight of the entire cube. **Pressure** is the force per unit area, so the pressure exerted on the bottom face by the weight of the cube is equal to the weight of the cube divided by the area of the bottom face, or

$$P = \frac{W}{A}$$

A cubic foot of steel weighs about 500 lb (we say it has a **density** of 500 lb per cubic foot). The first cube weighs 500 lb and has a bottom face with area 1 ft^2, so the pressure exerted on this face is 500 lb/ft^2.

The second cube is 3 ft on a side. The area of the bottom face increases with the square of the linear scaling factor, so it is $3^2 \times 1$ ft^2 = 9 ft^2. As we saw earlier, volume goes up with the cube of the linear scaling factor. So this larger cube has a volume of $3^3 \times 1$ ft^3 = 27 ft^3. Because both cubes are made of the same steel, the larger cube has 27 times as much steel as the smaller one. Hence, it weighs 27 times as much as the smaller cube, or 27×500 lb = 13,500 lb.

When we divide this weight by the area of the bottom face (9 ft²), we find that the pressure exerted on the bottom face is 1500 lb/ft², or three times the pressure on the bottom face of the original cube. This makes sense because over each 1 ft² area stands 3 ft³ of steel. In general, if the linear scaling factor for the cube is L, the pressure on the bottom face is L times as much. Using the notation of proportionality, we have $A \propto L^2$ and $W \propto V \propto L^3$, so

$$P = \frac{W}{A} \propto \frac{L^3}{L^2} \propto L$$

What About a 10-Foot Cube?

EXAMPLE 3

SOLUTION If we scale the original cube of steel up to a cube 10 ft on a side, then the dimensions are

$$10 \text{ ft} \times 10 \text{ ft} \times 10 \text{ ft}$$

The total volume is

$$V = \text{length} \times \text{width} \times \text{height}$$
$$= 10 \text{ ft} \times 10 \text{ ft} \times 10 \text{ ft} = 1000 \text{ ft}^3$$

The weight of the cube is

$$W = V \times \text{density}$$
$$= 1000 \text{ ft}^3 \times 500 \text{ lb/ft}^3 = 500{,}000 \text{ lb}$$

The area of the bottom face is

$$A = \text{length} \times \text{width}$$
$$= 10 \text{ ft} \times 10 \text{ ft} = 100 \text{ ft}^2$$

The pressure on the bottom face is

$$P = \frac{W}{A} = \frac{500{,}000 \text{ lb}}{100 \text{ ft}^2} = 5000 \text{ lb/ft}^2$$

This is 10 *times*—not "10 times *more* than"—the pressure on the bottom face of the original 1-ft cube. ∎

At some scale factor, the pressure on the bottom face will exceed the steel's ability to withstand that pressure—and the steel will deform under its own weight. That point for steel is reached for a cube about 3 miles (mi) on a side—the pressure exerted by the cube's weight exceeds the resistance to crushing (ability to withstand pressure, or **crushing strength**) of steel, which under Earth's gravity is about 7.5 million lb/ft². Because 3 mi = 3 × 5280 ft = 15,840 ft, a 3-mi-long cube of steel would be more than 15,000 times as long as the original 1-ft cube; that is, the linear scaling factor is more than 15,000. The pressure on the bottom face of the cube would therefore be more than 15,000 times as much as for the 1-ft cube, or more than $15,000 \times 500$ lb/ft^2 = 7.5 million lb/ft^2.

EXAMPLE 4   What About Taipei 101?

Taipei 101 in Taiwan, completed in 2004 and named for its 101 floors (see Spotlight 18.1), is the world's tallest skyscraper, at 1671 ft (how much is that in meters?), not counting radio and television antennas. What is the pressure at the bottom of its walls?

SOLUTION   The building is made of reinforced concrete, which weighs 160 lb/ft^3. Over each square foot of the foundation stands 1671 ft^3 of reinforced concrete weighing

$$1671 \text{ ft}^3 \times 160 \text{ lb/ft}^3 = 267{,}360 \text{ lb} \approx 270{,}000 \text{ lb}$$

So the pressure at the bottom of the wall from the wall's weight alone is about 270,000 lb/ft^2. That's not counting the contents of the tower, which also must be supported! ■

Could we have a Super Taipei Tower 10 times as high? The bottom of its walls would have to support about 2.7 million lb/ft^2. The crushing strength of reinforced concrete under Earth's gravity is 8.5 million lb/ft^3, which would leave some safety margin.

If built as planned, the Freedom Tower in New York City, whose cornerstone was laid July 4, 2004, and which was scheduled to be completed in 2009, would be taller still—1776 ft—but contain only 60 floors. Above would be windmills to generate power, topped by a 276-ft spire.

EXAMPLE 5   How High Can a Mountain Be?

The height of mountains too is limited to gravity, their composition, and their shape. Can a mountain be a mile high?

SOLUTION   We build a simple mathematical model of a mountain. Suppose that it is made of granite, a common material, with uniform density. Granite weighs 165 lb/ft^3 and has a crushing strength of about 4 million lb/ft^2.

In the interests of both realism and simplicity, we assume that the mountain is a solid cone whose width at the base is the same as its height. Let's model Mount Everest, the tallest mountain on Earth, at about 6 mi high. The base, then, is a circle with diameter (distance across) 6 mi. The radius (half the diameter) is 3 mi (Figure 18.4). Because we took round numbers (6 mi) for the height and width, we record as significant only the first two digits of the results of the calculations.

What does the model Everest weigh? The relevant formula is

$$\text{weight} = \text{density} \times \text{volume}$$

SPOTLIGHT 18.1   A Mile-High Building?

In 1956, the famous American architect Frank Lloyd Wright (1867–1959) proposed a mile-high tower for the Chicago lakefront. In the text, we focus on the problem of holding up the weight of such a structure, which would have to involve stronger materials and thicker walls and supporting columns.

But there are other limits to the height of a building. Some of these can also be addressed by similar strategies. For example, the bending of the building in the wind, which can go up dramatically with height, can be controlled by making the building stiffer.

The terrorist destruction of the World Trade Center towers in 2001—resulting not from the aircraft impacts but from the subsequent fires—revealed a vulnerability in the towers' structure.

Even if designed to better resist fires and impacts, however, a mile-high building might not be at all practical. For example, the enormous number of people (perhaps 100,000) living, working, or visiting in such a building would create enormous traffic problems (pedestrian, parking, deliveries) for blocks around.

Cost per square foot of usable area is an important economic consideration. Even if the building had the same width all the way up (as opposed to tapering near the top, like the planned Freedom Tower), the space in the upper floors might not justify their additional expense. With increasing height, an increasingly larger proportion of the cross-sectional area of all floors of the building must be devoted to services, such as elevators, plumbing, and conduits for heat and air-conditioning. It's not that there are more

people on the top floors (or that they use more water than people on other floors!). But everyone entering the building and going to any floor needs to start in an elevator on the ground floor, so there must be more elevators and more elevator shafts. In an emergency evacuation, the people must walk down!

Some architects, however, maintain that the main limit on height of a building is human physiology. Differences in air pressure between the top and bottom of a building impose a limit on how fast elevators can rise or drop without discomfort to passengers, thereby enforcing long travel times for "vertical commuters." Human psychology might also present some limits.

Empire State Building
Built 1931
Height 381 meters
New York

World Trade Center
Built 1972
Height 417 meters
New York
Destroyed 9/11/01 in terrorist attacks

Sears Tower
Built 1974
Height 443 meters
Chicago

Petronas Twin Towers
Built 1997
Height 452 meters
Kuala Lumpur, Malaysia

Taipei 101
Built 2004
Height 509 meters
Taipei, Taiwan

Freedom Tower
Proposed 2003
Height 541.3 meters
Planned for New York

Mile-High Tower
Proposed 1956
Never built
Height 1,609 meters
Planned for Chicago

FIGURE 18.4 Model of Mount Everest as a cone of granite.

We already know the density of granite (165 lb/ft³), so to find the weight we need the formula for the volume of a cone of radius r and height h:

$$\text{volume} = \frac{1}{3}\pi r^2 h$$

Using a radius of 3 mi and a height of 6 mi and π (pi) approximately 3.14, we find that the model Everest has a volume of about 57 mi³.

To find the weight of 57 mi³ of granite, we need to convert units, because the density is given in pounds per cubic foot (lb/ft³). Let's convert to units of feet:

$$1 \text{ mi}^3 = 1 \text{ mi} \times 1 \text{ mi} \times 1 \text{ mi}$$
$$= 5280 \text{ ft} \times 5280 \text{ ft} \times 5280 \text{ ft}$$
$$\approx 1.5 \times 10^{11} \text{ ft}^3$$

Thus

$$57 \text{ mi}^3 \approx 57 \times 1.5 \times 10^{11} \text{ ft}^3 \approx 8.6 \times 10^{12} \text{ ft}^3$$

So we have

$$\text{weight of mountain} = 165 \text{ lb/ft}^3 \times 8.6 \times 10^{12} \text{ ft}^3$$
$$\approx 1.4 \times 10^{15} \text{ lb}$$
$$\approx 1.4 \text{ quadrillion lb}$$

Now that we know the weight of the mountain, we want to find the pressure on the base of the cone and compare it with the crushing strength of granite. (Everest is standing, so if our model is any good, that pressure will be below the crushing strength.) Physics tells us that the weight of the mountain is spread evenly over the base of the cone (though we are oversimplifying the geology underlying mountains). Because

$$\text{pressure} = \frac{\text{weight}}{\text{area}}$$

we need to calculate the area of the base of the cone. The shape is a circle, and the familiar formula

$$\text{area} = \pi r^2$$

gives an area of 28 mi^2 for a radius of 3 mi.

Once again, we need to convert units to express the pressure in pounds per square foot, the units in which the crushing strength is expressed. We get

$$\text{area} = 28 \text{ mi}^2 = 28 \times 5280 \text{ ft} \times 5280 \text{ ft} \approx 7.8 \times 10^8 \text{ ft}^2$$

Then

$$\begin{aligned}
\text{pressure} &= \frac{\text{weight}}{\text{area}} \\
&= \frac{1.4 \times 10^{15} \text{ lb}}{7.8 \times 10^8 \text{ ft}^2} \\
&= 1.8 \times 10^6 \text{ lb/ft}^2
\end{aligned}$$

This number is below the crushing strength of granite under Earth's gravity, 4×10^6 lb/ft^2, with a safety factor of about 2.

For a mountain to come close to the limitation of the crushing strength of granite, it would have to be only about 10 mi high, not quite twice as high as Everest. Other physical considerations suggest a maximum height of at most 15 mi. That no current mountains are that high may be a consequence of Earth's high amount of volcanic activity and the structural deformation of Earth's crust. ■

What about mountains made of other materials—glass, ice, wood, old cars? They couldn't be nearly as high. The pressure would cause glass to flow, ice to melt, and old cars to compact. What about mountains on another planet? Or on an asteroid? Their potential height depends on the mass, hence the gravity, of the object.

18.4 Sorry, No King Kongs

Unfortunately, the resistance of bone to crushing is not nearly as great as that of steel or granite. This fact helps to explain why there couldn't be any King Kongs (unless they were made of steel or granite!). A King Kong scaled up by a factor of 20 would weigh $20^3 = 8000$ times as much. Though the weight increases with the cube of the linear scaling factor, the ability to support the weight—as measured by the cross-sectional area of the bones, like the area of the bottom face of the cube in Figure 18.3—increases only with the square of the linear scaling factor.

These simple consequences of the geometry of scaling apply not only to supermonsters but also to other objects, such as trees.

EXAMPLE 5   How Tall Can a Tree Be?

Galileo suggested that no tree could grow taller than 300 ft (see Spotlight 18.2). The world's tallest trees are giant sequoias, which grow only on the West Coast of the United States and hence were unknown to Galileo. The tallest today is 369 ft (Figure 18.5), and the tallest reliably measured was 413 ft.

What limits the height of a tree? If the roots do not adequately anchor it, a tall tree can blow over. (This happened in 1990 to the world's then-tallest tree, the Dyerville Giant, a giant sequoia in Humboldt Redwoods State Park in California.) The tree could buckle or snap under its own weight and the force of a strong wind. The wood at the bottom will crush if there is too much weight above. Finally, there is a limit to how far the tree can lift water and minerals from the roots to the leaves.

SOLUTION Could a tree be a mile high? To make a rough estimate of the pressure at the base of the tree due to gravity, let's model the tree as a perfectly vertical

FIGURE 18.5 Even these giant sequoias can grow no taller than their form and materials allow. *(Phil Schermeister/ CORBIS.)*

SPOTLIGHT 18.2   Galileo and the Problem of Scale

The Italian physicist Galileo Galilei (1564–1642) was the first to describe the problem of scale, in 1638, in his *Dialogues Concerning Two New Sciences* (in which he also famously discussed the idea of Earth revolving around the Sun):

You can plainly see the impossibility of increasing the size of structures to vast dimensions either in art or in nature; likewise, the impossibility of building ships, palaces, or temples of enormous size in such a way that their oars, yards, beams, iron-bolts, and, in short, all their other parts will hold together; nor can nature produce trees of extraordinary size because the branches would break down under their own weight, so also would it be impossible to build up the bony structures of men, horses, or other animals so as to hold together and perform their normal functions if these animals were to be increased enormously in height; for this increase in height can be accomplished only by employing material which is harder and stronger than usual, or by enlarging the size of the bones, thus changing their shape until the form and appearance of the animals suggests a monstrosity.

To illustrate briefly, I have sketched a bone whose natural length has been increased three times and whose thickness has been multiplied until, for a correspondingly large animal, it would perform the same function which the small bone performs for its small animal. From the figures shown here you can see how out of proportion the enlarged bone appears. Clearly then if one wishes to

One bone, with another three times as long and thick enough to perform the same function in a scaled-up animal.

maintain in a great giant the same proportion of limb as that found in an ordinary man he must either find a harder and stronger material for making the bones, or he must admit a diminution of strength in comparison with men of medium stature; for if his height be increased inordinately he will fall and be crushed under his own weight. Whereas, if the size of a body be diminished, the strength of that body is not diminished in proportion; indeed the smaller the body the greater its relative strength. Thus a small dog could probably carry on his back two or three dogs of his own size; but I believe that a horse could not carry even one of his own size.

Translated by Henry Crew and Alfonso De Salvo, and published by Macmillan, 1914, and Northwestern University, 1946.

cylinder. Over each square foot at the bottom, there is 5280 ft^3 of cells of wood, which weigh about half as much as water. To calculate the weight, we first translate 1 ft^3 into metric measurement:

$$1 \text{ ft}^3 = (12 \text{ in.})^3 = (12 \times 2.54 \text{ cm})^3 \approx 28,317 \text{ cm}^3$$

A reason to do so is the convenient fact that water weighs just about 1 gram (g) per cubic centimeter. So 1 ft^3 of water weighs about 28,300 g = 28.3 kg = 28.3 × 2.20 lb ≈ 62 lb. Consequently, 5280 ft^3 of water weighs about 5280 × 62 lb = 327,360 lb ≈ 327,000 lb. The pressure at the bottom of the tree would be half as much, or 327,000 lb/ft^2.

This is an overestimate, because we assumed that the tree does not taper. A tree that tapers steadily looks like an elongated cone; using a more realistic cone model (as we will do in the next section for a mountain), you would find that the pressure at the bottom of the tree would be one-third of 164,000 lb/ft^2, or about 55,000 lb/ft^2.

A biological organism needs a safety factor of at least two to four times the absolute minimum physical limits, so a tree a mile high would need from 110,000 to 220,000 lb/ft^2 of upward pressure for water and minerals. Tension in the string of water molecules from root to leaf ranges from 80,000 to 3.2 million lb/ft^2, for different kinds and heights of trees, so this consideration does not rule out mile-high trees.

At more than about 500 lb/in.^{2} = 70,000 lb/ft^2, though, the bottom of the tree would begin to crush under the weight above. On this point, a mile-high tree is barely feasible, with little margin of safety. However, researchers who hauled themselves up to the top of the tallest trees in 2004 found a much lower limit, at least for giant sequoias. With increasing height, leaves are smaller, dryer, and less efficient at photosynthesis. The researchers estimated that trees can't top out higher than 400 to 427 ft.

There are other considerations. The taller the tree, the greater the area from which it must draw water and minerals, for which nearby trees also compete. Moreover, for a tree to grow very tall, it would have to live for a very long time. Evolution and time may select against extremely tall trees, or maybe, for no reason at all, they have just never evolved. ■

18.5 Dimension Tension

A large change in scale forces a change in either materials or form. A major manifestation of the scaling problem is the tension between weight and the need to support it. For example, a real building or machine must differ from a scale model: The balsa wood or plastic of the model would never be strong enough for the real thing, which would need aluminum, steel, or reinforced concrete.

Another way to compensate is to redesign the object to distribute its weight better. Let's go back to the original cube. It supports all its weight on its bottom face. In the version scaled up by a factor of 3, each small cube of the bottom layer has a bottom face supporting that cube's weight plus the weight of the two cubes piled on top of it.

Let's redesign the scaled-up cube, concentrating for simplicity only on the front face, with its nine small cubes. We take the three cubes on top and move them to the bottom, alongside the three already there. We take the three cubes on the second level, cut each in half, and put a half-cube over each of the six

FIGURE 18.6 Nine small cubes rearranged to support greater weight.

ground-level cubes (see Figure 18.6). We have the same volume and weight that we started with, but now there is less pressure on the bottom face of each small cube. Of course, the new design is not geometrically similar to the object that we started with—it's no longer a cube. We have solved the scaling problem by changing the proportions.

We observe in nature both strategies for scaling: change of materials and change of form. Small animals (such as insects) do not have bony internal skeletons. Larger animals generally do. Animals made of similar materials but differing greatly in size, such as a mouse and an elephant, must differ in shape. If a mouse were scaled up to the size of an elephant, it would need the disproportionately thicker legs of the elephant to support its weight and the elephant's thick hide to contain its tissue.

Some dinosaurs, like *Supersaurus* (which weighed 30 tons), had special adaptations to lighten their weight, such as hollow bones, just as some birds have. Hollow bones are stronger, a paradox that Galileo analyzed. Of two bones of the same weight and length, the hollow one is wider across at its midpoint because of the air it contains; and the greater the width, the greater the resistance to fracture.

Falls, Jumps, and Flight

The need to support weight can be thought of as a tension between volume and area. As an object is scaled up, its volume and weight go up together, as long as the density remains constant (for example, no air bubbles introduced into the steel to make it into a Swiss cheese!). At the same time, the ability to support the weight goes up with the cross-sectional area, like the bottom face of the steel cube.

Area–Volume Tension

Area–volume tension is a result of the fact that as an object is scaled up, the volume increases faster than the surface area and faster than areas of cross sections.

Because volume V is proportional to the cube of the linear scaling factor L, we have $V \propto L^3$; taking each side to the one-third power, $L \propto V^{1/3}$. The fact that surface area A is proportional to the square of the linear scaling factor becomes

$$A \propto L^2 \propto (V^{1/3})^2 = V^{2/3}$$

so that surface area scales as the two-thirds power of volume.

In any crowded city, you can observe tension between length, area, and volume. Consider an apartment building that spans a city block. The area of parking spaces on the adjacent streets is proportional to the perimeter of (length around) the building. But the number of cars belonging to people in the building is proportional to the number of apartments, which is proportional to the volume of the building. So the higher the building, the greater the parking tension!

In some cities, zoning tries to help the situation by putting shops on the ground floor, which cuts out one floor of apartments. If the residents' cars are away during the day, customers and employees of the shops can park where the apartment dwellers do at night. A more common solution is an underground garage, usually with several levels (with an area for cars proportional to the volume of the building). However, garages that were designed for one car per apartment have proven inadequate now that families tend to have more than one car.

Other examples of dimensional tension solutions include the old-fashioned diner, with its serving counter in the form of S-shapes to expand its effective length, and your small intestine, which coils its 20 ft to fit into your abdomen.

Area–volume tension has many other practical consequences, some of them related to our childhood fantasies. We can forget about humans "leaping tall buildings in a single bound," "soaring like an eagle," free diving miles below the sea, and jumping from airplanes without parachutes. Consider the following examples.

EXAMPLE 6   Falls

Area–volume tension affects how animals respond to falling, another of gravity's effects. A mouse may be unharmed by a 10-story fall, and a cat by a two-story fall, but a human may well be injured by falling down while running, walking, or even just standing.

What is the explanation? The energy acquired in falling is proportional to the weight of the falling object, hence to its volume. This energy must be absorbed either by the object or by what it hits, or must be otherwise dissipated at impact—for example, as sound. The fall is absorbed over part of the surface area of the object, just as the weight of the cube was distributed over its base. With scaling up, volume—hence weight, hence falling energy—goes up much faster than area. As volume increases, the hazards of falling from the same height increase. ■

EXAMPLE 7   Jumps

A flea can jump about 2 ft vertically, many times its own height. Many people believe that if a flea were as large as a person, it could jump 1000 ft into the air.

Imagining—against our earlier arguments—that there could be so large a flea, we know its limits: A scaled-up flea could jump about the same height as a small flea. The strength of a muscle is proportional to its cross-sectional *area* (see Spotlight 18.3). A jump involves suddenly contracting the muscle through its length, so it turns out that the ability to jump is proportional to the *volume* of muscle. But the volume of the flea and the volume of its leg muscles go up in proportion.

Let's say that a real flea's leg muscles account for 1% of its body. If we scale the flea up to the size of a person (without any change in its form), the enlarged flea's leg muscles will still make up 1% of its body. For either flea, each bit of muscle has the same power: In a jump, it propels 100 times its own weight, and it can do so to the same height. Both the weight of the flea and the power of its legs go up proportionately. In fact, the maximum heights that people, fleas, grasshoppers, and kangaroos can jump are all within a factor of 3 of each other. ∎

Flight

EXAMPLE 8

Wouldn't it be nice to be able to fly? Well, you have to be able to stay up. The power necessary for sustained flight is proportional to the **wing loading,** which is the weight supported divided by the area of the wings. We know that in scaling up, weight grows with the cube of the length of the bird or plane, and wing area with the square of the length. So the wing loading is proportional to the length of the flying object.

For example, if a bird or plane is scaled up proportionally by a factor of 4, it will weigh $4^3 = 64$ times as much but will have only $4^2 = 16$ times as much wing area. So each square foot of wing must support 4 times as much weight.

Once you're up, you have to keep moving. To stay level, an airborne object must fly fast enough to maintain the lift on the wings. The minimum necessary speed is proportional to the square root of the wing loading. Combining this fact with the first consideration, we conclude that the minimum speed goes up with the square root of the length. A bird scaled up by a factor of 4 must fly $\sqrt{4} = 2$ times as fast. (Hovering helicopters, hummingbirds, and insects maintain lift by moving their wings directly rather than through forward motion.)

Take, for instance, a sparrow, whose minimum speed is about 20 miles per hour (mph). An ostrich is 25 times as long as a sparrow, so the minimum speed for an ostrich would be $\sqrt{25} \times 20 = 100$ mph. Have you seen any flying ostriches lately? Heavy birds have to fly fast or not at all!

Of course, ostriches are not just scaled-up sparrows, nor are eagles. Larger flying birds have disproportionately larger wings than a sparrow to keep the wing loading down. The largest animal ever to take to the air was *Quetzalcoatlus northropi,* a flying reptile of 65 million years ago, with a wingspan of 36 ft and a weight of about 100 lb.

You have to stay up, you have to keep moving—and you have to get up there. Here basic aerodynamics imposes further limits. Paleontologists originally thought that *Q. northropi* weighed 200 lb and had a 50-ft wingspan. Even though that works out to about the same wing loading as for 100 lb and a wingspan of 36 ft,

SPOTLIGHT 18.3 "Take That, King Richard!"

Shakespeare painted Richard III as a hump-backed Machiavellian monster. Did Richard have an advantage in armored combat because he was short? That suggestion was made some years ago by one of the leading modern historians of the Tudor era, Garrett Mattingly.

Between a short man and a tall man, height increases by the linear dimension–from 5 feet 2 inches, say, to 6 feet–while the surface of the body increases as the square. Because it's the surface of the body that the armorer must plate with steel, the armor of a short warrior, like Richard, would be lighter than a tall warrior's by a lot more than the few inches difference in height would indicate. So Richard's notorious deadliness in battle would have been possible at least in part because his armor, while protecting him as well as the big man's, left him less encumbered.

After the lecture, someone said to Mattingly that he had grasped the right idea–but by the wrong end. Muscle power, the listener claimed, is a matter of bulk–and physical volume goes up by the cube, whereas the surface to be protected goes up by the square. So the large warrior should have more strength left over than the little guy after putting on his armor. And the large warrior, swinging a bigger club, can deliver a far more punishing blow–because the momentum of the club depends on its weight, which goes up with its volume, which means by the cube. Richard was at a terrible disadvantage.

But wait a minute, a second listener said. That's true about the club–but not about the muscles. The strength of a muscle is proportional not to its bulk but to the area of its cross section. And because the cross section of muscles increases by the square, just as the surface of the body does, the big guy, plated out, has no more, or less, advantage over the little guy than if both were naked.

Did wearing armor give an advantage to the shorter warrior? *(Photodisc/Punchstock.)*

But hang on, a third person interjected–an engineer. That's right about the muscles, but it's not right about the armor. The weight of the armor increases not simply with the increase in the surface area that it must cover but slightly faster. There must be reinforcing ribs. Or else the metal must be significantly thicker overall. So maybe Richard had an advantage.

Armor was made as thin as possible. Thickness, reinforcement, and structural stiffening were concentrated where opponents' weapons were likely to hit. From these strong, shaped places, the metal tapered away, until the sheet steel was as thin as the lid of a coffee can at the sides of the rib cage beneath the arms, or across the fingers, or at the cheek of a helmet.

Source: Horace F. Judson, *The Search for Solutions* (Baltimore: Johns Hopkins University Press, 1987), pp. 54–56.

other considerations from aerodynamics show that at the larger size, the reptile couldn't have gotten off the ground. ■

Keeping Cool (and Warm)

Area–volume tension is also crucial to an animal's thermal equilibrium. Both warm-blooded and cold-blooded animals gain or lose heat from the environment in proportion to body surface area.

Warm-Blooded Animals

A warm-blooded animal's basal metabolism, or rate of food intake needed to maintain body heat, depends primarily on its surface area, the temperature of its environment, and the insulation provided by its coat or skin. Other factors being equal, a scaled-up mammal scales up its food consumption by *surface area* (proportional to the square of the linear scaling factor), *not by volume* (proportional to its cube). For example, a mouse eats about half of its weight in food every day, while a human consumes only about one-fiftieth of its own weight.

Thus, the metabolic rate should be proportional to the surface area. Using the symbolism of proportionality, we can find how the metabolic rate changes with the mass of the animal. We know that mass is proportional to volume, which in turn is proportional to the cube of length, or

$$M \propto V \propto L^3$$

Taking each side to the one-third power, we have

$$M^{1/3} \propto V^{1/3} \propto L \qquad \text{or} \qquad L \propto M^{1/3}$$

Meanwhile, the metabolic rate (call it P) is proportional to surface area, so

$$P \propto A \propto L^2 \propto (M^{1/3})^2 = M^{2/3}$$

So, based on area–volume tension, we would expect metabolic rate to scale as the two-thirds power of body mass. But it doesn't—instead, it scales as the *three-quarters* power of body mass, that is, $P \propto M^{3/4}$. The least-squares line (see section 6.4) through the points in the "mouse-to-elephant" curve of Figure 18.7 has a slope of 0.74, very close to three-quarters.

Why the difference from the two-thirds that area–volume tension would predict? And does the small difference between two-thirds and three-quarters matter? The answers lie in further considerations from geometry, physiology, and physics. A plant or animal needs a network of vessels (like the blood system) to transport resources to, and wastes away from, every part of the animal's tissues. The terminal branches (in the blood system, capillaries) tend to be just about the same size in all species, for reasons of the physics involved. To minimize the energy involved in transport, the network of vessels needs to be organized as a fractal-like tree, with smaller and smaller vessels branching off (see section 19.6 to learn about fractal patterns). With same-size smallest branches at the ends,

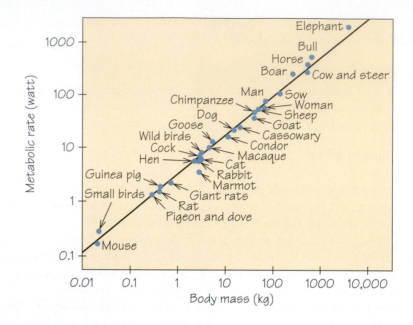

minimization of energy demands that the metabolic rate scale as the three-quarters power of body mass. Fractal branching makes it possible for the circulatory system of a whale, with 10^7 times the mass of a mouse, to have only 70% more branches than the mouse has.

EXAMPLE 9   Dives

Sperm whales (and some other species) regularly hold their breath and stay under water for an hour. Why can't we? In part, because we aren't as large as whales. A mammal's breath-holding ability depends on how much air it can hold in its lungs, which is proportional to its mass. It also depends on how fast it uses up air—in other words, on its metabolic rate, which is proportional to the three-quarters power of its mass. Hence, the limit of duration of a dive should be proportional to

$$\frac{M}{M^{3/4}} = M^{1/4}$$

For a 90,000-lb sperm whale, this limit is proportional to $90,000^{1/4} = 17.3$, while the corresponding figure for a 150-lb human is $150^{1/4} = 3.5$. So the sperm whale should be able to hold its breath for about $17.3/3.5 \approx 5$ times as long. However, humans cannot hold their breath for one-fifth of an hour (12 minutes)! This fact tells us that the whale must have some special adaptation—such as larger-than-proportional lungs, or a lower-than-predicted metabolism—to make its long dives possible. ■

Cold-Blooded Animals

Mammals and birds regulate their metabolism and maintain a constant internal body temperature. Cold-blooded animals, such as alligators and lizards, have a somewhat different problem. They absorb heat from the environment for energy, but they must also dissipate any excess heat to keep their temperatures below unsafe levels. The amount of heat that must be gained or lost is proportional to total volume, because the entire animal must be warmed or cooled. But the heat is exchanged through the skin, so the rate is proportional to surface area.

Dimetrodon was a large mammal-like reptile that roamed present-day Texas and Oklahoma 280 million years ago (see Figure 18.8). *Dimetrodon* had a great "sail," or fan, on its back. As an individual grew, and as the species evolved, the sail grew. But it did not grow according to *geometric similarity,* the kind of growth we refer to as *proportional growth.*

Proportional Growth

Proportional growth is growth according to geometric similarity, where the length of every part of the organism enlarges by the same linear scaling factor.

Instead, the area of *Dimetrodon*'s sail grew in proportion to the volume of the animal, a fact that strongly suggests to paleontologists that the sail was a

FIGURE 18.8
Dimetrodon may have evolved a sail to absorb and dissipate heat efficiently. *(Robert F. Walters.)*

temperature-regulating organ. Larger specimens of *Dimetrodon* didn't look like scaled-up smaller ones. We would say that the sail grew disproportionately compared to the rest of the animal. An individual twice as long would have eight (2^3) times as much weight and volume and a sail with eight times as much area. If it had grown according to geometric similarity, the sail would have been twice as high and twice as wide, and hence would have had only four times as much area.

Dimetrodon was a large animal, but heat regulation is even more important for small animals—like human babies, they can lose heat quickly because of their high ratio of surface area to volume. Leading paleontologists believe that birds evolved from dinosaurs and that feathers are modified reptilian scales. The wings of birds and insects may have evolved originally not for flight but as temperature-control devices.

Some scientists have speculated that African Pygmies are small in part because a small body can better lose heat in the hot, humid climate of the Ituri Forest in the Congo, where Pygmies live. The discovery announced in late 2004 of *Homo floresiensis*, a "hobbit-sized" people (just 1 m tall) who lived on the island of Flores in Indonesia 13,000 years ago, suggests another explanation. Being marooned on the island with a limited food supply (they hunted pygmy elephants) made large size—and a corresponding need for more calories—a disadvantage.

Other scientists have suggested that ancestors of human beings began walking on two legs in part to keep cool in a hot climate. Walking upright exposes less body area to the rays of the sun than walking on all fours and also reduces the amount of water needed by about one-half.

18.6 How to Grow

A large change of scale forces adaptive changes in materials or form. However, within narrow limits—in most cases, up to a factor of 2—creatures can grow according to geometric similarity. That is, they can grow proportionally, so that their shape is preserved. A striking example of such growth by a far greater factor is the chambered nautilus (*Nautilus pompilius*). Each new chamber that it adds to its shell is larger than, but geometrically similar to, the previous chamber and also similar to the shape of the shell as a whole—an *equiangular*, or *logarithmic*, spiral (see Figure 18.9).

Most living things grow over the course of their lives by a factor greater than 2. We've seen with *Dimetrodon* that a big specimen was not just a scaled-up small one. Nor is a human adult simply a scaled-up baby. Relative to the length of the body, a baby's head is much larger than an adult's, and its arms are disproportionately shorter than an adult's. In growth from baby to adult, the body does not scale up as a whole. But different parts of the body scale geometrically, each with a different scale factor. That is, a baby's eyes grow to perhaps twice their original size, while the arms grow to about four times theirs.

Although the laws for growth can be much more complicated than for proportional growth (or even for the allometric growth that we discuss in the next section), more sophisticated mathematics—for example, differential geometry, the

FIGURE 18.9 A chambered nautilus shell. *(Photodisc/Punchstock.)*

geometry of curves and surfaces—permits analysis of complex and interlocking scalings. For a model of the process in which a baby's head changes shape to grow into an adult head, we can use graph paper: First, we put a picture of the baby's skull on graph paper, then we determine how to deform the grid until the pattern matches an adult skull (see Figure 18.10 and Spotlight 18.4). The same idea lies at the heart of computerized "morphing," the process in which the face of one person can be changed smoothly into the face of another, with different scalings for different parts of the face.

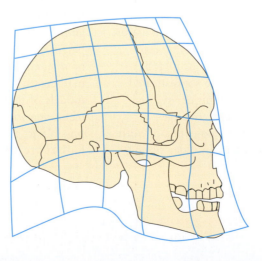

FIGURE 18.10 Modeling the changes in the shape of a human head from infancy to adulthood.

SPOTLIGHT 18.4   Helping to Find Missing Children

Photograph at age 7; photo showing one side of the child's face and one side of her mother's, to which the child's photo will be stretched.

Stretching, merging, and age progression to age 17. *(National Center for Missing and Exploited Children.)*

It can be valuable to be able to predict what a developing organism will look like in the future. For example, what does a child who was kidnapped at age 3 look like now, six years later?

At the National Center for Missing and Exploited Children (NCMEC) in Arlington, Virginia, a computer and a more sophisticated version of the graph-paper technique are used to answer such questions. Computer age-progression specialists scan photographs of both the missing child at age 3 and an older sibling

or a biological parent at age 9 into a computer. Then the face of the 3-year-old is stretched, depending on age, to reflect craniofacial growth and merged with the image of the sibling or parent at 9 years old. The result is a rough idea of what the missing child may look like. As mathematicians and biologists refine their models of how faces change over time, this technique will improve. It may even become possible to gain an idea of how a child may look at age 40 or 65.

Allometric Growth

If we measure the arm length or head size for humans of different ages and compare these measurements with body height, we observe that humans do not grow proportionally, that is, in a way that maintains geometric similarity. The head of a newborn baby may be one-third of the baby's length, but an adult's head is usually close to one-seventh of the individual's height. The arm, which at birth is one-third as long as the body, is by adulthood closer to two-fifths as long (see Figure 18.11a).

Graphing provides a way to test for differential growth. We plot body height on the horizontal axis and arm length on the vertical axis (see Figure 18.11b). A straight line would indicate proportional growth, that is, according to geometric similarity. We do get a straight line from 9 months (0.75 years) on up. But up to 9 months, we get a curve, which indicates that the ratio of arm length to height does not remain constant over the first year.

Is there an orderly law by which we can relate arm length to height? Let's plot again, this time using a different scale. For this **base-10 logarithmic scale,** we mark off equal units, as usual. But instead of labeling the marked points with 0, 1, 2, 3, and so on, we label them with the corresponding powers of 10: $10^0 = 1$, $10^1 = 10$, $10^2 = 100$, $10^3 = 1000$, and so on, which are also called **orders of magnitude.** Plotting a point on such a scale is not easy, because the point midway between 1 and 10 is not 5.5, but instead is closer to 3. Special graph paper (available in most college bookstores) marks smaller divisions and makes it easier to plot; paper marked with log scales on both axes is called **log-log paper,** while **semilog paper** has a logarithmic scale on just one axis. Also, many computer plotting packages can produce logarithmic scales.

We could use a logarithmic scale for either height or arm length, or for both. Using logarithmic scales for both, as in Figure 18.11c, the data plot closely to a straight line. Looking carefully, we can discern two different straight lines: a steeper one that fits early development (we will see shortly that it has slope 1.2), and a less steep one (with slope 1.0) that fits development after 9 months of age.

The change from one line to another after 9 months indicates a change in pattern of growth. The pattern after 9 months, characterized by the straight line with slope 1, is indeed proportional growth (sometimes called **isometric growth**). For the pattern before 9 months, the slope (1.2) being greater than 1 means that arm length is increasing relatively faster than height. This early growth also follows a definite pattern, called *allometric growth*.

∷ Allometric Growth

Allometric growth is growth of the length of one feature at a rate proportional to a power of the length of another.

FIGURE 18.11

(a) The proportions of the human body change with age. (b) A graph of human body growth on ordinary graph paper. The numbers shown beside the points indicate the age in years; they correspond to the stage of human development shown in part (a). (c) A graph of human body growth on log-log paper.

(a)

(b)

(c)

In geometric scaling, area grows according to the square (second power), and volume according to the cube (third power) of length, so they grow allometrically with length.

If we denote arm length by y and height by x, a straight-line fit on log-log paper corresponds to the algebraic relation

$$\log_{10} y = B + a \log_{10} x$$

where a is the slope of the line and B is the point where the graph crosses the vertical axis. If we raise 10 to the power of each side, we get

$$y = bx^a$$

where $b = 10^B$. This equation describes a **power curve:** y is a constant multiple of x raised to a certain power.

Finding the Power of Growth

EXAMPLE 10

How do we arrive at those slopes of 1.0 and 1.2?

SOLUTION You could use statistical software or your calculator to find the equation of the least-squares regression line (section 6.4) through the points on the log-log plots. Here we find approximate values for the slope a for each line from the coordinates of the points at the ends of the lines, for ages 0.42, 0.75, and 25.75. The observations and the corresponding logarithms are:

Age	Height	Log (Height)	Arm Length	Log (Arm Length)
0.42	30.0	1.48	10.7	1.03
0.75	60.4	1.78	25.1	1.40
25.75	180.8	2.26	76.9	1.89

The slope for the line from age 0.42 to age 0.75 is the vertical change over the horizontal change in terms of log units:

$$\frac{\log 25.1 - \log 10.7}{\log 60.4 - \log 30.0} = \frac{1.40 - 1.03}{1.78 - 1.48} = \frac{0.37}{0.30} \approx 1.2$$

The slope for the line from age 0.75 to age 25.75 is

$$\frac{\log 76.9 - \log 25.1}{\log 180.8 - \log 60.4} = \frac{1.89 - 1.40}{2.26 - 1.78} = \frac{0.49}{0.48} \approx 1.0$$

So $a = 1.2$ up to 9 months, and $a = 1.0$ after 9 months. Up to 9 months, arm length grows according to (height)$^{1.2}$. After 9 months, arm length grows according to (height)$^{1.0}$, and we get $y = bx^{1.0}$, which is a linear relationship describing proportional growth, that is, growth according to geometric similarity. On ordinary graph paper, proportional growth appears as a straight line, allometric growth as a curve. On log-log paper, both patterns appear as straight lines.

Allometry was used by paleontologists to determine that all specimens (just six!) of the earliest bird, *Archaeopteryx,* are indeed of the same species, and that the puzzling minute fossil fish *Palaeospondylus* (found only in Scotland) is probably just the larval stage of a better-known fish. ▪

We have examined the problem of scale and noted that a large change in scale forces a change in either materials or form. A particular instance of the problem of scale is area–volume tension, and we have seen how an animal's size and geometric shape affect its abilities to move and to keep itself warm or cool.

In this chapter we have explored the limitations on life imposed by dwelling in three dimensions. In Chapters 19 and 20, we will see that dimensionality also imposes surprising limits on artistic creativity in devising patterns.

:::::::::: REVIEW VOCABULARY

Allometric growth A pattern of growth in which the length of one feature grows at a rate proportional to a power of the length of another feature.

Area–volume tension A result of the fact that as an object is scaled up, the volume increases faster than the surface area and faster than areas of cross sections.

Base-10 logarithmic scale A scale on which equal divisions correspond to powers of 10.

Crushing strength The maximum ability of a substance to withstand pressure without crushing or deforming.

Density Mass per unit volume.

Dilation A linear scaling.

Geometrically similar Two objects are geometrically similar if they have the same shape, regardless of the materials of which they are made. They need not be the same size. Corresponding linear dimensions must have the same factor of proportionality.

Isometric growth Proportional growth.

Linear scaling factor The number by which each linear dimension of an object is multiplied when it is scaled up or down; that is, the ratio of the length of any part of one of two geometrically similar objects to the length of the corresponding part of the second.

Log-log paper Graph paper on which both the vertical and the horizontal scales are logarithmic scales; that is, the scales are marked in orders of magnitude 1, 10, 100, 1000, . . . , instead of 1, 2, 3, 4, ...

Orders of magnitude Powers of 10.

Power curve A curve described by an equation $y = bx^a$, so that y is proportional to a power of x.

Pressure Force per unit area.

Problem of scale As an object or being is scaled up, its surface and cross-sectional areas increase at a rate different from its volume, forcing adaptations of materials or shape.

Proportional growth Growth according to geometric similarity, where the length of every part of the organism enlarges by the same linear scaling factor.

Semilog paper Graph paper on which only one of the scales is a logarithmic scale.

Weight Force under gravity.

Wing loading Weight supported divided by wing area.

✔ SKILLS CHECK

1. A penny and a nickel are

(a) not geometrically similar, because they are made of different materials.

(b) not geometrically similar, because they are of different sizes.

(c) geometrically similar, because they have the same shape and proportional dimensions.

2. A scale model of a carillon stands 10 in. tall, and the actual carillon stands 100 ft tall. What is the linear scaling factor of the carillon compared to its model?

(a) 10

(b) 12

(c) 120

3. If a model car is built to a scale of 1 to 40, and the actual car has a turning circle of 37 ft, what should be the turning circle of the model?

(a) 11.1 in.

(b) 0.925 in.

(c) 1480 ft

4. You want to enlarge a small 3-in. by 5-in. photograph to a 12-in. by 20-in. copy. Assuming that the cost of photographic paper is proportional to its area, and that 3-in. by 5-in. reprints cost 40 cents each, how much would you expect to pay for the large copy?

(a) $1.60

(b) $3.20

(c) $6.40

5. If a medium 10-in. pizza costs $8 and a similar 14-in. pizza costs $14, which is the better buy?

(a) The 10-in. pizza

(b) The 14-in. pizza

(c) They are about the same price per square inch.

6. An artist plans to melt 100 pennies and re-form a larger penny proportional in all dimensions to an ordinary penny. What is the linear scaling factor of the large penny compared with the ordinary penny?

(a) 100

(b) 10

(c) 4.64

7. The actor Elijah Wood, who plays Frodo Baggins in the movie of Tolkien's *Lord of the Rings*, is 5 ft 6 in. tall, but his character is barely 4 ft tall. Put correctly, how much shorter is Frodo than Wood?

(a) 138% shorter

(b) 73% shorter

(c) 27% shorter

8. The distance for the marathon race was established in 1921 as 42.195 km. Converted to the U.S. Customary System, the distance is

(a) 26 mi plus 385 yd.

(b) 26.2 mi.

(c) 137,248 ft.

9. A kilometer is approximately equal in length to

(a) 5 mi.

(b) 3 mi.

(c) 3/5 mi.

10. A weight of 130 lb is approximately the same as

(a) 40 kg.

(b) 60 kg.

(c) 290 kg.

11. A 2-liter bottle contains approximately

(a) 2 quarts.

(b) 1 gallon.

(c) 10 pints.

12. A common speed limit in European neighborhoods is 30 km/h. About how much is that in miles per hour (try to answer with a rough mental estimate)?

(a) Below 20 mph

(b) Between 20 and 30 mph

(c) Above 30 mph

13. Coffee costs about $8 per pound in the United States. If a Canadian dollar (Cdn$) exchanges for U.S.$0.81, what is the approximate cost in Canadian dollars of 500 g of coffee?

(a) Cdn$2
(b) Cdn$7
(c) Cdn$11

14. A sculpture weighs 140 lb and is supported by three legs, each of which is 0.5 in. by 0.5 in. by 2 in. high. How much pressure do the legs exert on the floor?

(a) 47 lb/in.^{2}
(b) 105 lb/in.^{2}
(c) 187 lb/in.^{2}

15. If an object is scaled linearly so that its volume grows to 8 times its original volume, its surface area is scaled to _____ times its original surface area.

(a) 4
(b) 16
(c) 8/3

16. Assuming that a catfish maintains the same shape and proportions as its grows, and a catfish 8 in. long weighs about 1 lb, about how long is a 2-lb catfish?

(a) 10 in.
(b) 12 in.
(c) 16 in.

17. In comparing flight speeds of birds, an analysis of wing loading leads to the conclusion that

(a) light birds fly faster than heavy birds.
(b) heavy birds fly faster than light birds.
(c) heavy and light birds fly at about the same speed.

18. In comparing the heights that large and small animals jump, analysis of the impact on scaling leads to the conclusion that

(a) smaller jumping animals can jump much higher than larger jumping animals.
(b) larger jumping animals can jump much higher than smaller jumping animals.
(c) all jumping animals jump to about the same height.

19. The base-10 logarithm of 72 is approximately

(a) 2.
(b) 4.
(c) 7.

20. The population of Mexico grew by 2.4% in 2002, by another 2.4% (of the 2002 population) in 2003, and by another 2.4% (of the 2003 population) in 2004. The population is growing

(a) proportionally.
(b) allometrically.
(c) by a constant amount each year.

EXERCISES　　■ Challenge　◆ Discussion

Most of these exercises require a calculator; one with square roots will suffice.

Geometric Similarity

1. Suppose you are printing photographs from negatives of so-called 35-millimeter (35 mm) film, whose frames are just about 1 in. by $1\frac{1}{2}$ in.

(a) First you make some contact prints, which are exactly the same size as the negatives. What is the scaling factor of a contact print?
(b) One enlargement that you want to make is to be three times as high and three times as wide as the negative. What is the linear scaling factor for this print? How does its area compare with the area of the negative?
(c) What is the scaling factor for a 4-by-6 (4 in. by 6 in.) print?
(d) A common size of so-called 3-by-5 prints is about $3\frac{1}{16}$ in. by $4\frac{19}{32}$ in. Is such a print geometrically similar to the negative?
(e) The sizes and advertised costs of reprints available at Turtle Creek Bookstore (the Beloit College bookstore) are

3.5 × 5.25 ("standard" size)	$0.28
4 × 6 ("premium" size)	0.35
5 × 7	1.79
8 × 10	3.99
8 × 12	4.49
11 × 14	9.99
16 × 20	15.95
20 × 30	9.99

Assume that the prints have exactly the measurements indicated, in inches. Which sizes are geometrically similar to a 35mm negative? What do you think happens for sizes that aren't?

(f) The cost of photographic paper is very nearly proportional to the area of the paper. Compare the costs of a standard 4-by-6 enlargement and an 8-by-12 enlargement. What can you say about the relative costs of the two prints?

(g) Based on the amount of paper used and the cost of the 5-by-7 and 8-by-10 enlargements, how much would you expect an 11-by-14 enlargement to cost?

(h) The price for the 20-by-30 enlargement was a mistake on the price list. To scale with the other prices, what price should it have?

2. The area of a circle of radius r is πr^2. Expressed in terms of the diameter, $d = 2r$, the area is $\frac{1}{4}\pi d^2$. If we apply a linear scaling factor L to the diameter, then the area of the scaled circle—as in the case of the square that we considered in the text—changes with L^2, the square of the linear scaling factor. A natural application of this idea is with pizza. The prices at the pizza restaurant closest to Beloit College are $6.30, $7.35, $8.35, and $9.40, respectively, for small (10-in.), medium (12-in.), large (14-in.), and extra-large (16-in.) cheese pizzas.

(a) What is the linear scaling factor for an extra-large pizza compared with a small one?

(b) How many times as large in area is the extra-large pizza compared with the small one?

(c) How much pizza does each size give per dollar? What "hidden" assumptions are you making about how the pizzas are scaled up?

(d) The corresponding prices for a pizza with "the works" are $10.45, $12.55, $14.65, and $16.75. Is there any size of these for which you get more pizza per dollar than some size of the cheese pizzas?

(e) All of the prices are, to the nearest cent, exactly 5% higher than 3 years ago! What kind of scaling is that?

3. The human figures in Lego sets are 4 cm tall (without hats or helmets).

(a) What is the linear scaling factor of a Lego figure if it represents a human who is 160 cm tall?

(b) How does the volume of a real human compare with the volume of a Lego figure?

(c) The car in one Lego set is 10 cm long. Using the linear scaling factor in part (a), how long would a real car be?

4. Dollhouses and their furnishings are usually built to a scale of exactly 1 in. to 1 ft, meaning that an item 1 ft long in a real house is 1 in. long in a dollhouse.

(a) What is the linear scaling factor for a dollhouse?

(b) If a dollhouse were made of the same materials as a real house, how would their weights compare?

5. According to *Time* (March 7, 2005), men's brains on average are 10% larger than women's, even though men on average are only 8% taller. (The article mainly discusses the many differences in brain structure that likely outweigh any size differences.) If the brain scales linearly with height, and men are 8% taller, what percentage larger would you expect their brains to be?

6. At our house, we have some 10-in. frying pans and a 12-in. one; the 12-in. one seems to weigh a lot more, never get as hot, and cook food more slowly. Suppose that a 10-in. frying pan weighs 1 lb, apart from its handle. How much would a geometrically similar 12-in. frying pan weigh? How much would it weigh if it had the same thickness of metal as the 10-in. pan?

7. A famous geometry problem of Greek antiquity was *duplication of the cube*. Our knowledge of the history of the problem comes from the third century B.C. from Eratosthenes of Cyrene, famous for his estimate of the circumference of the earth. According to him, the citizens of Delos were suffering from a plague. They consulted the oracle, who told them that to rid themselves of the plague, they must construct an altar to a particular god that would be geometrically similar to the existing one but double the volume.

(a) How would the volume of the new altar compare with that of the old if each of its linear dimensions were doubled?

(b) What should the linear scaling factor be for the new altar? (The problem intended by the oracle was to construct with straightedge and compasses a line segment equal in length to this particular linear scaling factor. Not until the nineteenth century was the task shown to be impossible. Eratosthenes relates that the Delians interpreted the problem in this sense, were perplexed, and asked Plato about it. Plato told them that the god didn't really want an altar of double the volume but wished to shame them for their "neglect of mathematics and their contempt for geometry.")

8. The declining purchasing power of the dollar and the short life of a dollar bill in circulation suggest abolishing the paper dollar. Unfortunately, the Susan B. Anthony dollar coin (1979–1981, 1999) was a failure–the U.S. public found it too small and light. So was the gold-colored Sacagawea dollar (2000–2002), the same size and likely too late (a dollar today is worth one-third what it was in 1978). Suppose that you are to design a new *$5* coin (whom should it depict?). The sole requirement is that it be made of the same material as the quarter but weigh four times as much. A quarter can be described geometrically as a circular cylinder approximately $\frac{15}{16}$ in. in diameter and $\frac{1}{16}$ in. thick. Because your new dollar should weigh four times as much, it needs to have four times the volume of a quarter.

Susan B. Anthony dollar.   Sacagawea dollar.

Coins shown actual size (26.5 mm diameter). *(United States Mint.)*

[The formula for the volume of a cylinder is $\pi \times (\text{diameter}/2)^2 \times \text{height}$.]

(a) A member of your public advisory panel suggests just doubling the diameter and doubling the thickness. What do you tell this individual, in the most diplomatic terms?

(b) If you double the diameter, how thick does the coin need to be?

(c) Another member feels that the result of part (b) would be inconveniently large and proposes instead to scale up the quarter proportionality (she has studied an earlier edition of this book). What would the dimensions be for this coin?

9. Criticize the following statement (on sacks of the product) and write a correct version.

"Erin's Own Irish sphagnum moss peat. It enriches your soil and makes your growing easier. Compressed to $2\frac{1}{2}$ times normal volume."

10. Criticize the following claims, which were cited in *The New York Times* of September 25, 1987, and October 21, 1987:

(a) A new dental rinse "reduces plaque on teeth by over 300%."

(b) An airline working to decrease lost baggage has "already improved 100% in the last six months."

(c) "If interest rates drop from 10% to 5%, that is a 100% reduction."

■ **11.** The GMAT (Graduate Management Admission Test) is required for admission to most U.S. and Canadian M.B.A. programs. The following is a question on a sample test made up and published by a coaching firm (www.microedu. com/gmattest/freetest.htm):

The price of purchasing a car in Country Q is 120% less than the price of purchasing a car in Country Y. Even after transportation fees and tariff charges are added, it is still cheaper for a buyer to import a car from Country Q to Country Y than to buy a car in Country Y.

The statements above, if true, best support which of the following assertions?

A. Gasoline prices in Country Q are 120% below those in Country Y.

B. Importing cars from Country Q to Country Y will eliminate 120% of the sales of cars in Country Y.

C. The tariff on a car imported from Country Q to Country Y is less than 120% of the price of a car in Country Y.

D. The fee for transporting a car from Country Q to Country Y is more than 120% of the price of a car in Country Q.

E. It takes 120% less time to transport a car in Country Q than it does in Country Y.

See if you can figure out what the problem writer intended and express it in correct language. (*Tip:* According to Microedu.com, E gives a possible explanation, B is consistent with the passage, but the correct answer is C.)

12. Figure out what is correct and what is not in the following statement (Associated Press, November 12, 2003), and write a correct version.

"K.C. Moon, director of the Criminal Justice Resources Center, said Oklahoma's per capita expenditures for alcohol and drug abuse treatment programs are 117 percent less than the U.S. average. Oklahoma spends $6.06 per capita on treatment programs, compared with $13.17 nationally. On a regional basis, expenditures total $8.12, or 37 percent more than in Oklahoma." (www.mapinc.org/ccnews/v03/n1758/a04.html?212)

13. Abuses of the language of comparison aren't hard to find. For example, the phrase "ten times less than" occurs in more than 3000 Internet

documents. Search on the Internet and find either an abuse of *times* and *less than* together, or else an abuse of percentages. Figure out what the author meant to say, and write it in correct language.

14. Criticize the following claims:

(a) "A home near mine is on the market for $259,000, but the city says it is worth 141% less, just $107,500." [J. R. Ball, "Assessing the mess," The Greater Baton Rouge Business report.com (January 21, 2003) www.businessreport.com/cgi-bin/udt/im.display.printable?client.id=businessreport&story.id=3109.]

(b) "A veterinarian treating an animal is charged on average about 151% less than what the same drug is sold for human use. That doesn't make a lot of sense." [U.S. Rep. Bart Gordon (6th district of Tenn.), Press release: Gordon supports adding prescription drug benefit to Medicare (April 21, 2000), gordon.house.gov/HoR/TN06/Newsroom/Press+Releases/2000/April/04-21-00+Gordon+Supports+Adding+Prescription+Drug+Benefit+To+Medicare.htm.]

(c) "Let's look at our net direct debt as a percentage of gross domestic product. Our number in British Columbia is 7%. That is 100 percent less than the next-lowest province in Canada, Alberta, which sits at 14.7%." [Mr. Peterson, Official Report of Debates of the Legislative Assembly of British Columbia (May 28, 1991), p. 12260, www.legis.gov.bc.ca/hansard/34th5th/34p_05s_910528p.htm.]

How Much Is That in . . . ?

15. The cost of mailing a lightweight airmail letter from the United States to most of western Europe in 2005 was $0.80. How much was that in euros (€), the currency of the European Union, when the exchange rate was € 1 = $1.30? (For comparison, the cost then of an airmail letter to the United States varied from country to country in the European Union, ranging from € 1.09 to € 1.55.)

16. The cost of mailing a lightweight letter from the United States to Canada in 2005 was US$0.60. How much was that in Canadian dollars when the exchange rate was US$1 = Cdn$1.23? (The postage cost from Canada to the United States was Cdn$0.80.)

17. In Germany the fuel efficiency of cars is measured in liters of fuel per 100 km (L/100 km). A typical average in a 2005 diesel compact station wagon is 7.3 L/100 km. What is that in miles per gallon (mpg)?

18. According to Environmental Protection Agency ratings, the highest-mileage 2005 car seating four or more passengers was the gasoline/electric hybrid Toyota Prius at 60 mpg in the city. How many liters of gasoline does such a Prius use to travel 100 km in the city?

19. Consider a real locomotive that weighs 88 tons and an HO-gauge scale model of it, for which the linear scaling factor is 1/87.

(a) How much would an exact scale model weigh in tons?
(b) What assumptions are involved in your answer to part (a)?
(c) How much would an exact scale model weigh in pounds?
(d) In kilograms?
(e) In metric tonnes (1 metric tonne = 1000 kg)?

20. An ad for a software package for data analysis included a data set on tropical rain forests and deforestation. The data were given in hectares and were accompanied by the statement, "A hectare equals 10,000 mi² or 2471 acres." What conversion factors should have appeared instead?

21. Gasoline is sold in the United States by the U.S. gallon and in Europe by the liter (1 U.S. gal = 231 in.³; 1 L = 1000 cm³). What was the equivalent cost, in U.S. dollars per U.S. gallon, for gasoline in Germany priced in euros at € 1.42 per liter, when € 1 = $1.25, in September 2005?

22. In 1991, Edward N. Lorenz, a meteorologist who was an early researcher into chaos and dynamical systems (discussed in Chapter 23), received the Kyoto Prize in Basic Sciences,

consisting of a gold medal and 45 million Japanese yen (¥). If US$1 = ¥125 at the time, what was the value of the cash award in 1991 U.S. dollars? (In Chapter 21, we show how to convert such an amount to its value in today's dollars.)

23. In 200 B.C., Eratosthenes measured the circumference of the earth and expressed the result as 250,000 *stadia* (plural of *stadium*). In *The American Heritage Dictionary,* Second College Edition (Houghton Mifflin, Boston, 1982), we read for the second meaning of *stadium:* "An ancient Greek measure of distance . . . equal to about 185 kilometers, or 607 feet." The name of the unit came from the length of a racecourse that was a bit less than an eighth of a mile long. If the numbers in the definition are correct, what are the correct units that should have appeared?

Scaling Real Objects

24. The two fastest elevators in Taipei 101 go up (or down) at 55 ft/s.

(a) With no stops, how long would it take to get to the top of Taipei 101?
(b) With no stops, how long would it take to get to the top of the Petronas Towers, which are 1483 ft high and whose elevators run at 41 ft/s?
(c) The answers to parts (a) and (b) suggest that building designers find half a minute a reasonable standard for getting to the top of a building (or else elevators don't come any faster). How fast would the elevators in a mile-high building have to be to achieve that standard?

25. Calculate the speed, in miles per hour, of the elevators in the three buildings in parts (a), (b), and (c) of Exercise 24.

For Exercises 26–31, refer to the following.

The Canadian dollar for many years fell steadily in value in terms of the U.S. dollar, but that trend has been reversed in the past few years. In January 2002, a U.S. dollar was worth Cdn$1.60. In February 2005, US$1 = Cdn$1.23. How should you measure how much one currency has depreciated (lost value) against another? Let the *home currency* be the one whose change in value you are interested in, and let the *target currency* be the one in terms of which you will measure the change. In our example, we want to track the change in value of the U.S. dollar (home currency) in terms of the Canadian dollar (target currency).

There are two competing practices. Both begin by calculating the change as measured in the target currency, the new value minus the old value–here, Cdn$1.23 − Cdn$1.60 = Cdn − $0.37.

▶ Option A, used by the International Monetary Fund and the British periodical *The Economist,* divides this difference by the new trading value (Cdn$1.23) and multiplies by 100 to get a result in percent–here, −30%.

▶ Option B, sometimes called the "popular method," divides instead *by the old trading value,* Cdn$1.60, here getting −23% as the percentage change in value of the U.S. dollar in terms of the Canadian dollar.

So, depending on the method used, we can say that the U.S. dollar lost either 30% or 23% in value against the Canadian dollar.

26. To get a feeling for how the results of these two options come out, we consider an artificial example where the numbers are simple. Suppose that in January 2005 the imaginary currency of the imaginary country Middle Earth, the Middie ($\mathcal{M}$), traded at US$2 but in January 2006 its value was only $1.

(a) Using Option A, calculate how much percentage value the Middie lost against the dollar.
(b) Calculate how much percentage value the Middie lost against the dollar using Option B.

27. We use the same data as in Exercise 26 but look at matters from the perspective of an American entrepreneur importing rings of power from Middle Earth. In January 2005, $1 = $\mathcal{M}$0.50; in January 2006, $\mathcal{M}$1 = $1.

(a) Using Option A, calculate how much percentage value the dollar gained against the Middie.

(b) Using Option B, calculate how much percentage value the dollar gained against the Middie.

28. In January 2002, when European Union euro (€) currency coins and bills were introduced, the conversion to U.S. dollars was US$1 = € 1.160. By 3 years later, in March 2005, the dollar had declined severely, so that $1 = € 0.7435.

(a) Using Option A, calculate how much percentage value the dollar lost against the euro.
(b) Using Option B, calculate how much percentage value the dollar lost against the euro.

29. We use the same data as in Exercise 28 but look at matters from the perspective of a European considering a vacation in the United States. In January 2002, € 1 = $0.862; in March 2005, € 1 = $1.345.

(a) Using Option A, calculate how much percentage value the euro gained against the dollar.
(b) Using Option B, calculate how much percentage value the euro gained against the dollar.

◆ **30.** Using your results from either Exercises 26–27 or 28–29:

(a) Why don't the numbers agree? If the dollar loses a certain percentage against another currency, shouldn't that currency gain the same percentage against the dollar? Why or why not?
(b) Which option, A or B, seems to give a better sense of the effect of the change in relative values of currencies?

◆ **31.** For both Options A and B:

(a) Can either give a percentage loss that is more than 100%? Would it make sense to speak of a currency declining more than 100%?
(b) Is the percentage from Option A–loss or gain–always higher than the Option B percentage? Always lower? Neither, but do you see any pattern?
(c) Which option would you expect a person to use who wants to make a decline seem large? To make a gain seem large?

Sorry, No King Kongs

32. The weight of a 1-ft cube of steel is 500 lb. What is the pressure on the bottom face in

(a) pounds per square inch?
(b) atmospheres (1 atm = 14.7 lb/sq in.)?

33. In an article on adding organic matter to soil, the magazine *Organic Gardening* (March 1983) said, "Since a 6-inch layer of mineral soil in a 100-square-foot plot weighs about 45,000 pounds, adding 230 pounds of compost will give you an instant 5% organic matter."

(a) What is the density of the mineral soil, according to the quotation?
(b) How does this density compare with that of steel?
(c) How do you think the quotation should be revised to be accurate?

For Exercises 34 and 35, refer to the following.

A mature gorilla weighs 400 lb and stands 5 ft tall; its two feet combined have an area of about 1 ft^2.

34. (a) Give an estimate of the gorilla's weight when it was half as tall.

(b) What assumptions are involved in your estimate?

(c) When the gorilla is standing, what is the pressure on its feet in pounds per square inch?

35. Suppose King Kong is a gorilla scaled up with a linear scaling factor of 10.

(a) How much does King weigh?
(b) What is the pressure on King's feet in pounds per square inch?

36. You may want a waterbed but waterbeds are not allowed in your building. Apart from the danger of flood if the bed should puncture or leak, there is the weight.

(a) Suppose that a queen-size waterbed mattress is 80 in. long by 60 in. wide by 12 in. high, and water weighs 1 kg/L. How much does the water in the mattress weigh in pounds?

(b) If the weight of the mattress and frame is carried by four legs, each 2 in. by 2 in., what is the pressure, in pounds per square inch, on each leg?
(c) How does the pressure on the legs of the waterbed compare with the pressure that a person exerts on his or her feet—for example, a 130-lb person with a total foot area of about one-quarter of a square foot in contact with the ground?

37. If you aren't allowed to have a waterbed, how about a spa (hot tub)? Find the weight of the water in a spa that is in the shape of a cylinder 6 ft in diameter and 3.5 ft deep. (*Hint:* The volume of a cylinder is $\pi r^2 h$, where r is the radius and h is the height.)

38. What does the largest giant sequoia tree weigh? Model the tree as a (very elongated) cone. Assume that the tree is 360 ft high, has a circumference of 40 ft at the base, and that the density of the wood is 31 lb/ft^3. (The volume of a cone of height h and radius r is $\frac{1}{3}\pi r^2 h$.)

39. A 6-ft-tall indoor holiday tree needs four strings of lights to decorate it. How many strings of lights are needed for an outdoor tree that is 30 ft high? (Contributed by Charlotte Chell of Carthage College, Kenosha, Wisconsin.)

For Exercises 40 and 41, refer to the following.

An ancient measure of length, the *cubit*, was the distance from the elbow to the tip of the middle finger of a person's outstretched arm. So the length of a cubit depended on the person, though there was some attempt at standardization. Most estimates place the cubit between 17 and 22 in.

40. According to classical Greek sources, Pythagoras (sixth century B.C.) used geometric scaling to model the height of Hercules, the heroic figure of classical mythology. Pythagoras compared the lengths of two racecourses, one (according to tradition) paced off by Hercules and the other by a man of average height. Both were 600 "paces" long, but the one by Hercules was longer because of his longer stride. A normal man in the time of Pythagoras would have been about 5 ft tall.

(a) If the distance paced off by Hercules was 30% longer than the other racecourse, how tall was Hercules? What does your calculation assume?

(b) The sources give two conflicting answers, that Hercules was 4 cubits tall and 4 cubits 1 "foot" tall. What range does this give for his height in feet and inches? In centimeters? (Assume that a Greek "foot" was the length of a modern foot.)

41. Goliath [of David and Goliath, as related in the Bible (I Samuel 17:4)] was "six cubits and a span." A "span" was originally the distance from the tip of the thumb to the tip of the little finger when the hand is fully extended, about 9 in. What range of heights would this indicate for Goliath in feet and inches? In centimeters?

For Exercises 42–45, refer to the following.

The *body mass index* (BMI) is the basis for the National Heart, Lung and Blood Institute's weight guidelines. BMI is body weight (in kilograms) divided by the square of height (in meters). A BMI of 25 through 29 is considered "overweight"; a BMI of 30 or over is considered "obese." Some 55% of American adults have a BMI of 25 or above. (*Note:* BMI is not a useful measure for young children, pregnant or breastfeeding women, the frail elderly, or very muscular people.) For practice with this concept, calculate your own BMI.

42. Calculate the BMI for a woman 160 cm tall who weighs 65 kg. Is she overweight according to the institute's guidelines?

43. How much in kilograms must a man weigh who is 190 cm tall if he is not to be considered overweight according to the institute's guidelines?

44. Suppose that weight and height are measured instead in U.S. Customary units of pounds and inches. We can still calculate body weight divided by the square of height using these units. What

conversion factor is necessary to convert this number to the BMI?

◆ **45.** Because body weight is average density times body volume, BMI is average density times a quantity that has units of length. Discuss whether BMI makes sense as a measure of being overweight. Would dividing by a different power of height make for a better measure?

Dimension Tension

◆ **46.** Jonathan Swift's Gulliver traveled to Lilliput, where the Lilliputians were human-shaped but only about 6 in. tall. In other words, they were geometrically similar in shape to ordinary human beings but only one-twelfth as tall. What would a Lilliputian weigh?

Are Lilliputians ruled out by the size–shape and area–volume considerations in this chapter? If you think they are, what considerations do you find convincing? If not, why not?

47. (a) What would you expect an individual *Quetzalcoatlus northropi* to weigh if it had half the wingspan of an adult?

(b) If an individual weighed half as much as an adult, what would you expect its wingspan to be?

48. In the children's story *Peter Pan*, Peter and Wendy can fly. We may suppose that they are 4 ft tall, so they are about 12 times as tall as a sparrow is long. What should their minimum flying speed be?

49. Icarus of Greek legend escaped from Crete with his father, Daedalus, on wings made by Daedalus and attached with wax. Against his father's advice, Icarus flew too close to the Sun; the wax melted, the wings fell off, and he fell into the sea and drowned. What must have been his minimum cruising speed? What assumptions does your answer involve?

◆ **50.** Recent years have seen the beginnings of human-powered controlled flight, in the *Gossamer Condor* and other superlightweight planes, which have disproportionately large wings compared with geometric scaling up of birds. The *Gossamer Condor* is far longer than an ostrich but it flies at only 12 mph. How can it?

51. The cult movie *Them* (1954) features enormous ants (8 m long by 3 m wide). We can investigate the feasibility of such a scaled-up insect by considering its oxygen consumption. A common ant, 1 cm long, needs 24 milliliters (mL) of oxygen per second for each cubic centimeter of its volume. Because an ant has no lungs, it absorbs oxygen through its "skin" at a rate of 6.2 mL per second per square centimeter. Suppose that the tissues of a scaled-up ant would have the same need for oxygen for each cubic centimeter, and that its skin could absorb oxygen at the same rate, as a common ant.

(a) Compared with a common ant, how many times as large is an enormous ant's
  (i) length?
  (ii) surface area?
  (iii) volume?
(b) What proportion of such an ant's oxygen need could its skin supply?
(c) What can you conclude about the existence of such insects? (Adapted from George Knill and George Fawcett, Animal form or keeping your cool, *Mathematics Teacher*, May 1982, 395–397.)

For Exercises 52–55, refer to the following.

Maybe some trees could grow to a mile high, but they just don't live long enough to have the chance. In this problem, we try to determine how fast the height of a tree increases. We can measure indirectly how much mass the tree adds in a year by the area of the annual tree ring added. Here are two relevant facts:

▶ As you may have noticed from stumps, as a tree grows older, its annual rings get less wide. Although the width of the ring varies somewhat from year to year with the amount of rainfall and other factors, the total *area* of each annual ring is roughly the same over the years, meaning that *the tree adds roughly the same amount of mass each year*. Call that amount a; then the

mass M of the tree is $M = at$, where t is its age in years.

▶ Over a large range of tree sizes and tree species, the diameter d of a tree of a species is approximately proportional to the three-halves power of the height h of the tree (different species have different constants of proportionality). Thus, $d \propto h^{3/2}$ (this is shown in Exercise 52).

Now, if we assume that the bulk of the mass of the tree is in the trunk, and if we model the trunk either as a long cylinder or as a thin cone, the mass is proportional to the volume, so $M \propto d^2 h$. Then

$$at = M \propto d^2 h \propto (h^{3/2})^2 h = h^4$$

so $h \propto t^{1/4}$. In other words, *the tree grows in height as the fourth root of its age.*

52. Suppose that a tree grows to 20 m in 30 years. How tall will it be (if it lives long enough) when it is 60 years old?

53. How long would it take the tree in Exercise 52 to grow to be 40 m tall?

54. Giant sequoias can reach 100 m after about 1000 years. If it could keep on growing at the same rate of addition of mass, how long would it take a giant sequoia to grow to 200 m?

■ **55.** The branching of trees is similar to the branching of systems in the bodies of animals. For similar reasons, the area of the cross section of the tree at its base scales as the three-fourths power of the tree's mass, that is, $A \propto M^{3/4}$. Assume that most of the mass is in the trunk and model the tree either as a tall cylinder ($V = \pi r^2 / h$) or as a cone ($V = \pi r^2 h / 3$). Show that the diameter d of a tree is approximately proportional to the three-halves power of the height, that is, $d \propto h^{3/2}$.

◆ **56.** Some humans, such as the Bushmen of the Kalahari Desert in Africa, live in desert environments, where it is important to be able to do without water for periods of time. Would you expect such an environment to favor short people or tall ones? (Adapted from A. Zherdev, Horseflies and flying horses, *Quantum,* May–June 1994, 32–37, 59–60.)

◆ **57.** Smaller birds and mammals generally maintain higher body temperatures than do larger ones. Explain why you would expect this to be so. (Adapted from A. Zherdev, Horseflies and flying horses, *Quantum,* May–June 1994, 32–37, 59–60.)

How to Grow

58. Listed below are the numbers of species of reptiles and amphibians on some Caribbean islands, together with the approximate areas of the islands. (Suggested by Florence Gordon of the New York Institute of Technology, with contributions from Kevin Mitchell and James Ryan of Hobart and William Smith Colleges, Geneva, N.Y. This table is adapted from Tables 15 and 16 in P. J. Darlington, *Zoogeography: The Geographic Distribution of Animals,* Wiley, New York, 1957, pp. 483–484).

Island	Area (mi²)	Species
Redonda	1	3
Saba	4.9	5
Montserrat	40	9
Trinidad	2,000	80
Puerto Rico	3,400	40
Jamaica	4,500	39
Hispaniola	30,000	84
Cuba	40,000	76

(a) Plot number of species versus area on ordinary graph paper and then on log-log graph paper. If you don't have log-log paper available, use a calculator or spreadsheet to take the logarithms ($\log_{10}$) of all the numbers and graph logarithm of number of species versus logarithm of area on ordinary graph paper. (*Note:* Trinidad is an outlier from the general pattern—see Chapter 6).
(b) Is the relationship that you graphed in part (a) proportional? Allometric?
(c) What would be the expected number of species on an island of 400 mi²?

(d) For each 10-fold increase in the island's size, what happens to the number of species, approximately?

59. Listed below are the weights and wingspans of some birds and of some fully loaded airplanes. (Idea and data contributed by Florence Gordon of the New York Institute of Technology.)

Bird	Weight (lb)	Wingspan (ft)
Crow	1	2.9
Harris hawk	2.6	3.2
Blue-footed booby	4	3
Red-tailed hawk	4	4
Horned owl	5	5
Turkey vulture	6.5	6
Eagle	12	7.5
Golden eagle	13	7.3
Whooping crane	16.1	7.5
Vulture	18.7	9.3
Condor	22	9.9
Quetzalcoatlus northropi	100	36

Plane		
Boeing 737	117,000	93
DC9	121,000	93.5
Boeing 727	209,500	108
Boeing 757	300,000	156.1
Boeing 707	330,000	145.7
DC8	350,000	148.5
DC10	572,000	165.4
Boeing 747	805,000	195.7

(a) Use a calculator or spreadsheet to take the logarithms ($\log_{10}$) of all the numbers and then graph logarithm of weight versus logarithm of wingspan on ordinary graph paper.
(b) For the birds, is the relationship that you graphed in part (a) proportional? Allometric? How about for the planes?
(c) Does the same relationship of wingspan to weight seem to hold for birds and planes?

WRITING PROJECTS

1. A human infant at birth usually weighs between 5 and 10 lb and has a height (length) between 1 and 2 ft, with the shorter babies having the lesser weight. Considering the weight and height of an adult human, write a paragraph arguing that human growth must not be just proportional growth.

2. The principle that area scales with the square of length, and volume with the cube, has important consequences for the depiction and interpretation of data in graphic form. Suppose we wish to indicate in an artistic way that the weekly income of a U.S. carpenter is twice that of a carpenter in (mythical) Rotundia. We draw one moneybag for the Rotundian and another one "twice as large" for the American. (Illustration from Darrell Huff, *How to Lie with Statistics,* Norton, New York, 1954, p. 69.)

What's the problem? Well, first, people tend to respond to graphics by comparing areas. Because the larger moneybag is twice as high and twice as wide as the smaller one, its image has four times the area. Second, we are used to interpreting depth and perspective in drawings in terms of three-dimensional objects. Because the larger bag is also twice as thick as the smaller, it has eight times the volume. The graphic leaves the subconscious impression that the U.S. carpenter earns eight times as much, instead of twice as much. With these ideas in mind, evaluate—in a paragraph each—the following data depictions.

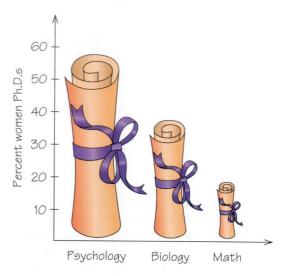

(a) Percentages of Ph.D.s earned by women in three fields. *[From* Science, *260 (April 16, 1993), 409, as reproduced in Jessica Utts,* Seeing Through Statistics, *Duxbury, Belmont, Calif., 1996, p. 142.]*

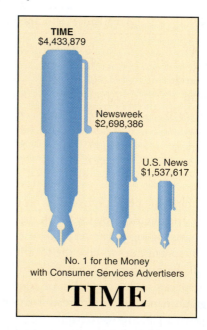

(b) Advertising spending in three prominent news-magazines. *(From* Time *magazine, as reproduced in David S. Moore,* Statistics: Concepts and Controversies, *4th ed., W. H. Freeman, New York, 1997, p. 207.)*

715
colleges

517
colleges

370
colleges

| Fewer than 2500 students | 2500 to 9999 students | 10,000 or more students |

(c) U.S. colleges as classified by enrollment. *(From David S. Moore, Statistics: Concepts and Controversies, 4th ed., W. H. Freeman, New York, 1997, p. 217.)*

3. Evaluate in a paragraph each of the following depictions (a–c). (Illustrations from Edward R. Tufte, *The Visual Display of Quantitative Information*, Graphics Press, 1983, pp. 70, 69, and 57)

1958—Eisenhower

1963—Kennedy

1968—Johnson

(a) Value of the dollar.

1973—Nixon

1978—Carter

1984—Reagan

1990—Bush

1994—Clinton

2006—Bush

THE SHRINKING FAMILY DOCTOR
In California

Percentage of Doctors Devoted Solely to Family Practice

1964	1975	1990
27%	16.0%	12.0%

1: 4,232
6,212

1: 3,167
6,694

1: 2,247 RATIO TO POPULATION
8,023 Doctors

(b) The shrinking family doctor.

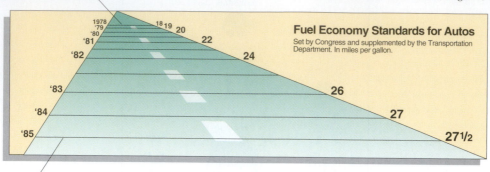

This line, representing 18 miles per gallon in 1978, is 0.6 inch long.

(c) Fuel economy standards for autos.
[*From* The New York Times, *August 9, 1978, p. D2.*]

Fuel Economy Standards for Autos
Set by Congress and supplemented by the Transportation Department. In miles per gallon.

1978 '79 '80 '81 18 19 20 22 '82 24 '83 26 '84 27 '85 27½

This line, representing 27.5 miles per gallon in 1985, is 5.3 inches long.

4. With the ideas of Writing Projects 2 and 3 in mind, collect and evaluate similar depictions of data from magazines and newspapers.

5. Dolls and human figures are usually scaled to be geometrically similar to actual humans. But are dolls designed to represent babies or adult humans? Go to a toy store and measure the height, the vertical height of the head, and the arm length of some dolls and other figures. Scale your measurements to compare them with Figure 18.11; from that comparison, try to estimate the ages of the humans that the figures resemble. Write up your procedure, data, calculations, and conclusions in a page or two.

SUGGESTED READINGS

DUDLEY, BRIAN A. C. *Mathematical and Biological Interrelations*, Wiley, New York, 1977. Excellent and gentle extended introduction to graphing, scale factors, and logarithmic plots.

GOULD, STEPHEN JAY. Size and shape. In *Ever Since Darwin*, Norton, New York, 1977, Chap. 21.

HALDANE, J. B. S. On being the right size. In *Possible Worlds and Other Papers*, Harper, New York, 1928. Reprinted in James R. Newman (ed.), *The World of Mathematics*, vol. 2, Simon & Schuster, New York, 1956, pp. 952–957. Also reprinted in John Maynard Smith (ed.), *On Being the Right Size and Other Essays by J. B. S. Haldane*, Oxford

University Press, Oxford, 1985, pp. 1–8. Succinctly surveys area–volume tension, flying, the size of eyes, and even the best size for human institutions.

McMAHON, T. A., and J. T. BONNER. *On Size and Life*, Scientific American Library, New York, 1983. Astonishingly beautiful and informative book on the effects of size and shape on living things.

SCHMIDT-NIELSEN, KNUT. *Scaling: Why Is Animal Size So Important?* Cambridge University Press, New York, 1984.

SUGGESTED WEB SITES

physics.nist.gov/cuu/Units/index.html In-depth information on SI, the modern metric system.

ftp://ftp.fifthwave.com/pub/Convert201.sit Convert (computer program for Macintosh) can convert many kinds of units; you can explore such units as turns (volume), blinks (time), and barns (area).

www.missingkids.com National Center for Missing and Exploited Children.

www.usmint.gov U.S. Mint.

www.thusness.com/bmi.t.html Body mass index calculator and further links.

Symmetry and Patterns

"The senses delight in things duly proportional." So said the famous philosopher-theologian Thomas Aquinas more than 700 years ago. In this chapter, we examine elements of esthetic appreciation, particularly *symmetry*.

What is symmetry and what does mathematics have to do with it? Symmetry, like beauty, is hard to define. Dictionaries talk about "correspondence of form on opposite sides of a dividing line or plane or about a center or an axis," "correspondence, equivalence, or identity among constituents of an entity," and "beauty as a result of balance or harmonious arrangement" (*American Heritage Dictionary*, 3rd ed.).

In the narrowest sense, symmetry refers to mirror-image correspondence between parts of an object. Crystals, in both their appearance and their atomic structure, provide examples of symmetry in this sense. Taken in a wider sense, though, symmetry includes notions of *balance, similarity,* and *repetition*.

Our wider sense of symmetry leads us to appreciate patterns. *Mathematics is the study of patterns,* and it gives important insights into symmetry.

Spirals

Patterns abound in nature. The successive sections of the beautiful chambered nautilus (Figure 18.9 and Figure 19.2b) grow according to a strict and specific spiral pattern, a kind of symmetry. This spiral has the same shape at any size: A photographic enlargement superimposed on it would fit exactly.

Botanists have long appreciated other spirals. In plant growth from a central stem, the shoots, leaves, and seeds often occur in a spiral pattern known as **phyllotaxis.** For instance, the scales of a pineapple or a pinecone are arranged in

FIGURE 19.1 Spirals of scales on a pinecone: 8 right, 13 left. *(From Verner E. Hoggatt, Jr., Fibonacci and Lucas Numbers, Houghton Mifflin, New York, 1969, p. 81.)*

FIGURE 19.2 (a) This sunflower has 55 spirals in one direction and 89 spirals in the other direction. *(Harvey Lloyd/The Stock Market.)* (b) A chambered nautilus shell. *(James Randkler/Tony Stone Images.)*

spirals (Figure 19.1), as are the seeds of a sunflower (Figure 19.2a) and the petals on a daisy. Like the chambers of the nautilus in Figure 19.2b, the spirals on these plants are geometrically similar to one another, and they are arranged in a regular way, with balance and "proportion." These plants have a **rotational symmetry.**

19.1 Fibonacci Numbers and the Golden Ratio

Fibonacci Numbers

Associated with the geometric symmetry of phyllotaxis is a kind of *numeric symmetry*, with a "proportion" in the sense of a ratio of numbers. Strangely, the number of spirals in plants with phyllotaxis is not just any whole number but always comes from a particular sequence of numbers called the *Fibonacci numbers* (see Spotlight 19.1).

SPOTLIGHT 19.1   Leonardo of Pisa ("Fibonacci")

Born in Pisa in 1170, Leonardo of Pisa has been known as "Fibonacci" for the past century and a half. This nickname, which refers to his descent from an ancestor named Bonaccio, is modern, and there is no evidence that he was known by it in his own time.

Leonardo was the greatest mathematician of the Middle Ages. His stated purpose in his book *Liber abbaci* (1202) was to introduce calculation with Hindu-Arabic numerals into Italy, to replace the Roman numerals then in use. Other books of his treated topics in geometry, algebra, and number theory.

We know little of Leonardo's life apart from a short autobiographical sketch in the *Liber abbaci*:

I joined my father after his assignment by his homeland Pisa as an officer in the customhouse located at Bugia [Algeria] for the Pisan merchants who were often there. He had me marvelously instructed in the Arabic-Hindu numerals and calculation. I enjoyed so much the instruction that I later continued to study mathematics while on business trips to Egypt, Syria, Greece, Sicily, and Provence and there enjoyed discussions and disputations with the scholars of those places. [L. E. Sigler,

Leonardo of Pisa ("Fibonacci")
A portrait of unlikely authenticity. *(From Columbia University, D. E. Smith Collection.)*

Leonardo Pisano Fibonacci, The Book of Squares: An Annotated Translation into Modern English, Academic Press, New York, 1987, p. xvi.]

The *Liber abbaci* contains a famous problem about rabbits, whose solution is now called the Fibonacci sequence. Leonardo did not write further about it.

Fibonacci Numbers (Fibonacci Sequence)

Fibonacci numbers occur in the sequence

$$1, 1, 2, 3, 5, 8, 13, 21, 34, 55, 89, 144, 233, 377, \ldots$$

This sequence begins with the numbers 1 and 1 again, and each next number is obtained by adding the two preceding numbers together.

Sometimes a sequence of numbers is specified by stating the value of the first term or first several terms and then giving an equation to calculate succeeding terms from preceding ones. This is called a *recursive rule,* and the sequence is said to be defined by **recursion.** Let's denote the *n*th Fibonacci number by F_n; then the Fibonacci sequence can be defined by

$$F_1 = 1,\ F_2 = 1,\qquad \text{and}\qquad F_{n+1} = F_n + F_{n-1}\qquad \text{for } n \geq 2$$

The recursive rule just expresses in algebraic form that the next Fibonacci number is the sum of the previous two.

Look at the sunflower in Figure 19.2a. You see a set of spirals running in the counterclockwise direction and another set in the clockwise direction. It is (just barely) possible to count the number of spirals in both directions. In the sunflower there are 55 in one and 89 in the other direction—two consecutive Fibonacci numbers. In the case of the pineapple, there are three sets of spirals, one each along the three directions through each hexagonally shaped scale. For the common grocery pineapple (*Ananas comosus*), there are always 8 spirals to the right, 13 to the left, and 21 vertically—again, consecutive Fibonacci numbers.

Why are the numbers of spirals in plants the same numbers that appear next to each other in a purely mathematical sequence? There is no easy answer. There are several intricate theories about the dynamics of the plant's growth.

The Golden Ratio

During the last several centuries, an attractive myth has arisen that the ancient Greeks considered a specific numerical proportion essential to beauty and symmetry. Known in modern times as the **golden ratio, golden mean,** or even **divine proportion,** this proportion was investigated by Euclid in Book II of his *Elements.* Recent research reveals little evidence connecting this proportion to Greek esthetics, but let's pursue the golden ratio briefly because of its intimate connection to the Fibonacci sequence and because it does have appeal as a standard for beautiful proportion.

Golden Ratio

The value of the **golden ratio,** which is usually denoted by the Greek letter phi (ϕ), is

$$\phi = \frac{1 + \sqrt{5}}{2} \approx 1.618034\ldots$$

The basic esthetic claim is that a **golden rectangle**—one whose height and width are in the ratio of 1 to ϕ—is the most pleasing of all rectangles. The Greeks treated lengths geometrically, so for them it was important to construct lengths using straightedge and compass. In Spotlight 19.2 we show how to construct a golden rectangle that is 1 unit by ϕ units.

SPOTLIGHT 19.2   How the Greeks Constructed a Golden Rectangle

In constructing a golden rectangle, the Greeks started from a one-by-one square (shown in black in the figure), which they made by constructing perpendiculars at the two ends of a horizontal segment of unit length. To extend the square to a golden rectangle, they bisected the original segment, getting a new point dividing it into two pieces of length one-half each. Using this new point and a compass opening equal to the distance from it to a far corner of the square (shown by the blue line in the figure), they could add the blue length to the length one-half to get an interval (in red at bottom) with total length ϕ.

A golden rectangle has the pleasing property that if you cut a square-shaped piece off one end of it, the rectangle that remains is again a golden rectangle.

Why would anyone think that this is an attractive ratio? And where did it come from? The answer lies not in Fibonacci numbers but in the Greeks' pursuit of balance in their study of geometry.

Given two line segments of different lengths, one way to find another length that "strikes a balance" between the two is to average them. For lengths l (the larger) and w (the smaller), their average, or *arithmetic mean*, is $m = (l + w)/2$, and it satisfies

$$l - m = m - w$$

The length m strikes a balance between l and w, in terms of a common difference from the two original lengths. More generally, the arithmetic mean of n numbers or lengths is their sum divided by n. (See Chapter 5 for its use in statistics.)

The Greeks, however, preferred a balance in terms of ratios rather than differences. They sought a length s, the *geometric mean*, that gives a common ratio

$$l \div s = s \div w \qquad \text{or} \qquad \frac{l}{s} = \frac{s}{w}$$

Hence $lw = s^2$, which expresses the geometric fact that s is the side of a square, the area of which is the same as the area of a rectangle that is l by w (the Greeks thought in terms of geometric objects). In geometry, the geometric mean s is called the *mean proportional* between l and w (see Figure 19.3).

The Greeks found symmetry and proportion in the geometric mean, but the geometric mean also has important practical applications (see Spotlight 19.3).

FIGURE 19.3 The line segment of length l is divided so that the length of s is the geometric mean between l and $w = l - s$. The dividing point divides the length l in the golden ratio.

Geometric Mean

The quantity $s = \sqrt{lw}$ is the **geometric mean** of l and w. More generally, the geometric mean of n numbers is the nth root of the product of all n factors: The geometric mean of $x_1, \ldots, x_n$ is $\sqrt[n]{x_1 \times \cdots \times x_n}$. For example, the geometric mean of 1, 2, 3, and 4 is $\sqrt[4]{1 \times 2 \times 3 \times 4} = \sqrt[4]{24} = 24^{1/4} \approx 2.213$.

The Greeks were interested in cutting a single line segment of length l into lengths s and w, where $l = w + s$, so that s would be the mean proportional between w and l. Surprisingly, the ratio ϕ arises, as we show. Denote by x the common ratio

$$\frac{l}{s} = \frac{s}{w} = x$$

Substituting $l = s + w$, we get

$$x = \frac{l}{s} = \frac{s + w}{s} = \frac{s}{s} + \frac{w}{s} = 1 + \frac{w}{s}$$

But w/s is just $1/x$, so we have

$$x = 1 + \frac{1}{x}$$

Multiplying through by x gives

$$x^2 = x + 1 \quad \text{or} \quad x^2 - x - 1 = 0$$

This is a quadratic equation of the form

$$ax^2 + bx + c = 0$$

The Bureau of Labor Statistics (BLS) uses the geometric mean—not the arithmetic mean—to calculate the Consumer Price Index (CPI), which tracks changes in the cost of the goods and services that people buy.

The geometric mean takes into account substitutions that consumers make when prices change. For example, if the price of beef goes up but the price of chicken doesn't, consumers may buy less beef and substitute the cheaper chicken for some beef.

Suppose that, overall, U.S. families consume equal dollar values of beef and chicken. A typical family might consume weekly 5 lb of beef at $4/lb and 10 lb of chicken at $2/lb, for $20 each and a total cost of $40. We say that beef and chicken each have a relative *market share* of 0.5 (50% beef, 50% chicken, by dollar value).

What if beef goes up to $6/lb but chicken stays at $2/lb? The *relative price change* in beef is $6/$4 = 1.5 and the relative price change in chicken is $2/$2 = 1.00 (no change). If the average family continues to eat just as much beef and chicken as before, the cost is now $50, an increase of 25%. Because $30 goes for beef and $20 for chicken, the relative market shares (0.6 and 0.4) have changed. The *relative price change* for the family's meat is $50/$40 = 1.25, which is just the arithmetic mean of the two relative price changes (1.50 and 1.00). A more general formulation is:

relative price change

$$= (\text{old market share of beef}) \frac{\text{new cost of beef}}{\text{old cost of beef}}$$

$$+ (\text{old market share of chicken}) \times$$

$$\frac{\text{new cost of chicken}}{\text{old cost of chicken}}$$

$$= 0.5 \times \frac{6.00}{4.00} + 0.5 \times \frac{2.00}{2.00}$$

$$= \frac{1.50 + 1.00}{2} = 1.25$$

A family that eats no beef sees no increase. A family that eats only beef sees an increase of 50%. The CPI is an average over *all* families, weighted by the dollar value that each consumes.

If instead we use the geometric mean, we get a relative price change of $\sqrt{1.50 \times 1.00} = 1.225$. The more general formulation is

relative price change

$$= \left(\frac{\text{new cost of beef}}{\text{old cost of beef}} \right)^{(\text{old market share of beef})}$$

$$\times \left(\frac{\text{new cost of chicken}}{\text{old cost of chicken}} \right)^{(\text{old market share of chicken})}$$

$$= \left(\frac{6.00}{4.00} \right)^{0.5} \times \left(\frac{2.00}{2.00} \right)^{0.5}$$

$$= \sqrt{1.50 \times 1.00} = 1.225$$

This relative price change, a 22.5% increase, is less than the 25% using the arithmetic mean.

The intention of the CPI is to measure the change in the cost of goods and services that still yield the same level of satisfaction to consumers. Use of the arithmetic mean presumes that a family buys the same amount of beef and chicken (5 lb beef, 10 lb chicken) as before. Use of the geometric mean presumes that a family buys the same *relative dollar value* of each meat as before, hence $24.50 (12.25 lb) of chicken and $24.50 (4.08 lb) of beef, for a total of $49 = 1.225 × $40. Buying 2.25 lb more chicken and 0.92 lb less beef is supposed to yield the "same satisfaction" as before.

Because the geometric mean is always less than or equal to the arithmetic mean (see Exercise 11), the geometric mean gives a lower figure for inflation than using the arithmetic mean would produce.

Social Security payments, some wage increases, and income tax rates are all automatically geared to the CPI, which we treat in detail in Chapter 21.

with $a = 1$, $b = -1$, and $c = -1$. We apply the famous quadratic formula,

$$x = \frac{-b \pm \sqrt{b^2 - 4ac}}{2a}$$

to get the two solutions

$$x = \frac{1 + \sqrt{5}}{2} \approx 1.618034 \ldots \quad \text{and} \quad \frac{1 - \sqrt{5}}{2} \approx -0.618034 \ldots$$

FIGURE 19.4 In a pentagon with equal sides, ϕ is the ratio of a diagonal to a side. The five-pointed star formed by the diagonals was the symbol of the followers of the ancient Greek mathematician Pythagoras.

We discard the negative solution because it does not correspond to a length. The first solution is the golden ratio ϕ. It occurs often in other contexts in geometry; for example, ϕ is the ratio of a diagonal to a side of a regular pentagon (see Figure 19.4).

Thanks to Roger Herz-Fischler (Wilfrid Laurier University) and George Markowsky (University of Maine), we know that the term *golden ratio* was not used in antiquity and that there is no evidence that the Great Pyramid was designed to conform to ϕ, nor that the Greeks used ϕ in the proportions of the Parthenon, nor that Leonardo da Vinci used ϕ in proportions for the human figure (Figure 19.5a). The area from the top of the head of the "Mona Lisa" to the

FIGURE 19.5 (a) Leonardo da Vinci's "Vitruvian Man" (ca. 1490), based on body proportions by Vitruvius (architect and engineer, first century B.C.). Despite claims on the Web and in the thriller *The Da Vinci Code*, neither Vitruvius nor Leonardo suggested using ϕ for human proportions or anything else. *(Accademia, Venice, Italy/Scala/Art Resource, New York)* (b) Le Corbusier, however, did use ϕ in his "Modulor" scale of proportions. *(Le Corbusier, "Le Modulor," 1945. © 2000 Artists Rights Society [ARS], New York/ADAGP, Paris/FLC.)*

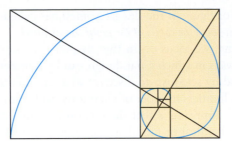

FIGURE 19.6 A logarithmic spiral determines a sequence of golden rectangles and corresponding squares.

top of her bodice may form a golden rectangle, as claimed by Bulent Atalay in his *Math and the Mona Lisa: The Art and Science of Leonardo da Vinci* (2004), but Leonardo left no documents saying that was his intention or design principle. Others have claimed that the impressionists Gustave Caillebotte (1848–1894) and Georges Seurat (1859–1891) used the golden ratio to design some of their paintings, but the painters themselves left no writings about it. Wolfgang Amadeus Mozart (1756–1791), who was fascinated by mathematics as a student, may have constructed the lengths of parts of some of his piano sonatas with an eye to the golden ratio. But we do not have evidence that this was his intention.

Moreover, experiments show that people's preferences for dimensions of rectangles cover a wide range, with golden rectangles not holding any special place.

It is true that human bodies exhibit ratios close to the golden ratio, as you can see by comparing your overall height to the height of your navel. The Swiss-born architect Le Corbusier (Charles-Edouard Jeanneret [1887–1965]) used the golden ratio (including a navel-height feature) as the basis for his "Modulor" scale of proportions (Figure 19.5b).

There are intriguing connections between the spiral of the nautilus and the spirals of the sunflower, and between the golden ratio and the Fibonacci sequence. The nautilus shape follows what is known as an *equiangular* or *logarithmic* spiral, which in its turning determines a sequence of golden rectangles (Figure 19.6). The spirals of the sunflower are approximations to a logarithmic spiral. The mathematical reason for this connection is that the ratios of consecutive Fibonacci numbers

$$\frac{1}{1} \qquad \frac{2}{1} \qquad \frac{3}{2} \qquad \frac{5}{3} \qquad \frac{8}{5} \qquad \frac{13}{8} \qquad \frac{21}{13} \, \cdots$$

$$1.0 \qquad 2.0 \qquad 1.5 \qquad 1.666 \ldots \qquad 1.6 \qquad 1.625 \qquad 1.615 \ldots$$

provide alternately under- and overapproximations to $\phi \approx 1.618034 \ldots$.

19.2 Symmetries Preserve the Pattern

The spiral distribution of the seeds in a sunflower head and the spiraling of leaves around a plant stem are instances of *similarity* and *repetition*, two key aspects of symmetry. They also illustrate *balance*, which refers to regularity in *how* the repetitions are arranged. In considering patterns with repetition, we distinguish the individual element or figure of the design (sometimes called the *motif*) from the *pattern* of the design—*how the copies of the motif are arranged.*

The problem that we focus on in this chapter is how to classify the fundamentally different ways in which a flat design can be symmetrical. The ideas that we discuss were used by scientists to discover what crystalline forms are possible. Although there is a limitless number of chemicals, and of motifs that people can make, what is quite surprising is that there is only a limited number of ways to arrange atoms in a chemical or motifs in a design in a symmetrical way.

How can we enumerate the ways that designs can be put together without counting all the actual designs themselves? The key mathematical idea is to look not at the motifs that make up the patterns but at what you can *do* to the pattern without changing its appearance. This is what we pursue in the next section.

Rigid Motions

Mathematicians describe various kinds of symmetry by using the geometric notion of a *rigid motion,* also known as an **isometry** (which means "same size"). A rigid motion is a specific kind of variation on the original pattern: We pick it up and move it, perhaps rotate it, possibly flip it over—but we *don't change its size or shape.* (The original figure and its image are not just geometrically similar, in the language of section 18.1, but also the same size.)

Figure 19.7 shows the results of various motions applied to the rectangle and its interior of Figure 19.7a. In Figure 19.7b, each side is shrunk by 50%—not a rigid motion, because the size of the rectangle changes. For Figure 19.7c, we shear the rectangle—again, this is not a rigid motion because the shape of the rectangle changes. In Figure 19.7d we rotate the rectangle 90° (a quarter-turn) clockwise

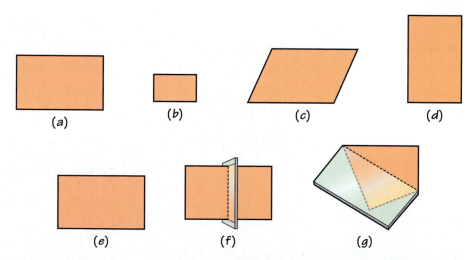

FIGURE 19.7 Results of various motions applied to a rectangle and its interior: (a) the original rectangle and interior; (b) 50% reduction (not a rigid motion); (c) shearing (not a rigid motion); (d) quarter-turn; (e) half-turn; (f) reflection along the vertical line down the middle; (g) reflection along a diagonal line.

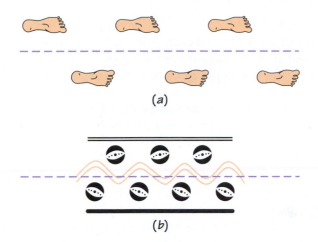

FIGURE 19.8 Glide reflection of (a) footprints; (b) design elements on a pot from San Ildefonso Pueblo.

around the center of the rectangle: This is a rigid motion. Similarly, in Figure 19.7c, rotating by 180° (a half-turn) is a rigid motion.

In Figure 19.7f we reflect the rectangle along a vertical mirror down the middle: Could you tell? The right and left halves exchange places.

Figure 19.7g shows the result of reflecting across a diagonal of the rectangle. All reflections and all rotations are rigid motions. So are all **translations,** which move every point in the plane a certain distance in the same direction.

The only remaining kind of rigid motion in the plane is a hybrid of reflection and translation. Known as a **glide reflection,** it is the kind of pattern that your footprints make as you walk along: Each successive element of the design (footprint) is a reflection of the previous one (Figure 19.8). The motion combines, in an integral way, translation ("glide") with a reflection across a line that is parallel to the direction of the translation.

Rigid Motion

A **rigid motion** is one that preserves the size and shape of figures. In particular, any pair of points is the same distance apart after the motion as before. Any rigid motion of the plane must be one of:

▶ Reflection (across a line)
▶ Rotation (around a point)
▶ Translation (in a particular direction)
▶ Glide reflection (across a line)

Performing one rigid motion after another results in a rigid motion that (surprisingly) must be one of the four types that we have just explored.

Preserving the Pattern

In terms of symmetry, we are especially interested in rigid motions like those of Figures 19.7e and 19.7f that **preserve the pattern**—that is, ones for which the pattern looks exactly the same, *with all the parts appearing in the same places,* after the motion is applied.

You might enjoy thinking of applying these motions as "The Pattern Game": You turn your back, I apply a transformation, then you turn back and see if you can tell whether anything is changed.

The 90° rotation of Figure 19.7a into Figure 19.7d does not preserve the pattern. The moved rectangle doesn't fit exactly over the original rectangle. On the other hand, the 180° rotation in Figure 19.7e does preserve the pattern. It's true that the top of the original rectangle is now on the bottom of the transformed version, but you can't tell that that has happened because you can't distinguish the two. If you had turned your back while the motion was applied, you would not be able to tell that anything had been done. A rotation by any multiple of 180° would also preserve the pattern.

Similarly, the reflection across the vertical line in Figure 19.7f preserves the pattern, while the one in Figure 19.7g, where the reflection line is along a diagonal, does not. Spotlight 19.4 discusses possible biological consequences of reflection symmetry or imperfections in it.

The pattern of footsteps in Figure 19.8a is not preserved under reflection along the direction of walking—there is not a left footprint directly across from a right footprint. The pattern is preserved under a glide reflection along the direction of walking, as well as by a translation of two steps, or one of four steps, and so on—but not by a translation of one step.

19.3 Rosette, Strip, and Wallpaper Patterns

Given a pattern, we analyze it by determining which rigid motions preserve the pattern. These are often referred to as the **symmetries of the pattern.** We then can classify the pattern by which rigid motions preserve it.

We may think of a pattern as a recipe for repeating a figure (motif) indefinitely. Of course, any pattern we see in nature or art has only finitely many copies of the figure. If the recipe for repetition is clear, we may imagine that we are looking at just a part of a pattern that extends indefinitely.

Patterns in the plane can be divided into those that have indefinitely many repetitions in

▶ no direction—the **rosette patterns**
▶ exactly one direction (and its reverse)—the **strip patterns**
▶ more than one direction—the **wallpaper patterns**

Rosette Patterns

A rosette pattern describes the possible symmetries for a single flower. The repetition aspect of asymmetry consists of the repetition of the petals around the

SPOTLIGHT 19.4   "Strive Then to Be Perfect"

Stand in front of a mirror and look at yourself. Are your left and right sides exactly symmetrical? Not on the inside, of course, but externally? What about the part in your hair, freckles on your face, evenness of your shoulders, outward bending of your ears?

According to studies, animals—including people—prefer bilateral symmetry in others. That preference is for as perfect a symmetry as possible, so that even slight deviations from exact symmetry can make an individual less desirable to others.

In some cases, symmetry may signal fitness. For example, the more symmetrical that a flower is, the more nectar it produces, making it a better food source for pollinating insects. And insects do prefer symmetrical flowers, thereby giving them a better chance of being pollinated than less perfect ones.

Symmetry, as a proxy for fitness, may affect mate selection by animals. Fruit flies and female barn swallows prefer males with symmetrical tails; a particular parasite can lead to an uneven tail. Symmetrical racehorses may run faster. Male lions with lopsided facial whisker-spot patterns die younger. Female zebra finches prefer males with symmetrical leg bands.

What about people? Computer-generated symmetrical female faces, generated from composites of individual photos, appear more attractive to men than photos of actual women. Studies also indicate that "symmetrical" men tend to have an earlier first sexual experience, more sexual partners, and more extramarital affairs. Women desiring fertile partners may

Michelangelo's David
(Roger Antrobus/Getty Images.)

want to consider a study that indicates that low sperm counts and poor sperm motility are associated with lack of symmetry of the hands. However, to what extent such symmetry in people signals individual or genetic fitness is yet to be determined.

stem. Translations and glide reflections do not come into play. The pattern is preserved under a rotation by certain angles corresponding to the number of petals. There may or may not be reflections that preserve it, depending on whether the petal itself has reflection symmetry. Most flowers do (Figure 19.9a), but some do not. An everyday example of the rosette pattern—a human-made one—that

FIGURE 19.9 (a) Flower; each petal has reflection symmetry. *(Gregory G. Dimijian/Photo Researchers, Inc.)* (b) Pinwheel with eight "leaves," each asymmetric symmetrical, hence pattern *c8* *(Travis Amos.)*

does not have reflection symmetry is a pinwheel (Figure 19.9b). If there is no reflection symmetry, the motif of the pattern (the element that is repeated) is an entire petal. If there is reflection symmetry, the motif is just half a petal, because the entire pattern can be generated by rotation and reflection of a half petal. The fact that these are the only possibilities is sometimes called *Leonardo's theorem,* after Leonardo da Vinci, who, in the course of planning the design of churches, needed to decide if chapels and niches could be added without destroying the symmetry of the central design.

Leonardo realized that there were two different classes of rosettes, the ones without reflection symmetry (*cyclic rosettes*) and those with it (*dihedral rosettes*) (see Figure 19.9). The respective notations for the patterns are *cn* and *dn*, where *n* is the number of times that the rosette coincides with its original position in one complete turn around the center. The rosette coincides with itself for every rotation of $360°/n$. A cyclic pattern has no lines of reflection symmetry, while the dihedral pattern *dn* has *n* different lines of reflection symmetry. The flower in Figure 19.9a has a dihedral pattern, because each petal has reflection symmetry. The pinwheel in Figure 19.9b has pattern *c8*.

Strip Patterns

We illustrate the different kinds of strip patterns, and their "ingredient" symmetries, with patterns in the art of the Bakuba people of the Democratic Republic of the Congo, who are noted for their fascination with pattern and symmetry (see Spotlight 19.5).

All the strip patterns offer repetition and **translation symmetry** along the direction of the strip. For simplicity, we always position the pattern so that its repetition runs horizontally.

SPOTLIGHT 19.5 Patterns Created by the Bakuba People

Among the Bakuba people of the Democratic Republic of the Congo (shaded area of map), it is considered an achievement to invent a new pattern, and every Bakuba king had to create a new pattern at the outset of his reign. The pattern was displayed on the king's drum throughout his reign and, for some kings, on his dynastic statue.

When missionaries first showed a motorcycle to a Bakuba king in the 1920s, he showed little interest in it. But the king was so enthralled by the novel pattern the tire tracks made in the sand that he had it copied and gave it his name.

Source: Adapted from Jan Vansina, *The Children of Woot*, University of Wisconsin Press, Madison, 1978, p. 221.

Two women with raffia cloths from the Bakuba village of Mbelo, July 1985. *Left:* Mpidi Muya with embroidered raffia (a kind of fiber) cloth. *Right:* Muema Kenye with plush and embroidered raffia cloth. *(Dorothy K. Washburn.)*

The pattern made by tire tracks fascinated the Bakuba people. *(Travis Amos.)*

It may be that the pattern has no other rigid motions that preserve it apart from translation, as in Figure 19.10a.

The simplest other rigid motion to check is reflection across a line. For a strip pattern, the center line of the strip may be a reflection line. If so, as in Figure 19.10b, we say that the pattern has symmetry across a horizontal line. There may instead be reflection across a *vertical* axis, such as the vertical lines through or between the V's in Figure 19.10c.

What kind of rotational symmetry can a strip pattern have? The only possibility for a strip pattern is a rotation by 180° (a half-turn), because any other angle won't even bring the strip back into itself. (We don't count rotations of 360° or integer multiples [full turns], because any pattern is preserved under these.) Figure 19.10d shows a strip pattern that is unchanged by a 180° rotation about any point at the center of the small crosshatched regions.

FIGURE 19.10 Bakuba patterns. (a) Carved stool; (b) pile cloth; (c) pile cloth; (d) embroidered cloth; (e) embroidered cloth; (f) carved back of wooden mask; (g) carved box.

What about glide reflections? A row of alternating p's and b's has glide reflection:

Glide:	p	p	p	p	p	p	p	p	p
Reflection:	p	p	p	p	p	p	p	p	p
	b	b	b	b	b	b	b	b	b
Glide reflection:	p	b	p	b	p	b	p	b	p

For glide reflection, a p is translated as far as the next b and is then reflected upside down. Figure 19.10e shows a Bakuba pattern whose only symmetry (except for translation) is glide reflection.

Having examined symmetries on strip patterns, we can ask: What *combinations* of the four are possible? It turns out that apart from the five kinds of patterns we have already seen, there are only two other possibilities: We can have vertical line reflection, half-turns, and glide reflection, either with horizontal line reflection (Figure 19.10g) or without (Figure 19.10f).

Mathematical analysis reveals the following.

There Are Only Seven Ways to Strip

There are only seven ways to repeat a pattern along a strip.

That this number is so small is quite surprising, because there are myriad different design elements (motifs). The key idea is that two designs may look entirely different yet share the same pattern of reproducing their design elements.

Wallpaper Patterns and Crystal Structures

So far we have classified the patterns that have no translation repetition (the rosette patterns) and those with repetition in one direction (the strip patterns). What about repetitions in more than one direction—say, in two different directions across a plane? It turns out that there are exactly 17 ways to do so, called **wallpaper patterns.** We give illustrations, notation, and a flowchart in Spotlight 19.6.

We emphasize again that "pattern" does not refer to the basic design but to how its repetition is structured across the plane. There is an infinite variety of possible designs that artists can devise. You should imagine that the artist has created one copy of the design and is contemplating how to place equal-sized copies of it in other parts of the (infinite) plane, in a way that is symmetrical. There are very few (17) strategies possible for doing so.

Mathematicians have further refined the classification of patterns of rosettes, strips, and wallpaper to take into account colors that are repeated in a symmetrical way. Crystallographers (physicists and chemists interested in the ways that crystals can occur or be built) in the nineteenth century classified three-dimensional crystal structures in terms of combinations of symmetry elements. They proved—after several years of coming up with different totals!—that there are exactly 230 patterns for crystals.

19.4 Notation for Patterns

It's useful to have a standard notation for patterns. Crystallographers' notation is commonly used. For the strip patterns, it consists of four symbols (an example is *pma2*):

1. The first symbol is always a *p*, which indicates that the pattern repeats (is "periodic") in the horizontal direction.
2. The second symbol is *m* if there is a vertical line of reflection. Otherwise, it is *1*.
3. The third symbol is
   ▶ *m* (for "mirror"), if there is a horizontal line of reflection (in which case there is also glide reflection)

SPOTLIGHT 19.6 — The 17 Wallpaper Patterns

As we noted earlier, there are exactly 17 wallpaper patterns. Here we give an example of each, together with a flowchart for identifying them. Crystallographers have standard notations and abbreviations for the patterns. The full notation consists of four symbols:

1. The first symbol is *c* (for "centered") if all rotation centers lie on the reflection lines, or *p* (for "primitive") otherwise.

2. The second symbol indicates rotational symmetry. It is either *1*, *2*, *3*, *4*, or *6*, corresponding to rotational symmetry of, respectively, 360°, 180°, 120°, 90°, or 60°. The symbol is the largest applicable number. For example, if symmetries of 360°, 120°, and 60° are present, the symbol if 6.

3. The third symbol is either *m*, *g*, or *1*, corresponding to the presence of "mirror," "glide," or no reflection symmetry.

4. The fourth symbol (*m*, *g*, or *1*) is for describing symmetry relative to an axis at an angle to the symmetry axis of the third symbol.

(*Note:* The patterns *p31m* and *p3m1* are exceptions to this scheme.)

Below each pattern illustration, we give both the standard abbreviation (on top) and the full notation (below).

| p1 | pm | pg | cm | p2 | pgg |
| p111 | p1m1 | p1g1 | c1m1 | p211 | p2gg |

| pmg | pmm | cmm | p4 | p4m | p4g |
| p2mg | p2mm | c2mm | p411 | p4mm | p4gg |

| | p3 | p31m | p3m1 | p6 | p6m |
| | p311 | | | p611 | p6mm |

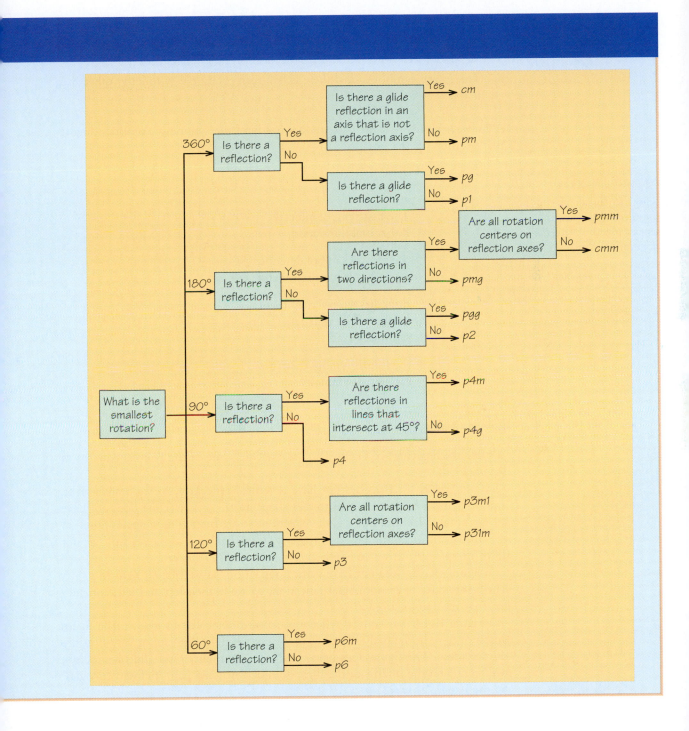

▶   *a* (for "alternating"), if there is a glide reflection but no horizontal reflection

▶   *1*, if there is no horizontal reflection or glide reflection

4.   The fourth symbol is *2*, if there is half-turn rotational symmetry; otherwise, it is *1*.

A *1* always means that the pattern does not have the symmetry corresponding to that position. In the notation:

FIGURE 19.11 A flowchart for identifying the seven ways patterns repeat, together with the notations for them.

The first symbol indicates horizontal translation,

the second symbol describes vertical reflection symmetry,

the third symbol describes horizontal or glide reflection symmetry, and

the fourth symbol describes rotational symmetry.

▦EXAMPLE 1   Bakuba Patterns

We use the flowchart of Figure 19.11 to analyze some of the Bakuba patterns of Figure 19.10.

SOLUTIONS Figure 19.10a does not have a vertical reflection, so we branch right, and the pattern notation begins to take shape as *p1_ _*. The figure does not have a horizontal reflection, nor a glide reflection, so we branch right again, filling in the third position in the notation to get *p11_*. A half-turn preserves part but not all of the pattern, so we conclude that we have a *p111* pattern.

Figure 19.10b does not have vertical reflection, so we branch right to *p1_ _*. The figure does have horizontal reflection, so we branch left and left, concluding that the pattern is *p1m1*.

Figure 19.10f has vertical reflection, so we branch left to *pm_ _*. The figure does not have horizontal reflection, so we branch right but cannot yet fill in the third symbol. The figure does have a half-turn symmetry (and glide symmetry), with center on the middle of the three lines between any pair of closest triangles. So the pattern is *pma2*. ■

Remember the Bakuba people's fascination with tire treadmarks? Apart from esthetic value, certain symmetries are important for practical purposes. The Museum of Transport in Glasgow, Scotland, includes all kinds of vehicle tires. How-

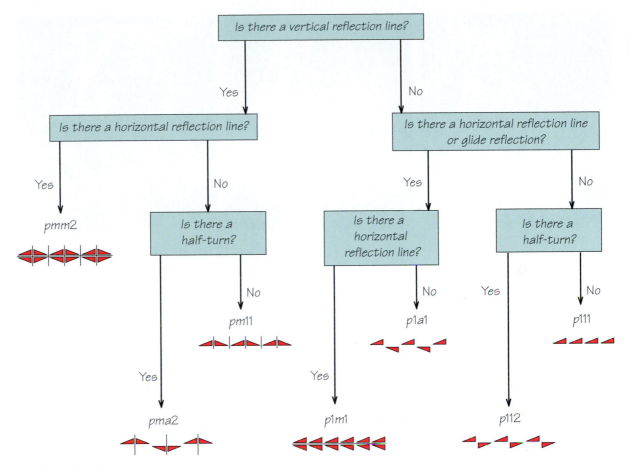

FIGURE 19.12 Flowchart for the seven strip patterns.

ever, only five of the seven strip patterns appear among treads of all the tires there. Examining Figure 19.12, can you guess which two patterns do not appear, what they have in common, and the practical reason why they are not used in tire treads?

Imperfect Patterns

In applying these classification schemes to patterns on real objects, we need to take into account that the pattern itself may not be perfectly rendered. Also, patterns that are not on flat surfaces—for example, the pattern around the rim of a bowl or around the body of a jar—require some latitude in interpretation.

Patterns on Pueblo Pottery   EXAMPLE 2

The pitchers in Figure 19.13 are from a thousand-year-old Pueblo site at Stark-weather Ruin near Reserve, New Mexico. Consider the patterns on the bodies of

FIGURE 19.13 Reserve black-on-white pitchers from the Pueblo II "horizon" (culture) (A.D. 900–1100), excavated 1935–1936 from Starkweather Ruin by Professor Paul H. Nesbitt and students from Beloit College. These pots are no longer available for study. In 2005 they were returned to representatives of Indian tribes who claim descent from the Pueblo culture and who requested return of the artifacts for reburial. *(Courtesy of Logan Museum of Anthropology, Beloit College, photos by Paul J. Campbell.)*

the pitchers, which continue on the back sides. Let's suppose that they could be unwrapped and continued as strip patterns, but we'll disregard the patterns on the spouts and handles.

We immediately come up against the question of the perfectness of the patterns. In Figure 19.13a the "teeth" (represented by the zigzagging of lightning bolts) in the left design element on the main body are "sharper" than those on the right. Is this lack of pattern, or just lack of perfection in executing one? For our analysis, we opt for the latter.

Similarly, what are we to make of the diagonal lines on the pitcher in Figure 19.13b? In the narrowest interpretation, these lines are part of the pattern, and rigid motion that is to qualify as a symmetry of the pattern must preserve them. More liberally, we may consider the lines as a kind of shading, a way to make the region appear gray. Indeed, to an observer at a distance, that is the effect of the lines.

For the pattern on the body of the pitcher in Figure 19.13c, we notice that the jagged white line in the design element on the left has three "steps," while that in the one on the right has four. If we were really strict, we would decide that the two are different design elements. But we do detect a similarity of the two that we do not want to deny totally. We attribute the variations in the jagged lines to artistic license and for our purposes consider the two jagged lines to be the same. ■

SOLUTIONS We follow the flowchart in Figure 19.12 and get the following:

▶ Figure 19.13a: Is there a vertical reflection? *No.* Is there a horizontal reflection or glide reflection? *No.* Is there a half-turn? *No.* Hence the pattern is *p111*.

▶   Figure 19.13b (narrow interpretation of the diagonal lines): Is there a vertical reflection? *No.* Is there a horizontal reflection or glide reflection? *No.* Is there a half-turn? *Yes* (around the center of each cross). The pattern is *p112.*

▶   Figure 19.13b (liberal interpretation–diagonal lines as shading, their direction doesn't have to be preserved): Is there a vertical reflection? *Yes* (on a vertical line through the center of a cross). Is there a horizontal reflection? *Yes* (through the center of a cross). The pattern is *pmm2.*

▶   Figure 19.13c: Is there a vertical reflection? *No.* Is there a horizontal reflection or glide reflection? *No.* Is there a half-turn? *Yes* (around the center of each jagged white line). The pattern is *p112.* (This pitcher has the interesting feature that the patterns on the neck and the body are mirror images of each other.)

Women made the pots at Starkweather. They strongly preferred the symmetry of half-turns. Very few of the pots have any reflection symmetry, either reflection or glide. The avoidance of reflection symmetry was a consistent feature of pottery of the indigenous peoples of the Western Hemisphere.

19.5 Symmetry Groups

We mentioned earlier that the key mathematical idea about detecting and analyzing symmetry is to look not at the motifs of a pattern but at its symmetries, the transformations that preserve the pattern.

The symmetries of a pattern have some notable properties:

▶   If we combine two symmetries by applying first one and then the other, we get another symmetry.

▶   There is an identity, or "null," symmetry that doesn't move anything, but leaves every point of the pattern exactly where it is.

▶   Each symmetry has an inverse, or "opposite," that undoes it and also preserves the pattern. A rotation is undone by an equal rotation in the opposite direction, a reflection is its own inverse, and a translation or glide reflection is undone by another of the same distance in the opposite direction.

▶   In applying a number of symmetries one after the other, we may combine consecutive ones without affecting the result.

The Symmetries of a Rectangle   EXAMPLE 3

Consider the rectangle of Figure 19.14. Its symmetries, the rigid motions that bring it back to coincide with itself (even as they interchange the labeled corners), are as follows:

▶   The identity symmetry *I,* which leaves every point where it is

▶ A 180° (half-turn) rotation R around its center

▶ A reflection V in the vertical line through its center

▶ A reflection H in the horizontal line through its center

You should convince yourself that the symmetries fulfill the four properties above.
SOLUTION

▶ Combining any pair by applying first one and then the other is equivalent
to one of the others. It's handy to have a notation for this combining; if
we apply V first and then H, we will write the result as $V \circ H$. You can
check that the result is the same as applying R, that is, $V \circ H = R$. Check
this by following where the corner A goes to under the symmetries.
Practice combining symmetries by making yourself a "multiplication table"
of them.

▶ The element I is an identity element.

▶ Each element is its own inverse.

▶ Try some examples to verify that associativity holds. For instance, check
that $R \circ H \circ V = (R \circ H) \circ V = R \circ (H \circ V)$. In other words, applying R
then H then V, we get the same result if we combine the first two and
then apply the third, or if we apply the first one and then apply the
combination of the second two. ■

The four properties of symmetries of an object are common to many kinds
of mathematical objects. The properties characterize what mathematicians call a
group. Various familiar collections of numbers, together with operations on them,
form groups.

EXAMPLE 4   A Group of Numbers

The positive real numbers form a group under multiplication:
SOLUTION

▶ Multiplying two positive real numbers yields another positive real number.

▶ The positive real number 1 is an identity element.

▶ Any positive real number x has an inverse $1/x$ in the collection.

▶ In multiplying several numbers together, it doesn't matter if we first multiply together some adjacent pairs of numbers; that is, it doesn't matter how we group or parenthesize the multiplication. For instance, $2 \times 3 \times 4 \times 5$ is equal to $2 \times (3 \times 4) \times 5 = 2 \times 12 \times 5$ and also to $(2 \times 3) \times 4 \times 5 = 6 \times 4 \times 5$. ▪

Group

A **group** is a collection of elements $\{A, B, \ldots\}$ and an operation $\circ$ between pairs of them such that the following properties hold:

Closure: The result of one element operating on another is itself an element of the collection ($A \circ B$ is in the collection).

Identity element: There is a special element I, called the identity element, such that the result of an operation involving the identity and any element is that same element ($I \circ A = A$ and $A \circ I = A$).

Inverses: For any element A, there is another element, called its inverse and denoted A^{-1}, such that the result of an operation involving an element and its inverse is the identity element ($A \circ A^{-1} = I$ and $A^{-1} \circ A = I$).

Associativity: The result of several consecutive operations is the same regardless of grouping or parenthesizing, provided the consecutive order of operations is maintained: $A \circ B \circ C = A \circ (B \circ C) = (A \circ B) \circ C$.

Symmetry Group of a Pattern

The symmetries that preserve a pattern form the **symmetry group of the pattern.**

Symmetry Groups of Strip Patterns

EXAMPLE 5

Each of the strip patterns of Figure 19.10 has a different group of symmetries. What do they have in common, and how do they differ?

SOLUTION The pattern of Figure 19.10a is preserved only by translations. If we let T denote the smallest translation to the right that preserves the pattern, then the pattern is also preserved by $T \circ T$ (which we write as T^2), by $T \circ T \circ T = T^3$, and so forth. Although the pattern looks the same after each of these translations by different distances, we can tell them apart if we number each copy of the motif and observe which other motif it is carried into under the symmetry. For instance, T^2 takes each motif into the motif two to the right. The symmetry T has an inverse T^{-1} among the symmetries of the pattern: the smallest translation to the *left* that preserves the pattern. Moreover, $T^{-1} \circ T^{-1}$ (which we write as T^{-2}), $T^{-1} \circ T^{-1} \circ T^{-1} = T^{-3}$, and so forth are also symmetries. The entire collection of symmetries of the pattern is

$$\{ \ldots, T^{-3}, T^{-2}, T^{-1}, I, T, T^2, T^3, \ldots \}$$

From this listing, you see that it is natural to think of the identity I as being T^0. All the strip patterns are preserved by translations, so the symmetry group of each includes the *subgroup* of all translations in this list. We say that the group is **generated by** T, and we write the group as $\langle T \rangle$, where between the angle brackets we list symmetries (**generators**) that, in combination, produce all of the group elements.

The symmetry group of Figure 19.10e includes in addition a glide reflection G and all combinations of the glide reflection with the translations. Doing two glide reflections is equivalent to doing a translation, which we express as $G^2 = T$. The glide is only "half as far" as the shortest translation that preserves the pattern. Check that $G \circ T = T \circ G$. The symmetry group of the pattern is

$$\{ \ldots, G^{-3}, G^{-2} = T^{-1}, G^{-1}, I, G, G^2 = T, G^3, \ldots \} = \langle G \rangle$$

The pattern of Figure 19.10c is preserved by vertical reflections at regular intervals. If we let V denote reflections at a fixed particular location, the other reflections can be obtained as combinations of V and T. To get a handle on what each of the symmetries does, it helps to make a "simplified" copy of the strip (we use V's), number fixed positions on the page, and identify individual copies of the V's with letters, as in the following:

The symmetries move the V's among the numbered positions. Let V be the reflection across the vertical line through the middle of position 3, and let T be the translation that moves each V one square to the right. To familiarize yourself with the symmetries, write out the result for each of V, T, and VT (V followed by T). (From now on, we will omit the operation sign between the two symmetries.)

The symmetry group of the pattern, the list of all of the symmetries, is

$$\{ \ldots, T^{-3}, T^{-2}, T^{-1}, I, T, T^2, T^3, \ldots ;$$
$$\ldots, T^{-3}V, T^{-2}V, T^{-1}V, V, TV, T^2V, T^3V, \ldots \}$$

This group is notable because not all of its elements satisfy the *commutative property* that $A \circ B = B \circ A$, which you are used to for numerical operations ($a + b = b + a$; $a \times b = b \times a$). In fact, we do not have $VT = TV$, but instead $VT = T^{-1}V$. Verify this fact by working out the effect of $T^{-1}V$, using your simplified strip from above, and compare with what you got for VT earlier.

We can express this group compactly as $\langle T, \ V | VT = T^{-1}V \rangle$, where we list the generators and indicate what relations hold among them. ∎

We have made a transition from thinking about patterns in geometrical terms to reasoning about them in algebraic notation—in effect, applying one branch of mathematics to another. This kind of cross-fertilization is characteristic of contemporary mathematics.

The concept of a group is a fundamental one in the mathematical field of abstract algebra. The generality ("abstractness") is exactly why groups and other algebraic structures arise in so many applications, in areas ranging from crystallography, quantum physics, and cryptography, to error-correcting codes (see Chapters 16 and 17) and anthropology (describing kinship systems).

19.6 Fractal Patterns and Chaos

We noted earlier that similarity and repetition are key aspects of symmetry, as are balance and proportionality. In most of our examples, the repetitions of a motif have been at the same size. Exceptions were the chambers of the nautilus in Figure 19.2b and the varying sizes of leaves and seeds in plants that feature the spiral pattern of phyllotaxis (Figures 19.1 and 19.2a). These exhibit a kind of "proportion," or numeric symmetry—symmetry with changes of scale. Another example of similarity with changes of scale are the nested dolls ("matrioshka") shown in Figure 19.15. They feature a linear scaling factor (see section 18.1) between one doll and another. Each part of one doll (face, arm, and so forth) has the same proportion (scaling factor) to the corresponding part of a second doll.

Fractals

Fractals are another example of symmetry in which linear scaling is used. The word *fractal* was invented in 1975 by Benoit Mandelbrot of IBM from the Latin word *fractus* meaning "broken into fragments" (of varied sizes), from which we get *fragment* and also *fracture* and *fraction*. Mandelbrot defined a fractal in strict mathematical terms but has never been satisfied that the definition captures all the examples that feature what he would consider "fractal-like" behavior. We sum up those features as follows:

> ### Fractal
>
> A **fractal** is a pattern that exhibits similarity at ever finer scales.

FIGURE 19.15 Nested "matrioshka" dolls from Russia exhibit symmetry at different scales.
(Photodisc Green/Getty Images.)

FIGURE 19.16
Various fractals. *(Courtesy of Noel Giffin and the Spanky Fractal Database.)*

(a) "Paradise."

(b) "Purgatory."

(c) "r-crest."

(d) "Scarab 2."

The scaling to different sizes is usually by a *linear scaling factor* and often is called a *dilation* (Chapter 18, p. 664).

We show various fractals in Figure 19.16. Figure 19.16a looks to us like an orchid with pronounced "bee guides" to the pollen. With its vertical mirror line, the overall pattern has *d1* rosette symmetry. However, the basic motif of the lacy wings is repeated at an infinite number of scales. In Figure 19.16b, the "suckers" on the "tentacles" appear in smaller and smaller sizes as the "tentacles" wind their way toward the point at the center. The pattern in Figure 19.16c has overall rosette symmetry of type *c2*, but the "seahorse" motif, with two large "seahorses" foot-to-foot in the center, is repeated in diminishing sizes throughout. Figure 19.16d features (to our imagination) "spikey snowmen," with smaller ones growing out of the sides of larger ones. What do they look like to you? And does the overall pattern as a rosette have symmetry *c1*, *c2*, *d1*, or *d2*?

A famous example of a fractal pattern is Maurits Escher's print "Circle Limit IV" (Figure 19.17). As you examine the angels and devils closer and closer to the boundary of the circle, you notice that they are not necessarily geometrically similar to the ones at the center. However, if you imagine that the print is the image of a hemisphere, then figures farther away from your viewpoint should indeed appear smaller.

Apart from their beauty and the opportunity that they offer as an art form (there is even "fractal music"!), fractals have two major applications:

▶ Fractals with very simple rules for replicating the motif mimic very well certain natural phenomena, such as the structure of a leaf, a tree, or a mountain (see Figure 19.18). This fact not only allows us to model leaves,

FIGURE 19.18
(a) Barnsley's fractal fern and (b) a snowy mountain landscape, both with templates showing how they were formed using reflections and linear "distortions" in addition to linear scaling. Each leaf of the fern and each mountain peak is in fact just a smaller copy of the entire image. *(Fred Solomon, Warren Wilson College.)*

FIGURE 19.19 A Sierpiński "triangle." Start with the big triangle, and remove its middle triangle. Then do the same for the three remaining smaller triangles. Recursively do the same for each subsequent smaller triangle. Can you guess the area of the resulting figure? (*Hint:* It may be less than you think.) (*Annalisa Crannell, Franklin and Marshall College.*)

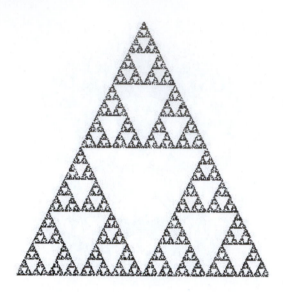

trees, and so on, using fractals but also suggests that such natural phenomena are produced by corresponding simple "rules of nature." Moreover, computer special effects in films can use fractals to mimic nature very closely, as in *Star Trek II: The Wrath of Khan,* for landscapes on the Genesis planet, and in *Return of the Jedi,* for the moons of Endor and the Death Star. In Chapter 23, we investigate in detail one particularly simple replication rule, called an **iterated function system (IFS);** this complicated-sounding name hides the fact that such a system is just a recursive rule, like the one for forming the Fibonacci numbers. Figure 19.19 shows a simple geometric IFS.

▶ Fractals form the basis for an important method of image compression. The key idea is to store not the millions of bits that make up an image but instead a much smaller number (maybe thousands) of rules for generating patterns that can be found in the image. A simple example (which doesn't use fractals) is that you can compress a checkerboard of a million pixels that alternate between black and white to just two simple rules: If the current pixel is white, the next one is black, and if the current one is black, the next one is white. A more realistic example is Microsoft's original 1992 Encarta Encyclopedia. It contained thousands of articles and photographs, plus color animations and hundreds of maps—all on one CD-ROM, thanks to fractal data compression.

Symmetry in Chaos

While the patterns in rosettes, strips, wallpaper patterns, and some fractals can be produced from very simple rules for the symmetries involved, you may be surprised to learn that symmetry can also arise from apparently random behavior.

SPOTLIGHT 19.7 The Father of Fractals

Benoit Mandelbrot (1924–) was born in Poland, grew up partly in France, and came to the United States to work for IBM. He found that many phenomena feature both repeating patterns and power curves (see section 18.6). The patterns are repeated at a change of scale governed by a power curve, as Barnsley's fern in Figure 19.18 and the Sierpiński triangle in Figure 19.19 show. Moreover, he found that the patterns could be described by very simple rules. In 1975 he coined the term *fractals*, and the systems of rules became known as iterated function systems. Now retired from IBM, he teaches a course in Fractal Geometry at Yale University. The Web site for the course (see Suggested Web Sites) lists 100 or so examples of fractal phenomena in nature and society. Mandelbrot's most recent book, *The*

Benoit Mandelbrot
(Roger Ressmeyer/CORBIS.)

(mis)Behavior of Markets: A Fractal View of Risk, applies ideas from fractals to describe how stock market prices vary.

We think of symmetry as referring to order, and chaos to disorder and randomness. Scientists use the word *chaos* to describe systems whose behavior over time is inherently unpredictable. We explore chaotic systems in Chapter 23 (section 23.5). Here we investigate how chaos can produce astonishingly beautiful designs on a computer screen.

One way to produce a graphic on the screen is to start with an initial pixel on the screen, apply a mathematical function (formula) to its coordinates to generate coordinates of a new pixel, light up the new pixel, then repeat the process with the new pixel. The process is recursive, in fact, an iterated function system.

Iterate the process a large number of times—millions or even hundreds of millions of times. Since the screen has many fewer pixels than that, by what mathematicians call the *pigeon-hole principle*, some pixels must be visited more than once—maybe even thousands of times.

The clue to producing art from this process is "color by number": Choose the color for each pixel according to how many times it is visited, and choose the colors with an eye to beauty. Figure 19.20 shows an example with D_5 symmetry that was produced by 30,000,000 iterations. The scale on the right in Figure 19.20 shows the colors for the number of times that pixels were hit; unhit pixels stay black. The order in which pixels are visited appears to be completely chaotic and is irrelevant to the final image.

FIGURE 19.20

"Emperor's Cloak," with D_5 symmetry. This work of art was produced by iterating a chaos-producing function, starting at one point and successively generating new points according to a fixed rule. The color bar shows the coloring of pixels according to how often they are visited by the iterations. *(Figure 1.13, p. 20, of Michael Field and Martin Golubitsky,* Symmetry in Chaos: A Search for Pattern in Mathematics, Art and Nature, *New York, Oxford University Press, 1992.)*

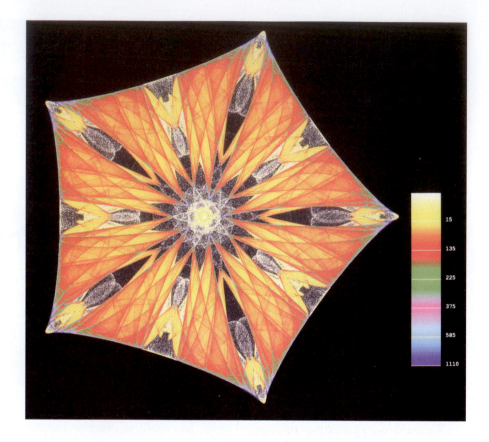

Other formulas can produce rosette, strip, or wallpaper patterns. Moreover,

▶ If you ignore the first thousand or so pixels visited, *it doesn't matter what pixel you start from—you get the same image!* But the pixels are visited in different orders.

▶ The formulas are variations on the *logistic map,* an iterated function system that we discuss in Chapter 23 in connection with biological populations.

REVIEW VOCABULARY

Divine proportion Another term for the golden ratio.

Fibonacci numbers The numbers in the sequence 1, 1, 2, 3, 5, 8, 13, 21, 34, ... Each number after the second is obtained by adding the two preceding numbers.

Fractal A pattern that exhibits similarity at ever-finer scales.

Generated, generators A group is generated by a particular set of elements (they are the generators) if composing them and their inverse in combinations can produce all elements of the group.

Geometric mean The geometric mean of two numbers a and b is $\sqrt{ab}$.

Glide reflection A combination of translation (= glide) and reflection in a line parallel to the translation direction. Example: pbpbpb.

Golden ratio, golden mean The number

$$\phi = \frac{1 + \sqrt{5}}{2} = 1.618034. \ldots$$

Golden rectangle A rectangle the lengths of whose sides are in the golden ratio.

Group A group is a collection of elements with an operation on pairs of them such that the collection is closed under the operation, there is an identity for the operation, each element has an inverse, and the operation is associative.

Isometry Another word for *rigid motion*. Angles and distances, and consequently shape and size, remain unchanged by a rigid motion. For plane figures there are only four possible isometries: reflection, rotation, translation, and glide reflection.

Itcrated function system (IFS) A sequence of elements (numbers or geometric objects) in which each successive element is determined recursively by applying the same function (rule) to the previous element.

Phyllotaxis The spiral pattern of shoots, leaves, or seeds around the stem of a plant.

Preserves the pattern A transformation preserves a pattern if all parts of the pattern look exactly the same after the transformation has been performed.

Recursion A method of defining a sequence of numbers, in which the next number is given in terms of previous ones.

Rigid motion A motion that preserves the size and shape of figures. In particular, any pair of points is the same distance apart after the motion as before. (Also called *isometry*.)

Rosette pattern A pattern whose only symmetries are rotations about a single point and reflections through that point.

Rotational symmetry A figure has rotational symmetry if a rotation about its "center" leaves it looking the same, like the letter S.

Strip pattern A pattern that has indefinitely many repetitions in one direction.

Symmetry of the pattern A transformation of a pattern is a symmetry of the pattern if it preserves the pattern.

Symmetry group of the pattern The group of symmetries that preserve the pattern.

Translation A rigid motion that moves everything a certain distance in one direction.

Translation symmetry An infinite figure has translation symmetry if it can be translated (slid, without turning) along itself without appearing to have changed. Example: AAA

Wallpaper pattern A pattern in the plane that has indefinitely many repetitions in more than one direction.

✔ SKILLS CHECK

1. What is the geometric mean of 4 and 36?

(a) 12

(b) 16

(c) 20

2. Which of the following rectangles is an approximate golden rectangle?

(a) 10 by 16

(b) 6 by 13

(c) 8 by 11

3. In the Fibonacci sequence, what number follows 13 and 21?

(a) 29

(b) 34

(c) 38

4. Recursion occurs in

(a) a sunflower blossom.

(b) a rosette pattern.

(c) the Fibonacci sequence.

5. The logarithmic spiral

(a) is used in Leonardo da Vinci's sketches.

(b) determines a sequence of golden rectangles.

(c) forms a symmetry pattern.

6. Which artist claimed to use the golden ratio in his work?

(a) Leonardo da Vinci
(b) Wolfgang Amadeus Mozart
(c) Neither

7. A rigid motion always moves any pair of points

(a) in the same direction.
(b) to another pair of points the same distance apart.
(c) to their mirror images.

8. Which of the following letters has a rotation isometry?

(a) S
(b) K
(c) Y

9. Assume the following two patterns continue in both directions. Which of these patterns has a reflection isometry?

ZZZZZZZZZ

UUUUUUUUU

(a) ZZZZZZZZZ only
(b) UUUUUUUUU only
(c) Neither

10. What isometries does this strip pattern have?

⌐Γ   ⌐Γ   ⌐Γ   ⌐Γ

(a) Translation and glide reflection only
(b) Translation and rotation only
(c) Translation, rotation, and glide reflection only

11. What isometries does this wallpaper pattern have?

(a) Translation and reflection only
(b) Translation and rotation only
(c) Translation, rotation, and reflection

12. What isometries does this wallpaper pattern have?

(a) Translation only
(b) Translation and reflection only
(c) Translation, rotation, and reflection

13. If a horizontal strip pattern has a glide reflection isometry, then

(a) it always has a horizontal reflection isometry.
(b) it may also have a horizontal reflection isometry.
(c) it cannot have a horizontal reflection isometry.

14. If a strip pattern has both vertical and horizontal reflection isometries, then

(a) it always has a half-turn rotation isometry.
(b) it may also have a half-turn rotation isometry.
(c) it cannot have a half-turn rotation isometry.

15. Consider the strip pattern in the raffia cloth held by the woman in the photo on the right. What isometries does it have?

(a) Vertical reflection
(b) Horizontal reflection
(c) Glide reflection

16. The symbol *p* indicates that a strip pattern has

(a) pretty symmetry.
(b) reflection symmetry.
(c) translation symmetry.

17. The symbol *2* indicates that a pattern has

(a) rotational symmetry.
(b) reflection symmetry.
(c) too much symmetry.

18. The symbol *m* indicates that a wallpaper pattern has

(a) marvelous symmetry.
(b) reflection symmetry.
(c) mathematical symmetry.

19. The symmetry group of a rectangle has how many elements?

(a) 4

(b) 6

(c) 8

20. The symmetry group of the strip pattern *pmm2* has how many elements?

(a) 4

(b) 8

(c) Infinitely many

EXERCISES   ■ Challenge   ◆ Discussion

Fibonacci Numbers

1. Examine the "scales" on the surface of a pineapple, which are arranged in spirals around the fruit. Note that there are spirals in three distinct directions. For each direction, how many spirals are there?

2. Repeat Exercise 1, but for a pinecone from your area.

3. Repeat Exercise 1, but for a sunflower.

4. Here are two primitive models of natural increase of biological populations, similar to those Fibonacci hypothesized around the year 1200. A pair of newborn male and female rabbits is placed in an enclosure to breed.

(a) Suppose that the rabbits start to bear young one month after their own birth. This may be unrealistic for rabbits, but we could substitute another species for which it is realistic. Fibonacci used rabbits. At the end of each month, they have another male–female pair, which in turn mature and start to bear young one month later. Assuming that none of the rabbits die, how many pairs of rabbits will there be at the end of six months from the start (just before any births for that month)? (*Hint:* Draw a month-by-month chart of the situation at the end of the month, just before any births.)

(b) Repeat part (a), but assume instead that the rabbits start to bear young exactly two months after their own birth.

The Golden Ratio

5. Put the golden ratio $\phi = (1 + \sqrt{5})/2$ into the memory of your calculator.

(a) Look at the value of ϕ. Now square it (either use the $\boxed{x^2}$ button or multiply it by itself). What do you observe?

(b) Back to ϕ. Now take its reciprocal (either use the $\boxed{1/x}$ button or divide it into 1). What do you observe?

(c) What formula explains what you saw in part (a)?

(d) What formula explains what you saw in part (b)?

6. The golden ratio satisfies the equation $x^2 = x + 1$. Show that $(1 - \phi)$ also satisfies the equation, so that $(1 - \phi) = (1 - \sqrt{5})/2$ is the other solution to $x^2 - x - 1 = 0$.

7. The geometric mean has interpretations in both arithmetic and geometry.

(a) Find the geometric mean of 3 and 27.

(b) Find the length of a side of a square that has the same area as a rectangle that is 4 by 64.

8. Here's further practice on arithmetic and geometric interpretations of the geometric mean.

(a) Find the geometric mean of 4 and 9.

(b) You are to make a golden rectangle with 6 inches of string. How wide should it be, and how high?

9. Another sequence closely related to the Fibonacci sequence is the *Lucas sequence*, which is formed using the same recursive rule but different starting numbers. The nth Lucas number L_n is given by

$$L_1 = 1, \; L_2 = 3, \text{ and } L_{n+1} = L_n + L_{n-1} \text{ for } n \geq 2$$

(a) Calculate L_3 through L_{10}.

(b) Calculate the ratio of successive terms of the Lucas sequence:

$$\frac{L_2}{L_1}, \frac{L_3}{L_2}, \ldots, \frac{L_{10}}{L_9}$$

What do you notice?

10. For a sequence specified by a recursive rule, finding an explicit expression for the nth term is not easy, nor is the form necessarily simple. An exact expression for the nth term of the Fibonacci sequence is given by the Binet formula:

$$F_n = \frac{1}{\sqrt{5}} \left(\frac{1 + \sqrt{5}}{2} \right)^n - \frac{1}{\sqrt{5}} \left(\frac{1 - \sqrt{5}}{2} \right)^n$$

(a) Verify the formula for $n = 1$ and $n = 2$ by multiplying out, not by using a calculator.
(b) Use the Binet formula and your calculator to find F_5.
(c) In fact, the second term on the right of the equation gets closer and closer to 0 as n gets large. Because we know that the Fibonacci numbers are integers, we can just round off the result of calculating the first term. Find F_{13} by calculating the first term with your calculator and rounding.

11. For two positive numbers x and y, show that the arithmetic mean $(x + y)/2$ is always greater than or equal to the geometric mean $x^{1/2}y^{1/2} = \sqrt{xy}$. [*Hint:* Suppose that the claim is false, so that $(x + y)/2 < \sqrt{xy}$.) Square both sides of the inequality, bring all terms to one side, factor, and observe a contradiction.]

■ **12.** You may remember having to work problems like, "If Joe can dig a ditch in 3 days, and Sam can dig it in 4, how long will it take the two of them working together?" The answer is related to the *harmonic mean* of 3 and 4. The formula for the harmonic mean of two numbers x and y is

$$\frac{2}{1/x + 1/y}$$

(a) Calculate the answer for Joe and Sam, which is *one-half* of the harmonic mean of 3 and 4. Explain why this is the correct answer.
(b) Show that the harmonic mean of two positive numbers is always less than or equal to the geometric mean. (Thus, in light of Exercise 11, we have the general conclusion that $H \leq G \leq A$, where H stands for the harmonic mean, G for the geometric mean, and A for the arithmetic mean.) (*Hint:* Suppose that the claim is false. Simplify the fraction that is the harmonic mean, square both sides of the inequality, and proceed as in Exercise 11.)
(c) Show once more that the harmonic mean of two positive numbers is always less than the geometric mean, but this time do it with less work: let $A = 1/x$ and $B = 1/y$, and discover one connection (equation) between the harmonic mean of x and y and the arithmetic mean of A and B, and a second connection between the geometric mean of x and y and the geometric mean of A and B. Then use Exercise 11 on A and B.
(d) What should be the formula for the geometric mean of three numbers? Of n numbers?
(e) Proceed as in part (d), but for the harmonic mean.

13. Here is a trick to "prove" that you can calculate faster than a person with a calculator. Turn your back and ask a friend to write down any two positive integers, then add them to get a third, then add the second and third to get a fourth, and so on, adding each time the last two integers until there are 10 numbers. Have your friend show you the list, whereupon you write down right away the total of all 10, while your friend begins to add them up on the calculator (to prove that you're right). The secret: The total is always 11 times the seventh number, and multiplying by 11 is pretty easy to do in your head—just add each pair of neighboring digits, carrying if necessary. Suppose that your friend writes down m and n as the first two numbers. Show that indeed the total of all 10 numbers is 11 times the seventh number. (Adapted from Martin Gardner, *Mathematical Circus*, Knopf, New York, 1979.)

14. The game of Fibonacci Nim begins with n counters. Two players take turns removing at least one counter, but no more than twice as many as the opponent just did. The winner is the player who takes the last counter. One other rule: The first player may not win immediately by taking all the counters on the first turn! (Adapted from

Martin Gardner, *Mathematical Circus*, Knopf, New York, 1979.)

(a) Play this game taking turns with an opponent and starting with different numbers n of counters and try to come up with a strategy for one player or the other to win. (*Hint:* The key is that any positive integer can be represented uniquely as a sum of Fibonacci numbers.)

(b) Proceed as in part (a), but with the rule changes that the player who takes the last counter loses and the first player may not take all but one counter.

15. Shari Lynn Levine, a high school student, published an article in *The Fibonacci Quarterly* that investigated the "Beta-nacci" sequence that results if instead of bearing one pair of baby rabbits per month, mature rabbits bear two pairs every month, starting when they reach two months of age. Here we ask you to rediscover some of Shari's results.

(a) How many rabbits will there be each month for the first 12 months?

(b) What is the recursive rule for the nth Beta-nacci number B_n?

(c) For the terms of the sequence in part (a), calculate the ratios B_{n+1}/B_n of successive terms. (*Motivating hint:* It's not the golden ratio this time.)

(d) Suppose that the ratio of successive terms approaches a number x. We show how to find x exactly. For very large n, we have $B_{n+1} \approx xB_n \approx x^2 B_{n-1}$. Substituting these values into the recursive rule for the sequence and dividing by B_{n-1} gives us the equation $x^2 = x + 2$. Solve this equation for x (you can use the quadratic formula). Make a table of values of $3B_n$ versus 2_n. From the evidence, can you suggest a formula for B_n?

16. Generalize Exercise 15, parts (a) through (d):

(a) to the case of each pair of rabbits having three pairs of rabbits (the "Gamma-nacci" sequence).

(b) to the case of each pair of rabbits having q pairs of rabbits.

For Exercises 17 and 18, refer to the following.

We have seen that the golden ratio is a positive root of the quadratic polynomial $x^2 - x - 1$. We

can generalize this polynomial to $x^2 - mx - 1$ for $m = 1, 2, 3, \ldots$ and consider the positive roots of those polynomials as generalized means—the "metallic means family," as they are sometimes known. In particular, for $m = 2, 3, 4,$ and 5, we have respectively the silver, bronze, copper, and nickel means. It is surely surprising that these numbers arise both in connection with quasicrystals (investigated in Chapter 20) and in analyzing the behavior of some dynamical systems (a topic investigated in Chapter 23) as the systems evolve into chaotic behavior.

17. Use the quadratic formula to find expressions in terms of square roots for the silver, bronze, copper, and nickel means, and approximate these to three decimal places. Find a general expression in terms of a square root for the mth metallic mean.

18. Just as the golden mean arises as the limiting ratio of consecutive terms of the Fibonacci sequence, each of the metallic means arises as the limiting ratio of consecutive terms of generalized Fibonacci sequences. A generalized Fibonacci sequence G can be defined by

$$G_1 = 1, \quad G_2 = 1, \quad \text{and} \quad G_{n+1} = pG_n + qG_{n-1}$$

where p and q are positive integers. The Fibonacci sequence itself is the case $p = q = 1$.

(a) Try various small values of p and q and determine which mean they lead to.

(b) Divide the equation for G_{n+1} by G_n. Assume that G_{n+1}/G_n and G_n/G_{n-1} both tend toward the same number x as n gets large, replace those quantities by x, and simplify the resulting equation. What must be the value of x?

(c) What happens to the sequence and to the mean if we allow one or both of p and q to be negative integers?

Symmetries Preserve the Pattern

19. Determine whether each of the following statements is always true or sometimes false. Drawing some sketches may be helpful.

(a) A line reflection preserves collinearity of points. That is, if the points A, B, and C are in a straight line (collinear), then their images reflected in some other line also lie in a straight line.
(b) A line reflection preserves betweenness. That is, if the collinear points A, B, and C (with B between A and C) are reflected about a line, then the image of B is between the images of A and C.
(c) The image of a line segment under a line reflection is a line segment of the same length.
(d) The image of an angle under a line reflection is an angle of the same measure.
(e) The image of a pair of parallel lines under a line reflection is a pair of parallel lines.

20. Determine whether each of the following statements is always true or sometimes false. Drawing some sketches may be helpful.

(a) The image of a pair of perpendicular lines under a line reflection is a pair of perpendicular lines.
(b) The image of a square under a line reflection is a square.
(c) Label the vertices of a square A, B, C, and D in a clockwise direction. Then their images A', B', C', and D' under a line reflection also follow a clockwise direction.
(d) The perimeter of a geometric figure is equal to the perimeter of its image under a line reflection.
(e) The image of a vertical line under a line reflection is always a vertical line.

21. Which of the capital letters of the alphabet, when drawn in the most symmetrical way, has the following symmetries? For example, assume that the upper and lower loops of B are the same size.

(a) A horizontal line of reflection symmetry
(b) A vertical line of reflection symmetry
(c) A rotational symmetry

22. Repeat Exercise 21 for the lowercase letters.

23. In *The Complete Walker III* (3rd ed., Knopf, New York, 1984, p. 505), Colin Fletcher's answer to "What games should I take on a backpacking trip?" is the game he calls "Colinvert": "You strive

to find words with meaningful mirror (or half-turn) images." Some of the words he found are

MOM   WOW   pod   MUd   bUM

(a) Which of his words reflect into themselves?
(b) Which of his words rotate into themselves?
(c) Find some more words or phrases of these various types—the longer, the better.

24. Repeat Exercise 23, but for words written vertically instead of horizontally.

25. For each of the following patterns, identify the rigid motions that preserve the pattern:

(a) AAAAAAAAAA
(b) BBBBBBBBBB
(c) 0000000000
(d) FFFFFFFFFF

26. Repeat Exercise 25, but for

(a) NNNNNNNNNN
(b) bdbdbdbdbd
(c) dbpqdbpqdbpq

Rosette, Strip, and Wallpaper Patterns

27. Give the notation (such as $d4$ or $c5$) for the symmetry patterns of the rosettes in hubcaps (a) through (c) on the next page, disregarding the logos in the centers. (Can you identify the make of car for each hubcap?)

28. Repeat Exercise 27 for hubcaps (d) through (f).

29. Repeat Exercise 27 for corporate logos (a) through (c) below. (Can you identify the corporations?)

(a)   *(b)*   *(c)*

30. Repeat Exercise 27 for automobile logos (d) through (f) below.

(d)   *(e)*   *(f)*

(a)

(b)

(c)

(d)

(e)

(f)

For Exercises 31–32, refer to the following.

Step patterns are found in Celtic illuminated manuscripts, metal work, and stone crosses. Square ones were constructed by first designing on a square lattice one quarter of the pattern (say, the top right), using horizontal and diagonal lines to produce a prototype such as the following:

Then three copies were added, either by (1) rotating the original successively by 90° [as in accompanying illustration (a)], or else by (2) reflecting it across its right and bottom edges [as in illustration (b)]. (Based on research by Mark A. M. Lynch of Glasgow Caledonian University, Scotland.)

(a)

(b)

31. Identify the rosette pattern for:

(a) step pattern (a).

(b) step pattern (b).

32. Which rosette pattern would result if the prototype, unlike the one above, has reflection symmetry across its diagonal from top left to lower right and

(a) strategy (1) is used.

(b) strategy (2) is used.

Hungarian needlework designs. (a) Edge decoration of table cover from Kalocsa, southern Hungary. (b) Pillow end decoration from Tolna County, southwest Hungary. (c) Decoration patched onto a long embroidered felt coat of Hungarian shepherds in Bihar County, eastern Hungary. (d) Embroidered edge decoration of bed sheet from the eighteenth century. (Note the deviations from symmetry in the lower stripes of the pattern.) (e) Shirt from Karád, southwest Hungary. (f) Pillow decoration pattern from Torockó (Rimetea), Transylvania, Romania. (g) Grape leaf pattern from the territory east of the river Tisza.

Strip Patterns

33. Use the flowchart in Figure 19.11 to identify the notation for the types of strip patterns from Hungarian needlework, shown in the illustrations above.

34. In each of the four accompanying examples, two adjacent triangles of an infinite strip are shown. (Contributed by Margaret A. Owens, California State University, Chico.)

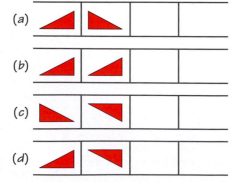

For each example:

(a) Determine a motion (translation, reflection, rotation, or glide reflection) that takes the first (= left) triangle to the second (= right) one.

(b) Draw the next four triangles of the infinite strip that would result if the second triangle is moved to the next space by another motion of the same kind, and so on.

(c) Identify (by notation) the resulting strip as one of the seven possible strip patterns.

Imperfect Patterns

35. Repeat Exercise 33 for the accompanying eight strip patterns, all of which appear on the brass straps for a single lamp from nineteenth-century Benin in West Africa. (From H. Ling Roth, *In Great Benin*.)

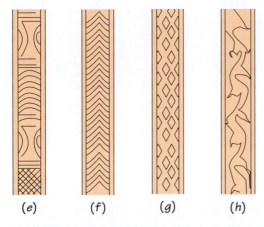

(e)   (f)   (g)   (h)

Note that the patterns are roughly carved, so you will need to discern the intent of the artist.

36. Repeat Exercise 33 for the accompanying patterns from San Ildefonso Pueblo, New Mexico.

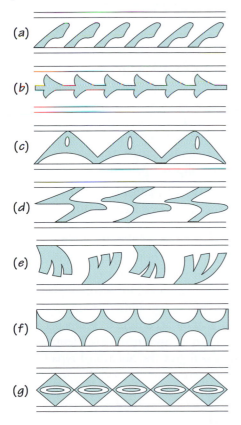

(a)

(b)

(c)

(d)

(e)

(f)

(g)

37. The following table shows comparative data about the frequency of occurrence of strip designs

of various types on pottery (Mesa Verde, United States) and smoking pipes (Begho, Ghana, Africa) from two continents.

Frequency of Strip Designs on Mesa Verde Pottery and Begho Smoking Pipes

| | Mesa Verde | |
Strip Type	Number of Examples	Percentage of Total
p111	7	4
p1m1	5	3
pm11	12	7
p112	93	53
p1a1	11	6
pma2	27	16
pmm2	19	11
Total	174	

| | Begho | |
Strip Type	Number of Examples	Percentage of Total
p111	4	2
p1m1	9	4
pm11	22	10
p112	19	8
p1a1	2	1
pma2	9	4
pmm2	165	72
Total	230	

(a) Which types of motions appear to be preferred for designs from each of the two localities?
(b) What other conclusions do you draw from the data of this table?
(c) On the evidence of the table alone, in which locality is each of the following strip patterns most likely to have been found?

(i)

(ii)

(iii)

(iv)

(v)

(vi)

(vii)

(viii)

(ix)

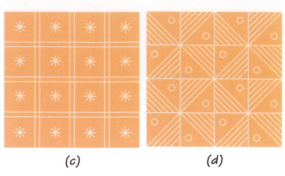

(c)   (d)

40. The triangles in the grid at the top of the following figure show beginning steps in forming instances of several of the wallpaper patterns by putting together a vertical motion and a horizontal motion.

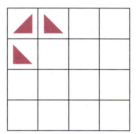

(a) Identify the horizontal motion.
(b) Identify the vertical motion.
(c) Fill in the remaining empty squares.
(d) Identify the wallpaper pattern.

■ **41.** Which of the 17 wallpaper patterns can be formed by the technique of Exercise 40?

For Exercises 42–47, refer to the following.

In Chapter 20, we study both repeating and nonrepeating plane patterns, from the point of view of their basic building blocks (tiles). Here we ask you to analyze the repeating patterns from figures in that chapter according to wallpaper type, using the flowchart of Spotlight 19.6. Identify all the symmetries and give the notational type for the wallpaper pattern of:

42. Figure 20.10.

43. Figure 20.11.

44. Figure 20.15.

38. For the Nigerian Yoruba cloths (a) and (b) in the following illustration, use the flowchart in Spotlight 19.6 to identify (by notation) the type of wallpaper pattern.

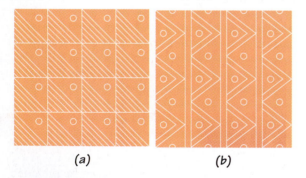

     (a)   **(b)**

Patterns on Yoruba (West Africa) *adire* cloth, made by starching a pattern onto white cloth, then dyeing the cloth before rinsing out the starch, so that the starched portion remains as a white design against a colored background.

39. For the Nigerian Yoruba cloths (c) and (d) in the accompanying illustration, use the flowchart in Spotlight 19.6 to identify (by notation) the type of wallpaper pattern.

(Courtesy of Dr. Ron Eglash, RPI. See www.ccd.rpi.edu/Eglash/csdt/na/loom/loom_symm4.html)

45. The figure in Spotlight 20.2.

46. The hexagonal regular tiling at upper right in Figure 20.5.

47. The convex hexagon tiling of type 3 in Figure 20.9.

48. Visit the Web site www-sphys.unil.ch/escher/, which features an interactive Java program called Escher Web Sketch. Experiment with choosing wallpaper patterns using crystallographic notation. For each, draw on the screen a colored design for the motif; the program will reproduce the motif using the pattern.

Symmetry Groups

49. For positive integers a and n, the expression a mod n means remainder when a is divided by n. Thus, 23 mod 4 = 3, because $23 = 5 \cdot 4 + 3$, and we say that "23 is equivalent to 3 modulo 4" (see Chapter 17 for further details about this *modular arithmetic*). Every positive integer is equivalent to 0, 1, 2, or 3 modulo 4. Consider the collection of elements {0, 1, 2, 3} and the operation $\oplus$ on them defined by $a \oplus b = (a + b)$ mod 4. Show that under this operation, the collection forms a group.

50. Explain, by referring to the properties of a group, whether the collection of all real numbers is a group under the operation of **(a)** addition; **(b)** multiplication.

51. For the traditional North American beadwork shown above:

(a) Which rosette pattern does it have?
(b) Specify two rigid motions that are generators of the group of the pattern.

(c) List the elements of the group.

52. Repeat Exercise 51 for the Plains Indian embroidery shown below.

53. What is the group of symmetries of a square?

54. What is the group of symmetries of:

(a) an equilateral triangle (all three sides equal)?
(b) an isosceles triangle (two equal sides) that is not equilateral?
(c) a scalene triangle (no pair of sides equal)?

55. (a) Give a numerical example to show that the operation of subtraction on the integers is not associative.
(b) Repeat part (a), but for division on the positive real numbers.

56. What are the elements of the group of symmetries of **(a)** Figure 19.10b? **(b)** Figure 19.10f?

57. What are the elements of the group of symmetries of **(a)** Figure 19.10d? **(b)** Figure 19.10g?

58. What are the elements of the group of symmetries of the dihedral pattern *d8?* (See the flower in Figure 19.9a.)

(Courtesy of Dr. Ron Eglash, RPI. See www.ccd.rpi.edu/Eglash/csdt.html)

59. What is the group of symmetries of the cyclic pattern *c8*?

■ **60.** What is the group of symmetries of a cube?

■ **61.** What is the group of symmetries of a general rectangular solid (its length, width, and height are all unequal)?

Fractal Patterns

◆ **62.** Explore Sprott's Fractal Gallery at sprott.physics.wisc.edu/fractals.htm, which features a "Fractal of the Day" and accompanying fractal music. There are various "rooms" in the gallery—including "Iterated Function Systems," "Natural Fractals" (I particularly like "Broccoli" and "Trees"), and "Publication Quality Attractors" ("SMKBNZQA" is our favorite)—together with PC programs for generating such fractals. What are your favorites, and why?

63. Explain how the pattern of the following illustration is fractal.

Exercises 64–68 use applets that require a computer with a Web browser equipped with Java and Flash plug-ins. These are available at links from the Web site www.rpi.edu/~eglash/csdt.html.

64. The Web site www.ccd.rpi.edu/Eglash/csdt/african/MANG_DESIGN/culture/mang_homepage.html has information about a fractal-patterned ivory hatpin from the Mangbetu culture in Africa. The site includes a tutorial on producing similar

designs using reflection, rotation, translation, and scaling. Work your way through the tutorial and then create a Mangbetu-style artifact.

65. Cornrow hairstyles are fractal in nature. At the Web site www.ccd.rpi.edu/Eglash/csdt/african/CORNROW_CURVES/, you can see how and why, including a tutorial on designing cornrow hairstyles using reflection, rotation, translation, and scaling. Work your way through the tutorial and then create a hairstyle. The Web site also includes instructions for actual braiding, with a short video.

66. The architecture of some African villages follows a fractal pattern—see www.ccd.rpi.edu/Eglash/csdt/african/archi/afractal/afarch.htm. At the Web site www.ccd.rpi.edu/Eglash/csdt/african/archi/mangbetuUpdate2_3_modified_final.swf is an applet, similar to the ones in Exercises 64–65 but without a tutorial. Decide on a plan and use the applet to create a simulated African village.

67. The Yupik (Western Eskimo, along the coast of Alaska) make their parka coats with black-and-white patterns. At the Web site www.ccd.rpi.edu/Eglash/csdt/na/yupik/yupik.html is an applet, similar to the ones in Exercises 64–65 but without a tutorial. Yupik designs emphasize reflection symmetry, but the applet allows you to use also rotation, translation, and scaling of a basic motif. Experiment with one of the motifs available and use the applet to create a pattern for a parka.

68. Download fractal-creation software and accompanying documentation and use the software to create your own fractal. Recommended software:

For DOS: Fractint, from spanky.triumf.ca/www/fractint/fractint.html

For Macintosh: FractaSketch, from ftp://ftp.info.ucl.ac.be/pub/fractasketch/FS203Archive.sea, with draft of manual (shows how to make fractal trees and leaves) at www.info.ucl.ac.be/people/PVR/FractaManual/index.html.

✎ WRITING PROJECTS

1. Generations of children have enjoyed the popular toy Spirograph®, which allows the user to trace out symmetric patterns. A pencil or pen is placed in a hole in one of several plastic circular disks with teeth on the outside rim. The disk is then meshed in the teeth of another plastic circle and rotated around its inside or outside. Each plastic piece is labeled with the number of teeth that it has on its circumference.

Either obtain a copy of Spirograph® or a closely related toy, or else visit the Web site www.wordsmith.org/~anu/java/spirograph.html#def, which offers an interactive Java application (which you can download) that mimics what the Spirograph® toy does.

(a) Experiment to determine, from the numbers of teeth on the rotating circular disk and the fixed circle, what symmetry pattern the result will have.

(b) Choose a rotating circular disk and a fixed circle for which the ratio of the number of teeth reduces to a whole number. For each of several "offsets" (holes to choose for the pencil or pen), trace overlapping designs. What symmetry pattern do you get for the design taken as a whole? Repeat this experiment for other pairs of pieces and try to reach a general conclusion.

(c) Write up, in a page or so, a description of your experiments and what conclusions you reached.

2. (Project for a team of 2 or 3) Explore your campus looking for symmetrical patterns in decorative elements of walls, floor, carpets, and ceilings. Find one example each of a rosette pattern, a strip pattern, and a wallpaper pattern. Take a digital photo of each and incorporate your photos into a document of three pages or so that explains to the reader where the pattern can be found, what symmetries (translation, rotation, reflection, glide reflection) it has, and how you identified the notation for it.

3. (Project for a team of 2 or 3) Visit a store that sells wallpaper and ask for a few old samples. Identify three that have different patterns according to the flowchart in Spotlight 19.6. Write in a page or two your explanations of how you identified the patterns, and attach the wallpaper samples to your report.

SUGGESTED READINGS

BELCASTRO, SARAH-MARIE, and THOMAS C. HULL. Classifying frieze patterns without using groups, *College Mathematics Journal,* 33 (March 2002): 93–98. Elementary analysis of why there are only seven ways to repeat a pattern along a strip.

CROWE, DONALD W. *Symmetry, Rigid Motions and Patterns,* High School Mathematics and Its Applications (HiMAP) Module 4, COMAP, Lexington, Mass., 1987. Reprinted in smaller format in *The UMAP Journal,* 8(3) (1987): 207–236. Instructional module on rigid motions of the plane, strip patterns, and wallpaper patterns, with worksheets.

LEE, KEVIN D. KaleidoMania!: Interactive Symmetry, Windows/Macintosh program, Key Curriculum Press, 1999. Lets the user construct rosette, strip, and wallpaper patterns.

LIVIO, MARIO. *The Golden Ratio: The Story of Phi, the World's Most Astonishing Number.* Broadway Books, New York, 2002.

WASHBURN, DOROTHY K., and DONALD W. CROWE. *Symmetries of Culture: Theory and Practice of Plane Pattern Analysis,* University of Washington Press, Seattle, 1988. An introduction to the mathematics of symmetry, splendidly illustrated with photographs of patterns from cultures all over the world. Includes a complete analysis of patterns with two colors, and proofs that there are only four rigid motions in the plane and exactly seven strip patterns.

SUGGESTED WEB SITES

www.geom.umn.edu/software/tilings Tessellation resources. Lists programs for various platforms that allow the user to create designs featuring the rosette, strip, and wallpaper patterns.

www.howe-two.com/nctm/wallpaperpatterns/ eschersketch/eschersketch.html/ Interactive Escher Web Sketch program that allows a user to design repeating patterns. Choose a wallpaper pattern using crystallographic notation and draw on the screen a colored design for the motif; the program then reproduces the motif using the pattern. The software (for Windows, Macintosh, and Unix) can also be downloaded.

www.geom.uiuc.edu/java/Kali/ Interactive Java Kali Web program that lets the user draw pictures under the action of rosette, strip, or wallpaper groups. Versions for various platforms can be downloaded.

ftp://ftp.uni-bielefeld.de/pub/math/tiling/ reptiles/RepTiles.2.0.1.sea.hqx Interactive Rep Tiles Macintosh application for designing wallpaper patterns, plus systematically generating all possible periodic tilings of the plane by applying "topological transformations" and "symmetry."

www.wordsmith.org/~anu/java/spirograph.html Interactive Spirograph Java application (which you can download) that lets you do electronically what the Spirograph toy does.

www.rpi.edu/~eglash/eglash.dir/afractal/afractal. htm African fractals site.

classes.yale.edu/fractals/index.htm Web site for Mandelbrot's course in fractals at Yale. Features many applets for different kinds of IFS (for example, incorporating randomness), including a fractal music composer.

Tilings

When our ancestors used stones to cover the floors and walls of their houses, they selected shapes and colors to form pleasing designs. We can see the artistic impulse at work in mosaics, from Roman dwellings to Muslim religious buildings (see Figure 20.1). The same intricacy and complexity arise in other decorative arts—on carpets, fabrics, baskets, and even linoleum.

Such patterns have one feature in common: They use repeated shapes to cover a flat surface, without gaps or overlaps. If we think of the shapes as tiles, we can call the pattern a *tiling*, or *tessellation*. Even when efficiency is more important than aesthetics, designers value clever tiling patterns. In manufacturing, for example, stamping the components from a sheet of metal is most economical if the shapes of the components fit together without gaps—in other words, if the shapes form a tiling.

::::: **Tiling (Tessellation)**

A **tiling (tessellation)** is a covering of the entire plane with nonoverlapping figures.

Some of the tilings in this chapter can be analyzed as wallpaper patterns, as in Spotlight 19.6. But in this chapter, we look at patterns "from the ground up." That is, we start with the basic ingredients of a type or types of tile and ask if or

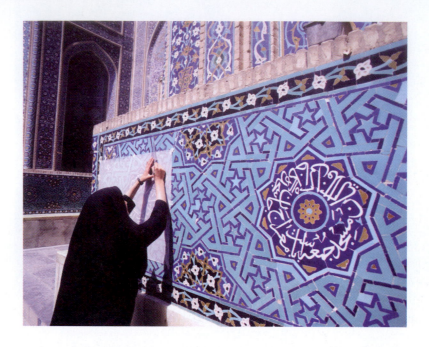

how they can be fit together into a pattern. If the result repeats in all directions, it must be one of the wallpaper patterns. We will see, though, that there are other surprising possibilities.

The major mathematical question about tilings is: Given one or more shapes (in specific sizes) of tiles, can they tile the plane? And, if so, how?

The surprising answer to the first question is that it is undecidable. That is, given an arbitrary set of tile shapes, there is no way to determine for certain if they can tile the plane or not. For some particular sets of tiles, we can exhibit tilings; for others, we can prove that there can't be any tiling. In this chapter we will see examples of both situations. But mathematicians have proved that there is no algorithm (mechanical step-by-step process) that can tell for every conceivable set of shapes which of the two situations holds. (See Chapter 9, pp. 356–359, for other examples of "unattainable ideals" in regard to voting.)

Given this sobering (and puzzling) limitation, we begin our investigation by considering the simplest kinds of tiles and tilings.

20.1 Tilings with Regular Polygons

The simplest tilings use only one size and shape of tile. They are known as *monohedral tilings*.

Monohedral Tiling

A **monohedral tiling** is a tiling that uses only one size and shape of tile.

In particular, we are interested especially in tiles that are **regular polygons,** figures all of whose sides are the same length and all of whose angles are equal. A square is a regular polygon with four equal sides and four equal interior angles; a triangle with all sides equal (an **equilateral triangle**) is also a regular polygon. A polygon with five sides is a pentagon, one with six sides is a hexagon, and one with *n* sides is an ***n*-gon.** Regular polygons are especially interesting because of their high degree of symmetry. Each has the reflection and rotation symmetries of a dihedral rosette pattern (see section 19.3). In three dimensions, the corresponding highly symmetrical figures are called *regular polyhedra* (see Spotlight 20.1).

By a convention dating back to the ancient Babylonians, angles are measured in degrees. An **exterior angle** of a polygon is one formed by one side and the extension of an adjacent side (Figure 20.2). Proceeding around the polygon in the same direction, we see that each **interior angle** (the angle inside a polygon formed by two adjacent sides) is paired with an exterior angle. If we bring all the exterior angles together at a single point, they add up to 360° (see Figure 20.2). If the polygon has *n* sides, then each exterior angle must measure 360/*n* degrees. For example, a square with *n* = 4 sides has 4 exterior angles, each measuring 90°; a pentagon with *n* = 5 sides has 5 exterior angles, each measuring 72°; a regular hexagon with *n* = 6 sides has 6 exterior angles, each measuring 60°. Each exterior angle plus its corresponding interior angle make up a straight line, or 180°. For a regular polygon with more than six sides, the interior angle is between 120° and 180°. This last consideration will prove crucial shortly.

FIGURE 20.2 The exterior angles of a regular pentagon, like those of any regular polygon, add up to 360°. Each exterior angle measures 72°.

Regular Tilings

:::::: **Regular Tiling**

A monohedral tiling whose tile is a regular polygon is called a **regular tiling.**

A square tile is the simplest case. Apart from varying the size of the square, which would change the scale but not the pattern of the tiling, we can get different tilings by offsetting one row of squares some distance from the next.

However, there is only one tiling that is edge-to-edge.

:::::: **Edge-to-Edge Tiling**

In an **edge-to-edge tiling,** the edge of a tile coincides entirely with the edge of a bordering tile.

Figure 20.3 shows one tiling that is not edge-to-edge and another that is.

SPOTLIGHT 20.1 Regular Polyhedra and Buckyballs

The three-dimensional analogue of a regular polygon is a regular polyhedron, a convex solid whose faces are regular polygons all alike (same number of sides, same size), with each vertex surrounded by the same number of polygons. Although there are infinitely many regular polygons, there are only five regular polyhedra, a fact proved by Theaetetus (414–368 B.C.). They were called the *Platonic solids* by the ancient Greeks.

If the restriction that the same number of polygons meet at each vertex is relaxed, five additional convex polyhedra are obtained, all of whose faces are equilateral triangles. If we allow more than one kind of regular polygon, 13 further convex polyhedra are obtained, known as the *semiregular polyhedra* or *Archimedean solids* (although there is no documented evidence that Archimedes studied them—but Kepler in the early 1600s catalogued them all). The truncated icosahedron, whose faces are pentagons and hexagons, is known throughout the world (once inflated) as a regulation soccer ball. Drawings of it appear in the work of Leonardo da Vinci.

The truncated icosahedron is also the structure of C_{60}, a form of carbon known as buckminsterfullerene and, more familiarly, the "buckyball." Sixty carbon atoms lie at the 60 vertices of this molecule, which was discovered in 1985. It is named after R. Buckminster Fuller

(1895–1983), inventor and promoter of the geodesic dome. The molecule resembles a dome.

The buckyball is part of a family of carbon molecules, the *fullerenes,* in which each carbon atom is joined to three others. Thirty years before the discovery of fullerenes, mathematicians had shown that a convex polyhedron in which every vertex has three edges must have 12 pentagon faces and may have any number of hexagon faces, from 0 on up, except for 1.

That there must be 12 pentagons follows from a famous equation due to Leonhard Euler (1707–1783). For any convex polyhedron, it must be true that $v - e + f = 2$, where v is the number of vertices, e is the number of edges, and f is the number of faces of the polyhedron.

In 2003 astronomers and mathematicians advanced a remarkable new theory about the shape of the universe, in an effort to explain why it does not show as much historic fluctuation in temperature as other models predict. This lack of fluctuation could be explained by the universe being in the shape of a dodecahedron (the figure shown here with 12 pentagonal sides), with opposite faces coinciding. This theory harks back to Kepler, who had conceived of the universe in terms of the five regular polyhedra nested within one another.

Tetrahedron

Cube

Octahedron

Dodecahedron

Icosahedron

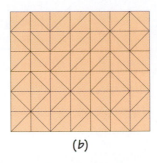

(a)   (b)

FIGURE 20.3 (a) A tiling that is not edge-to-edge. The horizontal edges of two adjoining squares do not exactly coincide. (b) A tiling by right triangles that is edge-to-edge.

For simplicity, from now on by *tiling* we mean "edge-to-edge tiling." The edges themselves may be curvy.

Any tiling by squares can be refined to one by triangles by drawing a diagonal of each square, but these triangles are not regular (equilateral). Equilateral triangles can be arranged in rows by alternately inverting triangles. As with squares, there is only one pattern of equilateral triangles that forms an edge-to-edge tiling.

What about tiles with more than four sides? An edge-to-edge tiling with regular hexagons is easy to construct (see the upper right pattern in Figure 20.5).

However, if we look for a tiling with regular pentagons, we won't find one. How do we know whether we're just not being clever enough or there really isn't one to be found? This is the kind of question that mathematics is uniquely equipped to answer. In the other sciences, phenomena may exist even though we have not observed them. Such was the case for bacteria before the invention of the microscope. In the case of an edge-to-edge tiling with regular pentagons, we can conclude with certainty that there is no edge-to-edge tiling with regular pentagons.

The proof is very easy. As we calculated earlier, the exterior angles of a pentagon are each 72°; each corresponding interior angle is thus 108° (see Figure 20.2). At a point where several pentagons meet, how many can meet there? The total of all of the angles around a point must be 360°. As you can see in Figure 20.4, four pentagons at a point would be too many (their angles would add to $4 \times 108° = 432°$, so they'd have to overlap), and three would be too few (their angles would add to $3 \times 108° = 324°$, so some of the area wouldn't be covered). Because 108 does not evenly divide 360, *regular pentagons can't tile the plane.*

With this argument, we can do something that is a favorite with mathematicians—we can *generalize* it to a criterion for when a regular polygon can tile the plane: when the size of its interior angles divides 360 evenly. We can apply this criterion to determine exactly which other regular polygons can tile the plane.

FIGURE 20.4
Polygons that come together at a vertex in a tiling must have interior angles that add up to 360°—no more, no less.

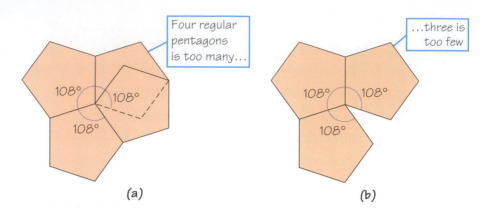

FIGURE 20.4
Polygons that come together at a vertex in a tiling must have interior angles that add up to 360°—no more, no less.

Four regular pentagons is too many...

...three is too few

(a)   (b)

⠿ EXAMPLE 1   Identifying the Regular Tilings

SOLUTION A regular hexagon has interior angles of 120°; 120 divides 360 evenly, and three regular hexagons fit together exactly around a point. A regular 7-gon (heptagon)—or any regular polygon with more than six sides—has interior angles that are larger than 120° but smaller than 180°. Now 360 divided by 120 gives 3, and 360 divided by 180 gives 2—and there aren't any other possibilities in between. Angles between 180° and 120° divided into 360° will give a result between 2 and 3, and consequently not an integer. So there are no regular tilings of the plane with polygons of more than six sides. ■

⠿ Only Three Regular Tilings

The only regular tilings are the ones with equilateral triangles, with squares, and with regular hexagons.

The follow-up question, of course, is which *combinations* of regular polygons of different numbers of sides can tile the plane edge-to-edge. The arrangement of polygons around a vertex in an edge-to-edge tiling is the **vertex figure** for that vertex.

⠿ Semiregular Tiling

A systematic tiling that uses a mix of regular polygons with different numbers of sides but in which all vertex figures are alike—the same polygons in the same order—is called a **semiregular tiling** (see Figure 20.5).

As before, the technique of adding up angles at a vertex (to be 360°) can eliminate some impossible combinations, such as "square, hexagon, hexagon"

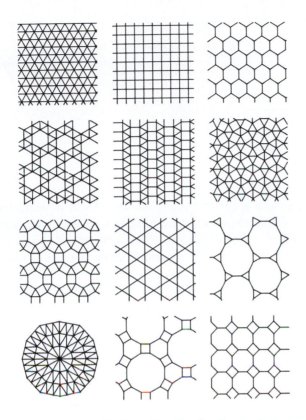

(Figure 20.4). Once we have found an arrangement that is numerically possible, we must confirm the actual existence of each tiling by constructing it (showing that it is geometrically possible). For example, even though a possible arrangement of regular polygons around a point is "triangle, square, square, hexagon," it is not possible to construct a tiling with that vertex figure at every vertex.

The result of such an investigation is that in a semiregular tiling, no polygon can have more than 12 sides. In fact, polygons with 5, 7, 9, 10, or 11 sides do not occur either. Figure 20.5 exhibits all of the semiregular tilings.

If we abandon the restriction about the vertex figures being the same at every vertex, then there are *infinitely many* systematic edge-to-edge tilings with regular polygons, even if we continue to insist that all polygons with the same number of sides have the same size.

20.2 Tilings with Irregular Polygons

What about edge-to-edge tilings with irregular polygons, which may have some sides longer than others or some interior angles larger than others? We will look just at monohedral tilings (in which all tiles have the same size and shape) and investigate in turn which triangles, **quadrilaterals** (four-sided polygons), hexagons, and so forth, can tile the plane.

FIGURE 20.6 (a) Two
scalene triangles form a
parallelogram. (b) Every
scalene triangle tiles the
plane.

(a)   (b)

The most general shape of triangle has all sides of different lengths and all interior angles of different sizes. Such a triangle is called a **scalene triangle,** from the Greek word for "uneven." We can always take two copies of a scalene triangle and fit them together to form a **parallelogram,** a quadrilateral whose opposite sides are parallel (Figure 20.6a). It's easy to see that we can then use such parallelograms to tile the plane by making strips and then fitting layers of strips together edge-to-edge (Figure 20.6b).

Tiling with Triangles

Any triangle can tile the plane.

What about quadrilaterals? We have seen that squares tile the plane, and rectangles certainly will, too. We have just noted that any parallelogram will tile. What about a quadrilateral (four-sided polygon) with its opposite sides not parallel, as in Figure 20.7a? The same technique as for triangles will work. We fit to-

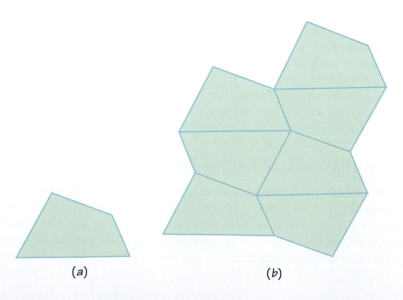

FIGURE 20.7 (a) A
general quadrilateral. (b)
Any quadrilateral tiles
the plane.

(a)   (b)

gether two copies of the quadrilateral, forming a hexagon whose opposite sides are parallel. Such hexagons fit next to each other to form a tiling, as in Figure 20.7b.

The quadrilaterals shown in Figure 20.7 are all **convex.** If you take any two points on the tile (including the boundary), the line segment joining them lies entirely within the tile (again, including the boundary). The quadrilateral of Figure 20.8a is not convex, but the same approach works for using it to form a tiling (Figure 20.8b).

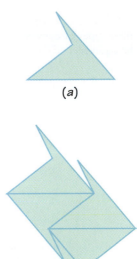

(a)

(b)

FIGURE 20.8 (a) A general nonconvex quadrilateral. (b) Any quadrilateral, convex or not, tiles the plane.

Tiling with Quadrilaterals

Any quadrilateral, even one that is not convex, can tile the plane.

We could hope that such success would extend to irregular polygons with any numbers of sides, but it doesn't. The situation for convex hexagons was determined by Karl Reinhardt in his 1918 doctoral thesis. He showed that for a convex hexagon to tile, it must belong to one of three classes, and that every hexagon in those classes will tile. Examples of the three classes are shown in Figure 20.9, together with their characterizations. Tilings with a hexagon of type 2 use both ordinary and mirror-image versions of the hexagon.

Tiling with Hexagons

Exactly three classes of convex hexagons can tile the plane.

Reinhardt also explored convex pentagons and found five classes that tile. For example, any pentagon with two parallel sides will tile. Reinhardt did not complete the solution, as he did for hexagons, by proving conclusively that no other pentagons could tile. He claimed that it would be very tedious to finish the analysis. Still, he felt that he had found them all. In 1968, after 35 years of working on the problem on and off, R. B. Kershner, a physicist at Johns Hopkins University, discovered three more classes of pentagons that will tile. Kershner was sure that he had found all pentagons that tile, but, like Reinhardt, he did not offer a complete proof, which "would require a rather large book."

When an account of the "complete" classification into eight types appeared in *Scientific American* (July 1975), the article provoked an amateur mathematician to discover a ninth type! A second amateur, Marjorie Rice, a housewife with no formal education in mathematics beyond high school "general mathematics" 36 years earlier, devised her own mathematical notation and found four more types over the next two years (see Spotlight 20.2). A fourteenth type was found by a

FIGURE 20.9 The three types of convex hexagon tile.

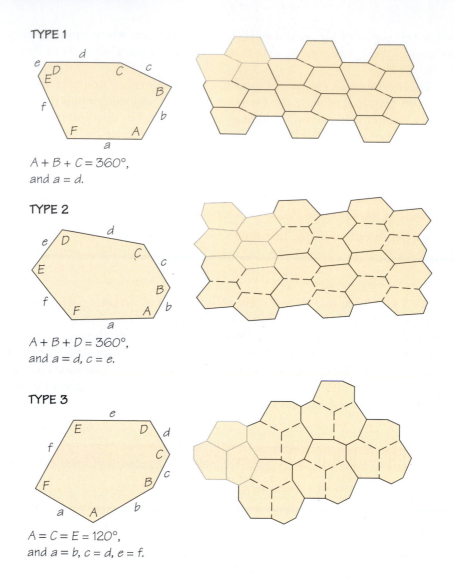

TYPE 1

$A + B + C = 360°$,
and $a = d$.

TYPE 2

$A + B + D = 360°$,
and $a = d$, $c = e$.

TYPE 3

$A = C = E = 120°$,
and $a = b$, $c = d$, $e = f$.

mathematics graduate student in 1985. Since then, no new types have been discovered, yet no one knows if the classification is complete.

With the situation so intricate for convex pentagons, you might think that it must be still worse for polygons with seven or even more sides. In fact, however, the situation is remarkably simple, as Reinhardt proved in 1927.

Tiling with Polygons with More Sides

A convex polygon with seven or more sides cannot tile.

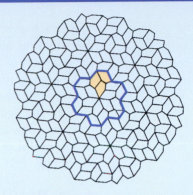

Marjorie Rice
(Courtesy Sharon Whittaker.)

R. B. Kershner's claim to have found all convex pentagons that tile was read by many puzzle enthusiasts, including Richard James III and Marjorie Rice. James found a tiling that Kershner had missed.

Rice, a San Diego housewife and mother of five, read about James's new tile. "I thought I would see if I could find still another type. It was a delightful new puzzle to me."

With no formal education in mathematics beyond a high school general mathematics course, she not only worked out her own method of attack but invented her own notation.

"I began drawing little diagrams on my kitchen counter when no one was there, covering them up quickly if someone came by, for I didn't wish to have to explain what I was doing. I was searching for a new type and a few weeks later, I found it." Over the next two years, she found three additional new tilings.

What makes a person pursue a problem so patiently and persistently? She was not trained for it nor paid, but she gained great personal satisfaction.

She was born in 1923 in St. Petersburg, Florida, and went to a one-room country school.

"When I was in the 6th or 7th grade, our teacher pointed out to us the Golden Section in the proportions of a picture frame. This immediately caught my imagination and I never forgot it. I've . . . been especially interested in

This tiling in the headquarters of the Mathematical Association of America in Washington, D.C., was discovered in 1995 by Marjorie Rice. The angles of each pentagon tile are 60°, 90°, 120°, and 150°—all multiples of 30°. The tiling is periodic, although not every pentagon is surrounded in the same way. Three pentagons form a fundamental block, and the outlined group of 18 pentagons tiles by translation.

architecture and the ideas of architects and planners such as Buckminster Fuller. I've come across the Golden Section again in my reading and considered its use in painting and design."

After high school, Rice worked until her marriage in 1945. She was drawn back into mathematics by her children, finding solutions to their homework problems "by unorthodox means, since I did not know the correct procedures." She became especially interested in textile design and the works of M. C. Escher. As she pursued the pentagonal tilings, she produced some imaginative Escher-like patterns (see Figure 20.19 and the figure here).

Intense spirit of inquiry and keen perception are the forte of all such amateurs. No formal education provides these gifts. Lack of a mathematical degree separates these "amateurs" from the "professionals," yet their curiosity and ingenious methods make them true mathematicians.

Source: Adapted from Doris Schattschneider, "In Praise of Amateurs," in David A. Klarner (ed.), *The Mathematical Gardner*, pp. 140–166, plus Plates I–III, Wadsworth, Belmont, Calif., 1981.

M. C. Escher and Tilings

The Dutch artist M. C. Escher (1898–1972) was inspired by the great variety of decoration in tilings in the Alhambra, a fourteenth-century palace built during the last years of Islamic dominance in Spain. He devoted much of his career of making prints to creating tilings with tiles in the shapes of living beings (a practice forbidden to Muslims). Those prints of interlocking animals and people have inspired awe and wonder among people all over the world. Figures 20.10–20.13 illustrate a few of his drawings and finished works. Like Marjorie Rice, he, too, developed his own mathematical notation for the different kinds of patterns for the tilings.

20.3 Using Translations

You may wonder just how much liberty can be taken in shaping a tile and how you might be able to design an Escher-like tiling yourself.

The simplest case is when the tile is just *translated* in two directions; that is, copies are laid edge-to-edge in rows, as in Figure 20.10. Each tile must fit exactly into the ones next to it, including its neighbors above and below. We say that

FIGURE 20.10 Escher No. 128 (*Bird*), from Escher's 1941–1942 notebook. *(© 1967 M. C. Escher Foundation, Baarn, Holland, all rights reserved.)*

FIGURE 20.11 (a) Escher No. 67 (*Horseman*), from Escher's 1941–1942 notebook. (© 1947 M. C. Escher Foundation, Baarn, Holland, all rights reserved.) (b) Sketch showing the tile design for the *Horseman* print. (© 1947 M. C. Escher Foundation, Baarn, Holland, all rights reserved. From the collection of Michael S. Sachs.)

each tile is a **translation** of each other one, because we can move one to coincide with another without doing any rotation or reflection.

When is it possible for a tile to cover the plane in this manner? The boundary of the tile must be divisible into matching pairs of opposing parts that will fit together. Figures 20.10 and 20.11 illustrate two basic ways that this can happen. In the first, two opposite pairs of sides match; in the second, three opposite pairs of sides match.

Translation Criterion

A tile can tile the plane by translations if either

1. There are four consecutive points A, B, C, and D on the boundary such that
   (a) the boundary part from A to B is congruent by translation to the boundary part from D to C, and
   (b) the boundary part from B to C is congruent by translation to the boundary part from A to D (see Figure 20.12a) or
2. There are six consecutive points A, B, C, D, E, and F on the boundary such that the boundary parts AB, BC, and CD are congruent by translation, respectively, to the boundary parts ED, FE, and AF (see Figure 20.12b).

FIGURE 20.12 Individual tiles traced from the Escher prints of Figures 20.10 and 20.11, with points marked to show they fulfill the criteria for tiling by translations. The two knights form a block that tiles by translation, although a single knight can tile by itself if we allow mirror-image reflections too.

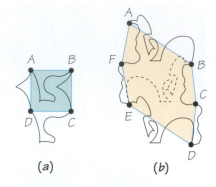

(a)   (b)

The tiles for Figures 20.10 and 20.11 are shown in outline form in Figure 20.12, together with points marked to show how the tiles fulfill the criterion.

In fact, alternative 1 of the criterion is a special case of alternative 2 (see Exercises 19–22 and 24). Moreover, alternative 2 *completely characterizes* tiles that can tile by translations. That is, not only is it true that *if alternative 2 is true, then the tile can tile by translations,* but also that the criterion works "in reverse": *If a tile can tile by translations, then alternative 2 must be true* (for some choice of six consecutive points).

A nice feature of the translation criterion is that if you can find points as required for alternative 2, then you can join them in order, as in Figures 20.12a and b, to see how to do the tiling.

To create tilings, though, you can proceed exactly as Escher did. His notebooks show that he designed his patterns in just the way that we now describe.

EXAMPLE 2   Tiling Starting from a Parallelogram

SOLUTION For the first alternative of the criterion, start from a parallelogram, make a change to the boundary on one side, then copy that change to the opposite side. Similarly, change one of the other two sides and copy that change on the side opposite it (Figure 20.13). Revise as necessary, always making the same change to op-

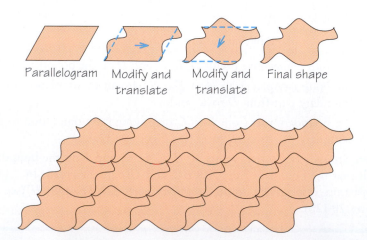

Parallelogram   Modify and   Modify and   Final shape
                 translate   translate

FIGURE 20.13 How to make an Escher-like tiling by translations, from a parallelogram base.

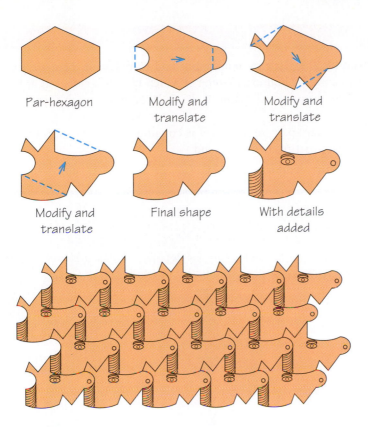

Par-hexagon

Modify and translate

Modify and translate

Modify and translate

Final shape

With details added

FIGURE 20.14 How to make an Escher-like tiling by translations, from a par-hexagon base.

posite sides. You might find it useful (as Escher did) to make your designs on graph paper, or you can work by cutting and taping together pieces of heavy paper. ■

Tiling Starting from a Hexagon

EXAMPLE 3

SOLUTION For the second alternative, start from a **par-hexagon,** one whose opposite sides are equal and parallel. This is one of the kinds of hexagons that tile the plane. Again, make a change on one boundary and copy the change to the opposite side, and do this for all three pairs of opposite sides (Figure 20.14). ■

20.4 Using Translations Plus Half-Turns

If the tiling is to allow half-turns, so that some of the figures are "upside down," the part of the boundary of a right-side-up figure has to match the corresponding part of itself in an upside-down position. For that to happen, that part of the boundary must be **centrosymmetric,** that is, symmetric about (unaltered by) a 180° rotation around its midpoint. The key to some of Escher's more sophisticated monohedral designs, and the fundamental principle behind some further easy recipes for making Escher-like tilings, is the **Conway criterion,** formulated by John H. Conway of Princeton University.

Conway Criterion

A tile can tile the plane by translations and half-turns if there are six consecutive points on the boundary (some of which may coincide, but at least three of which are distinct)—call them A, B, C, D, E, and F—such that

▶ the boundary part from A to B is congruent by translation to the boundary part from E to D, and

▶ each of the boundary parts BC, CD, EF, and FA is centrosymmetric.

FIGURE 20.15 Escher No. 6 (*Camel*), from Escher's 1941–1942 notebook. *(© 1937–1938 M. C. Escher Foundation, Baarn, Holland, all rights reserved.)*

(a)

(b)

FIGURE 20.17 Individual tiles traced from the Escher prints of Figures 20.15 and 20.16, with points marked to show they fulfill the Conway criterion for tiling by translations and half-turns (around the red dots).

The first condition means that we can match up the two boundary parts exactly, curve for curve, angle for angle. The second condition means that each of the remaining boundary parts is brought back into itself by a half-turn around its center. Either condition is automatically fulfilled if the boundary part in question is a straight-line segment.

The tiles for Figures 20.15 and 20.16 are shown in outline form in Figure 20.17, together with points marked to show how the tiles fulfill the Conway criterion.

Mathematicians do not know if Conway criterion completely characterizes tiles that can tile by translations and half-turns. Tiles that fulfill the Conway criterion must be able to tile by translations and half-turns, but not necessarily vice versa: There could be tiles that can tile that way but do not satisfy the criterion—however, nobody knows of any. (The Conway criterion does, however, completely characterize tiles that produce the wallpaper pattern *p2* of Spotlight 19.6: Any tile that satisfies the criterion can be used to make a *p2* pattern, and any tile that can produce that pattern must satisfy the criterion.)

Once again, you can make Escher-like tilings by starting from simple geometric shapes that tile. This time, the starting geometric tile can be any triangle or any quadrilateral.

FIGURE 20.18 How to make an Escher-like tiling by translations and half-turns, from a scalene triangle base.

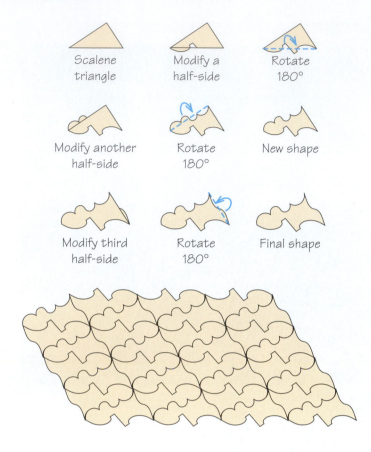

EXAMPLE 4   Tiling Using a Triangle

SOLUTION For a triangle, modify half of one side, then rotate that side around its center point to extend the modification to the rest of the side, thereby making the new side centrosymmetric. Then you may do the same to the second and third sides (Figure 20.18). ▪

EXAMPLE 5   Tiling Using a Quadrilateral

SOLUTION For the quadrilateral, do the same, modifying each of the four sides, or as many as you wish (Figure 20.19). ▪

The same approach will work with some of the sides of some pentagons and hexagons that tile. Because not all sides can be modified, there is less freedom for designing tiles, so it is more difficult to make the resulting tiles resemble intended figures. Figure 20.20 shows the beautiful results achieved by Marjorie Rice, using one of the unusual tilings by pentagons that she discovered.

Sketches in Escher's notebook indicate how he designed many of his prints. For the bird tiling of Figure 20.10, the single bird below the tiling shows that he modified the sides of a square. For the knights tiling of Figure 20.11a, the sketches

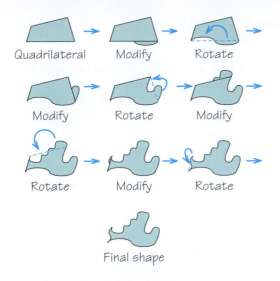

Quadrilateral Modify Rotate

Modify Rotate Modify

Rotate Modify Rotate

Final shape

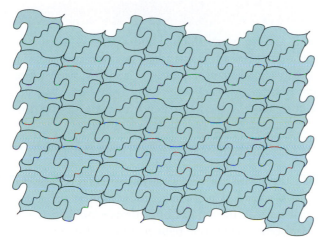

FIGURE 20.19 How to make an Escher-like tiling by translations and half-turns, from a quadrilateral base.

FIGURE 20.20 *Fish*, by Marjorie Rice, based on one of her unusual tilings by pentagons.

in Figure 20.11b show that he modified the pairs of sides of a par-hexagon. We redraw the two fundamental figures more clearly in Figure 20.12. (The knight tiling also has a reflection symmetry, taking a leftward-facing light knight to a rightward-facing dark knight; but we have not discussed criteria for producing a tiling with such a symmetry.)

As can be seen in faint lines in Figure 20.15, Escher used a parallelogram as a base for the camel tiling. In Figure 20.17a, the violet overlay shows how to make the tiling starting from a more general quadrilateral by modifying half of each side. For the seahorse tiling of Figure 20.16, Escher used a triangle base. However, once more he avoided modifying half of every side; instead, he treated the triangle ACF (Figure 20.17b) as a quadrilateral ACDF in which two adjacent sides (*CD* and *DF*) happen to continue on in a straight line.

Periodic Tilings

All the patterns that we have exhibited and discussed so far have been **periodic tilings**. If we transfer a periodic tiling to a transparency, it is possible to slide the transparency a certain distance horizontally, without rotating it, until the transparency exactly matches the tiling everywhere. We can also achieve the same result by moving the transparency in some second direction (possibly vertically) by a certain (possibly different) distance.

In a periodic tiling, you can identify a **fundamental region**—a tile, or a block of tiles—with which you can cover the plane by translations at regular intervals. For example, in Figure 20.10, a single bird forms a fundamental region. In Figure 20.15, two adjacent camels, one right side up and one upside down, form a fundamental region. In the terminology of Chapter 19, the periodic tilings are ones that are preserved under translations in more than two directions.

20.5 Nonperiodic Tilings

In Figure 20.3a, the second row from the bottom is offset one-half of a unit to the right from the bottom row, the third row from the bottom is offset one-third of a unit further, and so forth. Because the sum $\frac{1}{2} + \frac{1}{3} + \frac{1}{4} + \cdots + \frac{1}{n}$ never adds up to exactly a whole number, there is no direction (horizontal, vertical, or diagonal) in which we can move the entire tiling and have it coincide exactly with itself.

Nonperiodic Tiling

A **nonperiodic tiling** is a tiling in which there is no regular repetition of the pattern by translation.

A Random Tiling

SOLUTION Consider the usual edge-to-edge square tiling. For each square, flip a coin. Depending on the result, divide the square into two right triangles by adding either a rising or a falling diagonal (see Figure 20.3b). Because what happens in each individual square is unconnected to what happens in the rest of the tiling, the tiling by right triangles that is produced by this procedure has no chance of being periodic. ■

EXAMPLE 6

Penrose Tiles and Quasicrystals

For all known cases, if a single tile can be used to make a nonperiodic tiling of the plane, then it can also be used to make a periodic tiling. It is still an open question whether this property is true for every possible shape. In 1993, Conway discovered an example in three dimensions of a single convex polyhedron that tiles space nonperiodically but cannot be used to make a periodic tiling.

For a long time, mathematicians also tended to believe the more general assertion that if you can construct a nonperiodic tiling with a set of one *or more* tiles, you can construct a periodic tiling from the same tiles. But in 1964 a set of tiles was found that permits only nonperiodic tiling. It contains 20,000 different shapes! Over the next several years, smaller sets were discovered with the same property, with as few as 100 shapes. But it was still amazing when in 1975 Sir Roger Penrose, a mathematical physicist at Oxford, announced a set that tiles only nonperiodically–consisting of just two tiles! (See Figure 20.21 and Spotlight 20.3.)

Penrose called his tiles "darts" and "kites," and both of these *Penrose tiles* can be obtained from a single rhombus. A **rhombus** is a quadrilateral with four equal sides and equal opposite interior angles. The particular rhombus from which the Penrose tiles are constructed has interior angles of 72° and 108°. If we cut the

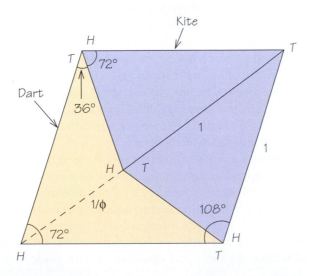

FIGURE 20.21
Construction of Penrose's "dart" (beige area) and "kite" (blue area). The length $1/\phi \approx$ 0.618 is the golden ratio.

Sir Roger Penrose
(Anthony Howarth/Photo Researchers.)

Sir Roger Penrose, a professor at the University of Oxford, received a doctorate in mathematics but has been seriously interested in physics for many years; he was one of the first to conjecture the existence of black holes. He discovered what are now called the Penrose pieces in 1973. His latest endeavor has been to try to establish that the mind is not a machine, that is, that the ideas and concepts of artificial intelligence cannot explain human consciousness.

A chance meeting with Maurits Escher resulted in Penrose sending him some of his grandfather's art, which helped inspire some of Escher's prints.

longer diagonal in two pieces so that the longer piece is the golden ratio, or $(1 + \sqrt{5})/2 \approx 1.618$ times as long as the shorter (see Chapter 19, p. 712), and connect the dividing point to the remaining corners, we split the rhombus into a dart and a kite (Figure 20.21).

Label the front and back vertices of the dart with H (for head) and its two wing tips with T (for tail), and do the reverse for the kite. Then the rule for fitting the pieces together is that only vertices with the same letter may meet: Heads must go to heads, tails must go to tails. Thus the rules don't allow the pieces to fit together as a rhombus (which would allow them to tile periodically).

A prettier method of enforcing the rules, proposed by Conway, is to draw circular arcs of different colors on the pieces and require that adjacent edges must join arcs of the same color. The result is the pretty patterns of Figure 20.22. In fact, Conway thinks of the darts as children, each with two hands. The rule for fitting the pieces together is that children are required to hold hands. Penrose patterns become dancing circles of children.

Figure 20.23 shows a tiling by a different pair of pieces, both rhombuses, that tile the plane only nonperiodically. Figure 20.24 shows a modification of the Penrose pieces into two bird shapes. Figure 20.25 shows a coloring of one particular tiling with the Penrose pieces so that no two adjacent pieces have the same color.

Although tilings with Penrose's pieces cannot be periodic, the tilings possess unexpected symmetry. As you recall, we have explored our intuitions of symmetry in terms of *balance, similarity,* and *repetition.* Patterns made with the Penrose

FIGURE 20.22 A Penrose tiling with specially marked tiles, forming what is known as the cartwheel tiling. *(From Sir Roger Penrose.)*

FIGURE 20.23 A Penrose nonperiodic tiling made with two rhombus shapes, one thin and one fatter. The fatter one has a yellow stripe across one end. *(Tiling by Sir Roger Penrose.)*

FIGURE 20.24 A modification of a Penrose tiling by refashioning the kites and darts into bird shapes. *(Tiling by Sir Roger Penrose.)*

FIGURE 20.25 A Penrose tiling by kites and darts, colored with five colors. A Penrose tiling can always be colored using four colors, in such a way that two tiles that share an edge have different colors. Whether a Penrose tiling can be colored in such a way using only three colors is an unsolved problem. However, we know that if even one Penrose tiling can be colored using three colors, all Penrose tilings can. *(Tiling by Sir Roger Penrose.)*

pieces certainly involve repetition, but it is the balance in the arrangement that we seek. What balance can there be in a nonperiodic pattern? It turns out that some Penrose patterns have a single line of reflection. But most surprising of all is that every Penrose pattern has a kind of fivefold rotational symmetry.

EXAMPLE 7   Fivefold Symmetry

SOLUTION Look again at Figure 20.21, which shows how to split a rhombus into the Penrose dart and kite pieces. Except in the recess of the dart and the matching part of the kite, all of the internal angles of the kite and of the dart are either 72° or 36°. Now, 72° goes into 360° five times, and 36° goes into 360° ten times. If we recall that it is the interior angles that matter in arranging polygons around a point, we see why it might be possible for a Penrose pattern to have fivefold or tenfold rotational symmetry.

A Penrose pattern with tenfold rotational symmetry is impossible, but there are exactly two Penrose patterns that tile the entire plane with fivefold rotational symmetry about one particular point. We show finite parts of these patterns in Figure 20.26. For each pattern, the center of rotational symmetry is at the center of the figure, surrounded by either five darts or five kites.

For any other Penrose pattern, the pattern as a whole does not have fivefold rotational symmetry. However, what is surprising is that the pattern must have arbitrarily large finite regions with fivefold rotational symmetry. You can see this feature in the regions of Figure 20.23 that are enclosed by yellow lines. In Conway's metaphor, whenever a chain of children (darts) closes, the region inside has fivefold symmetry. ▪

FIGURE 20.26
Successful deflation (that is, the systematic cutting up of large tiles into smaller ones) of patches of tiles of a Penrose nonperiodic tiling.

Conway invented a process called *inflation* that takes any Penrose pattern into a different Penrose pattern with larger darts and kites. The inflation operation (we don't give the details here) systematically cuts up the darts and kites into triangles and regroups the triangles into larger darts and kites.

We can use inflation to show that a Penrose pattern must be nonperiodic. Suppose (contrary to what we want to establish) that some Penrose pattern is periodic; that is, it has translation symmetry. Let d be the distance along the translation direction to the first repetition. Performing inflation does the same thing to each repetition, so the inflated pattern must still have translation symmetry and a distance d along the translation direction to the first repetition. Keep on performing inflation, time after time, until the darts and kites are so large that they are more than d across. The pattern, as we have just argued, must still have translation symmetry at a distance d, but it can't, because there's no repetition inside a single tile! We reach a contradiction. So what's wrong? Our initial supposition, that the pattern was periodic in the first place, must have been erroneous. We conclude that all Penrose tilings are nonperiodic.

Despite their being nonperiodic, all Penrose patterns are somewhat alike, in the following remarkable sense.

FIGURE 20.27
Penrose toilet paper
(Mario Ruiz/Time Magazine.)

Penrose Inside of Penrose

The subpattern of any finite region in one Penrose pattern is contained somewhere inside every other Penrose pattern. In fact, any subpattern occurs infinitely many times in every Penrose pattern.

The nonperiodicity of Penrose filings found a surprising application in 1997—to bathroom tissue. Quilted bathroom tissue needs to be embossed with a pattern to keep the layers together (Figure 20.27). If the pattern is regular, then the multiple layers on the roll can produce lumpy ridges and grooves. Using a nonrepeating Penrose pattern averts the lumpiness. However, the company used Penrose's pattern without his permission, and Penrose sued successfully.

Penrose tilings have another feature that allows us to characterize them as *quasiperiodic*, or somewhere between periodic and random. (Noting the precise definition of random would take us too far afield.) Robert Ammann introduced onto the two rhombic Penrose pieces used in Figure 20.23 lines now known as *Ammann bars*. In any Penrose tiling, these bars line up into five sets of parallel lines, each set rotated 72° from the next, forming a pentagonal grid (Figure 20.28). The distance between two adjacent parallel bars is one of only two values, either *A* or *B*. Do you want to guess what the ratio of the longer *A* is to the shorter *B*? You don't think it could possibly be anything but the golden ratio of Chapter 19, do you? And so it is.

Musical Sequences

EXAMPLE 8

What about the order in which the *A*'s and *B*'s occur, as we move from left to right in Figure 20.28? Is there any pattern to that?

SOLUTION From the limited part of the pattern that we can observe, we see the sequence as

ABAABABABAABABA

FIGURE 20.28

Penrose tilings with Ammann bars. Specially placed lines on the tiles produce five sets of parallel bars in different directions.

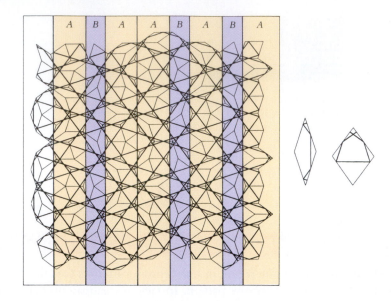

You might think from the figure that the pattern continues repeating the group

$$ABAAB$$

indefinitely; after all, there are five symbols in this group. But such is not the case. Known as a *musical sequence*, the sequence of intervals between Ammann bars is nonperiodic—it cannot be produced by repeating any finite group of symbols. We can think of it as a one-dimensional analogue of a Penrose tiling.

There is some regularity in musical sequences. Two *B*'s can never be next to each other, nor can we have three *A*'s in a row. Just as any finite part of any Penrose tiling occurs infinitely often in any other Penrose tiling, any finite part of any musical sequence appears infinitely often in any other one. The order of the symbols is neither periodic nor random, but between the two—quasiperiodic.

These sequences are called *musical* sequences because musicians represent the large-scale structure of songs in terms of the letters *A* and *B*. For example, a common pattern for popular songs is *AABA*, indicating that the first, second, and fourth verses have the same melody, but the third verse has a different melody. ■

The ratio of darts to kites in an infinite Penrose tiling, or of *A*'s to *B*'s in a musical sequence, is exactly the golden ratio, approximately 1.618. So if you are going to play with sets of Penrose pieces to see what kinds of patterns you can create, you will need about 1.6 times as many darts as kites.

As pointed out by geometers Marjorie Senechal (Smith College) and Jean Taylor (Rutgers University), Penrose tilings have three important properties:

▶ They are constructed according to rules that force nonperiodicity.

▶ They can be obtained from a substitution process (inflation and deflation) that features self-similarity at different scales (like the fractals in Chapter 19).

▶ They are quasiperiodic.

These properties are somewhat independent, meaning that one or two may be true of a tiling without all three being true.

Quasicrystals and Barlow's Law

Although Penrose's discovery was a big hit among geometers and in recreational mathematics circles in the mid-1970s, few people thought that his work might have practical significance. In the early 1980s some mathematicians even generalized Penrose tilings to three dimensions, using solid polyhedra to fill space nonperiodically. Like the two-dimensional Penrose patterns, these have orderly fivefold symmetry but are nonperiodic.

Yet in 1982 scientists at the U.S. National Bureau of Standards discovered unexpected fivefold symmetry while looking for new ultrastrong alloys of aluminum (mixtures of aluminum with other metals).

Manganese doesn't ordinarily alloy with aluminum, but the experimenters were able to produce small crystals of alloy by cooling mixtures of the two metals at a rate of millions of degrees per second. Following routine procedures, chemist Daniel Shechtman began a series of tests to determine the atomic structure of the special crystals. But there was nothing routine about what he found: The atomic structures of the manganese–aluminum crystals were so startling that it took Shechtman three years to convince his colleagues they were real.

Why did he encounter such resistance? His patterns—and the crystals that produced them—defied one of the fundamental laws of crystallography. Like our discovery that the plane cannot be tiled by regular pentagons, **Barlow's law,** also called the **crystallographic restriction,** says that a crystal must be periodic and hence can have only rotational symmetries that are twofold, threefold, fourfold, or sixfold. If there were a center of fivefold symmetry, there would have to be many such centers. Barlow proved this impossible.

Peter Barlow (1776–1862) was a British mathematician whose name survives today in the name of a book of mathematical tables. His argument was a very simple proof by contradiction, similar to Conway's proof in which we saw earlier that Penrose patterns are not periodic. Suppose (contrary to what we intend to show) that there is more than one fivefold rotation center. Let A and B be two of these that are closest together (see Figure 20.29). Rotate the pattern of Figure 20.28 by one-fifth of a turn clockwise around B, which carries A to some point A'. Because the pattern has fivefold symmetry around B, the point A', which is the image of the fivefold center A, must itself be a fivefold center. Now use A as a center and rotate the pattern by one-fifth of a turn clockwise, which carries B to some point B'. As we just argued in the case of A', B' must also be

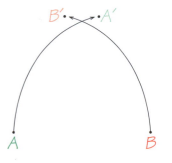

FIGURE 20.29
Barlow's proof that no pattern can have two centers of fivefold symmetry.

SPOTLIGHT 20.4 Quasicrystals

In 1984, working at the University of Pennsylvania, Paul Steinhardt and Don Levine did a computer simulation of what a three-dimensional Penrose pattern would be like. They decided to call such structures *quasicrystals*. Later that fall, their chemist colleague Daniel Shechtman showed that quasicrystals really exist. He produced images of an alloy of aluminum and manganese that were amazingly similar to images from the computer simulations. In short order, sevenfold, ninefold, and other symmetries were also shown to occur in real materials.

In 1991, Sergei Burkov showed that quasiperiodic tilings can be made using only a single kind of 10-sided tile, *provided the tiles are allowed to overlap*. With overlaps, the resulting patterns are no longer tilings. They are called *coverages*. In late 1998, scientists presented electron microscope photos that demonstrated that atoms really do form such coverages.

The current theory is that quasicrystals are packings of copies of a single type of atom cluster, with each cluster sharing atoms with its neighbors, that is, overlapping nearby clusters. The clusters form a quasiperiodic pattern that maximizes their density, thereby minimizing the energy of the atoms involved.

(a) A scanning electron microscope image of the quasicrystal alloy $Al_{5???}$ Li_3Cu (the question marks indicate uncertainty about how many aluminum atoms are involved). The fivefold symmetry can be seen in the five rhombic faces that meet at a single point in the center of the photograph, forming a starlike shape. (b) This image of the quasicrystal material $Al_{65}Co_{20}Cu_{15}$ was obtained with a scanning tunneling microscope. The resulting image has been overlaid with a nonperiodic tiling to display the local fivefold symmetry. *[Both adapted from Hans C. von Baeyer, "Impossible Crystals," Discover 11(2) (February 1990): 69–78, 84.]*

a fivefold center. But A' and B' are closer together than A and B, which is a contradiction. Hence our original supposition must be false, and a pattern can have at most one fivefold rotation center (as the patterns in Figure 20.26 in fact do) and so cannot be periodic.

Barlow's law, as a mathematical theorem, shows that fivefold symmetry is impossible in a *periodic* tiling of the plane or of space. Chemists, for good theoretical and experimental reasons, believe that crystals are modeled well by three-dimensional tilings. An array of atoms with no symmetry whatever would not be considered a crystal. Yet until Penrose's discovery, no one realized that nonperiodic tilings—or arrays of atoms—can have the regularity of fivefold symmetry.

Chemists could simply say that Shechtman's alloys aren't crystals. In the classic sense of being periodic, they aren't, but in other respects they do resemble crystals. It is scientifically more fruitful to extend the concept of crystal to include them rather than to rule them out. They are now known as *quasicrystals* (see Spotlight 20.4).

Once again, as so often happens in history, pure mathematical research anticipated scientific applications. Penrose's discovery, once just a delightful piece of recreational mathematics, has prompted a major reexamination of the theory of crystals. Barlow's law is not refuted, since it applies only to periodic crystals, not to quasicrystals.

REVIEW VOCABULARY

Barlow's law, or the **crystallographic restriction** A law of crystallography that states that a crystal may have only rotational symmetries that are twofold, threefold, fourfold, or sixfold.

Centrosymmetric Symmetric by 180° rotation around its center.

Convex A geometric figure is convex if for any two points on the figure (including its boundary), all the points on the line segment joining them also belong to the figure (including its boundary).

Conway criterion A criterion for determining whether a shape can tile by means of translations and half-turns.

Edge-to-edge tiling A tiling in which adjacent tiles meet only along full edges of each tile.

Equilateral triangle A triangle with all three sides equal.

Exterior angle The angle outside a polygon formed by one side and the extension of an adjacent side.

Fundamental region A tile or group of adjacent tiles that can tile by translation.

Interior angle The angle inside a polygon formed by two adjacent sides.

Monohedral tiling A tiling with only one size and shape of tile (the tile is allowed to occur also in "turned-over," or mirror-image, form).

n-gon A polygon with n sides.

Nonperiodic tiling A tiling in which there is no repetition of the pattern by translation.

Parallelogram A convex quadrilateral whose opposite sides are equal and parallel.

Par-hexagon A hexagon whose opposite sides are equal and parallel.

Periodic tiling A tiling that repeats at fixed intervals in two different directions, possibly horizontal and vertical.

Quadrilateral A polygon with four sides.

Regular polygon A polygon all of whose sides and angles are equal.

Regular tiling A tiling by regular polygons, all of which have the same number of sides and are the same size; also, at each vertex, the same kinds of polygons must meet in the same order.

Rhombus A parallelogram all of whose sides are equal—four equal sides and equal opposite interior angles.

Scalene triangle A triangle no two sides of which are equal.

Semiregular tiling A tiling by regular polygons; all polygons with the same number of sides must be the same size.

Tiling A covering of the plane without gaps or overlaps.

Translation A rigid motion that moves everything a certain distance in one direction.

Vertex figure The pattern of polygons surrounding a vertex in a tiling.

✔ SKILLS CHECK

1. How large is the exterior angle of a regular octagon?
(a) 30°
(b) 45°
(c) 135°

2. In a tiling of the plane, the tiles
(a) are allowed to overlap, as long as no area is left uncovered.
(b) are not allowed to overlap.
(c) must meet edge-to-edge.

3. In a tiling of the plane, the tiles
(a) have to be regular polygons.
(b) have to be polygons but need not be regular.
(c) can be any shape at all.

4. A regular tiling can be constructed using
(a) rectangles.
(b) hexagons.
(c) squares and octagons.

5. Regular octagons and squares can form a semiregular tiling of the plane with
(a) two octagons and one square at each vertex.
(b) two octagons and two squares at each vertex.
(c) a varying configuration at the vertices.

6. A semiregular tiling has a square, a regular dodecagon (12-gon), and another regular polygon at each vertex. What is this other polygon?

(a) Another square
(b) Another dodecagon
(c) A hexagon

7. Which of the following polygons *cannot* tile the plane?
(a) Regular hexagon
(b) Regular pentagon
(c) Nonrectangle parallelogram

8. A tiling of the plane can be formed using as a tile
(a) some but not all convex quadrilaterals.
(b) any convex quadrilateral, but no nonconvex quadrilaterals.
(c) any quadrilateral.

9. A tiling of the plane can be formed using as a tile
(a) some pentagons.
(b) any pentagon with at least two right angles.
(c) any pentagon with at least three right angles.

10. A tiling of the plane can be formed using as a tile
(a) some but not all convex hexagons.
(b) any convex hexagon, but no nonconvex hexagons.
(c) any hexagon.

11. Can the tile below be used to create a tiling of the plane?

(a) No
(b) Yes, using only translations
(c) Yes, using translations and half-turns

12. Can the tile below be used to create a tiling of the plane?

(a) No
(b) Yes, using translations
(c) Yes, using translations and half-turns

13. Any quadrilateral can tile the plane by

(a) translations.
(b) half-turns.
(c) translations and half-turns.

14. Which of the following is true?

(a) If a polygon fulfills the Conway condition, it can tile the plane by translations.
(b) If a polygon fulfills the Conway condition, it can tile the plane by translations and half-turns.
(c) If a polygon doesn't fulfill the Conway condition, it can't tile the plane at all.

15. In a nonperiodic tiling of the plane,

(a) the pattern never repeats.
(b) the pattern is not repeated by any translation.
(c) there must be at least three kinds of tiles.

16. Penrose tilings are

(a) periodic.
(b) random.
(c) quasiperiodic.

17. A rhombus always has the trait that

(a) opposite sides are unequal in length.
(b) opposite angles are congruent.
(c) it cannot be a square.

18. A Penrose dart always has the trait that

(a) opposite angles are congruent.
(b) it is nonconvex.
(c) the edges are all of different lengths.

19. In a Penrose tiling, the proportion of darts to kites is

(a) one-to-one.
(b) two-to-one.
(c) the golden ratio.

20. Barlow's law prohibits the existence of crystals with

(a) fivefold symmetry.
(b) sixfold symmetry.
(c) periodic tilings.

EXERCISES   ■ Challenge   ◆ Discussion

Hint: For the exercises about determining whether a shape will tile the plane, you should make a number of copies of the shape and experiment with placing them. One easy way to make copies is to trace the shape onto a piece of paper, staple half a dozen other blank sheets behind that sheet, and use scissors to cut through all the sheets along the edges of the traced shape on the top sheet.

Tilings with Regular Polygons

1. Determine the measure of an exterior angle and of an interior angle of a regular octagon (eight sides).

2. Determine the measure of an exterior angle and of an interior angle of a regular decagon (10 sides).

3. Discover a formula for the measure of an interior angle of a regular *n*-gon.

4. Using the formula from Exercise 3 and either your calculator or a short computer program, make a chart of the interior-angle measures of regular polygons with 3, 4, . . . , 12 sides.

■ **5.** Use the chart of interior-angle measures from Exercise 4 to determine all of the possible vertex figures of regular polygons (with at most 12 sides) surrounding a point.

■ **6.** Which of the vertex figures of Exercise 5 do not occur in a semiregular tiling?

■ **7.** In addition to the vertex figures of Exercise 5, exactly five others are possible, each involving one polygon with more than 12 sides. None of these vertex figures lead to a semiregular tiling. The five many-sided polygons involved in these five vertex figures have 15, 18, 20, 24, and 42 sides. Determine the other polygons in these vertex figures.

For Exercises 8 and 9, refer to the lower left corner of Figure 20.5, which shows a tiling by isosceles triangles.

8. Use the center vertex to determine the measures of the angles of the isosceles triangle tile.

9. Every vertex except the center vertex has the same vertex figure, in terms of the measures of the angles surrounding the vertex. What is that vertex figure?

10. Give a numerical reason why a semiregular tiling could not include both polygons with 12 sides and polygons with 8 sides (with or without any polygons with other numbers of sides).

Tilings with Irregular Polygons

11. You know that a regular pentagon cannot tile the plane. Suppose you cut one in half. Can this new shape tile the plane? (See Figure 19.4 for a regular pentagon that you can trace.)

12. For each of the tiles below, show how it can be used to tile the plane. (Adapted from *Tilings and Patterns,* by Branko Grünbaum and G. C. Shephard, Freeman, New York, 1987, p. 25.)

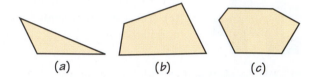

(a)   (b)   (c)

Using Translations

Refer to the following tiles (a) through (g) in doing Exercises 13 and 14.

13. For each of the tiles (a) through (c), determine whether it can be used to tile the plane by translations. (From *Tiling the Plane,* by Frederick Barber et al., COMAP, Lexington, Mass., 1989, pp. 1, 8, 9.)

(a)   (b)   (c)

(d)   (e)

(f)

(g)

14. Repeat Exercise 13, but for tiles (d) through (g).

15. Start from a par-hexagon of your choice and modify it to tile the plane by translations. (You will probably find it useful to do your work on graph paper. If you choose a regular hexagon, there is special graph paper, ruled into regular hexagons, that would be particularly useful.) Can you draw a design on the tile so as to make an Escher-like pattern?

16. Start from a parallelogram of your choice and modify it to tile the plane by translations. (You will probably find it useful to do your work on graph paper.) Can you draw a design on the tile so as to make an Escher-like pattern?

Refer to the following information in doing Exercises 17–20.

A particularly simple kind of polygon, called a *polyomino,* is one made of squares joined edge-to-edge. The name is a generalization of "domino"; indeed, there is only one kind of domino (two squares joined at an edge to form a rectangle). There are just two "trominos" (for "triominos"), the straight tromino and the L-tromino.

The straight tromino has the shape of a rectangle, so it can tile the plane by translations; and the L-tromino has the shape of a hexagon.

17. Is the L-tromino convex? Does the result about what hexagons can tile the plane (p. 763) give any information about whether the L-tromino can tile the plane or not?

18. Find a tiling of the plane using just the L-tromino and translations of it. Is there more than one way to do the tiling?

19. Show how alternative 2 of the translation criterion can be applied to the L-tromino.

■ **20.** Give an argument why alternative 1 of the translation criterion cannot be applied to the L-tromino. (*Hint:* Label each of the eight corners of the component squares of the tromino with the letters $S, T, \ldots, Z$. Let these be our candidates for the points $A, B, C,$ and D of the criterion.) Each of the sides of the tromino that is two units long has nowhere to go under a translation. Any application of the criterion must divide each side into two pieces, so their midpoints must be two of the points $A, B, C,$ and D. Make a similar argument about two corners of the tromino. Thus, we have four points, which can be labeled consecutively $A, B, C,$ and D, starting at any one of them. Show that none of the four possibilities "works." (This argument can be generalized to show that trying $A, B, C,$ and D at points other than the corners of the squares won't work either.)

Refer to the following information in doing Exercises 21–26.

Demonstrate to your own satisfaction that there are exactly five shapes of tetrominos (each made of four squares joined at edges)–plus differing mirror images of two of them–as shown below. In the order shown, they are called the square, straight, T, L, and skew tetrominos. The straight and the square tetromino certainly can tile the plane by translations.

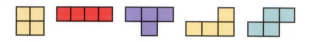

You will definitely find it useful to make yourself several copies of each of the polyominos mentioned below, by cutting them out of graph paper.

21. Apply alternative 1 of the translation criterion to the straight-tetromino and show how it can tile.

22. Show how alternative 2 of the translation criterion applies to the T-tetromino.

23. Apply alternative 1 of the translation criterion to the L-tetromino and show how it can tile.

24. Show how alternative 2 of the translation criterion applies to the skew-tetromino.

25. Show how alternative 2 of the translation criterion can be applied to the skew-tetromino, and show how it can tile.

26. In Exercises 23 and 25, we indulged in what appears to be "overkill," proving the same fact in two different ways. But those exercises should give you the idea that alternative 2 of the translation criterion can reduce to (and hence is more general than) alternative 1 if some points are allowed to coincide. For such a reduction, which pairs of points must coincide? (We are allowed to relabel the remaining four distinct points.)

Using Translations Plus Half-Turns

For Exercises 27 and 28, refer to tiles (a) through (g) on pp. 786–787.

27. For each of the tiles (a) through (c), determine whether it can be used to tile the plane by translations and half-turns.

28. Repeat Exercise 27, but for tiles (d) through (g).

29. Show how an arbitrary pentagon with two parallel sides can tile the plane.

30. The following is a pentagonal tile of type 13, discovered by Marjorie Rice. Show how it can tile the plane. (*Hint:* Carefully trace and cut out a dozen or so copies and try fitting them together.)

The parts of this pentagon satisfy the following relations: $A = C = D = 120°$, $B = E = 90°$, $2A + D = 360°$, $2C + D = 360°$, $a = e$, and $a + e = d$. [Adapted from "In Praise of Amateurs," by Doris Schattschneider, in David A. Klarner (ed.), *The Mathematical Gardner*, Wadsworth, Belmont, Calif., 1981, p. 162.]

31. Start from a triangle of your choice and modify it to tile the plane by translations and half-turns. (You will probably find it useful to do your work on graph paper.) Can you draw a design on the tile so as to make an Escher-like pattern?

32. Start from a quadrilateral of your choice and modify it to tile the plane by translations and half-turns. (You will probably find it useful to do your work on graph paper.) Can you draw a design on the tile so as to make an Escher-like pattern?

Refer to the information about polyominos preceding Exercise 17, and to the following, in doing Exercises 33–36.

We saw earlier that all the dominos, trominos, and tetrominos tile the plane by translations. Here we investigate the 12 pentominos, shown as follows with a letter notation for each (if you allow mirror images to count as different pentominos, there are

18). It will be useful for you to make several copies of each of the pentominos discussed below.

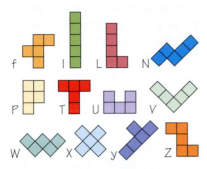

33. Just by experimenting, determine which of the pentominos can tile the plane by translations. (*Hint:* There are nine.)

34. Apply the Conway criterion to the f-pentomino, and show how it can tile by translations and half-turns.

35. Apply the Conway criterion to the U-pentomino, and show how it can tile by translations and half-turns.

36. Apply the Conway criterion to the T-pentomino, and show how it can tile by translations and half-turns.

37. In the text we discuss criteria and methods for generating Escher-like patterns that involve just translations or translations and half-turns. A slight variation on one of those methods allows construction of tilings that feature a tile and its mirror image.

Begin with a parallelogram made from two congruent isosceles triangles, as shown in the figure at right. Each of these triangles has two sides equal. Be sure that the two triangles are arranged so that they have one of the equal sides in common, forming a diagonal of the parallelogram.

Make any modification to half of the third side of one of the triangles. Mirror-reflect that modification across the side, then translate the reflection to become the modification of the other half of the side. Take the complete modification of this side, and translate it to become the

modification of the opposite side of the parallelogram.

Modify in any way one of the two remaining sides of the parallelogram, and make the same modification to the opposite side (that is, translate the modification, without rotation or reflection). Then make the mirror reflection of this modification the modification of the diagonal of the parallelogram.

The result is a modified parallelogram that tiles by translation and splits into two pieces that are mirror images of each other. Escher used a similar technique, but starting from a par-hexagon made from two quadrilaterals, in his *Horseman* print, as shown in his sketch in Figure 20.11b.

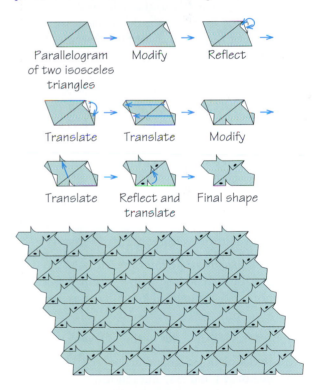

Use this technique to produce a tiling of your own design. Can you draw a design on the tile so as to make an Escher-like pattern?

38. Show that the modified parallelogram in Exercise 37 fulfills the Conway criterion, by identifying the six points of the criterion.

Nonperiodic Tilings

For Exercises 39–42, refer to the following.

The rabbit problem in Chapter 19 (Exercise 4) leads us directly into nonperiodic patterns and musical sequences. Let A denote an adult pair of rabbits and B denote a baby pair. We record the population at the end of each month, just before any births, in a particular systematic way—as a string of A's and B's. At the end of their second month of life, a rabbit pair will be considered to be adult and give birth to a baby pair. At the end of the first month, the sequence is just A, and the same is true at the end of the second month. When an adult pair A has a baby pair B, we write the new B immediately to the right of the A. So at the end of the third month, the sequence is AB; at the end of the fourth, it is ABA, because the first baby pair is now adult; at the end of the fifth month we have $ABAAB$.

Mathematicians and computer scientists call this manner of generating a sequence a *replacement system*. At each stage we replace each A by AB and each B by A.

39. What is the sequence at the end of the sixth month?

40. Why can't we ever have two B's next to each other?

41. Why can't we ever have three A's in a row?

42. Show that from the fourth month on, the sequence for the current month consists of the sequence for last month followed by the sequence for two months ago.

For Exercises 43–50, refer to the following.

We can define inflation and deflation of a sequence of A's and B's, and musical sequences themselves, without reference to Penrose patterns, and thereby arrive at an example of a nonperiodic pattern in one dimension. Inflation consists of replacing each A by AB and each B by A, and deflation consists of replacing each AB by A and each A by B; inflation and deflation undo each other on musical sequences. Call a sequence

musical if it results from applying inflation to the sequence consisting of a single B. Then inflation and deflation preserve musicality: If we inflate or deflate a musical sequence, we get another musical sequence. Another way to think of this relationship is that a musical sequence is self-similar under inflation and deflation.

■ **43.** Let the lone B be considered the first stage of inflation. Show that at the nth stage of inflation, for $n \geq 3$, there are F_n (the nth Fibonacci number, section 19.1) symbols in the sequence, of which F_{n-1} are A's and F_{n-2} are B's. (*Hint:* Check it for $n = 1, 2, 3,$ and 4.)

44. Show that no musical sequence contains AAA or BB.

45. Show that no musical sequence ends in AA or in $ABAB$.

46. Show that apart from the lone sequence B, every musical sequence is an initial subsequence of all the succeeding musical sequences.

47. Slightly modified, deflation can be used to check whether a finite block of A's and B's can belong to a musical sequence or not. First, if the block has length greater than one, we may suppose that it begins with an A (why?). So at any stage of the deflation with a block beginning with B, we may add an initial A. Second, we add the additional deflation rule to replace an ending AA with BA. If at any stage of this modified deflation we arrive at two or more B's in a row, or three or more A's in a row, then the original block could not be part of a musical sequence. Otherwise, the original block will eventually deflate to a single symbol, at which point we conclude that the original block is a part of a musical sequence. Check the two blocks $ABAABABAAB$ and $ABAABABABA$.

48. From Exercise 46 we know that each application of inflation to a musical sequence simply extends it. By successive inflation, then, we build an infinite sequence. Show that as we approach this limiting sequence, the ratio of B's to A's tends toward the golden ratio ϕ.

49. Conclude from Exercise 48 that the sequence cannot be periodic, nor settle into a period after a finite "burn-in" period. Thus, the sequence is nonperiodic. (*Hint:* ϕ is not a rational number; that is, it cannot be represented as a ratio m/n of whole numbers m and n.)

50. Show that any finite block of A's and B's that occurs in the infinite sequence must occur over and over again (just as any patch of tiles in a Penrose pattern occurs infinitely often in the pattern). Thus, the infinite sequence is self-similar.

 # WRITING PROJECT

Get computer software to make some tilings of your own. Check for software at www.geom.uiuc. edu/software/tilings/TilingSoftware.html. Print out your tilings and describe, in a sentence or two each, how you made them.

SUGGESTED READINGS

LEE, KEVIN. *TesselMania! Deluxe.* Computer program for producing Escher-like tilings for Macintosh or Windows. Available from www.worldofescher.com/store/compaccs.html. Free PC demo downloadable at www.worldofescher.com/down/tessdemo.exe.

RANUCCI, ERNEST, and JOSEPH TEETERS. *Creating Escher-Type Patterns,* Creative Publications, Oak Lawn, Ill., 1977.

SCHATTSCHNEIDER, DORIS. *Visions of Symmetry: Notebooks, Periodic Drawings, and Related Work of M. C. Escher,* W. H. Freeman, New York, 1990.

SEYMOUR, DALE, and JILL BRITTON. *Introduction to Tessellations,* Dale Seymour Publications, Palo Alto, Calif., 1989. An excellent introduction to tessellations, including how to make Escher-like tessellations.

TEETERS, JOSEPH L. How to draw tessellations of the Escher type, *Mathematics Teacher,* 67 (1974): 307–310.

SUGGESTED WEB SITES

www.geom.uiuc.edu/software/tilings Lists programs for various platforms that allow the user to design tilings.

www.geocities.com/SiliconValley/Pines/1684/Penrose.html A Java applet to play with Penrose tiles.

www.geom.uiuc.edu/apps/quasitiler Interactive Web program QuasiTiler 3.0 that draws Penrose tilings and their generalizations in higher dimensions.

ftp://geom.uiuc.edu/software/download/KaleidoTile.html Interactive Window and Macintosh program that lets the user design tilings on the plane, the sphere, and the hyperbolic plane. Spherical tilings can be realized as polyhedra.

Your Money and Resources

This part concentrates on numerical patterns of growth and decline in the realms of finance, resources, and biology. The unifying concept is a population, whether of dollars, barrels of oil, or tons of fish.

How much interest will your savings account earn in the next year? How much will the payment be on your credit card, your car loan, or a home mortgage? How much would you need to save to pay for a child's college education or for your retirement? What will inflation do to your savings? How much should you pay for a stock? These are problems of daily life for which mathematics provides custom-tailored models. In Chapter 21, Savings Models, and Chapter 22, Borrowing Models, you become familiar with the mathematics and terminology of situations that you will face repeatedly.

Good mathematical models are often versatile and flexible, and the financial models of these chapters apply broadly to important problems in other areas of life. Growth of money at interest is like growth of biological populations. Inflation of a currency or depreciation of an asset is like the decay of a radioactive substance. Finding out how long a retirement "nest egg" will last is similar to determining how long before a nonrenewable resource, such as oil or coal, may be exhausted. Managing a trust fund, such as the endowment of a college, presents problems similar to management of a renewable biological resource, such as a forest or a fishery. In Chapter 23, The Economics of Resources, we explore these similarities, together with the profound effect that economic conditions can have on natural resources.

Finally, you will see the surprisingly large and puzzling consequences that very small changes can produce in a physical system or biological population as a result of behavior that mathematicians call chaos. ∎

Savings Models

How much interest will your savings account earn in the next year? How much difference does it make how the interest is calculated? How much should you save for a comfortable retirement in the face of inflation? In this chapter, we consider questions such as these and show how the underlying mathematical models can explain such phenomena as the 2001 crash of "dot-com" stocks and the effect of interest-rate changes on stock prices.

21.1 Arithmetic Growth and Simple Interest

When you open a savings account, your primary concerns are the safety and the growth of the "population" of your savings. Suppose that you deposit $1000 in an account that "pays interest at a rate of 10% annually." (This is an unrealistic rate. We use it because it makes the calculations simple.) Assuming that you make no other deposits or withdrawals, how much is in the account after 1, 2, or 5 years?

The $1000 is the **principal,** the initial balance of the account. At the end of one year, **interest** is added. The amount of interest is 10% of the principal, or

$$10\% \times \$1000 = 10 \times 0.01 \times \$1000 = 0.10 \times \$1000 = \$100$$

("Percent" means "per 100," so you can think of the symbol "%" as standing for "× 0.01.") So the balance at the beginning of the second year is $1100.

With **simple interest,** interest is paid only on the original balance, no matter how much interest has accumulated. With simple interest, for a principal of $1000 and a 10% interest rate, you receive $100 interest at the end of the first year; so at the beginning of the second year, the account contains $1100. But at the end of the second year, you again receive only $100; so at the beginning of the third year, the account contains $1200. In fact, at the end of each year you receive just $100 in interest.

You no doubt find this method for interest rather strange, as you are accustomed from your savings account to a different system of awarding interest (which we will consider shortly). In Chapter 22 (pp. 834–836), we will look at the role of simple interest in consumer loans. However, simple interest also has an important application in the financing of corporations and governments through bonds.

Bonds

Governments and firms use bonds to borrow money. A bond is an obligation to repay a specified amount of money at the end of a fixed term, with simple interest usually paid annually. For instance, on February 15, 2005, you could buy a 10-year U.S. Treasury note ("T-note," a kind of bond) for $10,000 that pays 4.000% simple interest every subsequent February through February 15, 2015. (Actually, a T-note pays half the amount every six months, but for simplicity we lump a year's two payments together.)

The bond can be resold at a higher price than $10,000 if interest rates drop. If interest rates instead rise, the bond becomes less valuable (in terms of selling price) but still earns $400 every February 15 and returns the $10,000 principal on February 15, 2015.

The formulas involving simple interest are themselves simple. For a principal P and an annual rate of interest r, the interest earned in t years is

$$I = Prt$$

and the total amount A accumulated in the account is

$$A = P(1 + rt)$$

EXAMPLE 1   Earnings on That Bond

Suppose that you bought one of those 10-year T-notes on February 15, 2005, for $10,000. How much interest would it earn by February 15, 2015, and what would be the total amount accumulated?

SOLUTION We have $P = \$10{,}000$, $r = 4.000\% = 0.04000$, and $t = 10$ years:

$$I = Prt = \$10{,}000 \times 0.04000 \times 10 = \$4000$$

and

$$A = P(1 + rt) = \$10{,}000 \times (1 + 0.04000 \times 10)$$
$$= \$10{,}000 \times (1.4000) = \$14{,}000$$

Apart from bonds and some consumer loans, simple interest is little used. However, we frequently observe the corresponding kind of growth, called *arithmetic growth*, in other contexts.

Arithmetic Growth

Arithmetic growth (also called **linear growth**) is growth by a constant amount in each time period.

For example, the population of medical doctors in the United States grows arithmetically, because the fixed number of medical schools each graduates the same total number of doctors each year (and the number of doctors dying is also fairly constant). The concept of linear growth has appeared already in the discussions of linear programming (Chapter 4) and linear regression (Chapter 6).

21.2 Geometric Growth and Compound Interest

What you probably expected to happen to the savings account discussed in the last section is that during the second year the account would earn interest of 10% not on the *initial* balance of $1000 (as with simple interest) but on the *new* balance of $1100. Then, at the end of the second year, 10% of $1100, or $110, would be added to the account.

Thus, during the second year you would earn interest on both the principal of $1000 and on the $100 interest that was earned during the first year. Interest that is paid on both the principal and on the accumulated interest is known as **compound interest.** You receive more interest during the second year than during the first; that is, the account grows by a greater amount during the second year. At the beginning of the third year the account contains $1210, so at the end of the third year you receive $121 in interest. Again, this is larger than the amount you received at the end of the preceding year. Moreover, the increase during the third year,

$$\text{third-year interest} - \text{second-year interest} = \$121 - \$110 = \$11$$

is larger than the increase during the second year,

$$\text{second-year interest} - \text{first-year interest} = \$110 - \$100 = \$10$$

Thus, not only is the account balance increasing each year, but the amount added also increases each year.

Banks and savings institutions often compound interest more often than once a year—for example, quarterly (four times per year). With an interest rate of 10% per year and quarterly compounding, you get one-fourth of the rate, or 2.5%, paid in interest each quarter. The quarter (three months) is the **compounding period,** or the time elapsing before interest is paid.

Consider again a principal of $1000. At the end of the first quarter, you have the original balance plus $25 interest, so the balance at the beginning of the second quarter is $1025. During the second quarter, you receive interest equal to 2.5% of $1025, or $25.63, so the balance at the end of the second quarter is $1050.63. Continuing in this manner, the balance at the end of the first year is $1103.81. [You should "read" all calculations in this chapter by confirming them on your calculator. Here and in later calculations, we neglect rounding when interest is posted to the account. In the case at hand, standard rounding (half a cent or greater is rounded up) would result in a balance of $1103.82.]

Even though the account was advertised as paying 10% interest (the *nominal rate*), the interest of $103.81 for the year is 10.381% of the principal of $1000. The 10.381% is the *effective rate* or *equivalent yield*. (We give a precise definition below.)

If interest is compounded monthly (12 times per year) or daily (365 times per year), the resulting balance is even larger. A comparison of yearly, quarterly, monthly, and daily compounding for an interest rate of 10% is shown in Table 21.1. We will shortly summarize results in a general formula.

From now on, we express all interest rates as decimals. For example, 10% is 0.10. To convert a percentage to a fraction, divide the percentage by 100 by moving the decimal point two places to the left. An interest rate of 0.10 is the same as 10%, or 10/100. An interest rate i is the same as $100i\%$.

Terminology for Interest Rates

We have seen that an account at a particular annual rate of interest can produce different amounts of interest, depending on how the compounding is done. To help prevent confusion on the part of consumers, the Truth in Savings Act establishes specific terminology and calculation methods for interest.

A **nominal rate** is any stated rate of interest for a specified length of time, such as a 3% annual interest rate on a savings account or a 1.5% monthly rate on a credit-card balance. By itself, such a rate *does not indicate or take into account whether or how often interest is compounded.*

Effective Rate

The **effective rate** is the actual percentage rate of increase for a length of time, *taking into account compounding*. It equals the rate of simple interest that would realize exactly as much interest over that length of time.

We saw that $1000 at an annual interest rate of 10% (a nominal rate) compounded quarterly for one year yields $103.81 in interest, which is 10.381% of the principal. Hence the effective annual rate is 10.381%. In other words, $1000 at simple interest of 10.381% for one year would earn exactly the same amount of interest.

TABLE 21.1	Comparing Compound Interest: The Value of $1000, at 10% Annual Interest, for Different Compounding Periods*				
Years	Compounded Yearly	Compounded Quarterly	Compounded Monthly	Compounded Daily	Compounded Continuously
1	1100.00	1103.81	1104.71	1105.16	1105.17
5	1610.51	1638.62	1645.31	1648.61	1648.72
10	2593.74	2685.06	2707.04	2717.91	2718.28

*Without rounding at posting of interest and neglecting leap years; the difference in most cases is no more than one cent.

When stated per year ("annualized"), the effective rate is called the **effective annual rate (EAR)**. For savings, it is also called the **annual percentage yield (APY)** or *annual equivalent yield.*

To keep these different rates straight, we use i for a nominal rate for a specified **compounding period**—a day, a month, or a year—*within which no compounding is done.* Because no compounding is done for shorter intervals than this period, the effective rate and the nominal rate are the same for the compounding period.

Rate Per Compounding Period

For a nominal annual rate r compounded n times per year, the rate per compounding period is

$$i = \frac{r}{n}$$

For that $1000 in savings at 10% compounded quarterly, we have $r = 10\%$ and $n = 4$, so $i = 2.5\%$ per quarter.

We denote the number of compounding periods by n. We use r only for an annual rate (nominal or effective) and t for the number of years.

To avoid confusion, we don't use the terminology *annual percentage rate.* That term has a special meaning, and is used just for loans (see Chapter 22, p. 839).

Geometric Growth

Here we look for the underlying mathematical pattern of compounding. For quarterly compounding, you have at the end of the first quarter

initial balance + interest = $1000 + $1000(0.025) = $1000(1 + 0.025) and at the end of the second quarter

$$\begin{aligned}
\text{initial balance} + \text{interest} &= \$1000(1 + 0.025) \\
&\quad + [\$1000(1 + 0.025)](0.025) \\
&= [\$1000(1 + 0.025)] \times (1 + 0.025) \\
&= \$1000(1 + 0.025)^2
\end{aligned}$$

The pattern continues in this way, so that you have $\$1000(1 + 0.025)^4$ at the end of fourth quarter. You use the calculator button marked $\boxed{y^x}$ to evaluate expressions like $(1.0125)^2$.

More generally, with initial principal P and interest rate i ($= 100\ i\%$) per compounding period, you have at the end of the first compounding period

$$P + Pi = P(1 + i)$$

This amount can be viewed as a new starting balance. Hence, in the next compounding period, the amount $P(1 + i)$ grows to

$$P(1 + i) + P(1 + i)i = P(1 + i)(1 + i) = P(1 + i)^2$$

The pattern continues, and we reach the following conclusion.

Compound Interest Formula

If principal P is deposited in an account that pays interest at rate i per compounding period, then after n compounding periods the account contains the amount

$$A = P(1 + i)^n$$

The amount added each compounding period is proportional to the amount present. This type of growth is called *geometric growth*.

Geometric Growth (Exponential Growth)

Geometric growth (also called **exponential growth**) is growth proportional to the amount present.

EXAMPLE 2   Compound Interest

Suppose that you have a principal of $P = \$1000$ invested at 10% nominal interest per year. Using the compound interest formula $A = P(1 + i)^n$, you can determine the amount in the account after 10 years—once you know the compounding period.

SOLUTION

▶ *Annual compounding.* The annual rate of 10% gives $i = 0.10$, and after 10 years the account has

$$\$1000(1 + 0.10)^{10} = \$1000(1.10)^{10} = \$1000(2.59374)$$
$$= \$2593.74$$

▶ *Quarterly compounding.* Then $i = 0.10/4 = 0.025$, and after 10 years (40 quarters) the account contains

$$\$1000\left(1 + \frac{0.10}{4}\right)^{40} = \$1000(1.025)^{40} = \$1000(2.68506)$$
$$= \$2685.06$$

▶ *Monthly compounding.* Then $i = 0.10/12 = 0.008333$. The amount in the account after 10 years (120 months) is

$$\$1000\left(1 + \frac{0.10}{12}\right)^{120} = \$1000(1.00833333)^{120} = \$1000(2.70704)$$
$$= \$2707.04$$

These entries are found in the last row of Table 21.1.

In doing the calculations, use as many decimal places as possible. Don't round off until the final result. We will usually show intermediate results with enough decimal places to give the final result to the nearest cent. ■

Suppose that you want to make a one-time deposit of amount P that will grow to a specific amount A in n compounding periods from now by earning interest at a rate i per period. The quantities A, P, i, and n are related through the compound interest formula, $A = P(1 + i)^n$. The quantity P is called the *present value* of the amount A to be paid n compounding periods in the future.

Present Value

The **present value** P of an amount A to be paid in the future, after earning compound interest for n compounding periods at rate i per period, is

$$P = \frac{A}{(1 + i)^n}$$

EXAMPLE 3   Certificate of Deposit

A certificate of deposit (CD), like a bond, pays a fixed rate of interest for a term specified in advance. CDs are commonly sold for terms varying from one month to 10 years. (Unlike a bond, a CD cannot be resold. It can be cashed in, however, with a penalty for early withdrawal.) As I was researching this section, I realized that our son expects to enter college in five years. Based on current costs, we project that it could cost $15,000 for his first two years at the public university in our state capital and more if he goes to a private college. The best rate I could find for a five-year CD was 4.60%. How much would we need to set aside now in such a CD, compounded quarterly, to have $15,000 in five years?

SOLUTION We have $r = 0.0460$ and $i = 0.0115$. The compound interest formula $A = P(1 + i)^n$, with $n = 20$, gives

$$\$15,000 = P(1 + 0.0115)^{20}$$

Simplifying $(1 + 0.0115)^{20}$, we get

$$\$15,000 = P(1.25694925)$$

so that

$$P = \frac{\$15,000}{1.25694925} = \$11,993.66 \ \blacksquare$$

In some cases you know the principal, the current balance, and the interval of time, and you want to learn the interest rate. For example, money market funds typically report earnings to investors each month, based on interest rates that vary from day to day, but often do not report the average interest rate. We find the equivalent average *daily* effective rate, from which we calculate the annual yield.

The compound interest formula gives the end-of-month balance as $A = P(1 + i)^n$, where P is the balance at the beginning of the month, i is the average daily interest rate, and n is the number of days that the statement covers. So we have

$$\frac{A}{P} = (1 + i)^n$$

Taking the nth root gives

$$1 + i = \left(\frac{A}{P}\right)^{1/n} \qquad i = \left(\frac{A}{P}\right)^{1/n} - 1$$

EXAMPLE 4   Money Market Account

Suppose that the monthly statement from the fund reports a beginning balance (P) of $7373.93 and a closing balance (A) of $7382.59 for 28 days ($n$). What is the daily effective rate?

SOLUTION We thus have

$$i = \left(\frac{7382.59}{7373.93} \right)^{1/28} - 1 = (1.00117441)^{0.035714286} - 1 = 0.0000419194$$

Thus the average daily effective rate is 0.00419194%. Compounding daily for a year, we would have $(1 + 0.0000419194)^{365} = 1.01542$, for an annual yield of 1.54%. ■

Simple Interest Versus Compound Interest

The amounts in accounts paying interest at 10% per year with compound and simple interest are shown in Table 21.2 and in the graph in Figure 21.1, which dramatically illustrate the growth of money at compound interest compared with simple interest.

In some situations, the contrast is not so immediately dramatic at first glance. The amount of carbon dioxide in the atmosphere, which contributes to global warming, has been growing *superexponentially* since 1750 as a result of constantly increasing burning of fuels. This means that the amount is growing faster than exponentially. That is, the "interest rate," or growth rate, increases every year. The current growth rate is about 0.5% per year–a seemingly low rate of "interest." The international Kyoto Protocol that went into effect in early 2005 (without U.S. participation) aims to lower worldwide emissions. Even at a fixed lower level, though, the "population" of carbon dioxide atoms in the atmosphere would still increase–and global warming would intensify–just arithmetically, instead of super-exponentially. We are in effect "saving" carbon dioxide into the atmosphere, at an unknown future cost.

TABLE 21.2	The Growth of $1000: Compound Interest Versus Simple Interest	
Years	**Amount in Account from Compounded Interest**	**Amount from Simple Interest**
1	1100.00	1100.00
2	1210.00	1200.00
3	1331.00	1300.00
4	1464.10	1400.00
5	1610.51	1500.00
10	2593.74	2000.00
20	6727.50	3000.00
50	117,390.85	6000.00
100	13,780,612.34	11,000.00

FIGURE 21.1 The growth of $1000: compound interest and simple interest. The straight line explains why growth at simple interest is also known as linear growth.

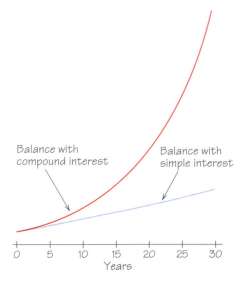

We noted earlier that the population of U.S. medical doctors grows as if it were at simple interest (arithmetic growth) because the same number of doctors graduate from medical school each year. On the other hand, general human populations tend to grow as if they were at compound interest (geometric growth), because the number of children born—the "interest"—increases as the population—the "balance"—increases.

The distinction between arithmetic growth and geometric growth is fundamental to the major theory of demographer and economist Thomas Robert Malthus (1766–1834). He claimed that human populations grow geometrically but food supplies grow arithmetically, so that populations tend to outstrip their ability to feed themselves (see Spotlight 21.1).

The situation of nuclear waste generated by a nuclear power plant is more complicated. The absolute volume of waste added each year depends on the fixed size and output of the power plant, not on the growing amount of waste in storage. Hence the volume of waste grows arithmetically. What about the total amount of radioactive material in the storage dump? The waste is a mixture of radioactive and nonradioactive substances. Over time, the radioactive ingredients decay slowly into nonradioactive ones. While the radioactivity of waste already in storage is decreasing, new amounts of radioactive material are being added each year. The situation requires a hybrid model that incorporates positive arithmetic growth (adding to the dump) accompanied by negative geometric growth (radioactive decay). The situation is like turning on the faucet in the bathtub while leaving the drain hole open a little. The faucet determines how fast water runs in, the height of the water determines how fast it runs out, and those two rates determine what happens to the volume of water in the tub.

SPOTLIGHT 21.1 Thomas Robert Malthus

Thomas Robert Malthus (1766–1834), a nineteenth-century English demographer and economist, based a well-known prediction on his perception of the different patterns of growth of the human population and growth of the "population" of food supplies.

He believed that human populations increase geometrically but food supplies increase arithmetically—so that increases in food supplies will eventually be unable to match increases in population. He concluded, however, that over the long run there would be restrictions on the natural growth of human populations, too, including war, disease, and starvation—hardly an optimistic forecast and doubtless responsible for the dreary image associated with his views.

Thomas Robert Malthus
(The Granger Collection, New York.)

21.3 A Limit to Compounding

The rows in Table 21.1 show a trend: More frequent compounding yields more interest. But as the frequency of compounding increases, the interest tends to a limiting amount, shown in the far right column.

Why is this so? Basically, because the extra interest from more frequent compounding is *interest on interest*. For example, in the first row of Table 21.1, the $3.81 extra interest from compounding quarterly is interest on the $100 yearly interest. The $3.81 is less than 10% of the $100 because the $100 interest is not on deposit for the whole year. Just part of it is credited to the account (and begins earning interest) at the end of each quarter. As compounding is done more and more often, smaller and smaller amounts of interest on interest are added.

How can we determine the limiting amount? Let's suppose that the initial balance is $1 and that we keep track at all stages of even the smallest fractions of a dollar.

For an interest rate i per compounding period, a principal of $1 grows to $(1 + i)^n$ in n periods, so the interest earned on that $1—which is the effective rate of interest for n periods—is given by:

Formula for Effective Rate

$$\text{effective rate} = (1 + i)^n - 1$$

Mostly, we will be interested in the effective rate on an annual basis. For a nominal *annual* interest rate r compounded n times, the interest rate per compounding period is $i = r/n$, and an amount of $1 grows in one year to

$$\left(1 + \frac{r}{n}\right)^n$$

The *annual* effective rate of interest (the annual percentage yield, or APY) is the amount of interest earned

$$\left(1 + \frac{r}{n}\right)^n - 1$$

divided by the original principal. Since that principal is $1, we have:

Formula for Annual Percentage Yield (APY)

$$\text{APY} = \left(1 + \frac{r}{n}\right)^n - 1$$

EXAMPLE 5   Finding the Annual Percentage Yield (APY)

With a nominal annual rate of 6% compounded monthly, what is the APY?
SOLUTION

$$\left(1 + \frac{0.06}{12}\right)^{12} - 1 = 0.0617 = 6.17\% \quad \blacksquare$$

Let's see what happens with the crazy interest rate of 100% per year compounded n times per year. This crazy rate makes the numbers simple (it also happened in Bolivia in the 1980s). Later we examine interest rates closer to those in stable economies. For an initial balance of $1, the amount at the end of one year—from the compound interest formula, with $P = \$1$ and $i = 100\%$—is

$$A = \$1 \times \left(1 + \frac{100\%}{n}\right)^n = \$\left(1 + \frac{1.00}{n}\right)^n$$

As n increases, this amount, which is just $(1 + 1/n)^n$, gets closer and closer to a special number called $e \approx 2.71828$ (see Spotlight 21.2). This is illustrated in Table

The number *e* is similar to the number π in several respects. Both arise naturally, π in finding the area and circumference of circles, and *e* in compounding interest continuously (*e* is also the base for the system of "natural" logarithms). In addition, neither is rational (expressible as the ratio of two integers, such as 7/2) or even algebraic (the solution of a polynomial equation with integer coefficients, such as $x^2 = 2$); we say that they are *transcendental* numbers. Finally, no pattern has ever been found in the digits of the decimal expansion of either number.

21.3, where the dots (ellipses) indicate that more decimal places follow. As *n* is made larger and larger, the limiting amount is e^r, and the interest method is called **continuous compounding**. The effective rate (APY, EAR) is ($e^r - 1$). (You can calculate powers of *e* using the $\boxed{e^x}$ button on your calculator. On some calculators, this button is the 2nd function of the button marked $\boxed{\text{LN}}$ or $\boxed{\ln\ x}$. For example, to calculate $e^{0.10}$, push $\boxed{\text{2nd}}$, push $\boxed{\ln\ x}$, and enter 0.10. You get 1.105170918.)

Continuous Compounding   EXAMPLE 6

For $1000 at an annual rate of 10%, compounded *n* times in the course of a single year, what is the balance at the end of the year?

TABLE 21.3	Yield of $1 at 100% Interest, Compounded *n* Times per Year
n	$\left(1 + \frac{1}{n}\right)^n$
1	2.0000000 . . .
5	2.4883200 . . .
10	2.5937424 . . .
50	2.6915880 . . .
100	2.7048138 . . .
1,000	2.7169239 . . .
10,000	2.7181459 . . .
100,000	2.7182682 . . .
1,000,000	2.7182818 . . .
10,000,000	2.7182818 . . .

SOLUTION It is

$$\$1000\left(1 + \frac{0.10}{n}\right)^{n}$$

This quantity gets closer and closer to $\$1000e^{0.1} = \$1105.17\ldots$ as the number of compoundings n is increased. No matter how frequently interest is compounded–daily, hourly, every second, infinitely often ("continuously")–the original $\$1000$ at the end of one year cannot grow beyond $\$1105.17$. The values after 5 and 10 years are shown in the lower rows of Table 21.1. ■

⋮⋮⋮⋮ Continuous Compound Interest Formula

For a principal P deposited in an account at a nominal annual rate r, compounded continuously, the balance after t years is

$$A = Pe^{rt}$$

We illustrate with $\$1000$ at 10%. For one year, we have $t = 1$ and

$$A = \$1000e^{0.10} = \$1000(1.10517) = \$1105.17$$

To find the amount in the account after 5 years, we have $t = 5$:

$$A = \$1000e^{(0.10)(5)} = \$1000e^{0.5} = \$1000(1.64872) = \$1648.72$$

exactly as shown in the rightmost column of Table 21.1.

It makes virtually no difference whether compounding is done daily or continuously over the course of a year. Most banks apply a daily periodic rate (based on compounding continuously) to the balance in the account each day and post interest daily (rounded to the nearest cent). The daily nominal rate is $r/365$, so each day the balance of the account is multiplied by $e^{r/365}$, the daily effective rate. Except for the rounding in posting interest, the effect is the same as continuous compounding throughout the year, because the compound interest formula gives $A = P(e^{r/365})^{365}$, which is the same as Pe^{r} from the continuous interest formula.

Also, it makes virtually no difference whether the bank treats a year as

▶ 365 days with daily nominal interest rate $r/365$ (the *365 over 365 method*)– this is the usual method for daily compounding; or

▶ 360 days with daily interest rate $r/360$ (the *360 over 360 method*)–this is the usual method for loans with equal monthly installments, because 360 is

TABLE 21.4	Comparing Methods of Compounding Interest				
Method	Compounding Periods per Year	Rate per Period	Formula per One Year	Effective Rate	Effective Rate for $r = 5\%$
360 over 360, daily	360	$r/360$	$(1 + \frac{r}{360})^3)^{360}$	$(1 + \frac{r}{360})^{360} - 1$	5.12674%
365 over 365, daily	365	$r/365$	$(1 + \frac{r}{365})^{365} - 1$	$(1 + \frac{r}{365})^{365} - 1$	5.12675%
Continuous			Pe^r	$e^r - 1$	5.12711%

evenly divisible into 12 equal "months" of 30 days. The daily interest is greater than for the 365 over 365 method, but there are fewer days in the year.

Table 21.4 gives a comparison of different interest methods.

21.4 A Model for Investment

The compound interest formula tells the fate over time of a single deposited amount, but another common question that arises in finance is: What size deposit do you need to make regularly in an account with a fixed rate of interest, to have a specified amount at a particular time in the future?

This question is important in planning for a major purchase in the future or accumulating a retirement nest egg. Later, in Chapter 22, we apply the same concepts and formula to paying off a mortgage and making installment payments on a car.

A Savings Plan

EXAMPLE 7

A graduate at her first job saves $100 per month, deposited directly into her credit union account on payday, the last day of the month. The account earns 1.8% per year, compounded monthly. How much will she have at the end of five years, assuming that the credit union continues to pay the same interest rate?

SOLUTION Note that she makes the first deposit at the end of the first month and the last deposit at the end of the sixtieth month. The monthly interest rate is $i = r/12 = 0.018/12 = 0.0015$.

It's easier to look at the deposits in reverse time order. The last deposit is made on the last day of the five years, so it earns no interest and contributes just $100 to the total.

The second-last deposit earns interest for one month, contributing $100(1 + i)$. Similarly, the third-last contribution is on deposit for two months, contributing $100(1 + i)^2$.

Continuing in the same way, we find that the first deposit earns interest for 59 months and contributes $\$100(1 + i)^{59}$. The total of all of the contributions is

$$\$100 + \$100(1 + i)^1 + \$100(1 + i)^2 + \cdots + \$100(1 + i)^{59}$$
$$= \$100[1 + (1 + i)^1 + (1 + i)^2 + \cdots + (1 - i)^{59}]$$

This expression is known as a **geometric series,** because the successive terms have geometric growth: Each succeeding term is a constant—in this case, $(1 + i)$—times the preceding term. For the sum of such a series with general ratio x, we have:

Formula for Sum of a Geometric Series

$$1 + x + x^2 + x^3 + \cdots + x^{n-1} = \frac{x^n - 1}{x - 1}$$

That this formula works for all x (except $x = 1$) can be confirmed by multiplying both sides $(x - 1)$ and watching terms on the left cancel. (You should do this confirmation for $n = 4$.)

In our example, we have $x = 1 + i$, and the formula becomes

$$1 + (1 + i)^1 + (1 + i)^2 + \cdots + (1 + i)^{n-1} = \frac{(1 + i)^n - 1}{i}$$

We have $n - 1 = 59$, or $n = 60$ months, and $i = 0.0015$, the interest rate per month. The total accumulation after five years is

$$A = \$100\left[\frac{(1 + 0.0015)^{60} - 1}{0.0015}\right] = \$6273.37$$

For a uniform deposit of d per compounding period (deposited at the end of the period) and an interest rate i per period, the amount A accumulated after n periods is given by the *savings formula*:

Savings Formula

$$A = d\left[\frac{(1 + i)^n - 1}{i}\right]$$

The savings formula involves four quantities: A, d, i, and n. If any three are known, the fourth can be found. A common situation is for A, i, and n to be known, with d (the regular payment) to be found, because the practical concern for most people is how much their monthly payment will be.

Sometimes the purpose of a savings plan is to accumulate a fixed sum by a certain date. Such savings plans are called **sinking funds,** because you sink money into them.

Sinking Fund

EXAMPLE 8

In the earlier example of putting aside money for our son's college education, we found that we would need to put away a hefty sum now to meet the needs of just his first two years of college five years from now. Well, we should have started saving earlier—for example, when he was born. Then we would have had 18 years to accumulate the money. How much would we have had to save regularly each month to accumulate $15,000 over the 18 years with an account earning a steady 5% per year, compounded monthly?

SOLUTION The monthly rate is $i = 5\%/12 = 0.004166667$. Applying the savings formula

$$ A = d\left[\frac{(1 + i)^n - 1}{i}\right] $$

with $A = \$15{,}000$, $i = 0.05/12 = 0.004166667$, and $n = 12 \times 18 = 216$, we get

$$ \$15{,}000 = d\left[\frac{(1 + \frac{0.05}{12})^{216} - 1}{\frac{0.05}{12}}\right] = d(349.20) $$

So $d = \$15{,}000/349.20 = \$42.96$. This sounds like a manageable amount to contribute, but it doesn't take into account inflation, nor costs beyond the first year, nor the higher cost of a private college. In the next section, we investigate how to take inflation into account. ■

Sinking Fund with Daily Interest

EXAMPLE 9

We illustrate how much—actually, how little—difference daily compounding makes compared with monthly compounding. What is the monthly contribution to the sinking fund with the same interest rate of 5% per year but compounded daily instead of monthly?

SOLUTION For simplicity, we calculate as if each month has 30 days and the savings institution uses the 360 over 360 method. The daily rate is $5\%/360 = 0.000138889$, so the effective monthly rate (for a 30-day month) is

$$ i = (1.000138889)^{30} - 1 = 0.004175072 $$

We have $A = \$15,000$, $i = 0.004175072$, and $n = 12 \times 18 = 216$. Applying the savings formula, we get

$$\$15,000 = d\left[\frac{1.004175072^{216} - 1}{0.004175072}\right] = d(349.56)$$

So $d = \$15,000/349.56 = \$42.91$. That is only a nickel less per month than with monthly compounding! ■

Saving for Retirement (It's Never Too Early to Start)

Financial advisers stress the importance of beginning early to save for retirement. Many firms offer a *401(k) plan* (named after a section of law regulating pensions), which allows an employee to make monthly contributions to a retirement account. The plan has the advantage that income tax on the contributions is deferred until the employee withdraws the money during retirement. That means, for example, that an employee making a $100 monthly contribution may see a reduction in take-home pay of only $75 or less, since taxes are not withheld on the contribution.

Sometimes a company's own pension plan consists of just contributing to the employee's individual 401(k) account either in cash or in the form of company stock. In early 2002, the bankruptcy of Enron Corporation resulted in thousands of its employees losing almost their entire retirement savings. Those savings consisted largely of Enron stock contributed by Enron, which fell from a value of $90 per share to under $1 per share in just a couple of months. The Enron bankruptcy illustrated how unwise it is for most of an employee's retirement fund to consist of stock in just one company, particularly if–as was the case for Enron–the employee is not free to sell the stock contributed by the employer. Even more people lost retirement savings and jobs when the stock of WorldCom declined more than 99% in value by midsummer 2002, after news emerged of prodigious financial fraud by its management.

EXAMPLE 10   Retirement Fund

Suppose that you start a 401(k) plan when you turn 23 and contribute $50 at the end of each month until you turn 65 and retire. Suppose (unlike some Enron employees) you put *your* contributions into a very safe long-term investment that returns a steady 5% annual interest compounded monthly. How much will be in your fund at retirement?

SOLUTION   Apply the savings formula with $d = \$50$, $i = 0.05/12 = 0.0041666667$, and $n = 12 \times (65 - 23) = 504$. We get

$$A = 50\left[\frac{(1.0041666667)^{504} - 1}{0.0041666667}\right] = \$50 \times 1711.348689 = \$85,567.43$$

At first glance, that may seem like a lot of money, but it is not so much if that's all you have to live off for the rest of your life (of course, there is also Social Security). In the exercises we explore the effects of saving more each month, getting a higher interest rate, saving on taxes, and—especially—having inflation erode the value of your savings. ■

A sinking fund is the reverse of an **annuity.** With a sinking fund, you make a specified number of payments at equal intervals of time. With an annuity, you receive such payments. Annuities are a common way for lotteries to pay out grand prizes (you probably won't have the pleasure of one of these!) and for retirees to receive funds saved up for retirement (you likely will have such an option). We examine an example of each in Chapter 22 (pp. 847–850).

21.5 Exponential Decay and the Consumer Price Index

Exponential Decay

In times of economic inflation, prices increase. When the rate of inflation is constant, the compound interest formula can be used to project prices.

Inflation

EXAMPLE 11

Suppose that there is constant 3% annual inflation from mid-2006 through mid-2010. What will be the projected price in mid-2010 of an item that costs $100 in mid-2006?

SOLUTION The compound interest formula applies with $P = \$100$, $r = 3\%$, and $n = 4$. The projected price is $A = P(1 + r)^n = \$100(1 + 0.03)^4 = \$112.55$. ■

During constant-rate inflation, prices grow geometrically (exponentially) and the value of the dollar goes down geometrically.

Exponential Decay

Exponential decay is geometric growth with a negative rate of growth.

Let a (for "additional") represent the rate of inflation; what costs $1 now will cost $\$(1 + a)$ this time next year. For example, if the inflation rate were $a = 25\%$, then what costs $1 now would cost $1.25 this time next year. A dollar next year would buy only 0.8 (= 1/1.25) times as much as a dollar buys today. In other words, a dollar next year would be worth only $0.80 in today's dollars—by next year, a dollar would have lost 20% of its purchasing power. We say that the present value (p. 801) of receiving a dollar next year is $0.80. Notice that although the inflation rate is 25%, the loss in purchasing power is 20%. For a general inflation rate a, a dollar a year from now will buy only a fraction of what a dollar today can buy.

> ## Present Value of a Dollar a Year from Now with Inflation Rate *a*
>
> $$\frac{\$1}{1 + a} = \$1 - \frac{\$a}{1 + a}$$

In other words, a dollar a year from now is worth $[1 − a/(1 + a)]$ today, and the loss in purchasing power is the fraction $a/(1 + a)$. (You should calculate what these expressions become for $a = 25\%$.) The quantity $i = −a/(1 + a)$ behaves like a negative interest rate. We can use the compound interest formula to find the value of P dollars n years from now as $A = P(1 + i)^n = P(1 − [a/(1 + a)])^n$.

The actual posted price of an item, at any time, is said to be in **current dollars.** That price can be compared with prices at other times by converting all prices to **constant dollars,** dollars of a particular year.

EXAMPLE 12   Deflated Dollars

Suppose that there is 25% annual inflation from mid-2006 through mid-2010. What will be the value of a dollar in mid-2010 in constant mid-2006 dollars? The inflation figure is unrealistic (we hope!), but it makes the calculations easy, so you can focus on the ideas.

We have $a = 0.25$, so $i = −a/(1 + a) = −0.25/1.25 = −0.20$. This, not 25%, is the negative interest rate, the rate at which the dollar is losing purchasing power. We have $n = 4$ years, so the value of $1 four years from mid-2006 in 2006 dollars will be

$$\$1(1 + i)^4 = \$(1 − 0.20)^4 = (0.80)^4 = \$0.41 \quad \blacksquare$$

In Example 11, we may think of the value of the dollar as "depreciating" 20% per year. Depreciation of the value of equipment is similar.

EXAMPLE 13   Depreciation

If you bought a car at the beginning of 2006 for $12,000 and its value in current dollars depreciates steadily at a rate of 15% per year, what will be its value at the beginning of 2009 in current dollars?

SOLUTION We have $P = \$12,000$, $i = −0.15$, and $n = 3$. The compound interest formula gives

$$A = P(1 + i)^n = \$12,000(1 − 0.15)^3 = \$7369.50$$

The Consumer Price Index

In our preceding model, we supposed that inflation stayed constant over a period of time. That is not generally the case. However, based on regular measures of inflation, we can determine the equivalent today of a price in an earlier year or how much a dollar in that year would be worth today.

The official measure of inflation is the Consumer Price Index (CPI), prepared by the Bureau of Labor Statistics. Here we describe and use the CPI-U, the index for all urban consumers, which covers about 80% of the U.S. population and is the index of inflation that is usually referred to in newspaper and magazine articles.

Each month, the Bureau of Labor Statistics determines the average cost of a "market basket" of goods, including food, housing, transportation, clothing, and other items. It compares this cost with the cost of the same (or comparable) goods in a base period. The base period used to construct the CPI-U is 1982–1984. The index for 1982–1984 is set to 100, and the CPI-U for other years is calculated by using the proportion

$$\frac{\text{CPI for other year}}{100} = \frac{\text{cost of market basket in other year}}{\text{cost of market basket in base period}}$$

For example, the cost of the market basket in 1976 (in 1976 dollars) was 0.569 times the cost in 1982–1984 (in 1982–1984 dollars), so the CPI for 1976 is 100×0.569, or 56.9.

Table 21.5 shows the average CPI for each year from 1913 through 2004, with estimates for 2005 and 2006. This table can be used to convert the cost of an item in dollars for one year to what it would cost in dollars in a different year, using the proportion

$$\frac{\text{cost in year A}}{\text{cost in year B}} = \frac{\text{CPI for year A}}{\text{CPI for year B}}$$

The Price of Our House and the Value of a Dollar

EXAMPLE 14

Where my family and I live, housing is relatively inexpensive. We bought our house in mid-1992 for $133,000 (close to the median price of U.S. housing at that time). What would be the equivalent cost in mid-2006 dollars?

SOLUTION We see from Table 21.5 that the CPI for 1992 is 140.3 and the CPI for 2006 is estimated to be 200.5. The table gives the average value for each year, which is very close to the value at midyear. Month-by-month values are available at the Bureau of Labor Statistics Web site.

Using the proportion, we have

$$\frac{\text{cost in 2006}}{\text{cost in 1992}} = \frac{\text{CPI for 2006}}{\text{CPI for 1992}}$$

or

$$\frac{\text{cost in 2006}}{\$133,000} = \frac{200.5}{140.3}$$

TABLE 21.5	U.S. Consumer Price Index (1982–1984 = 100)								
–	–	1931	15.2	1951	26.0	1971	40.5	1991	136.2
–	–	1932	13.7	1952	26.6	1972	41.8	1992	140.3
1913	9.9	1933	13.0	1953	26.7	1973	44.4	1993	144.5
1914	10.0	1934	13.4	1954	26.9	1974	49.3	1994	148.2
1915	10.1	1935	13.7	1955	26.8	1975	53.8	1995	152.4
1916	10.9	1936	13.9	1956	27.2	1976	56.9	1996	156.9
1917	12.8	1937	14.4	1957	28.1	1977	60.6	1997	160.5
1918	15.1	1938	14.1	1958	28.9	1978	65.2	1998	163.0
1919	17.3	1939	13.9	1959	29.1	1979	72.6	1999	166.6
1920	20.0	1940	14.0	1960	29.6	1980	82.4	2000	172.2
1921	17.9	1941	14.7	1961	29.9	1981	90.9	2001	177.1
1922	16.8	1942	16.3	1962	30.9	1982	96.5	2002	179.9
1923	17.1	1943	17.3	1963	30.6	1983	99.6	2003	184.0
1924	17.1	1944	17.6	1964	31.0	1984	103.9	2004	189.5
1925	17.5	1945	18.0	1965	31.5	1985	107.6	2005 (est.)	195.0
1926	17.7	1946	19.5	1966	32.4	1986	109.6	2006 (est.)	200.5
1927	17.4	1947	22.3	1967	33.4	1987	113.6		
1928	17.1	1948	24.1	1968	34.8	1988	118.3		
1929	17.1	1949	23.8	1969	36.7	1989	124.0		
1930	16.7	1950	24.1	1970	38.8	1990	130.7		

Note: This the CPI-U index, which covers all urban consumers, about 80% of the U.S. population. Each index is an average for all cities for the year. The basis for the index is the period 1982–1984, for which the index was set equal to 100. For each year, the figure is the average during the year, hence is usually close to the value at midyear.

Source: http://stats.bls.gov/

so that

$$\text{cost in 2006} = \$133{,}000 \times \frac{200.5}{140.3} = \$133{,}000 \times 1.4290805$$

$$= \$190{,}067.71 \approx \$190{,}000$$

That's what our house would sell for if its price exactly matched inflation. However, in many parts of the country, the cost of housing has far outstripped inflation; where we live, the reverse has been true—our house could not sell for nearly so much.

The 1.42908 is the *scaling factor* for converting 1992 dollars to 2006 dollars. What we are observing is a proportion, or *numerical similarity*, between 1992 dollars and 2006 dollars, analogous to the geometric similarity of Chapter 18 (pp. 664–665). To convert from 2006 dollars to 1992 dollars, we would multiply by $1/1.42908 = 0.699751 \approx 0.700$. ■

Spotlight 19.3 describes how the consumer price index is calculated, using the geometric mean (defined and introduced on p. 714).

21.6 Real Growth and Valuing Investments

It's natural to think that if your investment is growing at 6% per year and inflation is at 3% per year, then the real growth in the value (purchasing power) of your investment is $6\% - 3\% = 3\%$. Such, however, is not the case.

Let's suppose that you invest $500 for a year at 6%. At the beginning of the year, you have $500, which at $5 per pound could buy 100 pounds of steak. At the end of the year, you have $500(1 + 0.06) = \$530$, but steak now costs $\$5(1 + 0.03) = \$5.15$ per pound. How much steak would that buy? $530/$5.15 lb = 102.91 lb. In other words, in terms of purchasing power, or real gain, your investment has grown only 2.91%. This is not a great deal different from 3%, but it *is* different, and the difference is greater for higher rates of interest and inflation.

Consider an investment principal P and a market basket of goods of value m. Let the annual yield (rate of interest) of the investment be r and the rate of inflation be a. We calculate the rate of real growth g of the investment as follows.

At the beginning of the year, the investment would buy quantity $q_{old} = P/m$ of the market basket. At the end of the year, the investment would buy

$$q_{new} = \frac{P(1 + r)}{m(1 + a)}$$

market basket. Notice that the gain of r in the investment multiplies the principal by $(1 + r)$, while the erosion due to inflation divides the principal by $(1 + a)$. Here you see directly that the two influences on the investment have directly opposite effects.

The growth of the investment, relative to how many market baskets it could have bought originally, is

$$g = \frac{q_{new} - q_{old}}{q_{old}} = \frac{\frac{P(1 + r)}{m(1 + g)} - \frac{P}{m}}{\frac{P}{m}} = \frac{1 + r}{1 + a} - 1 = \frac{r - a}{1 + a}$$

In the last expression, the numerator is the difference of the two rates ($6\% - 3\%$ in our example), which is divided by a quantity greater than 1 if there is inflation. You should confirm that this formula gives 2.91% for $r = 6\%$ and $a = 3\%$.

One way to understand why this is the correct formula is to realize that the gain itself is not in original dollars but in deflated dollars.

The relationship between interest rate, inflation rate, and rate of real growth is called *Fisher's effect*, after the American economist Irving Fisher (1867–1947).

Valuing a Stock

A company sells shares of stock in the company to raise capital to make investments (such as in buildings and equipment) and expand the company. Buyers of the shares are part owners of the company. They can participate in selecting the management and can receive part of the profits of the company in the form of cash payments, called *dividends*. Shareowners can sell their shares to others, through traders or brokers or even over the Internet. The price for the shares is determined by current market forces.

Most individuals and business entities buy stocks as a way to invest; they try to pick stocks that will both pay dividends and rise in value. Some investors pay little attention to whether a company pays dividends or how much; they focus on "growth" stocks, ones whose prices will increase. However, a company that pays lower dividends than analysts expect usually sees the price of its stock go down. A stock is considered overvalued if the current price is higher than the potential for the company to earn profit, pay dividends, and expand the value of the business.

How can you tell if a stock is overvalued? The price of a stock depends on both logical factors ("fundamentals") and psychological factors.

We consider here logical factors, in what is known as the standard "rational" model of stock prices. This model does not take into account the possibility that herd psychology and frenzied trading may escalate the prices of stocks that never pay dividends to shareowners, as happened in 2000 with many "dot-com" stocks. Instead, the model visualizes the purchaser of a stock as buying a future "stream" of dividends (regular dividends each year or quarter). The formula for the sum of a geometric series figures importantly here.

The "rational" price to pay would be the present value of the future dividends, with their value "discounted" by a factor (the *discount rate*) that recognizes that getting a dollar a year from now is not as valuable as getting it today. The discount rate takes into account inflation, the rate of interest on alternative investments (such as government bonds, which involve no risk), and a measure of how risky the stock is (future dividends are not a sure thing, so the price of the stock should include a *risk premium* that reflects that uncertainty). This method of valuing a stock is called the *discounted cash-flow method*.

Let's buy a stock at the beginning of the year. How much should we pay? Suppose that last year's dividend was D, paid at the beginning of this year to those who owned it on a particular date last year (so, as new buyers, we don't get that dividend). Suppose that as the company expands and prospers, the annual dividend grows steadily at a rate g per year, so that the dividend paid next year is $D(1 + g)$. Suppose that the discount rate is r per year. This means that to find the present value of next year's dividend, we must divide it by $(1 + r)$, getting

$$D\,\frac{1 + g}{1 + r}$$

Present Value of Next Year's Dividend

EXAMPLE 15

Suppose that this year's dividend $D = \$5$, we expect it to go up by $g = 7\%$, and the discount rate is $r = 12\%$. So we expect next year's dividend to be $D(1 + g) = \$5.35$, but what was the present value of this future dividend when we bought the stock?

SOLUTION

$$D\frac{(1 + g)}{(1 + r)} = \$5\frac{(1 + 0.07)}{(1 + 0.12)} = \frac{\$5.35}{1.12} = \$4.78$$

Still, the price of the stock when we buy it should take the present value of future earnings into account—for next year and all later years.

Using the formula for a geometric series (on p. 810), we can calculate the total present value of all the dividends over the next n years. We spare you the algebra and simply tell you that the present value is

$$P = D(1 + g)\frac{x^n - 1}{g - r}$$

where

$$x = \frac{1 + g}{1 + r}$$

If $g > r$, then $x > 1$ and the value of P gets arbitrarily large as n increases. The stock would be a "gold mine," because its future earnings would outstrip inflation and any risk. ■

That is exactly what happened in the late 1990s with high-tech growth stocks. Buyers assumed that the growth rate g of dividends would remain forever higher than the prevailing discount rate r, so that they thought that $g > r$. Valuing high-tech stocks as gold mines led to a spiral of rising prices—a "bubble"—that collapsed in 2000–2001.

More conventionally, we have $g < r$, so $x < 1$ and x^n tends toward 0 as n gets large: Dividends from farther and farther in the future contribute less and less to the present value. For large n, we have the approximation

$$P \approx \frac{D(1 + g)}{r - g}$$

Let's examine this formula closely to see what it can tell us. First, a small change in the dividend D will lead to a correspondingly small change in the price P of the stock. For example, a 5% increase in D leads to a 5% increase in P.

However, small changes in either the anticipated rate of growth g or the discount rate r are *magnified* in the price of the stock, as we see in the next example.

⸬⸬⸬EXAMPLE 16   Higher Interest Rates

Suppose that the expected growth rate of dividends is $g = 7\%$, and the discount rate is $r = 12\%$, so that

$$P \approx D\frac{1+g}{r-g} = \$5 \times \frac{1.07}{1.05} = \$107$$

What might happen if the Federal Reserve raises interest rates by 0.25% (as happened repeatedly in 2005)?

SOLUTION The effect at first will be to raise r (loans become more expensive), but countereffects may diminish the rise (because decreased lending and spending leads to lesser inflation). If investors are confident that the economy will still grow and think that the net effect would be to increase r by, say, 0.125% (half the amount of the rise) without affecting g, then we have $r - g = 5.125\%$, $1/(r - g) =$ $1/0.05125 = 19.51$, and

$$P \approx D\frac{1+g}{r-g} = \$5 \times \frac{1.07}{1.05125} = \$104.39$$

That may not seem like much of a change from $107, but it means that the value of stocks should change by $(104.39 - 107)/107 = -2.44\%$. For the New York Stock Exchange with its Dow Jones Industrial Average at around 10,500 points (as through most of 2005), such a decrease would correspond to a dip of more than 256 points. Such decreases in value are one reason why investors are unhappy to see interest rates raised. ■

Of course, we cannot know precisely what future growth and an appropriate discount rate will be. The prices of stocks also depend on buyers' and sellers' expectations and psychology. However, the rational model of pricing can suggest when stocks may be overpriced or underpriced relative to earnings potential.

Other models lead to commonly used formulas for pricing bonds and other investments, such as financial derivatives.

A Model for Financial Derivatives

Financial markets trade everything from agricultural commodities (corn, orange juice, coffee) to precious metals (gold, silver), from national currencies (dollars, euros) to corporate stocks and bonds. In recent years, an increasing share of such trading has turned from the objects themselves to less tangible entities based on them, known as *financial derivatives*. Traditional derivatives include mutual funds, in which a purchaser owns shares in a company that itself owns and manages a portfolio of corporate stocks and bonds; commodities futures, in which an agricultural producer contracts to sell a commodity at a definite point in the future for a price agreed to in advance; and options.

An *option* allows (but does not require) an investor to acquire an asset (which may itself be a commodity or corporate stocks or bonds) during a particular interval at a price fixed in advance. Part of an employee's pay package may be options to acquire stock in the company. An option is valuable because if the ac-

The 1997 Nobel Memorial Prize in Economics was awarded to Robert C. Merton of Harvard University and Myron S. Scholes of Stanford University for their method of valuing financial derivatives. Together with the late Fischer Black in the 1970s, they formulated a mathematical model with appropriate assumptions and solved the resulting equation. At the time, Black was a mathematician with Arthur D. Little Consultants in Boston, Scholes was a professor of finance at MIT, and Merton was an assistant to the economist Paul Samuelson at MIT. All were under age 30 at the time.

The major achievement of Merton, Black, and Scholes was to incorporate variability of market prices into the formula for the value of an option. They realized that the risk involved in the market is already implicitly taken into account in the stock's current price and its volatility (tendency to vary), and they were able to find the right formulation for incorporating the risk into the value of the option. Black and Scholes actually modeled the rate of change of the option value and then used methods from calculus to work backward to calculate the value and the formula itself.

This fairly complicated formula is based on simplifying assumptions (no stock dividends, no transactions costs, fixed price volatility, fixed interest rate, efficient market) plus a major modeling assumption. That modeling assumption is that the change in the price of the stock, as a percentage of the price, has a fixed component proportional to elapsed time (price trends upward over time) plus a random component deriving from volatility (so the price also can jump around). The random component is modeled using the normal distribution of Chapter 5.

tual price A of the asset at some point during that interval is higher than the price C fixed in advance by the option, the investor could *exercise* the option by buying the asset at the option price C and selling it on the open market for A, making a profit of $A - C$.

Options may be bought and sold, but a vexing problem for many years was how to accurately assess the worth of an option. How can you determine the value of an option to buy 100 million euros (unit of European Union currency) at $1.45 per euro at any time between now and next Tuesday, or an option to buy 100 shares of Apple Computer at $50 per share before January 1?

One widely used pricing formula for valuing financial derivatives is the *Black–Scholes formula* (see Spotlight 21.3). In the next example, we examine a simpler model for valuing an option that still incorporates the main ideas.

Valuing an Option

EXAMPLE 17

Suppose that you are interested in an Internet company named Interjunk, but you're not sure whether or not to buy into it just now. You could buy a share of stock now for $100, or else buy an option to purchase a share of stock a year from now—no matter how high or low the price is then—at that same price of $100, called the *exercise price* (or *strike price*) of the option. The value of the option depends on the current price of the stock: The higher the price of the stock now, the more likely that it will be more than $100 a year from now, and hence the more valuable the option. As time goes by, if the price of the stock goes up, the value of the option does, too—but not quite as much, because there is still uncertainty about what will happen between then and expiry (when the option expires). What would be a fair price for such an option?

SOLUTION The answer depends on what you expect to happen to the price of the stock. Suppose (for simplicity, but unrealistically) that the consensus in the market is that there are only two possibilities: There is a 50% chance that the stock price a year from now will be $200, and a 50% chance that it will be $80.

If you buy a share of stock, the expected value of that share a year from now is $(0.50) \times \$200 + (0.50) \times \$80 = \$140$, for an expected profit of $\$140 - \$100 = \$40$.

If you buy an option for some cost C, you have a 50% chance of exercising it. Half the time you wind up with a share of stock that costs you only $100 but is worth $200, netting you $100; the other half of the time, the stock price is at $80 and the option is worth nothing. Either way, you paid $C for the option. Your expected profit is $(0.50) \times \$100 + (0.50) \times \$0 - \$C = \$50 - \$C$.

What should you be willing to pay for the option? If you pay less than $10, you expect on average to make a greater profit than the $40 you could make by buying the stock now; if you pay more than $10 for the option, you would be better off instead buying the stock now. So the value of the option is $10.

We could refine our model slightly to take into account alternative investments. Suppose that you can earn a risk-free 10% on U.S. Treasury obligations. Suppose further that you have $110 on hand and are going to invest it all in some combination of stock, "Treasuries" (T-notes), and options.

Strategy 1. Buy the stock ($100), put the rest into Treasuries ($10). You expect to realize $40 from the stock and $1 from the Treasuries, for $41.

Strategy 2. Buy an option ($10), put the rest into Treasuries ($100). You expect to realize $40 from the option and $10 from the Treasuries, for a total of $50.

Of course, the rest of the investors in the market can do the same calculations, and it is readily apparent that an option is worth more—hence, should be priced higher—than the $10 used in these calculations. In fact, an option is worth closer to $20. ■

⋮⋮⋮⋮⋮⋮ REVIEW VOCABULARY

Annual percentage yield (APY) The effective interest rate per year.

Annuity A specified number of (usually equal) payments at equal intervals of time.

Arithmetic growth Growth by a constant amount in each time period.

Compound interest Interest that is paid on both the original principal and the accumulated interest.

Compound interest formula Formula for the amount in an account that pays compound interest periodically. For an initial principal P and an effective rate i per compounding period, the amount after n compounding periods is $A = P(1 + i)^n$.

Compounding period The fundamental interval for compounding, within which no compounding is done. Also called simply *period*.

Constant dollars Costs are expressed in constant dollars if inflation or deflation has been taken into account by converting the costs to their equivalent in dollars of a particular year.

Continuous compounding Payment of interest in an amount toward which compound interest tends with more and more frequent compounding.

Current dollars The actual cost of an item at a point in time; inflation or deflation before or since then has not been taken into account.

e The base for continuous compounding, geometric (exponential) growth, and natural logarithms; $e = 2.71828. . . .$

Effective annual rate (EAR) The effective rate per year.

Effective rate The actual percentage rate, taking into account compounding. It equals the rate of simple interest that would realize exactly as much interest over the same period of time.

Exponential decay Geometric growth at a negative rate.

Exponential growth Geometric growth.

Geometric growth Growth proportional to the amount present.

Geometric series A sum of terms, each of which is the same constant times the previous term; that is, the terms undergo geometric growth.

Interest Money earned on a savings account or a loan.

Linear growth Arithmetic growth.

Nominal rate A stated rate of interest for a specified length of time; a nominal rate does not take into account any compounding.

Present value The value today of an amount to be paid or received at a specific time in the future, as determined from a given interest rate and compounding period.

Principal Initial balance.

Simple interest The method of paying interest only on the initial balance in an account, not on any accrued interest.

Sinking fund A savings plan to accumulate a fixed sum by a particular date.

✔ SKILLS CHECK

1. If a savings account pays 3% simple annual interest, how much interest will a deposit of $250 earn in 2 years?

(a) Less than $5

(b) Between $5 and $8

(c) Over $8

2. Simple interest is an example of

(a) linear growth.

(b) variable growth.

(c) constant growth.

3. If a bond matures in 3 years and will pay $10,000 at that time, what is the fair value of it today, assuming the bond has an interest rate of 6% compounded annually and there is no inflation?

(a) $8200.00

(b) $8396.19

(c) $8352.70

4. Which of the following pays more interest?

(a) 6% compounded annually

(b) 6% compounded monthly

(c) 6% compounded continuously

5. The 1.5% monthly rate on a credit-card balance is an example of

(a) an effective rate.

(b) a nominal rate.

(c) an adjusted rate.

6. If a single deposit is made into a compound interest certificate of deposit, the account

(a) earns interest only for the first period,

(b) earns the same amount of interest each period.

(c) earns more interest in each subsequent period.

7. If $800 is invested for one year at 6% compounded quarterly, the amount of interest earned is

(a) exactly $48.

(b) between $48 and $49.

(c) more than $49.

8. If you deposit $1000 at 6.2% simple interest, what is the balance after three years?

(a) $1186.00
(b) $1197.77
(c) $1224.63

9. Suppose you invest $250 in an account that pays 4.5% interest compounded quarterly. After 30 months, how much is in your account?

(a) $279.08
(b) $279.59
(c) $279.71

10. Suppose you deposit $15 at the end of each month into a savings account that pays 2.5% interest compounded monthly. After a year, how much is in the account?

(a) $184.50
(b) $182.50
(c) $182.08

11. What is the APY for 5.90% compounded monthly?

(a) 5.90%
(b) 6.06%
(c) 6.08%

12. Which of the following is the most generous interest rate for a one-year CD?

(a) 6% simple interest
(b) 5.9% compounded annually
(c) 5.9% compounded continuously

13. The effective rate for 5% compounded daily is

(a) exactly 5%.
(b) slightly more than 5%.
(c) slightly less than 5%.

14. The number e is

(a) irrational.

(b) irrelevant.
(c) irrotational.

15. When $1000 is invested at 8% compounded continuously for 5 years, the balance is approximately

(a) $1491.
(b) $1204.
(c) $1400.

16. The value of e is approximately

(a) 1.414.
(b) 2.718.
(c) 3.14.

17. An example of a sinking fund is

(a) buying a savings bond when a child is born.
(b) depositing $100 on a child's annual birth date.
(c) withdrawing $100 from a savings account at the beginning of each year.

18. An example of exponential decay is

(a) the depreciation of factory equipment.
(b) a retirement annuity.
(c) the Consumer Price Index.

19. If a new car costs $18,000 and loses value at a rate of 20% per year, what is its value after 3 years?

(a) About $7200
(b) About $9200
(c) Less than $5000

20. If your investment is growing at a rate less than the rate of inflation,

(a) you have a positive real growth in your investment.
(b) you do not have a positive real growth in your investment.
(c) you do not have enough information to determine whether real growth is positive or negative.

EXERCISES   ■ Challenge   ◆ Discussion

The exercises below require a scientific calculator with buttons for powers $\boxed{y^x}$, exponential $\boxed{e^x}$, and natural logarithm $\boxed{\ln x}$.

Geometric Growth and Compound Interest

1. An often heard claim is that "the amount of information in the world doubles every three days."

Presumably the claim refers to the amount of data, which can be quantified in terms of number of bits. (A bit is the smallest unit of storage in a computer.) Show that the claim is absolutely preposterous by doing a little arithmetic and comparing your result with the estimated number of particles in the universe (10^{70}). In particular:

(a) Start with one bit of data and double the number of bits every third day. How long does it take to get past 10^{70}? (*Hint:* Don't just keep multiplying by 2 over and over. Convince yourself that since the amount of data increases by a factor of 2 every 3 days, then it increases by a factor of $2^2 = 4$ every 6 days, a factor of $4^2 = 16$ every 12 days, a factor of $16^2 = 256$ every 24 days, and so forth.)

(b) Part (a) involves a lot of multiplying by 2, even if you do it efficiently. Another approach is to use the fact that $2^{10} = 1024 \approx 1000$. Thus, the amount of data increases by a factor of more than 1000 every $3 \times 10 = 30$ days, or every month (except February, but the 31-day months make up for it). By when will the total surely be past 10^{70}?

2. An old legend tells of a wizard who agreed to save a kingdom provided that the king would agree to a "modest" reward. The wizard asked to be given merely as much grain as would put one kernel on the first square of a chessboard, two kernels on the second, four on the third, eight on the fourth, and so forth, up through the 64th square. The king agreed and the wizard saved the kingdom, but the king was completely unable to honor the agreement. Why? [*Hints:* Notice that $1 = 2^1 - 1$, and $(1 + 2) = 2^2 - 1$, and $(1 + 2 + 4) = 2^3 - 1$. Generalize to arrive at a total for the number of kernels. A kernel of rice is about a quarter of an inch long and about a 16th of an inch wide and a 16th of an inch high. So about a thousand kernels will fit in a cubic inch (you should verify this calculation). Calculate the total volume of kernels.]

3. You deposit $1000 at 3% per year. What is the balance at the end of one year, and what is the annual yield, if the interest paid is

(a) simple interest?
(b) compounded annually?
(c) compounded quarterly?
(d) compounded daily?

4. Repeat Exercise 3, but for $1000 at 6% per year.

5. *Zero-coupon bonds* are securities that pay no current interest but are sold at a substantial discount from redemption value. The difference between purchase price and redemption value provides income to the bondholder at the time of redemption or resale. If the annual interest rate is 3%, what should be the price of a zero-coupon bond that will pay $10,000 eight years from now? (Use daily compounding.)

6. Repeat Exercise 5, but for a current interest rate of 4% and a zero-coupon bond that will pay $10,000 five years from now.

7. Suppose that on the statement for your money market account this month, the initial balance was $7744.70, the statement was for 34 days, and the final balance was $7770.84. Calculate the annual yield.

8. Repeat Exercise 7, but for the previous month, which had an initial balance of $7722.54, a period of 27 days, and a final balance of $7744.70.

9. *The rule of 72* is a rule of thumb for finding how long it takes money at interest to double: If r is the annual interest rate, then the doubling time is approximately $72/(100r)$ years.

(a) Calculate the balance at the end of the predicted doubling time for each $1000, with annual compounding, for the small growth rates of 3%, 4%, and 6%.
(b) Repeat part (a) for the intermediate interest rates of 8% and 9%.
(c) Repeat part (a) for the larger interest rates of 12%, 24%, and 36%.
(d) What do you conclude about the rule of 72?

10. More frequent compounding yields greater interest, but with diminishing returns as the frequency of compounding is increased. For small interest rates, there is little difference in yield for compounding annually, quarterly, monthly, daily, or continuously. Investigating doubling times with continuous compounding leads to understanding why the rule of 72 of Exercise 9 works. Recall that for continuous compounding at annual rate r, the balance A at the end of m years is Pe^{rm} for an initial principal of P. Let t be the number of years that it takes for the initial principal to double. Then we have $2P = A = Pe^{rt}$, so $e^{rt} = 2$. Taking the natural logarithm of both sides yields $rt = \ln 2$, where ln stands for the natural logarithm, represented on a calculator by a button marked either $\boxed{\ln}$ or $\boxed{\text{LN}}$ (not $\boxed{\log}$ or $\boxed{\log_{10}}$, which stands for a different kind of logarithm). Using the button gives $\ln 2 = 0.693$. So we have $rt = 0.693$, from which we can determine t if we know r.

Calculate the doubling times for continuous compounding at 3%, 6%, and 9%, and compare them with those predicted by the rule of 72. What do you conclude? Why do you think people prefer a rule of 72 over a rule of 69.3?

A Limit to Compounding

11. Use your calculator to evaluate for $n = 1, 10, 100, 1000,$ and $1,000,000$:

(a) $\left(1 + \dfrac{1}{n}\right)^n$

(b) $\left(1 + \dfrac{2}{n}\right)^n$

(c) As n gets larger, what numbers are the expressions in parts (a) and (b) tending toward?

12. (Contributed by John Oprea of Cleveland State University.) Use your calculator to evaluate for $n = 1, 10, 100, 1000,$ and $1,000,000$:

(a) $\left(1 - \dfrac{1}{n}\right)^n$

(b) $\left(1 - \dfrac{2}{n}\right)^n$

(c) As n gets larger, what numbers are the expressions in parts (a) and (b) tending toward?

13. You have $1000 on deposit at your bank at an annual rate of 3%. How much interest do you receive after one year if the bank compounds

(a) continuously?
(b) daily, using the 360 over 360 method?
(c) daily, using the 365 over 365 method?

14. Suppose that you have a bank account with a balance of $4532.10 at the beginning of the year and $4632.10 at the end of the year. Your bank advertises "continuous compounding," but in fact compounds continuously over each 24-hour day and posts interest to accounts daily.

(a) What effective rate did you receive?
(b) What nominal rate is the calculation based on?
(c) What difference is there between what the bank is doing and true continuous compounding?

15. We saw in Table 21.4 that a nominal rate of 5% compounded continuously yields an effective annual rate of 5.12711%.

(a) What effective annual rate does a nominal rate of 4% yield with continuous compounding?
(b) The difference between the effective rate under continuous compounding, $e^r - 1$ (see Table 21.4), and the nominal rate r is $e^r - 1 - r$. You can't calculate this formula in your head, but you can approximate it closely with one that you can: $e^r - 1 - r \approx \frac{1}{2}r^2$. Thus, for $r = 4\% = 0.04$, the difference is $\frac{1}{2}(0.04)(0.04) = \frac{1}{2}(0.0016) = 0.0008 = 0.08\%$. So the effective rate is about 4.08%. Apply this formula to approximate the difference for a nominal rate of 5%, and compare the result with the 5.12711% in Table 21.4.

16. [Suggested by Arthur R. Segal, University of Alabama at Birmingham (retired).] We approximate the smaller difference between regular compounding and continuous compounding. With a nominal annual rate r over t years, continuous compounding yields e^{rt}, while compounding n times per year yields $(1 + \frac{r}{n})^{nt}$, for a difference of $D = e^{rt} - (1 + \frac{r}{n})^{nt}$ that can be approximated by

$$D \approx \frac{r^2 t e^{rt}}{2n + \dfrac{4r}{3} + \dfrac{r^2 t}{2}}$$

For a $1000 initial investment, calculate both the true difference and the approximate difference between continuous compounding and

(a) quarterly compounding at a nominal rate of 4% for 10 years.

(b) daily compounding at a nominal rate of 18% for 10 years.

A Model for Investment

17. Suppose that you want to save up $2000 for a trip abroad two years from now. How much do you have to put away each month in a savings account that earns 5% interest compounded monthly?

18. Repeat Exercise 17, except that you have found a better deal–7% interest compounded monthly.

19. Parents struggle for the first few years after their child is born but are finally able to start saving toward the child's college education (because they stop paying for day care) when the child goes to school at age 6. If they save $400 per month in a credit union account paying 5.5% interest compounded monthly, how much will they have for college expenses 12 years later?

20. Suppose that you save for retirement by contributing the same amount each month from your 23rd birthday until your 65th birthday, in an account that pays a steady 5% annual interest compounded monthly.

(a) How much will be in your fund at age 65 if you save $100 a month?

(b) How much will be in your fund if you get a steady return of 7.5% compounded monthly?

(c) How much will be in your fund if you get a steady return of 10% compounded monthly? (This is comparable to the average annual return of about 11% for all stocks on the New York Stock Exchange from 1950 to 2000.)

21. A colleague of mine feels that he will need $1 million in savings to afford to retire at age 65 and still maintain his current standard of living. A younger colleague, age 30, decides to begin saving for retirement based on that advice. How much

does the younger colleague need to save per month to have $1 million at retirement if the fund earns a steady 5% annual interest compounded monthly?

22. Many young people do not start saving right away for retirement, although by the time that they do, they may be earning more and thus be able to afford to save more each month.

(a) How much will be in your fund at age 65 if you don't start saving until age 35 and at that age start saving $100 per month in an account paying a steady 6% annual interest compounded monthly?

(b) Suppose instead that you have children young, pay for their college expenses, and finally start saving for retirement at age 45. How much do you have to save per month, with a steady return of 7.5% compounded monthly, to accumulate $250,000 by age 65?

■ **23.** Suppose that you are 25 and in a 25% bracket for federal income tax and a 7% bracket for state and local income taxes (in 2004 this corresponded to an income, beyond exemptions and deductions, of $29,000–$70,000 for a single person). This means that you pay a total of 32% in income tax on part of your income but a lower rate on the rest. Assume that instead of paying 32% on some income, you put it into a tax-deferred retirement account (TDA) as follows:

(a) Suppose that you are willing to commit to $100 a month less take-home pay. You realize that you don't pay income tax on the money that you put into the retirement plan, so you can actually put in more than $100 per month while reducing your take-home pay by only $100. How much can you put into the retirement fund each month?

(b) How much will be in your fund at age 65 if you can get a steady return of 7.5% compounded monthly?

(c) Suppose that when you turn 65 you withdraw the entire amount in your account and pay the deferred taxes that are owed on it, say a total of 32% (federal, state, and local combined). How much do you net?

■ **24.** We continue the tax-deferral considerations of the previous exercise.

(a) Suppose that instead of contributing to a tax-deferred plan, you take the money as income, pay 32% income tax on it, and deposit what remains into a savings account or safe investment that pays a steady 7.5% compounded monthly. (Note that compared with not putting away any money, your paycheck is reduced by just what you contribute, since you still must pay income tax on the $147.06.) How much will be in your account at age 65?

(b) Under another alternative, you take $100 per month, pay 32% income tax on it, and deposit what remains into a *Roth IRA* (individual retirement account). For this kind of retirement account, the interest earned is not taxed. Assuming the same savings account or safe investment that pays a steady return of 7.5% compounded monthly, how much will be in your account, tax-free, at age 65?

25. (a) Use the geometric series formula to find the sum of $\frac{1}{2} + \frac{1}{4} + \frac{1}{8} + \frac{1}{16} + \frac{1}{32}$.

(b) Find the more general sum $\frac{1}{2} + \frac{1}{4} + \frac{1}{8} + \frac{1}{16} + \cdots + \frac{1}{2^n}$.

(c) As n gets larger and larger, what does the sum in part (b) tend to?

26. (a) Use the geometric series formula to find the sum of $1 - \frac{1}{3} + \frac{1}{9} - \frac{1}{27} + \frac{1}{81}$.

(b) Find the more general sum $1 - \frac{1}{3} + \frac{1}{9} - \frac{1}{27} + \frac{1}{81} + \cdots \pm \frac{1}{3^n}$, where the $\pm$ denotes $+$ if n is odd and $-$ if n is even (to fit the pattern of the earlier terms).

(c) As n gets larger and larger, what does the sum in part (b) tend to?

27. Apart from certificates of deposit, returns on investments are rarely the same from year to year, as they vary with prevailing interest rates. How should you calculate an "average" rate of return over several years? Consider a mutual fund that delivers 100% return one year and loses 50% the next year. Calculate just the ordinary average (the arithmetic mean) to get $(100\% + (-50\%))/2 = 25\%$. That sounds good, but check what happens to a $1000 investment: It grows to $2000, then halves back to $1000—for a 0% gain. The customary way used in finance to calculate the "average" return is to use the geometric mean. If the initial value of the portfolio was P, and its value after n years is A, then the average annual rate of return is the value of r that solves $(1 + r)^n = A/P$, or $r = (A/P)^{1/n} - 1$.

(a) Use this formula to determine the average annual rate of return for a portfolio with returns of 15%, 7%, and -20% in three consecutive years.

(b) Is the average rate that the formula finds a nominal rate or an effective rate?

■ **28.** We continue the theme of Exercises 19 and 20 by comparing three kinds of investments for retirement: an ordinary after-tax investment, a tax-deferred investment [such as a tax-deferred annuity or an individual retirement account (IRA)], and a Roth IRA. Let an investment earn interest at a steady annual yield i and let your income (in whatever year you receive it) be taxed at rate t.

(a) Ordinary after-tax investment: Explain why if you earn $E, pay taxes on it, let what remains earn interest, and pay tax each year on that year's interest, the $E grows after n years to $E(1 - t) \times [1 + r(1 - t)]^n$.

(b) Ordinary IRA: Explain why if you earn $E, defer taxes on it, let it earn interest, and defer taxes on all the interest, then the $E grows after n years to $E(1 + r)^n(1 - t)$.

(c) Roth IRA: Explain why if you earn $E, pay taxes on it, let what remains earn interest, and pay no taxes on all the interest, the $E grows after n years to $E(1 - t)(1 + r)^n$.

(d) Which investment gives the best return after n years?

(e) The assumptions of constant interest rate and stable tax rate won't necessarily hold, because interest rates fluctuate (though you can lock in a long-term constant interest rate by buying a long-term bond or certificate of deposit) and the tax rate may change (with your income, your state of residence, and changes in tax laws). If your marginal tax rate (the rate you pay on one more dollar of income) is lower in one year than the tax rate you expect to pay in retirement, what kind of retirement investment is better for you that year? If you have a windfall one year and your marginal tax rate is higher that year than the tax rate that you expect to pay in retirement, what kind of retirement investment is better for you that year?

Exponential Decay and the Consumer Price Index

29. Suppose inflation proceeds at a level rate of 4% per year from mid-2006 through mid-2009.

(a) Find the cost in mid-2009 of a basket of goods that cost $1 in mid-2006.
(b) What will be the value of a dollar in mid-2009 in constant mid-2006 dollars?

30. Suppose you bought a car in mid-2006 for $10,000. If its value (in current dollars) depreciates steadily at 12% per year, what will its value (in current dollars) be in mid-2009?

31. For the car in Exercise 30, suppose that there is also 3% annual inflation from 2006 through 2012. What will the value of the car be in mid-2012 in inflation-adjusted (mid-2006) dollars?

32. (a) I bought my first vinyl record in 1965, at list price, for $4.98. How much would that be in 2006 dollars? How does that compare with the list price of a CD today?

(b) My father bought a Royal portable typewriter in 1940 for $40 (I have the sales slip). What would be the equivalent price in 2006 dollars?

33. (a) My first-semester college mathematics book cost $10.75 in 1962. What would the equivalent price be in 2006 dollars? How does that compare with what you paid for this book? (My book had black-and-white text and figures, with no photographs, color or otherwise.)
(b) In 1970, before the OPEC oil embargo, gasoline cost about 25 cents per gallon. In 1974,

after the embargo, it cost about 70 cents per gallon. What would the equivalent prices be in 2006 dollars? How do they compare with the price of gasoline today?

34. From Table 21.5, you can determine the rate of inflation from one year to the next. For example, you find the rate of inflation from 1991 to 1992 by subtracting the two index numbers and dividing by the earlier one: $(140.6 - 136.2)/136.2 = 0.032 = 3.2\%$. Similarly, knowing the rate of inflation, you can compute one index number from another.

(a) What was the rate of inflation from 1980 to 1981?
(b) For a 3% rate of inflation per year from 2003 on, what would the Consumer Price Index be in 2006?

35. (Suggested by Ed Barbeau's column "Fallacies, Flaws, and Flimflam" in *The College Mathematics Journal*.) Suppose that you get a pay raise of 10%, but in the meantime, there has been inflation of 20%—so in effect you have suffered a pay decrease in terms of what your salary will buy. What is the percentage decrease?

36. What is the present value of a $2000 raise now, over the course of 40 years more of working, if inflation is a steady 3% per year?

Refer to the following in doing Exercises 37 and 38.

A new assistant professor at a typical American liberal arts college starts at age 30 with a salary of $42,000, while colleagues retiring at age 65 make about twice that. One college gives annual pay raises of inflation plus one percentage point, plus a promotion raise (to associate professor) of $1500 after (usually) 7 years and another promotion raise (to full professor) of $1500 after (usually) another 7 years.

37. (Requires spreadsheet) Can a new assistant professor who starts now expect to be making the equivalent of $90,000 in today's dollars when she retires 35 years from now if inflation holds steady at

(a) 3%?
(b) 5%?

38. (Requires spreadsheet) Repeat Exercise 37, but suppose that you are the vice president for academic affairs at the college. Suggest a salary policy that would result in the new assistant professor, when she retires in 35 years, making the equivalent of

(a) $90,000 in today's dollars.

(b) $135,000 in today's dollars. (She would prefer that!–and hence she would be more likely to accept an offer to come work at your college.)

39. Surprise! Just for fun, one of your friends wrote your name on an Illinois State Lottery ticket, and you are the sole winner of $40 million! You discover, however, that you don't get the $40 million all at once. In fact, it is paid in 20 equal annual installments of $2 million each. All you get right away is the first installment of $2 million (minus 20% withheld against federal income tax due and whatever you think your friend deserves for the favor). So, what is the prize really worth to you? That depends on the rate of inflation over the years. Assume a constant rate of 3% inflation over the 19 years until your last payment and calculate the present value of your prize winnings by using the formula for present value combined with the formula for the sum of a geometric series. Do the calculation for a rate of interest of 4%.

Actually, the checks will come not from the state of Illinois but from an insurance company from which Illinois purchases an annuity, whose price depends on current long-term interest rates.

40. Repeat Exercise 39, but for an interest rate of 6%.

41. (Suggested by Terence Blows, Northern Arizona University, Flagstaff, Ariz.) (Spreadsheet helpful) Your roommate (a business major) has already planned her retirement and started funding it in 2005. She plans to retire in 2040 at age 55 on $100,000 per year in 2040 dollars, living on just the interest on her investments. Assume that she realizes a steady 7.2% and assume a steady 4% annual inflation.

(a) What must the size of her nest egg be, and what should her quarterly investment be over the 35 years, to achieve this goal?

(b) What will be the value in 2005 dollars of her 2040 income of $100,000?

(c) What will be the value in 2005 dollars of her 2070 (when she is 85) income of $100,000?

42. (Spreadsheet helpful) You think what your roommate means is that she wants to retire in 2052 with a *steady* income of $50,000 a year in 2005 dollars, and that she should plan to receive that same value of income for 45 years in case she lives to 100. Having studied this book, you realize that one way to maintain the purchasing power of income is to devote part of the income to increasing the principal (this is what college endowments do). For example, if inflation is 4% and her investment earns 7.2%, she could reinvest 4% each year and draw the remaining 3.2% to live on.

(a) What is the present value in 2005 of the planned stream of 45 years of income?

(b) Following the reinvestment strategy, how big a nest egg does she need to have in 2040, and what should be her quarterly investment to achieve that goal?

Real Growth and Valuing Investments

43. On October 1, 2001, you could buy a U.S. Treasury 10-year note with a yield of 4.53%. At that time, the inflation rate was about 3.1%.

(a) What was the real growth rate of this investment?

(b) For an individual with a combined federal, state, and city income tax rate of 30%, what was the real growth rate of this investment?

44. Generalize the result in Exercise 43b by giving a formula for real growth rate g in terms of the interest rate r, the inflation rate i, and the tax rate t.

■ **45.** On the morning of October 2, 2001, in an effort to stimulate the economy after the attacks of September 11, 2001, the Federal Reserve ("the Fed") cut its federal funds interest rate (a rate paid by banks borrowing from the Fed) by 0.5%, to 2.5%, the ninth cut of the year. At the time, the rate of inflation was about 4% and the interest rate on long-term government bonds was 4.53%, but uncertainty about the economy in the wake of the attack on the United States had no doubt increased. Suppose then that the discount rate was $r = 15\%$ and that the rate cut by the Fed reduced r by 0.25%. Suppose also that the average stock dividend was $g = 3\%$. With this in mind, consider a stock with annual dividend $D = \$5$. By what percentage would you expect the price of that stock (and correspondingly the market as a whole) to rise as a result of the Fed's action? (In fact, on October 2, 2001, the Dow Jones Industrial Average rose by 114 points, from 8836.83 to 8950.59.)

■ **46. (a)** Let the symbol Δr (pronounced "delta r") denote a change in the discount rate, with Δr positive for an increase and negative for a decrease. Such a change could be produced by action of the Fed or by a change in the rate of inflation, in the rate on government bonds, or simply in investor confidence. Let $\Delta S/S$ denote the expected proportional change in stock prices. Find a simplified formula involving just r, g, and Δr for $\Delta S/S$.

(b) Apply the formula in part (a) with $g = 3\%$, $r = 15\%$, and $\Delta r = -0.25\%$ to find ΔS. You should get the same answer as in Exercise 45.
(c) Apply the formula "backward" with $g = 3\%$, $\Delta r = -0.25\%$, and ΔS (the change observed in the Dow Jones Industrial Average on October 2) to determine the corresponding value of r before the announcement by the Fed.

In the savings formula, the interest rate i appears twice. The particular ways in which i is involved make it impossible to solve it algebraically to get an explicit formula for i. However, with the help of a spreadsheet, you can find i approximately when the other quantities are given. Exercises 47–50 treat such situations.

47. (Requires spreadsheet) Suppose that you decide to lease a car. At the end of the 48-month lease period, you need to make a lump-sum payment of $5000 if you want to keep the car. You decide to save up, just in case you decide to

keep the car; if you don't keep this car, you will still have saved a good down payment on a new car. You feel comfortable with saving $70/month (over and above your lease payments). How high an annual nominal interest rate on savings do you need to accumulate $5000 in 48 months, with interest compounded monthly?

48. (Requires spreadsheet). A 1990 advertisement reads, "If you had put $100 per month into this fund starting in 1980, you'd have $37,747 today." Assume that deposits were made on the last day of the month, starting in January 1980, through December 1989, and that interest was paid monthly on the last day of the month (120 months).

(a) How much money was deposited during this period?
(b) What annual rate of interest, compounded monthly, would lead to the result described in the advertisement? What is the annual yield?

49. (Requires spreadsheet) The Powerball lottery drawing on December 25, 2002, resulted in the largest prize anywhere ever, with a jackpot of $314.9 million. The winner had the option of an annuity in 30 equal annual installments of $10.5 million (the first payment being right away) or else an instant lump sum of $170 million (actually, after taxes, about $112 million). Taking the lump sum (most lottery winners do) makes sense, particularly if the winner needs a large sum of money now or else feels that he or she can earn a higher rate of interest than the annuity is based on. The present value of the payment k years from now is given by the compound interest formula $A = P(1 + i)^k$, where $P = 10.5$ million and i is the rate of interest built into the annuity. Hence, we have $P_k = A/(1 + i)^k$ for the kth payment. The complete stream of 30 payments has present value (in millions of dollars)

$$170 = 10.5\left[1 + \frac{1}{1 + i} + \frac{1}{(1 + i)^2} + \cdots + \frac{1}{(1 + i)^{29}}\right]$$

We use the geometric series formula with $x = 1/(1 + i)$, finding that the right-hand side is also equal to

$$10.5\left[\dfrac{1 - \dfrac{1}{(1 + i)^{30}}}{1 - \dfrac{1}{1 + i}}\right]$$

Enter $i = 0.04$ (annual rate) in the preceding expression; the result is larger than 170. For $i = 0.05$, the result is smaller than 170. Make changes in the value of i until you determine to two decimal places the rate i that gives the closest value to 170. This is the rate on which the annuity is based. Is it a nominal rate or the effective rate?

50. (Requires spreadsheet) Suppose that your parents are willing to lend you $20,000 for part of the cost of your college education and living expenses. They want you to repay them the $20,000, without any interest, in a lump sum 15 years after you graduate, when they will be about to retire and move. Meanwhile, you will be busy repaying federally guaranteed loans for the first 10 years after graduation. But you realize that you can't repay the lump sum without saving up. So you decide that you will put aside money in an interest-bearing account every month for the five years before the payment is due. You feel comfortable with the idea of putting aside $275 a month (the amount of the payment on your government loans). How high an annual nominal interest rate on savings do you need to accumulate the $20,000 in 60 months, if interest is compounded monthly? Enter into a spreadsheet the values $d = 275$, $r = 0.05$ (annual rate), and $n = 60$, and the savings formula with r replaced by $r/12$ (the monthly interest rate). You will find that the amount accumulated is not enough. Change r to 0.09—it's more than enough. Try other values until you determine r to two decimal places.

 ## APPLET EXERCISE

To do these exercises, go to www.whfreeman.com/fapp7e.

How important is it to begin a retirement fund at an early stage of one's career? In the *Saving for* *Retirement* applet, you will discover that early funding of a retirement plan can make a huge difference in the ultimate amount that will be available when a person retires.

 ## WRITING PROJECTS

1. Exercises 23, 24, and 28 asked you to look at various forms of tax-deferred and ordinary savings and compare them on the basis of amount of tax-free income accumulation at age 65.

Ordinary savings have the important advantage that, at any time, you can do anything you want with the money accumulated so far (buy a car, put down money on a house, and so on). A second advantage is that the money is free and clear, in that taxes have already been paid on it.

A tax-deferred 401(k) retirement fund has the disadvantage that for any funds withdrawn before age $59\frac{1}{2}$, you must pay income tax in the year of withdrawal and, in addition, pay a 10% penalty for "early withdrawal." (These plans were given the advantage of tax deferral to encourage individuals to save for retirement—hence the penalty for withdrawing the money earlier.)

A third option, the Roth IRA, has some of the advantages and disadvantages of each of the other plans.

Look into the details of the rules for 401(k) plans and Roth IRAs, compare your answers in Exercises 23, 24, and 28, and devise and describe your own plan for how you will save for retirement. Your report should run two to three pages.

2. Plan your own retirement. Decide on a retirement age and desired income (in today's dollars). Estimate yield on investments, inflation rate, and Social Security benefits. (*A few notes:* By the time of your retirement, a woman retiring in her mid-60s will likely live an average of 20 years more, a man 18 years more. Social Security income goes up with inflation. Annuities are available whose income grows to keep up with inflation. Various

other financial products, such as a *life annuity*, can make sure that you don't outlive your retirement income. We treat life annuities in Chapter 22. Also, we neglect here any consideration of any tax deferral of income or earnings, as considered in Exercises 23, 24, and 28.) Write up your assumptions, justifications for them, calculations, and conclusions in three to four pages. Be sure to note any additional factors that you think should be taken into account but which your analysis does not include.

SUGGESTED READINGS

KASTING, MARTHA. *Concepts of Math for Business: The Mathematics of Finance.* UMAP Modules in Undergraduate Mathematics and Its Applications: Module 370–372. COMAP, Inc., Arlington, Mass., 1980.

LINDSTROM, PETER A. *Nominal vs. Effective Rates of Interest.* UMAP Modules in Undergraduate Mathematics and Its Applications: Module 474. COMAP, Inc., Arlington, Mass., 1988. Reprinted in Paul J. Campbell (ed.), *UMAP Modules: Tools for Teaching 1988,* COMAP, Inc., Arlington, Mass., 1989, pp. 21–53. A learning module, requiring no more background than this chapter, that teaches about nominal and effective rates of interest and how to calculate them. Gives real examples of banks using different options for calculating interest.

MILLER, CHARLES D., VERN E. HEEREN, and JOHN HORNSBY. Consumer mathematics. In *Mathematical Ideas,* 9th ed., Addison Wesley Longman, Boston, 2001, pp. 786–845.

VEST, FLOYD, and REYNOLDS GRIFFITH. The mathematics of bond pricing and interest rate risk. *Consortium (COMAP),* 59 (Fall 1996): HiMAP Pullout Section 1–6.

SUGGESTED WEB SITES

www.bls.gov/cpi/ Home page for the inflation tables prepared by the Bureau of Labor Statistics.

www.dol.gov/dol/esa/public/minwage/ Department of Labor data on the U.S. federal minimum wage over the years, including a graph of its inflation-adjusted value.

www.westegg.com/inflation/ Inflation calculators for the United States (1800–2001), Canada, and Italy, by S. Morgan Friedman, with links to sites about the current purchasing power of amounts of currencies of other countries in the past.

CHAPTER 22

Borrowing Models

In the previous chapter, we looked at consumer financial models for saving and formulas for calculating the amount accumulated. Savings or investments would not earn interest unless they could be loaned to someone to make productive use of the money.

In this chapter, we examine the other side of consumer finance, borrowing. You are likely to borrow to buy a car, you will almost certainly borrow if you buy a house or apartment, and you are borrowing if you use a credit card. You will pay interest (and perhaps finance charges) in addition.

But there are different ways to structure loans that can make a lot of difference in how much you pay for the use of the money that you borrow. We investigate and compare some common kinds of loans.

We briefly (re)acquaint you with compound interest and a few formulas from Chapter 21. If you have a grasp of the ideas behind compound interest and can use the formulas, you can proceed with this chapter without first reading Chapter 21.

22.1 Simple Interest

The amount of **interest** charged on a loan is determined by the **principal,** by the amount borrowed, and by the method used to calculate the interest. With **simple interest,** the borrower pays a fixed amount of interest for each period of the loan. The interest rate is usually quoted as an annual rate.

In the previous chapter (p. 796), we cited the example of simple interest paid on a bond. We do not repeat the example here, but proceed to consumer loans. First, we remind you of the relevant formula, now in the context of loans rather than savings.

For a principal P and an annual rate of interest r, the interest owed after t years is

$$I = Prt$$

and the total amount A due on the loan is

$$A = P(1 + rt)$$

Add-On Loan

A common type of consumer loan is the *add-on* loan. You borrow an amount P to be repaid in t years. The interest is simple interest at an annual rate $r\,(= 100r\%)$, for a total of $I = Prt$. You must pay $P + I = P(1 + rt)$ in installments (usually monthly). With n payments, each payment is $d = P(1 + rt)/n$.

Add-on Loan

By the terms of an **add-on loan** over n installments, you pay $\frac{1}{n}$th of the principal and $\frac{1}{n}$th of the total interest with each payment. The total interest is "added on" to the principal, and the sum is paid in equal installments.

Add-On Loans

EXAMPLE 1

Suppose that you have to borrow $8000 to buy a used car. The dealer offers you a 5% add-on loan to be repaid in monthly installments over four years. This sounds like a much better deal than the 8% loan that you can get at the credit union. How much is your payment d on the dealer's add-on loan?

SOLUTION To find the total amount to be paid over the four years, first calculate the interest on the add-on loan:

$$I = Prt$$
$$I = \$8000(0.05)(4)$$
$$I = \$1600$$

The total amount that must be repaid is then $P + I = \$8000 + \$1600 = \$9600$, and the monthly payment over 4 years (48 months) is $\$9600 \div 48 = \$200$. ∎

With an add-on loan, everything sounds simple and straightforward (even the calculation of the payment!). The interest is calculated on the entire principal. Because you slowly pay back the principal, however, you do not have the use of the whole amount for the entire duration of the loan. In fact, you have the use of the full principal for just one month. You do, however, have the use of the car! But the net value that you have at any point is the cost of the car minus the amount of principal already repaid (here we neglect depreciation of the car). It is on this net value, and the amount of interest, that the interest rate should be calculated. In effect, the "true" interest rate (we will make this concept more precise) is much higher than the rate r quoted.

Discounted Loan

Another type of consumer loan is a *discounted loan*. The interest is computed as simple interest, just as for an add-on loan, but it is subtracted from the amount given to the borrower. In other words, instead of getting the principal P, the borrower gets the *proceeds $P - I = P - Prt = P(1 - rt)$* but must pay back the amount P over the term of the loan. In effect, the interest is paid in advance. What is discounted is not the cost of the loan but how much the borrower gets! The interest is based on the entire P, but the borrower never has the use of that much and, as with an add-on loan, is paying interest on the entire amount P but has the use of less and less of it as the loan is repaid.

> ### Discounted Loan
>
> With a **discounted loan** over n installments, you pay $\frac{1}{n}$th of the principal with each payment. The amount that you receive from the lender is the principal "discounted by" (minus) the total interest, but you pay back the entire principal in equal installments.

EXAMPLE 2   Discounted Loans

Suppose again that you have to borrow $8000 to buy a used car. Your neighborhood loan office offers you a 5% discounted loan to be repaid in monthly installments over four years. Is this a better deal than the dealer's offer of a 5% add-on loan over four years (see Example 1)?

SOLUTION   The total amount to be paid over the four years is $P = \$8000$ over 48 months, for a monthly payment of $166.67. This is indeed a lower payment, but the problem is that the lender loans you only $\$8000 \times (1 - 0.05 \times 4) = \$8000 \times 0.80 = \$6400$, which isn't enough to buy the car. How big a discounted loan do you need? Call the amount x. You need $x(0.80) = \$8000$, or $x = \$8000/0.80 = \$10{,}000$. Your monthly payment is $\$10{,}000/48 = \$208.33$, so this loan is *not* cheaper than the add-on loan; for the same terms, this is always so. ■

22.2 Compound Interest

Compounding is the calculation of interest on interest. A common example is the balance on a credit card. As long as there is an outstanding balance owed, the interest owed is calculated on the entire balance, including any part of it that was interest calculated and added to the balance in earlier months.

Credit-Card Interest

EXAMPLE 3

Suppose that you owe $1000 on your credit card, the company charges 1.5% interest per month, and you just let the balance ride. How much interest do you pay in the first year?

SOLUTION Your interest the first month is 1.5% of $1000, or $0.015 \times \$1000 = \$15$. The new balance owed is $(1 + 0.015) \times \$1000 = \$1015$. Your interest the second month is not 1.5% of $1000, or $15 (as would be the case for simple interest), but 1.5% of $1015, or $0.015 \times \$1015 = \$15.23$, so the new balance is

$$(1 + 0.015) \times \$1015 = \$(1 + 0.015) \times (1 + 0.015) \times \$1000 = \$1030.23$$

(We neglect the extra charges for your failure to make minimum payments.) After 12 months of letting the balance ride, it has become

$$(1.015)^{12} \times \$1000 = \$1195.62$$

In other words, the actual interest for the year comes to $195.62, which is 19.562% of $1000. So, although the quoted rate of interest is 1.5% per month, which seems as if it should amount to $12 \times 1.5 = 18\%$ per year, the interest owed is actually more. ■

We apply two formulas from Chapter 21: the **compound interest formula** (p. 800) and the **savings formula** (p. 810), here phrasing them for loans.

Compound Interest Formula

If a principal P is loaned at interest rate i per compounding period, then after n compounding periods (with no repayment) the amount owed is

$$A = P(1 + i)^n$$

This formula just generalizes what we saw happen with the credit-card balance.

Borrowing from Joe, the Patient Loan Shark

EXAMPLE 4

Suppose that you arrive on campus without enough money to buy your textbooks (including this one), not to mention other supplies and necessities (such as late-night pizzas). So you go to Joe's Friendly Loan Service at a nearby off-campus location, where Joe offers to loan you $1000 at an incredible interest rate of 1% per week, compounded weekly. Joe is such a good sport that he doesn't want the money back until a year from now, which gives you a chance to earn the repayment over the summer. How much will you have to pay back at the end of a year?

SOLUTION We have $P = \$1000$, $i = 1\% = 0.01$ per compounding period, and $n = 52$ compounding periods:

$$A = P(1 + i)^n = \$1000 \times (1 + 0.01)^{52} = \$1677.69 \ \blacksquare$$

EXAMPLE 5   Saving Up to Pay Joe Back

Summer comes and it's time to save up to pay Joe back. Suppose that you work for three months and at the end of each month deposit an amount d to a savings account that pays 3% interest per year, or 0.25% per month. How much does d need to be to pay off the $1667.69 that you owe?

SOLUTION We have $A = \$1667.69$, $i = 0.25\% = 0.0025$ per compounding period, and $n = 3$ compounding periods:

$$d = \frac{Ai}{(1 + i)^n - 1} = \frac{\$1667.69 \times 0.0025}{(1 + 0.0025)^3 - 1} = \$554.51 \ \blacksquare$$

Terminology for Loan Rates

We have seen that the rate of interest for a loan depends on whether or not compounding is done and how the interest is calculated. Just like the Truth in Savings Act mentioned in Chapter 21 (p. 798), the Truth in Lending Act establishes terminology and calculation methods for interest. We remind you of it here briefly.

A **nominal rate** is any stated rate of interest for a specified length of time. For instance, a nominal rate could be a 1.5% monthly rate on a credit-card balance. By itself, such a rate does not indicate or take into account whether or how often interest is compounded.

Similarly, from Chapter 21 (p. 806) you are familiar with the **effective rate,** the actual percentage rate of increase for a length of time, *taking into account compounding.* It equals the rate of simple interest that would realize exactly as much interest over that length of time.

We saw that $1000 at a yearly interest rate of 18% (a nominal rate), or 1.5% per month, compounded monthly for one year yields $195.62 in interest owed, which is 19.562% of the original principal. Hence, the effective annual rate is 19.562%. In other words, a $1000 loan at simple interest of 19.562% for one year would owe exactly the same interest.

Finally, when stated per year ("annualized"), the effective rate is called the **effective annual rate (EAR).**

To keep the rates straight, we use i for a nominal rate for the specified **compounding period**–such as a day, month, or year–*within which no compounding is done;* this rate is the effective rate for that length of time. For a nominal rate compounded n times per year, we have $i = r/n$. For that $1000 credit-card balance at 18% compounded monthly, we have $r = 18\%$ and $n = 12$, so $i = 1.5\%$ per month.

A new term that the Truth in Lending Act introduces is the *annual percentage rate (APR).*

Annual Percentage Rate (APR)

The **annual percentage rate (APR)** is the rate of interest per compounding period times the number of compounding periods per year.

$$APR = i \times n$$

In the example of the credit-card balance, the interest is compounded monthly, or $n = 12$ times per year, and the interest rate for the compounding period is $i = 1.5\%$, so the APR is $12 \times 1.5\% = 18\%$. The APR is the rate that the Truth in Lending Act requires the lender to disclose to the borrower. *The APR is not equal to the EAR* (as we have already seen in the credit-card example). Spotlight 22.1 explains how and why.

22.3 Conventional Loans

A common situation that you are likely to encounter is a loan—for a house, a car, or college expenses—to be paid back in equal periodic installments. Your payments are said to **amortize** (pay back) the loan. In these so-called **conventional loans,** each payment pays the current interest and also repays part of the principal. *As the principal is reduced, there is less interest owed, so less of each payment goes to the interest and more toward paying off the principal.*

Buying a House

EXAMPLE 6

Let's suppose that you buy a house with a $100,000 loan to be paid off over 30 years in equal monthly installments. Suppose that the interest rate for the loan is 6.00%. How much is your monthly payment?

SOLUTION Imagine changing the setup slightly so that instead of making monthly payments, you are supposed to pay off the entire principal and interest at the end. Meanwhile, you make payments to a savings fund that you're building up to pay off the loan, and the savings fund earns the same rate of interest as the loan costs. The interest rate of 6.00% on the loan is compounded monthly, so the monthly rate is 0.5%. At the end of 30 years, the principal and interest on the loan will (by the compound interest formula) amount to

$$\$100,000 \times (1 + 0.005)^{30 \times 12} = \$602,257.52$$

On the other hand, saving $\$d$ each month for 30 years at 6.00% interest compounded monthly, we know from the savings formula that you will accumulate

$$d\left[\frac{(1 + 0.005)^{360} - 1}{0.005}\right]$$

Financial experts agree that the real, "true" rate of interest for savings or loans is the effective annual rate (EAR).

The 1991 Federal Truth in *Savings* Act requires that savers be told the annual percentage yield (APY), which is the EAR.

The 1968 Federal Truth in *Lending* Act, however, requires that borrowers be told the *annual percentage rate (APR)*, which is *not* the same as the EAR. The APR is the rate of interest per compounding period times the number of compounding periods per year. Thus, a credit-card rate of 1.5% per month translates to an APR of 18%. The APR does not take into account compounding. Hence it is not equivalent to—indeed, it understates—the true cost of borrowing, that is, the EAR. For the credit-card loan, with monthly compounding, the EAR is

$$(1 + 0.015)^{12} - 1 \approx 19.6\%$$

The APR also ignores costs that are sometimes involved in borrowing, such as a flat charge for making the loan in the first place ("loan-processing fee"), charges for late payments, and charges for failing to make a minimum payment.

For mortgage loans, the Truth in Lending Act requires that lenders include in the APR some of the upfront costs referred to as *closing costs:* any "loan origination" fee, "loan-processing" fee, and "points" (additional charges calculated as percentage points of the purchase price). The APR does not include title insurance, appraisal, credit-report fees, or transaction taxes.

Closing costs are paid at the closing of the sale, while interest is paid over the life of the loan. However, the APR treats the closing costs included in it as if they were amortized over the term of the mortgage, despite the fact that they are paid beforehand. Here, too, the APR understates the true costs.

But very few people hold a mortgage to its maturity. The median life of a 30-year mortgage is only about 5 years; that is, half of all mortgage holders pay off their mortgage before 5 years are up, usually because they sell the home and move elsewhere. Thus, for almost all home loans, the APR also includes interest that will never be paid.

Finally, we must take into account inflation. One advantage of buying a home with a fixed-rate mortgage is that your payment stays the same but your earnings and the value of your home are likely to go up with inflation: You are thus paying back the loan with dollars of lesser value. For any loan in a time of inflation, *Fisher's effect* (Chapter 21, p. 817) comes into play: If your loan has an EAR of 7% but inflation is running at 3.5% per year, the true cost to you of the loan is not exactly $7\% - 3.5\% = 3.5\%$. Instead, for an EAR of r and an inflation rate of a, the cost of the loan at the beginning of the first year is indeed $r - a$ ($= 3.5\%$ in our example), but at the end of the first year it is

$$c = \frac{r - a}{1 + a}$$

For $r = 7\%$ and $a = 3.5\%$, we get $c = 3.38\%$. The reason this is less than the expected 3.5% is that at the end of the first year you are paying back the loan with dollars that have been inflated for a year. As inflation mounts over the term of a mortgage, the cost c goes down steadily each year. For example, at the end of five years of steady inflation at 3.5%, the total inflation has been $a = (1 + 0.035)^5 - 1 = 18.8\%$, and we have $c = 2.95\%$.

A final—and major—consideration is that interest paid on your home mortgage is deductible from taxable income on federal, state, and some local income tax returns. Thus, your home ownership is subsidized by other taxpayers (just as you help subsidize home buyers among them), and the cost to you of the loan is reduced further.

To make d just the right amount to pay off the loan exactly, we need to solve the equation

$$d\left[\frac{(1 + 0.005)^{360} - 1}{0.005}\right] = \$100{,}000 \times (1 + 0.005)^{30 \times 12} = \$602{,}257.52$$

for the value of d, getting $d = \$599.55$ as your monthly payment. The total of the payments is "only" $360 \times \$599.55 = \$215{,}838.00$—on a loan of just $\$100{,}000$. ∎

We put this idea into a more general setting.

Paying Off a Conventional Loan Is Like Saving

Let the principal be A, the effective interest rate per period i, the payment at the end of each period d, and let there be n periods. Then we have

$$A(1 + i)^n = d\left[\frac{(1 + i)^n - 1}{i}\right]$$

The compound interest formula is on the left and the savings formula is on the right. You can think of paying off the loan as making payments to a savings account that earns interest at the same rate as the loan. The savings balance will exactly equal the principal plus the interest on the loan at the end of the loan term.

Thus, the quantity A is sometimes called the *present value of an annuity* of n payments of d, each at the end of an interest period, with interest compounded at rate i in each period. This terminology agrees with that used in some calculators, such as the TI-83, which have a financial mode with the option to calculate A. Solving for A gives the **amortization formula.**

Amortization Formula

$$A = d\left[\frac{1 - (1 + i)^{-n}}{i}\right]; \qquad d = \frac{Ai}{1 - (1 + i)^{-n}}$$

Making Weekly Payments to Joe

EXAMPLE 7

You're worried about having to come up with the lump sum ($\$1677.69$) that you will owe Joe after a year, so you ask him if you can make weekly payments instead. He is delighted at the idea but says that the interest will have to be higher,

1.5% per week. He agrees to let you amortize the loan over 52 weeks, so that each payment covers the week's interest on the amount still outstanding and pays off some of the principal. How much is your weekly payment?

SOLUTION We have $A = \$1000$, $i = 1.5\% = 0.015$ per compounding period (one week), and $n = 52$ compounding periods; we want to find d:

$$d = \frac{\$1000 \times 0.015}{1 - (1 + 0.015)^{-52}} = \$27.83 \quad \blacksquare$$

EXAMPLE 8   Buying a Car

You decide to buy a new Wheelmobile car. After a down payment, you need to finance (borrow) $12,000. Comparing interest rates offered by the car dealership, local banks, and your credit union, the best deal you can find is 4.9% compounded monthly over 48 months. What is your monthly payment?

We have $A = \$12,000$, monthly interest rate $i = 0.049/12$, and $n = 48$. Using the amortization formula, we have

$$d = \frac{\$12,000 \times \frac{0.049}{12}}{1 - (1 + \frac{0.049}{12})^{-48}} = \$275.81$$

How much interest do you pay? You make payments totaling $48 \times \$275.81 = \$13,238.88$, so the interest is $\$13,238.88 - \$12,000 = \$1,238.88$.

If you had bought a Plushmobile instead, with $24,000 to finance, you would have borrowed twice as much and your monthly payment would have been twice as much. $\blacksquare$

EXAMPLE 9   APR Versus EAR for an Add-On Loan

We return to the earlier 5% add-on loan for $8000 to be repaid over four years. The effective interest rate per month is i in the amortization formula. As hard as you try, though, you cannot solve the amortization formula for i in terms of A, d, and n. Mathematicians have shown that it cannot be done! What can you do?

SOLUTION You can use a spreadsheet and try successive values of i until you find one that makes the two sides of the equation close enough to being equal. If your spreadsheet or calculator has a Solve feature, you can use that. In this case, you find $i = 0.0077014725$.

Here we now see a distinction. The APR (see Spotlight 22.1), which the dealer is required to tell you, is 12 times the monthly rate, or $12 \times 0.0077014725 = 9.24\%$. The EAR, which takes into account monthly compounding, is $(1 + i)^{12} - 1 = 9.64\%$.

Either way, is this a better rate than the 8% credit union loan? Before you jump to conclusions, better find out the details of that loan! $\blacksquare$

A car loan is usually for 48 or 60 months, but when you buy a home, you usually borrow a great deal more money and hence pay it off over a much longer period. The usual term for a home mortgage is 30 years. We disregard here some of the closing costs connected with buying a home and getting the loan, to focus on the loan itself.

As a simple example, consider a mortgage for $184,000 (the median cost of a U.S. home in mid-2004) at an interest rate of 5.4% over 30 years. Payments are monthly. The monthly interest rate is $5.4\%/12 = 0.45\% = 0.0045$ and there are $30 \times 12 = 360$ months. We use the amortization formula with $A = \$184,000$, $i = 0.0045$, and $n = 360$:

$$d = \frac{\$184,000 \times \frac{0.054}{12}}{1 - (1 + \frac{0.054}{12})^{-360}} = \$1033.22$$

When you decide to buy a home, a key question that you need to ask yourself is how much home you can afford.

Thirty-Year Mortgage on Median-Priced Home

EXAMPLE 10

Let's suppose that you are a family with the U.S. median income of about $65,000 for a family of four, that you want to buy a median-priced home ($206,000 in April 2005) with a 30-year fixed-rate mortgage at 5.4%, and that you can make a down payment of only about 5%, or $10,000. Can you afford such a home?

SOLUTION Lenders have "affordability" guidelines that suggest that a family can afford to spend about 28% of its monthly income on housing. Thus, by their guidelines, you can afford $0.28 \times \$65,000/12 = \$1516.67$ per month.

What is the monthly payment on the loan? The principal is $A = \$196,000$, the monthly interest rate is $i = 0.054/12 = 0.0045$, and $n = 360$ months. The amortization formula gives

$$d = \frac{\$196,000 \times \frac{0.054}{12}}{1 - (1 + \frac{0.054}{12})^{-360}} = \$1100.60$$

Well, that sounds good. But unfortunately there is more to the mortgage than just the amount needed to amortize the loan. Your payment to the bank must also cover real estate taxes and homeowner's insurance on the property. On a $206,000 home, these may add $450 to the monthly payment, which will then total about $1550.

So, no, the median family can't afford the median-priced home, at least not without a bigger down payment or a lower-interest loan. ■

(Norbert Schwerin/The Image Works.)

EXAMPLE 11 Home Equity

My wife's parents sold their house in rural Minnesota to move to the town where we live. They had bought their house in 1980 for $100,000 with a 30-year mortgage at an 8% interest rate. After 22 years, how much **equity** did they have in the house—that is, how much of the principal had been repaid? And how much did they still owe on the house?

SOLUTION What may shock you is that when they sold their house in May 2002—after 269 months of payments, almost exactly three-quarters of the 30 years of the mortgage—they had only $50,000 in equity (hence still owed $50,000 on the house) but had already paid $147,000 in interest. *Three-quarters of their payments had gone to interest.*

We can use the amortization formula to determine just how much equity they had after 269 months of payments, but first we need to determine their monthly payment. We see $A = \$100,000$, $n = 360$ months, and $i = 0.08/12$ monthly interest, getting $d = \$733.76$.

Now we use the formula again, this time "in reverse." Knowing $i = 0.08/12$ and $d = \$733.76$, we find out how much of a loan A is paid off by the remaining $n = 360 - 269 = 91$ payments:

Cumulative equity, 30-year mortgage for $100,000
at 8% interest

FIGURE 22.1 Equity
grows almost exponentially,
especially in the later
years of a mortgage.

$$A = d\left[\frac{1 - (1 + i)^{-n}}{i}\right] = \$733.76\left[\frac{1 - (1 + \frac{0.08}{12})^{-91}}{\frac{0.08}{12}}\right] = \$49{,}940.03$$

This is how much my parents-in-law had yet to pay, so their equity was $100,000 −
$49,940.03 = $50,059.97.

Figure 22.1 shows equity versus time for such a loan. The equity grows almost exponentially, but that means that the rate of growth is very slow in the
early years of the mortgage. ■

When you buy a home, you have several options for the mortgage: a conventional 30-year mortgage, a conventional 15-year mortgage, or a mortgage for
either length of time but with an interest rate that can vary.

You might expect the payment on a 15-year mortgage to be double that of
a 30-year mortgage. On the contrary, the payment is only 47% more (for a 5%
mortgage) to 26% more (for a 9% mortgage). This range includes the prevailing
mortgage rates over the past 20 years. Moreover, over the course of a $100,000
mortgage at 6%, you would pay $164,000 in interest over 30 years but only $72,000
over 15 years. At 9%, the interest totals are $190,000 versus $83,000. (Some financial counselors advise taking a 30-year mortgage and making extra payments
when you can afford them, rather than incurring the higher payment obligation
of a 15-year loan, which, if you encounter tight personal financial circumstances,
you might not be able to afford.) In Spotlight 22.2, we tell what we did in our
own circumstances and mention other options.

Since very few mortgages are held for the full term, it is useful to compare
the status of mortgages after five years. Table 22.1 shows the equity after five years
for a variety of interest rates. For a 30-year mortgage, the equity after five years

SPOTLIGHT 22.2 — What We Did with Our House, and What Else You Could Do

We bought our house in 1992. We were offered a choice between an 8.375% fixed-rate 30-year mortgage and an adjustable-rate 30-year mortgage (ARM) at 6.875% whose rate could be raised (or lowered) by up to 2% every year. When we asked, we were also quoted slightly lower rates for corresponding 15-year mortgages.

We were planning to stay in the house much longer than the median of five years, and we were concerned that inflation might force the ARM considerably higher. Also, we did not want the obligation of the higher payments of a 15-year mortgage, in case our circumstances changed (such as through job loss or death). There was no penalty for making extra payments (if we could afford them).

We chose the 8.375% fixed-rate 30-year mortgage (and made some extra payments). Others in other circumstances, or with a different tolerance for risk, would no doubt have decided otherwise. Had we been sure then that interest rates would not go higher in the 1990s, we would have gone for the ARM. But hindsight is always better than foresight. During the early years of this decade, most homeowners with mortgage interest rates such as ours refinanced at much lower prevailing rates, near 5% for a fixed-rate 30-year mortgage.

Currently, about a third of people take ARMs rather than fixed-rate mortgages as one way to respond to soaring real estate prices in some parts of the country. Newer mortgage "products" include interest-only mortgages and shared appreciation mortgages (SAMs). With an interest-only ARM, payments are (just slightly) lower than for a conventional 30-year mortgage, but you accumulate no equity (still, the market value of the house may rise). After five to seven years, you start also paying off the principal—which means that your payments go up then. In some such loans, the interest rate—and your payments—fluctuates as frequently as every month.

In a SAM, interest payments are lower or absent, but the lender receives a portion of any appreciation (rise in value) when the house is sold. In a nationally reported instance in 2003, a single mother received a no-interest SAM loan to finance the $30,000 down payment on a $223,000 house in Pleasanton, California, through the city's affordable housing program. Four years later, she sold the house for $385,000, and the "affordable housing" lenders got 60% of the $162,000 appreciation, or $97,000. She herself realized $65,000 (minus the cost of the sale) but complained bitterly, saying that she would have been better off to have put the loan on her credit card! Critics have termed SAMs an urban form of sharecropping.

TABLE 22.1 Equity on a $100,000 Mortgage After Five Years					
Mortgage Term (years)	5%	6%	7%	8%	9%
15	25,700	24,000	22,600	21,200	19,900
30	8,200	6,900	5,900	4,900	4,100

may be less than the cost of selling the home through a realtor. Of course, the resale value of the home may also be higher after five years.

A mortgage with an interest rate that can vary is called an **adjustable-rate mortgage (ARM).** Often such mortgages have a substantially lower interest rate (hence a lower payment) than a fixed-rate mortgage. The ARM's interest rate may go up or down with interest rates in the economy. Usually the rate can be raised or lowered only every year or two, and then by a limited percentage. An ARM may be attractive if you plan to pay off the mortgage after only a few years or because it allows lower payments, it facilitates buying a more expensive home, or you do not plan to keep the home long (hence, you would be selling before the interest rate could rise substantially).

22.4 Annuities

An **annuity** is a specified number of (usually equal) payments at equal intervals of time. We restrict our discussion to *ordinary annuities,* for which payments are made at the end of each interval and the interval is also the compounding period.

An annuity can be interpreted as involving borrowing. For example, winners of lotteries are often offered the choice of receiving either the jackpot amount paid as an annuity over a number of years or else a smaller lump sum to be paid immediately. The cost to the lottery administration is the same. If the winner wants an annuity, the administration buys one from an insurance company for the lump sum. You can think of the insurance company as borrowing the lump sum in exchange for making the payments of the annuity. In effect, the insurance company is amortizing the lump sum over the duration of the annuity.

Winning the Lottery

EXAMPLE 12

On April 16, 2002, three winning tickets shared an almost-record Big Game (since renamed Mega Millions) jackpot of $331 million. Each ticket's share was one-third of the total, or $110.333 million, to be paid as an annuity in 26 equal annual installments of $4.244 million each, the first payment being right away. However, each winner instead chose an instant lump sum of $58.9 million. What was the interest rate of the annuity?

SOLUTION The insurance company that sold the annuity to the lottery administration regarded $58.9 million as the present value of the annuity. To consider the annuity as an ordinary annuity, with payments made at the end of each interval, we must subtract the first payment, leaving $110.333 − $4.244 = $106.089 million to be paid in 25 equal annual installments at the end of each year, and $58.9 − $4.244 = $54.656 million as the present value.

In the formula for the present value of an annuity,

$$A = d\left[\frac{1 - (1 + i)^{-n}}{i}\right]$$

we have $A = \$54.656$, $d = \$106.089/25$, and $n = 25$. To solve for i, the interest rate used by the insurance company, we must use either a calculator with financial mode or a spreadsheet. Either way, we find $i = 5.92\%$. ■

As you save for retirement, it is probably wise to save part of your funds in the form of a tax-deferred annuity (employees of Enron who kept their retirement savings in Enron stock lost big when the company went bankrupt in 2002). If you do not, at retirement you can still sell all of your holdings in other forms and purchase an annuity.

Such annuities differ in a crucial way from the lottery annuity in Example 12. If you retire at 65 and purchase an ordinary annuity, you would be in trouble if you live longer than the term of the annuity (past age 89), because the payments would stop and you would have no further income from the annuity. (About 2% of U.S. children born today can expect to live to age 100.) Similarly, if you die sooner, your estate would still get the payments due after your death, but they wouldn't have helped you meet your living expenses while you were alive.

An approach that avoids these two disadvantages is the *life income annuity*: You receive a fixed amount of income per month for as long as you live. How much you receive per month is based on the life expectancy of people your age, as determined from population data. There are many variations on life annuities, such as payments that increase with anticipated cost-of-living increases, or payments that last until both you and your life partner die (see Spotlight 22.3). But we focus on a simple one-life annuity.

The insurance company that sells you the annuity makes money if you die younger than average and loses money if you die older than average. As in any kind of insurance, over a large number of people, the company can expect gains to balance losses. This is a manifestation of the law of large numbers of Chapter 8. Also, the company's profits vary with the prevailing interest rate during the annuity as compared with the rate built into the annuity.

How much you receive per month depends on your gender. Because women on average live longer than men, the monthly payment to a woman is lower.

EXAMPLE 13   Life Income Annuity

Suppose that you are a 65-year-old male retiring with $250,000 in a life income annuity. According to the table from one particular insurance company, you would receive $6.54 per month for every $1000, so your monthly income would be $1635. According to the Social Security Administration, your life expectancy at age 65 is about 16.6 years = 199 months. If you lived exactly that long, you would receive a total of $199 \times \$1635 = \$325,365$. The rate of interest that your annuity would need to earn to last that long can be calculated from the amortization formula. For example, using $=\text{RATE}(199,1635,-250000)$ in Excel gives a monthly rate of 0.276%, for an effective annual rate of 3.32%.

SPOTLIGHT 22.3 What Actuaries Do

The Truth in Savings Act and the Truth in Lending Act specify that the APY for savings and the APR for loans must be calculated "according to the actuarial method." A loan might not be repaid, and the risk of that happening must be taken into account in setting the interest rate.

Actuaries are financial experts who assess the costs of risks and investigate the probability of various contingencies—for example, death or cancellation—that might occur. Actuaries are crucially involved in setting premiums. Their calculations take into account historical rates—such as the percentage of female 85-year-olds who live to be 86, or the percentage of unmarried male drivers under age 25 who have auto accidents—and project those rates and the accompanying costs into the future.

Other actuaries concentrate on setting up and evaluating pension and fringe benefit plans. For example, the city of Beloit, Wisconsin, hired a consulting actuary to estimate the current and future costs of free lifetime medical benefits to families of police and firefighters.

Another major activity of actuaries is managing return on investment. Contrary to

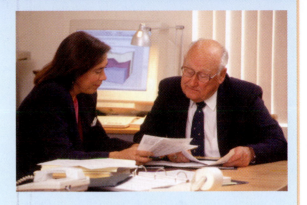

(David Young-Wolff/PhotoEdit.)

popular belief, insurance companies (particularly life insurance companies) do not earn all of their money from premiums paid. In fact, a substantial portion of their income comes from return on investment of financial *reserves*, funds that they are required to have to meet current and future insurance obligations.

Becoming an actuary requires training in mathematics, statistics, economics, and finance, and includes a sequence of professional exams taken over several years.

If you are female and retire now at the same age with the same $250,000 savings in a life income annuity, you would receive $6.30 per month for every $1000, or $1575 per month. Your life expectancy would be about 19.6 years = 235 months. If you lived exactly that long, you would receive a total of 235 × $1575 = $370,125. The rate of interest that your annuity would need to earn to last that long can be calculated from the amortization formula; using =RATE(235, 1575, −250000) in Excel gives a monthly rate of 0.358%, for an effective annual rate of 4.29%. The difference of this figure from that for a man probably reflects the company's use of different values for life expectancy, which vary with region of the country. ■

Notice that a man and a woman who save the same amount receive different monthly incomes at retirement: The woman receives less but for longer—

about 90% as much for 25% longer. Yet their living expenses are likely to be the same. That consideration has resulted in some companies offering "merged gender" rate schedules for annuity payments, so that the individual receives the same monthly payment regardless of gender.

REVIEW VOCABULARY

Add-on loan A loan in which you borrow the principal and pay back principal plus total interest with equal payments.

Adjustable-rate mortgage (ARM) A loan whose interest rate can vary during the course of the loan.

Amortization formula Formula for installment loans that relates the principal A, the interest rate i per compounding period, the payment d at the end of each period, and the number of compounding periods n needed to pay off the loan:

$$A = d\left[\frac{1 - (1 + i)^{-n}}{i}\right]$$

Amortize To repay in regular installments.

Annual percentage rate (APR) The rate of interest per compounding period times the number of compounding periods per year.

Annuity A specified number of (usually equal) payments at equal intervals of time.

Compound interest formula Formula for the amount in an account that pays compound interest periodically. For an initial principal A and effective rate i per compounding period, the amount after n compounding periods is $A = P(1 + i)^n$

Compounding period The fundamental interval for compounding, within which no compounding is done. Also called simply *period*.

Conventional loan A loan in which each payment pays all the current interest and also repays part of the principal.

Discounted loan A loan in which you borrow the principal minus the interest but pay back the entire principal with equal payments.

Effective annual rate (EAR) The effective rate per year.

Effective rate The actual percentage rate, taking into account compounding.

Equity The amount of principal of a loan that has been repaid.

Interest Money earned on a loan.

Nominal rate A stated rate of interest for a specified length of time; a nominal rate does not take into account any compounding.

Principal Initial balance.

Savings formula Formula for the amount in an account to which a regular deposit is made (equal for each period) and interest is credited, both at the end of each period. For a regular deposit of d and an interest rate i per compounding period, the amount A accumulated is

$$A = d\left[\frac{(1 + i)^n - 1}{i}\right]$$

Simple interest The method of paying interest on only the initial balance in an account and not on any accrued interest. For a principal P, an interest rate r per year, and t years, the interest I is $I = Prt$.

✔ SKILLS CHECK

1. With a $5000 discounted loan,

(a) you get a discount on payments.

(b) you get less than $5000.

(c) you get a discount on your next loan.

2. An add-on loan

(a) is a traditional simple-interest loan.

(b) uses compound-interest calculations.

(c) computes interest on the principal for the total loan period.

3. Your car dealer offers to finance a $6000 add-on loan at 3% to be repaid in four years of monthly payments. What is the monthly payment?

(a) $140
(b) $150
(c) $180

4. Repeat Skills Check 3, but the dealer offers a discounted loan instead under the same terms. How much is the loan for if you need to have proceeds (the amount that you get) of $6000?

(a) Exactly $6000
(b) Approximately $6818
(c) Approximately $7200

5. For the same proceeds from a loan (the amount that the borrower gets), the same number of months, and the same rate of interest,

(a) a discounted loan is always cheaper than an add-on loan.
(b) an add-on loan is always cheaper than a discounted loan.
(c) a discounted loan may or may not be cheaper than an add-on loan.

6. Credit-card interest

(a) is computed using compound interest.
(b) is computed using simple interest.
(c) is included in the late fees.

7. If a store credit account charges 1.5% interest each month, what is the effective annual rate?

(a) 18%
(b) More than 18%
(c) Less than 18%

8. If a store credit account charges 1.5% interest each month, what is the annual percentage rate?

(a) 18%
(b) More than 18%
(c) Less than 18%

9. Your credit union offers to finance a $6000 conventional loan at 4% to be repaid in four years

of monthly payments. What is the monthly payment?

(a) Approximately $135
(b) Approximately $138
(c) Approximately $145

10. The nominal rate of interest for a loan is

(a) the same as the effective rate.
(b) less than the effective rate.
(c) never greater than the effective rate.

11. If you finance $15,000 for 3 years at 6% compounded monthly, the monthly payments will be

(a) about $456.
(b) about $492.
(c) about $560.

12. If you establish a 30-year mortgage, most of the initial payments

(a) go toward reducing the balance.
(b) go toward paying the interest.
(c) pay insurance costs.

13. After 15 years of payments on a 30-year mortgage, the balance remaining is

(a) about one-third of the original balance.
(b) about one-half of the original balance.
(c) about two-thirds of the original balance.

14. An adjustable-rate mortgage

(a) has variable interest rates but maintains a fixed payment amount.
(b) has variable payment amounts.
(c) is always a better alternative to fixed-rate mortgages.

15. Payments for a 15-year mortgage

(a) are double the payments for a 30-year mortgage of the same amount.
(b) are about 50% to 80% more than payments for a 30-year mortgage of the same amount.
(c) are about 25% to 50% more than payments for a 30-year mortgage of the same amount.

16. Equity in a 30-year conventional mortgage grows

(a) linearly.
(b) logarithmically.
(c) exponentially.

17. A convenient rule of thumb is that for a 30-year mortgage at 6%, the monthly payment is about 0.6% of the loan. So, on a $100,000 mortgage, the monthly payment is about $600. How much of the first payment goes toward interest?

(a) All of it
(b) About $100
(c) About $500

18. Which of the following arrangements could be an ordinary annuity?

(a) Monthly payments, annual compounding
(b) Annual payments, monthly compounding
(c) Annual payments, annual compounding

19. If you just won a lottery jackpot paid in 25 equal annual installments of $1 million each, what is the present value for the jackpot?

(a) $25 million
(b) More than $25 million
(c) Less than $25 million

20. A life income annuity is designed to pay a fixed amount each period until

(a) the annuity runs out of money.
(b) you die.
(c) you reach your life expectancy.

EXERCISES   ■ **Challenge**   ◆ **Discussion**

Simple Interest

1. On August 13, 2001, you could buy $10,000 U.S. Treasury bonds that pay simple interest of

▶ 5.51% per year in February each year through maturity in February 2031, at a cost of $9,802 each; or

▶ 11.25% per year in February each year through 2015, at a cost of $15,529 each.

Each bond also returns $10,000 at maturity.

(a) What is the annual yield of each investment?
(b) Compare these two investment opportunities, which differ vastly in interest rate but also in cost.

2. In late September 2001, you could buy a U.S. Treasury bond offering 5.75% interest and maturing one year later. The face value of the bond was $10,000; this is the amount of the principal that the Treasury pays back to the owner at maturity. But resellers of this particular bond wanted $10,318. On this bond, what was the *current yield*—the annual percentage return to the purchaser at the current price?

3. You need to buy a car and need to finance $5000 of the cost. The dealer offers you a 5.9% add-on loan to be repaid in monthly installments over four years. How much is your monthly payment?

4. You have to make some home improvements— well, to be honest, they're really maintenance that you can't defer any longer!—and need to borrow $3000 to pay for them. You can get an 8.5% add-on loan from a savings and loan association to be repaid in installments over two years. How much is your monthly payment?

5. You are in the same situation as in Exercise 4, except that you find that you can get a 9% discounted loan from a loan company to be repaid in monthly installments over four years. What is the monthly payment on this loan?

6. You are in the same situation as in Exercise 4, except that you find that you can get an 8.5% discounted loan from a loan company to be repaid in monthly installments over five years. What is the monthly payment on this loan?

7. Suppose that you need $1000 (no less) and have available to you either an add-on loan or a discounted loan, both at the same interest rate and for the same period. Which will have the lower monthly payment?

8. Show algebraically that your claim in Exercise 7 is true in general for a loan of any amount.

Compound Interest

9. A credit-card bill of mine showed $500 due, with a minimum payment of $10 and daily interest rate of $r = 0.04932\%$. If I make no more charges on the card and pay $10 a month as soon as I get each bill, how long will it take to pay off the total? (*Hint:* The amortization formula can be changed algebraically into the form

$$(1 + i)^n = \frac{d}{d - iP}$$

Note that making the first payment immediately will reduce the principal P to be amortized to $500 - 10 = 490$. If I delay payment, I incur additional daily interest on the amount due. Evaluate the right-hand side. Then, using either a spreadsheet or the power key $\boxed{y^x}$ on your calculator, raise the value of $(1 + i)$ to higher and higher powers n until you find the smallest value for n that makes the left-hand side larger than the right-hand side.)

10. (Requires spreadsheet) Regarding the credit-card bill in Exercise 9: By paying $10 each month, approximately how much interest would I have paid by the time I pay off the original $500? (*Hint:* You won't be off by much if you estimate the last payment to be $10.)

11. (Requires spreadsheet) According to the regulations for the credit card discussed in Exercises 9 and 10, the minimum payment is supposed to be the greater of $10 or 2% of the balance (rounded down to the next higher dollar amount). Suppose that you have such a card and the balance is $1500. Neglect the rounding down and assume that you pay exactly 2% of the current balance each month until the balance reaches $500. Notice that if you make a payment of 2% and the bank charges daily interest of 0.04932% in each 30-day month, you in effect reduce the balance to $(1 - 0.02)(1.0004932)^{30} = 99.46043\%$ of what it was the previous month. How many 30-day months will it take to reduce the balance of

$1500 to $500? (Again, assume that you make payments right away.)

12. By making the minimum payment each month in the situation that exists in Exercise 11, approximately how much interest will you have paid by the time you pay off the original $1500?

13. Adding your answers from Exercises 9 and 11, determine how long it will take to pay off a balance of $1500 by making minimum payments.

14. (Requires spreadsheet) Assume the same situation as in Exercise 11, but instead of paying exactly 2% of the balance due, you make a minimum payment of the greater of $10 or 2% of the balance (rounded to the next higher dollar amount).

Conventional Loans

15. In September 2004 you could buy a new Kia Amanti car for $24,995 with a loan from the manufacturer at 0% (!) interest over 60 months. Assume that you had cash or a trade-in worth $2000. What was the monthly payment?

16. You can't afford the new car in Exercise 15 but need a better-than-junk car, and you need to borrow the entire $7000 of its cost. You can get a 48-month conventional loan—not an add-on loan or a discounted loan—from your credit union at 8.5% compounded monthly. What is your monthly payment?

17. A television ad that ran in February 2002 advertised a new 2002 Pontiac for 16% down with a five-year loan at 5.9%, with a payment of $19.09 per month for each $1000 financed. Check whether the monthly payment is correct (I may have miscopied it from the fine print that showed for three seconds at the end of the ad!).

18. For the same car mentioned in Exercise 17, a car dealer offered a 7.9% APR conventional loan over 48 months but a 6.8% APR conventional loan over 60 months.

(a) What is the payment per $1000 under each loan?
(b) With the 60-month loan, it takes longer for the dealer to be paid in full. So what is the

advantage to the dealer in offering a lower percentage loan for 60 months, instead of a higher percentage loan for 48 months? (Thanks to Terence Blows of Northern Arizona University.)

19. Suppose that you have good credit and can get a 30-year mortgage for $100,000 at 6.5%. What is your monthly payment?

20. Assume the same situation as in Exercise 19, except that your credit is not as good and the rate that you are offered is 7.125%.

21. Assume the same situation as in Exercise 19, but you inquire about a 15-year loan instead. You are offered 6.125%. What is your monthly payment?

22. Assume the same situation as in Exercise 21, but your credit is not as good, and you are offered 6.75%. What is your monthly payment?

23. For the mortgage in Exercise 19, how much equity would you have after five years?

24. For the mortgage in Exercise 20, how much equity would you have after five years?

25. For the mortgage in Exercise 21, how much equity would you have after five years?

26. For the mortgage in Exercise 22, how much equity would you have after five years?

27. Despite a filter, lots of spam gets into my email. For a while I was getting mortgage offers, such as "$160,000 for less than $735 per month" (for a 30-year mortgage). What would be the corresponding interest rate?

28. At about the same time as the offer in Exercise 27, I was getting spam offers for 3.6% mortgages. In terms of the monthly payment, would this be a better deal than the mortgage in Exercise 27? The 3.6% was about 2% below the going rate at local banks. What would you guess might explain the difference?

29. Suppose that you and two friends decide to live off-campus in your senior year. One of them (who has wealthy parents) suggests that instead of renting an apartment, you could buy a house

together, live in it for your senior year, then rent it out or else sell it. Assuming that (with the help of her parents and their good credit rating) you could get a mortgage for $180,000 to buy a house near the campus, what would be the monthly mortgage payment on a 30-year mortgage at 6.75%?

30. For the add-on loan in Exercise 3, what is the APR? What is the EAR?

31. For the add-on loan in Exercise 4, what is the APR? What is the EAR?

32. For the discounted loan in Exercise 5, what is the APR? What is the EAR?

33. For the discounted loan in Exercise 6, what is the APR? What is the EAR?

34. For the credit card in Exercises 9 and 10, what is the APR? What is the EAR? What annual rate must the credit-card company tell its customers applies?

For Exercises 35 and 36, refer to the following.

Payday lenders provide short-term loans, usually for two weeks, until the borrower's next payday, in one of two forms:

▶ The borrower writes a postdated check to a lender and the lender advances the cash, minus a finance charge; this is in effect a discounted loan.

▶ The borrower signs an authorization for the lender to debit the borrower's bank account on payday for the amount of the loan plus the finance charge. This is in effect an add-on loan with a single installment payment.

The industry grew in the 1990s out of check-cashing services to approximately 12,000 firms in 31 states and the District of Columbia, with a revenue of over $2 billion by 2000. Comptroller of the Currency John D. Hawke, Jr., noted that "California alone has more payday loan offices—nearly 2,000—than it does McDonalds and Burger Kings."

The average loan is $300, for two weeks, with an average fee of $54.

35. For the discounted loan format, and average loan size and fee, what is the APR?

36. For the add-on loan form, and average loan size and fee, what is the APR?

For Exercises 37 and 38, refer to the following.

Many income tax preparation services, including the large national chains, offer refund anticipation loans (RALs), or "rapid refunds." These are similar to payday loans in providing an advance on anticipated income—in this case, a tax refund. The loan is repaid when the IRS pays the refund, usually about 7 to 17 days after the loan is made. The cost of the loan is deducted from the proceeds to the client, so this is a discounted loan. The RAL business takes in about $2 billion each year (including tax preparation and check-cashing fees) to arrange payment of the earned-income tax credit to working parents, about 7% of the total of this aid to poor families. A RAL for an anticipated refund usually is issued for a flat fee, often $88, and the average loan is $1500.

37. Suppose that the RAL speeds the refund by 7 days. What is the APR for the average RAL?

38. Suppose that the RAL speeds the refund by 17 days. What is the APR for the average RAL?

39. When interest rates drop, it may become attractive to refinance your home. Refinancing means that you acquire a new mortgage to borrow the current principal due on your home and use the proceeds to pay off your old mortgage. You then begin a new 15- or 30-year mortgage at the new, lower interest rate. A second factor that reduces your monthly payment is that the equity you accumulated under the old mortgage reduces the amount that you have to borrow under the new mortgage. Suppose that you have an existing 30-year $100,000 mortgage at 8.375%, on which you have been paying for five years, and you are considering refinancing at 7.0%.

(a) What is your payment under the old mortgage?

(b) How much equity do you have in the home?
(c) If you use all your equity to reduce the amount of the new mortgage, how much will your monthly payment be?
(d) How long is the payback period for the $2000 loan charge—that is, how many months will it take before you have saved $2000 in monthly payments?

40. (Requires spreadsheet) Even if interest rates go up, you may find it beneficial to refinance if your financial circumstances suddenly worsen. Consider an actual individual (as recounted in the *Wall Street Journal* of October 30, 2001), a 39-year-old plumber whose income was cut in half by worsening economic conditions. She had missed eight monthly mortgage payments totaling $14,000, still owed $160,000 on the mortgage, and was in danger of foreclosure and losing her home. Data on her loan were not given in the *Journal*, but data that fit are a 15-year mortgage for $180,000 at 8.375%.

(a) What was her monthly payment?
(b) How long had she been paying on the mortgage?
(c) Under a "loan modification" program, the bank added the $14,000 in missed payments to her existing debt of $160,000. Suppose that the bank offered her a new 30-year loan for the total of $174,000 at 8.75% (despite interest rates in late 2001 being generally lower than when she got her first loan, her delinquency on that loan made her a worse risk and hence she incurred a higher interest rate). How much was her new monthly payment?

■ **41.** Fisher's effect (Chapter 21, p. 817) gives the real rate for borrowing, sometimes called the *real cost of capital*. Suppose that you buy a house with a $100,000 mortgage at an interest rate of 6.75% and that inflation remains at 3% for the duration of the mortgage. You make monthly payments in constantly deflating dollars. Simplify the situation by supposing that you sell the house one month after you buy it (the neighbors really drove you crazy, right from the start).

(a) What would the house sell for if it kept up with inflation of 3%/12 = 0.25% per month?

(b) Calculate your mortgage payment at the end of the month. How much is interest and how much is principal?

(c) After making the mortgage payment, paying off the balance of the $100,000 cost of the house, and selling the house for the amount in part (a), how much did it cost you to own the house for the month?

(d) The cost in part (c) is in deflated dollars. Using a formula from Chapter 21, convert this to constant (month-before) dollars and divide by $100,000 to express the result as a percentage rate. This is the real rate of interest.

42. One of the advantages of buying a home with a fixed-rate mortgage is that your payment stays the same but your earnings and the value of your home are likely to go up with inflation. You are paying back the loan with dollars of lesser value.

Consider the following scenario. Suppose that you buy a "starter" two-bedroom home for $105,000 under a special program for first-time home buyers that requires a down payment of only $5000. You have a 30-year fixed-rate mortgage for $100,000 at 7%, on which the monthly payment is $665.30. You also have a $2000 one-time expense in closing costs and annual costs of $200 for insurance and $2000 for property taxes.

You live in the home for five years and spend $10,000 on maintenance, upkeep, and improvements. You then sell the home for $125,000, pay a realtor $9000 to sell it, and pay closing costs of $500 (for title insurance and other costs). Finally, it costs $3000 to move.

(a) Make out a balance sheet of revenue and expenses. How did you make out on owning the home?

(b) Remember, you also got to live in the home without paying rent! Translate the cost of owning the home into an equivalent monthly rent.

Annuities

43. The largest amount won by an individual in the Florida Lotto lottery was $81.6 million on March 29, 2000 (other jackpots have been higher but were shared). The advertised jackpot amount is paid in an annuity of 30 annual payments, including one immediate payment.

Instead of the 30-year payment plan, the winner may choose to receive a smaller lump sum right away in cash. If not, that sum (minus the initial payment) is invested by the State Board of Administration in U.S. government securities on behalf of the player. The cash invested, plus all interest earned, goes to the winner.

Suppose that in April 2000 the state of Florida could buy U.S. securities paying an interest rate of 7%. What was the present value of the jackpot annuity—that is, how much could the winner have received in cash instead? (We say that $81.6 million is the *future value* of the annuity.)

44. Suppose that you retire at age 65 and in addition to Social Security need $2000 per month in income. Based on an expected lifetime of 16.6 more years (for men) or 19.6 more years (for women), how much would you have to invest in a life income annuity earning 4% to pay you that much per year?

45. (Spreadsheet helpful) We return to your forward-thinking roommate of Exercises 41 and 42 in Chapter 21, who has already planned her retirement. She now wants to retire at 55 in 2040 with a steady income of $50,000 a year in 2005 dollars. Her plan is to amass a fortune and live off just the interest. But why not live it up and spend every last dollar? You suggest to her that instead she plan to buy in 2040 a special kind of 45-year annuity (in case she lives to 100!) whose payments increase 4% each year. (Recall, she assumes 7.2% APR on her investment and 4% inflation.) What will such an annuity cost? How much will she have to invest quarterly between 2005 and 2040? (*Hint:* What is the present value in 2040 dollars of the annuity payments?) (Under this plan she celebrates her 100th birthday by going broke!)

46. (Spreadsheet helpful) Your other roommate also started a retirement fund in 2005, but she took a few other factors into account, too (she plans to be an actuary). She aims to start retirement with $50,000 a year in 2005 dollars in 2052 when she is 67, the age at which she can receive government health insurance (Medicare) and full Social Security payments. She consulted the Social Security Administration, which estimates that a medium-earning person retiring then can expect to receive $1628 per month in 2001 dollars. She estimates (more conservatively) a steady 6% APY on her investment and a steady 3% annual inflation. She plans to live to 100 (why not try?) and to buy a 33-year annuity in 2052

(also paying 6%). The annuity would supplement Social Security payments to give a total the first year of $50,000 in 2005 dollars, and the same *current* dollar amount each year of the 33 anticipated years of retirement. What must the size of her nest egg be, and what should be her monthly investment over the 47 years, to achieve her goal?

47. Plan as we may, the future is not certain. The roommate in actuarial studies knows that various plans are afoot to "bail out" Social Security. One such plan would raise the age for full Social Security benefits to 68 and reduce those benefits by 32.5% for those retiring in 2053. As in Exercise 46, calculate how much she should accumulate (over 48 years) to meet her financial goal for the remaining 32 years until she reaches 100.

APPLET EXERCISES

To do these exercises, go to www.whfreeman.com/fapp7e.

There are two ways to buy a car: save up and pay cash or borrow the money from a bank. In the

Buying a Car: Cash vs. Loan applet, you can explore just how much more expensive it is to borrow the money.

WRITING PROJECTS

1. In recent years, incentives from auto manufacturers to potential customers have taken the form of offering either a reduced interest rate on the loan for the car or else a rebate (reduction in price) on the cost itself. In fall 2004 you could buy a 2005 Toyota Camry for $13,570, with one of the following options for payment:

▶ $750 rebate
▶ 1.9% APR over 24 months
▶ 1.9% APR over 36 months
▶ 2.9% APR over 48 months
▶ 3.9% APR over 60 months

Suppose that you could afford a $2000 down payment and could get a loan from a credit union at 6.0% over 60 months if you opt for the $750 rebate.

(a) What is your monthly payment under each option?
(b) Suppose that the rate of inflation over the course of the loan is a steady 3% per year. How do the various options compare in terms of present value of the loan?
(c) Locate current advertised incentives for a car that you would like to buy and compare them in an essay of two to three pages.

2. A substantial proportion of new cars today are not sold but leased. Contact a local car dealer about a car that you are interested in and find out the details on leasing. Compare the cost of the lease and associated expenses with the cost of purchasing and owning the car. Include estimated maintenance, repair, and insurance costs for each option. Which seems like a better deal, and why? Write two to three pages describing and comparing the two options.

3. Banks typically offer mortgages with various combinations of interest rates and points. Loans at a higher interest rate tend to come with more points, and vice versa. Like interest, points are deductible on income taxes, but in the year in which the loan is made. In recent years, home buyers have (when offered the option) generally favored paying fewer points. Suppose that you have a choice between a mortgage at 6% with 2 points (2%) and a mortgage at 8% and no points. Which would you choose, and why? Does it make a difference how long you are planning to own the home? Or how expensive the home is? Write a page justifying your decision.

4. Explore actual costs of homes in your area, mortgages with local banks (including closing costs), and property taxes and insurance. Come up with data such as those in Exercise 40 and make out a corresponding balance sheet for five-year ownership.

SUGGESTED READINGS

KASTING, MARTHA. *Concepts of Math for Business: The Mathematics of Finance* (UMAP Modules in Undergraduate Mathematics and Its Applications: Module 370–372), COMAP, Inc., Arlington, Mass., 1980.

MILLER, CHARLES D., VERN E. HEEREN, and JOHN HORNSBY. Consumer mathematics. In *Mathematical Ideas,* 9th ed., Addison Wesley Longman, Boston, 2001, pp. 786–845.

VEST, FLOYD, and REYNOLDS GRIFFITH. The mathematics of bond pricing and interest rate risk, *Consortium* (COMAP) no. 59 (Fall 1996): HiMAP Pullout Section 1–6.

YAREMA, CONNIE H., and JOHN H. SAMPSON. Just say "Charge it!" *Mathematics Teacher* 94 (7) (October 2001), 558–564. Shows how to apply the savings formula and the amortization formula and graph the results on the TI-83 calculator. Notes that the 78% of undergraduates in the United States who have credit cards carry an average debt of more than $2700, with 10% owing more than $7000.

SUGGESTED WEB SITES

http://www.lendingtree.com/stmrc/calculators1.asp Java applet calculators (for any platform) to calculate payments and amortization schedules for conventional loans, adjustable-rate mortgages, auto loan vs. home-equity loan, and credit-card payoff. (*Note:* Lending Tree, Inc., is a loan broker; mention here of calculators at its Web site does not imply endorsement of its other services by this book's authors, editors, or publisher.)

http://www.udayton.edu/sba/wf/wksht/wksht5.htm Excel spreadsheet templates for comparing consumer loans, tax-deferred retirement plans, mortgage loans, present value, lease vs. buy, and more.

http://www.edmunds.com/apps/calc/Calculator Controller?pmtcalAction=apr_cash_calc Commercial site offering a calculator to compare rebate vs. interest-rate offers for car purchase. (*Note:* Edmunds is a loan broker; mention here of calculators at its Web site does not imply endorsement of its other services by this book's authors, editors, or publisher.)

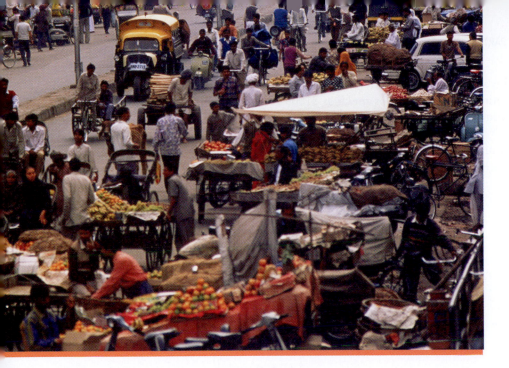

The Economics of Resources

We use resources all the time: food, money, natural resources, labor, time. Time and labor are by definition consumed in their use. Some resources, such as annual flowers, are perishable, while others, such as money and standing timber, can be used now or saved for later use.

Our use of resources involves a complex intermixture of biological, ethical, practical, technical, and economic issues:

▶ How many people will there be in the world in another 20 or 40 years, and how will those numbers affect resources available to you and to them?

▶ What should we consume for our own use, give to others more needy, or leave for future generations, in terms of wealth, well-being, and wilderness?

▶ Will our standard of living keep getting better, or is it not maintainable, even at current levels, in the long run?

▶ How long will it be, at current patterns of use, until we exhaust some particular resources?

▶ Can we develop more efficient technology, so that we can get more out of the resources that we use?

▶ How do we balance economics, the needs and desires of individuals, with other important considerations? How much would it cost—how much is it worth—to ensure that we do not let tigers, elephants, or rhinos go extinct?

On a personal level, we face similar but more immediate questions about the balance between consumption and conservation. How much must you save to

859

be able to afford to retire? How do you take into account that $1000 after you retire won't be worth as much as $1000 today? Once you retire, how much can you spend without exhausting your nest egg?

In Chapters 21 and 22, we explored mathematical models for saving, accumulating, and borrowing—the building up of resources. From Chapter 21, we use here only two formulas, specialized to annual interest rate r:

Compound interest formula: If a principal P is deposited into an account that pays interest at rate r per year, then after n years the account contains the amount

$$A = P(1 + r)^n$$

Savings formula: For a uniform deposit of d per year (deposited at the end of the year) and an interest rate r per year, the amount A accumulated after n years is

$$A = d\left[\frac{(1 + r)^n - 1}{r}\right]$$

A reader who can use these formulas can proceed in this chapter without first reading Chapters 21 and 22.

In this chapter, we consider models for processes in the other direction—the decay, depletion, or spending down of resources. Some resources, however, replenish themselves regularly. Our models of them will lead us into the mathematics of dynamical systems and the mathematical concept of chaos. The models of this chapter provide important insights into the consequences of answers to the preceding questions.

23.1 Growth Models for Biological Populations

The economic models for savings are flexible and can be applied to other arenas. We now use a geometric growth model to make rough estimates about sizes of human populations. In addition to the **rate of natural increase**—the annual birth rate minus the annual death rate—we must take into account net migration. The sum of the two, in the terminology of financial models, is the effective rate.

Birth, death, and migration rates rarely remain constant for long, so projections must be made with care. In the short run, however, predictions based on the model may be useful. Let's apply this model to two questions about the population of the United States.

EXAMPLE 1   Predicting the U.S. Population

The U.S. population is increasing at an average growth rate of 1.0% per year to an expected total of 301 million by mid-2007. What is the anticipated population for mid-2011? What is it if the rate of natural increase changes to 1.2% per year or to 0.8% per year?

SOLUTION We apply the compound interest formula with initial population size ("principal") 301 million. Using a year as the compounding period and the formula $A = P(1 + r)^n$, where $n = 4$, the projected population size in 2011 for a rate $r = 0.010$ is

$$\text{population in 2011} = (\text{population in 2007}) \times (1 + \text{growth rate})^4$$
$$= 301,000,000 \, (1 + 0.01)^4$$
$$= 301,000,000 \, (1.04060401)$$
$$\cong 313,000,000$$

Because the original estimates of population and growth rate are approximations, we round off the final answer. The result of a calculation can't be more precise than the ingredients.

In the same way, with a growth rate of 1.2% per year, we predict a population of 316 million, while a growth rate of 0.8% per year yields a prediction of 311 million.

So an uncertainty of two-tenths of one percentage point, or 0.002, in the growth rate has major implications, even over fairly short time horizons. The presence or absence of 5 million people would have a significant impact on our social and economic systems! Indeed, much of the concern over long-range funding of Social Security programs results from uncertainties over birth and immigration rates. Figure 23.1a gives a graph of the U.S. population in 2000, structured by age and sex. Figure 23.1b shows possible futures for India. ▪

Rates of increase in most developing nations are much higher than in industrialized nations. With a growth rate of 2.8%, Nigeria, Africa's most populous country, will grow from 135 million in mid-2007 to 231 million in mid-2030, an increase of more than two-thirds. Such projections raise concern over providing sufficient food and resources for all people.

It is not just the number of people that is crucial, but also the **population structure.** In poorer countries, the proportion of the population over 60 years of age will be 20% by 2050, compared with 8% now; in Japan, it will be 40%.

Limitations on Growth

A population that keeps adding a fixed percentage each year, like a bank account accumulating compound interest, would eventually grow to astronomical numbers. But no biological population can continue to increase without limit (see Spotlight 23.1). Its growth is eventually constrained by the availability of resources such as food, shelter, and psychological and social "space." There may be a maximum population size that can be supported by the available resources, the **carrying capacity** of the environment.

FIGURE 23.1 Graphs of the population of the United States in 2007 grouped by age and gender, and of scenarios for the population of India in 2030.

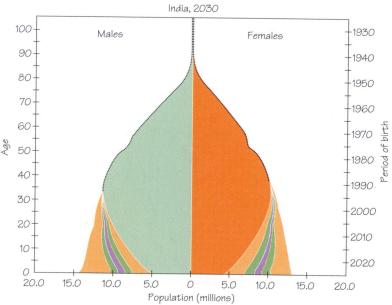

The **logistic model** for population growth takes carrying capacity into account by reducing the annual increase rP by a factor of how close the population size P is to the carrying capacity M:

$$\text{growth rate } P' = rP\left(1 - \frac{\text{population size}}{\text{carrying capacity}}\right) = rP\left(1 - \frac{P}{M}\right)$$

How many people can the world hold? Are developing countries heading for a population disaster? Will falling fertility play havoc with Social Security in the United States? Will aging result in 50% of Japanese being over 60 in 2100?

Answers come from mathematical modeling of the future from predicted trends. The best analyses suggest a probability distribution over a range of estimates. They project separately by age, gender, education, and other characteristics. They try to factor in improvements in agriculture, spread of diseases (especially HIV), changes in urbanization, increases in economic aspirations, and the potential for climate change (for example, from global warming).

A basic concept is *total fertility rate* (TFR), the average number of births per woman. Absent catastrophes (such as war or disease), a rate of 2.1 continues a population at the same size. A model that assumes a value above 2.1 will predict an ever-increasing population; one that assumes a value below 2.1 will predict an ever-dwindling one. Most of the world's population growth will occur in the lesser-developed countries, whose TFR values are well above 2.1 (for example, it is 4.6 for Africa). Many countries in Europe have TFR values near 1; without immigration, they will lose population and struggle with fewer workers to provide social benefits to the elderly.

The situation with China illustrates *demographic momentum*. Even though its fertility rate has been below replacement level for 20 years, the number of women in the childbearing years was (and still is) so large that China's population will continue to grow until 2040.

The most sophisticated models try to assess how the TFR will vary with changing social circumstances. The most important single factor impacting fertility is education of women. There is a strong negative association between level of education and TFR, and the effect can be very large: In India and China, women with some college education have on average only half as many children as women with no education. Hence, a country's policies about

(Will & Deni McIntyre/CORBIS.)

education, and their success, may directly impact future population levels.

As we have revised this book for successive editions, we have seen population projections change. The estimates have decreased, because fertilities have declined. The key questions are how to model such declines, whether they will continue, and how they will adapt to other world changes. In Spotlight 21.1, we saw Thomas Robert Malthus's over-simplified prediction that mere arithmetic growth in food supplies would limit the geometric increase of human populations. Some demographers now think that population growth will remain a serious problem in some parts of the world but that global population may stabilize or even decline after 2050.

How many people the world will have depends on how well we as a world conserve the environment, distribute food, provide jobs, produce and consume energy, and make other critical decisions about our money and resources. The key concern is the quality of life of *all* people. Political and economic events in far corners of the world, and even natural disasters such as the tsunami at the end of 2004, impact us all. Neglecting problems faced by increasing numbers of poor people provides no security, peace, or moral refuge for anyone.

As the population increases toward M, the growth rate decreases. For a population equal to the carrying capacity ($P = M$), the growth rate is zero.

If the population ever exceeds the carrying capacity, the growth rate becomes negative (because $P > M$) and the population decreases. The carrying capacity is the long-range capacity to support the population, so the population could exceed it for short periods of time. This could happen either because the population grows very rapidly and surges above the carrying capacity or because of a sudden decrease in the food supply, thus temporarily lowering the carrying capacity, as happens to deer and other animals in winter. The logistic model is a simple model, but it provides excellent predictions for some populations.

EXAMPLE 2 ⁞⁞⁞⁞⁞   Logistic Model for the U.S. Population

How well does the historical U.S. population fit a logistic model?

SOLUTION The U.S. population from 1790 to 1950 closely followed a logistic model with $r = 0.031$, P = population in 1790 = 3,900,000, and $M = 201$ million. In the first decades after 1790, the population was a small fraction of this carrying capacity, and it grew at close to the rate r of 3.1% per year (a rate higher than in many developing nations today). By 1920, the U.S. population had reached 106 million, a little more than half of the carrying capacity, and the growth rate had slowed by about one-half, to 1.5% per year (see Figure 23.2).

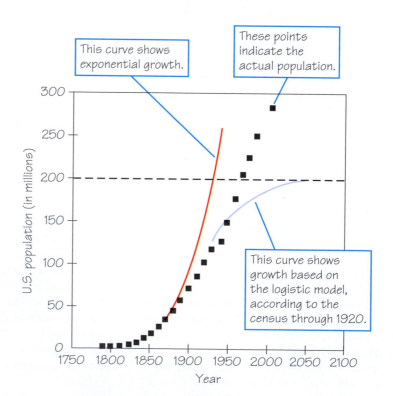

FIGURE 23.2 U.S. population by year, showing actual growth, exponential (geometric) growth, and logistic growth.

The mid-2007 U.S. population of 301 million far exceeds the hypothesized carrying capacity of 201 million. The structure of the U.S. population changed, from a large proportion of people making their livings on family farms to a highly urbanized society. The average number of children per family shrank. As the structure changed, the model based on the prior structure gradually became invalid. ∎

The logistic model applies not only to population growth limited by carrying capacity but also to modeling the spread of a technology or product, such as DVD players or flat-screen plasma TVs. Initially, sales are slow. Then they begin to climb rapidly. Finally, as the market gets saturated, new sales slow. We will see in the next section that the logistic model also can apply to exhaustion of nonrenewable resources.

23.2 How Long Can a Nonrenewable Resource Last?

People use resources, some of which are renewable, but others are not. In this section, we model depletion of nonrenewable resources. In the next, we treat renewable resources.

A **nonrenewable resource** is one that does not tend to replenish itself. Gasoline, coal, and natural gas are important examples, while lottery winnings and inheritances could be examples from personal affairs. There is no practical way to recover or reconstitute these resources after use. Some substances, such as aluminum or the sand used to make glass, are potentially recyclable, but to the extent that we do not recycle them, they, too, are nonrenewable.

For a nonrenewable resource, only a fixed supply S is available. Even without human population increases, we face dwindling nonrenewable resources. We are interested in the question: How long will the supply of a resource last?

As long as the rate of use of the resource remains constant, the answer is easy. If we are using U units per year and continue using U units per year, then the supply will last S/U years. This kind of calculation is the basis for statements such as those claiming that at the current rate of consumption, U.S. recoverable coal reserves will last 250 years or that the U.S. strategic reserve of gasoline (stored in underground salt domes in the South) will last 60 days.

However, the rate of use of resources tends to increase with population and with a higher standard of living. For example, projections for use of electric power often assume that use will increase by a fixed percentage each year. This is the simplest situation (apart from constant usage) and one that we can easily model.

Suppose that $U_1 = U$ is the rate of use of the resource in the first year (this year), and that usage increases $r = 0.05 = 5\%$ each year. Then the usage in the second year is

$$U_2 = U_1 + 0.05U_1 = 1.05U$$

and usage in the third year is

$$U_3 = U_2 + 0.05U_2 = 1.05U_2 = 1.05(1.05U) = (1.05)^2U$$

Generalizing, we see that usage in year i will be $(1.05)^{i-1}U$. Total usage over the next five years, for example, will be

$$U + (1.05)^1 U + (1.05)^2 U + (1.05)^3 U + (1.05)^4 U$$

This situation should remind you of the accumulation of regular deposits plus interest (see Chapter 21). Here the usage U corresponds to a deposit, and the increasing rate of use r corresponds to the interest rate. We may think of the situation as making regular withdrawals (with interest) from a fixed supply of the nonrenewable resource. The savings formula gives

$$A = d\left[\frac{(1 + r)^n - 1}{r}\right]$$

In the resource situation, A is the accumulated amount of the resource that has been used up at the end of n years, and U is the initial rate of use. We have

$$A = U\left[\frac{(1 + r)^n - 1}{r}\right]$$

To find out how long the supply S will last, we set the supply S equal to the cumulative use A over n years and then determine what n has to be. We have

$$S = U\left[\frac{(1 + r)^n - 1}{r}\right]$$

We perform some algebra to isolate the term involving n, to get

$$(1 + r)^n = 1 + \frac{S}{U}r$$

At this point, to isolate n, we have to take the natural logarithm of both sides.

$$\ln[(1 + r)^n] = n \ln(1 + r) = \ln\left(1 + \frac{S}{U}r\right)$$

thus gives the final expression

$$n = \frac{\ln[1 + (S/U)r]}{\ln(1 + r)}$$

which may look complicated but is quite easy to evaluate on a calculator. The expression S/U is called the *static reserve*, and n is called the *exponential reserve*.

Static Reserve and Exponential Reserve

The **static reserve** is how long the supply S will last at a particular constant annual rate of use U, namely, S/U years. The **exponential reserve** is how long the supply S will last at an initial rate of use U that is increasing by a proportion r each year, namely

$$\frac{\ln[1 + (S/U)r]}{\ln(1 + r)} \text{ years.}$$

U.S. Coal Reserves

EXAMPLE 3

Coal accounts for 30% of U.S. energy use. We noted earlier that recoverable reserves of U.S. coal would last about 250 years at the current rate of use, so the static reserve is 250 years. How long would the supply last if the rate of use increases 2.25% per year, as it did from 2002 to 2003?

SOLUTION The corresponding exponential reserve is

$$n = \frac{\ln[1 + (250)(0.0225)]}{\ln 1.0225} = \frac{\ln 6.625}{\ln 1.0225} = 85 \text{ years}$$

That's quite a difference! ■

   We must not take such projections as exact predictions. Estimates of supplies of a resource may underestimate how much is available, and previously unknown sources may be discovered or the technology improved to extract previously unavailable supplies. In addition, as supplies dwindle, the economic considerations of supply, demand, and price come into play. We will never completely run out of oil. It will always be available "at a price."

   We must not take such projections lightly, either, because we are discussing resources that, once used, are gone forever. In any projection, it is very important to examine the assumptions, because small differences in the rate of increase of use can make big differences in the exponential reserve.

Using Up Retirement Savings

EXAMPLE 4

Suppose that you begin retirement with $1 million in savings, and you don't trust banks or the stock market, so you keep it all under your mattress. Suppose that it costs you $50,000 per year to live at your accustomed standard of living and there is no inflation. How long will your retirement nest egg last?

SOLUTION The static reserve is $1,000,000/$50,000 per year = 20 years. If, however, there is constant 5% per year inflation, then it will cost you increasingly more per year to live, so you should realize that your savings will last only for the length of the exponential reserve, which is

$$ n = \frac{\ln(1 + 20(0.05))}{\ln 1.05} = 14.2 \text{ years} $$

You have a fine strategy if you expect to live just 14 more years and want to die broke! ■

In our examples so far, we have assumed that the resource is just sitting there, waiting to be used up. For many natural resources, however, we have to find and develop new sources. As doing that becomes more difficult and more costly, at some point the exponentially increasing demand outstrips the ability to meet that demand.

Such a situation is modeled well by the logistic model famously applied to oil by M. King Hubbert, director of Shell Oil Company's research laboratory. Figure 23.3 shows data for cumulative U.S. oil production through 2001 compared with the logistic curve for the ultimate production of $M = 240$ gigabarrels (240 billion barrels). In 1956, Hubbert predicted that U.S. production would peak in the early 1970s (it did) and decline steadily thereafter (it did that, too, except for a blip from Alaska in the 1980s) (see Figure 23.4, whose curve is similar to but "heavier in the tails" than the normal distribution curve of Chapter 5).

If we rearrange the logistic equation on p. 862 by dividing both sides by P and doing a little algebra, we get

$$ \frac{P'}{P} = r - \frac{r}{M} P $$

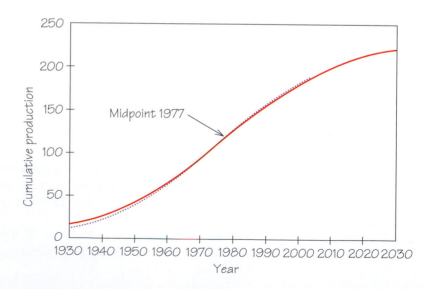

FIGURE 23.3 Logistic model (dashed line), assuming ultimate production of 240 billion barrels, and actual data (points) for cumulative U.S. oil production, in billions of barrels.

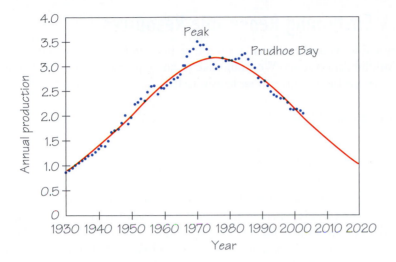

FIGURE 23.4 U.S. oil production, in billions of barrels versus year, and production as predicted by the logistic model, assuming ultimate production of 240 billion barrels.

You can recognize this as the equation of a straight line $y = a - bx$, where P'/P takes the role of y and P takes the role of x. In other words, for a logistic model, if we graph P'/P against P, we get a straight line. Figure 23.5 shows that the data fit the Hubbert model well.

The world's oil and gas are running out far faster than most people realize or than their governments are willing to acknowledge (see Exercises 11 and 12). The need for "affordable" fuels will likely soon dominate the political agendas of the entire world, particularly since experts have long predicted that world oil output would peak around 2006.

FIGURE 23.5 Display of fit of data to the logistic model (dashed line) for U.S. oil production. The line follows the equation $y = 0.054 + 0.000225x$.

23.3 Sustaining Renewable Resources

A **renewable natural resource** is a resource that tends to replenish itself, such as fish, wildlife, and forests. We would like to know how much we can harvest and still allow for the resource to replenish itself.

EXAMPLE 5   College Endowment

Donors sometimes give money to colleges and universities to fund prizes, scholarships, or professorships, which they would like to see continue in perpetuity. The college invests the donor's money, so that this resource renews itself, and "harvests" the earnings to fund the project. Suppose that the college can earn 8% on its endowed funds. Then an individual who wants to make a one-time donation to fund an annual full scholarship (tuition, books, room and board) of $25,000 has to donate enough so that the fund earns that much; the donation has to be $25,000/0.08 = $312,500. But what about inflation?

SOLUTION   The current cost of $25,000 will increase with inflation. If the donor wants each year's future interest to cover that year's full cost, the donation has to be larger. If the college endowment earns 8%, but there is 3% inflation in the annual cost, the net interest is only about 5% (actually, 4.85%; see Fisher's effect in section 21.6, p. 817), and the amount of the donation has to be $25,000/0.05 = $500,000 (more accurately, about $25,000/0.0485 ≈ $515,000). ■

Other renewable resources are biological populations. We concentrate on the subpopulation with commercial value. For a forest, this might be trees of a commercially useful species and appropriate size. We measure the population size as its **biomass,** the physical mass of the population. For example, we measure a fish population in pounds rather than in number of fish, and a forest not by counting the trees but by estimating the number of board feet of usable timber.

Reproduction Curves

Our models for growth include many simplifications. Real populations may behave according to one of our models or another known model. Complicated factors that can affect populations, such as climatic or economic change, may mean that the only way to understand a population is to plot a graph of its size over time. Either from such a graph or from a model, we can construct a **reproduction curve,** which predicts next year's population size (biomass) based on this year's size. Although the precise shape of the curve varies from one population to another, the shape in Figure 23.6 is typical. For all possible sizes, it shows for the size this year (on the horizontal axis), the size next year (on the vertical axis), taking into account growth in size of continuing members and addition of new members, minus losses due to death and other factors.

Let x on the horizontal axis be a typical size of the population in the current year. The size *next* year is given by the height of the curve above the point marked x. This value is denoted by $f(x)$. (You can think of f as standing for "function of," or even as "forthcoming.")

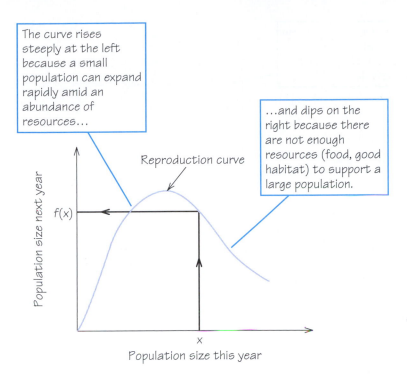

The curve rises steeply at the left because a small population can expand rapidly amid an abundance of resources...

...and dips on the right because there are not enough resources (food, good habitat) to support a large population.

Reproduction curve

$f(x)$

Population size next year

x

Population size this year

FIGURE 23.6 A typical reproduction curve.

Figure 23.7a shows the same reproduction curve, plus the broken line $y = x$ (which makes a 45° angle with the horizontal axis). You can trace what happens for various choices for x. For an x for which the curve is above the broken line, next year's size, $f(x)$, is larger than this year's, x. In Figure 23.7b, the **natural increase,** or gain in population size, is shown as the length of the green vertical line from the broken line to the curve, which in algebraic terms is $f(x) - x$. For an x for which the curve is below the broken line, next year's size is smaller than this year's and $f(x) - x$ is negative. For the size labeled x_e, for which the curve crosses the broken line, the size is the same next year as it was this year. This is the *equilibrium population size.*

Equilibrium Population Size

An **equilibrium population size** does not change from year to year.

Sustained-Yield Harvesting

In the case of fishing, for example, the harvest **yield** h depends on both the population x of fish and the amount of fishing effort (number of boats, hours of fishing). If the fishing fleet fishes the same banks with the same effort in a year when there are only half as many fish as usual, we can expect the fleet to catch only half as many fish. A simple model is that for any particular level of fishing

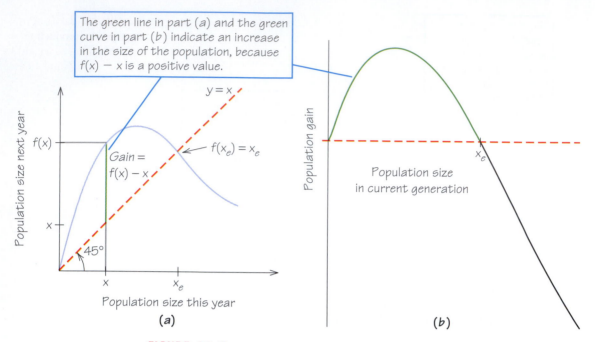

The green line in part (a) and the green curve in part (b) indicate an increase in the size of the population, because $f(x) - x$ is a positive value.

FIGURE 23.7 Depiction of the natural increase (gain) in population from one year to the next. The population size x_e is the equilibrium population size, for which the population one year later is the same, or $f(x_e) = x_e$.

effort, the harvest is proportional to the fish population, so that a graph of harvest versus population would be a straight line with a steeper slope corresponding to greater effort.

For each level of fishing effort, there is a level of **sustainable yield,** one that could be sustained year after year because the fish population would recover to the same level after each harvest. Figure 23.8 shows the gain curve and two possible harvest lines, with gain and harvest on the vertical axis and current population along the horizontal axis. We focus first on comparing the gain curve with the black harvest line. Where the curve lies above the harvest line, the gain exceeds the number harvested. Where the curve lies below the line, the harvest exceeds the annual growth. Where the curve and the line intersect, harvest equals growth *and the population will return to the same size next year.*

The reason is that we harvest all of the gain and bring the population back to the same level as before the current season's growth, from which the population should generate the same gain next year. In the figure, the curve and the black harvest line intersect at a harvest of about 5 million pounds (vertical axis) for a preharvest population this year and next of about 13.5 million pounds (horizontal axis) and a postharvest population of 8.5 million pounds.

However, at a higher level of fishing effort, shown by the steeper orange harvest line, the fleet could take a harvest of 11 million pounds on a continuing

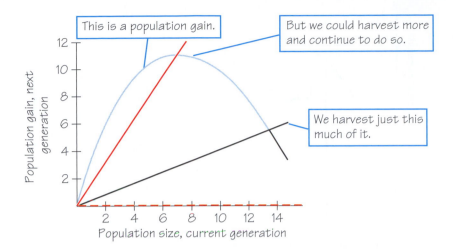

FIGURE 23.8 This population curve shows the effects of harvesting on a fish population. The scales for population and gain sizes are in millions of pounds.

basis, leaving a population of 7 million pounds, which should reproduce next year again to a preharvest level of 18 million pounds.

Sustained-Yield Harvesting Policy

A **sustained-yield harvesting policy** is a policy that if continued indefinitely will maintain the same yield.

For a sustainable yield, the same amount is harvested every year and the population remaining after each year's harvest is the same. To achieve this stability, the harvest must exactly equal the natural increase each year, the length of the green vertical line in Figure 23.7a.

Each value of x between 0 and x_e determines a different green vertical line and corresponding sustained-yield harvest (Figure 23.9). This harvest can vary from 0 (for $x = 0$ or $x = x_e$) up to some maximum value (for some x between 0 and x_e). A goal for a timber company or a fishery could be to harvest the **maximum sustainable yield:** to select an x whose colored vertical line is as long as possible, marked as x_M in Figure 23.9b. Such a yield level is difficult to determine, and we will see that economic and environmental factors—plus the inevitable involvement of mathematical chaos—also come into play.

Decline of a Fishery

EXAMPLE 6

A successful fishing effort attracts more fishers and more boats. A lower price for fish causes fishers to fish longer hours to maintain their incomes. Whichever is the case, the long-run effect can be catastrophic. How so?

FIGURE 23.9 The reproduction curve, with the population size x_M corresponding to the maximum sustainable yield.

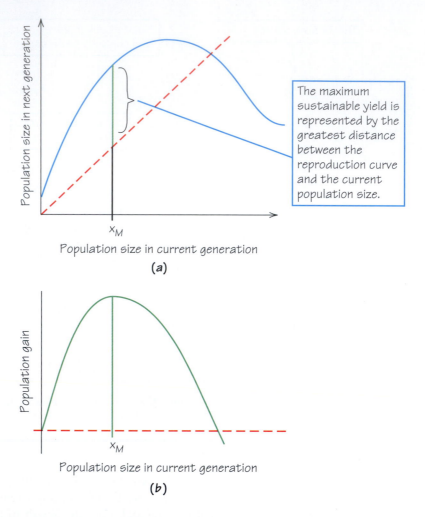

The maximum sustainable yield is represented by the greatest distance between the reproduction curve and the current population size.

Population size in next generation

Population size in current generation

(a)

Population gain

Population size in current generation

(b)

SOLUTION Figure 23.10 shows the sudden decline of a fishery with only a 20% increase beyond the original fishing effort. Extinction of the resource could result from additional fishing effort. ■

Dynamics of a Population over Time

The line $y = x$ provides a convenient way to trace the evolution of the population over several years (see Figure 23.11a), by alternating steps vertically to the curve and horizontally to the line $y = x$. Begin with the first year's population on the horizontal axis and go up vertically to the curve. The height is the population in the second year. Proceed horizontally from the curve over to the line $y = x$. Proceeding vertically from there to the curve yields a height that is the population in the third year. The result is a **cobweb diagram,** so-called because it can resemble a spider web. A cobweb diagram gives a convenient visual representation of the evolution of the population over time.

Figure 23.11 shows several traces for the same reproduction curve, each starting from a different initial population on the horizontal axis. The resulting

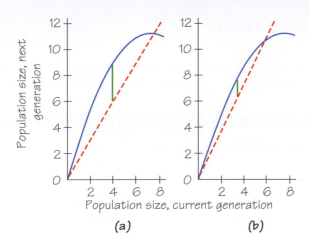

FIGURE 23.10 Deterioration of a fishery with increasing fishing pressure. Yield is shown in green. (a) Shows fishing effort geared to maximum sustainable yield. (b) Shows a 20% increase beyond the original fishing effort (the slope of the harvest line increases). The yield declines from the maximum sustainable yield of 2.4 million pounds to less than 1 million pounds.

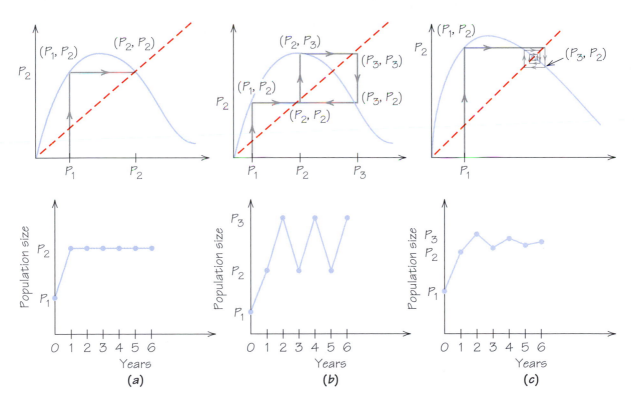

FIGURE 23.11
Examples of the dynamics, over time, for the same reproduction curve but different starting populations.
(a) The population goes in one year to the equilibrium population and stays there year after year. (b) After initial adjustment, the population cycles between values over and under the equilibrium population. (c) The population spirals in toward the equilibrium population.

variation is surprising and can be "chaotic" in a very specific mathematical sense, showing how apparently random behavior can result from strict rules.

23.4 The Economics of Harvesting Resources

We consider two models: one for a cattle ranch and one for either a fishing boat or a tree farm.

We assume that the price p received is the same for each harvested unit and does not depend on the size of our harvest. In effect, we assume that our operation is a small part of the total market, not substantially affecting overall supply and hence price.

We want to stay in business, so we do not extinguish the resource for quick profits. For any given population size, we harvest just the natural increase.

EXAMPLE 7   Cattle Ranching

We assume that the cost of raising and bringing a steer to market is the same for every steer and does not depend on how many steers we bring to market. What should our sustainable-yield harvesting policy be?

SOLUTION   Because the cost does not depend on the population size, the cost curve is a horizontal line (Figure 23.12).

As long as the selling price per unit is higher than the harvest cost per unit, we make a profit. The points of view of economics and biology agree, because the maximum profit occurs for the maximum sustainable yield. ◾

EXAMPLE 8   Fishing and Logging

In this model, we assume that the cost of harvesting a unit of the population decreases as the size of the population increases. This is the familiar principle of **economy of scale.** For example, the same fishing effort yields more fish when

FIGURE 23.12 The unit cost, unit revenue, and unit profit of harvesting one unit, as a function of population size, for the cattle ranch.

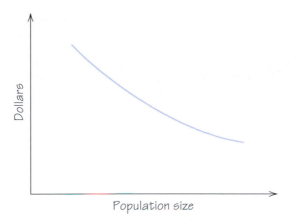

FIGURE 23.13 The unit cost, as a function of population size, for fishing or logging.

fish are more abundant. Similarly, a logger's harvest costs per tree are less when the trees are clumped together. This is the logger's motivation to clear-cut large stands. What should our sustainable-yield harvesting policy be?

SOLUTION The cost curve slopes downward and to the right, as in Figure 23.13. The size of the population from which one unit is harvested is shown on the horizontal axis. The cost of harvesting a single unit is measured on the vertical axis.

An optimal sustainable-yield harvesting policy depends on the relation between price and costs. There are two cases, as shown in Figure 23.14.

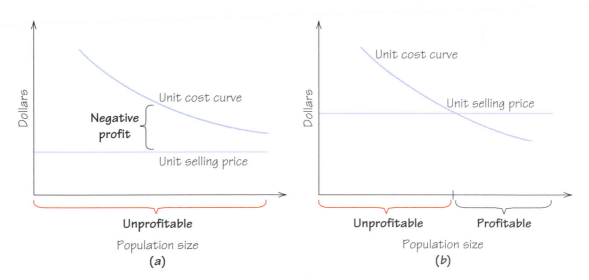

FIGURE 23.14 The unit cost, unit revenue, and unit profit of harvesting one unit, as a function of population size, for fishing or logging. (a) The market price is below the harvesting cost for all population sizes. (b) The operation is profitable for populations above a certain minimum size.

FIGURE 23.15
Regions of profitability for sustained-yield policy, with the economically optimal population size x_Q marked.

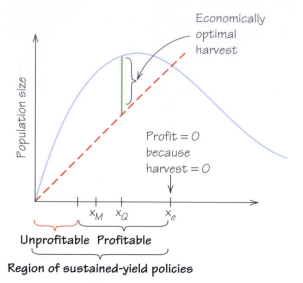

Economically optimal harvest

Population size

Profit = 0 because harvest = 0

x_M x_Q x_e

Unprofitable Profitable

Region of sustained-yield policies

▶ *The unit cost curve lies entirely above the unit price line.* The price for a harvested unit is less than the cost of harvesting it, no matter how large the population. It is impossible to make a profit.

▶ *The unit cost curve intersects the unit price line.* Above a certain population size, the price for a harvested unit is more than the cost of harvesting it, so profit is possible (Figure 23.15). Some population size, call it x_Q, gives a maximum net profit. Using calculus, it can be shown that x_Q is actually larger than x_M, the population that gives the maximum sustainable yield. Compared to harvesting the maximum sustainable yield, economic considerations result in harvesting less, but also in maintaining a larger stock of the population. ■

Our simple models fail to take into account a very critical feature of a modern economy: the *time value of money,* as measured by the interest that capital can earn. We now investigate why biological populations are susceptible to overexploitation and even extinction.

Why Eliminate a Renewable Resource?

Some species, such as the passenger pigeon, were harvested to extinction. In other cases, an entire ecosystem has been destroyed (see Spotlight 23.2). Why would anyone eliminate a renewable resource? Our approach helps explain why.

Sustained-yield policies involve revenues that will be received, year after year, in the future. The value of these revenues should be discounted to reflect the lost investment income that we could earn if instead we had the revenues today. For funds invested at a return of $100r\%$ per year, compounded annually, the present value P of an amount A to be received in n years in the future is related to A by the compound interest formula $A = P(1 + r)^n$.

SPOTLIGHT 23.2   The Tragedy of Easter Island

Easter Island
(Tom Till/Tony Stone Images.)

Easter Island is famous for its isolation—1400 miles to the nearest island—and for its hundreds of huge stone statues. For 30,000 years before the arrival of people in about A.D. 400, Easter Island maintained a lush forest, with several species of land birds. By the time of the first visit by Europeans in 1722, the island was barren, denuded of all trees and bushes over 10 feet high, and with no native animals larger than an insect. The 2000 or so islanders had only three or four leaky canoes made of small pieces of wood.

What happened? Careful analysis of pollen in soil samples tells the sad story. The settlers and their descendants cut wood to plant gardens, build canoes, make sledges and rollers to move the huge statues, and burn for cooking and warmth in the winter. In addition to crops they raised and chickens they had brought to the island and cultivated, they ate palm fruit, fish, shellfish, the meat and eggs of birds, and the meat of porpoises that they hunted from seagoing canoes. The population of the island grew to 7000 (or perhaps even 20,000).

By 1500, the forest was gone. Most tree species, all land birds, half of the seabirds, and all large and medium-sized shellfish had been extinguished. There was no firewood, no wood for sledges and rollers to transport hundreds of statues at various stages of completion, and no wood for seaworthy canoes. Without canoes, fishing declined and porpoises could not be taken. Stripping the trees exposed the soil, which eroded, so crop yields fell. The people continued raising chickens, but warfare and cannibalism ensued. By 1700, the population had crashed to 10–25% of its former size.

Why didn't the people notice earlier what was happening, imagine the consequences of keeping on as they had been, and act to avert catastrophe? After all, the trees did not disappear overnight.

From one year to the next, changes may not have been very noticeable. The forests may have been regarded as communal property, with no one charged with limiting exploitation or ensuring new growth. There was no quantitative assessment of the resources available and need for conservation versus the long-term needs of the "public works" program of erecting statues. Moreover, the religion of the people, the prestige of the chiefs, and the livelihood of hundreds depended on the statue industry. There was no perceived need to limit the population and no technology for birth control. Once the large trees were gone, there was no means for excess population to emigrate.

Adapted from Jared Diamond, "Easter's end," *Discover,* 16(8) (August 1995): 63–69.

The economic goal is to maximize the sum of the present values of all future receipts from harvesting. The optimal harvesting policy thus must depend on the expected rate of return r. We don't delve into the details of the calculations here but instead just give the results of the analysis.

Again, there are several cases to consider:

1. The unit cost of harvesting exceeds the unit price received, for all population sizes. Then it is impossible to make a profit.

2. For some population size x, the unit cost of harvesting equals the unit price received. Then there is a size between x and x_e (the equilibrium population size) for which the present value of the total return is maximized and the population and its yield are sustained.

3. The unit price exceeds the unit cost for *all* population sizes.

   ▶ For small r, the situation is the same as in the second case just cited.

   ▶ For larger r, *the economically optimal policy may be to harvest the entire population, immediately extinguish the resource, and invest the proceeds.*

Let's put this in the simplest and starkest terms. Suppose that you own a resource, such as a forest, whose cost of harvesting is small relative to its value. If the rate at which the forest population grows is greater than what you can earn on other investments, it pays to let the forest keep on growing.

On the other hand, if the forest grows more slowly than the rate of return on other investments, the economically optimal harvesting policy is to cut down all the trees now and invest the money. You could then start raising cattle on the land—and right there you have the scenario that is resulting in deforestation all over the world.

The sobering fact is that *very few economically significant renewable resources can sustain annual growth rates over 10%.* Many, like whales and most forests, have growth rates in the 4% to 5% range. These values—even a growth rate of 10%— are far below the return that many investors expect on their investment. For example, over the long run, the U.S. stock market has yielded an average 11% return, but venture capital firms expect to exceed a 25% profit.

The concept of maximum sustainable yield is an attractive ideal if the expectations of investors are low enough. However, there are still difficult problems:

   ▶ One problem is "the tragedy of the commons," discussed by ecologist Garrett Hardin. Several hundred years ago, English shepherds would graze their flocks together on common land. The grass of the commons could support only a fixed number of sheep. Each shepherd could reasonably think that adding just one or two more sheep to his flock would not overtax the commons. Yet if each did so, there could be disaster, with all the sheep starving. Many natural-products industries, such as fisheries, are a form of commons. Small overexploitation by each harvester can produce disastrous results for all. Global warming may be a tragedy of a worldwide commons.

▶ How, in the presence of human needs or greed, can we anticipate and prevent overexploitation and possible extinction of a resource? By and large, it has been politically impossible to force a harvesting industry to reduce current harvests to ensure stability in the future.

▶ In some industries, such as a fishery, growth of the population may be abundant one year but meager another, so that a steady yield cannot be sustained without damaging the resource. A few good years in a row may provoke increased investment in fishing capacity. Then attempting to harvest at the same levels in succeeding normal or below-normal years results in overfishing. This exact scenario destroyed the California sardine fishery in the 1930s, the Peruvian anchovy fishery in 1972, and much of the North Atlantic fishery in the 1980s.

Were the fishers and regulators mentioned above at fault for extinguishing these fisheries by overexploiting a dependable resource? Or were the extinctions due to chance variations of the fish stocks? In the next section, we examine a third possibility—that the fish stocks followed simple rules that nevertheless produced "chaotic" behavior of stocks, that is, wide variation from one year to the next. When we do not see the pattern, we interpret such behavior as randomness, much as the moves in a chess game may appear random and inexplicable to someone who does not know the rules of the game.

23.5 Dynamical Systems and Chaos

Throughout this and the two previous chapters, we have considered systems that change over time: bank accounts, the amount due on a loan, and the size of a population. Other examples are a dripping faucet, a playground swing, a pinball play, the solar system, the business cycle, epidemics, the passage of a drug through the human body, and the weather. Some of these are very predictable (interest on a bank account), while others are notoriously unpredictable (the weather). Some involve no outside influences (the amount due on a loan, assuming that you don't get behind on payments!), while others are the result of many contributing factors (the business cycle).

In some systems (the population of a country), the state of the system may depend largely on its states at previous times, while in other systems (epidemics) connection to past behavior may be harder to discern. A system whose state depends only on its states at previous times is called a **dynamical system.** Various dynamical systems provide useful models for physical, biological, chemical, financial, and other phenomena.

Mathematical Chaos

We think of **chaos** as referring to general confusion, unpredictability, and apparent randomness. Mathematicians and other scientists use the word to describe systems whose behavior over time is inherently unpredictable.

There are three main features of mathematical chaos:

► **Determinism**—future behavior of the system is completely determined by its present state, its past history, and known laws; chance is not involved
► Complex behavior
► Sensitivity to initial conditions—a small change now can make a big difference later

The last feature is sometimes known as the **butterfly effect,** from the title of a 1979 talk by meterologist E. N. Lorenz: "Predictability: Does the Flap of a Butterfly's Wings in Brazil Set Off a Tornado in Texas?" (The phrase probably traces to a 1953 science fiction story by Ray Bradbury, "A Sound of Thunder," in which history is changed by a time-traveler who steps on and kills a prehistoric butterfly.)

We can get a feel for chaotic systems by playing with some **iterated function systems (IFS).** The fancy name just means that we take an initial value, apply a function to it, then repeat over and over. This is exactly what we did earlier, geometrically, with reproduction curves for populations. (See section 19.6, pp. 735–740, for more about IFS and their connection to fractals.)

EXAMPLE 9   Doubling on a "Stone Age" Calculator

Imagine that you have a calculator that keeps only the last two digits of a number. It has a special key marked DBL that doubles the number in the display and keeps *only* the last two digits. For example, DBL applied to 52 gives 04 (*not* 104).

Let's start with two numbers that are as close together as can be on this calculator, such as 37 and 38. As we push the DBL key over and over again, will the result stay close?
SOLUTION
37, 74, 48, 96, 92, 84, 68, 36, 72, 44, 88, 76, 52, 04, . . .
38, 76, 52, 04, 08, 16, 32, 64, 28, 56, 12, 24, 48, 96, . . .

Already, by the fourth iteration, the two sequences are far apart. ▪

EXAMPLE 10   The Solar System

The American moon landings in 1969 and later, as well as all other space missions, were possible because of the predictability, or *determinism,* of the solar system. The moon and planets follow their orbits like clockwork. So how could the solar system be chaotic?
SOLUTION Over tens of millions of years, the orbit of each planet is chaotic, meaning that the slightest change in its position or velocity—due to, say, a comet passing nearby—could produce a huge difference later. ▪

More down-to-earth examples of physical systems that can exhibit chaos include the fluttering of a falling autumn leaf, heart arrhythmias, and the Tilt-A-Whirl amusement park ride.

Chaos in Biological Populations

If we measure this year's population as a fraction x of the carrying capacity, and do the same for next year's population as a fraction $f(x)$, the logistic model takes the form

$$f(x) = \lambda x(1 - x)$$

where the Greek letter lambda $\lambda = 1 + r$ is the amount by which the population is multiplied each year. When expanded, the equation has the familiar form of a quadratic in x:

$$f(x) = -\lambda x^2 + \lambda x$$

The Logistic Population Model

EXAMPLE 11

What behaviors can occur in the logistic model?

SOLUTION For different values of the parameter λ and different starting values for the population fraction, each of the behaviors of Figure 23.11 can occur:

▶ $\lambda = 2.8$ and starting population fraction $x = 0.36$ produces Figure 23.11a.
▶ $\lambda = 3.1$ and starting population fraction $x = 0.235$ produces Figure 23.11b.
▶ $\lambda = 3.0$ and starting population fraction $x = 0.4$ produces Figure 23.11c.

In other words, for population growth rates (values of λ) that are fairly close together (2.8, 3.1, 3.0), the population evolves very differently. This is a surprising and nonintuitive conclusion.

But there is more. For $\lambda = 4$ and any starting population fraction, the population does not settle down into any of the patterns of Figure 23.11; year after year it wanders "unpredictably" all over the place (Figure 23.16). This is *chaotic*

FIGURE 23.16
Chaotic behavior of a population.

behavior: It is deterministic, complex, and—in the long run—unpredictable. In the short run, the behavior is completely predictable. For example, from this year's population fraction, the equation tells us exactly what next year's will be. Repeating the use of the equation, we can determine what it will be the following year. But as the years pass, any sense of pattern gets lost in the complexity. ■

This potentially chaotic behavior of a biological population is bad news for those who manage a ranch or any biological population in the wild or in captivity. In recent years, the lobster catch in Maine has been much higher than in previous years, reaching record levels, for no discernible reason. On the other hand, in the late 1950s the annual harvest of Dungeness crabs off the central California coast declined from 12 million pounds to less than 1 million pounds without any evidence of disease, heightened predation, or increased crabbing effort. Researchers who modeled that population in 1994 found that booms and busts are the rule. The population can remain nearly level for generations before suddenly exploding or crashing without warning.

Searching for an environmental cause for these fluctuations could be futile, because there may not be one. Moreover, observing the population over a few generations provides no help in predicting future behavior.

EXAMPLE 12   Childhood Disease Epidemics

The incidence of childhood diseases such as chickenpox and measles varies greatly from year to year. Why?

SOLUTION   There are three plausible explanations for the fluctuations:

▶ There is an underlying regular cycle that is perturbed and occasionally overwhelmed by random events ("noise").

▶ There is no discernible pattern, because the fluctuations are due solely to chance.

▶ There is no discernible pattern, because such fluctuations are inherent in the epidemiology of the disease, as it is a chaotic system.

The first explanation, a perturbed cycle, fits chickenpox, with a cycle of one year. For measles, either the second or the third explanation may be correct, depending on the size of the community. For small communities, chance is an adequate explanation. For large communities, historical data from before the era of mass immunization suggest that measles cases were chaotic. That doesn't mean that they occurred at random but rather that they were unpredictable. Research also shows that there is a critical community size above which a disease will not die out solely by chance. For measles, this size is about 250,000. ■

What you need to realize about chaos is that behavior that appears to be random can be produced even with very simple systems that are completely governed by deterministic rules. Just because the behavior appears chaotic does not mean a lack of underlying order and structure, though discovering that structure may be difficult.

Even if we discover the structure, prediction may elude us because of chaos. If we had an absolutely correct model of how weather behaves and measurements at every location on and above the Earth, we might still not be able to forecast the weather accurately a week ahead.

What about the fishery extinctions? Perhaps the fishers and the fish were victims not of greed or chance but of the chaotic nature of the reproduction curve for the fish.

REVIEW VOCABULARY

Biomass A measure of a population in common units of equal value.

Butterfly effect A small change in initial conditions of a system can make an enormous difference later on.

Carrying capacity The maximum population size that can be supported by the available resources.

Chaos Complex but deterministic behavior that is unpredictable in the long run.

Cobweb diagram A kind of graphical portrayal of the evolution of a dynamical system, such as a population.

Compound interest formula Formula for the amount in an account that pays compound interest periodically. For an initial principal P and effective rate r per year, the amount after n years is $A = P(1 + r)^n$.

Deterministic A system is deterministic if its future behavior is completely determined by its present state, past history, and known laws.

Dynamical system A system whose state depends only on its states at previous times.

Economy of scale Costs per unit decrease with increasing volume.

Equilibrium population size A population size that does not change from year to year.

Exponential reserve How long a fixed amount of a resource will last at a constantly increasing rate of use. A supply S, as an initial rate of use U that is increasing by a proportion r each year, will last

$$\frac{\ln\left(1 + \dfrac{S}{U}\,r\right)}{\ln(1 + r)} \text{ years}$$

Iterated function system (IFS) A sequence of elements (numbers or geometric objects) in which the next element is produced from the previous one according to a function (rule).

Logistic model A particular population model that begins with near-geometric growth but then tapers off toward a limiting population (the carrying capacity).

Maximum sustainable yield The largest harvest that can be repeated indefinitely.

Natural increase The growth of a population that is not harvested.

Nonrenewable resource A resource that does not tend to replenish itself.

Population structure The division of a population into subgroups.

Rate of natural increase Birth rate minus death rate; the annual rate of population growth without taking into account net migration.

Renewable natural resource A resource that tends to replenish itself; examples are fish, forests, wildlife.

Reproduction curve A curve that shows population size in the next year plotted against population size in the current year.

Savings formula Formula for the amount in an account to which a regular deposit is made (equal for each period) and interest is credited, both at the end of each period. For a regular deposit of d and an effective interest rate r per year, the amount A accumulated is

$$A = d\left[\frac{(1 + r)^n - 1}{r}\right]$$

Static reserve How long a fixed amount of a resource will last at a constant rate of use; a supply S used at an annual rate U will last S/U years.

Sustainable yield A harvest that can be continued at the same level indefinitely.
Sustained-yield harvesting policy A harvesting policy that can be continued indefinitely while maintaining the same yield.
Yield The amount harvested at each harvest.

 # SKILLS CHECK

1. The carrying capacity of a population is

(a) the largest recorded population.
(b) the largest supportable population.
(c) the change in population.

2. The shape of the reproduction curve reflects the fact that a small population has abundant resources by

(a) rising steeply at the left.
(b) falling to the right.
(c) crossing the diagonal.

3. The logistic curve is a model for a population that is growing

(a) linearly.
(b) exponentially.
(c) with a ceiling.

4. U.S. oil use has grown

(a) linearly.
(b) exponentially.
(c) logistically.

5. Management of a nonrenewable resource can be modeled by

(a) an annuity.
(b) a savings account.
(c) an add-on loan.

6. If we have enough reserves of a product to last 200 years at the current rate of use, but the rate of use increases by 10% per year, how long will the supply last?

(a) About 30 years
(b) About 40 years
(c) About 50 years

7. If we have enough reserves of a product to last 1000 years at the current rate of use, but the rate of use increases by 1% per year, how long will the supply last?

(a) About 700 years
(b) About 480 years
(c) About 240 years

8. The equilibrium population size is the same as

(a) the carrying capacity.
(b) the intersection point of the reproduction curve and the diagonal.
(c) the natural increase of the population.

9. If the starting population for a reproduction curve is changed, the subsequent population pattern

(a) will still drift to the same pattern.
(b) will change to a different pattern.
(c) will sometimes change to a different pattern.

10. Economic considerations

(a) always work against the conservation of a resource.
(b) never interfere with the conservation of a resource.
(c) always affect in some way the conservation of a resource.

11. The population size that leads to the maximum net profit under sustainable-yield harvesting is always

(a) the same as the maximum sustainable yield.
(b) smaller than the maximum sustainable yield.
(c) larger than the maximum sustainable yield.

12. A system whose current state depends solely on its previous states is called

(a) a dynamical system.
(b) a stable system.
(c) an optimal system.

13. The "butterfly effect" refers to which feature of chaos?

(a) Predictability
(b) Complexity
(c) Sensitivity to initial conditions

14. For the logistic model $f(x) = 3x(1 - x)$, if the starting population fraction is 0.5, what is the next population fraction?

(a) 0.75
(b) 0.6
(c) 0.25

15. For the logistic model $f(x) = 3x(1 - x)$, if the starting population fraction is 0.4, what is the next population fraction?

(a) 0.116
(b) 0.416
(c) 0.720

16. For the logistic model $f(x) = 4x(1 - x)$, if the starting population fraction is 0.1, what is the next population fraction?

(a) 0.4
(b) 0.36
(c) 0.09

17. For the logistic model $f(x) = 4x(1 - x)$, are there any starting population fractions that will immediately lead to 0 population?

(a) No
(b) Yes, 1.0
(c) Yes, 0.5

18. Chaotic behavior appears to be random

(a) but is actually not random.
(b) and is random.
(c) and is sometimes random.

19. Pressing a digit and then repeatedly pressing the $\boxed{\text{SIN}}$ key is a model of

(a) an iterated function system.
(b) chaos.
(c) randomness.

20. Elimination of a natural resource

(a) always results from human greed.
(b) happens just by chance.
(c) may be caused by chaotic variation in the population.

EXERCISES   ■ Challenge   ◆ Discussion

Growth Models for Biological Populations

1. (Requires spreadsheet) For many years, China has been the world's most populous country. However, India has been catching up, with 1110 million in mid-2007 and growing at 1.7% per year, versus China then with 1322 million and growing at only 0.6% per year. If these rates continue, when will India have more people than China?

2. The population of the less-developed countries (excluding China) in mid-2007 of 4.056 billion is expected to grow at 1.8% per year (this is an annual yield, so you may think of it as compounded annually). If this growth rate continues until mid-2025, what will be the size of the population then?

3. If the growth rate of the less-developed countries of Exercise 2 changed suddenly to 1.7% in mid-2007, what would be the size of the population in mid-2025?

4. (Requires spreadsheet) If the growth rate of the less-developed countries of Exercise 2 decreased by $\frac{1}{25}$ of a percentage point (0.04%) per year from

2007 through 2011, beginning in mid-2007, what would be the size of the population in mid-2012?

5. An advertisement for Paul Kennedy's book *Preparing for the Twenty-First Century* (Random House, 1993) asked: "By 2025, Africa's population will be: 50%, 150% or 300% greater than Europe's?" The population of Europe in mid-2007 of 729 million is expected to stay constant through 2025. The population of Africa in mid-2007 of 925 million is expected to increase at about 2.4% per year. What answer would you give to the question?

6. In its estimates for doubling times for populations in the world, the Population Reference Bureau uses a rule of 70, similar to (but slightly more accurate than) the rule of 72 used in banking and explained in Exercises 7 and 8 in Chapter 22. The rule of 70 says that if a country's population continues to grow at a constant rate of $r\%$ per year, then it will double in size every $70/r$ years. (As noted in Chapter 22, a rule of 69.3 would be even more accurate, but the difference between that and the rule of 70 is only 1%.) Apply the rule of 70 to estimate the doubling times for the following populations (figures are for mid-2007):

(a) Africa, 925 million, 2.4%

(b) United States, 301 million, 0.8%

7. Do the calculations as in Exercise 6, but for:

(a) China, 1.321 billion, 0.6%

(b) The world as a whole, 6.593 billion, 1.3%

8. (Requires spreadsheet) **(a)** Wisconsin's electricity demand increased 2.8% per year for the 10 years from 1991 through 2000. If that trend continues, when will Wisconsin need to have twice as much generating capacity as it did in 1991?

(b) Wisconsin's electricity demand in 1991 was about half from business use and half from residential use. Over the 10 years through 2001, business demand increased at 4.2% per year and residential demand increased at 1.5% per year. Assuming that these rates of growth continue, calculate business and residential demands separately for the year that you calculated in part (a). What do you observe?

For Exercises 9 and 10, refer to the following.

Is Warren Sanderson right about world population growth slowing down (see Spotlight 23.1)? How much difference does it make in projections if we look at the world as a whole or break it down by countries or regions? In Exercises 9 and 10, we investigate this question, first projecting as a whole and then projecting by regions and adding.

9. The population of the world in mid-2007 of 6.593 billion is expected to increase 1.3% per year.

(a) Project the population to mid-2025 (by then you may have finished having children, if you have any) and to mid-2050 (by then you may be thinking about retiring).

(b) What are the assumptions involved in your projections?

10. Divide the countries of the world into three groups with differing rates of increase (see the table). (Why is this useful?)

Group	Population Mid-2007 (billions)	Rate of Growth (%)
More-developed countries	1.216	0.1
Less-developed countries (excluding China)	4.056	1.8
China	1.321	0.6

(a) Redo the projections in Exercise 9a for the years 2025 and 2050 by projecting each group separately and adding the totals. Is there a major difference from the results in Exercise 9a?

(b) Will the world be able to support the numbers of people that you project? What problems will these greater numbers of people cause? What could be done to avert those problems? Do you think that anything will be done before there is some kind of worldwide crisis?

   (In mid-1995, the world population was 5.7 billion and growing at 1.5% per year. Those figures led to projections as in Exercise 9 of 8.3 billion for 2020 and 11.1 billion for 2040. The corresponding growth rates for the groups of Exercise 10 were

0.2%, 2.2%, and 1.1%, which led to projections of 8.9 billion for 2020 and 12.7 billion for 2040.)

How Long Can a Nonrenewable Resource Last?

11. In 2004, world oil "resources" (which include estimates of undiscovered oil) totaled 2934.8 billion barrels, while annual consumption was 77.9 billion barrels. The U.S. Geological Survey (USGS) projected then that world consumption would increase 1.9% per year through 2025.

(a) What was the static reserve for oil in 2004?
(b) What was the exponential reserve for oil in 2004?
(c) What considerations may affect the answers to parts (a) and (b) over time?

12. In 2004, world natural gas "resources" (which include estimates of undiscovered gas) totaled 6076.5 trillion cubic feet, while annual consumption was 90 trillion cubic feet. The USGS projected then that world consumption would increase 2.2% per year through 2025.

(a) What was the static reserve for natural gas in 2004?
(b) What was the exponential reserve for natural gas in 2004?
(c) What considerations may affect the answers to parts (a) and (b) over time?

13. Can our energy problems be solved by increasing the supply? [Thanks for the idea to Evar D. Nering of Arizona State University, in "The mirage of a growing fuel supply," *The New York Times* (June 4, 2001) Op-Ed page.]

(a) Suppose that we have a 100-year supply of a resource (such as oil, for which known world reserves will last less than 100 years at the current world rate of use). That is, the resource would last 100 years at the current rate of consumption. Suppose that the resource is consumed at a rate that increases 2.5% per year (this is the average increase in consumption for oil in the United States since 1973). How long will the resource last?
(b) Suppose that we underestimated the supply and actually have a 1000-year supply at the current rate of use. How long will that last if consumption increases 2.5% per year?

(c) Let's think big and suppose that there is 100 times as much of the resource as we thought—a 10,000-year supply. How long will that last if consumption increases 2.5% per year?

14. In this problem we explore the consequences of reducing the rate of growth of oil use. Suppose that we halve the growth rate from the 2.5% per year given in Exercise 13 to 1.25% per year. [Thanks for the idea to Evar D. Nering of Arizona State University, in "The mirage of a growing fuel supply," *The New York Times* (June 4, 2001) Op-Ed page.]

(a) How long will the 100-year supply last?
(b) How long will the 1000-year supply last?
(c) How long will the 10,000-year supply last?

15. We continue the ideas of Exercises 13 and 14, but with a more radical hypothesis.

(a) How long would the 100-year supply last if we reduced our consumption by just $\frac{1}{2}$% per year— that is, if we used $\frac{1}{2}$% less each year instead of 2% more?
(b) If we used 1% less each year?

16. By the time there is concern about using up a nonrenewable resource, it may be too late. Suppose that a resource has a static reserve of 10,000 years, but consumption is growing at 3.5% per year.

(a) How long will the resource last?
(b) How long before half the resource is gone?
(c) How much longer will the resource last if after half of it is gone, consumption is stabilized at the then-current level?
(d) What implications do you see to your answers?

17. Do a calculation to criticize the claim in the following quotation: "The United States holds 437 billion tons of known (coal) reserves, enough energy to keep 100 million large electric generating plants going for the next 800 years or so." [*Forbes* (December 15, 1975), p. 28; thanks to Albert A. Bartlett.]

For Exercises 18–20, refer to the following.

The formula for the average growth rate over a period of time is even simpler than the one for the exponential reserve of a resource. If usage at the beginning is N_0 and at the end of an interval of t years it is N, then the average annual rate of growth is

$$\frac{1}{t} \ln \left(\frac{N}{N_0} \right)$$

18. In Exercise 11, we note that the average increase in oil consumption in the United States during 1993–2000 was "nearly" 2%. In fact, consumption in 1993 was 6.291 billion barrels and consumption in 2000 was 7.211 billion barrels. What is a more accurate (two-decimal-place) estimate of the average annual percentage increase in oil consumption?

19. The U.S. population was 3.93 million in 1790 and 62.95 million in 1890. What was the average annual rate of growth, to two decimal places?

20. The U.S. population was 62.95 million in 1890 and 301 million in 2007 (at roughly the same times of the year). What was the annual rate of growth, to two decimal places?

Sustaining Renewable Resources

For Exercises 21–25, refer to the following.

We suppose that a population has the reproduction curve shown in the figure at top right, with units of thousands of tons of biomass. The mathematical description is that the population in the following year, x_{n+1}, depends on the population x_n in the current year (after any harvest) according to

$$x_{n+1} = f(x_n) = \tfrac{1}{5} x_n (20 - x_n) = 4x_n(1 - 0.05x_n)$$

for x_n between 0 and 10, in units of millions of pounds. We start with a population this year of x_1 whose value we vary. (For these exercises, a spreadsheet or a programmable calculator is useful.)

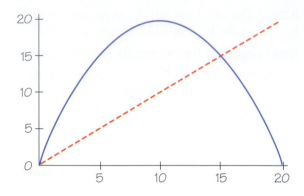

21. Start with $x_1 = 5$. Calculate numerically the population in the first few years, draw a cobweb diagram, and briefly describe the qualitative behavior of the population.

22. Repeat Exercise 21 with the starting value $x_1 = 10$.

23. Repeat Exercise 21 with the starting value $x_1 = 7$, going at least as far as x_{10}.

24. Try to find a starting value (besides 0, 5, 10, and 15) that leads to a stable population over time.

25. What is the equilibrium population size?

For Exercises 26–36, refer to the following.

We suppose that the population has the reproduction curve shown in the following figure, with units of thousands of tons of biomass, whose mathematical description is

$$x_{n+1} = f(x_n) = 3x_n(1 - 0.05x_n)$$

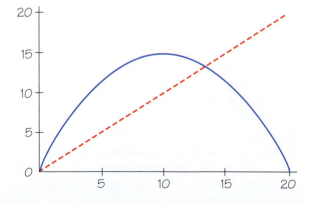

26. Start with $x_1 = 5$. Calculate numerically a population in the first 10 years, draw a cobweb diagram, and briefly describe the qualitative behavior of the population.

27. Repeat Exercise 26 with the starting value $x_1 = 10$.

28. Try to find a starting value (besides 0 and 20) that leads to extinction of the population.

29. What is the equilibrium population size?

◆ **30.** Which of the reproduction curves–the one for Exercises 21–25 or the one for Exercises 26–29–seems to you more realistic as a model of a biological population, and why?

31. What is the significance of the red dashed line in the preceding figures and its intersection with the blue curve?

32. (Spreadsheet helpful) The following figure shows the annual population gain in the absence of any harvesting. Determine the maximum sustainable yield to one decimal point.

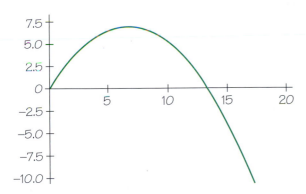

33. Suppose that the population of Exercises 26–32 starts with $x_1 = 11$ and each year we harvest half of the population. For example, in year 1, we harvest 5.5 million pounds, leaving the remaining 5.5 million pounds to reproduce (according to the reproduction curve) for the next year. Calculate numerically the population in the first 10 years, draw a cobweb diagram, and briefly describe the qualitative behavior of the population.

34. Repeat Exercise 33 but for a starting population $x_1 = 5$.

35. Harvesting a set proportion of a population is unrealistic for some situations, such as fishing, in which we can't know the size of the population or when we have harvested half of it. A more realistic situation for fishing is that increasing harvests attract increasing fishing effort (e.g., more boats). Repeat Exercise 33 with $x_1 = 11$ and a harvesting strategy that harvests 1 million tons the first year and every year harvests an extra 1 million tons (over the harvest of the previous year).

36. Suppose that the population of Exercises 26–35 has been overharvested to the point that only 1 million tons remain at the end of a particular year. If there is no harvesting at all until a year after a year with a population of 11 million tons, when can harvesting resume?

37. A reproduction curve for a population is shown in the following figure. Estimate the equilibrium population size and the maximum sustainable yield. (The units are in millions of pounds.)

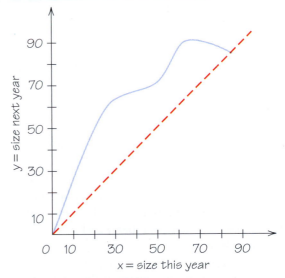

38. Suppose that a reproduction curve for a certain population is as in the following figure, where the units are in millions of pounds.

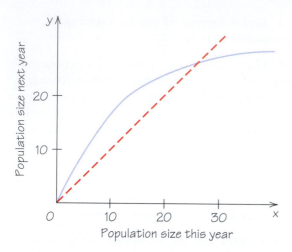

(a) Estimate the sustainable yields corresponding to a population of size 10 remaining after the harvest.

(b) Estimate the maximum sustainable yield.

Mathematical Chaos

39. In doubling on a "Stone Age" calculator (see Example 9):

(a) What do you notice about the two sequences that were produced?

(b) Suppose that we had started with a different seed, say 39. What would happen as we iterate the doubling?

(c) Explain what you observe in parts (a) and (b) and give a general argument about why it is true.

40. Explain why the logistic model on p. 864 is the same as the one on p. 883, where we have $f(x) = \lambda x(1 - x)$.

41. You saw in Figure 23.11 that a logistic model can result in a stable value, produce cycling between several values, or result in chaos. Other dynamical systems can exhibit similar behavior. Here we examine the system in which we start with a positive whole number n and iterate the following function:

$$f(n) = \text{sum of the squares of the digits of } n$$

For example, we have $f(133) = 1^2 + 3^2 + 3^2 = 19$.

(a) Calculate what happens as f is applied repeatedly, starting with 133. What do you observe?

(b) Pick a number different from 133 and different from 1, and iterate f repeatedly. What do you observe?

(c) Why did we exclude 1 in part (b)?

(d) Try some other values. Can you offer a general conclusion? Can you offer an argument why your conclusion is correct?

42. The behavior of some very simple dynamical systems is still not completely known. Consider the system that starts with a positive whole number n and gives as the next number

$$f(n) = \begin{cases} 3n + 1 & \text{if } n \text{ is odd} \\ n/2 & \text{if } n \text{ is even} \end{cases}$$

[This iterative function system was devised by Lothar O. Collatz, later of the University of Hamburg, during his student days before World War II. It is sometimes called the "$3n + 1$ problem" or the "Syracuse problem" (because it became popular at the Mathematics Department of Syracuse University), and the sequences generated are sometimes called "hailstone numbers."]

(a) Start with $n = 1$. What happens?

(b) Start with $n = 13$. What happens?

(c) Start with $n = 12$. What happens?

What you observe is known to happen for all $n < 10^{40}$, but after more than 60 years mathematicians have been unable to show that it happens for every n whatsoever.

43. (Requires programmable calculator, spreadsheet, or BASIC programming) A population model slightly different from the logistic model is given by the iterative function system

$$g(x) = x + rx(1 - x)$$

where x is a fraction of the limiting population and r is a growth rate.

(a) Set $r = 3$, start with $x = x_1 = 0.01$, and calculate the first 20 values $x_1, \ldots, x_{20}$.

(b) In part (a), you should have found $x_{10} = 0.722914$. Replace this value with the rounded-up value $x_{10} = 0.723$ and continue on to calculate x_{20}.

(c) Now replace x_{10} with the rounded-down value $x_{10} = 0.722$ and continue on to calculate x_{20}.

44. A dynamical system expressed as an iterated functional system $f(x)$ has an *equilibrium point* at a value x_0 if, once the system reaches x_0, it always stays at that value. In terms of an equation, an equilibrium point exists at x_0 if $f(x_0) = x_0$.

(a) For the dynamical system of Exercise 43, find all equilibrium points.

(b) For the logistic population model of Exercise 40, find all equilibrium points.

■ **45.** (Requires programmable calculator, spreadsheet, BASIC programming, or preferably use of software available under Suggested Web Sites) The behavior of the logistic population model $f(x) = \lambda x(1 - x)$ depends on the value of the positive parameter λ. As λ increases from 0 to 4, the system changes from one behavior to another, through the following possible states:

▶ The population simply dies out.
▶ The population tends toward a nonzero equilibrium point.
▶ The population oscillates between 2 points.
▶ The population oscillates between 4 points, then 8 points, then 16 points, and so on.
▶ The population oscillates between numbers of points that are not powers of 2, until at last . . . the population oscillates between 3 values.
▶ The population behaves chaotically.

Explore what happens for various values of λ between 0 and 4, trying to identify where the shifts in the system's behavior take place.

 WRITING PROJECTS

1. Based on the calculations you did in Exercises 9 and 10, writes a one- to two-page guest editorial for a newspaper. Describe your projections and how you arrived at them, how serious a problem you think population growth is, what problems it is likely to cause, what you think needs to be done, and what the implications are for your own life.

2. Identify a particular regional, national, or world nonrenewable primary resource (such as coal) or a secondary resource (one, such as electric power, that is produced from primary resources). Research how much of it is available now and what the current rate of consumption is. Determine the static reserve. Estimate the growth rate in consumption, taking into account human population increase, and determine the exponential reserve.

What social and technological factors contribute to the increasing rate of consumption? Brainstorm how those factors could be changed. Write an essay of three to five pages.

3. Identify a particular regional, national, or world renewable resources (such as timber or clean drinking water). Research how much of it is produced now, how much is harvested now, and what the current rate of consumption is. Estimate the growth rate in consumption, taking into account human population increase. For how long can this resource continue to meet the demand? What social and technological factors contribute to the increasing rate of consumption? Brainstorm how those factors could be changed. Write an essay of three to five pages.

SUGGESTED READINGS

BARTLETT, ALBERT A. *The Essential Exponential! For the Future of Our Planet.* Center for Science, Mathematics, & Computer Education, University of Nebraska–Lincoln (126 Morrill Hall, Lincoln, NE 68588-0350), 2004.

COHEN, JOEL. *How Many People Can the Earth Support?* Norton, New York, 1995.

GLEICK, JAMES. *Chaos: Making a New Science,* Viking, New York, 1987.

PETERSON, IVARS. *Newton's Clock: Chaos in the Solar System,* W. H. Freeman, New York, 1993.

SCHWARTZ, RICHARD H. *Mathematics and Global Survival,* 4th ed., Ginn Press, Needham Heights, Mass., 1998.

SUGGESTED WEB SITES

www.popin.org/ UN Population Division population statistics and estimates for all countries.

www.prb.org/ Population Reference Bureau population statistics and rates of growth by regions.

www.maths.anu.edu.au/~briand/chaos Downloadable Java applets to illustrate chaos.

http://www.ac.wwu.edu/~stephan/Animation/ pyramid.html Animated display of population structures of individual countries from 1950 to 2050.

math.bu.edu/DYSYS/applets/ Downloadable Java applets designed to accompany Robert L. Devaney's *A Toolkit of Dynamics Activities,* but can be used independently.

http://staff.science.uva.nl/~alejan/tour.html Java applet for the logistic population model of Figures 23.11 and 23.16.

http://www.census.gov/ipc/www/idbnew.html International data base (IDB) of demographic data for 227 countries, with projections.

Answer to Skills Check Exercises

Chapter 1

1. b
2. c
3. c
4. c
5. b
6. b
7. c
8. c
9. a
10. a
11. b
12. a
13. c
14. a
15. c
16. b
17. a
18. a
19. c
20. b

Chapter 2

1. c
2. b
3. c
4. b
5. b
6. b
7. c
8. c
9. a
10. c
11. b
12. a
13. b
14. b
15. c
16. a
17. b
18. c
19. a
20. c

Chapter 3

1. c
2. c
3. b
4. b
5. c
6. b
7. c
8. b
9. b
10. c
11. b
12. b
13. c
14. a
15. b
16. a
17. c
18. a
19. a
20. c

Chapter 4

1. a
2. c
3. b
4. b
5. b
6. c
7. c
8. c
9. b

10. c
11. b
12. c
13. c
14. c
15. c
16. c
17. c
18. a
19. a
20. c

Chapter 5

1. a
2. a
3. b
4. b
5. c
6. a
7. b
8. c
9. b
10. b
11. c
12. c
13. b
14. a
15. b
16. a
17. a
18. a
19. b
20. c

Chapter 6

1. a
2. a
3. c
4. b

5. c
6. c
7. a
8. b
9. c
10. b
11. c
12. b
13. b
14. b
15. a
16. a
17. a
18. c
19. a
20. b

Chapter 7

1. b
2. a
3. c
4. b
5. a
6. c
7. a
8. b
9. a
10. a
11. c
12. b
13. b
14. b
15. c
16. b
17. b
18. c
19. a
20. a

Chapter 8

1. a
2. b
3. b
4. c
5. a
6. b
7. b
8. c
9. b
10. a
11. b
12. c
13. a
14. c
15. c
16. b
17. a
18. a
19. c
20. c

Chapter 10

1. c
2. c
3. c
4. c
5. a
6. c
7. c
8. b
9. a
10. b
11. c
12. a
13. c
14. c
15. a
16. a
17. d
18. b
19. a
20. b

Chapter 12

1. b
2. b
3. c
4. a
5. c
6. c
7. a
8. c
9. c
10. b
11. b
12. b
13. a
14. a
15. b
16. b
17. c
18. a
19. b
20. a

Chapter 14

1. c
2. a
3. b
4. a
5. a
6. b
7. c
8. a
9. a
10. b
11. a
12. b
13. c
14. c
15. a
16. b
17. c
18. a
19. c
20. c

Chapter 9

1. b
2. c
3. b
4. b
5. c
6. d
7. b
8. b
9. c
10. a
11. a
12. c
13. a
14. b
15. c
16. a
17. b
18. b
19. c
20. c

Chapter 11

1. b
2. c
3. c
4. b
5. b
6. b
7. a
8. a
9. b
10. b
11. c
12. a
13. b
14. c
15. c
16. a
17. a
18. c
19. c
20. b

Chapter 13

1. a
2. b
3. a
4. b
5. d
6. c
7. b
8. c
9. b
10. c
11. b
12. c
13. a
14. a
15. b
16. c
17. a
18. b
19. b
20. a

Chapter 15

1. c
2. c
3. b
4. c
5. a
6. c
7. a
8. a
9. c
10. a
11. a
12. b
13. c
14. b
15. b
16. b
17. b
18. c
19. b
20. b

Chapter 16

1. a
2. c
3. b
4. b
5. b
6. c
7. b
8. a
9. a
10. a
11. b
12. b
13. c
14. b
15. b
16. c
17. c
18. c
19. c
20. a

Chapter 17

1. b
2. b
3. b
4. c
5. a
6. b
7. a
8. c
9. b
10. b
11. a
12. b
13. c
14. a
15. c
16. b
17. c
18. c
19. a
20. a

Chapter 18

1. c
2. c
3. a
4. c
5. b
6. c
7. c
8. b
9. c
10. b
11. a
12. a
13. c
14. c
15. a
16. a
17. b
18. c
19. a
20. a (or b)

Chapter 19

1. a
2. a
3. b
4. c
5. b
6. c
7. b
8. a
9. b
10. b
11. c
12. b
13. b
14. a
15. c
16. c
17. a
18. b
19. a
20. c

Chapter 20

1. b
2. b (or c)
3. c
4. b
5. a
6. c
7. b
8. c
9. a
10. a
11. c
12. b
13. c
14. b
15. b
16. c
17. b
18. b
19. c
20. a

Chapter 21

1. c
2. a (or c)
3. b
4. c
5. b
6. c
7. c
8. a
9. b
10. c
11. b
12. c
13. b
14. a
15. a
16. b
17. b
18. c
19. b
20. b

Chapter 22

1. b
2. c (or a)
3. a
4. b
5. b
6. a
7. b
8. a
9. a
10. c
11. a
12. b
13. c
14. b
15. c
16. c
17. c
18. c
19. c
20. b

Chapter 23

1. b
2. a
3. c
4. b
5. a
6. a
7. c
8. b
9. c
10. c
11. b
12. a
13. c
14. a
15. c
16. b
17. b
18. a
19. a
20. c

Answers to Odd-Numbered Exercises

Chapter 1

1. *E* has valence 0; *A* has valence 1; *H, D,* and *G* have valence 2; *B* and *F* have valence 3; *C* has valence 5. *E* is "isolated." *E* might have valence 0 because it is on an island with no road access.

3. (a) This diagram fails to be a graph because a line segment joins a single vertex to itself. The definition being used does not allow this.

(b) The edge *EC* crosses edges *AD* and *BD* at points that are not vertices; edge *AC* crosses *BD* at a point that is not a vertex.

(c) This graph has 5 vertices and 5 edges.

5. (a) *FDCBF*

(b) (i) *BD; BCD*

    (ii) *CBF; CDF; CDBF*

(c) *CDFBC*

7. (a) 4 vertices; 4 edges

(b) 7 vertices; 6 edges

(c) 10 vertices; 14 edges

9. Remove the dotted edges in the figure below and the remaining graph will be disconnected.

11. (a)

(b)

13. Yes, a disconnected graph can arise. One possible example is shown below.

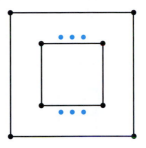

This gives rise to the disconnected graph here.

15. (a)

(b)

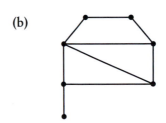

17. The supervisor is not satisfied because all of the edges are not traveled by the postal worker. The worker is unhappy because the end of the worker's route wasn't the same point as where the worker began. The original job description is unrealistic because there is no Euler circuit in the graph.

19. There is such an efficient route. The appropriate graph model has an additional edge joining the same pair of vertices for each of the edges shown in the graph of Exercise 17. Since this graph is connected and even-valent, it has many Euler circuits, any one of which will provide a route for the snowplow. Routes without 180-degree turns are better choices.

21.

23. (a) The largest number of such paths is 3: *AF, ABEF,* and *ACF*.
(b) This task is simplified when we note that there are many symmetries in this graph.
(c) In a communication system, such a graph offers redundant ways to get messages between pairs of points, even when some of the communication links (edges removed) fail.

25. Do not choose edge 2, but edge 1 or 10 could be chosen.

27.

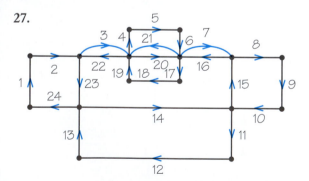

29. Two edges need to be dropped to produce a graph with an Euler circuit. Persons who parked along these stretches of sidewalk without putting coins in the meters would not need to fear that they would get tickets.

31.

(a)   (b)

33. (a) There are four 3-valent vertices. By properly removing two edges adjacent to these four vertices, one can make the graph even-valent.
(b) Yes, because the resulting graph is connected and even-valent.
(c) It is possible to remove two edges and have the resulting graph be even-valent.
(d) No, because the resulting graph is not connected, even though it is even-valent.

35.

(a)   (b)

(c)

(d)

Nine is the best one can do.

37.

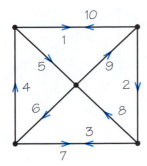

39. There are many different circuits that will involve three reuses of edges. These are the edges that connect the six 3-valent vertices in pairs.
41. There are many circuits that achieve a length of 44,000 feet. The number of edges reused is 8, because a shorter tour can be found by repeating more shorter edges than fewer longer edges.
43. Both graphs (b) and (c) have Euler circuits. The valences of all of the vertices in (a) are odd, which makes it impossible to have an Euler circuit there.

(b)

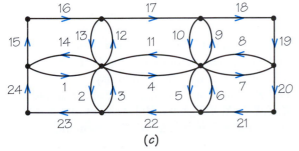

(c)

45. If the graph G is connected, the newly constructed graph will be even-valent and thus will have an Euler circuit. If G is not connected,

the new graph will not have an Euler circuit because it will not be connected either.

47. (a)

(b) The best eulerization for the four-circle four-ray case adds two edges.
(c) *Hint:* Consider the cases where r is even and odd separately.

49. Yes. A graph with six vertices where each vertex is joined to every other vertex will have valence 5 for each vertex.

51. When you attach a new edge to an existing graph, it gets attached at two ends. At each of its ends, it makes the valence of the existing vertex go up by 1. Thus the increase in the sum of the valences is 2. Therefore, if the graph had an even sum of the valences before, it still does; if its valence sum was odd before, it still is.

53. The graph is connected.

55. When chemists say, for example, that hydrogen has valence 1, they mean that hydrogen forms one chemical bond with other elements. This usage has similarities to the graph theory concept of valence.

57. A tour that begins and ends at vertex *A* and respects the traffic directions would be *ABDEFBEBFEDBACDCA*. The cutting machine has to make "sharp turns" at some intersections.

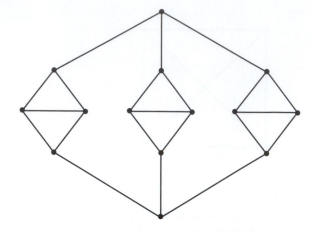

Chapter 2

1. (a) $X_5X_6X_1X_3X_4X_2X_5$
(b) $X_5X_4X_3X_2X_1X_6X_7X_8X_9X_{10}X_{11}X_{12}X_5$
(c) $X_5X_4X_3X_1X_2X_7X_6X_9X_8X_5$
(d) $X_5X_8X_3X_4X_7X_6X_1X_2X_5$
(e) $X_5X_4X_3X_2X_8X_1X_{10}X_7X_6X_9X_5$

3. (a) A Hamiltonian circuit will remain for graphs (a) and (b), but there will be no Hamiltonian circuit for graphs (c), (d), and (e).
(b) The removal of a vertex might correspond to the failure of the equipment at that site.

5. Other Hamiltonian circuits include *ABIGDCEFHA* and *ABDCEFGIHA*.
7. (a) a. Add edge *AB*.
　　　　b. Add edge X_1X_3.
(b)

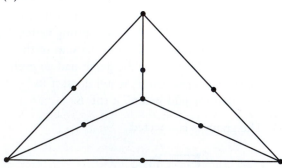

9. The following graph has no Hamiltonian circuit and every vertex of the graph has valence 3.

11. (a) Any Hamiltonian circuit would have to use both edges at the vertices X_5, X_4, and X_2. This would cause a problem in the way a Hamiltonian circuit could visit vertex X_1. Thus, no Hamiltonian circuit exists.
(b) If there were a Hamiltonian circuit, it would have to use the edges X_4 and X_5 and X_6 and X_7. This would make it impossible for the Hamiltonian circuit to visit X_8 and X_9. Thus, no Hamiltonian circuit exists.
13. (a) No Hamiltonian circuit.
(b) No Hamiltonian circuit.
(c) No Hamiltonian circuit.
15. (a) There is a Hamiltonian path from X_3 to X_4.
(b) No. There is a Hamiltonian path from X_1 to X_8 in graph (b).
(c) Here are two examples. A worker who inspects sewers may begin at one garage at the start of the workday but may have to report to a different garage for the afternoon shift. A school bus may start at a bus garage and then pick up students to take them to school, where the bus sits until the end of the school day.
17. (a) Hamiltonian circuit, yes; one example is $X_1X_3X_7X_5X_6X_8X_2X_4X_1$. Euler circuit, no.
(b) Hamiltonian circuit, yes. Euler circuit, yes.
(c) Hamiltonian circuit, yes. Euler circuit, no.

(d) Hamiltonian circuit, no. Euler circuit, yes; one example is $U_1U_2U_5U_6U_{16}U_{15}U_{11}U_4U_5U_{12}$ $U_{11}U_{10}U_{14}U_{13}U_7U_3U_8U_{10}U_9U_3U_1$.

19. (a) Hamiltonian circuit, yes; Euler circuit, no.
(b) Hamiltonian circuit, yes; Euler circuit, no.
(c) Hamiltonian circuit, yes; Euler circuit, no.
(d) Hamiltonian circuit, no; Euler circuit, no.

21. (a)

(b)

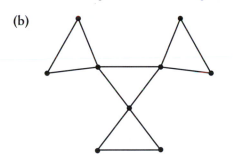

(c) A graph has an Eulerian path if for two different vertices u and v of the graph there is a path from u to v that uses each edge of the graph once and only once.

23. The new system is an improvement since it codes 676 locations compared with 504 for the old system. This is 172 more locations.

25. (a) $9 \cdot 8 \cdot 7 \cdot 6 \cdot 5 = 15,120$
(b) $(26)(26)(26) = 17,576$

27. (a) $7 \times 6 \times 5 \times 4 \times 3 = 2520$
(b) $7 \times 7 \times 7 \times 7 \times 7 = 16,807$
(c) $(7)^5 - 7 = 16,800$

29. (a) $26(26)(26)(10)(10)(10) - (26)(26)(26) =$ $(26)^3(10^3 - 1)$
(b) Answers will vary.

31. With no other restrictions, 10^7. With no other restrictions, 9×10^2.

33. These graphs have 6, 10, and 15 edges, respectively. The n vertex complete graph has $[n(n - 1)]/2$ edges. The number of TSP tours is 3, 12, and 60, respectively.

35. (a)

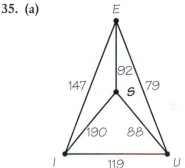

(b) (1) Tour $UISEU$; mileage $= 119 + 190 + 92 + 79 = 480$
(2) Tour $USIEU$; mileage $= 88 + 190 + 147 + 79 = 504$
(3) Tour $UIESU$; mileage $= 119 + 147 + 92 + 88 = 446$
(c) $UIESU$ (Tour 3)
(d) No.
(e) Starting from U, one gets [see answer (b)] Tour 1. From S one gets Tour 2; from E one gets Tour 2; and from I one gets Tour 1.
(f) Tour 2. No.

37. $FMCRF$ gets her home in 36 minutes.

39. $MACBM$ takes 344 minutes to traverse.

41. A traveling salesman problem.

43. A sewer drain inspection route at corners involves finding a Hamiltonian circuit, and there is such a circuit. If the drains are along the blocks, a route in this case involves solving a Chinese postman problem. Since there are 18 odd-valent vertices, an optimal route would require at least nine reuses of edges. There are many such routes that achieve nine reuses.

45. The complete graph shown has a different nearest-neighbor tour that starts at A ($AEDBCA$), a sorted-edges tour ($AEDCBA$), and a cheaper tour ($ADBECA$).

47. The optimal tour is the same, but its cost is now $4200 + 10(30) = 4500$.
49. (a) a. Not a tree because there is a circuit. Also, the wiggled edges do not include all vertices of the graph.
(b) The circuit does not include all the vertices of the graph.
(b) a. The tree does not include all vertices of the graph.
   b. Not a circuit.
(c) a. Not a tree.
   b. Not a circuit.
(d) Not a tree.
   b. Not a circuit.
51. (a) 1, 2, 3, 4, 5, 8
(b) 1, 1, 1, 2, 2, 3, 3, 4, 5, 6, 6
(c) 1, 1, 1, 2, 2, 2, 2, 2, 3, 3, 3, 3, 4, 4, 4, 5, 5, 6, 7
(d) 1, 2, 2, 3, 3, 3, 4, 5, 5, 5, 6, 6
The cost is found by adding the numbers given.
53. The spanning tree will have 26 vertices. H also will have 26 vertices. The exact number of edges in H cannot be determined, but H has at least 25 edges and no more edges than the complete graph having 26 vertices.
55. Yes.
57. Yes. Change all the weights to negative numbers and apply Kruskal's algorithm. The resulting tree works, and the maximum cost is the negative of the answer you get. If the numbers on the edges represent subsidies for using the edges, one might be interested in finding a maximum-cost spanning tree.

59. A negative weight on an edge is conceivable, perhaps a subsidization payment. Kruskal's algorithm would still apply.
61. Two different trees with the same cost are shown:

63. (a) True
(b) False (unless all the edges of the graph have the same weight)
(c) True
(d) False
(e) False
65. (a) Answers will vary for each edge, but the reason it is possible to find such trees is that each edge is an edge of some circuit.
(b) The number of edges in every spanning tree is 5, one less than the number of vertices in the graph.
(c) Every spanning tree must include the edge joining vertices C and D, since this edge does not belong to any (simple) circuit in the graph.
67.

	A	B	C	D
A	0	14	13	5
B	14	0	17	9
C	13	17	0	8
D	5	9	8	0

69. (a) The earliest completion time is 22 because the longest path, the unique critical path $T_3T_2T_5$, has length 22.
(b) The earliest completion time is 30 because the longest path, the unique critical path $T_3T_5T_7$, has length 30.

71. The only tasks that, if shortened, will reduce the earliest completion time are those on the critical path; in this case, these are the tasks T_1, T_5, and T_7. If T_5 is shortened to 7, then the longest path will have length 28, which becomes the earliest completion time. The tasks on this critical path are T_1, T_4, and T_7.

73. Different contractors will have different times and order-requirement digraphs. However, in any sensible order-requirement digraph, the laying of the foundation will come before the erection of the side walls and the roof. The fastest time for completing all the tasks will be the length of the longest path in the order-requirement digraph.

75. One example is given below:

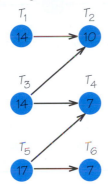

Chapter 3

1. (a) Scheduling final examinations, the times for courses being offered, cleaning of classrooms, etc.
(b) Scheduling the buses, the selling of tickets, cleaning, etc.
(c) Scheduling regular staff and supervisory staff, deliveries of items for sale, etc.
(d) Scheduling mealtimes, shopping, etc.
(e) Scheduling staff, maintenance of backup equipment, etc.
(f) Scheduling nurses, doctors, operating rooms, medical imaging, etc.
(g) Scheduling maintenance and repair of hoses and equipment, personnel to respond to alarms.

(h) Scheduling store personnel, inventory work, shelving new books, and restoring books to proper order on the shelves.
(i) Scheduling ground and squad car patrols around the clock, officers on surveillance assignments, general work involving nonemergency interaction with the public; arrangements must be made regarding who is on call for emergency duty.

3. Jocelyn must perhaps launder her clothes, arrange care for her cat, pack, arrange for a taxi to the airport, and get to the airport. Unless she can get a friend to help her with some of these tasks, she must do all the tasks herself. She can launder her clothes during the time she arranges for a taxi, but most of the tasks cannot be done simultaneously.

5. (a) i. Processor 1: T_1 from 0 to 13, T_3 from 13 to 25, T_6 from 25 to 45.
  Processor 2: T_2 from 0 to 18, T_4 from 18 to 27, T_5 from 27 to 35, idle from 35 to 45.
  ii. Processor 1: T_1 from 0 to 13, T_3 from 13 to 25, T_4 from 25 to 34, T_5 from 34 to 42.
  Processor 2: T_2 from 0 to 18, T_6 from 18 to 38, idle from 38 to 42.
(b) The schedule produced in (ii) is optimal, because the sum of the task times is 80 and no set of tasks can be arranged that will feasibly sum to 40 on each processor.
(c) The critical path is T_2T_6 and it has length 38. No schedule can be completed by time 38 on two processors because the sum of the task times divided by 2 is 40.

7. (a) Processor 1: T_1, T_2, T_3, T_5, T_7
  Processor 2: Idle 0 to 2, T_4, T_6, idle 4 to 5
(b) Processor 1: T_1, T_2, T_3, T_6 T_7
  Processor 2: Idle 0 to 2, T_4, T_5, idle 4 to 5
(c) Yes
(d) No
(e) T_3 and T_5

9. (a) Processor 1: T_1, T_6, idle 15 to 21, T_7, idle 27 to 31
  Processor 2: T_2, T_5, T_8
  Processor 3: T_3, T_4, idle from 13 to 31

(b) Processor 1: T_1, T_6, idle 15 to 21, T_7, idle 27 to 31

   Processor 2: T_3, T_4, idle from 13 to 21, T_8

   Processor 3: T_2, T_5, idle from 21 to 31

(c) Processor 1: T_4, idle 10 to 11, T_6, idle 18 to 21, T_8

   Processor 2: T_2, T_5, T_7, idle 27 to 31

   Processor 3: T_1, T_3, idle 11 to 31

11. Examples include inserting identical mirror systems on different models of cars and vaccinating different children against polio.

13. T_1, T_2, T_3, T_4, T_8, T_9, T_{10}, T_{11}, T_5, T_6, T_7, T_{12}.

15. (a) No. Consider the tasks that begin after the stretch where all machines are idle. Pick one of these tasks T and say machine 1 was the machine that it was given to. This task was ready for machine 1 just prior to when it began T because no task was just being completed on any other machine at this time because they were all idle. Thus, T should have begun earlier on machine 1.

(b) This schedule cannot arise using the list-processing algorithm, because T_2 should have been scheduled at time 0.

(c) Use the digraph with no edges and the list: T_2, T_1, T_3, T_4, T_5.

17. (a) T_1, T_2, T_3, and T_6 are ready at time 0.

(b) No tasks require that T_1 and T_6 be done before these other tasks can begin.

(c) The critical path consists only of T_6 and has length 20.

(d) Processor 1: T_1, T_6; Processor 2: T_2, T_4, idle from 18 to 30; Processor 3: T_3, T_5, idle from 12 to 30.

(e) No

(f) Processor 1: T_6, idle from 20 to 22; Processor 2: T_3, T_5, T_1; Processor 3: T_2, T_4, idle from 18 to 22.

(g) Yes

(h) Another list leading to the same optimal schedule is T_6, T_3, T_2, T_4, T_5, T_1.

19. (a) $5! = 120$

(b) No. Whatever list is used T_1 must be assigned to the first machine at time 0 because it is the only task ready at time 0.

(c) No. First, while Processor 1 works on T_1. Processor 2 must be idle. Second, the task times are integers with sum 31. If there are two processors, one of the processors must have idle time since when 2 divides 31 there is a remainder of 1.

(d) No

21. Using the order-requirement digraph shown and any list with one or more processors yields the same schedule:

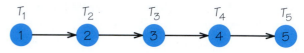

23. (a) One reasonable possibility is (time in min):

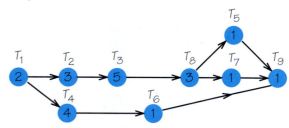

The earliest completion time is 15.

(b) The decreasing-time list is T_3, T_4, T_2, T_8, T_1, T_5, T_6, T_7, T_9. The schedule is Processor 1: T_1, T_4, T_6, idle 7 to 10, T_8, T_5, T_9; Processor 2: idle 0 to 2, T_2, T_3, idle 10 to 13, T_7, idle 14 to 15.

25. No. At time 11, T_4 should have been assigned to Machine 1 because Machine 1 was free at this time and T_4 was ready.

27. (a) Task times: $T_1 = 3$, $T_2 = 3$, $T_3 = 2$, $T_4 = 3$, $T_5 = 3$, $T_6 = 4$, $T_7 = 5$, $T_8 = 3$, $T_9 = 2$, $T_{10} = 1$, $T_{11} = 1$, and $T_{12} = 3$. This schedule would be produced from the list: T_1, T_3, T_2, T_5, T_4, T_6, T_7, T_8, T_{11}, T_{12}, T_9, T_{10}.

(b) Task times: $T_1 = 3$, $T_2 = 3$, $T_3 = 3$, $T_4 = 2$, $T_5 = 2$, $T_6 = 4$, $T_7 = 3$, $T_8 = 5$, $T_9 = 8$, $T_{10} = 4$, $T_{11} = 7$, $T_{12} = 9$, $T_{13} = 3$. This schedule would be produced from the list: T_1, T_5, T_7, T_4, T_3, T_6, T_{11}, T_8, T_{12}, T_9, T_2, T_{10}, T_{13}.

29. (a) (i) Processor 1: T_1, T_3, T_5, T_7, idle from 16 to 20; Processor 2: T_2, T_4, T_6, T_8.

(ii) Processor 1: T_8, T_5, T_4, T_1; Processor 2: T_7, T_6, T_3, T_2.

(b) The schedule in (ii) is optimal.

31. Such criteria include decreasing length of the times of the tasks, order of size of financial gains when each task is finished, and increasing length of the times of the tasks.

33. (a) Machine 1: T_1, T_6, T_{10}, idle from 8 to 9; Machine 2: T_3, T_4, T_{11}, T_{12}; Machine 3: T_2, T_7, idle from 8 to 9; Machine 4: T_5, T_8, T_9, idle from 8 to 9.

(b) Machine 1: T_1, T_8, T_{10}; Machine 2: T_5, T_6, T_2, T_{13}; Machine 3: T_7, T_{12}; Machine 4: T_4, T_{11}, idle from 9 to 12; Machine 5: T_3, T_9, idle from 11 to 12.

The schedule in part (a) had to have idle time because the total task time was 33 and 33 is not exactly divisible by 4. The schedule in part (b) had to have idle time because the total task time was 56 and 56 is not exactly divisible by 5.

35. (a) List for (i) yields (with items coded by task time): Machine 1: 12, 9, 15, idle from 36 to 50; Machine 2: 7, 10, 13, 20. List for (ii) yields: Machine 1: 12, 13, 20; Machine 2: 7, 9, 15, 10, idle from 41 to 45. List for (iii) yields: Machine 1: 20, 12, 9, idle from 41 to 45; Machine 2: 15, 13, 10, 7.

(b) These schedules complete earlier than those where precedence constraints hold. An optimal schedule is possible, however: Machine 1: 20, 10, 13; Machine 2: 12, 15, 9, 7. The associated list is: $T_6, T_1, T_5, T_7, T_2, T_4, T_3$.

(c) The critical path list is $T_6, T_5, T_4, T_1, T_7, T_2, T_3$ using the first processor. It finishes at time 41 and is idle until 45, when Processor 2 finishes.

37. (a) The tasks are scheduled on the machines as follows: Processor 1: 12, 13, 45, 34, 63, 43, 16, idle 226 to 298; Processor 2: 23, 24, 23, 53, 25, 74, 76; Processor 3: 32, 23, 14, 21, 18, 47, 23, 43, 16, idle 237 to 298.

(b) The tasks are scheduled on the machines as follows: Processor 1: 12, 24, 14, 34, 25, 23, 16, 16,

76; Processor 2: 23, 23, 21, 63, 43, idle 173 to 240; Processor 3: 32, 23, 53, 74, idle 182 to 240; Processor 4: 13, 45, 18, 47, 43, idle 166 to 240.

(c) The decreasing-time list is 76, 74, 63, 53, 47, 45, 43, 43, 34, 32, 25, 24, 23, 23, 23, 23, 21, 18, 16, 16, 14, 13, 12.

The tasks are scheduled on three machines as follows: Processor 1: 76, 45, 43, 24, 23, 18, 16, 13; Processor 2: 74, 47, 34, 32, 23, 21, 14, 12, idle 257 to 258; Processor 3: 63, 53, 43, 25, 23, 23, 16, idle 246 to 248.

The tasks are scheduled on four machines as follows: Processor 1: 76, 43, 24, 23, 16, idle 182 to 194; Processor 2: 74, 43, 25, 23, 16, 13; Processor 3: 63, 45, 32, 23, 18, 12, idle 193 to 194; Processor 4: 53, 47, 34, 23, 21, 14, idle 192 to 194.

(d) The new decreasing-time list is 84, 82, 71, 61, 55, 45, 43, 43, 34, 32, 25, 24, 23, 23, 23, 23, 21, 18, 16, 16, 14, 13, 12.

The tasks are scheduled as follows: Processor 1: 84, 45, 43, 25, 23, 23, 16, 12; Processor 2: 82, 55, 34, 32, 23, 18, 14, 13; Processor 3: 71, 61, 43, 24, 23, 21, 16, idle 259 to 271.

39. Examples include jobs in a videotape copying shop, data entry tasks in a computer system, and scheduling nonemergency operations in an operating room. These situations may have tasks with different priorities, but there is no physical reason for the tasks not to be independent, as would be the case with putting on a roof before a house had walls erected.

41. Each task heads a path of length equal to the time to do that task.

43. The times to photocopy the manuscripts, in decreasing order, are 120, 96, 96, 88, 80, 76, 64, 64, 60, 60, 56, 48, 40, 32. Packing these in bins of size 120 yields Bin 1: 120; Bin 2: 96; Bin 3: 96; Bin 4: 88, 32; Bin 5: 80, 40; Bin 6: 76; Bin 7: 64, 56; Bin 8: 64, 48; Bin 9: 60, 60. Nine photocopy machines are needed to finish within 2 minutes using FFD. The number of bins would not change, but the placement of the items in the bins would differ for worst-fit decreasing.

45. (a) Using the next-fit algorithm, the bins are filled as follows: Bin 1: 12, 15; Bin 2: 16, 12; Bin 3: 9, 11, 15; Bin 4: 17, 12; Bin 5: 14, 17; Bin 6: 18; Bin 7: 19; Bin 8: 21; Bin 9: 31; Bin 10: 7, 21; Bin 11: 9, 23; Bin 12: 24; Bin 13: 15, 16; Bin 14: 12, 9, 8; Bin 15: 27; Bin 16: 22; Bin 17: 18.
(b) The decreasing list is 31, 27, 24, 23, 22, 21, 21, 19, 18, 18, 17, 17, 16, 16, 15, 15, 15, 14, 12, 12, 12, 12, 11, 9, 9, 9, 8, 7.

The next-fit decreasing schedule is Bin 1: 31; Bin 2: 27; Bin 3: 24; Bin 4: 23; Bin 5: 22; Bin 6: 21; Bin 7: 21; Bin 8: 19; Bin 9: 18, 18; Bin 10: 17, 17; Bin 11: 16, 16; Bin 12: 15, 15; Bin 13: 15, 14; Bin 14: 12, 12, 12; Bin 15: 12, 11, 9; Bin 16: 9, 9, 8, 7.
(c) The worst-fit schedule using the original list is Bin 1: 12, 15, 9; Bin 2: 16, 12; Bin 3: 11, 15; Bin 4: 17, 12; Bin 5: 14, 17; Bin 6: 18, 7; Bin 7: 19, 9; Bin 8: 21, 15; Bin 9: 31; Bin 10: 21, 9; Bin 11: 23, 8; Bin 12: 24; Bin 13: 16, 12; Bin 14: 27; Bin 15: 22; Bin 16: 18.
(d) The worst-fit decreasing schedule would be Bin 1: 31; Bin 2: 27, 9; Bin 3: 24, 12; Bin 4: 23, 12; Bin 5: 22, 14; Bin 6: 21, 15; Bin 7: 21, 15; Bin 8: 19, 17; Bin 9: 18, 18; Bin 10: 17, 16; Bin 11: 16, 15; Bin 12: 12, 12, 11; Bin 13: 9, 9, 8, 7.
47. The bins have a capacity of 120. (First fit): Bin 1: 63, 32, 11; Bin 2: 19, 24, 64; Bin 3: 87, 27; Bin 4: 36, 42; Bin 5: 63. This schedule would take five station breaks; however, the total time for the breaks is under 8 minutes.

The decreasing list is 87, 64, 63, 63, 42, 36, 32, 27, 24, 19, 11. (First-fit decreasing): Bin 1: 87, 32; Bin 2: 64, 42, 11; Bin 3: 63, 36, 19; Bin 4: 63, 27, 24. This solution uses only four station breaks.
49. (a) There are theoretical results that show that best fit "usually" performs better than worst fit.
(b) Try 8, 7, 5, 3, 3, 2 in bins of capacity 14.
51. The total performance time exceeds what will fit on four discs. Using FFD, one can fit the music on five discs.
53. For problems with few weights to be packed, small integer weights, and a small integer as bin

capacity, this method can work well. However, when these special conditions are not met, it is very time-consuming to carry out this method. For example, imagine trying to use this method for 2000 random real numbers of the form .*xyz*, where *x*, *y*, and *z* are decimal digits and the bin capacity is 1.

55. It makes sense to leave bins open as more items arrive to be packed if the cost of having many bins open at once is reasonable and there is room to have many partially full bins open without incurring great inconvenience or cost. One such example might be a company that has room for many identical trucks to park as they are loaded with goods to be delivered. There may be complex cost trade-offs between sending off fewer trucks because we wait to pack as much into each truck as possible and sending out more partially filled trucks.

57. (a) The schedule with four secretaries is as follows: Processor 1: 25, 36, 15, 15, 19, 15, 27; Processor 2: 18, 32, 18, 31, 30, 18; Processor 3: 13, 30, 17, 12, 18, 16, 16, 16, 14; Processor 4: 19, 12, 25, 26, 18, 12, 24, 9.

The schedule with five secretaries is as follows: Processor 1: 25, 25, 31, 12, 16, 14; Processor 2: 18, 12, 17, 12, 15, 30, 9; Processor 3: 13, 32, 26, 16, 15, 18; Processor 4: 19, 36, 18, 19, 24; Processor 5: 30, 18, 15, 18, 16, 27.
(b) The decreasing-time list is 36, 32, 31, 30, 30, 27, 26, 25, 25, 24, 19, 19, 18, 18, 18, 18, 18, 17, 16, 16, 16, 15, 15, 15, 14, 13, 12, 12, 12, 9.

The schedule using this list on four processors would be Processor 1: 36, 25, 19, 18, 17, 16, 13, 9; Processor 2: 32, 26, 25, 18, 16, 15, 12; Processor 3: 31, 27, 24, 18, 16, 15, 12, 12; Processor 4: 30, 30, 19, 18, 18, 15, 14.

The schedule using this list on five processors would be Processor 1: 36, 24, 18, 16, 14, 12; Processor 2: 32, 25, 18, 18, 15, 12; Processor 3: 31, 25, 19, 18, 15, 9; Processor 4: 30, 27, 18, 17, 15, 13; Processor 5: 30, 26, 19, 16, 16, 12.

(c) The five-processor decreasing-time schedule is optimal (time 120), but the four-processor decreasing-time schedule is not. One can see this, since when the task of length 17 scheduled on processor 1 and the task of length 18 on processor 3 are interchanged, the completion time is reduced to 154 from 155 for the four-processor decreasing-time schedule.

(d) As a bin-packing problem, each bin will have a capacity of 60. Using the decreasing list, we obtain the following packings:

(First-fit decreasing): Bin 1: 36, 24; Bin 2: 32, 27; Bin 3: 31, 26; Bin 4: 30, 30; Bin 5: 25, 25, 9; Bin 6: 19, 19, 18; Bin 7: 18, 18, 18; Bin 8: 18, 17, 16; Bin 9: 16, 16, 15, 13; Bin 10: 15, 15, 14, 12; Bin 11: 12, 12.

(e) NFD uses 13 bins. Bin 1: 36; Bin 2: 32; Bin 3: 31; Bin 4: 30, 30; Bin 5: 27, 26; Bin 6: 25, 25; Bin 7: 24, 19; Bin 8: 19, 18, 18; Bin 9: 18, 18, 18; Bin 10: 17, 16, 16; Bin 11: 16, 15, 15; Bin 12: 15, 14, 13, 12; Bin 13: 12, 12, 9. WFD uses 11 bins. Bin 1: 36, 24; Bin 2: 32, 26; Bin 3: 31, 27; Bin 4: 30, 30; Bin 5: 25, 25; Bin 6: 19, 19, 18; Bin 7: 18, 18, 18; Bin 8: 18, 17, 16; Bin 9: 16, 16, 15, 13; Bin 10: 15, 15, 14, 12; Bin 11: 12, 12, 9.

(f) An optimal packing with 10 bins exists. Bin 1: 36, 24; Bin 2: 32, 16, 12; Bin 3: 31, 17, 12; Bin 4: 30, 30; Bin 5: 27, 18, 15; Bin 6: 26, 18, 16; Bin 7: 25, 19, 16; Bin 8: 25, 19, 15; Bin 9: 18, 18, 12, 9; Bin 10: 18, 15, 14, 13.

59. (a) Packing boxes of the same height into crates: packing want ads into a newspaper page.

(b) We assume, without loss of generality, $p \geq q$. One heuristic, similar to first-fit, orders the rectangles $p \times q$ as in a dictionary (i.e., $p \times q$ listed prior to $r \times s$ if $p > r$ or $p = r$ and $q \geq s$). It then puts the rectangles in place in layers in a first-fit manner; that is, do not put a rectangle into a second layer until all positions on the first layer are filled. However, extra room in the first layer is "wasted."

(c) The problem of packing rectangles of width 1 in an $m \times 1$ rectangle is a special case of the two-dimensional problem, equivalent to the bin-packing problem we have discussed.

(d) Two 1×10 rectangles cannot be packed into a 5×4 rectangle, even though there would be an area of 20 in this rectangle.

61. There is an example of a bin-packing problem for which a given list takes a certain number of bins, and when an item is deleted from the list, more bins are required. In this example, the deleted item is not first in the list.

63. (a) Graphs (a), (d), (e), and (f) can be colored with three colors, but graphs (b) and (c) cannot.

(b) Graphs (a), (b), (d), (e), and (f) can be colored with four colors, but graph (c) cannot.

(c) The chromatic numbers for graphs (a) through (f) are, respectively, 3, 4, 5, 2, 3, and 2.

65. (a) Construct the graph shown below:

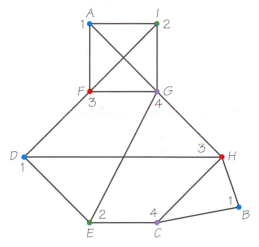

The vertices of this graph can be colored with no fewer than four colors (1, 2, 3, 4 are used to denote the colors in the figure). Hence, four tanks can be used to display the fish.

(b) The coloring in (a) shows that one can display two types of fish in three of the tanks, and three types of fish in one tank. Since 4 does not divide 9, one cannot do better.

67. (a) The graph for this situation is shown on the following page. The vertices can be labeled with the colors 1, 2, 3 as shown.

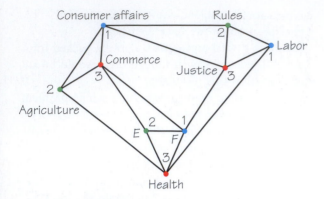

(b) Since the vertices can be colored with three colors (and no fewer), the minimum number of time slots for scheduling the committees is three.

(c) The committees can be scheduled in three rooms during each time slot. This might be significant if there were only three rooms that had microphone systems.

69. (a) Three time slots. To solve this problem, draw a graph by joining the vertices representing two committees if there is no x in the row and column of the table for these two committees.
(b) It is possible to three-color this graph so that each of the three colors is used three times. This means that one needs three rooms to arrange the scheduling of the nine committees.

71. Start at any vertex of the tree and label this vertex with color 1; color any vertex attached by an edge to this vertex with color 2. Continue to color the vertices in the tree in this manner, alternating the use of colors. If some vertex were attached to both a vertex colored 1 and another vertex colored 2, at some stage this would imply the graph had a circuit (of odd length), which is not possible, since trees have no circuits of any length.

73. The edge-coloring numbers for graphs (a) through (f) of Exercise 63 are, respectively, 6, 8, 6, 3, 3, and 4. The minimum edge-coloring number of any graph is either the maximal valence of any vertex in the graph or one more than the maximal valence. (This fact was discovered by the Russian mathematician Vizing.)

75. (a) Graph (a) four colors; graph (b) two colors; graph (c) four colors; graph (d) four colors; graph (e) two colors; graph (f) three colors.
(b) Coloring the maps of countries in an atlas would be one application of face colorings of graphs.

77. 3

Chapter 4

1. (a)

(b)

(c)

(d)

(e)

(f)

(b)

3. (a)

(c)

(b)

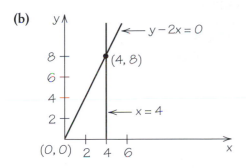

5. *Note:* The situation shown is for the first quadrant only.

(a)

(d)

7. (a) $6x + 4y \leq 300$
(b) $30x + 72y \leq 420$

9.

11.

13.

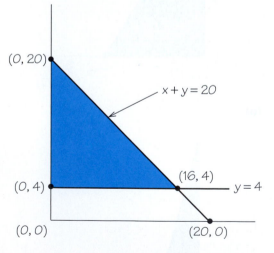

15. For Exercise 9, (2, 4) is feasible, but (10, 6) is not. For Exercises 11 and 13, both (2, 4) and (10, 6) are feasible.

17. Optimal production policy: Make 0 skateboards and 30 dolls for a profit of $111.

19. (a)

(b)

21.

23.

25. The maximum value is 28.

27. The maximum value is 38.

29. (a) The coordinates of $Q = (2, 0)$.
(b) $(2, 0)$ is not a feasible point.
(c) The profit at $(2, 0)$ is greater than the profit at $(13/7, 0)$.
(d) R is feasible. Profit at R is 33, which is less than the profit at Q.
(e) Maximization problems involving linear constraints, but where the variables are required to take on integer values, cannot be solved by first solving the associated linear-programming problem and rounding the answer to the nearest integer.

31. Schedule 400 oil changes and no tune-ups. With nonzero minimums of $x \geq 50$ and $y \geq 20$, schedule 300 oil changes and 20 tune-ups.

33. Schedule 360 routine visits. With nonzero minimums of $x \geq 20$, $y \geq 30$, schedule 210 routine visits and 30 comprehensive visits.

35. Take four math courses and no other courses. With nonzero minimums of $x \geq 2$, $y \geq 2$, either take two math and three other or find extra study time and take three math and two other.

37. Make 2 grade A and 5 grade B batches. With nonzero minimums $x \geq 1$ and $y \geq 1$, there will be no change, because the optimal production policy already obeys these nonzero minimums.

39. Make 3000 cartons of regular and 2000 cartons of diet. With zero minimums, there is no change in the optimal production policy, because the corresponding corner point does not touch either line from a minimum constraint.

41. Make no desk lamps and 1200 floor lamps. With nonzero minimum of $x \geq 150$, $y \geq 200$, make 150 desk lamps and 1080 floor lamps.

43. Optimum production policy: Make 50 chairs, 10 tables, and no beds each month.

45. Optimum production policy: Make 470 pounds of Excellent, none of Southern, 2400 of World, and 320 of Special.

47. 39

49. Make 60 business and no charity calls. Make 45 business and 10 charity calls.

51. Make 3 bikes and 2 wagons. With nonzero minimums of $x \geq 2$ and $y \geq 2$, there will be no change, because the optimal production policy obeys these minimums.

53. (a)

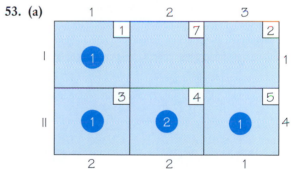

(b) The cost is 17.
(c) The indicator value for cell (I, 2) is 5 and for cell (I, 3) is −1.

55. (a) The graph is a tree because it is connected and has no circuit.
(b) If we add the edge joining vertex I to vertex 2, we get the circuit 2, I, 1, II, 2, while if we add the edge from vertex I to vertex 3, we get the circuit 3, I, 1, II, 3.
(c) Circuit 2, I, 1, II, 2 corresponds to the circuit of cells (I, 2), (I, 1), (II, 1), (II, 2), (I, 2). Circuit 3, I, 1, II, 3 corresponds to the circuit of cells (I, 3), (I, 1), (II, 1), (II, 3), (I, 3).

57. (a)

(i)

(ii)

(iii)

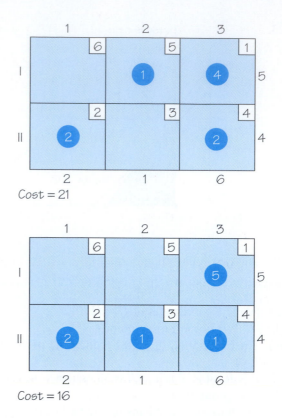

Cost = 21

Cost = 16

(b) For (i) the tableau shown is optimal. However, there are also other optimal tableaux. For (ii) the tableau shown is optimal, although there are also other optimal tableaux. For (iii) the tableau shown is not optimal. The optimal tableau is shown below and was obtained by two improvement steps from the starting tableau.

Chapter 5

1. (a) The vehicle models.
(b) Type, Transmission, Cylinders, City MPG, and Highway MPG. Histograms would be helpful for Cylinders (maybe) and the two MPGs (certainly).
3. Draw a histogram with a peak at the right and lower bars trailing out to the left of the peak. Most coins were minted in recent years, producing a peak at the right (highest-numbered years, like 2004 and 2005). There are few coins from 1990 and even fewer from 1980.
5. (a) Otherwise, big countries would top the list even if they had low emissions for their sizes.
(b) The distribution is skewed to the right. There appear to be three high outliers: Canada, Australia, and the United States.
7. (a) Alaska, 5.7%; Florida, 17.6%.
(b) The distribution is single-peaked and roughly

symmetrical. The center is near 12.7%. The spread is from 8.5% to 15.6%.

9. There is one high outlier, 200. The center of the 17 observations, other than the outlier, is 137 (9th of 17). The spread is 101 to 178.

11. (a) $\bar{x} = 2539/18 = 141.1$
(b) Without the outlier, $\bar{x} = 137.6$. The high outlier pulls the mean up.

13. The distribution of incomes is strongly right-skewed, so the mean is much higher than the median. Thus, $57,852 is the mean.

15. 1, 2, 2, 2, 3, 3, 4, 17

17. 5.7, 11.7, 12.75, 13.5, 17.6

19. 13, 18, 23, 28, 66

21. 0.0, 0.75, 3.2, 7.8, 19.9; the third quartile and maximum are much farther from the median than the first quartile and minimum.

23. The income distribution for bachelor's degree holders is generally higher than for high school graduates: The median for bachelor's is greater than the third quartile for high school. The bachelor's distribution is very much more spread out, especially at the high income end, but also between the quartiles.

25. (a) Either a histogram or a stemplot.
(b) $\bar{x} = 48.25$ and $M = 37.8$. The long right tail pulls the mean up.
(c) 2.0, 21.5, 37.8, 60.1, 204.9; the third quartile and maximum are much farther above the median than the first quartile and minimum are below it, showing that the right side of the distribution is much more spread out than the left side.

27. Values less than -7.8 or greater than 18.6 are suspected outliers: Arizona, California, Nevada, New Mexico, and Texas.

29. (a) $\bar{x} = 5.448$ and $s = 0.221$
(b) $M = 5.46$. The mean and median are close, but the one low observation pulls $\bar{x}$ slightly below M.

31. $s^2 = 225$

33. For both data, $\bar{x} = 7.50$ and $s = 2.03$ (to two decimal places). Data A have two low outliers and Data B have one high outlier.

35. Depends on your calculator.

37. (a) 1, 1, 1, 1
(b) 0, 0, 10, 10
(c) Answer (b) is unique; (a) is not.

39. $A =$ mean and $B =$ median

41. (a) 327 to 345 days
(b) 16%

43. $Q_1 = 886$ and $Q_3 = 1166$

45. (a) -32.3% to 48.9%
(b) A loss of 32.3% or greater.

47. (a) Median $= 10\%$, by symmetry
(b) 9.6% to 10.4%
(c) 9.866% to 10.134%

49. (a) 50%, 2.5%
(b) 0.37 to 0.43

51. Use whole millimeters as stems and tenths of millimeters as leaves. The lengths of red flowers are somewhat right-skewed, with no outliers. The lengths of yellow flowers are quite symmetrical, with no outliers.

53. Red: $\bar{x} = 39.71$, $s = 1.799$; Yellow: $\bar{x} = 36.18$, $s = 0.975$. The mean and standard deviation are better suited to the symmetrical yellow distribution.

55. Between 16% and 2.5% of yellow flowers are longer that 37.4 millimeters.

Chapter 6

1. (a) Study time explanatory, exam grade response.
(b) Relationship only.
(c) Rainfall explanatory, corn yield response.
(d) Relationship only.

3. (a) Life expectancy increases with GDP in a curved pattern. The increase is very rapid at first, but levels off for GDP above roughly $5000 per person.
(b) Richer nations have better diets, clean water, and better health care, so we expect life expectancy to increase with wealth. But once food, clean water, and basic medical care are in place, greater wealth has only a small effect on life span.

5. (a) See Exercise 29 for the scatterplot.
(b) There is a strong positive straight-line relationship.

7. (a) Here is the scatterplot, with speed as the explanatory variable:

Speed (km/hr)

We have added the regression line from Exercise 31.
(b) The relationship is curved; fuel usage first decreases as speed increases, and then increases at high speeds).
(c) There is no overall direction.
(d) The relationship is quite strong—there is little scatter about the overall curved pattern.
9. The estimated slope is 2.31.
11. (a) Choose two values of weeks, preferably near 1 and 150. Find pH from the equation given, plot the two points (weeks horizontal), and draw the line through them.
(b) 5.42 and 4.64 (rounded to two places)
(c) The slope -0.0053 says that on average pH declined by 0.0053 per week during the study period.
13. Line $y = x + 2$; slope $b = 1$
15. Correlation $r = 0.9757$, matching a very strong straight-line pattern.
17. Correlation $r = 0.9934$, stronger when the Insight is added, because that point extends (strengthens) the straight-line pattern.
19. See Exercise 7 for the scatterplot. The correlation is $r = -0.17$. Correlation measures the strength of only linear (straight-line) relationships. This relationship is strong but curved.

21. The correlation would be $r = 1$, because there is a perfect straight-line relationship, $y = x + 2$.
23. (a) Negative: Older cars will in general sell for lower prices.
(b) Negative: Heavier cars will (other things being equal) get fewer miles per gallon.
(c) Positive: Taller people are on average also heavier than shorter people.
(d) Small: There is no reason to expect that height and IQ are related.
25. (a) Dividend Growth, $r = 0.98$; Small Cap Stock, $r = 0.81$; Emerging Markets, $r = 0.35$.
(b) No: It just says that they tend to move in the same direction, whether up or down.
27. (a) Predicted mpg $= 4.87 + 1.11x$
(b) Predicted mpg $= 4.87 + (1.11)(18) = 24.85$ mpg
(c) The points in the plot closely follow a straight line, so we expect quite accurate predictions.
29. Choose two city mpg values, such as $x = 10$ and $x = 30$, and use the equation of the line to find y. Plot the two points and draw the line between them. Here is the plot.

Highway miles per gallon

The predicted highway mileage is about 25 mpg.
31. See Exercise 7 for the plot. The predicted values from the equation given are 10.9, 10.0, and

8.85. The observed values are 21.0, 6.3, and 12.88. The least-squares line gives the best straight-line fit, which is of little value here.

33. The slope of the least-squares line is $b = 0.54$. The intercept is $a = 33.67$. So for $x = 67$, we predict 69.85 inches.

35. In $y = a + bx$, substitute $a = \bar{y} - (rs_y/s_x)\bar{x}$, $b = rs_y/s_x$, and $x = \bar{x}$.

37. Open-ended. Suggest boxplots to compare the distributions for the two years. Stemplots show that the 2003 returns are roughly single-peaked and symmetrical, and that the 2002 returns are left-skewed with an extreme high outlier. The median returns are -12.8% in 2002 and 32.3% in 2003. Because of the influence of outliers on correlation, it is better to report the correlation without the outlier, $r = -0.838$. That is, the funds that went down most in 2002 tended to go up most in 2003. A scatterplot confirms this.

39. (a) All four sets of data have $r = 0.816$ and regression line $y = 3 + 0.5x$ to a close approximation.
(c) Only A is a "normal" regression setting in which the line is useful for prediction. Plot B is curved, C has an extreme outlier in y, and D has all but one x identical.

41. Heavier people who are concerned about their weight may be more likely than lighter people to choose artificial sweeteners in place of sugar.

43. Higher income generally means better water and sewage utilities, better diet, and better medical care, which will produce better health. But better health means more children can go to school and more workers are able to work and can stay on the job, which raises national income.

45. Explanatory: parents' income. Response: amount of college debt. We expect a negative association: Children of richer parents do not need to borrow as much to pay for college.

47. (a) There is a positive association, so r will be positive. The pattern is a bit irregular so that r won't be close to 1.
(b) $r = 0.5653$

49. (a) The slope is $b = 0.68$. For each additional inch of women's height, the height of the next man dated goes up by 0.68 inch on the average.
(b) The prediction is 69.56 inches.

Chapter 7

1. Population: U.S. residents aged 18 and older. Sample: the 1002 who responded.
3. Students passing the student center may not fairly represent all students. For example, they may underrepresent commuters or students whose classes are far from the center. In addition, a woman student may be reluctant to stop men, thus underrepresenting male students.
5. (a) This isn't clear: possibly its readers, possibly all adults in its circulation area.
(b) Larger. People with strong opinions, especially negative opinions, are more likely to respond. This is bias due to voluntary response.
7. Label the complexes 01 to 33 in alphabetical order. Line 117 gives 16 = Fairington, 32 = Waterford Court, and 18 = Fowler.
9. (a) 001 to 371
(b) Area codes labeled 214, 235, 119, 033, 199
11. If you always start at the same point in the table, your sample is predictable in advance. Repeated samples of the same size from the same population will always be the same—that's not random.
13. (a) Because $200/5 = 40$, we divide the list into 5 groups of 40. Label the first 40 rooms 01 to 40. Line 120 chooses room 35. The sample is rooms 35, 75, 115, 155, and 195.
(b) Each of the first 40 rooms has chance 1 in 40 to be chosen, so every room has equal chance, 1 in 40. An SRS gives all samples of 5 rooms an equal chance to be chosen.
15. (a) All people aged 18 and over living in the United States.
(b) The nonresponse rate is $669/1800 = 0.37 = 37\%$.

(c) It is hard to remember exactly how many movies you saw in exactly the past 12 months.

17. People are more reluctant to "change" the Constitution than to "add to" it. So the wording "adding to" will produce a higher percent in favor.

19. No treatment was imposed on the subjects. This observational study collected unusually detailed information about the subjects but made no attempt to influence them.

21. (a) It is an observational study that gathers information (e.g., through interviews) without imposing any treatment.
(b) "Significant" means "unlikely to be due simply to chance."
(c) Nondrinkers might be more elderly or in poorer health than moderate drinkers.

23. The design resembles Figure 7.3. Be sure to show randomization, two groups and their treatments, and the response variable (change in obesity). Label the 29 classes 01 to 29; the treatment group contains classes 17, 09, 22, 13, 07, 02, 27, 01, 18, 25, 29, 19, 14, 15, 08.

25. This is a randomized comparative experiment with four branches, similar to Figure 7.4 with one more branch. The "flow chart" outline must show random assignment of subjects to groups, the four treatments, and the response variable (healthcare spending). We can't show the group sizes because we don't know how many people or households are available to participate.

27. (a) There are six treatments:

	Discount Level		
	20%	**40%**	**60%**
50% on sale	1	2	3
100% on sale	4	5	6

(b) The outline randomly assigns 10 students to each of the 6 treatment groups, then compares the attractiveness ratings. It resembles Figure 7.4, but with 6 branches. Label the subjects 01 to 60. The first group contains subjects labeled 54, 58, 08, 15, 07, 27, 10, 25, 60, 55.

29. The design resembles Figure 7.3:

(b) Label the rats 01 to 18. The tea group contains 15, 06, 13, 09, 18, 03, 05, 16, 17.

31. (a) This is a randomized comparative experiment with four branches, similar to Figure 7.4 with one more branch. The "flow chart" outline must show random assignment of subjects to groups, the group sizes and treatments, and the response variable (colon cancer). It is best to use groups of equal size, 216 people in each group.
(b) With labels 001 to 864, the first five chosen are 731, 253, 304, 470, and 296.
(c) Those working with the subjects did not know the contents of the pill each subject took daily.
(d) The differences in colon cancer cases in the four groups were so small that they could easily be due to the chance assignment of subjects to groups.
(e) People who eats lots of fruits and vegetables may eat less meat or more cereal than other people. They may drink less alcohol or exercise more.

33. During the experiment, only the experimental cars had center brake lights, so they attracted attention. Once most cars had them, they were less noticed and so did a poorer job of preventing collisions. This is an example of an experiment that could not be completely realistic.

35. (a) What percent of college students say that being very wealthy is one of their goals in life? Consider themselves conservatives? Have had five drinks in one sitting in the past week?
(b) Will offering nightly tutoring sessions improve performance in a math course? Will showing videos of drunken students reduce binge drinking?

37. Both are statistics.

39. (a) Approximately normal with mean 0.14 and standard deviation 0.0155.
(b) $0.14 \pm (2)(0.0155)$, or 0.109 to 0.171

41. (a) Each digit in the table has one chance in 10 to be any of the 10 possible digits. So in the long run, 60% of the digits we encounter will be 0, 1, 2, 3, 4, or 5.
(b) Line 101 contains 29 digits 0 to 5. Use lines 101 to 110 to simulate 10 samples; the counts of "yes" responses are 29, 24, 23, 23, 20, 24, 23, 19, 24, and 18. Thus, three samples are exactly correct (24/40 = 60%), one overestimates, and six underestimate.

43. The sample proportion is 0.42. The approximate 95% confidence interval is 0.42 ± 0.023, or 0.397 to 0.443.

45. (a) The sample proportion is 0.194. The approximate 95% confidence interval is 0.194 ± 0.027, or 0.167 to 0.221.
(b) It is likely that more than 171 ran a red light, because some people are reluctant to admit illegal or antisocial acts.

47. If Harris takes a very large number of poll samples using the same methods, the poll result will be within ± 3 percentage points of the truth about the population in 95% of the samples.

49. (a) 11.3%
(b) The response rate is so low that it is likely that those who responded differ from the population as a whole. That is, there is a bias that the margin of error does not include.

51. (a) No. The number of e-filed returns in all states is much larger than the sample size. When this is true, the margin of error depends only on the size of the sample, not on the size of the population.
(b) The sample sizes vary from 970 to 49,000, so the margins of error will also vary.

53. The margin of error for 90% confidence comes from the central 90% of a normal sampling distribution. We need not go as far out to cover 90% of the distribution as to cover 95%. So the margin of error for 90% confidence is smaller than for 95% confidence.

55. We are 95% confident that the true proportion of articles that discuss the success of blinding is between 0.02 and 0.124 (that is, 2% to 12.4%).

57. The probability is about 0.16.

Chapter 8

1. Results will vary.
3. 0.105
5. (a) $S = \{0, 1, 2, 3, 4, 5, 6, 7, 8, 9, 10\}$
(b) $S = \{0, 10, 20, 30, 40, 50, 60, 70, 80, 90, 100\}$
(c) $S = \{Yes, No\}$
7. (a) $S = \{$HHHH, HHHM, HHMH, HMHH, MHHH, HHMM, HMMH, MMHH, HMHM, MHHM, MHMH, HMMM, MMMH, MHMM, MMHM, MMMM$\}$
(b) $S = \{0, 1, 2, 3, 4\}$
9. (a) 0.19
(b) 0.287
11. Any two events that can occur together will do.
13. (b) 0.64
15. (a) Yes
(b) No
(c) Yes
17. The probability model is equally likely; outcomes 1 to 12 each with probability 1/12.
19. 42/90
21. (a) 1024
(b) 2/1024
23. There are 46,656 different codes. The probability of no x is 0.919. The probability of no digits is 0.377.
25. The possibilities are *ags, asg, gas, gsa, sag, sga,* of which *gas* and *sag* are English words. The probability is $2/6 = 0.33$.
27. There are 44,916,768 possible IDs. Of these, 12,355,928 have no numbers. The probability is 0.275.
29. (a) The area is 1.

(b) Probability 1/2.

(c) The area representing this event is 0.125.

31. The area is the half of the square below the $y = x$ line, probability 1/2.

33. The mean is 2.78. The variance is 0.7516, so the standard deviation is 0.8669.

35. The mean for owner-occupied units is 6.248. For rented units, $\mu = 4.321$.

37. Both models have mean 1, because both density curves are symmetrical about 1.

39. (a) $\mu = 3.5$

(b) $\mu = 7$

(c) We could roll two dice separately and add the spots later.

41. (a) 4.5

(b) 2.5

(c) The mean count for the two dice is 7. This is the same as for rolling two standard dice, with mean 3.5 for each.

43. (a) 0.99749

(b) $623.218

45. Between 0.11 and 0.19.

47. (a) 5.77 mg

(b) $n = 4$ measurements. Averages of several measurements are less variable than individual measurements, so an average is more likely to give about the same result each time.

49. (b) 4580 to 4620

(c) 4588.46 to 4611.54

51. (a) About 0.16

(b) Mean 20.8, Standard deviation 1.6

(c) About 0.0015

53. (a) 45,697,600

(b) 2600

(c) 0.0000569

55. (a) 0.15

(b) 0.32

57. (a) 1.7689 days

(b) 0.676 days and 1.384 days

Chapter 9

1. Minority rule satisfies condition (1): An exchange of marked ballots between two voters leaves the number of votes for each candidate unchanged, so whichever candidate won on the basis of having fewer votes before the exchange still has fewer votes after the exchange. Minority rule also satisfies condition (2): Suppose candidate X receives n votes and candidate Y receives m votes, and candidate X wins because $n < m$. Now suppose that a new election is held, and every voter reverses his or her vote. Then candidate X has m votes and candidate Y has n votes, and so candidate Y is the winner. Minority rule, however, fails condition (3): Suppose, for example, that there are 3 voters and that candidate X wins 1 out of the 3 votes. Now suppose that one of the 2 voters who voted for candidate Y reverses his or her vote. Then candidate X would have 2 votes, and candidate Y would have 1 vote, thus resulting in a win for candidate Y.

3. A dictatorship satisfies condition (2): If a new election is held and every voter (in particular, the dictator) reverses his or her ballot, then certainly the outcome of the election is reversed. A dictatorship also satisfies condition (3): If a single voter changes his or her ballot from being a vote for the loser of the previous election to a vote for the winner of the previous election, then the single voter could not have been a dictator (since the dictator's ballot was not a vote for the loser of the previous election). Thus, the outcome of the new election is the same as the outcome of the previous election. A dictatorship, however, fails to satisfy condition (1): If the dictator exchanges his or her marked ballot with any voter whose marked ballot differs from that of the dictator, then the outcome of the election is certainly reversed.

5. With an odd number of voters, each one-on-one score will have a winner because there cannot be a tie. If one of the candidates, say A, is the winner by Condorcet's method, then A would have beaten every other candidate in a one-on-one competition.

7. (a) Yes.

(b) Alfonse D'Amato (D) is the winner by plurality voting.

9. (a) C

(b) A

(c) *D*
(d) *D*
11. (a) They tie.
(b) *C*
(c) They all tie.
(d) *E*
13. (a) *C*
(b) *E*
(c) *E*
(d) *E*
15. (a) *E*
(b) One possible scenario is having candidates *A*, *B*, and *C* with the following preference lists.

Rank	Number of Voters (2)	
	1	1
First	*A*	*C*
Second	*B*	*B*
Third	*C*	*A*

17. (a) Plurality voting satisfies the Pareto condition, because if everyone prefers *B* to *D*, for example, then *D* has no first-place votes at all. Thus, *D* cannot be among the winners in plurality voting.
(b) Plurality voting satisfies monotonicity, because if a candidate wins on the basis of having the most first-place votes, then moving that candidate up one spot on some list (and making no other changes) neither decreases the number of first-place votes for the winning candidate nor increases the number of first-place votes for any other candidate. Hence, the original winner remains a winner in plurality voting.
19. (a) Sequential pairwise voting satisfies the Condorcet winner criterion because a Condorcet winner always wins the kind of one-on-one contest that is used to produce the winner in sequential pairwise voting.
(b) Sequential pairwise voting satisfies monotonicity because moving a candidate up on some list only improves that candidate's chances in one-on-one contests.
21. With the plurality runoff method, assume everyone prefers *B* to *D*. Then *D* has no first-place votes. So, if *D* is involved in the runoff,

then some other alternative has *all* the first-place votes. This other alternative will emerge as the sole winner.
23. Answers will vary.
25. One possible scenario is having candidates *A*, *B*, and *C* with the following preference lists.

Rank	Number of Voters (5)	
	3	2
First	*A*	*B*
Second	*B*	*C*
Third	*C*	*A*

27. (a) Since *D* has the least number of first-place votes, *D* is eliminated. Then since *B* has the least number of first-place votes, *B* is eliminated. Thus, *A* is the unique winner by the Hare system.
(b) *B*
29. Answers will vary.
31. Answers will vary.
33. *A*
(a) *A*
(b) *B*
(c) Answers will vary.

Chapter 10

1. One example of two such elections is the following:

Rank	Election 1 Number of Voters (3)		
First	*A*	*A*	*B*
Second	*B*	*B*	*A*

Rank	Election 2 Number of Voters (3)		
First	*B*	*A*	*B*
Second	*A*	*B*	*A*

With the voting system in which the candidate with the fewest first-place votes wins, *B* is the winner in the first election. However, if the leftmost voter changes his or her ballot as shown in the second election, then *A* becomes the winner. Taking the ballots in the first election to be the sincere preferences of the voters, the leftmost voter (who prefers A to B) has secured a more favorable outcome by the submission of a disingenuous ballot.

3. One example of two such elections is the following:

	Election 1		
Rank	**Number of Voters (3)**		
First	A	B	B
Second	B	A	A

	Election 2		
Rank	**Number of Voters (3)**		
First	B	B	B
Second	A	A	A

With the voting system in which the candidate receiving an even number of first-place votes wins, B is the winner in the first election. However, if the leftmost voter changes his or her ballot as shown in the second election, then A becomes the winner. Taking the ballots in the first election to be the sincere preferences of the voters, the leftmost voter (who prefers A to B) has secured a more favorable outcome by the submission of a disingenuous ballot.

5. (a) The voting system does not treat all voters the same.
(b) A dictatorship in which voter 1 is the dictator.
(c) A dictatorship in which voter 2 is the dictator and a dictatorship in which voter 3 is the dictator.
7. With the given ballots, the winner using the Borda count is A. But the winner becomes C if the leftmost voter changes his or her ballot to

$$C$$
$$B$$
$$D$$
$$A$$

9. With the given ballots, the winner using the Borda count is B. But the winner becomes A if the leftmost voter changes his or her ballot to

$$A$$
$$C$$
$$D$$
$$B$$

11. The following is one such example:

A	B	B	A	D	A	F	A	F
B	A	A	B	C	B	E	B	E
C	C	C	C	B	C	D	C	D
D	D	D	D	A	D	C	D	C
E	E	E	E	E	E	B	E	B
F	F	F	F	F	F	A	F	A

13. With the given ballots, the winner using the weak Condorcet rule is C. But the winner becomes B if the leftmost voter changes his or her ballot to

$$B$$
$$A$$
$$D$$
$$C$$

15. With the given ballots, the winner using Coombs's rule is C. But the winner becomes A if the leftmost voter changes his or her ballot to

$$A$$
$$C$$
$$B$$

17. With the given ballots, the winner using the plurality runoff rule is C. But the winner becomes B if the leftmost voter changes his or her ballot to

$$B$$
$$A$$
$$C$$

19. (a) For B to win, consider the agenda $DACB$.
(b) For C to win, consider the agenda $BDAC$.
(c) For D to win, consider the agenda $BACD$.
21. If the voters in the 7% group all change their ballots to

$$H$$
$$J$$
$$D$$

then the winner switches from D to H, which they all prefer.

23. Properties 1, 2, and 3

25. Properties 1, 2, and 4

27. Consider the following scenario: The chair votes for A and I vote for B. If you vote for C, the winner is A (your least preferred outcome), while the winner is B if you vote for B. This shows that voting for C does not weakly dominate your strategy of voting for B.

Chapter 11

1. (a) A winning or blocking coalition would be 50 senators plus the vice president, or more than 50 senators.

(b) A winning or blocking coalition requires 50 or more senators.

(c) A winning coalition requires at least 67 senators. A coalition of 34 or more senators can block.

3. (a) No

(b) The weight-5 and weight-4 voters have veto power.

(c) The weight-3 voter is a dummy.

5. No

7. Let's call the voters A, B, C, and D in declining order of voting weight.

(a) There are 12 permutations: $BACD$, $BADC$, $CABD$, $CADB$, $DABC$, $DACB$, $BCAD$, $CBAD$, $BDAC$, $DBAC$, $CDAB$, $DCAB$.

(b) B is pivotal in the following four permutations: $ABCD$, $ABDC$, $CDBA$, and $DCBA$.

(c) $\left(\dfrac{1}{2}, \dfrac{1}{6}, \dfrac{1}{6}, \dfrac{1}{6} \right)$

9. (a) $\left(\dfrac{5}{12}, \dfrac{1}{4}, \dfrac{1}{4}, \dfrac{1}{12} \right)$

(b) $\left(\dfrac{1}{3}, \dfrac{1}{3}, \dfrac{1}{6}, \dfrac{1}{6} \right)$

(c) $\left(\dfrac{1}{4}, \dfrac{1}{4}, \dfrac{1}{4}, \dfrac{1}{4} \right)$

11. (a) We can represent a "yes" with 1, and a "no" with 0. Then the voting combinations are the 16 four-bit binary numbers 0000, 0001, 0010, 0011, 0100, 0101, 0110, 0111, 1000, 1001, 1010, 1011, 1100, 1101, 1110, 1111.

(b) {}, {D}, {C}, {C, D}, {B}, {B, D}, {B, C}, {B, C, D}, {A}, {A, D}, {A, C}, {A, C, D}, {A, B}, {A, B, D}, {A, B, C}, {A, B, C, D}.

(c) If the first bit of a given permutation is 1, then A votes yes. If the second bit is 1, B votes yes in the corresponding coalition. The third bit tells us how C votes, and the fourth indicates the vote of D.

(d) i. 1

    ii. 4

    iii. 6

13. (a) (10, 6, 6, 2)

(b) (8, 8, 4, 4)

(c) (6, 6, 6, 6)

(d) (8, 4, 4, 4)

(e) (6, 6, 2, 2)

(f) (4, 4, 4, 0)

(g) (2, 2, 2, 2)

15. (a) 1001001001

(b) 10101010101

(c) 11111010101

17. (a) 20

(b) 4950

(c) 4950

(d) 126

19. (a) {A, C, D} and {A, B}.

(b) A belongs to each winning coalition, so if A opposes a motion it will not pass. {B, C} and {B, D} are also minimal blocking coalitions.

(c) (10, 6, 2, 2).

(d) [5 : 3, 2, 1, 1] is one set of weights that works, but there are many other solutions.

(e) $\left(\dfrac{7}{12}, \dfrac{1}{4}, \dfrac{1}{12}, \dfrac{1}{12} \right)$

21. (28, 4, 4, 4, 4)

23. Let's look at the minimal winning coalitions. We call the faculty members $F_1, \ldots, F_4$ and the administrators A_1, A_2, and A_3. Here are two winning coalitions: { $F_1, F_2, F_3, F_4, A_1, A_2$}, in which the administrators are critical but the faculty members aren't; and $\{F_1, F_2, F_3, A_1, A_2, A_3\}$, in which the faculty members are critical but the administrators are not. In any weighted voting system, the critical voters in a coalition must have more weight than those who are not critical. The first coalition that we cited indicates that the administrators should have more weight, while the second indicates that the faculty members have more weight. These contradictory requirements cannot be satisfied, so the system is not equivalent to a weighted voting system.

25. The Shapley–Shubik index of this voting system is (in lowest terms)

$$\left(\frac{11}{70}, \frac{11}{70}, \frac{11}{70}, \frac{11}{70}, \frac{13}{105}, \frac{13}{105}, \frac{13}{105}\right)$$

By this measure, each faculty member is more powerful than any administrator. A faculty member has about 15.7% of the power, and an administrator has about 12.4%.

27. All four-voter systems can be presented as weighted voting systems:

Minimal Winning Coalitions	Weights
$\{A, B, C, D\}$	[4:1, 1, 1, 1]
$\{A, B\}, \{A, C, D\}$	[5:3, 2, 1, 1]
$\{A, B, C\}, \{A, B, D\}$	[5:2, 2, 1, 1]
$\{A, B\}, \{A, C\}, \{A, D\}$	[4:3, 1, 1, 1]
$\{A, B\}, \{A, C\}, \{B, C, D\}$	[5:3, 2, 2, 1]
$\{A, B\}, \{A, C, D\}, \{B, C, D\}$	[4:2, 2, 1, 1]
$\{A, B, C\}, \{A, B, D\}, \{A, C, D\}$	[4:2, 1, 1, 1]
$\{A, B\}, \{A, C\}, \{A, D\}, \{B, C, D\}$	[4:3, 2, 1, 1]
$\{A, B, C\}, \{A, B, D\}, \{A, C, D\},$ $\{B, C, D\}$	[3:1, 1, 1, 1]

29. (b)

31. (a) [8 : 6, 1, 1, 1, 1, 1, 1, 1, 1]
(b) (492, 16, 16, 16, 16, 16, 16, 16, 16)
(c) $\left(\frac{2}{3}, \frac{1}{24}, \frac{1}{24}, \frac{1}{24}, \frac{1}{24}, \frac{1}{24}, \frac{1}{24}, \frac{1}{24}, \frac{1}{24}\right)$
(d) In this system, the chairperson is 30.75 times as powerful as an ordinary member according to the Banzhaf index, but only 16 times as powerful according to the Shapley–Shubik power index.

33. The weighted voting system is [9 : 4, 4, 4, 1, 1, 1, 1, 1].
35. The three weight-3 voters or two weight-3 voters and one weight-1 voter form minimal winning coalitions. The Banzhaf power index of this system is (30, 30, 30, 6, 6, 6).
37. If there are three weight-1 voters, the Shapley–Shubik power index of this system is $\left(\frac{17}{60}, \frac{17}{60}, \frac{17}{60}, \frac{1}{20}, \frac{1}{20}, \frac{1}{20}\right)$. The weight 3 voter is times as powerful as the weight-1 voter.

With four weight-1 voters, the Shapley–Shubik index changes to

$$\left(\frac{9}{35}, \frac{9}{35}, \frac{9}{35}, \frac{2}{35}, \frac{2}{35}, \frac{2}{35}, \frac{2}{35}\right)$$

Each weight-3 voter is now $4\frac{1}{2}$ times as powerful as a weight-1 voter.
39. Consider Maine, which has two congressional districts. Ignoring the rest of the country, Maine would be a 3-voter system, in which the state has 2 votes, and each congressional district has 1. There are 3! permutations of voters, but only 2 are possible: $1M2$ and $2M1$, where the congressional districts are identified by numerals, and the state is M. The reason is that each entity (district or statewide) is given a score, which is the number of votes recorded for the Bush–Cheney ticket, divided by the number of votes cast in that entity for the Kerry–Edwards ticket. The entity's position in the permutation is determined by that ratio.

Let r_1, r_2, and r_M denote the ratios for the two districts and the state as a whole, respectively, and let y_1, y_2, and y_M be the number of votes cast for the Kerry–Edwards ticket in each entity. The reason that some electoral permutations are impossible is that $y_M = y_1 + y_2$. Each vote is actually counted twice: once for the elector representing the voter's congressional district, and once for the two statewide electors. The numbers of votes for the Bush–Cheney ticket in the three entities were $r_1 y_1$, $r_2 y_2$, and $r_M(y_1 + y_2)$. Because the number of statewide votes for the

Bush–Cheney ticket can also be determined by adding the votes in the two districts:

$$r_M(y_1 + y_2) = r_1 y_1 + r_2 y_2$$

Dividing by $(y_1 + y_2)$, we can obtain a formula for r_M:

$$r_M = \frac{r_1 y_1 + r_2 y_2}{y_1 + y_2}$$

Suppose that $r_1 > r_2$. Then $r_1 y_2 > r_2 y_2$, and thus

$$r_1 = \frac{r_1 y_1 + r_1 y_2}{y_1 + y_2} > \frac{r_1 y_1 + r_2 y_2}{y_1 + y_2} = r_M$$

Also, $r_1 y_1 > r_2 y_1$, so

$$r_M = \frac{r_1 y_1 + r_2 y_2}{y_1 + y_2} > \frac{r_2 y_1 + r_2 y_2}{y_1 + y_2} = r_2$$

These inequalities taken together show that $r_1 > r_M > r_2$ if $r_1 > r_2$. Of course, if $r_2 > r_1$, the same argument would show $r_2 > r_M > r_1$.

Any permutation of the states in which the statewide electors for Maine do not fall between the electors for the two congressional districts is therefore impossible. The same is true for Nebraska: At least one district elector must come before the statewide electors, and one must come after them.

The Shapley–Shubik power index should be computed by using only the possible permutations. A given entity's Shapley–Shubik power index would be the number of possible permutations in which it is pivotal, divided by the total number of possible permutations.

Chapter 12

1. Assume a distribution is skewed to the left. The heavier concentration of voters on the right means that fewer voters are farther from the median. Because there are fewer voters "pulling" the mean rightward, it will be to the left of the median. Likewise, a distribution skewed to the right will have a mean to the right of the median.

3. While there is no median position such that half the voters lie to the left and half the voters lie to the right, there is still a position where the middle voter (if the number of voters is odd) or the two middle voters (if the number of voters is even) are located, starting from either the left or right. In the absence of a median, less than half the voters lie to the left and less than half to the right of the middle voter's (voters') position(s). Hence, any departure by a candidate from a position of a middle voter to the position of a nonmiddle voter on the left or right will result in that candidate's getting less than half the votes—and the opponent's getting more than half. Thus, the middle position(s) is (are) in equilibrium, making it (them) the extended median.

5. When the four voters on the left refuse to vote for a candidate at 0.6, his opponent can do better by moving to 0.7, which is worse for the dropouts.

7. The voters are spread from 0.1 to 0.9, so it is a position at 0.5 that minimizes the maximum distance (0.4) a candidate is from a voter. If the candidates are at the median of 0.6, the voter at 0.1 would be a distance of 0.5 from them. In this sense, the median is worse than the mean of 0.56, which would bring the candidates closer to the farthest-away voter and, arguably, be a better reflection of the views of the electorate.

9. The middle peak will be in equilibrium when it is the median or the extended median. Yes, it is possible that, say, the peak on the left is in equilibrium, as illustrated by the following discrete-distribution example, in which the median is 0.2:

Position i	1	2	3	4	5	6	7
Location l_i of position i	0.1	0.2	0.3	0.5	0.6	0.8	0.9
Number of voters n_i at position i	7	8	1	2	1	2	1

11. If the population is not uniformly distributed and, say, 80% live between $\frac{3}{8}$ and $\frac{5}{8}$

and only 10% live to the left of $\frac{3}{8}$ and 10% to right of $\frac{5}{8}$, then the bulk of the population will be well served by two stores at $\frac{1}{2}$. In fact, stores at $\frac{1}{4}$ and $\frac{3}{4}$ will be farther away for 80% of the population, so it can be argued that the two stores at $\frac{1}{2}$ provide a social optimum.

13. Presumably, the cost of travel would have to be weighed against how much lower more competitive prices are.

15. Since the districts are of equal size, the mayor's median or extended median must be between the leftmost and rightmost medians or extended medians; otherwise, at least $\frac{2}{3}$ of the voters would be on one side of the mayor's position, which would preclude it from being the median or extended median. This is not true of the mean, however, if, say, the left-district positions are much farther away from the mayor's median or extended median than the right-district positions. In such a case, the mayor's mean would be in the interval of the left-district positions.

17. If, say, A takes a position at M and B takes a position to the right of M, C should take a position just to the left of M that is closer to M than B's position, giving C essentially half the votes and enabling him or her to win the election. If neither A nor B takes a position at M, C should take a position next to the player closer to M; the position that C takes to maximize his or her vote may be either closer to M (if the candidates are far apart) or farther from M (if the candidates are closer together), but this position may not be winning. For example, assume the voters are uniformly distributed over [0, 1]. If $\frac{3}{16}$ of the voters lie between A (to the left of M) and M, and $\frac{3}{16}$ of the voters lie between M and B (to the right of M), then C does best taking a position just to the left of A or just to the right of B, obtaining essentially $\frac{5}{16}$ of the vote. To be specific, assume C moves just to the left of A. Then A will obtain $\frac{3}{16}$ of the vote, but B will win with $\frac{1}{2}$ of the vote (that to the right of M), so C's maximizing position will not always be sufficient to win.

19. Following the hint, C will obtain $\frac{1}{3}$ of the vote by taking a position at M, as will A and B, so there will be a three-way tie among the candidates. Because a nonunimodal distribution can be bimodal, with the two modes close to M, C can win if he or she picks up most of the vote near the two modes, enabling C to win with more than $\frac{1}{3}$ of the vote.

21. B should enter just to the right of $\frac{3}{4}$, making it advantageous for C to enter just to the left of A, giving C essentially $\frac{1}{4}$ of the vote. With C and A almost splitting the vote to the left of M and a little beyond, B would win almost all the vote to the right of M. (If C entered at $\frac{1}{2}$, he or she would get slightly more than $\frac{1}{4}$ of the vote but lose to A, who would get $\frac{3}{8}$.)

23. If the distribution is uniform, these positions are $\frac{1}{6}$, $\frac{5}{6}$, and $\frac{1}{2}$ for A, B, and C, respectively, making D indifferent between entering just to the left of A, just to the right of B, or in between A and C at $\frac{1}{3}$ or between C and B at $\frac{2}{3}$, which would give D $\frac{1}{6}$ of the vote in any case.

25. No, because Gore would get 49%, the same as Bush; so instead of winning, Gore would tie with Bush.

27. It seems far too complicated a "solution" for avoiding effects caused by the Electoral College. Why not just abolish the Electoral College?

29. By definition, more voters prefer the Condorcet winner to any other candidate. Thus, if the poll identifies the Condorcet winner as one of the top two candidates, he or she will receive more votes when voters respond to the poll by voting for one or the other of these candidates. The possibility that the Condorcet winner might not be first in the poll, but win after the poll is announced, shows that the plurality winner may not be the Condorcet winner. Some argue that the Condorcet winner is always the "proper" winner, but others counter that a non-Condorcet winner who is, say, everybody's second-most-preferred candidate is a better social choice than a 51%-Condorcet winner who is ranked last by the other 49%.

31. D is the Condorcet winner. It is strange in the sense that a poll that identifies either the top two or the top three candidates would not include D.

33. A would win with 4 votes to 3 votes for B and 3 votes for C. It is strange that the number of top contenders identified by a poll can result in opposite outcomes (A in this exercise, whereas B defeats A when only two top contenders are identified by a poll, as in Exercise 32).

35. Assume a voter votes for just a second choice. It is evident that voting for a first choice, too, can never result in a worse outcome and may sometimes result in a better outcome (if the voter's vote for a first choice causes that candidate to be elected).

37. Following the hint, the voter's vote for a first and third choice would elect either A or C. If the voter also voted for B, then it is possible that if A and B are tied for first place, then B might be elected when the tie is broken, whereas voting for just A and C in this situation would elect A.

39. No. Voting for a first choice can never hurt this candidate and may help elect him or her.

41. No. If class I and II voters vote for all candidates in their preferred subsets, they create a three-way tie among A, B, and C. To break this tie, it would be rational for the class III voter to vote for both D and C and so elect C, whom this voter prefers to both A and B. But now class I voters will be unhappy, because C is a worst choice. However, these voters cannot bring about a preferred outcome by voting for candidates different from A and B.

43. Without polling, A in case (i), D in case (ii), and B and D in case (iii); with polling, B in case (i), D in case (ii), and D in case (iii).

45. Exactly half the votes, or 9.5 votes each.

47. Substitute into the formula for r_i in Exercise 46 $d_i = (n_i/N)(D)$ and $D = R$. The proportional rule is "strategy-proof" in the sense that if one player follows it, the other player can do no better than to follow it. Hence, knowing that an opponent is following the proportional rule does not help a player optimize against it by doing anything except also following it.

49. To win in states with more than half the votes, any two states will do. Thus, there is no state to which a candidate should not consider allocating resources. In the absence of information about what one's opponent is doing, all states that receive allocations should receive equal allocations since all states are equally valuable for winning.

51. The Democrat can win the election by winning in any two states or in all three. The first three expressions in the formula for PWE_D give the probabilities of winning in the three possible pairs of states, whereas the final expression gives the probability of winning in all three states.

53. Yes, but in a complicated way. Intuitively, the large states that are more pivotal, and whose citizens therefore have more voting power (as shown in Chapter 11), are more deserving of greater resources (as shown in this chapter).

Chapter 13

1. Donald receives the Trump Tower triplex and $\frac{13}{15}$ (about 87%) ownership of the Palm Beach mansion for a total of about 72.7 of his points, and Ivana gets the rest (for about 72.7 of her points).

3. Phil gets his way on the stereo level issue, the smoking rights issue, the phone time issue, the visitor policy issue, and $\frac{13}{15}$ (about 87%) of his way on the alcohol issue for a total of 68 points. Mike gets his way on the rest.

5. Answers will vary.

7. *Allocation 1:*
(a) Not proportional
(b) Not envy-free
(c) Not equitable
(d) Example: Give Bob X, Carol Y, and Ted Z.
   Allocation 2:
(a) Not proportional
(b) Not envy-free
(c) Not equitable
(d) Example: Give Bob Y, Carol X, and Ted Z.

Allocation 3:
(a) Not proportional
(b) Not envy-free
(c) Not equitable
(d) It is Pareto-optimal.
  Allocation 4:
(a) Not proportional
(b) Not envy-free
(c) Not equitable
  Allocation 5:
(a) It is proportional.
(b) Not envy-free
(c) It is equitable.

9. Mary gets both the car and the house and pays John $15,081.25 + $28,750 = $43,831.25.

11. *A*, in total, receives the farm plus $7333. Similarly, *B* receives $132,334 and *C* receives both the house and the sculpture, while paying $139,667.

13. Bob first chooses the investments, and the final allocation has him also receiving the car and the CD player.

15. Mark first chooses the tractor, and the final allocation has him also receiving the truck and the tools.

17. Donald first chooses the Palm Beach mansion, and the final allocation has him also receiving the Trump Tower triplex and the cash and jewelry.

19. The chooser.

21. (a) Bob gets a piece whose value to him is 9 units (assuming that Bob is the divider), and Carol gets a piece whose value to her is 12 units.
(b) Carol gets a piece whose value to her is 9 units (assuming that Carol is the divider), and Ted gets a piece whose value to him is 15 units.

23. (a)

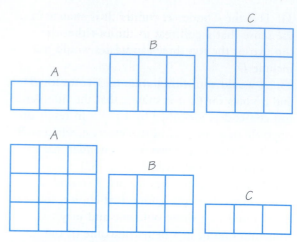

(b) Player 2 finds *B* acceptable (6 square units) and *C* acceptable (9 square units). Player 3 finds *A* acceptable (9 square units) and *B* acceptable (6 square units).
(c) Player 3 chooses *A* (9 square units). Player 2 chooses *C* (9 square units). Player 1 chooses *B* (6 square units). Yes, there is another order. Player 2 chooses *C* (9 square units). Player 3 chooses *A* (9 square units). Player 1 chooses *B* (6 square units).

25. (a)

(b) Player 2 will further trim the piece:

in player 2's eyes.
(c) Player 3 will further trim the piece:

in player 3's eyes.

(d) Player 3 receives it, and thinks it is 6 units of value. The one leaving with the first piece always thinks it is one-nth of the value with n players.

(e) Assume player 1 is the divider. He sees it as 16 units of value, and he divides it as follows:

$\frac{1}{3}$ $\frac{2}{3}$

Player 2 chooses the piece on the left, which he sees as follows:

(f) Assume player 2 is the divider. He sees it as 14 units of value, and he divides it as follows:

$\frac{1}{2}$ $\frac{1}{2}$

Player 1 chooses the piece on the right, which he sees as follows:

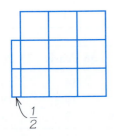

$\frac{1}{2}$

27. Bob, Carol, and Ted each divide the piece he or she has in four parts (equal in his or her own estimation). Alice then chooses one of Bob's four pieces, one of Carol's four pieces, and one of Ted's four pieces.

29.

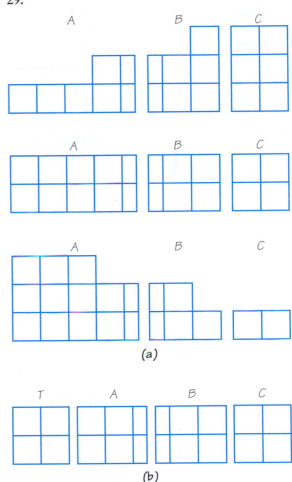

(a)

(b)

(c) Player 3 will choose A (which he thinks is of size 6 square units).

Player 2 will choose B (which he thinks is of size 5 square units).

Player 1 will receive C (which he thinks is of size 6 square units).

The proviso does not come into play (since player 3 took the trimmed piece).

Chapter 14

1.

	Rounded Percentage
Rent	44%
Food	23%
Transportation	10%
Gym	17%
Miscellaneous	7%

These percentages add up to 101%.

3.

Level	Quota	Rounded
I	9.69	10
II	1.71	2
III	5.71	6
IV	2.90	3

This calls for too many teaching assistants, so the numbers must be adjusted.

5. $0 + 2 + 1 + 2 + 3 + 2 = 10$

7. The total population is 510,000, so the standard divisor is $510,000 \div 102 = 5000$. The quotas are obtained by dividing each state's population by this divisor, obtaining 50.8, 30.6, and 20.6, respectively. The lower quotas add up to 100, so we must increase the apportionment of two states to their upper quotas. The first state, whose quota of 50.8 has the largest fractional part, gets an increase. The fractional parts of the quotas of the remaining two states are both equal to 0.6: They are tied for priority in receiving the last seat. A coin toss is probably the fairest way to settle this dispute.

9. There will be three sections of Geometry, and one section each of Algebra and Calculus.

11. The apportionments before and after the tax was paid are

	Before Tax	After Tax
Abe	36	36
Beth	19	19
Charles	23	22
David	20	20
Esther	2	3
Total	100	100

So, Aunt Esther not only got a dollar back, but Charles had to give her one of his rare coins! At least it's still in the family. The cause of this confusion is, of course, the population paradox.

13.

State	Population	Apportionments					
A	5,576,330	25	26	26	27	28	28
B	1,387,342	6	6	6	7	7	7
C	3,334,241	15	15	16	16	17	17
D	7,512,860	34	34	35	37	37	38
E	310,968	2	2	1	2	1	1
Total	18,121,741	82	83	84	89	90	91

The Alabama paradox occurs when the apportionment for the smallest state decreases from 2 to 1 as the house size increases from 83 to 84, and it occurs again as the house size increases from 89 to 90.

15. There will be four Geometry sections, one Calculus section, and no Algebra section.

17.

Jefferson method:		36 Pearls	37 Pearls
	Abe	14	15
	Beth	19	19
	Charles	3	3

Webster method:		36 Pearls	37 Pearls
	Abe	14	15
	Beth	19	19
	Charles	3	3

	36 Pearls	37 Pearls
Abe	15	15
Beth	18	19
Charles	3	3

Hill–Huntington method:

With 36 pearls, there is a difference between the Hill–Huntington apportionment and the others, but with 37, the three methods produce the same results. If there is a principle on which to choose a method, it would probably be to choose the method by which the cost per pearl is as close as possible to the same for each of the friends. The cost per pearl is the district size. The method that minimizes relative differences in the cost per pearl is the Hill–Huntington method. If the friends would prefer to minimize absolute differences, they would have to use the Dean method, which was not covered in this chapter. Charles might want to study up on it, though, because it allocates the 36th pearl to Beth and the 37th to him!

19. Hamilton method:

$$88 + 2 + 2 + 1 + 1 + 1 + 1 + 1 +$$
$$1 + 1 + 1 = 100\%$$

The first three percentages are rounded to upper quotas, and the remaining percentages are rounded to lower quotas. The quota condition is satisfied.

Jefferson method:

$$90 + 1 + 1 + 1 + 1 + 1 + 1 + 1 + 1 +$$
$$1 + 1 = 100\%$$

This apportionment rounds 87.85% to 90%, more than the upper quota. The quota condition is violated.

Webster method: The apportionment is the same as the Jefferson apportionment, so it, too, violates the quota condition.

21. (a) $\sqrt{0 \times 1} = 0$

(b) $\sqrt{1 \times 2} = 1.4142$

(c) $\sqrt{2 \times 3} = 2.4495$

(d) $\sqrt{3 \times 4} = 3.4641$

23.

	Webster	Hill–Huntington
Algebra	2	2
Geometry	3	2
Calculus	Cancelled!	1

It's likely that the principal would prefer the Webster method, because classes as small as the calculus class, with 3 students, should be cancelled. Notice that the Hill–Huntington apportionment gives Geometry less than its lower quota in order to accommodate Calculus.

25. 2241

27.

	Before Tax	After Tax
Abe	36	36
Beth	19	19
Charles	22	23
David	20	20
Esther	3	2
Total	100	100

Esther must give one of her three rare coins to her nephew.

29. With the Hamilton and Webster methods, the Liberals get 49 votes and the Tories get 50. With the Jefferson method, the split is 48 for the Liberals, 50 for the Tories, and a tie for the last seat.

31. A seat that had been assigned to Vermont by the Hill–Huntington method moves to Pennsylvania. There are no other changes.

33. 10.77%

35. (a) California, 640,204; Utah, 745,571

(b) Absolute difference: 105,367; relative difference: 16.46%

(c) The absolute difference would be 93,336.5; the relative difference would be 16.69%.

(d) The absolute difference in district populations would be less if California had 52 seats and Utah had 4. With that revised apportionment, the relative differences would be greater. Thus, the Hill–Huntington method, which was used in apportioning Congress after the 2000 census, did not minimize absolute differences in district population. It minimized relative differences.

37. The Webster method would allocate 11 seats to Massachusetts, and 5 to Oklahoma.

39. (a) Lowndes favors small states.
(b) Yes, because like the Hamilton method, the Lowndes method presents a way to decide, for each state, if the lower or upper quota should be awarded.
(c) Yes. Since the method is not a divisor method, the population paradox is inevitable.
(d)

State	p_i	a_i
VA	630,560	18
MA	475,327	14
PA	432,879	12
NC	353,523	10
NY	331,589	10
MD	278,514	8
CT	236,841	7
SC	206,236	6
NJ	179,570	5
NH	141,822	4
VT	85,533	3
GA	70,835	2
KY	68,705	2
RI	68,446	2
DE	55,540	2
Total	3,615,920	105

41. (a) Let $n = \lfloor q \rfloor$. If q is between n and $n + 0.4$, then the Condorcet rounding of q is equal to n. Since $q + 0.6 < n + 1$ in this case, it is also true that $\lfloor q + 0.6 \rfloor = n$. On the other hand, if $n + 0.4 \leq q < n + 1$, then the Condorcet rounding of q is $n + 1$, and also $n + 1 \leq q + 0.6 < n + 1.6$, so $\lfloor q + 0.6 \rfloor = n + 1$.
(b) The method favors small states, since numbers will be rounded up more often than down. This makes it more likely that the quotas will be adjusted downward.

(c) If the sum of the tentative apportionments is less than the house size, the critical divisor for state i, with population p_i, is $d_i = \dfrac{p_i}{n_i + 0.4}$. The state with the largest critical divisor gets the next seat, and then its critical divisor is recomputed. The process stops when the house is full.

If the total apportionment is more than the house size, then the critical divisor for state i is $d_i = \dfrac{p_i}{n_i - 0.6}$. The state with the least critical divisor of all has its tentative apportionment decreased by 1. Its critical divisor is then recomputed. The process stops when enough seats have been removed so that the number of seats apportioned is equal to the house size.

43. Let $f_i = q_i - \lfloor q_i \rfloor$ denote the fractional part of the quota for state i. Since the Hamilton method assigns to each state either its lower or its upper quota, each absolute deviation is equal to either f_i (if state i received its lower quota) or $1 - f_i$ (if it received its upper quota). For convenience, let's assume that the states are ordered so that the fractions are decreasing, with f_1 the largest and f_n the smallest. If the lower quotas add up to $h - k$, where h is the house size, then states 1 through k will receive their upper quotas. The maximum absolute deviation will be the larger of $1 - f_k$ and f_{k+1}.

The maximum absolute deviation for the Hamilton method is less than 1, because each fractional part f_i and its complement, $1 - f_i$, is less than 1. If a particular apportionment fails to satisfy the quota condition, then for at least one state, the absolute deviation exceeds 1–hence the maximum absolute deviation is greater than that of the Hamilton apportionment.

If an apportionment satisfies the quota condition then–as with the Hamilton method–k states receive their upper quotas and $n - k$ states receive their lower quotas.

If a state j, where $j \leq k$, receives its lower quota, then–to compensate–a state l, where $l > k$, must get its upper quota. The absolute deviations for these states would be f_j and $1 - f_l$, respectively. Because of the way the fractions have been ordered, we have $f_j \geq f_{k+1}$ and $1 - f_l \geq 1 - f_k$. Therefore, the absolute deviation for one of states j and l will be equal to or exceed the maximum absolute deviation of the Hamilton apportionment. We conclude that no apportionment is better than Hamilton's, if what we mean by "better" is "smaller maximum absolute deviation."

Chapter 15

1. (a), (b) Saddlepoint at row 1 (maximim strategy), column 2 (minimax strategy), giving value 5. **(c)** Row 2 and column 1
3. (a) No saddlepoint. **(b)** Rows 1 and 2 are both maximin strategies; column 1 is the minimax strategy. **(c)** None
5. (a), (b) Saddlepoint at row 3 minimum that is the same as column 3 maximum, giving value -20. **(c)** Columns 1 and 2
7. Batter's optimal mixed strategy is $\left(\frac{1}{4}, \frac{3}{4}\right)$ and pitcher's is $\left(\frac{1}{4}, \frac{3}{4}\right)$, giving value 0.275.
9. Saddlepoint is "not cheat" and "audit," giving value $-\$100$.
11. (a)

	Officer Does Not Patrol	Officer Patrols
You park in street	0	-40
You park in lot	-32	-16

(b) You: $\left(\frac{2}{7}, \frac{5}{7}\right)$; officer: $\left(\frac{3}{7}, \frac{4}{7}\right)$; value: $-\$22.86$
(c) It is unlikely that the officer's payoffs are the opposite of yours—that she always benefits when you do not.
(d) Use some random device, such as a die with seven sides.
13. (a) Move first to the center box; if your opponent moves next to a corner box or to a side box, move to a corner box in the same row or column. There are now six more boxes to fill, and you have up to three more moves (if you or your opponent does not win before this point). But the

rest of your strategy becomes quite complicated, involving choices like "move to block the completion of a row/column/diagonal by your opponent."
(b) Showing that your strategy is optimal involves showing that it guarantees at least a tie, no matter what choices your opponent makes.
15. Player I should play H, winning 1 on average.
17. (a) Player II should avoid "call" because "fold" dominates it.
(b) Player I: $\left(\frac{1}{3}, \frac{2}{3}, 0\right)$; player II: $\left(\frac{2}{3}, 0, \frac{1}{3}\right)$; value: $-\frac{1}{12}$
(c) Player II. Since the value is negative, player II's average earnings are positive and player I's are negative.
(d) Yes. Player I bets first while holding L with probability $\frac{2}{3}$. Player II raises while holding L with probability $\frac{1}{3}$, so sometimes player II raises while holding L.
19. (a) Leave umbrella at home if there is a 50% chance of rain; carry umbrella if there is a 75% chance of rain.
(b) Carry umbrella in case it rains
(c) Saddlepoint at "carry umbrella" and "rain," giving value -2
(d) Leave umbrella at home
21. The Nash equilibrium outcomes are (4, 3) and (3, 4). [It would be better if the players could flip a coin to decide between (4, 3) and (3, 4).]
23. The Nash equilibrium outcome is (2, 4), which is the product of dominant strategies by both players.
25. The players would have no incentive to lie about the value of their own weapons unless they were sure about the preferences of their opponents and could manipulate them to their advantage. But if they do not have such information, lying could cause them to lose more than 10% of their weapons, as they value them, in any year.
27. The sophisticated outcome, x, is found as follows: Y's strategy of y is dominated; with this strategy of Y eliminated, X's strategy of x is dominated; with this strategy of X eliminated, Z's

strategy of z is dominated, which is eliminated. This leaves X voting for xy (both x and y), Y voting for yz, and Z voting for zx, creating a three-way tie for x, y, and z, which X will break in favor of x.

29. The payoff matrix is as follows:

		Even		
		2	4	6
	1	(2, 1)	(2, 1)	(2, 1)
Odd	3	(2, 4)	(6, 3)	(6, 3)
	5	(2, 4)	(4, 8)	(10, 5)

Odd will eliminate strategy 1, and Even will eliminate strategy 6, because they are dominated. In the reduced 2×2 game, Odd will eliminate strategy 5. In the reduced 1×2 game, Even will eliminate strategy 4. The resulting outcome will be (2, 4), in which Odd chooses strategy 3 and Even chooses strategy 2. The outcome (2, 1), in which Odd chooses strategy 1 and Even chooses strategy 2, is also in equilibrium.

31. If the first player shoots in the air, he will be no threat to the two other players, who will then be in a duel and shoot each other. If a second player fires in the air, then the third player will shoot one of these two, so the two who fire in the air will each have a 50–50 chance of survival. Clearly, the third player, who will definitely survive and eliminate one of her opponents, is in the best position.

33. In a duel, each player has incentive to fire—preferably first—because he or she does better whether the other player fires (leaving no survivors, which is better than being the sole victim) or does not fire (you are the sole survivor, which is better than surviving with the other player). In a truel, if you fire first, then the player not shot will kill you in turn, so nobody wants to fire first. In a four-person shoot-out, if you fire first, then you leave two survivors, who will not worry about you because you have no more bullets, leading them to duel. Thus, the incentive in a four-person shoot-out—to fire first—is the same as that in a duel.

35. Nobody will shoot.

37. The possibility of retaliation deters earlier shooting.

Chapter 16

1. Since $3 + 9 + 5 + 3 + 9 + 8 + 1 + 6 + 4 + 0 = 48 = 9 \times 5 + 3$, the check digit is 3.

3. Since $873345672 = 7 \times 124763667 + 3$, the check digit is 3.

5. Since $30860422052 = 7 \times 4408631721 + 5$, the check digit is 5.

7. Since $3 \cdot 3 + 8 + 3 \cdot 1 + 3 + 3 \cdot 7 + 0 + 3 \cdot 0 + 9 + 3 \cdot 2 + 1 + 3 \cdot 3 = 69$, the check digit is 1.

9. Since $10 \cdot 0 + 9 \cdot 6 + 8 \cdot 6 + 7 \cdot 9 + 6 \cdot 1 + 5 \cdot 9 + 4 \cdot 4 + 3 \cdot 9 + 2 \cdot 3 = 265 = 11 \times 24 + 1$, the check digit is X.

11. Since $7 \cdot 0 + 3 \cdot 9 + 9 \cdot 1 + 7 \cdot 9 + 3 \cdot 0 + 9 \cdot 2 + 7 \cdot 0 + 3 \cdot 4 = 129$, the check digit is 9.

13. Since $4 + 6 + 1 + 2 + 1 + 2 + 0 + 2 + 3 = 21$, the check digit is 6.

15. Since $7 \cdot 3 + 8 + 7 \cdot 1 + 3 + 7 \cdot 7 + 0 + 7 \cdot 0 + 9 + 7 \cdot 2 + 1 + 7 \cdot 3 = 133$, the check digit is 7. This check-digit scheme will detect all single-digit errors.

17. In the odd-numbered positions, if a digit a is replaced by the digit b, where $a - b$ is even, the error is not detected.

19. We begin with $(3 + 0 + 2 + 6 + 0 + 9 + 4 + 1) \times 2 = 50$. Adding 2, we obtain 52. Then $52 + 0 + 1 + 5 + 0 + 1 + 6 + 3 = 68$. So, the check digit is 2.

21. **(a)** Since $1 \cdot 0 + 1 \cdot 1 + 3 \cdot 2 + 3 \cdot 1 + 1 \cdot 6 + 3 \cdot 9 + 1 \cdot 0 = 43$, the check digit is 7.
(b) Since $1 \cdot 0 + 1 \cdot 2 + 3 \cdot 7 + 3 \cdot 4 + 1 \cdot 5 + 3 \cdot 5 + 1 \cdot 1 = 56$, the check digit is 4.
(c) Since $1 \cdot 0 + 1 \cdot 7 + 3 \cdot 6 + 3 \cdot 0 + 1 \cdot 0 + 3 \cdot 2 + 1 \cdot 2 = 33$, the check digit is 7.
(d) Since $1 \cdot 0 + 1 \cdot 4 + 3 \cdot 9 + 3 \cdot 6 + 1 \cdot 5 + 3 \cdot 8 + 1 \cdot 0 = 78$, the check digit is 2.

23. First observe that the given number 0669039254 results in a weighted sum that has a remainder of 5 after division by 11. So all we need do is check for successive pairs of digits of this number that results in a contribution to the weighted sum of 5 less or 6 more, since either of these will make the weighted sum divisible by 11. Checking each pair of consecutive digits we see that 39 contributes $5 \cdot 3 + 4 \cdot 9 = 51$, whereas 93 contributes $5 \cdot 9 + 4 \cdot 3 = 57$. So, the correct number is 0669093254.

25. Notice that when we add the weighted sum used for the actual check digit $7a_1 + 3a_2 + 9a_3 + 7a_4 + 3a_5 + 9a_6 + 7a_7 + 3a_8$ and the weighted sum $3a_1 + 7a_2 + a_3 + 3a_4 + 7a_5 + a_6 + 3a_7 + 7a_8$, we obtain $10a_1 + 10a_2 + 10a_3 + 10a_4 + 10a_5 + 10a_6 + 10a_7 + 10a_8$, which always ends with 0. So, the actual check digit and the check digit calculated with the weighted sum $3a_1 + 7a_2 + a_3 + 3a_4 + 7a_5 + a_6 + 3a_7 + 7a_8$ are both 0 or their sum is 10.

27. Replacing Z by 9, or vice versa, is not detected.

29. Since the value of the weighted sum determines whether or not a number is valid, the position of the check digit is not relevant.

31. Since the remainder after dividing by 9 is between 0 and 8, 9 never can be a check digit.

33. If the remainder of the sum of the noncheck digits after dividing by 7 is k, then the check digit is $7 - k$ if $k \neq 0$, and 0 if $k = 0$. So, 7, 8, and 9 never can be a check digit.

35. Yes. The ISBN scheme detects all transposition errors.

37. For the transposition to go undetected, it must be the case that the difference of the correct number and the incorrect number is divisible by 11. That is,

$$(10a_1 + 9a_2 + 8a_3 + \cdots + a_{10}) -$$
$$(10a_3 + 9a_2 + 8a_1 + \cdots + a_{10})$$

is divisible by 11. This reduces to $2a_1 - 2a_3 = 2(a_1 - a_3)$ is divisible by 11. But $2(a_1 - a_3)$ is divisible by 11 only when $a_1 - a_3$ is divisible by 11, and this happens only when $a_1 - a_3 = 0$. In this case, there is no error. The same argument works for the fourth and sixth digits.

39. The combination 72 contributes $3 \cdot 7 + 2 = 23$ or $7 + 3 \cdot 2 = 13$ (depending on the location of the combination) toward the total weighted sum, while the combination 27 contributes $3 \cdot 2 + 7 = 13$ or $2 + 3 \cdot 7 = 23$. So, the total weighted sum resulting from the number with the transposition is still divisible by 10. Therefore, the error is not detected. When the combination 26 contributes $3 \cdot 2 + 6 = 12$ toward the total weighted sum, the combination 62 contributes $3 \cdot 6 + 2 = 20$ toward the total weighted sum. So the new weighted sum will not be divisible by 10. Similarly, when the combination 26 contributes $2 + 3 \cdot 6 = 20$ to the total, the combination 62 contributes $6 + 3 \cdot 2 = 12$ to the total. So, the total for the number resulting from the transposition will not be divisible by 10 and the error is detected. In general, an error that occurs by transposing ab to ba is undetected if and only if $a - b = \pm 5$.

41. The error $\cdots abc \cdots \rightarrow \cdots cba \cdots$ is undetected if and only if $a - c = \pm 5$. To see this in the case where the weights for abc are 7, 3, 9, notice that a and c contribute $7a + 9c$ toward the weighted sum, whereas in the case of cba, the c and a contribute $7c + 9a$. Thus the error is undetected if and only if $7a + 9c$ and $7c + 9a$ contribute equal amounts to the last digit of the weighted sum. This means that they differ by a multiple of 10. That is, $-2a + 2c = 2(c - a)$ is a multiple of 10. This occurs when $c - a = 0$, or $c - a = \pm 5$. When $c - a = 0$, there is no error.

43. Since any error in the position with weight 10 does not change the last digit of the weighted sum, no error in that position is detected. In the position with weight 5, replacing an even digit by any other even digit is not detected. In positions with weights 12, 8, 6, 4, or 2, replacing a by b is undetected if $a - b = \pm 5$. In positions with weights 11, 9, 7, 3, or 1, all errors are detected.

45. Since both numbers are valid, the difference of their weighted sums is divisible by 10. That is, $(7w + 3 + 2w + 1 + 5w + 6 + 7w + 4) - (7w + 3 + 2w + 1 + 5w + 6 + 6w + 1)$ is divisible by 10. The difference simplifies to $w + 3$. So, $w = 7$.

47. (a) The code is 51593-2067; since $5 + 1 + 5 + 9 + 3 + 2 + 0 + 6 + 7 = 38$, the check digit is 2.
(b) The code is 50347-0055; since $5 + 0 + 3 + 4 + 7 + 0 + 0 + 5 + 5 = 29$, the check digit is 1.
(c) The code is 44138-9901; since $4 + 4 + 1 + 3 + 8 + 9 + 9 + 0 + 1 = 39$, the check digit is 1.

49. (a) Since the sixth block of five bars (ignoring the first bar) has one long bar and four short bars, that block is incorrect. Call the digit corresponding to that block x. Then the code is $20782x9607$. Since the sum of the digits is $x + 41$, $x = 9$.
(b) Since the eighth block of five bars (ignoring the first bar) has three long bars and two short bars, that block is incorrect. Call the digit corresponding to that block x. Then the code is $5543599x2$. Since the sum of the digits is $x + 42$, $x = 8$.
(c) Since the tenth block of five bars (ignoring the first bar) has one long bar and four short bars, that block is incorrect. Since this is the check digit, the nine-digit ZIP code is 52735-2101.

51. If a double error in a block results in a new block that does not contain exactly two long bars, we know this block has been misread. If a double error in a block of five results in a new block with exactly two long bars, the block now gives a digit different from the original one. If no other digit is in error, the check digit catches the error, since the sum of the 10 digits will not end in 0. So, in every case an error has been detected. Errors of the first type can be corrected just as in the case of a single error. When a double error results in a legitimate code number, there is no way to determine which digit is incorrect.

53. If you replace each short bar in the bar code table by an a and replace each long bar in the bar code table by a b, the resulting strings are listed in alphabetical order.

55. Since there is an even number of 1's in 1000100, the scanner is reading from right to left.

57. Wyoming, Nevada, and Alaska
59. The size of the population
61. The Canadian scheme detects any transposition error involving adjacent characters. Also, there are $26^3 \times 10^3 = 17{,}576{,}000$ possible Canadian codes but only $10^5 = 100{,}000$ U.S. five-digit ZIP codes. Hence the Canadian scheme can target locations more precisely.
63. S000, S200, L550, L300, E663, O451
65. A person born in 1999 is too young for a driver's license.
67. For a woman born in November or December, the formula $40(m - 1) + b + 600$ gives a number requiring four digits.
69. The 58 indicates that the year of birth is 1958. Since 818 is larger than $12 \cdot 31 = 372$, we subtract 600 from 818 to obtain 218. Then $218 = 7 \cdot 31 + 1$ tells us that the person was born on the first day of the eighth month. So, the birth date is August 1, 1958.
71. Since $248 = 3 \cdot 63 + 58 + 1$, the number 248 corresponds to a female born on March 29; since $601 = 9 \cdot 63 + 34$, the number corresponds to a male born on September 17.
73. Likely circumstances could be twins; sons named after their fathers (such as John L. Smith, Jr.); common names such as John Smith and Mary Johnson; and states that do not include year of birth in the code.
75. Because of the short names and large population, there would be a significant percentage of people whose names would be coded the same.

Chapter 17

3. (a) 6 **(b)** 3
5. 1001101
7. 000000, 100011, 010101, 001110, 110110, 101101, 011011, 111000
9. 0000000, 1000001, 0100111, 0010101, 0001110, 1100110, 1010100, 1001111, 0110010, 0101001, 0011011, 1110011, 1101000, 1011010, 0111100, 1111101. No, since 1000001 has weight 2.

11. 000000, 100101, 010110, 001011, 110011, 101110, 011101, 111000. 001001 is decoded as 001011; 011000 is decoded as 111000; 000110 is decoded as 010110; 100001 is decoded as 100101.
13. 00000000, 00010111, 00101110, 01001011, 10001101, 11000110, 10100011, 10011010, 01100101, 01011100, 00111001, 11101000, 11010001, 10110100, 01110010, 11111111. The code will detect any three errors or correct any single error.
15. $2^5 = 32$; $2^8 = 256$
17. 0000, 1012, 2021, 0111, 0222, 1120, 2210, 2102, 1201
19. $3^4 = 81$; $3^6 = 729$
21. AATAAAGCAA
23. 111101000111001010; *AABAACAEADB*
25. *t*, *n* and *r*; *e*
27. In the Morse code, a space is needed to determine where each code word ends. In a fixed-length code of length *k*, a word ends after each *k* digits.
29. 2015 42 22 -19 -10 3 12 -35 -5 -23; we go from 49 characters to 34 characters, a reduction of 30.6%.
31. *BCEEFCDDCFF*
33. *B* is the least likely letter, *J* is the second least likely, and *G* is the third least likely letter.
35. UHWUHDW; ADVANCE
37. 13
39. ROLLING STONES
41. I'll
43. I am a man.
45. could've or would've
47. and
49. Oh, I love your magazine. Especially the "Enrich Your Word Power" section. I think it's really, really, really good.
51. **(a)** 11000000; **(b)** 10011001
53. 23, 49, 16
55. 13
57. *N* converts to 14 and *O* converts to 15, but 14 and 77 have a greatest common divisor of 7. On the other hand, using blocks of length 4, NO converts to 1415 and the greatest common divisor of 77 and 1415 is 1.

59.

P	Q	$P \wedge Q$	$P \vee (P \wedge Q)$
T	T	T	T
T	F	F	T
F	T	F	F
F	F	F	F

Since the entries in the column for the variable P are exactly the same as the entries in the column for $P \vee (P \wedge Q)$, the two expressions are logically equivalent.

61. First we construct the truth table for $\neg(P \wedge Q)$:

P	Q	$P \wedge Q$	$\neg(P \wedge Q)$
T	T	T	F
T	F	F	T
F	T	F	T
F	F	F	T

Next we construct the truth table for $\neg P \vee \neg Q$:

P	Q	$\neg P$	$\neg Q$	$\neg P \vee Q$
T	T	F	F	F
T	F	F	T	T
F	T	T	F	T
F	F	T	T	T

Since the last columns of the two truth tables are identical, we have shown that $\neg(P \wedge Q)$ is logically equivalent to $\neg P \vee \neg Q$.

63. First we construct the truth table for $P \wedge (Q \vee R)$:

P	Q	R	$Q \vee R$	$P \wedge (Q \vee R)$
T	T	T	T	T
T	T	F	T	T
T	F	T	T	T
T	F	F	F	F
F	T	T	T	F
F	T	F	T	F
F	F	T	T	F
F	F	F	F	F

Next we construct the truth table for $(P \wedge Q) \vee (P \wedge R)$:

P	Q	R	$P \wedge Q$	$P \wedge R$	$(P \wedge Q) \vee (P \wedge R)$
T	T	T	T	T	T
T	T	F	T	F	T
T	F	T	F	T	T
T	F	F	F	F	F
F	T	T	F	F	F
F	T	F	F	F	F
F	F	T	F	F	F
F	F	F	F	F	F

Since the last columns of the two truth tables are identical, we have shown that $P \wedge (Q \vee R)$ is logically equivalent to $(P \wedge Q) \vee (P \wedge R)$.

65. The truth table for $\neg P \vee Q$ is:

P	Q	$\neg P$	$\neg P \vee Q$
T	T	F	T
T	F	F	F
F	T	T	T
F	F	T	T

Because the last column of this truth table is identical to the one for $P \rightarrow Q$, we conclude that the two expressions are logically equivalent.

67. Let P denote "it snows" and let Q denote "there is school." Then the statement "If it snows, there will be no school" can be expressed as $P \rightarrow \neg Q$. Similarly, the statement "It is not the case that it snows and there is school" can be expresed as $\neg (P \wedge Q)$. We now constuct the truth tables for each of these expressions. A truth table for $P \rightarrow \neg Q$ is:

P	Q	$\neg Q$	$P \rightarrow \neg Q$
T	T	F	F
T	F	T	T
F	T	F	T
F	F	T	T

A truth table for $\neg (P \wedge Q)$ is:

P	Q	$P \wedge Q$	$\neg (P \wedge Q)$
T	T	T	F
T	F	F	T
F	T	F	T
F	F	F	T

Because the two tables have identical last columns, the two expressions are logically equivalent.

Chapter 18

1. (a) 1 **(b)** 3; nine times as large **(c)** 4 **(d)** Yes **(e)** The 5×7, 8×10, 11×14, and 16×20 are not geometrically similar to the negative; in printing one of those, the negative must be cropped. **(f)** Based just on paper used, the 8×12 would cost $2^2 = 4$ times as much as the 4×6. **(g)** The 5×7 and the 8×10 cost about $0.05/in.^2$, so you could expect the 11×14 to cost about $11 \times 14 \times \$0.05 = \$7.70$, probably \$7.79. The 20×30 should probably be priced at \$29.99.

3. (a) $\frac{1}{40} = 0.025$ **(b)** 64,000 times as large **(c)** 400 cm = 4.00 m ≈ 13.1 ft

5. 26% larger

7. (a) The volume would be 8 times as large–not "8 times greater than" or "8 times larger than," and definitely not twice as large, as the old altar. **(b)** $\sqrt[3]{2} \approx 1.26$

9. The writer meant that the volume was 2.5 times as much before packaging. The new volume is $1/2.5 = 0.4$ "times as much as" before. We could also correctly say that the peat moss has been compressed "to 40% of its original volume" or "by 60%," or that the compressed volume is "60% less than" the original volume.

11. If the original statement read instead "a car costs 120% more in Country Y than in Country Q," then C is correct.

13. Answers will vary.

15. € 0.59

17. 32 mpg

19. (a) 0.000134 tons **(b)** That all parts of the scale model are made of the same materials as the real locomotive. **(c)** 0.267 lb **(d)** 0.121 kg **(e)** 0.000121 metric tonnes

21. US \$6.88/gal

23. 185 m, or 607 ft

25. (a) 37.5 mph **(b)** 28 mph **(c)** 120 mph

27. (a) 50% **(b)** 100%

29. (a) 35.9% **(b)** 56.0%

31. (a) For Option A, yes; for Option B, no. **(b)** For either a loss or a gain, the absolute value of the percentage is higher for Option B. **(c)** Either way, use Option B.

33. (a) 900 lb/ft^3 **(b)** Almost twice as dense as steel **(c)** The mineral soil weighs about 4500 to 4600 lb.

35. (a) 400,000 lb **(b)** 28 lb/in.^{2}

37. 6100 lb, or 2800 kg

39. The lights are strung around the outside of the tree branches, so in effect they cover the outside "area" of the tree (thought of as a cone). Hence, the number of strings needed grows in proportion to the square of the height: A 30-ft tree will need $5^2 = 25$ times as many strings as a 6-ft tree. However, you could also argue that a 30-ft tree is meant to be viewed from farther away, so that stringing the lights farther apart on the 30-ft tree would produce the same effect as with the shorter tree.

41. 9 ft 3 in. to 11 ft 9 in. (in modern times, there have been men over 9 ft tall); 282 cm to 358 cm

43. Less than 90.25 kg

45. Answers will vary.

47. Answers will vary.

49. Answers will vary.

51. (a) 800 **(b)** 640,000 **(c)** 512,000,000 **(d)** One eight-hundredth. There couldn't be any such giant ants.

53. 480 years

55. $A \propto d^2$ and $A \propto M^{3/4} \propto (d^2h)^{3/4} = d^{3/2}h^{3/4}$, so $d^2 \propto d^{3/2}h^{3/4}$, hence $d^{1/2} \propto h^{3/4}$ and $d \propto h^{3/2}$

57. A small, warm-blooded animal has a large surface-area-to-volume ratio. Pound for pound, it loses heat more rapidly than a larger animal, hence it must produce more heat per pound, resulting in a higher body temperature.

59. (a)

Log Weight by Log Wingspan

(b) Both relationships are allometric, since the results are good fits to straight lines whose slopes are not 1. **(c)** The slope for birds is less steep than for planes.

Chapter 19

1. 5, 8, and 13

3. Answers will vary but will be Fibonacci numbers.

5. (a, b) The digits after the decimal point do not change. **(c)** $\phi^2 = \phi + 1$ **(d)** $1/\phi = \phi - 1$

7. (a) 9 **(b)** 16

9. (a) 4, 7, 11, 18, 29, 47, 76, 123 **(b)** 3, 1.333, 1.75, 1.571, 1.636, 1.611, 1.621, 1.617, 1.618. The ratios approach ϕ.

11. Answers will vary.

13. The seventh number is $5m + 8n$, and the total is $55m + 88n$.

15. (a) 1, 1, 3, 5, 11, 21, 43, 85, 171, 341, 683, 1365 **(b)** $B_n = B_{n-1} + 2B_{n-2}$ **(c)** 1, 3, 1.667, 2.2, 1.909, 2.048, 1.977, 2.012, 1.994, 2.003, 1.999 **(d)** $x = 2, -1$; we discard the -1 root. **(e)** $B_n = [2^n - (-1)^n]/3$. We omit the table.

17. Silver mean: $1 \pm \sqrt{2} \approx 2.414$
Bronze mean: $\frac{1}{2}(3 \pm \sqrt{13}) \approx 3.303$
Copper mean: $2 \pm \sqrt{5} \approx 4.236$
Nickel mean: $\frac{1}{2}(5 \pm \sqrt{29}) \approx 5.193$
General expression: $\frac{1}{2}(m + \sqrt{m^2 + 4})$.

19. All are true.

21. (a) B, C, D, E, H, I, K, O, X **(b)** A, H, I, M, O, T, U, V, W, X, Y **(c)** H, I, N, O, S, X, Z.

23. (a) MOM, WOW; MUd and bUM reflect into each other, as do MOM and WOW. **(b)** pod rotates into itself; MOM and WOW rotate into each other. **(c)** Here are some possibilities: NOW NO; SWIMS; ON MON; CHECK BOOK BOX; OX HIDE.

25. For all parts, translations. **(a)** Reflection in vertical lines through the centers of the A's or between them. **(b)** Reflection in the horizontal midline, glide reflections. **(c)** Reflection in the horizontal midline, reflections in vertical lines through the centers of the O's or between them, 180° rotation around the centers of the O's or the midpoints between them, glide reflections. **(d)** None other than translations.

27. (a) *c5* **(b)** *c12* **(c)** *c22*

29. (a) *c6* **(b)** (CBS) *d2* **(c)** (Dodge Ram) *d1*

31. (a) *c4* **(b)** *d2*

33. *p111, p1a1, p112, pm11, p1m1, pma2, pmm2*

35. (a) *pmm2* **(b)** *p1a1* **(c)** *pma2* **(d)** *p112*
(e) *pmm2* (perhaps) **(f)** *p1m1* **(g)** *pma2* **(h)** *p111*

37. (a) Half-turns are preferred on Mesa Verde pottery, while reflections predominate on Begho smoking pipes.
(b) Neither culture completely excludes any strip type. Begho designs are heavily concentrated (almost 90%) in *p1m1*, *p112*, or *pmm2*, while Mesa Verde designs are more evenly distributed over the seven patterns.
(c) (i) *pm11* or *pma2*: Mesa Verde (ii) *p112*: Mesa Verde (iii) *pmm2*: Begho (iv) *pm11*: Begho (v) *p1m1*: Difficult to say (vi) *pmm2*: Begho (vii) *pmm2*: Begho (viii) *pma2*: Mesa Verde (ix) *p1a1*: Mesa Verde

39. (c) Smallest rotation is 90°, there are reflections, there are reflections in lines that intersect at 45°: *p4m*. **(d)** Smallest rotation is 90°, there are no reflections: *p4*.

41. None of the five patterns with hexagonal symmetry can be realized, nor *p4g*, *p4*, *cm*, and *cmm*. The remaining eight can all be formed by the technique.

43. *pg*, if color is disregarded; otherwise, *p1*.

45. *p6*

47. *p3*

49. Answers will vary.

51. (a) *d2* **(b)** Any two of: R (180° rotation around the center), V (reflection in vertical line through its center), H (reflection in horizontal line through its center). **(c)** $\{I, R, V, H\}$

53. There are four rotational symmetries (including the identity), two reflection symmetries, and two reflections across diagonal lines: $\{I, R, R^2, R^3, H, V, RH = VR, RV = HR\}$.

55. Answers will vary.

57. As in Example 5, number fixed positions, label with letters copies of the pattern elements in

the positions, and pick a fixed position about which to make a half-turn R.
(a) $\langle T, R \mid R^2 = I, T \circ R = R \circ T^{-1}\rangle = \{ \ldots, T^{-1}, I, T^1, \ldots; \ldots, R \circ T^{-1}, R, R \circ T^1, \ldots \}$.
(b) $\langle T, R, H \mid R^2 = H^2 = I, T \circ H = H \circ T, R \circ H = H \circ R, (R \circ T)^2 = I\rangle = \{ \ldots, T^{-1}, I, T, \ldots; \ldots, R \circ T^{-1}, R, R \circ T, \ldots; \ldots, H \circ T^{-1}, H, H \circ T, \ldots; \ldots, R \circ H \circ T^{-1}, R \circ H, R \circ H \circ T, \ldots \}$.

59. $\langle R \mid R^8 = I\rangle = \{I, R, R^2, R^3, R^4, R^5, R^6, R^7\}$, where R is a rotation by 45°.

61. There are four rotational symmetries (including the identity), three reflection symmetries, and an inversion through the center.

63. The carved head is reproduced in the same shape at different scales.

65. Answers will vary.

67. Answers will vary.

Chapter 20

1. Exterior: 45°; interior: 135°

3. $180° - \dfrac{360°}{n}$

5. The usual notation for a vertex figure is to denote a regular *n*-gon by *n*, separate the sizes of polygons by periods, and list the polygons in clockwise order starting from the smallest number of sides, so that, e.g., 3.3.3.3.3.3 denotes six equilateral triangles meeting at a vertex. The possible vertex figures are 3.3.3.3.3.3, 3.3.3.3.6, 3.3.3.4.4, 3.3.4.3.4, 3.3.4.12, 3.4.3.12, 3.3.6.6, 3.6.3.6, 3.4.4.6, 3.4.6.4, 3.12.12, 4.4.4.4, 4.6.12, 4.8.8, 5.5.10, and 6.6.6.

7. 3.7.42, 3.9.18, 3.8.24, 3.10.15, and 4.5.20

9. At each of the vertices except the center one, six triangles meet, with angles (in clockwise order) of 75°, 75°, 30°, 30°, 75°, and 75°.

11. Yes

13. (a) No **(b)** No **(c)** No

15. Answers will vary.

17. No; no

19. The only way to tile by translations is to fit the outer "elbow" of one tile into the inner "elbow" of another. Labeling the corners as follows works: the corners on the top A and B, those on the rightmost side C and D, the middle of the bottom E, and the middle of the leftmost side F.

21. Just label the four corners consecutively: A, B, C, and D.

23. Place the skew-tetromino on a coordinate system with unit length for the side of a square and with the lower left corner at (0, 0). Then $A = (1, 2)$, $B = (3, 2)$, $C = (2, 0)$, and $D = (0, 0)$ works.

25. Place the skew-tetromino on a coordinate system with unit length for the side of a square and with the lower left corner at (0, 0). Then $A = (0, 1)$, $B = (1, 2)$, $C = (3, 2)$, $D = (3, 1)$, $E = (2, 0)$, and $F = (0, 0)$ works.

27. **(a)** Yes **(b)** No **(c)** No

29. See figure below.

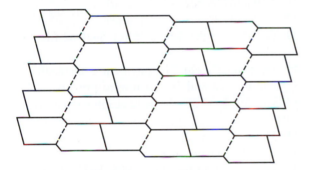

31. Answers will vary.

33. N, Z, W, P, y, I, L, V, X. See www.srcf.ucam. org/~jsm28/tiling/5-omino-trans.ps.gz.

35. Place the U on a coordinate system, with unit length for the side of a square and with the lower left corner at (0, 0). Then $A = (2, 2)$, $B = (3, 2)$, $C = (3, 0)$, $D = (1, 0)$, $E = (0, 0)$, and $F = (0, 1)$ works. See www.srcf.ucam.org/~jsm28/tiling/ 5-omino-rot.ps.gz.

37. Answers will vary.

39. $ABAABABA$

41. The two leftmost A's would have had to come from two B's in a row in the preceding month.

43. Let S_n, A_n, and B_n be the total number of symbols, the number of A's, and the number of B's at the nth stage. We note that the only B's at the nth stage must have come from A's in the previous stage, so $B_n = A_{n-1}$. Similarly, the A's at the nth stage come from both A's and B's in the previous stage, so $A_n = A_{n-1} + B_{n-1}$. Using both of these facts together, we have $A_n = A_{n-1} + A_{n-2}$. We note that $A_1 = 0$, $A_2 = 1$, $A_3 = 1$, $A_4 = 2, \ldots$. The A_n sequence obeys the same recurrence rule as the Fibonacci sequence and starts with the same values one step later; in fact, it is always just one step behind the Fibonacci sequence: $A_n = F_{n-1}$. Consequently, $B_n = A_{n-1} = F_{n-2}$, and $S_n = A_n + B_n = F_{n-1} + F_{n-2} = F_n$.

45. If a sequence ends in AA, its deflation ends in BB, which is impossible for a musical sequence. Similarly, if a sequence ends in $ABAB$, its deflation ends in AA, which we just showed to be impossible.

47. The first is part of a musical sequence, but the second is not.

49. If the sequence were periodic, the limiting ratio of A's to B's would be the same as the ratio in the repeating part, which would be a rational number, contrary to the result of Exercise 48.

Chapter 21

1. **(a)** 698 days **(b)** After 24 months

3. **(a)** \$1030.00; 3.000% **(b)** \$1030.00; 3.000% **(c)** \$1030.34; 3.034% **(d)** \$1030.45; 3.045%

5. Using 365-day years: The daily interest rate $i = 0.03/365$ is in effect for $n = 8 \times 365 = 2920$ days. We have in the compound interest formula $A = \$10,000 = P(1 + i)^n$, so we get $P = \$10,000/1.2712366 = \$7866.36$. (Fine point: In fact, the eight years must contain two February 29 days. Calculating interest for $6 \times 365 = 2190$ days at $i = 0.03/365$ and for $2 \times 366 = 732$ days at $i = 0.03/366$ gives a result that differs by less than one one-hundredth of a cent.)

7. 3.68%

9. **(a)** \$2032.79; \$2025.82; \$2012.20 **(b)** \$1999.00; \$1992.56 **(c)** \$1973.82; \$1906.62; \$1849.60 **(d)** For small and intermediate interest rates, the rule of 72 gives good approximations to the doubling time.

11. **(a)** 2, 2.59, 2.705, 2.7169, 2.718280469 **(b)** 3, 6.19, 7.245, 7.3743, 7.389041321 **(c)** $e =$ 2.718281828 . . . ; $e^2 =$ 7.389056098. . . . Your calculator may give slightly different answers, because of its limited precision.
13. In all cases, $30.45, not taking into account any rounding to the nearest cent of the daily posted interest.
15. **(a)** $(e^{0.04} - 1) \times 100\% = 4.08108\%$ **(b)** The approximation for the effective rate is $r + \frac{1}{2}r^2 = 0.05 + \frac{1}{2} \times (0.05)^2 = 0.05 + 0.00125 = 0.05125$ or 5.125%, very slightly less than the true effective rate.
17. $79.41
19. $81,327.45
21. $880.21
23. **(a)** $147.06 **(b)** $444,683.29 **(c)** $302,384.64
25. **(a)** 31/32 **(b)** $\frac{2^n - 1}{2^n}$ **(c)** 1
27. **(a)** $A = P(1.15)(1.07)(0.80) = 0.98440P$, so $r = (0.98440)^{1/3} - 1 = -0.00523 = -0.523\%$. **(b)** It is the effective rate.
29. **(a)** $(1.04)^3 \approx $1.12 **(b)** $1/1.12 \approx $0.89
31. Approximately $3900
33. **(a)** About $70; answers will vary. **(b)** 1970: $1.29; 1974: $2.85
35. 8.3%
37. Nowhere close. **(a)** $59,339.44 **(b)** $57,394.20
39. $29.8 million; with 4% interest, $33.4 million through last payment.
41. **(a)** $161.39 **(b)** $25,341.55 **(c)** $7813.27
43. **(a)** 1.39% **(b)** 0.97% (however, some states and cities do not tax interest earned on U.S. government securities)
45. 2.13%, or 188 points. The answer does not depend on the value of D.
47. 1.60% per month, or 19.2% annual rate
49. 4.97%; it is the effective rate.

Chapter 22

1. **(a)** 5.62%; 7.24% **(b)** Answers will vary; the first bond locks in the interest rate much farther into the future.
3. $128.75
5. $97.66

7. Answers will vary. The add-on loan always has a lower payment.
9. All but a tiny amount after 91 months.
11. After 203 months (more than 16 years!), the balance is $500.16.
13. 294 months (24.5 years!), plus a few cents in the 295th month
15. $383.25
17. $19.28 per month for each $1000 financed
19. $632.07
21. $850.62
23. $6388.73
25. $23,813.20
27. 3.68%
29. $1167.48
31. APR = 15.55%; EAR = 16.71%
33. APR = 24.37%; EAR = 27.29%
35. 571% (572% for 365-day year)
37. 324% (325% for 365-day year)
39. **(a)** $760.07 **(b)** $4612.20 **(c)** $634.62 **(d)** 16 months
41. **(a)** $100,250 **(b)** The payment is $648.60: $583.33 interest and $65.27 toward principal. **(c)** $333.33 **(d)** $332.50; the annual interest rate is 3.99%.
43. $36.1 million.
45. $d = $994.71 per quarter.
47. Using the CPI for 2001–2005, $624.32 per month; using 3% inflation, $618.65. If the 6% APR is interpreted as 6% APY, then $658.84 (CPI) or $652.85 (3% inflation).

Chapter 23

1. In late summer 2023
3. 5.49 billion
5. Almost 100% greater than Europe's population.
7. **(a)** 117 years **(b)** 54 years
9. **(a)** 8.3 billion, 11.5 billion **(b)** No change in growth rate, no change in death rates, no global catastrophes, etc.
11. **(a)** 38 years **(b)** 29 years **(c)** Answers will vary.
13. **(a)** 51 years **(b)** 132 years **(c)** 224 years
15. **(a)** 138 years **(b)** Forever!

17. That's about 5.5 tons per plant per year ≈ 30 lb per plant per day, which is unreasonable.
19. 2.77%
21. After the first year, the population stays at 15.
23. 7, 18.2, 6.6, 17.6, 8.4, 19.5, 2.0, 7.3, 18.6, 5.3
25. We must have $f(x_n) = x_n$, or $4x_n(1 - 0.05x_n) = x_n$. The only solutions are $x_n = 0$ and $4(1 - 0.05x_n) = 1$, or $x_n = 15$.
27. 10, 15.0, 11.3, 14.8, 11.6, 14.6, 11.8, 14.5, 11.9, 14.4. The population is oscillating but slowly converging to $40/3 \approx 13.3$.
29. $x_n = 0$, $40/3 \approx 13.3$
31. The red dashed line indicates the same size population next year as this year; where it intersects the blue curve is the equilibrium population size.
33. The population sizes are 11, 12.0, 12.6, 12.9, 13.1, 13.2, 13.3, 13.3, 13.3, 13.3.
35. The population sizes are 11, 15.0, 13.7, 14.9, 14.9, 15.0, 14.8, 14.3, 13.0, 9.5–and the following year the population is wiped out.
37. About 15 million pounds. Maximum sustainable yield is about 35 million pounds for an initial population of 25 million pounds.
39. (a) The last entry shown for the first sequence is the fourth entry of the second sequence, so the first "joins" the second and they then both end up going through the same cycle (loop) of numbers over and over.
(b) 39, 78, 56, and we have "joined" the second sequence. However, an initial 00 stays 00 forever; and any other initial number ending in 0 "joins" the loop sequence 20, 40, 80, 60, 20, ...
(c) Regardless of the original number, after the second push of the key, we have a number divisible by 4 and all subsequent numbers are divisible by 4. There are 25 such numbers between 00 and 99. You can verify that an initial number either joins the self-loop 00 (the only such numbers are 00, 50, and 25); joins the loop

20, 40, 80, 60, 20, . . . (the only such numbers are the multiples of 5 other than 00, 50 and 25); or joins the big loop of the other 20 multiples of 4.
41. (a) 133, 19, 82, 68, 100, 1, 1, ... The sequence stabilizes at 1.
(b) Answers will vary. **(c)** That would trivialize the exercise! **(d)** For simplicity, limit consideration to three-digit numbers. Then the largest value of f for any three-digit number is $9^2 + 9^2 + 9^2 = 243$. For numbers between 1 and 243, the largest value of f is $1^2 + 9^2 + 9^2 = 163$. Thus, if we iterate f over and over–say 164 times–starting with any number between 1 and 163, we must eventually repeat a number, since there are only 163 potentially different results. And once a number repeats, we have a cycle. Thus, applying f to any three-digit number eventually produces a cycle. How many different cycles are there? That we leave you to work out. *Hints*: (1) There aren't very many cycles. (2) There is symmetry in the problem, in that some pairs of numbers give the same result; for example, $f(68) = f(86)$.
43. (a) 0.0397, 0.15407173, 0.545072626, 1.288978, 0.171519142, 0.59782012, 1.31911379, 0.0562715776, 0.215586839, **0.722914301**, 1.32384194, 0.0376952973, 0.146518383, 0.521670621, 1.27026177, 0.240352173, 0.78810119, 1.2890943, 0.171084847, **0.596529312**
(b) 0.723, 1.323813, 0.0378094231, 0.146949035, 0.523014083, 1.27142514, 0.236134903, 0.777260536, 1.29664032, 0.142732915, **0.509813606 (c) 0.722**, 1.324148, 0.0364882223, 0.141958718, 0.507378039, 1.25721473, 0.287092278, 0.901103183, 1.16845189, **0.577968093**
45. Period 2 begins at $\lambda = 3$, period 4 at $1 + \sqrt{6} \approx 3.449$, period 8 at 3.544, period 3 at $1 + 2\sqrt{2} \approx 3.828$, and chaotic behavior onsets at about 3.57. See www.answers.com/topic/logistic-map.

Index

Probability, 296–323
  area under a curve as, 310,
    311–312
  assignment of, 301
  Benford's law and, 304–305, 306
  cancer clusters and, 317
  central limit theorem and,
    316–323
  for coin tossing, 298–300
  definition of, 298
  for dice rolling, 300–304
  equally likely outcomes and,
    305–309
  in gambling, 298–304, 319–322
  law of large numbers and, 315
  randomness and, 296–297
  risk and, 302
  in roulette, 319–322
Probability histogram, 304
Probability models, 297–305
  continuous, 309–312
  definition of, 299
  discrete, 304–305, 313–316
  expected value of, 313
  for first digits, 304–305, 306,
    313–316
  mean of, 313–315
  sample space S in, 299–300
  standard deviation of, 315–316
  uniform, 310
  variance of, 316
Problems of scale, 663–664,
  678–681
Processors, in scheduling, 80, 81
Profit formula, 136–137
  corner points and, 134
  for one-resource mixture,
    136–137
  optimal production policy and,
    134–138
  for two-resource mixture,
    138–141
Profit line, 142–143
Programming
  integer, 125–126
  linear, 124–163. *See also* Linear
    programming
  mathematical, 125–126
Proportion
  divine, 712–717
  golden ratio and, 712–717

Modular scale of, 717, 718
sampling distribution of,
  322–323
Proportional allocation, in divide-
  and-choose procedure, 476,
  487–490
Proportional growth, 687–690. *See
also* Growth and form
Proportional rule, 457, 461
Proportional symmetry, 735
Prospective studies, 269
Pseudo-random numbers, 259
Public key cryptography, 640–644
Pueblo pottery, patterns on,
  729–731
Pukelsheim, Friedrich, 345
Pure strategy, 543, 544–550
Pygmies, 688

Q

Quadrilaterals, 761, 762–763
Quarterly compounding, 801
Quartiles, 192–195
  in boxplots, 194–195
  calculation of, 193
  definition of, 192
  five-number summary of, 193–194
  of normal distributions, 203–204
Quasicrystals, 782, 783
Quasiperiodic tiling, 779
Quetzalcoatlus northropi, 683–685
Quota(s)
  in apportionment, 506, 507, 508,
    509–510, 518–520
  adjusted, 514, 519
  lower, 509
  upper, 509
  in weighted voting, 391
Quota condition, 518–520
Quota method, 519

R

Random numbers, 259
  table of, 258–262
Random phenomena
  definition of, 296
  equally likely outcome of, 305–309
  examples of, 297
  probability of, 298. *See also*
    Probability
  sample space S in, 299–300

Random sample. *See* Sample;
  Sampling
Random tiling, 775
Randomized comparative
  experiment, 266–270
Randomized strategy, 543, 550–560
Rank voting method, 347–353
Rate of natural increase, 860
Rational choice, 542
Ready task, 83–84
Reagan, Ronald, 433
Real growth, 817–820
Rectangle, symmetries of, 731–732
Rectangular networks, 17
Recursion, 712
Reed, Irving, 621
Reed–Solomon codes, 621
Reflection, rigid motion and, 719
Reflection symmetry, 719, 721
Regression lines, 226–228
  best-fitting, 232–233
  guidelines for, 235–239
  intercept of, 228
  interpretation of, 235–238
  least-squares, 233–235
  slope of, 228
Regular pentagons, tiling with, 759
Regular polygons, 757
Regular polyhedra, 758
Regular tiling, 757
Reinhardt, Karl, 763, 764
Relationship between variables
  causal, 237–238
  description vs. interpretation of,
    236
  direction of, 224–225
  form of, 224
  negative, 225
  patterns and deviations in, 224
  positive, 225
  strength of, 224, 225
Relative difference, 529
Relative dollar value, 715
Relative price change, 715
Renewable resources, 870–876
  elimination of, 878–881
  overexploitation of, 880–881
  sustained-yield harvesting of,
    871–874, 876–881
Repetition, in symmetry, 709,
  717–718